Dynamics of Civil Engineering and Transport Structures and Wind Engineering (DYN-WIND 2017)

MATEC Web of Conferences Volume 107 (2017)

Trstena, Slovakia
21 - 25 May 2017

Editors:

Jozef Melcer
Kamila Kotrasova

ISBN: 978-1-5108-4114-7

Printed from e-media with permission by:

Curran Associates, Inc.
57 Morehouse Lane
Red Hook, NY 12571

Some format issues inherent in the e-media version may also appear in this print version.

Printed by Curran Associates, Inc. (2017)

For additional information, please contact EDP Sciences – Web of Conferences at the address below.

EDP Sciences – Web of Conferences
17, Avenue du Hoggar
Parc d'Activité de Courtabœuf
BP 112
F-91944 Les Ulis Cedex A
France

Phone: +33 (0) 1 69 18 75 75
Fax: +33 (0) 1 69 28 84 91

contact-edps@webofconferences.org

Additional copies of this publication are available from:

Curran Associates, Inc.
57 Morehouse Lane
Red Hook, NY 12571 USA
Phone: 845-758-0400
Fax: 845-758-2633
Email: curran@proceedings.com
Web: www.proceedings.com

TABLE OF CONTENTS

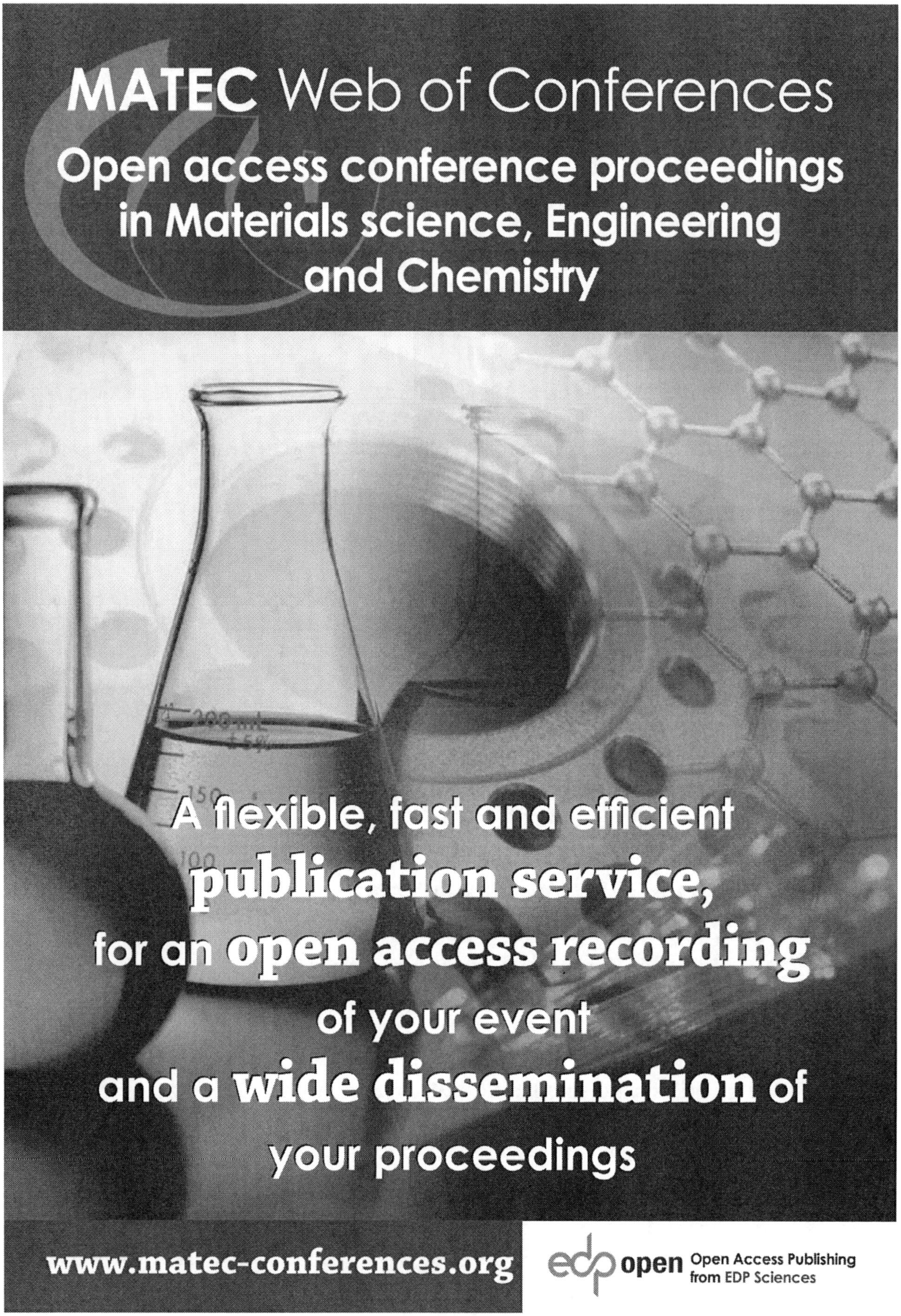
MATEC Web of Conferences
Open access conference proceedings in Materials science, Engineering and Chemistry
A flexible, fast and efficient publication service, for an open access recording of your event and a wide dissemination of your proceedings
www.matec-conferences.org
edp open Open Access Publishing from EDP Sciences

ISSN (electronic edition): 2261-236X

Statement of Peer review

In submitting conference proceedings to MATEC Web of Conferences, I certify to the Publisher that I adhere to the **Policy on Publishing Integrity** of the journal in order to safeguard good scientific practice in publishing.

1. All articles have been subjected to peer review administered by the proceedings editors.
2. Reviews have been conducted by expert referees, who have been requested to provide unbiased and constructive comments aimed, whenever possible, at improving the work.
3. Proceedings editors have taken all reasonable steps to ensure the quality of the materials they publish and their decision to accept or reject a paper for publication has been based only on the merits of the work and the relevance to the journal.

Title, date and place of the conference DYN-WIND 2017. May 21-25,2017, Oravský Háj, Slovakia

Prof. Ing. Jozef Melcer, DrSc., University of Žilina
Assoc. Prof. Kamila Kotrasova, PhD., University of Košice
jozef.melcer@fstav.uniza.sk; kamila.kotrasova@tuke.sk
Proceedings editor(s):

Date and editor's signature May 5, 2017

Structural condition assessment of the bridge in Ostrava

Jindřich Čech[1,*], *Petr* Tej[1], *Jiří* Kolísko[1] and *Jaromír* Král[1]

[1]Klokner institute, Czech Technical University in Prague, Department of mechanics, Czech Republic

Abstract. This paper deals with the comparison of results of dynamic test and numerical modelling of a road bridge across the Dr. Kudela street in the street Rudná in Ostrava. The bridge is a reinforced concrete structure, made of post-tensioned KA-beams. On the bridge were verified material properties. Calculation model to validate the static and dynamic behavior of structures was created. On the bridge were measured dynamic properties (frequency, mode shapes and attenuation) during excitation construction by hydraulic vibration exciter. Article will focus on comparing the results of the calculation of dynamic structures and properties measured on real structure upon excitation.

1 Description of structure

The bridge object consists of two bridge structures, which are composed of three fields (11 m + 16 m + 11 m). Each bridge structure is formed by 14 prefabricated KA-61 beams (in each field). The width of bearing structure is 13.25 m. The bridge is placed on the supports and elastomeric bearings. Bridge foundations are monolithic reinforced concrete. Outer supports are massive reinforced concrete. Spans of individual fields of the bridge are 11 m (short field) and 16 m (long field). During the reconstruction was made a coupling plate of thickness 160 to 300 mm.

Fig. 1. View of the bridge structure.

* Corresponding author: jindrich.cech@cvut.cz

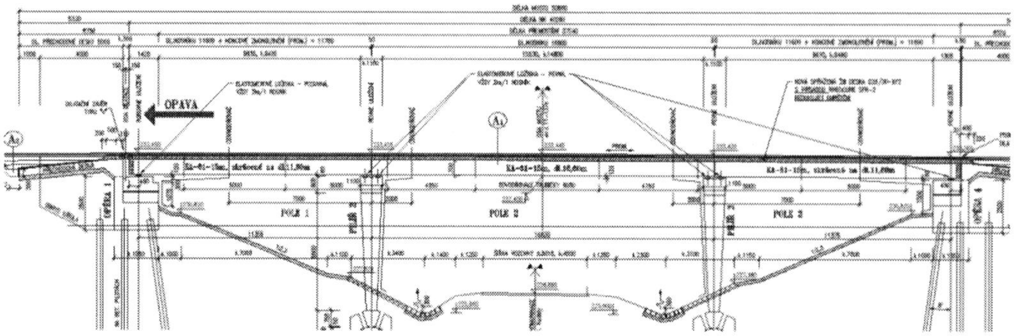

Fig. 2. Longitudinal section of the bridge structure.

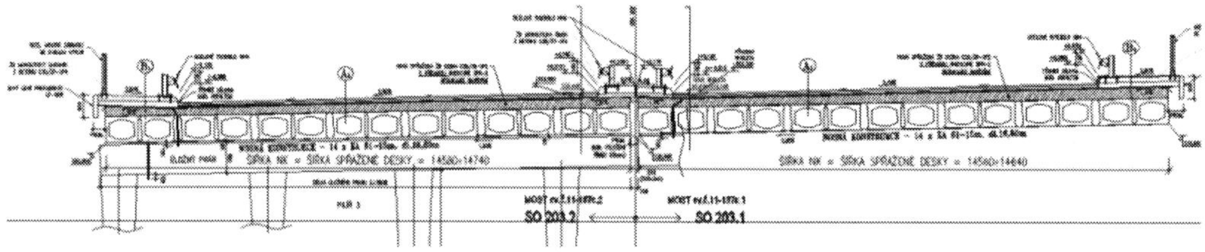

Fig. 3. Cross-section of the bridge structure.

2 Diagnosis and load tests

2.1 Diagnosis

The lower surface of the bearing structure is locally disrupted by leaking through the bridge deck waterproofing and longitudinal cracks in lower slabs of beams. Two outer cable canals are exposed with exposed corroded prestressing wires at the bridge structure ev. Nr. 11-137c.1. Outer beams of the bridge structure is further disrupted by longitudinal cracks. Near the cracks occur coatings/extracts of bright colour. The above mentioned failures were identified mainly as a result of alkali-silica reaction of aggregate [1].

Tensile strength of concrete was tested in the corner areas of reinforced concrete blocks. Tests were performed on the upper and lower surfaces of the sample. Core drill, which could affect measured values negatively, was not used for thinning of test sites with respect to anticipated concrete structure defects.

Tests of compressive strength were performed on cube samples, which were carved from the central part of the block. For strength tests were excluded samples that had been disrupted by pruning from tensile strength tests and drilled (dismantling) holes. Further parts of the concrete with visible cracks from alkali-silica reaction or samples with a higher rate of reinforcement rods have been omitted.

2.2 Load tests

Dynamic loading test of the bridge was performed using electrohydraulic vibration exciter with a movable mass of 500 kg and electronics FlexTest (MTS Systems Corp.). Modal analysis was performed for each field separately with broadband excitation in the frequency band 7 Hz to 20 Hz. From the transfer functions were estimated lowest natural frequency of individual fields and the response was recorded on the bridge excitation by harmonic forces with these frequencies (Table 1). The course of transfer functions and the response of the bridge after the end of excitation by harmonic force show a large attenuation of the bridge structure. The response was measured in 9 sections (over the supporting blocks and columns, in the middle of the short fields and in the middle and quarters of span of middle field) [2]. To measure of the response of the bridge was in each section used 7 acceleration sensors BK 4379 with amplifiers of charge BK2635 (Bruel&Kjaer). PULSE system (Bruel&Kjaer) was used for the record response and evaluation of complex transfer functions. Evaluation of modal characteristic was performed using MEScope (Vibrant Technology, Inc.).

3 Numerical modeling

The numerical computational model of individual fields of the bridge was made in an environment of software Scia Engineer 15.3. As the material of the walls and slabs forming in the model KA beams was chosen material model C40/50 from the catalog of program Scia. As the material of the coupling plate was chosen C30/37. Individual KA beams were supported in the model at the edges of the bottom plates, always on one side of a fixed joint and on the second side of a displaceable joint in the length direction of the beam. Slabs and walls of the model were meshed of planar finite elements with side 1 m. On the thus modelled structure were identified shapes and values of relevant frequencies.

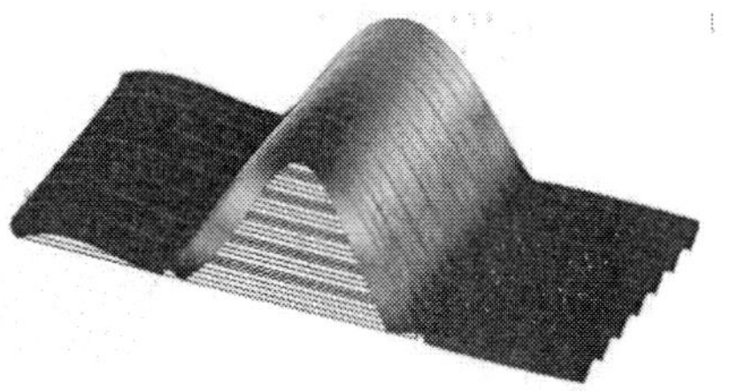

Fig. 4. The first natural shape of the bridge structure – frequency 8.84 Hz.

Fig. 5. The second natural shape of the bridge structure – frequency 11.07 Hz

Fig. 6. The third natural shape of the bridge structure – frequency 14.95 Hz.

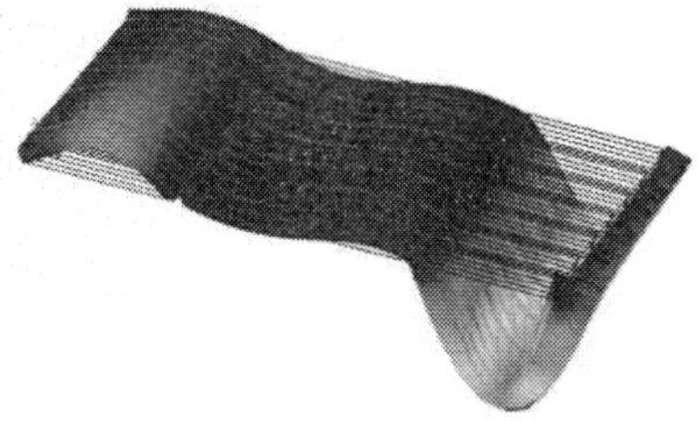

Fig. 7. The fourth natural shape of the bridge structure – frequency 15.76 Hz

DOI: 10.1051/matecconf/201710700001

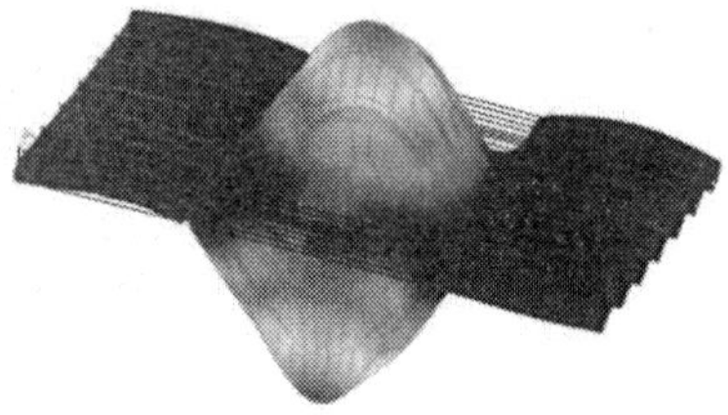

Fig. 8. Fifth natural shape of the bridge structure – frequency 16.42 Hz.

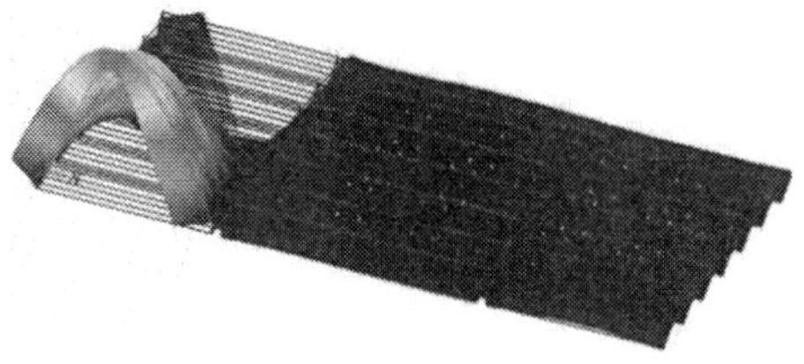

Fig. 9. Sixth natural shape of the bridge structure – frequency 17.24 Hz

Fig. 10. Seventh natural shape of the bridge structure – frequency 18.54 Hz

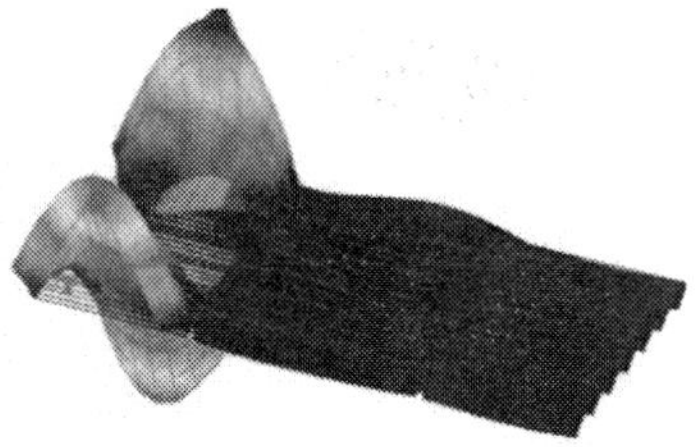

Fig. 11. Eighth natural shape of the bridge structure – frequency 23.39 Hz

4 Conclusions

Table 1. The natural frequency and damping in the middle span.

f [Hz] (dynamic load test)	f [Hz] (numerical modelling)	[%]	Description of vibration shape
9.1	8.84	4.8	B1
11.4	11.07	5.1	B2
13.9	14.95	2.3	C1
15.4	15.76	1.1	A1
17.2	16.42	4.4	C2, B3, A2
18.4	17.24	3.7	B3,C2
21.0	18.54	4.5	A1, B4
23.4	23.39	1.0	B4; A3 a C3 in antiphase
23.9	-	2.2	A3 a C3 in antiphase

Shorter spans are marked A and C, the middle span is B. The number B2 in the specification in Table indicates the second shape of vibration of middle span. The bridge at this level of amplitude vibrates as continuous beam with three spans. I.e. artificially created joints between each field do not show the shapes of the vibration considerably (they are bypassed with a plate probably). It is impossible to separate the second and higher shapes of vibration with a large attenuation as they occur simultaneously in pairs or threes.

The measured natural frequencies correspond roughly wall-slab computational model, shapes of vibration not. The maximum amplitudes of oscillation at mid-span, and in the outer fields are identical in all cases regardless of whether the driver in the middle span or in one of the outer spans.

This project is supported by grant GAČR P105/12/G059

References

1. J. Čech, P. Tej, P. Kněž, M. Blank, Static load test and numerical analysis of the bridge ev. Nr. 11-137C.2 Road I/11, Ostrava-Poruba, Expert report, Klokner Institute CTU in Prague (2017)
2. J. Čech, P. Tej, P. Kněž, J. Král, Dynamic load test and numerical analysis of the bridge ev. Nr. 11-137C.2 Road I/11, Ostrava-Poruba, Expert report, Klokner Institute CTU in Prague (2017)
3. A. Mecke, I. Lee, J.R. Baker jr., M.M. Banaszak Holl, B.G. Orr, Eur. Phys. J. E **14**, 7 (2004)
4. M. Ben Rabha, M.F. Boujmil, M. Saadoun, B. Bessaïs, Eur. Phys. J. Appl. Phys. (to be published)
5. F. De Lillo, F. Cecconi, G. Lacorata, A. Vulpiani, EPL, **84** (2008)
6. L. T. De Luca, *Propulsion physics* (EDP Sciences, Les Ulis, 2009)

Assessment of structural condition of Libeň Bridge

Petr Kněž[1,*], *Petr* Tej[1], *Jindřich* Čech[1] and *Marek* Blank[1]

[1]Klokner institute, Czech Technical University in Prague, 16608 Šolínova 7, Prague, Czech Republic

Abstract. The paper presents diagnostic and load tests of the Inundation bridge which is part of a group of bridges called The Libeň bridge group in Prague. The Libeň bridge group consists of two arched and several framed bridges spanning the Vltava river. One of the vaulted bridges consists of 5 arches and the other vaulted bridge (called Inundation bridge) consists of only one arch. Arched bridges are extraordinary structures with both technical and historical value. Since the inundation bridge has the largest arch of whole group, it was selected for testing purposes. The bridge is assembled with three-hinged arch made of concrete. The hinges are made of reinforced concrete and lead contact slabs. Detailed measurements of geometry and material properties were made on the bridge. Based on these measurements a computer model was created to verify the behavior of the structure. Both static and dynamic calculations were performed. Measurements of dynamic characteristics were made during normal operation and with hydraulic vibration exciter. This article will focus on comparing the results of dynamic calculation of the modeled structure and properties measured on real structure excited by hydraulic vibration exciter.

1 Description of construction

The Libeň bridge was built between 1924-1928 according to a project by architect Pavel Janak. The arched parts of the bridge were designed by engineer František Mencl. The bridge was opened for traffic on October 29, 1928 at the 10th anniversary of the Czechoslovakia in the presence of the former president Thomas G. Masaryk. The bridge arches with three hinges made of concrete are considered of an immense technical value. The arch of the inundation bridge is the largest arch of its kind in Prague. The overlap of arches on consoles are of 3 to 4.5 meters and the use of prefabricated concrete blocks for the hinges are a creative contribution of engineer Mencl. The thickness of the arch at the top hinge is 670 mm at the smallest arch, 830 mm at the largest arch, and 800 to 950 mm at base hinges of arches.

The arch of the Inundation bridge (Fig. 1, 2) has a span of 48 m. The bridge carries urban traffic across an inundation area. The width of the roadway on the bridge is 14.5 meters (traffic lanes + tram line), walkways on both sides have a width of 3.25 m. The driveway is covered with asphalt concrete, pavement surface consists of cast asphalt, and

* Corresponding author: petr.knez@cvut.cz

tramway lane is made of a special type of reinforced concrete panels with canals called BKV panels.

The supporting structure of the arches consist of a cast-in-place, normal strength concrete segments with embedded reinforced concrete hinges. The arches in cross section are divided into 4 segments. The segments are 4.85 m wide and their thickness varies between 0.9 and 1.1 meters. Each arch was designed as a three-hinged arch with one hinge at the top and two hinges positioned on cantilevers roughly 3 meters of a distance from pillars.

Fig. 1. The construction of Inundation Bridge.

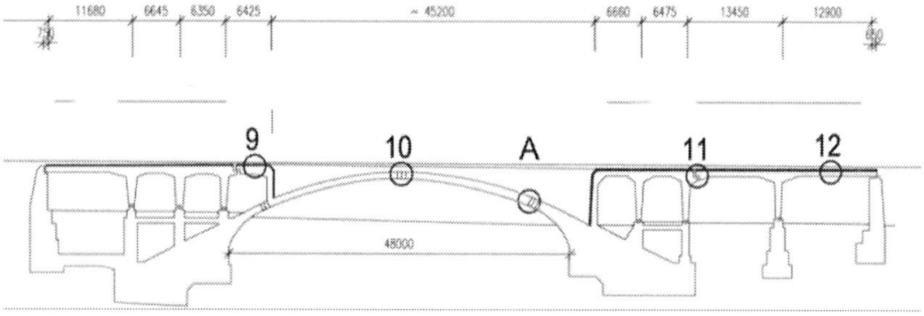

Fig. 2. Longitudinal section of Inundation Bridge.

2 Diagnostics and load tests

2.1 Diagnostics

Within the project, control tests of the concrete specimens taken from the bridge arch were carried out. To determine the mechanical properties and a frost resistance, tests were performed on these specimens.

The following statements were made based on the results obtained from the tests:

- The cement paste has been leached for an extended period of time. Concrete degraded due to the alternation of freezing and thawing and due to the technology of casting used during construction. The technology of casting of concrete induced very high inhomogeneity manifested as caverns, macropores and gaps in the concrete structure. The inhomogeneity due to casting is also linked to high variability of properties. Completely disintegrated regions with strength between 0 and 5 MPa were found in the

samples. These sections could not be mechanically tested; thus, they were not included in the evaluation of the strength by conventional methods according to standards.
- when evaluating the concrete samples according to EN 13791, the characteristic strength of concrete for the group of samples was determined with regard to the presence of gaps, caverns and overall visual assessment. The lowest characteristic strength in the concrete mass of the arch was determined as 21.9 MPa therefore the corresponding strength class C 20/25.
- The average compressive modulus of elasticity of all examined arches was set at 27.3 GPa. This value corresponds to strength class C 16/20.
- Embedded concrete hinges were found to be stronger than the rest of the structure, with a corresponding strength class C 35/45.
-Water absorption of the examined concrete structure was found to be of a relatively low value, ranging from 2.5% to 6.1%. Such low percentage should indicate sufficient frost resistance. However, during the frost resistance testing the opposite was revealed. The test was carried out on 19 concrete samples according to CSN 731322. Out of the 19 specimens, 4 were completely disintegrated after 25 freeze-thaw cycles. After additional 50 freezing cycles, another 7 samples were completely disintegrated. Therefore after 75 freeze-thaw cycles 58% of the concrete samples disintegrated. The concrete in its current state is not resistant to frost.

2.2 Load tests

2.2.1 Static load test

The load of the static load test was represented by four trucks Iveco Trakker 8×4 weighing 25 t symmetrically placed at the top of arch (load condition 1) or three trucks Iveco Trakker 6×4 weighing 25 t placed asymmetrically on one side of the arch (load condition 2 and 3).

Tested bridge complied the static load test in terms of permanent deformation to conditions of CSN 73 6209 "Load testing of bridges" in 1996.

The measured deflection was 0.76 mm at the top hinge in the vertical direction for the first load condition.

2.2.2 Dynamic load test

Dynamic load test was executed by a hydraulic vibration exciter type INSET in Mode for harmonic excitation (frequency sweep) in the frequency range of 1-20 Hz. Two locations of the exciter were proposed for the dynamic load test. The first location (B1) was in one third of the span of the arch above the first transverse arch at 2/3 of its width. The second location (B2) was in the middle of the span over the second transverse in the middle of the arch. In the first position, the excitation was induced in the vertical and horizontal direction. For the second position, the excitation was induced only vertically. The measured eigenfrequencies are shown in Table 1. The values in parentheses were not accepted.

Response of the structure was measured using sensors of speed of oscillation. First group of sensors was fastened to the lower surface of the arch. Second group of sensors was placed on the roadway approximately in the same position as the first group, but on the upper surface of the arch. The sensors were placed on all four arches of the arch. Sensors were always in the place of the hinges and in the distance corresponding to approximately 1/10 of the span. The eigenfrequencies, the corresponding mode shapes (Fig. 3, 4) and the attenuation of structure were examined.

The load tests revealed that the structure is eligible for operation even in its current state.

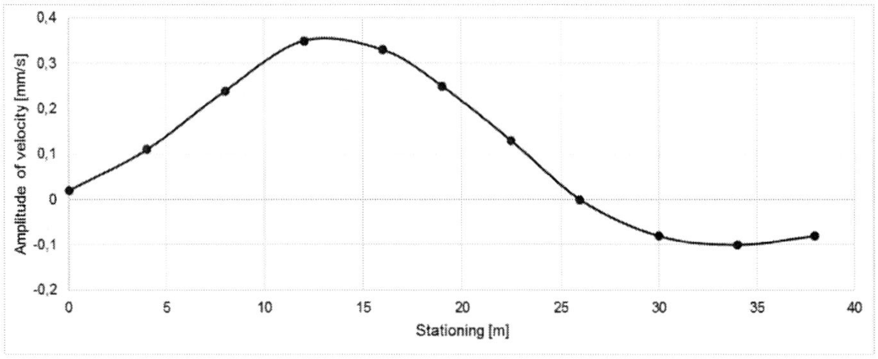

Fig. 3. Mode shapes during excitation in the vertical direction in the first position, frequency of forced vibration 4.63 Hz.

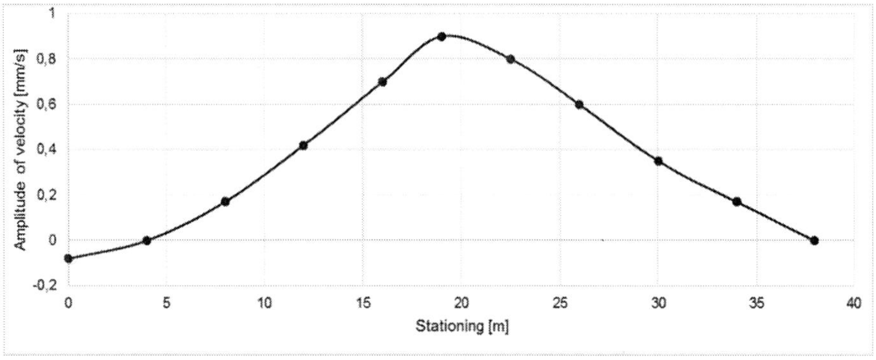

Fig. 4. Mode shapes during excitation in the vertical direction in the second position, frequency of forced vibration 4.98 Hz.

DOI: 10.1051/matecconf/201710700002

Table 1. Measured eigenfrequencies for vertical (V) and horizontal (H) direction.

Mode shape	Frequency [Hz]		
	Exciter B1 (V)	Exciter B1 (H)	Exciter B2
1	4.63	5.00	(4.63)
2	(4.98)	6.73	4.98
3	(7.03)	7.40	(7.03)
4	-	7.76	8.03
5	(9.08)	-	9.08
6	(10.30)	-	10.30
7	(12.20)	-	(12.20)
8	(13.40)	-	13.40
9	(16.60)	-	(16.60)

3 Numerical modeling

A model of a three-hinged arch bridge KL6 was created in Scia Engineer 15 as a slab-wall construction with serrated top hinge and sloping base hinges. The modeled shape of the bridge was based on the measured geometry and historical documents.

Stiffness of the model was adjusted for the deformation to match the deflection of the structure determined in a static load test (Fig. 5). Maximum deflection of the structure during the static load test was measured in a top hinge and had a value of 0.76 mm.

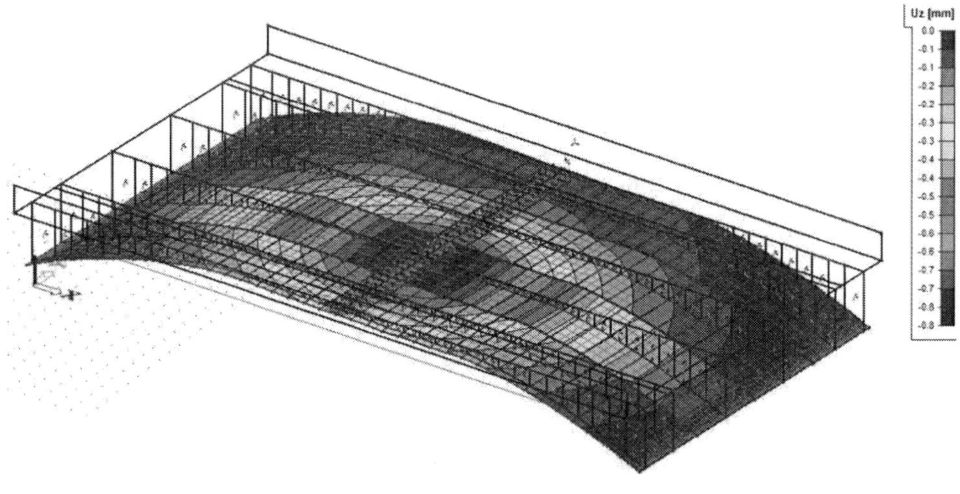

Fig. 5. Displacement in the vertical direction with modified stiffness.

Material characteristics of the concrete arches were taken from the diagnostics. Classes C16/20 and C20/25 of concrete were used. Material properties for embankment, roadway and wall bearing sidewalk bracket were altered to obtain adequate density and stiffness.

The resulting modulus of elasticity: E_{embank} = 7.5 GPa, $E_{roadway}$ = 10 GPa E_{wall} = 27 GPa. Rigidity of the central hinge was set at 150 MNm /rad/m.

Eigenfrequencies and mode shapes were calculated using the modal analysis (Fig. 6, 7).

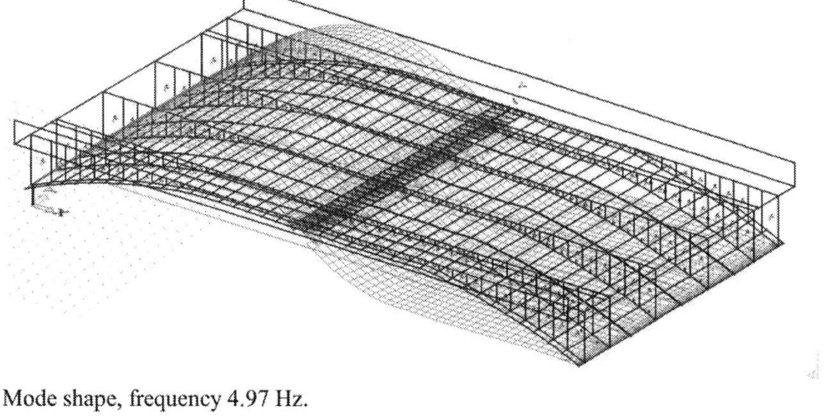

Fig. 6. Mode shape, frequency 4.97 Hz.

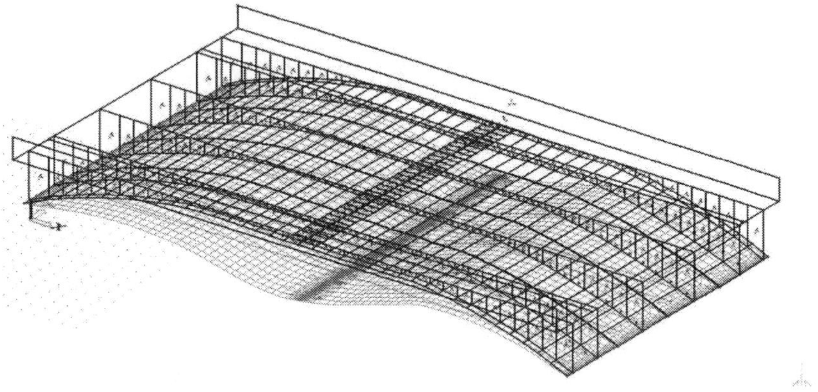

Fig. 7. Mode shape, frequency 5.78 Hz.

4 Conclusion

Static and dynamic load tests showed that the structure of the inundation bridge in its current state can handle the required load transfer induced by traffic.

The stiffness of the numerical model was adjusted according to results from the static load test. After the adjustment of the model, a relatively good agreement of the results was found for the first and fourth eigenfrequency. The differences between the other eigenfrequencies and mode shapes was most likely caused by the degradation of concrete at some parts of arches and the due to discrepancies between the modeled and real nature of the connection between arches. In the SCIA model the connection between arches was provided only by top slab of roadway. The model depicts an arch in perfect condition and made of a homogeneous concrete, whereas the real structure is at least partly degraded. Comparison between calculation and measurement with mode shape assignment is shown in Table 2.

Table 2. Mode shapes and eigenfrequencies assignment.

Mode shape	Dynamic calculation [Hz]	Modal analysis [Hz]
1	4.97	4.63 (B1, V)
2	5.78	4.98 (B2)
3	6.29	5.00 (B1, H)
4	6.71	6.73 (B1, H)
5	8.72	8.03 (B2)
6	11.39	9.08 (B2)

This project is supported by grant GAČR 15-05791S

References

1. Z. Bittnar, J. Šejnoha,: *Numerical Methods in Sructural Mechanics*, ASCE Press, (1996)
2. Dynamics of Bridges, Vol. 5, Proceedings of the 28th IMAC, A Conference on Structural Dynamics (2010)
3. Czech standard ČSN 73 6209, *Load testing of bridges* (1996)
4. Czech standard ČSN 73 6201 – *Desing of bridge structures*
5. Czech standard ČSN 73 6203 – *Bridge loads*
6. Czech standard ČSN EN 1991-2 – *Traffic loads on bridges*"
7. Czech standard ČSN EN 1992-2 – *Concrete bridges - Design and detailing rules*
8. P. Tej, P. Kněž, J. Kolísko, P. Bouška: *Libeň Bridge, Prague 7 A 8. innundation bridge X-656 - arch KL 6 and subsequent frame constructions,* Expert report (only in Czech), Klokner institute, (2016)
9. P. Tej, J. Kolísko, P. Bouška, V. Vacek: *Libeň bridge, Prague 7 A 8, N. A. 999 984. Analysis and assessment of the current technical condition of motifs and possibilities of repairs or construction of a new bridge based on submitted diagnostic examinations and project documentation*, Expert report (only in Czech), Klokner institute, (2015)
10. M. Krejcar, P. Chlopčíková, L. Záleský: *Static load test, report* (only in Czech), Inset, (2016)
11. M. Krejcar, J. Machač, J. Formánek: *Dynamic load test of bridge, report* (only in Czech), Inset, (2016)

 DOI: 10.1051/matecconf/201710700003

Assessment of structural condition of steel bridge in Brandýs nad Orlicí

Jan Mourek[1,*], *Petr* Tej[1], *Jaromir* Kral[1] and *Jiří* Kolísko[1]

[1]Klokner Institute, Department of Mechanics, 16608 Šolínova 7 Prague 6, Czech Republic

Abstract. This paper deals with structural condition assessment of historical steel through arch bridge in Brandýs nad Orlicí. Structure was dynamically tested to obtain mode shapes and eigenfrequencies on non-loaded structure and to estimate logarithmic damping decrement after traffic loading. Material properties were verified. Calculation model was created to validate its statical and dynamical behaviour. Article focuses on comparison of calculated results and properties of real structure upon excitation.

1 Description of a bridge

Bridge was built (according to bridge control protocol from 2014 [1]) between 1905 and 1929. Superstructure consists of single span through arch bridge made out of riveted steel sections. Compressed upper part of the arch is strutted using diagonal struts in outer profile of the cross section. Base of an arch is connected by steel tie. Original deck was rebuilt in 2014 using modern rolled "I" beams with steel ribbed plate. Structure is stiffened by longitudinal and transversal diaphragms. Main arches with ties are simply supported using cast iron bearings. Longitudinal deck beams are resting on abutments, which are made out of stone blocks.

Fig. 1. Tested bridge over Orlice river.

* Corresponding author: mourekj@gmail.com

1.1 Description of bridge surroundings

Bridge is located in active industrial area with nearby railroad and a weir. These surroundings are affecting the bridge itself and the dynamic response of the structure in idle non-loaded state is fairly measurable, while train passage had a significant effect on a structure.

2 Dynamic load testing

Structure was tested in two states [2]. Dynamic response was recorded in idle non-loaded state, where the dynamic response of the structure is caused by natural and technical seismicity without the need of using vibration exciter. For this state the operational modal analysis was performed. In second state the bridge was opened for traffic and dynamic response due vehicular and pedestrian loading was recorded to estimate logarithmic damping decrement.

Fig. 2. Example of sensor assessment during modal analysis.

Dynamic behaviour was measured using 8 piezoelectric acceleration sensors in vertical and horizontal direction. Sensor's placement is shown in Fig. 3.

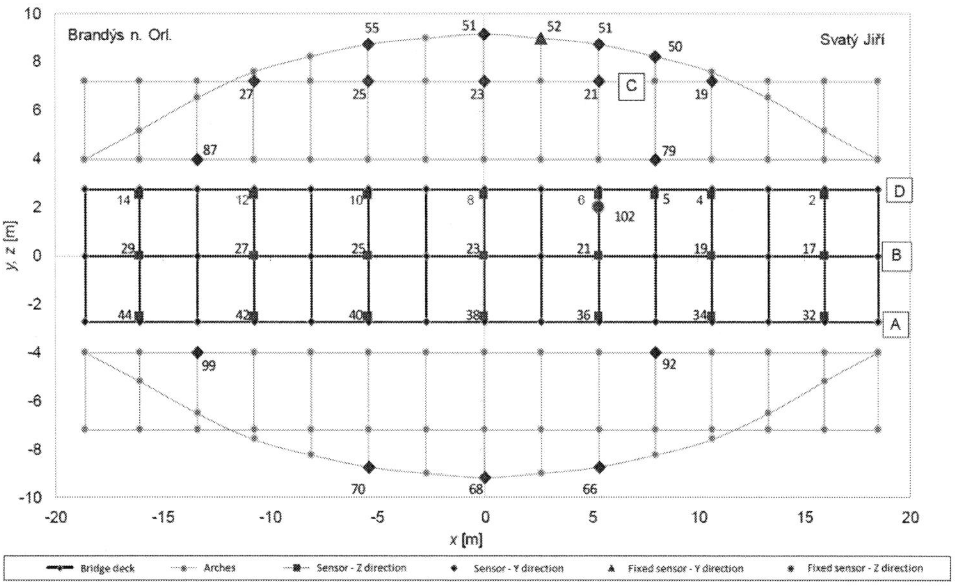

Fig. 3. Sensor's placement along the structure.

2.1 Results

In Table 1, the results of operational modal analysis are shown. Certain doubts exist for the values in brackets. These might be caused by chosen method of measurement or the nonstationarity of external conditions, or overall condition of a bridge and hidden failures. These doubts were aroused by anomalies in corresponding mode shapes.

Table 1. Measured eigenfrequencies.

Mode shape	Frequency [Hz]	Damping [%]
1	5.31	0.90
2	5.99	0.52
3	6.43	0.61
4	(7.57)	0.49
5	7.69	0.38
6	(8.11)	0.10
7	(8.25)	0.50
8	8.43	0.79
9	8.80	0.67
10	11.32	0.16
11	11.77	0.29

Maximal peaks of acceleration in vertical direction recorded during vehicular loading were ±0.4 m/s^2 for cars and ±1.0 m/s^2 for bus. Acceleration in horizontal direction was recorded as ±0.6 m/s^2 for car and ±1.5 m/s^2 for bus. Logarithmic damping decrement was calculated as $\delta = 0.05$.

2.1.1 Idle non-loaded state

Bridge permanently vibrates with low acceleration amplitudes. Cause of this behaviour is technical and natural seismicity of the surroundings. Eigenfrequnecies can be found in acceleration spectra (Fig. 4) as sharp peaks. These frequencies are systematically lower than frequencies calculated in FEA model. Both results are various around 8.0 Hz area, where significant peaks are missing in measurement. Vertical lines in upper part are representing calculated eigenfrequencies.

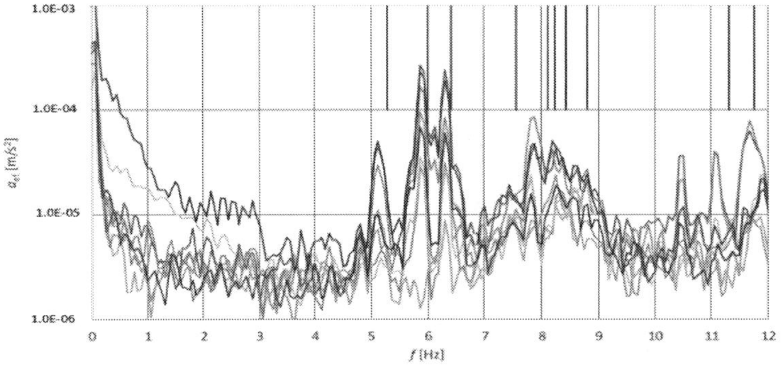

Fig. 4. Acceleration spectra in idle state of the structure.

3 Result comparison

In Table 2, measured mode shapes are assigned to calculated mode shapes in FEA model [3]. Assignment was based on characteristic common signs of each mode shape.

As can be seen in Table 2 the lowest mode shape wasn't found during modal analysis. It is possible that the lowest calculated self-frequency exists but could not be measured because the excitation in this part of the spectrum was small and the response measured by the acceleration sensor was lost in the noise of the used electronics. Most common mode shape occurred at 8.80 Hz, but commonalities may only be found for vertical direction. Horizontal mode shape is different for this frequency. First calculated eigenmode is a simple oscillation of the structure in vertical direction. Also mode shapes No. 5 and 8 could not be matched to any measured mode shapes.

Table 2. Mode shape assignment.

Mode shape	Dynamic calculation [Hz]	Modal analysis [Hz]
1	2.84	-
2	4.63	5.31
3	5.29	5.99
4	6.77	6.43
5	7.04	-
6	7.64	7.68
7	7.67	8.43
8	8.64	-
9	-	(7.57)
10	-	(8.11)
11	-	(8.25)
12	-	11.32
13	-	11.77

4 Conclusion

Operational modal analysis performed on the bridge has shown, that 5 of 11 mode shapes are comparable to calculated mode shapes using FEA model of the bridge. Three of the measured mode shapes are incomparable. While the number of irregularities is fairly high it is convenient to find the cause of such behaviour. Two mode shapes are indicating on non-specified anomalies around one of its abutment. Causes of these anomalies might be hidden failures of the structure. While no information explaining these irregularities can be found the bridge should stay closely monitored and should be further analysed to found the answer on deviations in dynamic behaviour. Standardized tolerance [4] are slightly exceeded if we take 2.84 Hz as the lowest eigenfrequency.

This project is supported by grant GAČR 17-22796S

References

1. L. Bystřický, Bridge control protocol (2014)
2. J. Král, Dynamic load testing protocol (2017)
3. L. Černík, Modal analysis of a bridge (2016)
4. Czech standard ČSN 73 6209, *Load testing of bridges* (1996)

Probabilistic and sensitivity analysis of Botlek Bridge structures

Juraj Králik[1,*] and *Peter* Rosko[2]

[1]STU Bratislava, Faculty of Civil Engineering, Radlinského 11, 813 68 Bratislava, Slovakia
[2]Institute for Building Construction and Technology Centre of Mechanics and Structures, TU Vienna, Austria

Abstract. This paper deals with the probabilistic and sensitivity analysis of the largest movable lift bridge of the world. The bridge system consists of six reinforced concrete pylons and two steel decks 4000 tons weight each connected through ropes with counterweights. The paper focuses the probabilistic and sensitivity analysis as the base of dynamic study in design process of the bridge. The results had a high importance for practical application and design of the bridge. The model and resistance uncertainties were taken into account in LHS simulation method.

1 Introduction

During the structural design process, an engineer has to consider problems of the safety, reliability and durability of a single structural element as well as the entire structure from the point of view of its planned life cycle. Randomness in the loading and the environmental effects, the variability of the material and geometric characteristics of structures and many other "uncertainties" affecting errors in the computing model lead to a situation where the actual behavior of a structure is different from the modeled one [1 - 9]. Recent advances and the general accessibility of information technologies and computing techniques give rise to assumptions concerning the wider use of the probabilistic assessment of the reliability of structures through the use of simulation methods [5].

2 Simulation Methods

In the case of simulation methods the failure probability is calculated from the evaluation of the statistical parameters and theoretical model of the probability distribution of the reliability function $Z = g(\boldsymbol{X})$. The failure probability is defined as the best estimation on the base of numerical simulations in the form

$$p_f = \frac{1}{N}\sum_{i=1}^{N} I\left[g(X_i) \le 0\right] \le p_d \tag{1}$$

* Corresponding author: juraj.kralik@stuba.sk

where N in the number of simulations, $g(.)$ is the failure function, $I[.]$ is the function with value 1 if the condition in the square bracket is fulfilled, otherwise is equal to 0.

Variation of the failure function can be defined by Melchers [10] in the form

$$s_{p_f}^2 = \frac{1}{(N-1)}\left\{\frac{1}{N}\left[\sum_{i=1}^{N} I^2\left[g\left(X_i\right)\leq 0\right]\right] - \left[\frac{1}{N}\sum_{i=1}^{N} I\left[g\left(X_i\right)\leq 0\right]\right]^2\right\} \qquad (2)$$

The various forms of analyses (statistical analysis, sensitivity analysis, probabilistic analysis) can be performed. Most of these methods are based on the integration of Monte Carlo (*MC*) simulations. Three categories of methods have been presently realized - Direct methods, approximation methods or modified methods.

The most effective method is the modified *LHS* method based on the same number of simulations of the safety function as in the *MC* method [5]. Using the *LHS* strategy, we get values like the reliability reserve parameter – the mean value of safety function.

The modified LHS method is based on the simulations of the function $g(X)$ so thus *MC* method, but the definition domain of the distribution function $\Phi(X)$ is divided to N intervals with the identical probability $1/N$. Characteristic values of the simulations as calculated randomly on the base of K permutation integer number 1, 2,...N. The reliability function $g(X)$ must be determined from the N simulation. The interval of the random simple is used one time only.

This method gives us the best estimation of the statistical parameters of the structure in the comparison with *MC* method.

3 Botlek Bridge structure

The new Botlek Bridge recently built in Rotterdam (Holland) with two lifting spans of approximately 100 m and pylons reaching over 60 m above water level is the largest vertical - lift bridge in the world.

Fig. 1. Botlek Bridge.

The value of this project was approximate equal to 1.5 billion EUR. The design analysis of sophisticated structures of high importance require more detailed and expensive analyses, provide an opportunity for theory development, comparison of different analysis procedures and gaining experiences [11]. Because of dynamic loading (traffic loads in "down position" of decks, movement of decks, wind loads and seismic loads), the movable bridge model for deterministic and consistently the stochastic analysis was developed. The paper focuses the probabilistic analysis as the base of dynamic study in design process of the bridge.

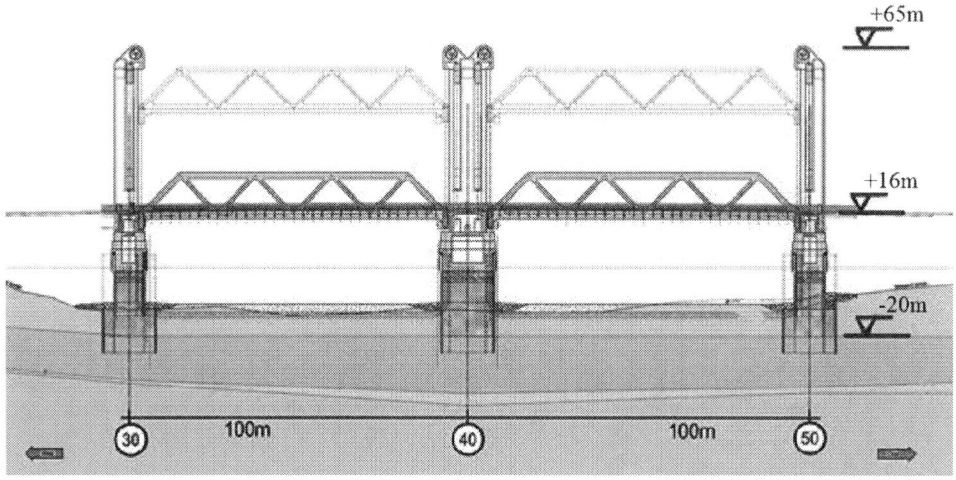

Fig. 2. Botlek Bridge. Section plan.

Two steel decks, 4000 tons weight each, are independently vertically movable on six concrete towers – the lifting height is 31 m. The length of steel decks is 96 m, the width 49 m (each deck has approximately the area of soccer stadium). The lifting system is fully balanced. The four concrete counterweights are guided on pylon towers and are connected through ropes with decks at their ends. The three main bridge piers (from left to right in Figure 1: Pier 30, Pier 40 and Pier 50) are founded on rigid concrete blocks with footing dimensions of 15 × 60 m, at 8 m below river bed at the top of the first dense (Pleistocene) sand layer. For the geotechnical design the foundation was essentially treated as a shallow foundation [12]. At a depth of approximately 16 m below the foundation footing a relatively soft clay layer is present with varying thicknesses between 0 and 4 m. This stratum complicated the design, especially with respect to the settlement behavior which has a major impact on the performance of the total bridge and influences the different design disciplines (e.g. mechanical, electrical and structural). This article considers the probabilistic analysis with respect to the deformation behavior of the subsoil which was undertaken as part of the foundation design.

4 Soil-structure interaction

For the determination of the soil parameters an extensive soil investigation has been performed [12]. A relatively dense grid of Cone Penetration Tests (CPT's) with a mutual distance of about 15 m was executed to a depth of about 3 times the foundation width. In addition, a number of boreholes were drilled and undisturbed samples were taken at regular intervals for geotechnical laboratory tests by means of light percussion drilling in combination with thin-walled samplers. From the CPT's and borehole logs the soil stratigraphy is determined, see Table 1. The foundation is modelled as an infinitely stiff foundation block, supported by linear elastic (stochastic) springs at a spacing of about 3 m. Since the foundation consists of a massive concrete block with a thickness of about 20 m the assumption of a stiff foundation is considered reasonable.

In the case of the dynamic loads the soil material characteristics depend on compression wave (P - wave) and shear wave (S - wave) velocities in the soil

$$v_P = \sqrt{\frac{E_d\left(1-\nu\right)}{\left(1+\nu\right)\left(1-2\nu\right)\rho}} \cong \sqrt{\frac{E_d}{\rho}}, \qquad v_S = \sqrt{\frac{E_d}{2\left(1+\nu\right)\rho}} = \sqrt{\frac{G_d}{\rho}}, \tag{3}$$

where v_p is the compression wave velocity, v_s is the shear wave velocity, E_d is the dynamic Young modulus, G_d is the dynamic shear modulus, ν is the Poissonov number, ρ is the soil density. The typical ranges of P- and S-wave velocities for different subsoil conditions were published in [13] (see Fig. 3).

Table 1. General soil stratigraphy [12].

Top of layer [m NAP]	**Soil description**	**Soil layer**
From -7 to -14	SAND, clayey	cover layer
From -14 to -20	SAND, (medium) dense	1st sand layer
From -33 to -39	CLAY, stiff	deep clay layer
From -34 to -42	SAND, (medium) dense	2nd sand layer
At -60	Max. investigation depth	

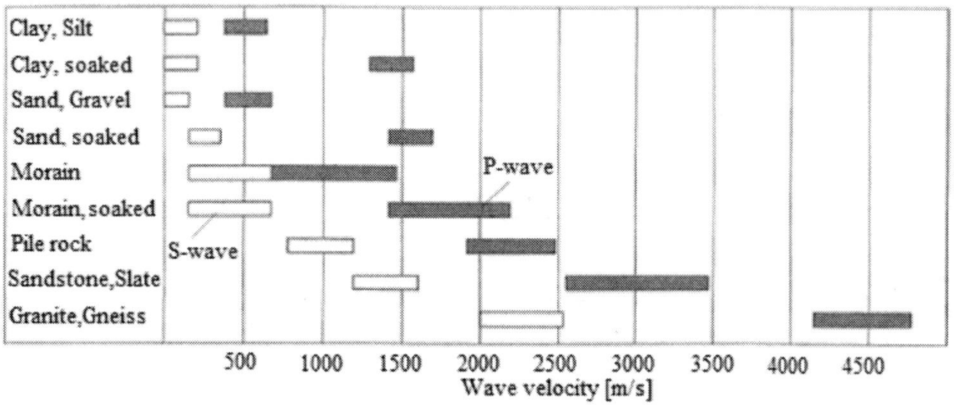

Fig. 3. Typical ranges of P- and S-wave velocities for different subsoil conditions [13].

The consideration of SSI effects is very important. The influences of stiffness and damping characteristics of the soil on the structure are not negligible. There are many ways of mathematical representation of the soil. The soil can be represented by a set of equivalent springs or by continuum. For FE modelling, it is well known that a narrow domain with fixed boundaries is not likely to represent realistic soil behavior, whereas a very large domain would result in a too complicated problem. It is, therefore, necessary to find an optimum value that reflects the realistic behavior of soil without significant loss in accuracy. The spring stiffness of the soil under the concrete block foundation can be determined in accordance with the [14] as follows

$$K_v = \frac{4G_d r_o}{1-\nu}, \quad K_h = \frac{32(1-\nu)G_d r_o}{7-8\nu}, \quad K_\psi = \frac{8G_d r_o^3}{3(1-\nu)}, \quad K_\theta = \frac{16G_d r_o^3}{3} \tag{4}$$

$$r_o = \sqrt{W\frac{L}{\pi}} \text{ for translation, } r_o = \left(\sqrt{W\frac{L^3}{3.\pi}}\right)^{1/2} \text{ for rotation, } r_o = \left(\sqrt{WL\frac{W^2+L^2}{6.\pi}}\right)^{1/2} \text{ for torsion,}$$

where W is the width of foundation, L – length of foundation.

5 Calculation model

The finite element software ANSYS [15] have been used for all models and analyses. The model of the bridge was for deterministic and consistently for probabilistic analysis prepared.

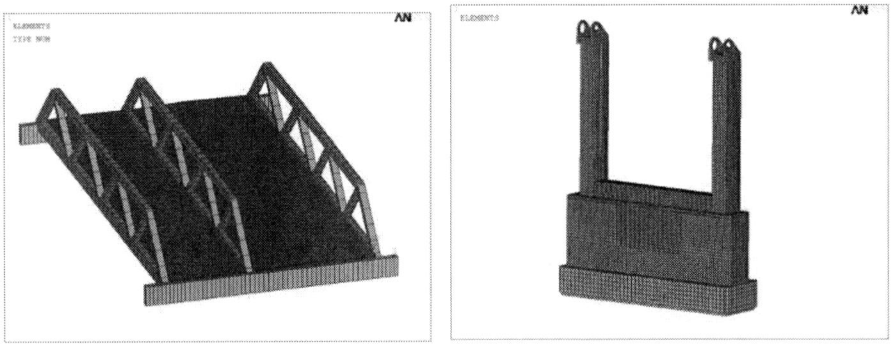

Fig. 4. FEM model of the bridge deck and pylon.

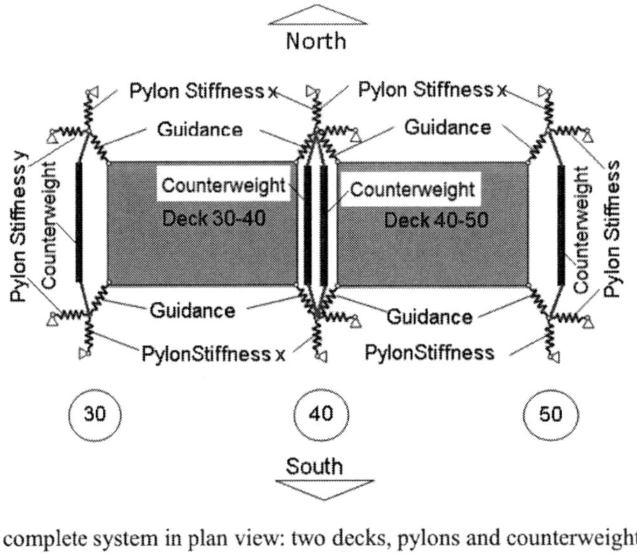

Fig. 5. Scheme of complete system in plan view: two decks, pylons and counterweights.

The elements SOLID186, SOLID187 and SURFACE154 were used in this detailed FEM model. The preliminary analyses have shown, that the vertical displacements of the tops of the pylons are neglectable in comparison to the horizontal displacements. For that reason was the simplification of the model in 2D possible. The model consist of two approximately rigid plates (decks) with weight G and substituted thickness H, models of pylons - springs with stiffness K. SHELL63 and LINK8 finite elements were applied.

Let consider situation when the both decks are in upper position. The mass more than 8000 tons is in the height of more than 30 m supported by six pylons. The system is sensitive to horizontal loading e.g. wind loading, seismic loading. The natural frequencies and modes of the bridge system influence the dynamic behavior of the bridge system. The dynamic component of the loading depends on the base of Eurocode [16, 17] and on natural frequency of the structure and can be calculated according to Eurocode procedure [18].

6 Probability and sensitivity analysis

The deterministic eigenvalue analysis tested the model used for the probabilistic analysis [5]. The uncertain parameters K, G and H were defined as random input variables (Table 2.) as follows:

$$K = K_d.\Phi(\boldsymbol{k_{var}}),\ G = G_k.\Phi(\boldsymbol{g_{var}}),\ H = H_k.\Phi(\boldsymbol{h_{var}}), \tag{5}$$

where K_d, G_k and H_k are the characteristic values of the input data, $\boldsymbol{k_{var}}$, $\boldsymbol{g_{var}}$, $\boldsymbol{h_{var}}$ are the variable parameters and $\Phi(.)$ is the distribution function of the variables (Table 2).

Special macro has been written in APDL ANSYS [15] using LHS simulation method to the probabilistic analysis of this calculation model. The uncertainties are coming from the standard requirements and the other publications [3, 5, 7, 18 and 19]. The mean values and standard deviations were defined in accordance of the experimental test and design values of the material properties and the action effects (see Table 2). Based on the results from the simulated modal analysis of the Boltek bridge and the variability of the input parameters 10^4 LHS simulations were performed in the software ANSYS. The sensitivity analyses give us the informations about the influences of the variable properties of the input data to the output data (Fig. 6). These analyses are based on the correlations matrixes.

Table 2. Variability of input parameters [19].

Quantity	**Charact. value**	**Variable**	**Histog. type**	**Mean** μ	**Deviat.** σ**[%]**	**Minim. value**	**Maxim. value**
Stiffness	K_d	k_{var}	N	1	20	0.451	1.490
Plate thickness	H_k	h_{var}	N	1	5	0.808	1.195
Dead load	G_k	g_{var}	N	1	5	0.808	1.195
Action	E_k	e_{var}	N	1	5	0.813	1.190
Resistance	R_k	r_{var}	N	1	5	0.812	1.201

We used four models considering the various numbers of the input data (see Table 3). The uncertainties of the pylon stiffness considering the soil-stiffness interaction to the modal characteristic of the Boltek Bridge were investigated. In the case of the model 1 two independed characteristics were considered. In addition 18 independed input characteristics were considered in the model 4 (see Table 3). The extrem values of the interval of the mode distribution functions are in the model 2 with 3 independed input characteristics (pylon stiffnes *K*, masses *G* and model uncertainties action *E*). The interval of the response distribution function diminishes with the increase number of the input variables.

Table 3. Analysis of the influence of the input data uncertainties to the Bridge frequencies.

Number of		First mode			Second mode			Third mode		
Model	**Variab. data**	**Min**	**Mean**	**Max**	**Min**	**Mean**	**Max**	**Min**	**Mean**	**Max**
1	2	0.809	1.202	1.676	0.964	1.432	1.996	1.353	2.010	2.802
2	3	0.738	1.202	1.711	0.880	1.432	2.038	1.235	2.010	2.861
3	14	0.993	1.108	1.227	1.287	1.428	1.589	1.612	1.850	2.101
4	18	0.898	1.111	1.367	1.155	1.429	1.718	1.613	2.016	2.594

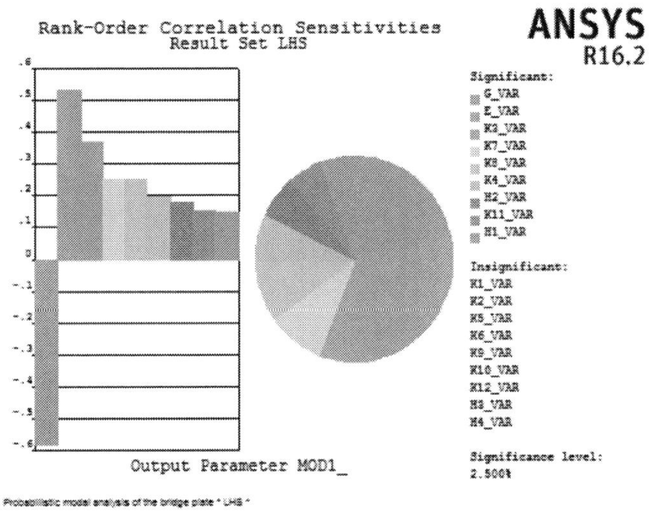

Fig. 6. Sensitivity analysis of the influences of the number of the variable input data to the first mode- Model 4 with 18 variable parameters .

7 Conclusion

The paper presents the probability and sensitivity analysis results of the movable bridge. For the design of the new Botlek Bridge soil deformations can potentially have a major effect on one of the most critical design requirements, which is a limited rotation of the large foundation footing. Application of a simplified stochastic subsoil model enabled a quantitative risk analysis in order to deal with the uncertainties described in this paper. The decks were considered in top position. The stiffness of the pylons and weight of decks were as random input variables defined. The first, second and the third natural frequencies were as output parameters defined. The results were useful in dynamic analysis of wind impact.

The project was performed with the financial support of the Grant Agency of the Slovak Republic (VEGA 1/0265/16).

References

1. C. Bucher, *Computational Analysis of Randomness in Structural Mechanics* (Taylor & Francis Group, London, UK, 2012)
2. A. Haldar and S. Mahadevan, *Probability, Reliability and Statistical Methods in Eng. Des.* (J.W. & S., New York, 2000)
3. Holický M. et al. *Basics for Assessment of Existing Structures, Base of Reliability Theory and Risk Evaluation* (Ed. K.I. CTU Prague, 2013)
4. Z. Kala, Applications of Global Sensitivity Analysis Methods in Civil Engineering, In Proc. *ESREL 2015,* Zürich (Switzerland), pp. 2541-2545 (2015)
5. J. Králik, *Reliability Analysis of Structures Using Stochastic Finite Element Method*, (Pub. by STU Bratislava, 143pp, 2009)
6. J. Králik, J. Králik, jr. Deterministic and Probabilistic Analysis of NPP Communication Bridge Resistance due to Extreme Loads, *Trans. VŠB TU Ostrava*, **14**, 54-63, DE GRUYTER OPEN (2014)
7. M. Krejsa, J. Králik, Probabilistic Computational Methods in Structural Failure Analysis, *Jour. of Mult. Model.* **6**, 2, Imperial College Press (2015)
8. V. Salajka, J. Kala, Z. Čada, Modification of response spectra by probabilistic approach. *ESREL 2014* (2015)
9. J. Melcer, Dynamic analysis of the new bridge across the Danube in Bratislava. *Proc. of the 4th Int. Conf. on Bridges acr. the Danube 2001*, Bratislava, 193-198 (2001)
10. R. E. Melchers, *Structural Reliability: Analysis and Prediction* (J.W&S., Chichester, U.K. 1999)
11. Dynamic Analysis, *Nieuwe Beweegbare Botlekbrug*, Tech. Rep., VCE Consult ZT-GmbH, pp. 1-219, August (2012)
12. J. A. Jacobse et al. Probabilistic Settlement Analysis for the Botlek Lifting Bridge Design, *Proc. of the 18th Int. Conf. on Soil Mech. and Geot. Eng.* Paris (2013)
13. H. Bachman et al., *Vibration Problems in Structures. Practical Guidelines* (Birkhäuser Verlag Basel, Berlin, 1995)
14. G. B. Baecher, and T. S. Ingra, Stochastic FEM in Settlement Prediction, J. of the Geot. Eng. Div. ASCE, **107**, GT4, 451-463 (1981)
15. K. G. Bhatia , *Foundation for Industrial Machines, Handbook for Practical Engineers*, (D-CAD Publisher, New Delhi, 2009)
16. P. Kohnke, *ANSYS, Theory*, SAS IP Inc. Canonsburg (2008)
17. Eurocode 1990, *Eurocode – Basis of structural design* (CEN/TC250 Bruxelles 2002)
18. HANBOOK 5, *Implem. of Euroc. Reliab. Back. Design of Buildings for the Fire Situation.* L. Da V. P. Project CZ/02/B/F/PP-134007. Prague (2005)
19. JCSS 2011. *JCSS Probabilistic Model Code.* Zurich: Joint Committee on Structural Safety. www.jcss.byg.dtu.dk (2011)

Modal parameter identification of all-GFRP composite cable-stayed footbridge in Denmark

Piotr Górski[1,*], *Beata* Stankiewicz[1] and *Marcin* Tatara[1]

[1]Opole University of Technology, Faculty of Civil Engineering and Architecture, Department of Roads and Bridges, Katowicka 48, Opole 45-061, Poland

Abstract. The aim of this paper is to investigate of dynamic characteristics of cable-stayed Fiberline Bridge in Kolding, Denmark, made entirely of Glass Fiber Reinforced Polymer (GFRP) composite. During examination based on in situ free-decay measurements and using accelerometers under human jumping the primary five natural frequencies, corresponding mode shapes and damping ratios of the footbridge were identified. The Peak Picking (PP) and Frequency Domain Decomposition (FDD) approaches were applied to identify the natural frequencies and mode shapes. The corresponding damping ratios were extracted by a linear regression on the extremes of modal decays. The estimated damping ratios were compared with published data for selected footbridges made of various conventional materials. The obtained experimental results provide a relevant data regarding the dynamic response prediction or structural health monitoring of all-GFRP composite footbridges.

1 Introduction

Nowadays, Glass Fiber Reinforced Polymer (GFRP) composite material is one of the most innovative material being slowly introduced in a bridge structures. The decisive meaning has several key advantages of the material such as relatively high strength, low self-weight, ease of installation, corrosion resistance, electromagnetic transparency and ease of creation an innovative shapes of pultruded profiles as well as whole structure. Besides the advantages mentioned above, GFRP has also some limitations like a relatively low Young's modulus, low alkaline resistance and low long-term strength due to stress rupture [1-2]. For this reason, the examinations of mechanical properties and durability of GFRP material are still being performed, e.g. [3], in order to improve this technology as well as to increase the use of this material in civil engineering structures.

Currently, the GFRP profiles are willingly combined with some conventional materials, like steel or concrete, in a modern hybrid footbridges and are usually used as a bridge deck system. Some examples of such structures are given in the papers [4-5]. There are only few all-composite bridges reported in the literature. Among them the following footbridges can be mentioned: the cable-stayed Scripps Bridge in La Jolla, USA [6], Aberfeldy Bridge in Scotland, UK [7], or Fiberline Bridge in Kolding, Denmark (considered in this paper), as

* Corresponding author: p.gorski@po.opole.pl

well as the Pontresina Bridge in Switzerland [8] with a truss structural system or the girders footbridge constructed in Tainan, Taiwan [9].

The knowledge of an actual value of modal characteristics of the structure is very useful for the procedure of calculation model verification, calculation model updating, dynamic response prediction, assessment of the technical state or structural health monitoring of the structures. It is very important issue, especially, in the case of all-GFRP footbridges for which the lack of specific design codes and insufficient number of studies in existing literature in this matter (e.g. [10-11]) can be observed. At a design stage, the natural frequencies and corresponding mode shapes are usually obtained from a solution of a free vibration response for an idealized calculation model. Unfortunately, the damping ratio, which is one of the most important and uncertain parameter significantly reducing the reliability of dynamic response prediction, could not be estimated basing on theoretical method. Thus, it is very difficult to predict damping parameters with a great certainty. The most appropriate way to investigate the actual dynamic characteristics of existing structures is to analyze their dynamic response recorded in a full scale. Nowadays, it is very advisable and valuable to conduct some studies in order to recognize the fundamental dynamic characteristics of all-GFRP footbridges.

The paper concerns with the investigation of dynamic characteristics of cable-stayed Fiberline Bridge in Kolding, Denmark, made entirely of GFRP composite material. During investigation based on in situ free-decay measurements, using accelerometers under human jumping, the primary five natural frequencies, corresponding mode shapes and damping ratios of the footbridge were identified. The Peak Picking (PP) and Frequency Domain Decomposition (FDD) approaches were applied to identify the natural frequencies and mode shapes. The corresponding damping ratios were extracted by a linear regression on the extremes of modal decays. The estimated damping ratios were compared with data published in the literature for selected footbridges made of various conventional materials. The obtained experimental results provide relevant data for the dynamic response prediction and structural health monitoring of all-GFRP composite footbridges.

2 Description of the Fiberline Bridge in Kolding

The cable-stayed Fiberline Bridge, located in Kolding city in Denmark, was constructed in 1997 using 12 different pultruded profiles all made of GFRP material. The footbridge is crossing an overhead main railway line near a salt water fjord. In spite of these disadvantageous conditions, the expected life time of the structure is minimum 100 years. The general view of the footbridge and its longitudinal section with the most important dimensions are shown in Fig. 1.

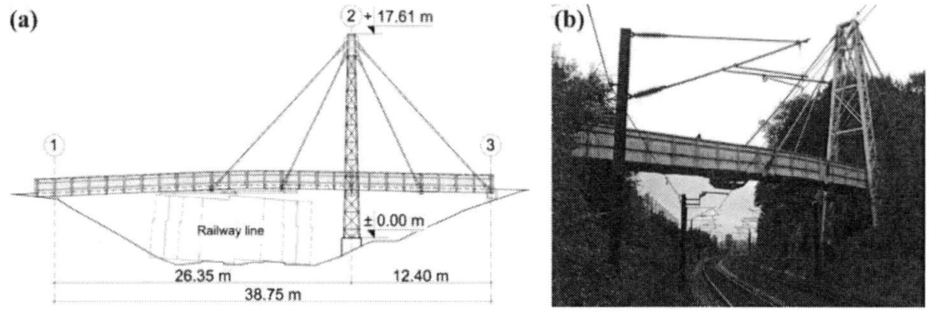

Fig. 1. Fiberline Bridge in Kolding: (a) longitudinal section and (b) general view.

The 40.3 m long and 3.21 m width footbridge comprises two continuous spans with lengths of 27.9 and 12.4 m, supported by a single A-shaped 18.5 m tall pylon. Four pairs of cables with a square cross-section and a length of about 17.8, 17.6, 13.6 and 13.4 m are secured to the top of the pylon. The structural system of the deck comprises two I-shaped pultruded profiles with a height of 1.4 m used as main girders. Total weight of the bridge is just 12 tonnes. The additional information about the footbridge is given in the Fiberline Composites Website [12].

3 Experimental set-up and measurement results

In order to identify the modal parameter of the considered footbridge a free structural response for excitation under human jumping was measured using accelerometers. Since no significant lateral vibrations of the bridge were exhibited only the vertical acceleration component was measured. Due to the necessity of the natural modes determination, two accelerometers were used simultaneously in the field test. In total 28 measurement points were chosen directly on the upper edge of two girders, namely girders denoted as "A" and "B", mainly within 3.1 m distance along the longitudinal axis of the girders. The measurement points are shown in Fig. 2. During the field measurement one reference accelerometer was fixed at the reference point, while the second rover accelerometer was moved step by step from a measurement points No. 1 to 27. With every rearrangement of the rover accelerometer, i.e. at each measurement point, the series of two human jumps were performed at the deck center near the reference point and the free vibration responses of the footbridge were recorded during at least 25 s with the sampling rate of 200 Hz. The reference point was chosen at a point where it was assumed that the most important modes have the most significant modal component. Fig. 3(a) depicts vertical accelerations of the footbridge girder "A" at the reference point and measurement point No. 23 surveyed due to two human jumps.

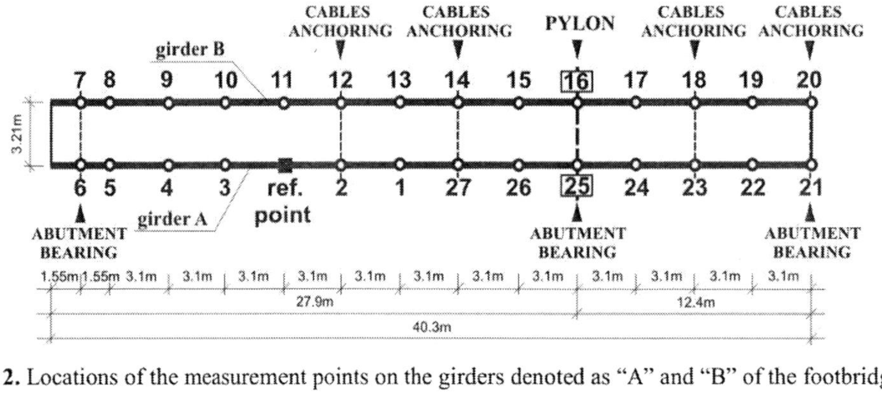

Fig. 2. Locations of the measurement points on the girders denoted as "A" and "B" of the footbridge.

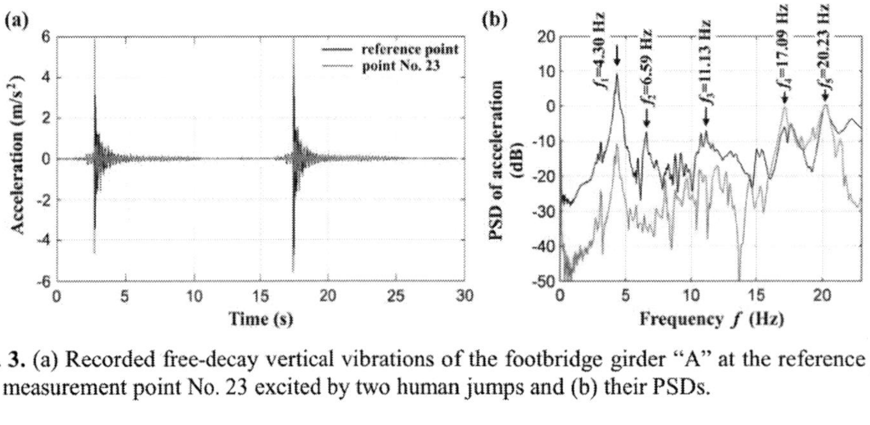

Fig. 3. (a) Recorded free-decay vertical vibrations of the footbridge girder "A" at the reference point and measurement point No. 23 excited by two human jumps and (b) their PSDs.

In the field test two uniaxial high sensitivity accelerometers type PCB 3711E112G, based on the Micro-Electro-Mechanical System (MEMS) technology, were used. The measurement range of the sensors was ±2 g, the frequency range 0–400 Hz, the broadband resolution 0.1 mg rms in 0.5 to 100 Hz bandwidth, and the spectral noise 15 μg/√Hz in 1 to 100 Hz bandwidth. Such sensitivity is adequate for studying behaviour of the footbridge.

All measured free-decays were analysed by the Fast Fourier Transform. The Power Spectral Density (PSD) functions were estimated for the total duration of the time interval of all free vibration responses. The PSDs of measured vertical accelerations of the footbridge girder "A" at the reference point and measurement point No. 23 are presented in Fig. 3(b). These PSDs are dominated by five well-separated peaks with the values above 10 dB, representing the lowest five natural frequencies, i.e. f_1 = 4.30 Hz, f_2 = 6.59 Hz, f_3 = 11.13 Hz, f_4 = 17.09 Hz, and f_5 = 20.23 Hz. In accordance with further analysis performed using the FDD method, presented in Section 4, it was found that these five frequencies correspond with specific natural mode shapes of the considered footbridge.

4 Modal parameter estimation

4.1 Mode shapes identification using FDD technique

The natural mode shapes of the footbridge were evaluated using the FDD method based on the free decays time series measured for all arrangement of two accelerometers at the reference point and measurement points denoted from No. 1 to 27 (see Fig. 2).

The FDD technique was introduced by Brincker in 2000 [13] as one of non-parametric frequency domain signal processing methods used in an output-only modal analysis. The method is appropriate to extract the natural frequencies and corresponding mode shapes using the Singular Value Decomposition (SVD) of the PSD matrices of a structural responses at each discrete frequency calculated with the assumption that the structural damping is light. Such decomposition is performed in order to identify a structure as a single degree of freedom (SDOF) system for each singular value corresponding to the resonant frequency of the structure and corresponding to the singular vector containing the information about the mode shape of the structure.

The first step of FDD is the estimation of a set of the PSD matrices of the structural response at each discrete frequency. All the matrices are Hermitian, i.e. the matrices are symmetric with complex conjugate elements around the diagonal. For example, in case of the considered field measurement of the footbridge response performed simultaneously in two measuring points the PSD matrix $\mathbf{G}_{xx}$ of the response at an i-th discrete frequency f_i is expressed as follows

$$\mathbf{G}_{xx}\left(\mathrm{i}f_i\right)=\begin{bmatrix} PSD_{11}\left(\mathrm{i}f_i\right) & CSD_{12}\left(\mathrm{i}f_i\right) \\ CSD_{21}\left(\mathrm{i}f_i\right) & PSD_{22}\left(\mathrm{i}f_i\right) \end{bmatrix} \tag{1}$$

where $PSD(\mathrm{i}f)$, as the elements on the diagonal, are the magnitudes of the PSD of the structural response at a chosen point of the structure, which for a real stochastic process $x(t)$ are real elements, $CSD(\mathrm{i}f)$, as the elements on the off-diagonal, are the cross PSD of the structural response at two various points of the structure and are the complex conjugate, Hermitian's PSDs, which fulfill equation $CSD_{qr}(\mathrm{i}f) = CSD^*_{rq}(\mathrm{i}f)$, whereas $q \neq r$, the symbol * denotes a complex conjugate value, and $\mathrm{i}=\sqrt{-1}$ is the imaginary unit.

Since the FDD technique is well-known in the existing literature, e.g. [13], the further explanations are omitted. It should be noted that the final form of the PSD matrix of the structural response could be decomposed at each discrete frequency f_i into a set of singular values and corresponding singular vectors using SVD according to the following expression [13]

$$\mathbf{G}_{xx}(\mathrm{i}f_i) = \mathbf{U}_i\, \mathbf{S}_i\, \mathbf{U}^{\mathrm{H}}{}_i \qquad (2)$$

where $\mathbf{S}_i = \mathrm{diag}[s_1(f_i), s_2(f_i),\ldots, s_j(f_i),\ldots, s_n(f_i)]$ is the diagonal matrix, it corresponds to the i-th discrete frequency f_i containing a set of discrete singular values $s_j(f_i)$, $\mathbf{U}_i = [\{u_{i1}\},\{u_{i2}\},\ldots,\{u_{ij}\},\ldots,\{u_{in}\}]$ is the unitary matrix containing column singular vectors $\{u_{ij}\}$ it corresponds to the i-th discrete frequency f_i, $j = 1, 2,\ldots, n$, and n denotes the dimension of $\mathbf{G}_{xx}(\mathrm{i}f_i)$ and is equal to number of measuring points in which the response of the structure was measured.

The decomposition given in Eq. (2) allows the modal parameter identification of a SDOF systems, i.e. the modal frequencies and mode shapes, at each singular value corresponding to the resonant frequency. Each set of the discrete singular values $s_j(f)$, for $j = 1, 2,\ldots, n$, can be presented in the frequency domain as n independent plots. The plot of the first singular values $s_1(f_i)$ is dominated by maximum values which, on the frequency-axis, indicate a set of the modal frequencies of a structure f_k. According to the FDD theory, the first singular vectors $\{u_{i1}\}$ of the unitary matrix $\mathbf{U}_i$ corresponding to the first maximum singular vales $s_1(f_k)$ are the k-th unscaled mode shape vectors corresponding to the f_k, thus

$$\phi_k = \{u_{k1}\}. \qquad (3)$$

The FDD procedure was adopted to write a computer program in Matlab, which was used for the numerical calculations. The PSDs and CSDs of the discrete digital signal, representing the free response of the considered footbridge, were estimated using the modified averaged periodogram method, called the Welch's technique. According to this technique the time series of the signal are split and then windowing procedure are overlapped before averaging them together. This technique minimizes the bias noise caused by the spectral leakage. As an example, the plots of the singular values $s_1(f)$ and $s_2(f)$ of the decomposed PSD matrices $\mathbf{G}_{xx}(\mathrm{i}f)$ of the free vibration responses of the footbridge girder "A" measured at the reference point and measurement point No. 23 are given in Fig. 4. Due to the field test results it was possible to identify first five specific vibration modes of the analysed footbridge. These normalised mode shapes and their theoretical approximation are shown in Fig. 6.

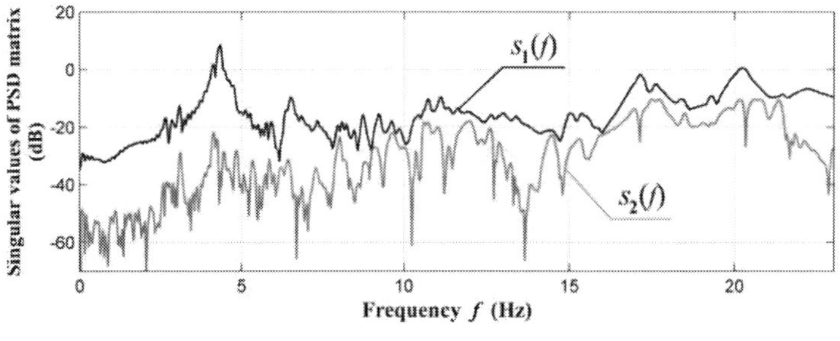

Fig. 4. Singular values of the decomposed PSD matrices of the free vibration responses of the footbridge girder "A" measured at the reference point and measuring point No. 23.

4.2 Damping estimation

The damping ratios of the footbridge were analysed on the basis of the free decays recorded only at the reference point. The analysed free vibration response was decomposed into five modal decays by isolating the contributions of the primary five of the most important modes of the footbridge. For this purpose, the filtering procedure using the eighth-order Type 1 Chebyshev band-pass digital filters with passband ripple 1 dB was applied. The following pass-bands in the filtering procedure were chosen: 3.5–5.0 Hz, 6.0–6.8 Hz, 11.0–11.5 Hz, 16.8–17.3 Hz, and 19.5–20.7 Hz. Fig. 5 (left side) depicts two modal decays provided by the filtering procedure with pass-band 3.5–5.0 Hz and 6.0–6.8 Hz, regarding the natural frequencies $f_1 = 4.30$ Hz and $f_2 = 6.59$ Hz, respectively.

Basing on the filtered data, containing only the contribution of a single mode, the modal damping ratios were estimated by a linear regression on the extremes of these modal decays (using the least square approximates). In this case, each m-th extreme value of the modal decay, denoted as $\hat{x}_m$, was found and the corresponding values A_m were calculated according to the formula

$$A_m = 2 \cdot \ln\left(\frac{\hat{x}_0}{|\hat{x}_m|}\right) \tag{4}$$

where $\hat{x}_0$ is the initial value of the modal decay.

Then, all A_m values were plotted as separate points and were compared with a straight line obtained by a linear regression on $\Delta \cdot m$ regarding unknown logarithmic decrement Δ. The damping ratio value was calculated according to the equation

$$\xi = \frac{\Delta}{\sqrt{\Delta^2 + 4\pi^2}} . \tag{5}$$

Since the stability of the modal decays is limited, the use of the linear regression is helpful for assessment of the stability as well as for more precise estimation of the damping ratio. Fig. 5 (right side) depicts the linear regression on the extremes of two modal decays provided by the filtering procedure with pass-band 3.5–5.0 Hz and 6.0–6.8 Hz. The obtained damping ratio values, corresponding to five mode shapes of the Fiberline Bridge presented in Fig. 6, are $\xi_1 = 2.87\%$, $\xi_2 = 1.63\%$, $\xi_3 = 2.59\%$, $\xi_4 = 2.09\%$, and $\xi_5 = 2.57\%$.

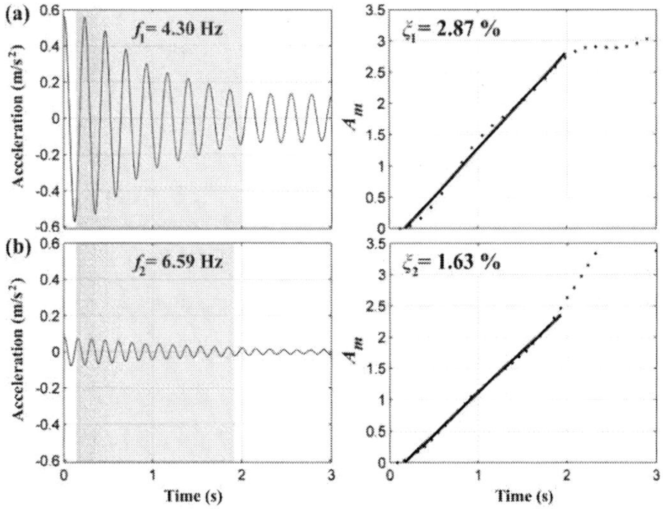

Fig. 5. Modal decays (left side) and linear regression on their extremes (right side) provided by filtering procedure using the eighth-order Type 1 Chebyshev band-pass digital filter with passband ripple 1 dB and pass-band (a) 3.5-5.0 Hz and (b) 6.0-6.8 Hz.

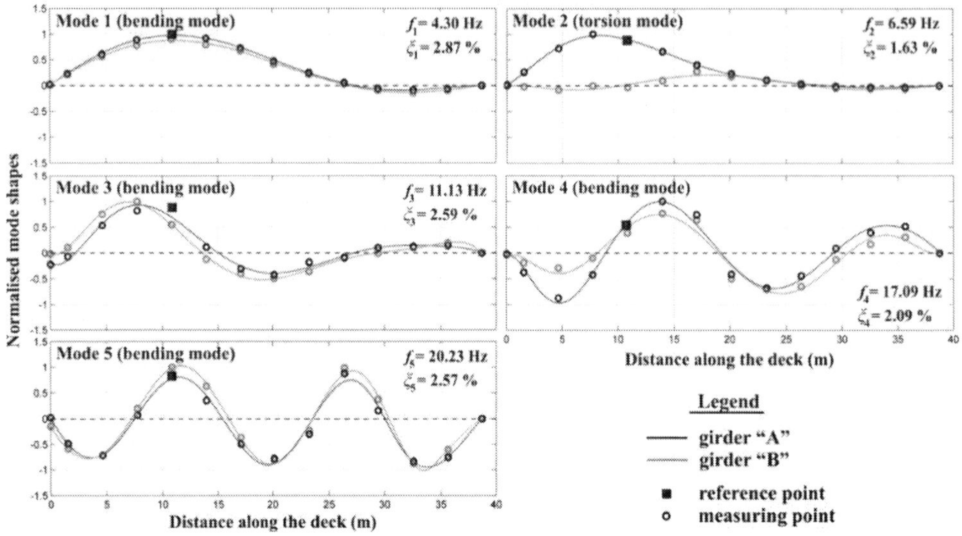

Fig. 6. Normalized mode shapes, corresponding to natural frequencies and damping ratios estimated from measured free vibration responses of the Fiberline Bridge.

5 Comparison of the damping ratios of selected footbridges

The comparison of the measured damping ratio values of the Fiberline Bridge with data published in the literature [14-16] for several selected footbridges made of various conventional materials without a control devices regarding their natural frequencies is presented in Fig. 7.

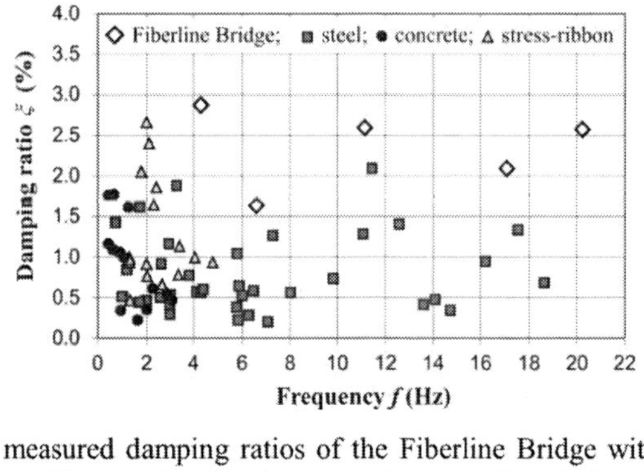

Fig. 7. Comparison of measured damping ratios of the Fiberline Bridge with data published in the literature for selected footbridges made of various materials regarding their natural frequencies.

6 Conclusions

The aim of this paper was to present methods and results of modal parameter identification of the all-GFRP composite cable-stayed Fiberline Bridge in Kolding, Denmark, based on in situ free-decay measurements using accelerometers. The application of an impulse due to human jumping in a single free vibration test was sufficient to excite the specific five vibration modes in the vertical direction of the analysed footbridge. In case of the used approaches the primary five natural frequencies, corresponding mode shapes and modal damping ratios were identified. The modes No. 1, 3, 4, and 5 are the vertical bending type of modes. The mode No. 2 is the untypical torsion mode of the footbridge deck as the result of vibrations mainly the girder "A", probably due to a relatively low bending-torsional stiffness of the deck system. In order to explain this problem the numerical analysis of free vibrations of the analysed footbridge will be made in the next stage.

The comparison of the estimated damping ratios of the Fiberline Bridge with data published in the literature for selected footbridges with comparable length made of various conventional materials showed that the modal damping of the cable-stayed footbridge made entirely of GFRP composite is relatively high. The values of the damping ratios of the analysed footbridge were within the range between 1.6% and 2.9% for all identified natural frequencies. The damping ratios for concrete footbridges were within the range of 0.2 to 1.8%, for steel footbridges – within the range of 0.2 to 2.1%, and for stress-ribbon footbridges – within the range of 0.5 to 2.7%.

Nowadays, it is very advisable and valuable to conduct study in order to recognize the fundamental dynamic characteristics of all-GFRP footbridges. The collected information would be very useful for the procedure of the calculation model verification, the calculation model updating, the dynamic response prediction and the assessment of the technical state or structural health monitoring of this type of footbridges.

References

1. B. Stankiewicz, Procedia Engineer **40**, 423 – 427 (2012)
2. A. Kabiri Ataabadi, S. Ziaei-Rad, H. Hosseini-Toudeshky, Steel Compos. Struct. **12**, 53-72 (2011)
3. B. Stankiewicz, J. Compos. Mater. doi: 10.1177/0021998316669857 (to be published)
4. B. Stankiewicz, M. Tatara, J Civ. Eng. Architecture **9**, 791-797 (2015)
5. A.B.S Santos Neto, H.L. La Rovere, Compos. Struct. **92**, 2554-2564 (2010)
6. S. Robert, *Bridging the World* (Bridge Ink, Wilsonville, USA, 2003)
7. W.J. Harvey, Struct. Eng. Int. **4**, 93 (1993)
8. M. Kutz, *Handbook of materials selection* (John Wiley & Sons, 2002)
9. L. Yeou-Fong, B. Sainey, W.C. Walter, C. Yu-Tsung, Case Stud. Construction Mater. **1**, 83-95 (2014)
10. Y. Bai, T. Keller, Compos. Struct. **82**, 1 (2008)
11. R. Burgueño, V.M. Karbhari, F. Seible, R.T. Kolozs, Compos. Struct. **54**, 4 (2001)
12. Fiberline Composites Website, https://fiberline.com/fiberline-bridge-kolding (accessed January 2017)
13. M. Batel, Sound Vib. **36**, 8 (2002)
14. C. Butz, M. Feldmann, C. Heinemeyer, G. Sedlacek, B. Chabrolin, A. Lemaire, M. Lukic, P.O. Martin, E. Caetano, Á. Cunha, A. Goldack, A. Keil, M. Schlaich, *Project RFS-CR-03019, Final Report* (RFCS, Brussels, Belgium, 2007)
15. J.M.W. Brownjohn, M. Bocian, D. Hester, A. Quattrone, W. Hudson, D. Moore, S. Goh, M.S. Lim, J. Sound Vib. **384**, 339–355 (2016)
16. K. Maes, K. Van Nimmen, E. Lourens, A. Rezayat, P. Guillaume, G. De Roeck, G. Lombaert, Mech. Syst. Signal Pr. **75**, 245–260 (2016)

Estimation of cables' tension of cable-stayed footbridge using measured natural frequencies

Przemysław Jakiel[1*] and *Zbigniew* Mańko[2]

[1]Opole University of Technology, Katowicka Street, No 48, 45-061 Opole, Poland
[2]International University of Logistic and Transport, Soltysowicka, No. 19B, 51-168 Wrocław, Poland

Abstract. This paper presents analysis of cables' tension of steel cable-stayed footbridge using their field-test natural frequencies. A vibration method is usually used for the measured cable tension during the construction of cable systems stiffened with inclined cables. Practical formulas for the vibration method applied herein, mainly based on cable-sag and vertical angle effects (a survey measurement), have been verified on the one-tower steel cable-stayed bridge. The bridge is situated in Sieradz (Poland) and it was the structure with the longest span concerning all the cable-stayed bridges in Poland until 1999. The obtained cable axial forces for estimated natural frequencies of low- and high-order modes are verified using FEM models. The final conclusions drawn on the basis of conducted studies can be useful for technical diagnosis, monitoring programs and repair works of similar class of cable-stayed bridges.

1 Introduction

A knowledge of cable tensions is very important concerning suitable geometry of the cable-stayed bridges and it allows to create their detailed calculation models [1]. The cable axial force can be estimated by the method using measured natural frequencies, which depend not only on axial force but on flexural stiffnes, cable sag and inclination of the cable chord as well. According horizontal cable configuration with relatively small sag the cable tension F we can obtain from well known formula $F = A\rho L^2 f^2/g$ (A - cross section area, ρ - material density, L - length, f – natural frequency, g - gravitational acceleration) applicable for first-order mode only, but the result may be burden with error concerning longer tendons [2].

In this paper, verification of the accurate method which allows to estimate the cable axial force using measured its natural frequencies, taking into consideration aforementioned material and geometrical parameters is presented.

Considering the very slender tendons, i.e. those that are the structural members in the cable-stayed bridges for instance is difficult to excite the cable artificially to first or second-order mode oscillation. In this case, one should use the results obtained from stationary vibrations, in which modes of high order are usually dominant.

* Corresponding author: p.jakiel@po.opole.pl

So called vibration method proposed by Zui, Shinke and Namita [3] is herein briefly described with unification of practical formulas, and applied to the results of natural frequencies obtained from the measurement conducted on a cable-stayed pedestrian bridge in Sieradz (Poland) and the accuracy is confirmed using FEM models [4].

2 Short description of footbridge structure

The cable-stayed footbridge in Sieradz was built in 1978 for pedestrian traffic and emergency vehicle, e.g. fire engine. The total length of this bridge is 142 m with effective spans equal to 9.12+75.88+3x19.00 m, respectively. The steel superstructure consists of one portal tower, consisted of closed, rectangular cross-section, cable system arranged in harp layout and an orthotropic plate deck (Fig. 1a). Total height of the tower transversally braced in the locations of cable joints is 43.48 m. The cables are fixed to the tower and anchored to the deck. The tower is rigidly fixed to the main girders and supported by the steel hinge bearings situated on a pier P2 (Fig. 1b).

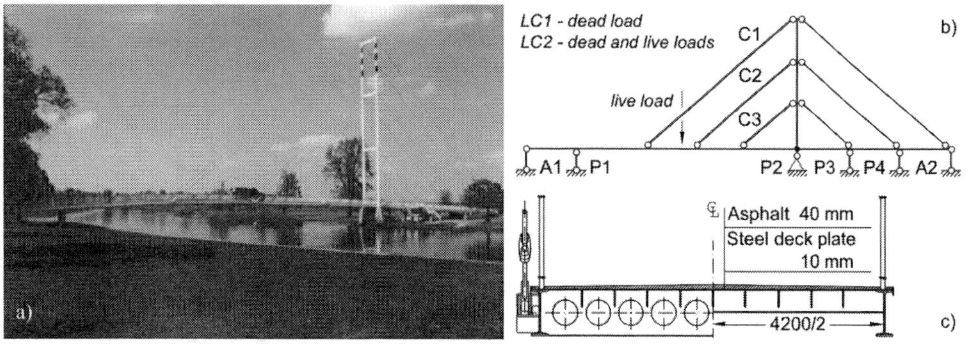

Fig. 1. Sieradz cable-stayed footbridge considered in this investigation: (a) side view, (b) calculation scheme and (c) cross-sections.

The steel deck structure, which is 4.10 m wide, is composed of two I-beams 550 mm main girders, the openwork and plain girder double-tee cross-beams of 525 mm and 250 mm high respectively (Fig. 1c). The cross-beams are alternately spaced: the major ones 7.40 m and the minor ones 1.90 m. The steel plate orthotropic deck consists of 10 mm thick plate and the longitudinal stiffening ribs (12×200). The plate deck is directly topped with a 50 mm thick layer of asphalt [5].

The cables are made of helical wires closed by the z-profile wires in the cross-section in diameter of 48 mm. A peculiar feature of this bridge are the rocker bearings at each support except the pier P2, which enable limited longitudinal and vertical displacements of the spans at their supports.

3 Vibration method's formula

A simplified algorithm of presented herein the vibration method is used for free-oscilation frequencies of the cable with its span length denoted as L, sag d and angle of cable inclination α (Fig. 2a) [3].

Introducing a dimensionless parameter χ (1):

$$\chi = L \times \sqrt{F / EJ}\ , \tag{1}$$

where EJ is flexural rigidity of cable, the range of application of the method is specified as any region of χ, so far as first or second-order vibration' modes obtained on the basis of the field measurement. Moreover, Shimada, using high natural frequency, discovered a formula useful for very slender cables ($\chi \geq 200$), which often appear in the cable-stayed bridges [6].

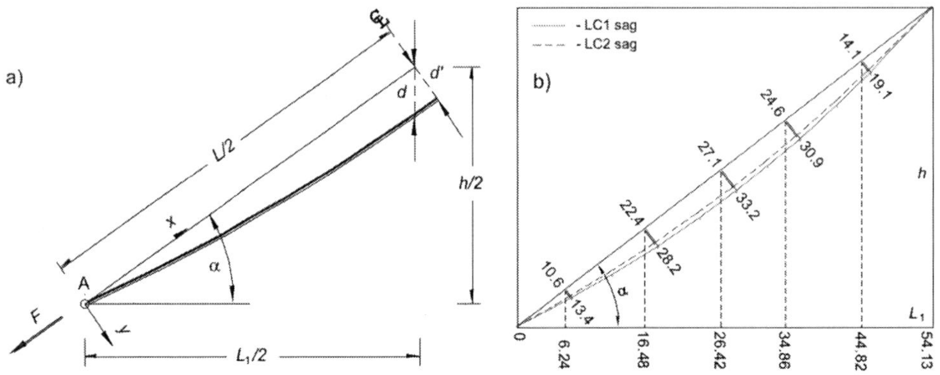

Fig. 2. Scheme of inclined cable (a) and analyzed bridge C1 cable with the sags under load case 1 and load case 2 (b).

The following assumptions have been made in the analysis:
– a second-order parabola is the geometric shape of the cable,
– the sag-to-span ratio is sufficiently small ($e = d/L_1 << 1$),
– the cable vibrates in x-y plane only and its motion in the x-direction is neglibly small.

The method states the basic equation of motion in the y-direction of a beam with axial tension F as follows [3]:

$$EJ\frac{\partial^4 v(x,t)}{\partial x^4} - F\frac{\partial^2 v(x,t)}{\partial x^2} + \frac{q_c}{g}\frac{\partial^2 v(x,t)}{\partial t^2} = 0, \tag{2}$$

where $v(x, t)$ - dynamic deflection in the y-direction and q_c - weight of cable per unit length.

By means of variable separation and after transformation of above equation a transcendental equation can be obtained and thus, assuming the dynamic characteristic of a cable is similar to that of a string, i.e. parameter χ is large, a nondimensional parameter η_n was introduced (3):

$$\eta_n = \frac{f_c}{f_n^s}, \tag{3}$$

where $f_c = \omega/2\pi$, and f_n^s is theoretical value of the n-th order natural frequency of a string expressed in the form [3]:

$$f_n^s = \frac{n}{2L}\sqrt{\frac{Fg}{q_c}}, \tag{4}$$

By substituting obtained values into equation of free vibration of the string clamped at both ends [3], after transformation we can obtain also transcendental equation and the solution η_n can be obtained by one of an iterative methods.

In a case when a cable has relatively large sag, there are introduced an approximate solutions for the first-order mode in a form:

$$f_n^{\mathrm{c}} = \frac{\beta}{\pi}\sqrt{\frac{g}{8d}}, \tag{5}$$

where $\beta = n\,\pi$ ($n = 1, 2, \ldots$) for unsymmetric mode, and $\beta = \Lambda_0$ for symmetric one [3]:

$$\frac{\beta - \mathrm{tg}\beta}{\beta^3} = \Lambda_0 \quad \text{and} \quad \Lambda_0 = \frac{q_{\mathrm{c}} L}{128 E A e^3 \cos^5\alpha} \tag{6}$$

The effect of a cable elongation and/or contraction is not concerned in above equations, except included cable shape changes, what appears in solutions as coincidence with those of a string for unsymmetric modes. Finally, the effect of extension and/or contraction of a cable is significant and derivative cable forces F due to vibrations cannot be neglected for symmetric modes [3]. This natural frequency can be determined by parameter Λ_0 (6), as a function of length, sag, weight, flexural rigidity and inclination angle of cable.

The range of applicability of considered statements is precised by parameter Λ' and χ, which are both introduced in a new parameter Λ in the form:

$$\Lambda = \Lambda' \frac{0{,}31\chi + 0{,}5}{0{,}31\chi - 0{,}5} \quad \text{where} \quad \Lambda' = \sqrt{\Lambda_0}\,. \tag{7}$$

In a case of $\Lambda < 3$, two values of cable force F give the same first-order symmetric natural frequency. We can observe that cable force is very sensitive to the change of first-order natural frequency in the area of large values of cable sag and its derivative axial force and from this reason even small measurement error of free-vibration frequency may generate a large error of cable force [3]. Thus, it is desirable to use second-order natural frequency (or higher-order one) for the measurement of cable force with insignificant error, i.e. when effect of cable sag and derivative of cable force is relatively small.

Applicable region of the formula depends on parameter χ, which is influenced by cable sag in relation with the natural modes n, what is specified as $\Lambda \geq 3$ (for cable with sufficiently small sag, $n = 1$) and $\Lambda < 3$ (for cable with relatively large sag, $n = 2$) and also as $\chi \geq 200$ for very long cables, $n \geq 2$ (Table 1) [3].

Table 1. Basic formulas allowing to calculate axial force in a cable.

Mode	Equations	Parameters		Coefficients		
		χ	Λ	A	B	C
1	$F = \frac{4q_{\mathrm{c}}}{\mathrm{g}}(f_1 L)^2\left[\mathrm{A} + \mathrm{B}\frac{\Psi}{f_1} - \mathrm{C}\left(\frac{\Psi}{f_1}\right)^2\right]$	0–6	≥ 3	0.828	0	10.65
		6–17		0.865	0	11.60
		≥ 17*)		1	–2.20	0.55
2	$F = \frac{q_{\mathrm{c}}}{\mathrm{g}}(f_2 L)^2\left[\mathrm{A} + \mathrm{B}\frac{\Psi}{f_2} - \mathrm{C}\left(\frac{\Psi}{f_2}\right)^2\right]$	0–17	< 3	0.882	0	85.00
		17–60		1.030	–6.33	1.58
		≥ 60*)		1	–4.40	1.10
$n > 2$	$F = \frac{4q_{\mathrm{c}}}{n^2\mathrm{g}}(f_n L)^2\left(1 - 2{,}20\frac{n\Psi}{f_n}\right)$*)	≥ 200	–	–	–	–
$\Psi = \sqrt{\frac{EJ\mathrm{g}}{q_{\mathrm{c}} L^4}}$ and $\Lambda = \sqrt{\frac{q_{\mathrm{c}} L}{128 E A e^3 \cos^5\alpha}} \times \frac{0{,}31\chi + 0{,}5}{0{,}31\chi - 0{,}5}$						
*) the formulas applied herein						

4 Tension of cables using field test free-vibration measurement

In order to determine axial forces in the selected bridge cables (C1 – C3) natural frequencies obtained from the measurements were used [4], [5]. Selected investigated results of C3 cable's natural frequencies are illustrated in Fig. 3.

What is indispensable for the application of the given method is knowledge of the cable sag, which – although does not directly impact the axial force being calculated – is used indirectly for classifying it in the relation of the Λ factor: sag size – cable chord length. It was justified in the paper [4] that the lack of measurement results for the sag of cables shorter than C1 (e.g. C2 and C3) is of no significant influence on the error value.

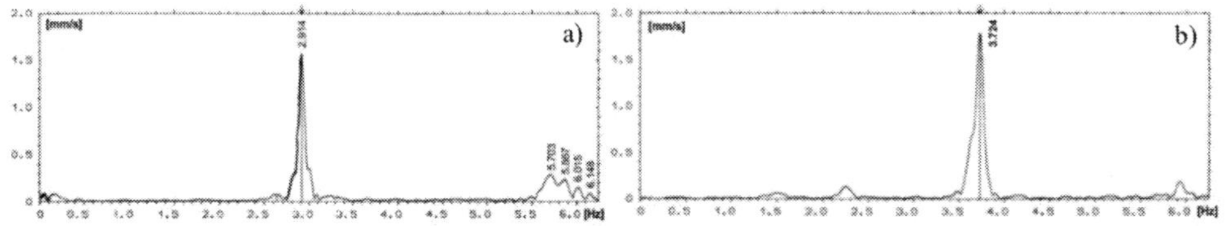

Fig. 3. Spectrum of manually excited first-order free-vibrations of C3 cable under (a) dead load (LC1) and (b) live load (LC2).

The following were assumed for the analyzed slender cables: experimentally determined value of Young's modulus $E_c = 172$ GPa, a unit dead load $q_c = 134.8$ N/m, a cross-section area $A = 1.635\times10^{-3}$ m^2, a moment of inertia $J = 2.606\times10^{-7}$ m^4 and g = 9.81 m/s^2 [4].

Formula (1) indicates the need to apply a preliminary estimated value (e.g. a design value) of the F_0 – axial tensile force in the cable for further calculations, which leads to the determination of the χ parameter. It was justified that high accuracy while determining F_0 is of no particular significance for the final result and thus the results obtained from the preliminary FEM analysis of the bridge model were applied for further calculations of cable tensions. In order to determine the F_1 preliminary cables' tensions, it was assumed that their effort balances the bridge spans' dead load (LC1), whereas in order to determine the design value of the F_2 axial force (LC2), an additional live load provided by a vehicle (weighing 250.6 kN) in the middle of the longest bridge span was applied (Fig. 1a, b) [4].

While the difference between the assumed $F_{1\text{-}0}$ cable tension and the F_1 real value might be quite large, the difference in cable tensions under load case 1 and load case 2, determined with the use of the suggested method and FEM, should not be significant.

Parameters used in the calculations of values of the F axial forces in C1 – C3 cables under dead load only (LC1), as well as under live load has been applied (LC2), were compiled in Table 2, where detailed C1 cable sags are showed on Figure 2b.

Table 2. Assumed values and parameters necessary for the determination of the F axial force in the selected cables of the cable-stayed bridge under analysis.

Cable	F_o	L	χ	d	α	e	Parameters Λ	Parameters Ψ	f_n	F
	(kN)	(m)	–	(m)	(° ' ")	×10^{-3}	–	×10^{-3} (s^{-1})	(Hz)	(kN)
C1	84.70	67.822	93	0.416	37 02 40	7.685	1.36	12.417	1.72 (f_2)	180.612
	186.34		138				1.35			
	125.00	67.831	113	0.340	37 04 24	6.282	1.83	12.413	2.80 (f_3)	213.233
	228.19		153				1.82			
C2	145.90	46.140	83	0.193	36 27 33	5.201	1.99	26.828	3.21 (f_2)	290.322
	297.00		118	0.157		4.231	2.68		3.90 (f_2)	431.004
C3	142.90	22.667	41	0.047	36 43 02	2.587	4.16	111.161	2.91 (f_1)	219.482
	303.10		59	0.038		2.091	5.59		3.72 (f_1)	365.120

Additional calculations were carried out for the longest C1 cable, taking into account the initially determined (design) value of the axial force, which amounted to 84.7 kN for LC1 and, with the use of its value determined on the basis of considerations concerning the changes of geometry of cable (sag), amounting to 186.3 kN for the actual load condition [4]. What can be noticed is the fact that although the calculation axial force F_o has more than doubled it is of no substantial influence on the final value which in both cases amounts to $F_{1(C1)} = 180.6$ kN for load case 1 and $F_{2(C1)} = 213.2$ kN for load case 2. The only values that have changed are the ones of χ i Λ parameters which – being of the utmost importance in the phase of selecting a proper algorithm for the δ relation – did not have an impact on the change of classification of a given element (e.g. for cables with a relatively small sag that is at $\Lambda > 3$, Table 1).

5 Tension of cables using MES model

On the basis of our own experience included in the paper [4] it was assumed that, with the knowledge of specific geometrical characteristics of a bridge cable, it is possible to apply the FEM model of a single (isolated) cable without the risk of making a serious mistake, as well as to determine the axial force on this basis.

Calculations under the static loads were made for the C1 cable discretized with 24 two-node linear elements (of the cable type) in the *2D* set for the geometrical and material data given in item 4. Due to the strong nonlinearity of such an element, calculations were carried out taking into consideration geometrical nonlinearity with the use of Newton-Raphson iterations (Fig. 4a).

The analysis was carried out in three stages assuming a single span cable clamped at both ends and the cable axial force almost equal to zero ($F = 1$ N ≈ 0 N) in the first one, with one roller support and a maximum expected vertical displacement (sag) in the middle of the span of a macroelement $y_{max} = 332.3\times10^{-3}$ m (LC1) in the second stage, and finally $y_{max} = 271.2\times10^{-3}$ m (LC2) in the third one (Fig. 4b, c) [4].

In order to establish the approximate values of cable tensile forces for the known d' sags, a load bigger than the expected one was applied in the initial phase, while the relation of displacement – load ($y - \lambda$) parameters was being controlled. It was found that the following tensile forces of a cable refer to its specific d' values: $F_1 = 182.75$ kN and $F_2 = 223.95$ kN.

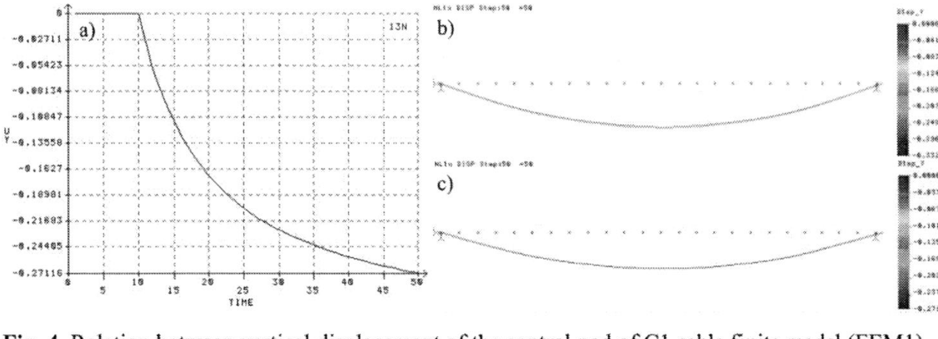

Fig. 4. Relation between vertical displacement of the central nod of C1 cable finite model (FEM1) and live load, LC2 (a); deflection of discrete the same cable model under the cable dead load and the F_1 axial force, LC1 (b) and the F_2 axial force, LC2 (c), scale 20:1.

6 Results and analysis

Relationships between natural frequencies in the function of the obtained axial forces in C1 – C3 cables under two load cases were presented in Figure 5.

The analysis of the obtained results allows to claim that the increase in the F force, similar for each of the cables, in case of C1 occurred for the smallest region of natural frequencies due to the fact that the third mode vibration was assumed for consideration.

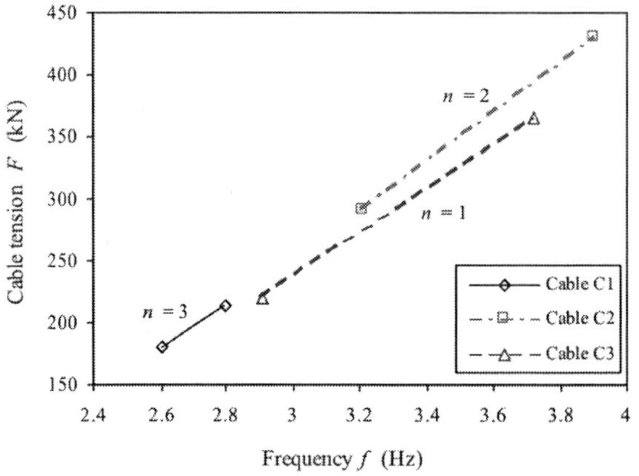

Fig. 5. Relation between tension F and natural frequency f (for n = 3, 2 i 1) for the bridge C1-C3 cables.

On the basis of FEM calculations (the course of y vertical displacements), it was noted that the cable model displays a strong nonlinearity [2]. Comparing the values of the axial tensile forces in the C1 cable obtained from the FEM (model of an isolated cable), it was concluded that they were in agreement with the results obtained pursuant to the considered method determining the cable axial force by measuring its natural frequency (VM) with the difference not exceeding 5% (Table 3). On the other hand, if we take into account the increase in the axial force $\Delta F_{C1} = F_2 - F_1$ in the same cable, the difference is slightly bigger and amounts to 20.8%.

Table 3. Comparison of the axial forces in C1-C3 cables resulting from vibration method and FEM calculations.

Cable No	**Load case**	**Axial force** (kN)			**VM / FEM** (%)
		VM	FEM1	FEM2	
C1	F_1 (LC1)	180.61	182.75	–	98.8
	F_2 (LC2)	213.23	223.95	–	95.2
	$\Delta F_{C1}=F_2-F_1$	32.62	41.20	36.07	79.2 / 90.5
C2	ΔF_{C2}	140.68	–	137.95	102.4
C3	ΔF_{C3}	145.64	–	141.32	103.1

VM – vibration method, FEM1 / FEM2 – discrete models of the single cable (2D) and the bridge (3D) respectively

Certain discrepancies in the obtained results might be due to, among others, the fact that two distinct initial tensions of a given element were assumed, as well as the fact that cable sag was not included in the calculations according to the second order theory (FEM) in the given comparative example for the calculation full bridge model [4].

7 Final conclusions

Verification of the method allowing to assess the tension level of cables in cable-stayed bridges taking into account the geometric and material characteristics of these first ones was presented in the paper. The following conclusions have been formulated on the basis of our own observations and after the results obtained with the use of the FEM have been verified:

- The suggested method allows to determine with quite an ease the current condition of axial forces in cables, however, what is indispensable to do so is a precise measurement of their actual geometry and at least estimated identification of their initial axial tension. Also, a change of the cross-section of a cable due to corrosion or an additional load of the anticorrosion protection might be of certain influence here.
- Particular attention should be paid to the manner of inducing cable vibrations, i.e. solely in the plane of its sag, as well as to the register of dynamic parameters and interpretation of the vibration spectrum. The aforementioned is due to the sensitivity of this method to the measurement error, in particular for lower natural frequencies.
- The obtained results of axial forces in the selected cables of the bridge point to certain discrepancies between both applied methods, which might be caused by the above named reasons.
- The presented algorithm along with the analysis methodology of the isolated cable element (FEM) taking into consideration geometrical nonlinearity allows to prospectively apply them in the course of research and calculation analysis not only of cable-stayed bridges but also other systems in which cables constitute bearing structural members.
- The need to verify the suggested algorithm within the scope of determining the changes of cable tension in laboratory conditions is also pointed out. This shall allow to, among others, determine the influence of damping on the obtained cable tensions.

References

1. R. Walther, B. Houriet, W. Isler, P. Moia, *Cable stayed bridges* (London, Thomas Telford Ltd, 1988)
2. J. Hajduk, J. Osiecki, *Cable structures. Theory and calculations* (WNT, Warsaw, 1970) (*in Polish*)
3. H. Zui, T. Shinke, Y. Namita, Practical formulas for estimation of cable tension by vibration method. J. Str. Eng., ASCE, **122**, 6, 651–656 (1996)
4. P. Jakiel, Evaluation of nonlinear effects in steel cable-stayed bridges. Ph.D. Thesis (Wrocław University of Technology, Wrocław 2005) (In Polish)
5. P. Jakiel, Z. Mańko, Dynamic response of steel cable-stayed bridge in Sieradz. XLVI Scientific Conf., Wrocław-Krynica, **4**, 34–40 (2000)
6. T. Shimada, Estimating method of cable tension from natural frequency of high mode. Proc. of JSCE, 501/1-29, 163-171 (1994)

Modal analysis of cable-stayed UHPC bridge

Petr Tej[1,*], *Petr* Kněž[1], *Lukáš* Vráblík[2] and *Jiří* Kolísko[1]

[1]Klokner Institute, Czech Technical University in Prague
[2]Faculty of Civil Engineering, Czech Technical University in Prague

Abstract. This paper deals with the dynamic analysis of cable-stayed UHPC bridge over the Vltava river near town Melnik in Czech Republic, Europe. Bridge serves for pedestrians and cyclists. This work aims to familiarize the reader with dynamic calculations carried out and the results obtained, describing the dynamic properties of proposed bridge. The construction of bridge is designed as a cable-stayed structure with prestressed bridge deck consisting of prefabricated UHPC panels and reversed "V" shaped steel pylon with height of approximately 40 meters. The deck is anchored using 24 steel hangers in one row in a steel pylon - 17 ropes in the main span and 7 cables on the other side. Range of the main span is 99.18 meters and the secondary span is 31.9 m. Deck width is 4.5 meters with 3.0 meters passing space. The bridge is designed for the possibility of passage of vehicles weighting up to 3.5 tonnes. Deck panels are made of UHPC with reinforcement. At the edge of the bridge on the side of the shorter span the bridge deck is firmly connected with abutment and on the other deck it is stored using a pair of sliding bearings.

1 Structural design

Bridge transfers the bike path and the pedestrian trail over the river Vltava, connecting the village Luzec nad Vltavou with surrounding municipalities on the other side of the river. The axis of the bridge leads over the river bed in approximately perpendicular direction and at the end of the bridge, on the shorter span side, passes in earth ramp located partly in a left curve of radius 30.0 meters for the connection to the existing road leading to the current ferry that the footbridge will permanently replace. The ramp is made of soil with grassed slopes. The bridge deck height is kept in top curved surface of radius 777.5 meters. This ensures enough camber of the footbridge over the rivers canaliculated profile measuring 35.0 meters with height of 7.0 meters, which was necessary due to the requirement of waterway administrator. The rise of an arch is chosen so that the ends of the supporting structure would still have the maximum allowed slope for wheelchair use of the building.

The UHPC bridge deck consists of panels stapled to the contact joint by prestressed rods and cables. Pretension leads in outer beams with the width of 0.75 meters and a thickness of approximately 0.3 meters. These beams are connected by desk with at least 60 mm thickness, which forms a walk-bearing surface.

* Corresponding author: petr.tej@cvut.cz

The surface of this area is fitted with sprayed anti-slip insulation finish in natural color of the original UHPC material. Each panel is reinforced by a pair of ribs located in about ¼ the length of the panel from each end.

Each panel contains an embedded steel weldment at a distance of 1.25 meters from the end of the panel, in which will be anchored the hanging rope. The total length of the panels is 5.65 m. End panels of the deck will be accompanied by a monolithic part for the realization of the restraint (connection of the bridge deck and abutments) and for the installation of bridge bearings and the expansion joint on the opposite side of the bridge. The deck therefore dilates its entire length to the one abutment. Dewatering of the bridge is provided by a free water drain from the surface. The deck will be illuminated by spotlights mounted in the construction of the railing, which has the height of 1.3 meters.

2 Materials

Ultra High Performance Concrete - UHPC is a new cementitious composite material, which is mainly used for reducing or optimizing a weight and dimensions of structures. Its outstanding features, mainly high compressive strength (>150 MPa), high tensile strength (>10MPa) and high durability, are generally known. UHPC is a homogenous material whose mixture consists of very fine aggregates, cement, fine fillers (silica fume, fly ash), superplasticizers and randomly distributed fibres (steel, glass, and polymer). A very low water-cement ratio and a fine grained mixture provide extremely low porosity. [2-7]

The bridge will be built using these materials: Bridge deck is made of ultra-high performance concrete which is designed with compressive strength on columns 150 MPa with dispersed steel fibres. Prestressing steel of cables is Ø15.7 - 1770/1570 MPa with very low relaxation, steel S355 K2+Nc, concrete C30/37 and reinforcement B500B.

3 Modal analysis

Basic static scheme of the footbridge is formed by a beam suspended by rope elements (able to transmit only tension force) into the steel pylon, while the last hanger on the side of the shorter span is anchored to the abutment outside the supporting structure. On this side is the desk cantilevered to the abutment, which positively influences the dynamic properties of the system. On the opposite side is the structure deposited on the sliding joints (bearings) that allow the dilation of the structure. Computing model was created in Scia Engineer 2014, Civil Midas and Midas FEA. For the making of the model have been used 2D shell finite elements.

The analysis was based on the basic principle of the hangers acting as elements with a sagging which axial stiffness is affected by the magnitude of a tensile force. The calculation is, given the nature of the construction, materially and physically nonlinear. The structure is loaded by permanent and variable load effects. Permanent loads include the self-weight of the construction and all additionally built parts. Variable loads represent loads caused by traffic on the bridge - pedestrians, vehicles up to 3.5 tons, and climatic loads - wind, snow and temperature. Pedestrian movement on the structure is simulated by moving surface load with size of 4.0 kN / m^2 and vehicle load is represented by moving two axle vehicles of total weight of 3.5 tons. Both loads are considered in the model so that their effect would be maximal due to the monitored variable - thus unloading effects are neglected. Within the primary static calculation was mainly controlled the extreme normal stress at crucial parts of the structure to maintain the pressure reserve and tension in the walls of the pylon. Furthermore, the stability of the controlled compression of the steel pylon and two fundamental methods – Euler's concept of stability and the introduction of production inaccuracies into the shape the pylon. The result of the calculation was the need to fill the foot of the pylon with concrete to reduce the otherwise high number of reinforcements and prevent buckling of the walls. Further was also observed the deformation of the structure, taking into consideration that all shifts and deflections caused by the self-weight of the structure will be removed by the camber and simultaneously by prestressing of the hangers. At the very hangers occurs reducing of their axial stiffness due to their elongation. Introduced normal force is partially depleted by changing the geometry of the element until the moment when the cable is activated.

From this point it can be assumed that the element behaves linearly and the full stiffness of the rope applies. This behavior is known as tension stiffening. Static design effort was to ensure such axial forces in the cables (by permanent loads, variable loads, but also the actual final tightening of the ropes) that during their minimal stress still remained sufficient tensile reserve in ropes, which would guarantee their linear behavior and thus the linear behavior of the structure. The model takes into account the influence of the construction process and the gradual tensioning hangers. Tensioning of the cables, and the normal force in general applied in them has a beneficial effect of reducing tensile stresses arising in the deck of the footbridge from the extreme positions of the live load. Minimum required tensile strength, which ensures its linear behavior, was set for each rope by using the calculation based on its geometric characteristics. The effort is within the construction of the object to strain the ropes so that it acts almost linearly by minimizing their sag. From the static model, it is possible to determine that the ropes at the edge of the bridge on the side of his longer range will need to be prestressed, because of their angle does not allow the transfer of sufficient tensile normal force of self-weight of the structure to ensure their linear behavior. Given the nature of the structure, it was made its basic modal analysis for the determination of natural frequencies and associated mode shapes. From the results it is possible to conclude the behaviour of the structure and its response to load by pedestrians. With regard to the fact that it is a very lightweight construction, it is possible to predict from the modal analysis results, that after the dynamic load test (to determine the logarithmic decrement of attenuation) it will be necessary to fit a passive vibration damper on the construction, which will change the mass distribution on the structure.

Table 1. Modes of shape and frequency.

Mode shape number	Frequency [Hz]	Mode shape description
1	0.632	Deck bend
2	0.945	Deck torsion
3	1.098	Deck and pylon bend
4	1.508	Deck and pylon bend
5	1.864	Pylon bend
6	1.949	Deck and pylon bend
7	1.955	Deck torsion and longitudinal bend of the pylon
8	2.389	
9	2.758	Deck and pylon bend
10	2.981	Deck torsion and longitudinal bend of the pylon

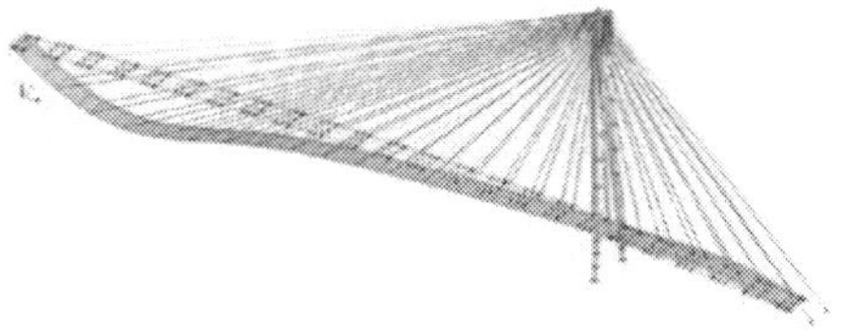

Fig. 1. First mode shape.

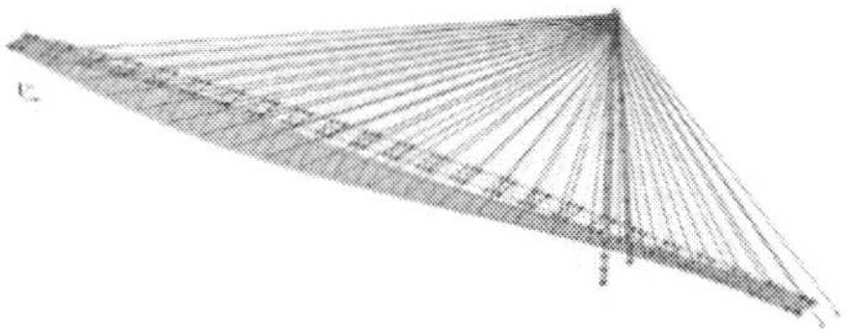

Fig. 2. Second mode shape.

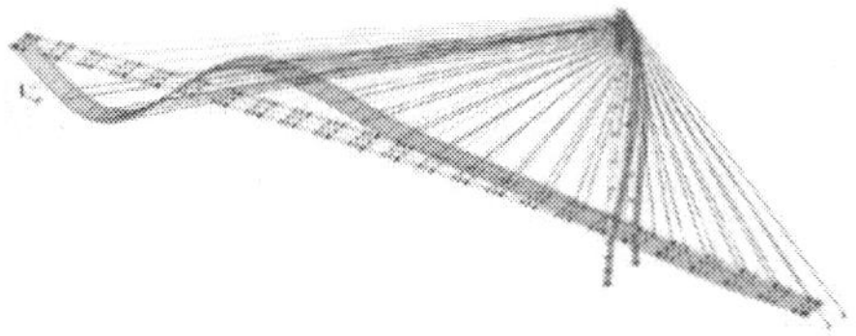

Fig. 3. Third mode shape.

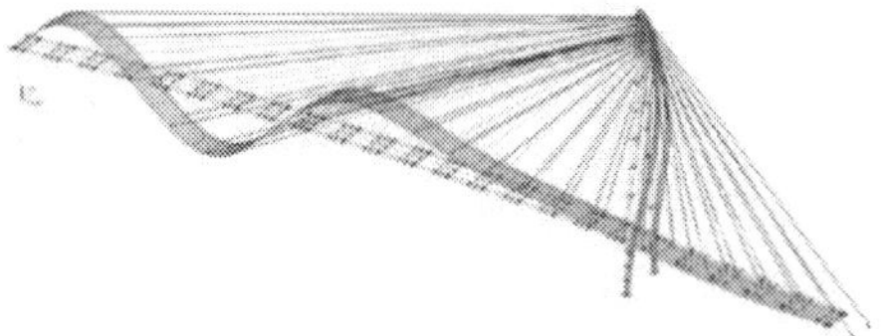

Fig. 4. Fourth mode shape.

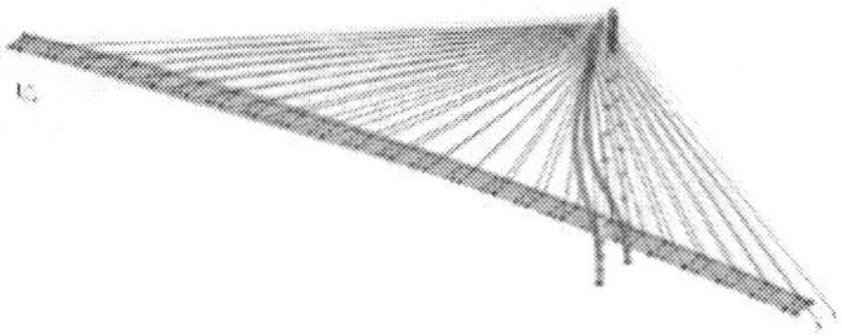

Fig. 5. Fifth mode shape.

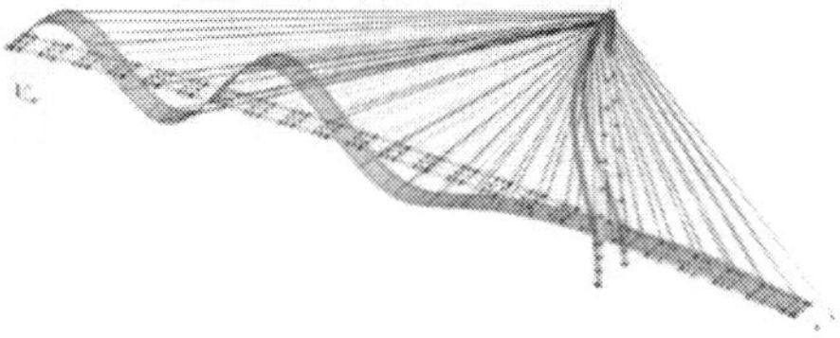

Fig. 6. Sixth mode shape.

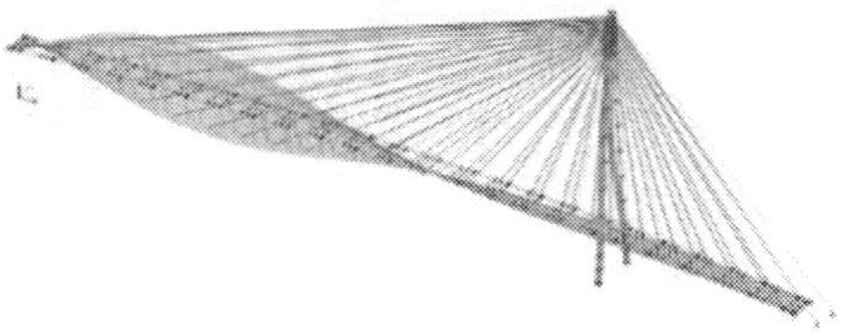

Fig. 7. Seventh mode shape.

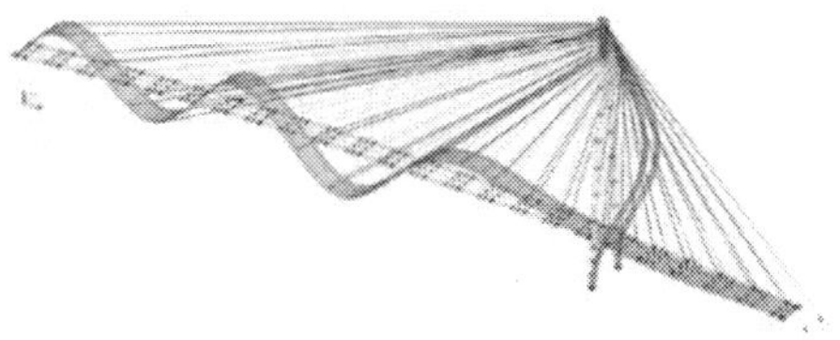

Fig. 8. Eight mode shape.

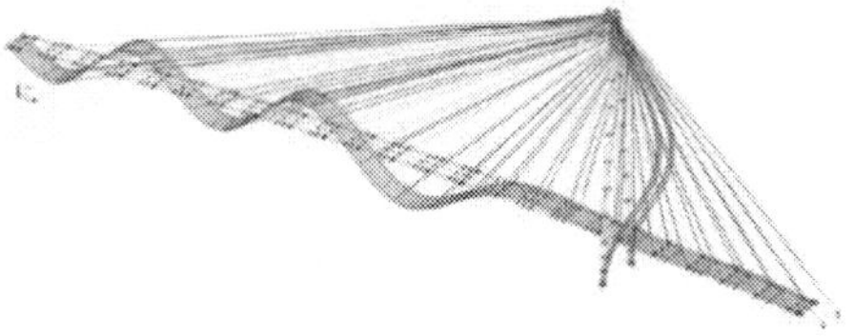

Fig. 9. Ninth mode shape.

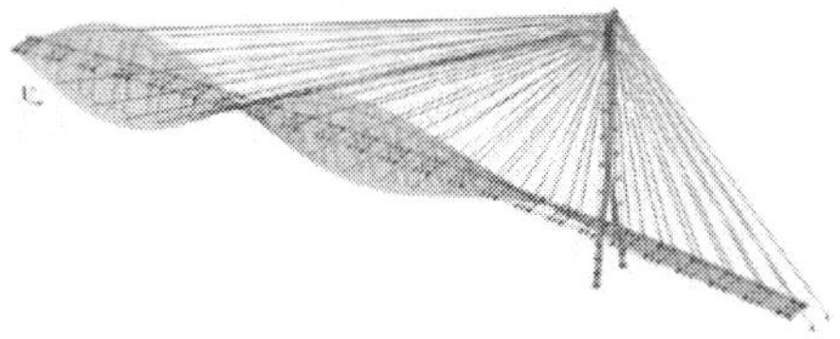

Fig. 10. Tenth mode shape.

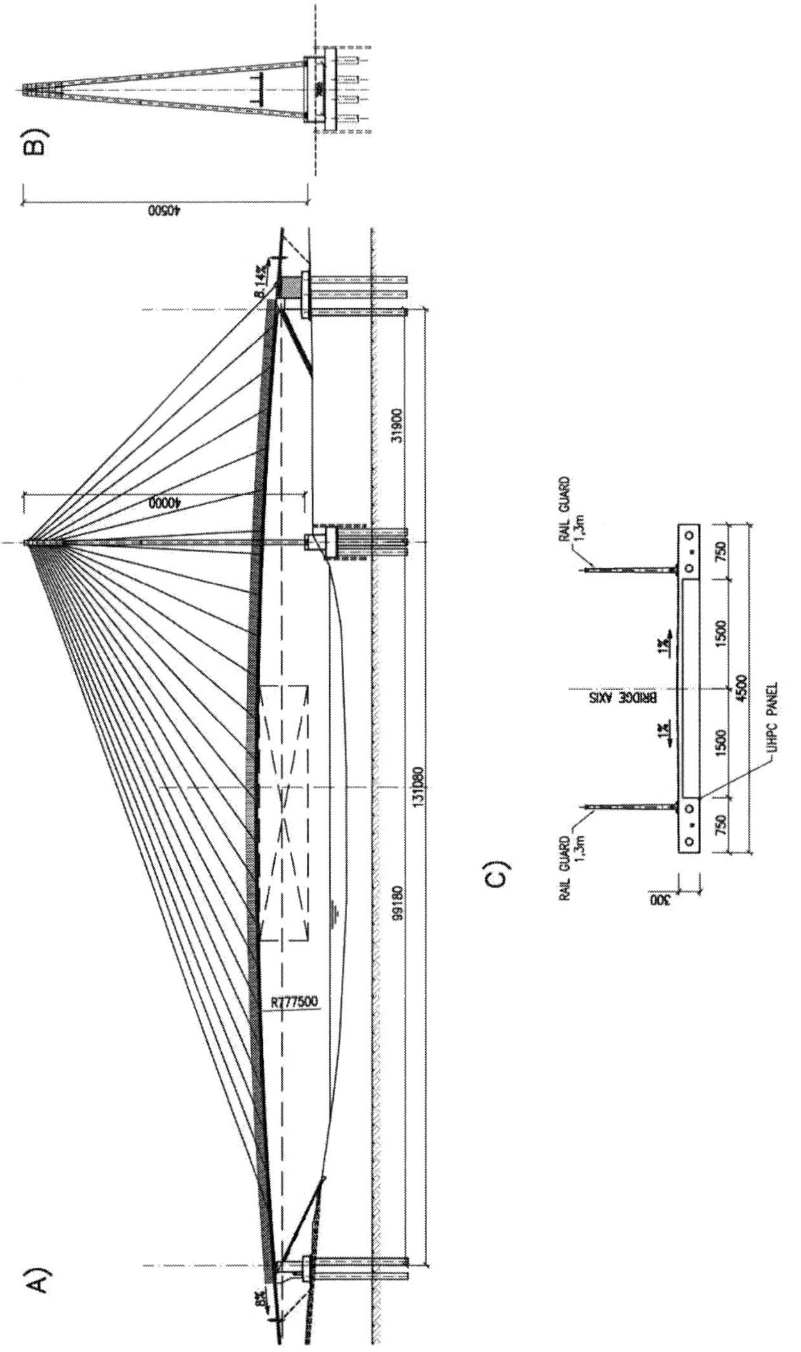

Fig. 11. A) Longitudinal section, B) Cross-section, C) Typical cross-section of the deck.

4 Conclusion

The basic modal analysis of the structure was performed and all kinds of load regarding the nature of the structure were considered. Due to the significant nonlinear behaviour of components of the structure it was necessary to deal with the construction as a geometrically and physically nonlinear task. Given the nature of the structure, it was made its basic modal analysis for the determination of natural frequencies and associated mode shapes. From the results it is possible to conclude the behaviour of the structure and its response to load by pedestrians. With regard to the fact that it is a very lightweight construction, it is possible to predict from the modal analysis results, that after the dynamic load test (to determine the logarithmic decrement of attenuation) it will be necessary to fit a passive vibration damper on the construction, which will change the mass distribution on the structure. The construction met all the criteria and it is possible to conclude, that the design is feasible and meets the requirements for this type of construction.

This research is supported by grant GACR P105/12/G059.

References

1. Z. Bittnar, J. Šejnoha, *Numerical Methods in Sructural Mechanics* (ASCE Press, 1996)
2. JSCE-USC: *Recommendations for Design and Construction of Ultra-High Strength Fiber-Reinforced Concrete Structures – Draft.*
3. SETRA/AFGC: *Ultra High Performance Fibre-Reinforced Concretes: Interim Recommendations* (2002)
4. AFGC/SETRA, *Ultra High Performance Fibre-Reinforced Concretes, Recommendations*. Paris CEDEX (2013)
5. Processings from HIPERMAT 3rd International symposium on UHPC and Nanotechnology for High Performance Material, Kassel (2012)
6. P. Paulík: The effect of curing conditions (in situ vs. laboratory) on compressive strength development of high strength concrete. Procedia Engineering **65**, 113-119, (2013)
7. FIB Model Code for Concrete Structures 2010, fib, Ernst & Sohn, (2013)

Identification of the moving vehicle presence on the bridge with the time-frequency analysis

Ján Kortiš[1,*], *Ľuboš* Daniel[1] and *Michaela* Holešová[1]

[1]University of Žilina, Faculty of Civil Engineering, Univerzitná 8215/1, 010 26 Žilina, Slovakia

Abstract. The article presents the application of common techniques of signal processing as the Wavelet Transformation and Short-time Fourier Transformation to detect the position of the moving vehicle on the bridge. The measured acceleration in the middle of the span of the investigated bridge structure are used for this purpose. The aim of the analysis is to determine the time when the vehicle comes to the bridge and when it leaves. This can be important to know if there is required to investigate free and forced vibration of the bridge separately. It is done for two cases that differ in presence of the obstacle which is positioned on the road in the middle of the span. The obstacle is used to increase the dynamic excitation of the bridge. It is recommended by Slovak standard STN 73 6209 for measurement of dynamic characteristics of bridges as a source of excitation.

1 Introduction

The measurement of dynamic characteristics of bridges is an essential practice in the assessment of these structures [1]. The measurement techniques that are used to identify natural frequencies and natural shapes are based on the techniques of operational modal analysis [2]. However, the operational modal analysis is effective technique with high accuracy, it has got a few practical issues. One of the biggest is to find the suitable and sufficiently powerful source of excitation. The vehicle moving through the obstacle on the bridge is a good alternative how to excite the bridge. In this case, the presence of the heavy vehicle on the bridge could influence the results obtained from the measurement [3]. Because of this, there is important to know the position of the vehicle to recognise if the bridge oscillate without the mass of the vehicle or with that and how they differ. For this purpose, there is required to know the time period in the signal while the vehicle is on the bridge and while it is not. The direct detection of the moment when the vehicle comes to the bridge and when it leaves is one possible way. The special measured equipment like steel plates with accelerometers that are located at the beginning and at the end of the bridge was used for this purpose (Fig. 1).

* Corresponding author: jan.kortis@fstav.uniza.sk

 DOI: 10.1051/matecconf/201710700008

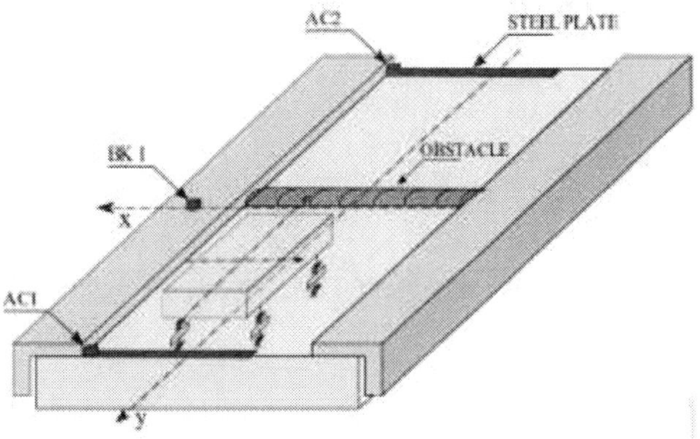

Fig. 1. The scheme of experimental measurement with the location of the steel plates with accelerometers (AC1, AC2), accelerometer BK1 measured response of the bridge.

The measured signals of the accelerometers AC1 and AC2 significantly change when the wheels of the vehicle reach the steel plates. This is used to detect the moment when the vehicle reaches the places where they are positioned (Fig. 2). They give good results but there is a problem that they only give information in time about the moment when the vehicle comes to the bridge and when it leaves. There is no information about the real position of the vehicle on the bridge.

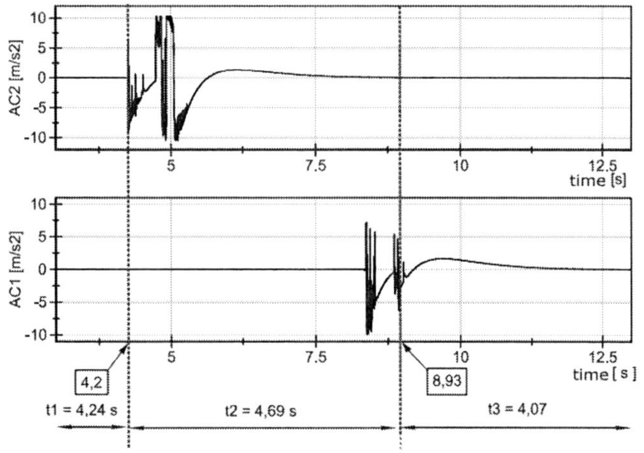

Fig. 2. The identification of the position of the vehicle on the bridge trough the response of the accelerometers AC1 and AC2

The next possible way how to detect the position of the vehicle is to investigate the changes in the frequency spectrum of measured acceleration that arise when the vehicle is moving through the bridge. The Fast Fourier Transformation gives information just about the changes in the frequency domain of the whole signal. In the article, there are presented two different transformations Short Fourier Transformation and Continuous Wavelet Transformation which can show the changes of frequency domain in time [4], [5].

2 Mathematical background of the analysis

The Fourier Transformation is used to transform the signal from time domain to the frequency domain. There is assumed that the signal has an infinite length and it does not change in time. Nonetheless, the real signals are not infinite and they are nonstationary. Thus, the transformation that is useful for practical purposes desires also the possibility to provide the information about the changes of frequency components of the spectrum with respect to time. Determining the time localization of frequency components of nonstationary signals requires time-frequency transformation such as Short Time Fourier Transformation or Continuous Wavelet Transformation.

Let the Hilbert space

$$L^2(R) = \left\{g: R \to C,\ \int_{-\infty}^{\infty} |g(x)|^2 dx < \infty\right\} \tag{1}$$

and the scalar product of functions $f,\ h \in L^2(R)$ is

$$\langle f, h\rangle = \int_{-\infty}^{\infty} f(x)\overline{h(x)}dx, \tag{2}$$

where $\overline{h(x)}$ complex conjugation of the function $h(x)$. If the signal represented by a function f of the Hilbert space $L^2(R)$, then Short Time Fourier Transform (STFT) of signal $f(t)$ is based on the fact that we analyse this function using of some window function $h \in L^2(R)$ with constant length. Then there is computed the Fourier Transformation separately on each shorter segment of a continuous signal $f(t)$, hence STFT is defined as follows

$$STFT\{f(t)\}(\omega, \tau) = \int_{-\infty}^{\infty} f(t)\overline{h(t-\tau)}\, e^{-i\omega t}\, dt, \tag{3}$$

Where $\overline{h(t-\tau)}$ complex conjugation of the function $h(t\text{-}\tau)$. We used for our computation Gaussian window with length $N = 256$.
STFT decomposes (localizes) this function of signals in time-frequency domain $(t, \omega) \subset R \times R$. Spectrogram $SPEC\ (t,\ \omega)$ is used for graphical representation of STFT as a complex function. It is the function of two variables identified as a square of modulus of the STFT, i.e.

$$SPEC(t, \omega) = |STFT\{f(t)\}(\omega, \tau)|^2. \tag{4}$$

Although STFT gives information between time and frequency representation of signals, it cannot reach the high resolution in time and frequency. Selecting of a long window allows good frequency resolution, but worse time resolution. Short window leads to a good resolution in time but poor frequency resolution. Therefore, the compromise between precision of the determination of frequency and time have to be find by the optimal width of the window. However, this can cause mistakes if results with high accuracy are required. For example, good resolution helps to detect the moment when vehicle enters the bridge precisely. In this case, Continuous Wavelet Transformation (CWT) seems to be a better alternative. CWT can be used to discover difference of the signal properties at various scales in time [6], [7]. This transformation is suitable for the analysis of continuous signal that considerably varies. It gives good time resolution at higher frequencies.
Let $\psi \subset L^2(R)$ a mother (base) wavelet function which is usually $\|\psi(t)\| = 1$.

We define

$$\psi_a(t) = \frac{1}{\sqrt{a}}\psi\left(\frac{t}{a}\right) \text{ and } \psi_{a,b}(t) \equiv \psi_a(t-b) = \frac{1}{\sqrt{a}}\psi\left(\frac{t-b}{a}\right). \quad (5)$$

Continuous wavelet transform (CWT) of a signal $f \in L^2(R)$ is

$$CWT\{f(t)\}(a,b) = \int_{-\infty}^{\infty} f(t)\overline{\psi_{a,b}(t)}\,dt = \frac{1}{\sqrt{a}}\int_{-\infty}^{\infty} f(t)\overline{\psi\left(\frac{t-b}{a}\right)}\,dt = \langle f, \psi_{a,b}(t)\rangle, \quad (6)$$

Where $\overline{\psi_{a,b}(t)}$ complex conjugation of the function $\psi_{a,b}(t)$. The number $a>0$ is the scaling parameter and the mother wavelet ψ for $a>1$ expands in the time domain that corresponds to a compression in the frequency domain. For $0<a<1$, the wavelet compressed in the time domain that corresponds to expansion in the frequency domain. The number $b \in R$ determines the size of displacement of wavelet in the timeline. This increases the ability to distinguish signals in time. If the signal contains, for example, some discontinuity at the time $t = t_1$ then for small values of the parameter a in the time translation $b = t_1$ this discontinuity correlates with compressed wavelet and $CWT\{f(t)\}(a,b)$ for these values of parameters a, b will be relatively high. It leads to precise location of the discontinuity in the signal. The mother wavelet functions Morlet is used in the presented CWT.
CWT decomposes (localizes) this function of signals in time-scale domain $(b,a) \subset R \times R$ and the area of time-scale windows is always constant and independent of the parameters a, b. In contrast to the Short-time Fourier transformation, the time-scale window changes its shape depending on the scale a. When displaying, we must remember that the parameter a is inversely proportional to frequency, i.e. with the increase of the scale we have lower frequencies. The scalogram $SCAL(b,a)$ is used to present CWT graphically. It is the function in two variables identified as square of modulus of the CWT.

$$SCAL(b,a) = |CWT\{f(t)\}(a,b)|^2. \quad (7)$$

3 Experimental measurement and analysis of results

The measured data presented in the article were acquired by the transducer which measures acceleration of the bridge in the middle of the span. It is labelled as BK1 and it was positioned on the surface of the side-walk (Fig. 1). The responses were measured while the vehicle was moving through the bridge with the constant speed. The moments when the vehicle comes to the bridge and when it leaves were detected through the steel plates with accelerometers fastened on them (Fig. 2).

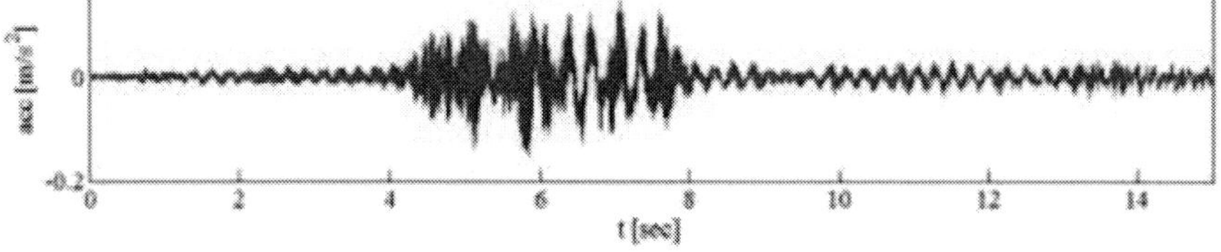

Fig. 3. The measured signal BK1 without the obstacle on the bridge.

The analysed signals were measured for two cases. In the first case the vehicle moves with the speed 9.1 ms^{-1}. There is no obstacle on the bridge. In the second case, the obstacle is positioned in the middle of the span to increase the excitation of the bridge. The obstacle is made of wood with special half-sine wave shape that is recommended by STN 73 6209 standard. The measured amplitude of acceleration on the bridge increases from 0.2 ms^{-2} to 1.2 ms^{-2}. This significant increase in the response of the bridge is achieved by the obstacle. The speed of the vehicle 7.1 ms^{-1} is a little bit smaller than in the previous case. It is influenced with the fact that the driver of the vehicle intuitively slow down in front of the obstacle. This small difference in the speed is not considered in the analysis of the measured data as something that can dramatically change the results.

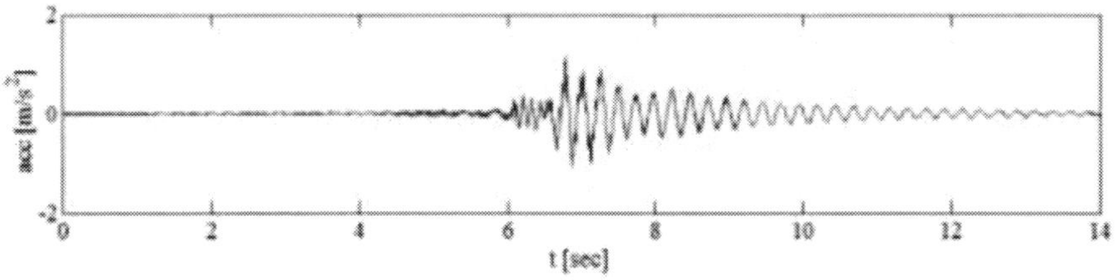

Fig. 4. The measured signal BK1 with the obstacle on the bridge.

The two measured signals are analysed with the two techniques of time-frequency analysis. Firstly, the STFT was used to analyse the changes in the frequency spectrum during the time. The Gaussian window with the length of 256 samples is used. The sample rate of the signal is 1000 Hz so the range of investigated frequency is from 0 to 500 Hz. The frequency step in the results is 3.91 Hz with regards to the length of window (Fig. 5).

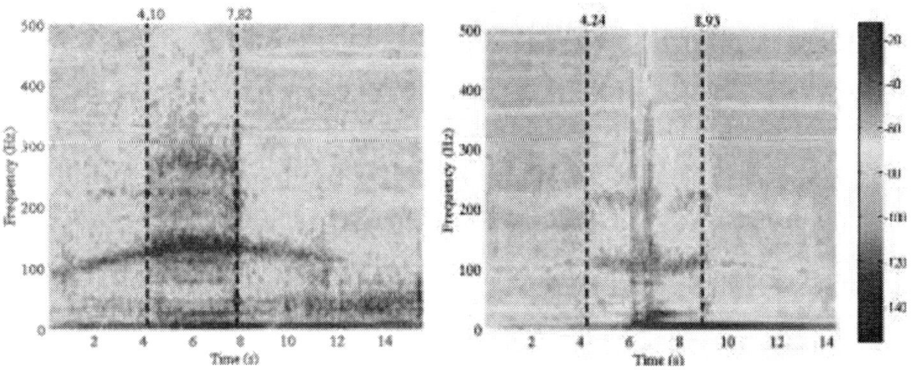

a) Response of the bridge without the obstacle b) Response of the bridge with the obstacle

Fig. 5. Spectrogram of acceleration amplitude in the middle of the bridge (Gaussian window with length 256 samples).

The moments in time that were detected with steel plates when the vehicle came to their location are used to show how the STFT changes when the vehicle is on the bridge and when it is not. The amplitudes arises for the whole range of investigated frequency domain for both cases if the vehicle is on the bridge. There is possible to recognise the time when it starts very clearly. It means that this moment when the vehicle comes to the bridge is possible to detect. But the moment when the vehicle leaves the bridge is not so apparent because the bridge continues to oscillate with higher amplitudes even after the vehicle is not on the bridge. This time moment would be better identified if the damping characteristics of the bridge were considered but this has been not done till now.

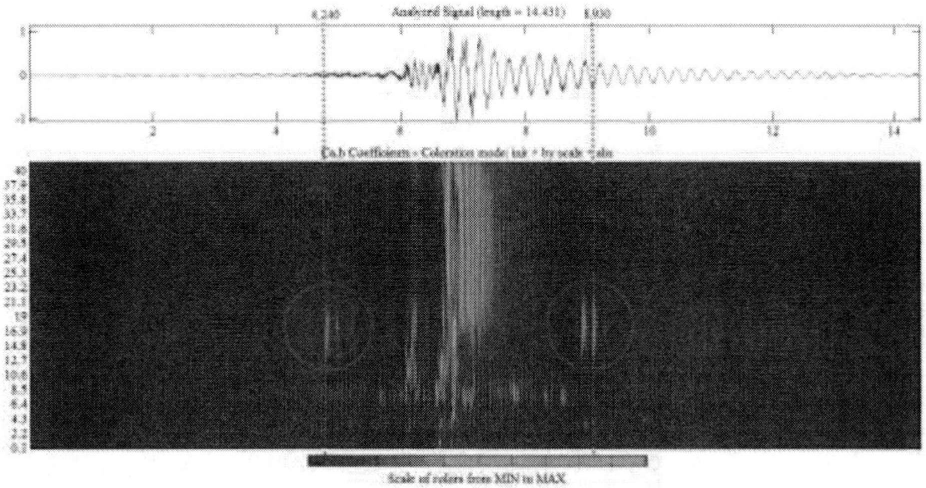

Fig. 6. Wavelet transform of acceleration in the middle of the bridge, Morlet wavelet, with the highlighted discontinuities.

The first one when there is an obstacle on the bridge there are detected the discontinuities in the signal in time when the vehicle comes and lives the bridge (Fig. 6). They can be used to recognise this time moment but it is not clear which is the source of these discontinuities in the signal. There is supposed that the steel plates are a source of these discontinuities as they are not fasten on the surface of the pavement and they can vibrate freely. They also produce a noise when the wheels moves through them during the measurement. However, there is no proof which could be used to confirm it.

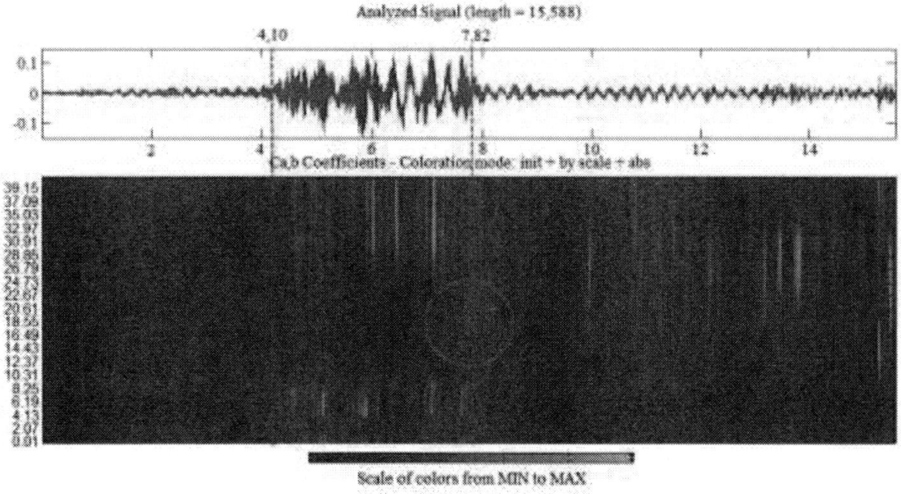

Fig. 7. Wavelet transform of acceleration in the middle of the bridge without obstacle, Morlet wavelet, with the highlighted discontinuities.

The next approach focused on the identification of the time period when the vehicle is on the bridge is based on the CWT with the application of the Morlet (mother) wavelet. There are again investigated both cases with and without the obstacle on the bridge. The signal which was measured for the case when the obstacle is not on the bridge does not contain so big sources of discontinuities in the signal (Fig. 7). It is evident as there is not an obstacle. There is also a problem to identify the time period when the vehicle is on the bridge. In the moment when the vehicle comes to the bridge, there are not identified any noticeable discontinuities. Surprisingly, they are identified in this time moment in the previous case. There is identified the discontinuity only at the moment when the vehicle leaves the bridge.

Conclusions

The response of the bridge to the moving vehicle was measured by the accelerometer. The common techniques of time-frequency analysis of measured signals are used to recognise the time period in the signal when the vehicle is on the bridge. The results are compared with the identification of the moment when the vehicle comes to the bridge and when it leaves by steel plates positioned on the road of the bridge. It shows that both transformation from time domain to the frequency domain presented in the article can be used for this purpose. The STFT shows differences in the signal in time better for higher frequencies. Thanks this it can be detect the presence of the vehicle on the bridge. In the case, when there is an obstacle, it is also possible to detect the moment when the wheels bump to this obstacle. The CWT does not give so good results, but it could be better if the other type of mother wavelet were used. So the next investigation of this transformation should be focused on application of other types of wavelets to show if it can improve the results.

This paper was supported by the Grant National Agency VEGA of the Slovak Republic (grant No. 1/0005/16) and by the Grant Research Centre of The University of Zilina (grant No. 26220220183)

References

1. Nuno Manuel Mendes Maia, Julio Marins Manalvaou a Silva, *Theoretical and experimental modal analysis* (Research Studies Press, 1997)
2. Peeters Bart, C. E. Ventura, Mech. Syst. Signal Pr. **17**, 5, 965-988 (2003)
3. A. Tesar, J. Melcer, Int. J. Numer. Meth. Eng. **74**, 11, 1670-1678 (2008)
4. Yhong Hong Yan, Anyaho Miyamato, Zhongwei Jiang, Mech. Syst. Signal Pr. **23**, 1474-1489 (2009)
5. Daniel Cantero, Mahir Ülker-Kaustell, Raid Karoumi, Mech. Syst. Signal Pr. **76**, 518-530 (2016)
6. K. Nayzar, *Základy teorie waveletů* (Univerzita Karlova v Praze, Karolinum, 2004)
7. D. Lee Fugal, *Conceptual Wavelets in digital signal processing* (Space & Signals Technologies LLC, 2009)

Pedestrian load models of footbridges

Jiří Máca[1,*] and *Jan* Štěpánek[1]

[1]Czech Technical University in Prague, Faculty of Civil Engineering, Department of Mechanics, Thákurova 7, Praha 6, CZ-166 29, Czech Republic

Abstract. The increase of vibration problems in modern footbridges shows that footbridges should no longer be designed for static loads only. Not only natural frequencies but also damping properties and pedestrian loading determine the dynamic response of footbridges and design tools should consider all of these factors. In this paper the pedestrian load models for serviceability verification of footbridges, which are missing in the current European codes, are presented. For simplicity reasons the proposed pedestrian load models are based on stationary pulsating loads instead of moving pulsating loads. It is shown that simplified procedure can be used in verification of the serviceability limit state related to vibration due to pedestrians. Footbridge vibrations don't cause usually structural problems, but if the vibration behaviour does not satisfy the comfort criteria, changes in the design or damping devices could be considered. The most popular external damping devices are viscous dampers and tuned mass dampers (TMD). The efficiency of TMD is demonstrated on the example of a footbridge prone to vibrations induced by pedestrians. It is shown that if the TMD is tuned quite precisely the reduction of accelerations can be very significant.

1 Introduction

Modern footbridges are very often lightweight and flexible structures, where the first natural frequencies of vibration may fall close to dominant frequencies of the dynamic excitation due to walking or running. Such bridges are susceptible to vertical as well as to horizontal vibrations leading to a resonant response characterized by high levels of vibration and a dynamic design is necessary.

In this paper, the different models of dynamic loads caused by pedestrian crossing the bridge, which can be used in serviceability verification, are presented. Although footbridge vibrations do not cause usually structural problems, they can induce some uncomfortable sensation, and so many codes establish maximum acceptable values of acceleration. Provided that the vibration behaviour due to expected pedestrian traffic is checked with dynamic calculations and satisfies the required comfort, any type of footbridge can be designed and constructed. If the vibration behaviour does not satisfy some comfort criteria, changes in the design or damping devices could be considered.

[*] Corresponding author: maca@fsv.cvut.cz

2 Pedestrian loads

In order to verify serviceability limit state related to vibration due to pedestrians it is necessary to define dynamic pedestrian load. Numerous studies deal with determination of human walking, running or jumping force over the years, cf. e.g. [1].

The possible loading scenario can be divided into five categories:

- Single person loading;
- Normal traffic - spatially unrestricted traffic where each individual can move freely without having to change walking pattern to avoid contact with others;
- Crowd loading - spatially restricted traffic where the walking of each individual is restricted due to limited space;
- Group loading - a number of persons is walking closely together;
- Vandal loading - a person, or a group of people, tries to excite the structure by moving in a correlated harmonic way in response-sensitive areas.

In addition to these five groups three different types of human motion are commonly considered to model the dynamic loads applied by pedestrians, namely walking, running and rhythmic jumping. All these load models can often be categorized into deterministic and probabilistic models. In this paper only the deterministic models of a single pedestrian, group and crowd loading will be considered.

Two types of analytical force models can be found in the literature: time-domain models (deterministic and probabilistic force models) and frequency-domain models – for a detailed review cf. [1] and [2]. The suitable model of mutual interaction between human gait and elastic bridge has been developed in [3].

As an example the deterministic force model for walking is given. The vertical force component is greater than the horizontal one, but the lateral and longitudinal horizontal components can also cause vibration related problems of slender bridges. Frequency of lateral movement, which occurs as a result of moving the centre of mass from one foot to the other, is equal to half of the step frequency of vertical or longitudinal movement.

General shapes of the temporal evolution of the pedestrian loads - assuming a perfect periodicity of the force - can be performed using appropriate load-time functions, for a vertical periodic force $F_{p,ver}(t)$, lateral periodic force $F_{p,lat}(t)$ and longitudinal periodic force $F_{p,long}(t)$:

$$F_{p,ver}(t) = G + \sum_{i=1}^{3} Ga_i \sin(2\pi i f_p t - \varphi_i) \tag{1}$$

$$F_{p,lat}(t) = 0.05G \sin 2\pi \frac{f_p}{2} t \tag{2}$$

$$F_{p,long}(t) = 0.20G \sin 2\pi f_p t \tag{3}$$

where G is the weight of the person (usually G = 700 N), f_p is the pacing frequency, a_1 = 0.4, $a_2 = a_3$ = 0.1 are the Fourier coefficients of the i-th harmonic for vertical, lateral and longitudinal forces, φ_1 = 0 and $\varphi_2 = \varphi_3 = \pi/2$ are the phase shifts of the ith harmonic contributions.

The pacing frequency f_p and the pedestrian forward speed v_p are two parameters that play a fundamental role in terms of the characterisation of the excitation. The corresponding average values are presented in Table 1 for walking and running.

Table 1. Data on walking and running [2].

	Pacing frequency f_p [Hz]	Forward speed v_p [ms^{-1}]	Stride length $L=v_p/f_p$ [m]	Vertical fundamental frequency f_{vert} [Hz]	Horizontal fundamental frequency f_{lat} [Hz]
Slow walk	1.7	1.1	0.65	1.7	0.85
Normal walk	2.0	1.5	0.75	2.0	1.0
Fast walk	2.3	2.2	0.96	2.3	1.15
Slow running (jogging)	2.5	3.3	1.32	2.5	1.25
Fast running (sprinting)	>3.2	5.5	1.72	>3.2	>1.6

A general proposal as to the typical frequency ranges for different human activities is given for walking 1.6-2.4 Hz and 3.5-4.5 Hz (first and second walking harmonics), for running 2.0-3.5 Hz, for jumping 1.8-3.4 Hz and for bouncing 1.5-3.0 Hz. Commonly adopted mean value frequency for running and jumping is 2.5 Hz [10].

3 Proposed load models

The current European standard for determination of traffic loads on bridges [4] does not recommend the load models for serviceability limit verification due to pedestrians. The Guidelines for the design of footbridges [5] gives the certain pedestrian load models. The load models are divided into three categories: Single pedestrian load model (DLM1), Group of pedestrians load model (DLM2) and Continuous pedestrian stream load model (DLM3). Instead of pulsating forces in vertical and lateral direction which move with the speed of $0.9\,f_p$, the stationary pulsating forces applied at the most adverse position on the bridge are defined.

DLM1 defines vertical $F_{p,v}(t)$ and horizontal (lateral) components $F_{p,h}(t)$ as:

$$F_{p,v}(t) = 180 \sin 2\pi f_v\, t \quad [\text{N}] \tag{4}$$

$$F_{p,h}(t) = 70 \sin 2\pi f_h\, t \quad [\text{N}] \tag{5}$$

DLM2 defines the effect of a group of 8 ÷ 15 persons walking across the bridge by vertical $F_{g,v}(t)$ and horizontal components $F_{g,h}(t)$ as:

$$F_{g,v}(t) = 180\, k_v \sin 2\pi f_v\, t \quad [\text{N}] \tag{6}$$

$$F_{g,h}(t) = 70\, k_h \sin 2\pi f_h\, t \quad [\text{N}] \tag{7}$$

The effect of synchronisation of step frequencies and the phase shift between pedestrians is taken into account by coefficients k_v and k_h (Fig. 1).

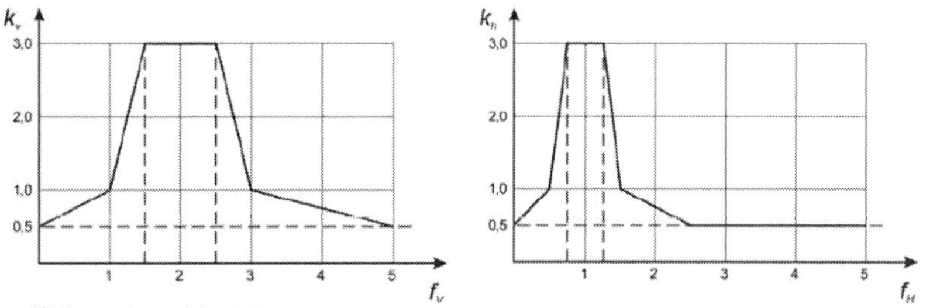

Fig. 1. Coefficients k_v and k_h [5].

DLM3 defines the continuous pedestrian stream is modelled as a uniformly distributed pulsating load with a vertical component $q_{p,v}(t)$ and a lateral component $q_{p,h}(t)$ as:

$$q_{p,v}(t) = 12.6\, k_v \sin 2\pi f_v\, t \quad [\mathrm{Nm}^{-2}] \tag{8}$$

$$q_{p,h}(t) = 3.2\, k_h \sin 2\pi f_h\, t \quad [\mathrm{Nm}^{-2}] \tag{9}$$

The load should be applied in the way to produce the most unfavourable loading case (depending on the mode shape) and a uniformly distributed mass of 400 kN/m^2 (if unfavourable) should be applied at the same location.

4 Example - Footbridge in Čelákovice

The footbridge is a cable-stayed structure with 3 spans 43.0 + 156.0 + 43.0 meters made of Ultra-High Performance Concrete. The height of the steel pylons is 36 meters (Fig. 2 and Fig. 3). Natural frequencies are summarized in Table 2 and important modes of vibration are shown in Fig. 4 to 6.

4.1 Results

Pedestrian loading was modelled using Eq. (6) and Eq. (7) for three principal load cases – a) horizontal excitation with pacing frequency corresponding to the fundamental lateral frequency; b) vertical excitation with pacing frequency corresponding to the fundamental vertical frequency; c) vertical excitation with pacing frequency corresponding to the commonly adopted mean value frequency for walking 2.0 Hz. The results of the analysis are given in Table 3. It can be seen that for the load case c) the accelerations are higher than the limit values taken from Eurocode EN 1990. In such that case the changing of vibration characteristics of the footbridge (natural frequencies) or damping devices should be considered.

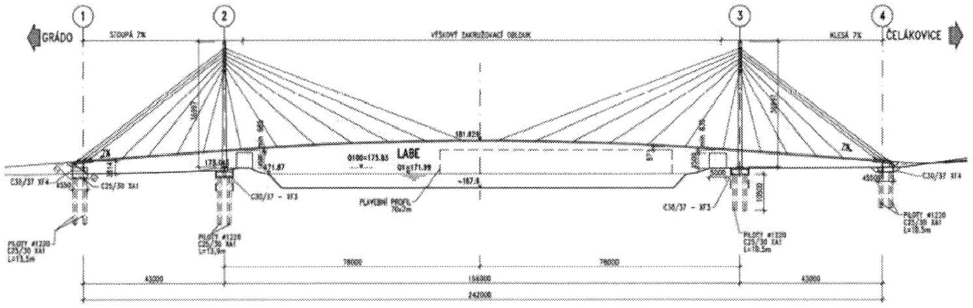

Fig. 2. Footbridge Čelákovice – view [6].

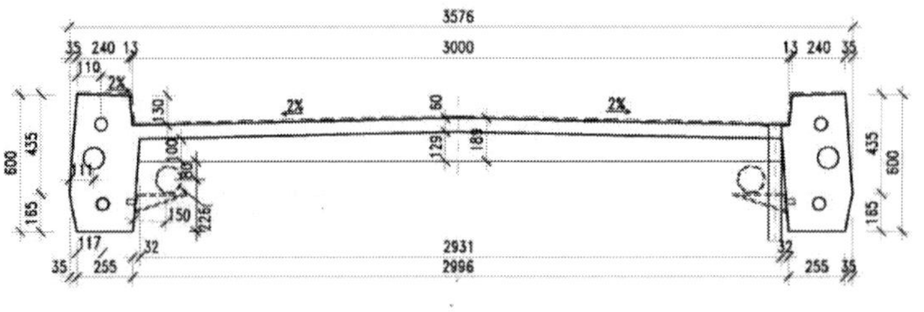

Fig. 3. Footbridge Čelákovice – cross section [6].

Table 2. Footbridge Čelákovice – natural frequencies and modes of vibration.

Mode of vibration	Natural frequency f [Hz]
1 – horizontal bending	0.59
2 – vertical bending	0.63
3 – vertical bending	0.81
4 – vertical bending	1.16
5 – vertical bending	1.44
6 – torsional	1.47
7 – horizontal bending	1.58
8 – vertical bending	1.63
9 – vertical bending	1.72
10 – vertical bending	1.82
11 – pylon	1.87
12 – pylon	1.99
13 – vertical bending	2.04

Table 3. Footbridge Čelákovice – response due to pedestrian loading

Mode of vibration	Pacing frequency f_p [Hz]	Maximum displacement d_{max} [10^{-3} m]	Maximum acceleration a_{max} [ms^{-2}]	Limit acceleration a_{lim} [ms^{-2}]
1 – horizontal bending	0.59	7.0	0.10	0.2
2 – vertical bending	0.63	15.3	0.24	0.7
13 – vertical bending	2.04	4.5	0.74	0.7

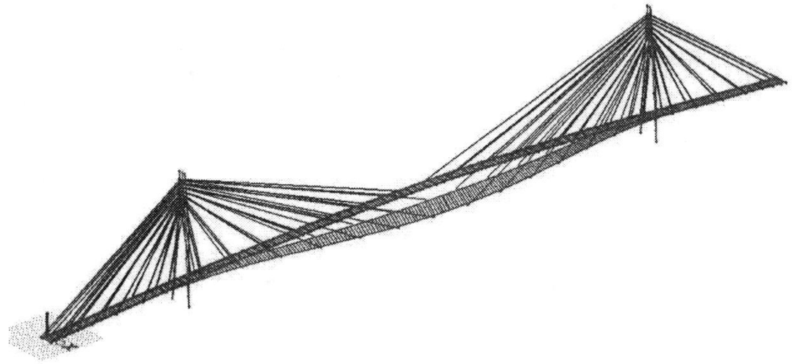

Fig. 4. Footbridge Čelákovice – horizontal bending mode of vibration – f = 0.59 Hz.

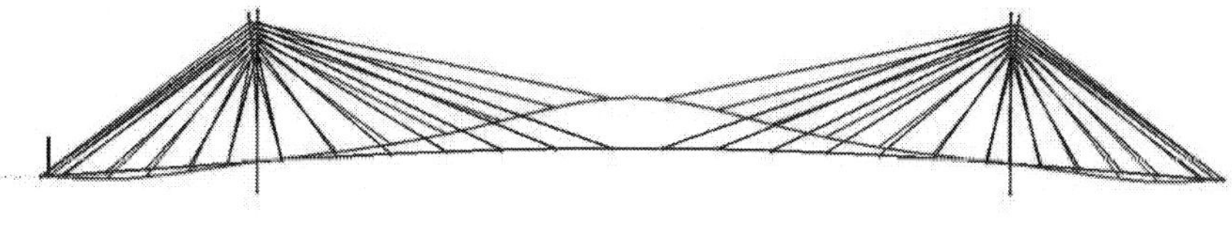

Fig. 5. Footbridge Čelákovice – vertical bending mode of vibration – f = 0.63 Hz.

Fig. 6. Footbridge Čelákovice – vertical bending mode of vibration – f = 2.04 Hz.

4.2 Footbridge with TMD

To avoid undesirable vibrations of the structure it is a good idea to install tuned mass dampers (TMDs) on the footbridge to dissipate the energy from one or more modes. A TMD is often a much more lucrative solution when compared to changing the natural frequencies of the structure.

The theory of how a TMD works, and how to determine the optimal characteristics are summarized in [7]. With respect to the antisymmetric shape with natural frequency of vertical bending 2.04 Hz (cf. Fig. 5) two TMDs were mounted at the points with maximum displacements close to the midspan.

Parameters of the footbridge:

Mass m_s = 440 000 kg

Natural frequency f_s = 2.04 Hz

Parameters of the TMD:

Mass m_d = 2 200 kg

Mass ratio $\mu = 2m_d / m_s = 0.01$

Natural frequency $f_d = f_s /(1+\mu) = 2.02$ Hz

Stiffness $k_d = (2\pi f_d)^2\, m_d = 354$ kN/m

As a result of the increased mass of the footbridge with two TMDs) the corresponding natural frequency changed to the value of 1.83 Hz. The response was calculated for the pacing frequency 1.83 Hz and results are given in Table 4.

Table 4. Footbridge Čelákovice – response due to pedestrian loading

Mode of vibration	Pacing frequency f_p [Hz]	Maximum displacement d_{max} [10^{-3} m]	Maximum acceleration a_{max} [ms^{-2}]	Limit acceleration a_{lim} [ms^{-2}]
vertical bending	2.04	4.5	0.74	0.7
vertical bending with TMDs	1.83	2.2	0.29	0.7

5 Conclusions

In this paper the pedestrian load models for serviceability verification of footbridges, which are missing in the current European codes, are presented. For simplicity reasons the proposed pedestrian load models are based on stationary pulsating loads instead of moving pulsating loads. It is shown that simplified procedure can be used in verification of the serviceability limit state related to vibration due to pedestrians.

Not only natural frequencies but also damping properties and pedestrian loading determine the dynamic response of footbridges and design tools should consider all of these factors. Footbridge vibrations don't cause usually structural problems, but if the vibration behaviour does not satisfy the comfort criteria, changes in the design or damping devices could be considered. The most popular of these are viscous dampers and TMDs. The efficiency of TMD is demonstrated on the example of a footbridge prone to vibrations induced by pedestrians. Is has been shown that if the TMD is tuned quite precisely (especially its frequency) the reduction of accelerations can be very significant.

The authors gratefully acknowledge support from Technology Agency of the Czech Republic (TAČR), project TE01020168 *Centre for Effective and Sustainable Transport Infrastructure* (CESTI) and CTU project SGS17/043/OHK1/1T/11 *Development and application of advanced algorithms for numerical analysis and modeling of structures and materials.*

References

1. S. Živanovič, A. Pavic A, P. Reynolds, J. Sound Vib. **279**, 1-74 (2005)
2. H. Bachman, W. Ammann, *Vibrations in Structures Induced by Man and Machines* (IABSE Structural Engineering Document 3e, 1987)
3. J. Máca, M. Valášek, *8th International Conference on Structural Dynamics EURODYN 2011*, 1083-1089 (2011)
4. *Eurocode 1: Actions on structures - Part 2: Traffic loads on bridges*, (CEN, 2003)
5. *Guidelines for the design of footbridges*, (FIB Bulletin 32, 2005)
6. *Lávka přes Labe - statický výpočet,* (Pontex, 2013)
7. J. Máca, Mater Sci Eng (to be published)

Requirements of technical standards for the dynamic analysis of the load-bearing structures of footbridges

Petr Hradil[1,*], Vlastislav Salajka[1] and *Jiří Kala*[1]

[1]Brno University of Technology, Faculty of Civil Engineering, Department of Structural Mechanics, Veveří 95, Brno, Czech Republic

Abstract. The load-bearing structures of footbridges are designed to be slender and feature spans of considerable length. It can be expected that the natural frequencies of such load-bearing structures range from 0.5 Hz to 5.0 Hz. These low natural frequencies are problematic as regards the effects caused by the dynamic component of wind or the movement of persons. Increased acceleration values can lead to the serviceability limit state of structures being exceeded or the heightening of stress which can result in damage to the structure mainly in the area of details prone to fatigue. The contribution deals with the requirements concerning the execution of dynamic analyses which are listed in the relevant technical standards. It will present procedures for the determination of dynamic loading, methods of solving dynamic tasks as well as design criteria enabling delicate bridge structures to be designed correctly.

1 Introduction

In the Czech Republic, the evaluation of comfort levels for structures under dynamic loading caused by the movement of pedestrians is primarily covered by the ČSN EN 1990 Standard [12] and other related standards. The standards mainly specify the necessity of meeting comfort criteria and keeping within limit acceleration values. The standards do not deal with the determination of the load which affects the structures.

2 Dynamic loading

Loads caused by the movement of pedestrians along structures are interpreted as forces which are a function of position and time. The characteristics of the loading functions are affected by the physiology of the person, their clothes, the stiffness of the structure and other similar factors. Experimental measurements show that the loading function is periodic and is mainly characterized by the pacing frequency. According to performed measurements, standard walking can be described using Gaussian distribution with an

* Corresponding author: hradil.p@fce.vutbr.cz

average pacing frequency of 2.0 Hz and a standard deviation of 0.2 Hz. The loading function is usually defined using a Fourier series:

$$F(t) = G_0 + G_1 \cdot \sin(2 \cdot \pi \cdot f_m \cdot t) + \sum_{i=2}^{n} G_i \cdot \sin(2 \cdot \pi \cdot i \cdot f_m \cdot t - \varphi_i) \tag{1}$$

where

G_0 the weight of one pedestrian (static value), 700 N
G_1 the first harmonic amplitude
G_i the i-th harmonic amplitude
f_m the pacing frequency
ϕ_i the phase displacement of the i-th harmonic member

In engineering practice, only the first member of the Fourier loading function is used to determine load in calculations, as it does this with sufficient accuracy. The loading function is distributed in the vertical direction, F_v and in two horizontal directions, lateral F_{ht}, and longitudinal F_{hl}. Function (1) for one freely walking person can be written using the following formulae (2):

$$F_v(t) = G_0 + 0.4 \cdot G_0 \cdot \sin(2 \cdot \pi \cdot f_m \cdot t) \tag{2a}$$

$$F_{ht}(t) = 0.05 \cdot G_0 \cdot \sin\left(2 \cdot \pi \cdot \frac{f_m}{2} \cdot t\right) \tag{2b}$$

$$F_{hl}(t) = 0.2 \cdot G_0 \cdot \sin(2 \cdot \pi \cdot f_m \cdot t) \tag{2c}$$

It needs to be noted that when dealing with a pacing frequency identical to the frequency of transverse loading, half of the value is considered in comparison with the frequencies of vertical and longitudinal loading. This assumption is derived from human walking, during which a transverse frequency results from steps taken alternately with the left and right leg. Footbridge structures are usually exposed to the concurrent movement of several pedestrians. As the movement of any pedestrian can be specified using individual characteristics (weight, pacing frequency, speed of walking, etc.), the loading caused is more or less synchronized. As a result, initial phase displacement of loading arises due to the movement of various pedestrians along the structure of the bridge. In addition, it is also necessary to consider the possibility of modifications to the movement of individual pedestrians. It is very difficult to analyze the stated characteristics of pedestrian movement along structures and implement them in computer software.

It is also very difficult to carry out complete simulations of the effects of the movement of pedestrians. Assumptions can be made and simplifying hypotheses created based on executed studies which assume that loading should be determined via the multiplication of the effects of the movement of one pedestrian using a selected coefficient. Loading from a crowd of pedestrians is therefore determined using two different models. It is a traditional and comprehensive way of dealing with this complex situation. The traditional loading model is used for pedestrians moving independently along bridges where no synchronization is expected. The loading is determined by multiplying the loading caused by one pedestrian by a coefficient, k. Coefficient k is determined using formula (3)

$$k = \sqrt{\lambda \cdot T} \tag{3}$$

where

λ expresses the number of persons entering the bridge per second,
T is the time the pedestrian takes to cross the bridge.

A comprehensive model of pedestrian loading is able to take the rigidity and natural frequency of the structure into account during calculations. It introduces loading models derived from the number of persons crossing the structure. The equivalent number of persons is added to the calculation as a quantity (N_{eq}) for the case of a scattered or compact stream of persons on the structure (4a) or a very dense stream of persons (4b)

$$N_{eq} = 10.8\sqrt{N \cdot \xi} \tag{4a}$$

$$N_{eq} = 1.85\sqrt{N} \tag{4b}$$

where
N the number of persons on the structure,
ξ the damping ratio of the structure.
Pedestrian-induced dynamic loading used for the calculation of the response is determined for each significant natural frequency. The determination of dynamic loading is also related to the number of persons moving along the structure. A distinction is made between cases when the bridge is crossed by solitary pedestrians, small groups and crowds.

3 Parameters influencing the design, frequency, comfort criteria, damping

The swaying of a footbridge caused by vibration has a negative effect on the movement of pedestrians as they cross it. When designing such structures, it is important to check the parameters that cause vibration, particularly the natural frequencies and natural modes of the vibration, in common the damping value. In the majority of cases, it is sufficient to examine the response of the structure for the first several natural frequencies. This simple method leads to a problem concerning the response of the lowest natural frequencies of the monitored structure. Various rules recommend the examination of the vertical vibration of a structure in a certain frequency range which is given in Table 1, with an emphasis on the characteristics of the movement of persons across the structure.

Table 1. Risk frequency range values noted in literature.

EN 1992	1.60 Hz – 2.40 Hz
EN 1995	0.00 Hz – 5.00 Hz
BS 5400	0.00 Hz – 5.00 Hz
ISO/DIS 10137	1.70 Hz – 2.30 Hz
CEB 209 Bulletin	1.65 Hz – 2.30 Hz

The comfort criteria for the use of a footbridge are usually defined by the owner of the structure. There are three levels comfort level with regard to structural vibration. The first level is the maximum comfort level for pedestrians, who find that the effects of vibration while walking are imperceptible. The second level is one where slight vibration is felt, and the third involves structural vibration that a person can definitely feel, though walking along the structure does not become impossible. The level of comfort is strongly influenced by the users of the bridge and can differ for various groups of inhabitants. A different comfort level is acceptable for children or adults than for older persons or the infirm. It is generally known that there is a difference between the vibrations affecting a structure and the vibrations actually perceived by pedestrians. Nevertheless, the existing knowledge in this area is still inaccurate and insufficient. This is why, in the case of projects, it is necessary to consider recommended values and use parameters which can be determined using calculations and measurements. Threshold acceleration values are defined which should not be exceeded. Eurocode [2] requires the evaluation of the pedestrian comfort criterion for footbridges. An acceptable level of acceleration is stated as being a_{lim} = 0.7 m/s^2 in the vertical direction and 0.2 m/s^2 in the horizontal direction. The lock-in phenomenon must be mentioned in connection with the effect of pedestrian movement across a structure. When persons begin crossing the bridge, the frequencies of the steps they take exhibit random distribution and differing initial phases. However, the pacing frequency of bridge users eventually synchronizes with the natural frequency of the structure. Amplitudes that should not be exceeded are therefore recommended for vertical (10 mm) and transverse (2 mm) vibration.

The lowering of vibration in built structures is achieved via damping. The most frequent method is via internal damping provided by the construction material itself. Another way of damping structures is through friction between individual structural elements, and their mutual displacement. Such damping is interpreted as a force which limits the movement of the structure, and is referred to as viscous damping. A large proportion of the results of experiments conducted on viscous damping in structures suggest that vibration damping is independent of the natural frequency of the structure but is related to the amplitude of movement. Generally, the critical damping value rises with increasing vibration. Studies have been produced which use damping values ranging from 0.1% to 2.0%. The values used for damping are comprehensively summarised in the CEB bulletin and are listed in Table [2]. It should be mentioned that damping can only be determined via measurements at a previously completed structure.

Table 2. Damping of structures.

Type of bridge deck	**Damping ratio ξ**	
	Minimum value	Average value
Reinforced concrete structure	0.8	1.3
Prestressed concrete structure	0.5	1.0
Steel structure	0.2	0.4
Composite structure	0.3	0.6
Wooden structure	1.5	3.0

4 Vibration limits

When acceleration limit values have been exceeded, it is necessary to lower vibration to an acceptable limit. One mechanical means of limiting excessive vibrations is to change the natural frequency of the structure, which involves a radical change in its stiffness or mass, e.g. by adding an auxiliary support. The purpose is to alter the natural frequency of the structure so that it is not near the human pacing frequency. One method of reducing excessive structural vibrations that is considered to be reliable is to increase damping using vibration absorbers in places where significant movement of the structure is expected. Dynamic passive absorbers appear to be suitable for the load-bearing structures of footbridges. These take the form of a mechanical mass-spring-damper set connected to the original structure. A well-designed absorber causes the energy of the vibrating structure to be absorbed by the additional mass of the damper. Passive dynamic absorbers are suitable for damping the movement of structures with low relative damping and one prevailing frequency.

The first step in designing the parameters of such a vibration absorber is the determination of its weight. A good value is taken as being approximately 1/30 to 1/10 of the vibrating mass of the bridge structure. The next stage involves finding the natural frequency and the relative damping value of the absorber. To this end, use is made of formulae (5), (6) and (7), which can be found in the relevant literature, e.g. [1], [7], [8] or [9]

$$f_{TMD} = f_{str} \Big/ \left(1 + \frac{m_{TMD}}{m_{str}}\right) \tag{5}$$

$$\xi_{TMD} = \sqrt{\left(3 \cdot m_{TMD} / m_{str}\right) / \left(8 \cdot \left(1 + m_{str} / m_{TMD}\right)^3\right)} \tag{6}$$

$$B_{TMD} = \xi_{TMD} \cdot \left(2 \cdot m_{TMD} \cdot \omega_{TMD}\right) \tag{7}$$

where f_{TMD} is the natural frequency of the absorber, m_{TMD} is the weight of the absorber, f_{str} is the natural frequency of the bridge and m_{str} is the size of the vibrating mass of the bridge for which the absorber is designed. The value ξ_{TMD} is the damping of the vibration absorber, ω is the circular frequency and B_{TMD} is the damping force proportional to speed.

Formulas (5) to (7) are design formulas for the parameters of the vibration absorber, which must always be specified using dynamic analyses. These analyses are conducted to find optimum parameters. An example of the characteristics of an optimized vibration absorber is shown in Figure 1. It presents the frequency characteristics of the acceleration of a load-bearing structure and a vibration absorber in the monitored frequency.

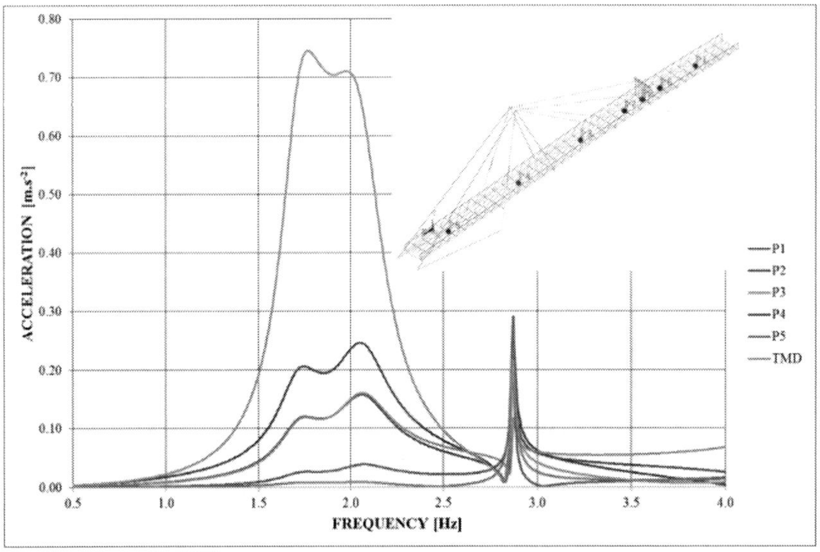

Fig. 1. Characteristics of absorber in a monitored area.

5 Conclusion

The contribution describes issues relevant to the calculation of the dynamic response of footbridge structures. Methods of calculating harmonic load imposed by the movement of pedestrians are shown, as well as the parameters affecting the design and evaluation of a structure. These are mainly the natural frequency and relative damping values for the given structure. In the case that permitted or contractually agreed and pertinent structural design criteria are exceeded, vibration absorbers can be required: their function is described. Vibration absorbers are installed in the load-bearing structures of footbridges in order to reduce excessive vibration to an acceptable level.

This paper was created with the financial support of project GACzR 17-23578S "Damage assessment identification for reinforced concrete subjected to extreme loading"

References

1. H. Bachmann, Lively footbridges – A real challenge. *AFGC and OTUA Footbridge Conference* (Paris, 2002)
2. H. Kreuzinger, Dynamic design strategies for pedestrian and wind actions. *AFGC and OTUA Footbridge Conference* (Paris, 2002)
3. British Standards Institutions, BS 5400, Part 2, Appendix C: Vibration Serviceability Requirements for Foot and Cycle Track Bridges (1978)
4. ČSN EN 1992-2 Eurocode 2 - Design of concrete structures - Concrete bridges - Design and detailing rules (2005)
5. ČSN EN 1995–2 Eurocode 5: Design of timber structures - Part 2: Bridges (2004)
6. J. Stráský, Stress ribbon and cable-supported pedestrian bridges, *Thomas Telford Publishing* (London, 2005)
7. H. Bachman, Vibration problems in structures: *Practical Guidelines, Birkhauser* (Basel, 1995)
8. M. Pirner, O. Fischer, Wind loading of structures, *Information centre of CKAIT* (Prague 2003)
9. J. Kala, P. Hradil , V. Salajka, V. Juttner, Dynamic response of cable stayed steel bridge with tuned mass damper installed - Numerical and experimental approach, *10th International Conference on Modern Building Materials, Structures and Techniques* (Vilnius, 2010)
10. V. Salajka, J. Kala, Z. Cada, P. Hradil, Modification of response spectra by probabilistic approach, *Proceedings of the European safety and reliability conference* (Wroclaw, 2004)
11. J. Kralik, J. Kralik, Probability Assessment of Analysis of High-Rise Buildings Seismic Resistance, Advances in manufacturing science and engineering, Advanced Materials Research (2013)
12. ČSN EN 1990, Eurocode - Basis of structural design, ed. 2 (2015)

MATEC Web of Conferences 107, 00011 (2017) DOI: 10.1051/matecconf/201710700011
DYN-WIND'2017

Experimental study of horizontal forces of pedestrian dynamics

Shota Urushadze[1,*] and *Miroš* Pirner[1]

[1]Institute of Theoretical and Applied Mechanics, AS CR, v. v. i., Prosecká 76, Prague, 190 00, Czech Republic

Abstract. At present similar detailed examination of pedestrian dynamics has been brought about chiefly by the requirements of determining the loads on footbridges. In their case – contrary to bridges – the magnitude of dynamic response is more important than that of static response, as it determines also the serviceability of the footbridge with reference to pedestrian comfort. The forces frequencies of step or strides for different walking velocities is the most important for the further analysis and analysis of the mechanisms.

1 Dynamic response of footbridge in the horizontal direction

The excessive lateral sway motion of the footbridge by the people walking across has great attention in literature due to some examples: especially the very flexible Millennium footbridge in London.

The human organism is more sensitive on horizontal vibration than to the vertical one. The warning example was the London Millennium Footbridge, which after pedestrian traffic showed insufficient stiffness in the horizontal direction and had to be closed. Followed suppress horizontal vibration using the dampers [2], which cost a city about 5 mil. Pounds (original price of 18.2 mil. Pounds). That was the case when the absurd proposal by architect the designer agreed to the project.

1.1 One pedestrian

For an approximate solution of acceleration, assuming that the footbridge is simple beam and the pedestrian is in resonance with the frequency of the footbridge, in this case Kreuzinger recommends [6]

$$a_1 = \frac{\alpha_1 \cdot 700}{M \cdot \xi \cdot k_1} \qquad (1)$$

where $\alpha_1 = 0{,}10$, when $f_{str} < 1{,}25$ Hz a $f_p = 2\,f_{str}$,

---

* Corresponding author: urushadze@itam.cas.cz

$\alpha_1 = 0.05$, when $f_{str} > 1.25$ Hz and $f_p = f_{str}$,
$k_1 = 0.75$
M – mass of footbridge [kg],
700 – weight of pedestrians [N],
f_p – step frequency,
f_{str} – frequency of construction

1.2 The group and the continuous stream of pedestrians

Approximate solution of acceleration a_{hor} recommends Kreuzinger [6] by the expression

$$a_{hor} = a_1 \cdot N_{ef} \cdot k_{hor} \tag{2}$$

where $N_{ef} \approx 0.2\ N$, $k_{hor} = 1{,}0$, when $f_{str} \leq 1.25$ Hz,
$k_{hor} = 0.5$, when $f_{str} = 1.75$ Hz,
k_{hor} approaching 0, when f_{str} approaching 2.5 Hz.
Intermediate values k_{hor} can be interpolated linearly.

For the construction in resonance with the evenly distributed mass along its length and uniformly loaded, with the shape of the excited vibration of a sine wave, calculates the maximum acceleration in the j-th shape by the Stoyanoff [8]

$$a_{j,\max}^{hor} = \frac{2 \cdot c_R \cdot N \cdot \alpha_j^H \cdot w_p}{\pi \cdot m_{deck} \cdot l \cdot \xi_j} \tag{3}$$

where c_R is a correlation coefficient 0,2 (for vandals 1,0),
N – the number of pedestrians,
α_j^H – 0.125,
w_p – 700 N,
$m_{deck} \cdot l$ – mass of footbridge,
ξ_j – damping (j-th shape of vibration).

1.3 Sensitivity of pedestrians

Reaction of the human body to the pad movement in the case of footbridges, instead of bridges, is important for evaluating of pedestrian comfort. In most cases is evaluated acceleration of motion [1, 2], (in the vertical and horizontal direction), as well as a velocity or displacement. From the results of many experiments [3] have been recommended limits of dynamic responses. In general confirmed, that the man is less sensitive to the vibration on the footbridge than the vibrations in residential or in other buildings. Some researchers have found that the recommended limit of pedestrian comfort depends on the time of crossing over the footbridge - with increasing duration decreases its value. Other researchers declare, on the contrary, that the pedestrian "get accustomed" to the movement and limit comfort is unchanging.

In the assessment and determination of boundaries between acceptable and unacceptable dynamic response (boundary comfort) from the viewpoint the human body, it is important to realize, that it depends on the degree of subjectivity, based only on experience†.

2 Lateral horizontal load

The lateral horizontal force depends on the weight of pedestrian on the speed of the walking and on the length of the stride.

Typical time histories of horizontal lateral loads (right, left, right leg) are on Fig. 1.

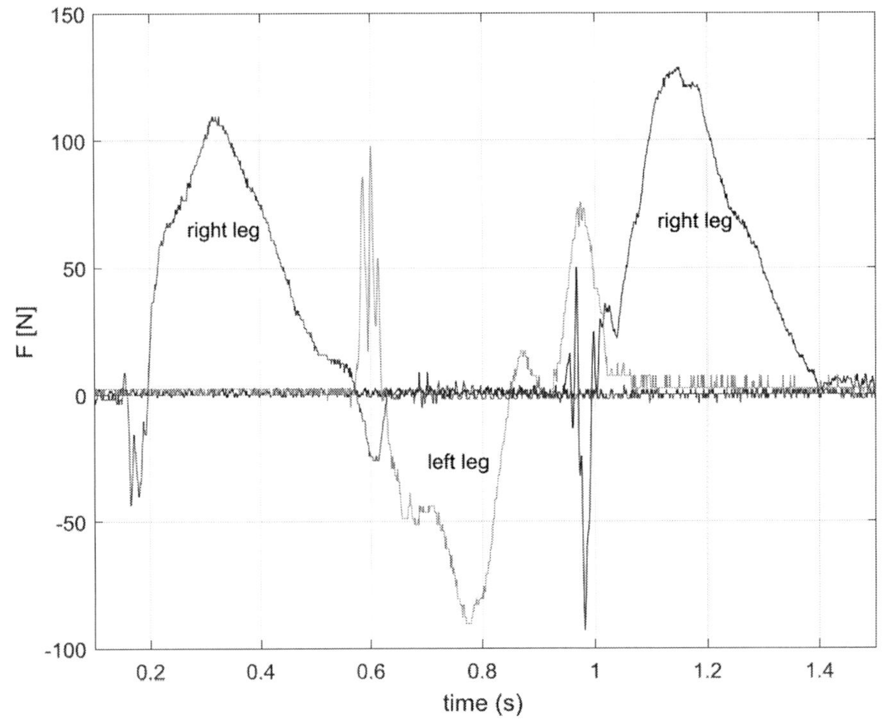

Fig. 1. Typical time histories of horizontal lateral loads.

3 Sensors of the lateral horizontal load

The sensor is the series of three steel strips, 370 mm long, 30 mm wide, supported by boundary box. The defection stress of steel strips was measured by strain gauges which register their horizontal deflection. On the top of steel strips is the plate from soft material, which guarantees the participation of all strips. The sensor is shown on Fig. 2. The position of sensors in plain view is drawn on Fig. 3.

† The human body is very sensitive : the second step of Mercalli–Cancani–Sieberg scale (it is the acceleration 2.5 up to 5.0 mm/s^2) is the start of the human feeling.

Fig. 2. The sensor.

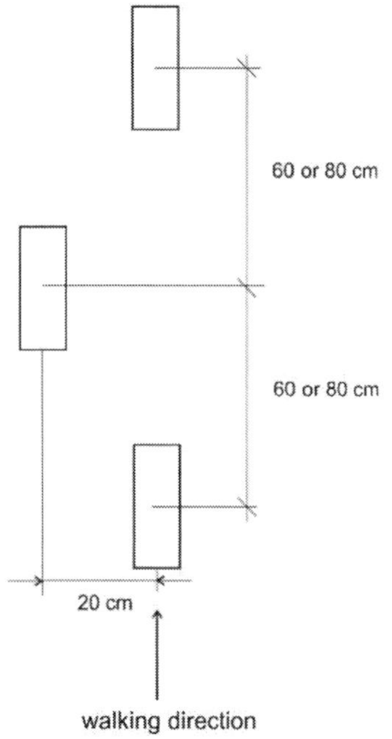

Fig. 3. Sensors positions for lateral horizontal loads.

In Fig. 4 are results of all our experiments. The weights of pedestrians was from 700 N up to 1125 N and the walking speed from 0.45 m/s up to 1.44 m/s (from 1.6 km/h up to 5.1 km/h).

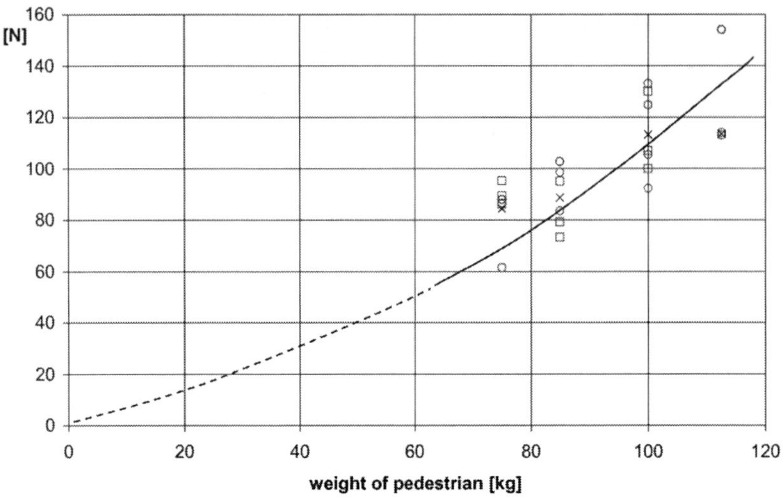

Fig. 4. Relation between weight of pedestrian and lateral horizontal load.
○– stride length 60cm, □– stride length 80cm, ×– the mean value.

4 The dynamic load in the horizontal lateral direction

Stoyanoff [8] gave the formula for the load

$$F(t) = c_R \cdot N \cdot \alpha \cdot w_P \cdot \cos \Omega t \tag{4}$$

where c_R (correlation coefficient ≈ 1)
N number of pedestrians (20 ÷ 25 persons)
α dynamic coefficient (0.125)
w_P weight of the pedestrian
Ω dominant walking circular frequency (commonly $f = 1$ Hz)

5 Power spectral density of footbridge displacement

Pedestrian(s) - induced forced vibrations have two components - transient and stationary. In case of standard values of damping ($\xi = 0.015$) and stride frequency ($f_s = 2$ Hz) the excited amplitude attains about 60% of its maximum after 10 strides, about 85 % of its maximum after 20 strides and the absolute maximum only after 60 strides. (i.e. after 30 seconds of walking). Hence it follows that the first component, i.e. transient vibrations, is in-significant for long footbridges.

After this verification let us derive the power spectral density for stationary displacement. The power spectral density of the consolidated displacement of a point s is

$$G_{v_s v_s}(f,s) = \sum_{n=1}^{\infty} \left| H_n(f,s) \right|^2 \int_0^l \int_0^l v_n(x_1) \cdot v_n(x_2) \cdot G_{p_1 p_2}(f, x_1, x_2) \cdot dx_1 \cdot dx_2 \tag{5}$$

where we have neglected the influence of correlation among natural vibration modes, where:

$$H_n(f,s) = \frac{v_n(x_s)}{M_n\left[\left(f_n^2 - f^2\right) + 2if_b f\right] \cdot 4\pi^2} \tag{6}$$

And G_{p1p2} is the cross spectral density of loads in points x_1 and x_2.

The cross spectral density of loads can be expressed by means of the power spectral density of loads G_{pp} and the function $g_{1,2}(f)$, i.e.

$$\left|G_{p_1p_2}(f)\right| = g_{1,2}(f) \cdot G_{pp}(f) \tag{7}$$

So far there are very few experimental verifications of the first quantity of the righ-hand side of Eq. (7). According to our measurements and according to [7] it is possible to express $g_{1,2}(f)$, by means real coherence

$$g_{1,2}(f) = \frac{\left|G_{p_1p_2}(f)\right|}{G_{pp}(f)} = \sqrt{coh(f)} \tag{8}$$

in the form of

$$\sqrt{coh(f)} = \exp\left(-\frac{\kappa f \cdot |x_1 - x_2|}{N}\right) \tag{9}$$

where k is so called the damping coefficient and N the number of pedestrians filling the whole deck area.

According to our measurements, see e.g. [7][‡], the value of the damping coefficient is approximately $k \approx 60$ to 80

If we express the power spectral density of loads for the correlation function of one process $x(t)$, see [5], with mean values

we obtain

$$R(\tau) = \lim \frac{1}{T}\int_0^T x(t) \cdot x(t+\tau)\,dt, \text{ for } T \to \infty \tag{10}$$

Power spectral density is related to it by the expression

$$G(f) = 4\int_0^\infty R(\tau)\cos(2\pi f \tau)\,d\tau \tag{11}$$

The correlation function of loads can be determined with sufficient accuracy from the formula for the amplitude of the dynamic component of the vertical force. From the formula for the power spectral density of point as we obtain the variance of the displacement

$$\sigma_v^2 = \int_0^\infty G(f)\,df \tag{12}$$

[‡] footbridge length $l = 51$ m, $N = 408$ pedestrians, $f = 1.5$ Hz

6 Conclusion

The time histories of lateral forces were registered and statistical analyzed: the authors received the lateral force dependence on the walking velocity.

The important results of measurements done in the ITAM laboratory and on the footbridges of various supportive systems follow:

1) Mutual relations among the stride frequency, step frequency, step length, dynamic coefficient and the striding velocity depend on individual body characteristics of a pedestrian.
2) The dynamic coefficient for a given pedestrian can be larger than for a group of pedestrians, if they do not move in a synchronous way.
3) The obtained dynamic coefficients are of use for computations of load exerted by a single pedestrian, a group of pedestrians, a connected stream of pedestrians, and vandals, and for the computation of responses of footbridges with different supportive systems.

The supports of grants GA CR 17-26353J and MOST 105-2221-E-032-005 are gratefully acknowledged. Identification code of research project of the Institute of Theoretical and Applied mechanics is AVOZ 6838297.

References

1. C. Barker, Some observations on the nature of the mechanism that drives the self-excited lateral response of footbridges, Int. Conf. „Footbridge 2002“, IABSE, Paris (2002)
2. Footbridge 2002, Int. Conf. „*Footbridge 2002*“, IABSE, Paris (2002)
3. M. J. Griffin, *Handbook of Human Vibration*. Sydney, Academic Press 1990.
4. V. Koloušek, et. all.: *Wind effects on civil engineering structures* (Academia – Elsevier, 1983)
5. V. Koloušek, et. all.: *Aeroelasticity of engineering structures* (Prague, Academia, 1977, in Czech)
6. H. Kreuzinger, *Dynamic design strategies for pedestrian and wind actions*. In: "Footbridge 2002, Proceedings of the International Conference on the Design and dynamic behaviour of footbridges" Paris, 20. - 22. Nov. 2002. Editor: AFGC and OTUA. 129-141.
7. M. Pirner, Sh. Urushadze, Pedestrian dynamics – footbridge loads, Acta Technica CSAV **52**, 269-283 (2007)
8. S. Stoyanoff, M. Hunter, D.D. Byuers, Human-induced vibrations on footbridges, in: *Proceedings of the International Conference on the Design and Dynamic Behaviour of Footbridges*, Paris, France, November 20–22, (2002)

MATEC Web of Conferences **107**, 00012 (2017) DOI: 10.1051/matecconf/201710700012
DYN-WIND'2017

Dynamics effects on a wooden footbridge

Veronika Vašková[1,*], *Lenka* Poništová[1] and *Roman* Fojtík[1]

[1] VŠB-TU Ostrava, Department of Civil Engineering, Ludvíka Podéště 1875/17, 708 33 Ostrava-Poruba, Czech Republic

Abstract. The timber is the current trend for the construction of many footbridges because of many reasons as price, aesthetics and ecology. Most of these structures are designed using simple static models and massive elements. However, there are implemented more complicated constructions including suspended footbridge in Příbor in Czech Republic. This construction with efficient use of material is more susceptible to dynamic effect. The article describes monitoring of dynamics effects at the construction with result of installation dynamics dampers.

1 Introduction

The bridge construction put into service must meet the criteria both in terms of load carrying capacity but also in terms of usability perspective, which includes vibration limits on the structure. Exceeding these limits may lead to reduced mobility (comfort) of people on the construction caused by concerns of pedestrian while passing through construction. In addition, the dynamics effects may also lead to a significant damage of the structure. Especially the excessive vibrations may cause damages on the construction and destruction of a part or the whole construction. Adverse influence of vibrations can cause a substantial reduction of lifetime of the structure or its part due to cyclic fatigue.

These negative impacts are necessary to be removed or suppressed to the values that ensure safe operation of the structure and prescribed comfort, as well. It is necessary to perform the prescribed dynamic tests because of order purpose of appropriate measures and determining the status of the dynamic characteristics.

The footbridge (see Fig.1), located in Příbor in Czech Republic was designed as a subtle structure, which implies a high risk of experiencing negative dynamic impacts. Because of this, the designers of this construction prescribed dynamic load test, which had to confirm or disprove the need to use vibration dampers or determine the values that will serve for tune this damper.

* Corresponding author: veronika.vaskova.st@vsb.cz

Fig. 1. A timber footbridge in Příbor in Czech Republic.

2 Description of construction

The subject of this article is the cable-stayed footbridge with a length of 43 m and a total width of 3.44 m. This is a direct construction with curved overhanging. The suspension is realized by using twelve steel bar rods with diameters of 28 mm and 45 mm. The rods are anchored to the steel pylons with diameter of 457 mm. Rods and pylons are made of steel S355.

The main structure (see Fig.2) consists of two main beams, which are made from glued laminated timber grade GL 28h with the dimensions 220×700 mm. On the main beams, there are placed crossbeams of glued laminated timber grade GL24h and also steel crossbeams grade S355. Timber and steel crossbeams have the dimensions of 200×240 mm. On the crossbeams, there are placed three longitudinal beams of glued laminated timber grade GL24h with cross section of 120×160 mm. The footbridge pavement is made of solid oak timber grade D35. There is bracing solidification below the bridge, made of steel rods grade S355 with diameter of 18 mm.

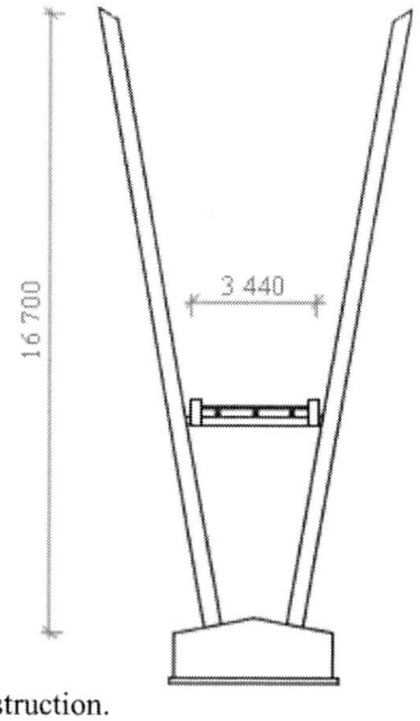

Fig. 2. Cross section of construction.

3 Dynamic load tests

On the monitored footbridge there was performed a dynamic loading test to ensure that comfort of users will not be influenced during normal operation. This test was performed according to ČSN 73 6209. For deducing the dynamic effects at the footbridge, movement of one person and then movement of group of persons were used.

These load models were applied for loading by one person moving in one lane through:

- walk with frequency of about 1 Hz;
- walk with frequency of about 2 Hz;
- trot;
- run.

These load models were applied for loading by group of persons moving side by side in the through lane:

- walk with frequency of about 1 Hz;
- walk with frequency of about 2 Hz;
- trot;
- run;
- demonstrating response in the middle of the span of the footbridge
- demonstrating response in one third of the span.

The load models realized by using the walk, trot and run were always performed in both directions to ensure the relevance of results. The speed of the people movement was chosen per the adverse condition according to the standard ČSN 73 6209. The aim of the test was to demonstrate the maximum acceleration amplitudes for the verification of reduced comfort of pedestrians.

3.1 Load tests weather conditions

All parameters of the weather conditions were measured at the beginning and at the end of the dynamic loading tests. There have been no major fluctuations in the weather conditions during measurement. The average air temperature has been measured 20.5 °C. The average wind speed varied from 0 to 1.8 m/s. Direction of the wind was South-to-Southwest.

3.2 Used devices

The temperature was recorded by optical thermometer. Wind speed was obtained using manual anemometer.

For dynamical measurements the seismic acceleration sensors KB12VD were used. This sensor works on the piezoelectric principle. These sensors have a great dynamical and frequency range and they do not contain the moving parts, which could cause inaccuracy of measurement.

Parameters of sensor:

- sensitivity: 10 000 mV/g
- measuring range: ±0.6g
- the resonant frequency >0.35 kHz

The record was conducted by using the control panel Spider 8 with sampling rate 100 Hz to obtain the desired values. The sensors (see Fig. 3) were attached to the footbridge on the both sides, in the middle and in the third of the span (see Fig. 4).

Fig. 3. Acceleration sensors.

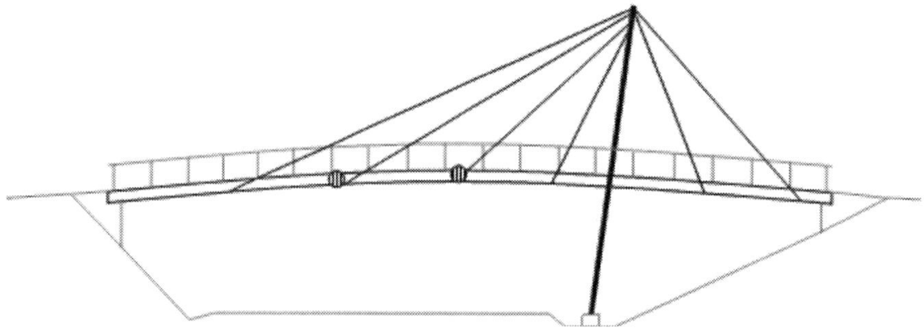

Fig. 4. Chart placement of sensors.

3.3 Measurement data

According to ČSN 73 6209 include determination of the natural frequencies of oscillation of the construction, which were evaluated from measured data (sample is in fig. 5) for each of the excitation load. Dominant frequencies were counted using specific load model because of suspension construction on the ropes.

As the most significant measured (excitation) natural frequency in vertical direction were evaluated these values: 2.5 Hz; 2.8 Hz; 5.1 Hz; 7.5 Hz.

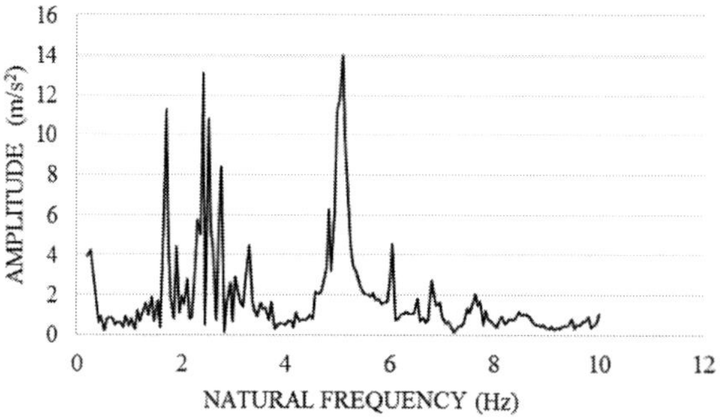

Fig. 5. One of records the measured data.

3.4 Evaluation of dynamical load test

Due the fact that the measured values of natural frequency were higher than 5 Hz, it was necessary to perform the analysis of another dynamic parameters which are values of acceleration. The most unfavourable acceleration values are presented in Table 1.

Table 1. Amplitudes of acceleration obtained for each state.

load models	amplitude	maximum acceleration	effective acceleration
	[m/s²]		
walking one person by frequency 1Hz	1.0228	0.59	0.4172
walking group of persons by frequency 2Hz	1.7224	1.17	0.8273
trot group of persons	1.2324	0.95	0.6717
run group of persons	3.1482	1.95	1.3788

The limits of permissible effective value of acceleration according to ČSN 73 6209 is shown in Fig. 6. The values listed in Figure 6 correspond to the maximum amplitude acceleration acquired in dynamic load test. The response to excitation of the bridge structure is at maximum amplitude of the range 1.6 Hz to 2.1 Hz.

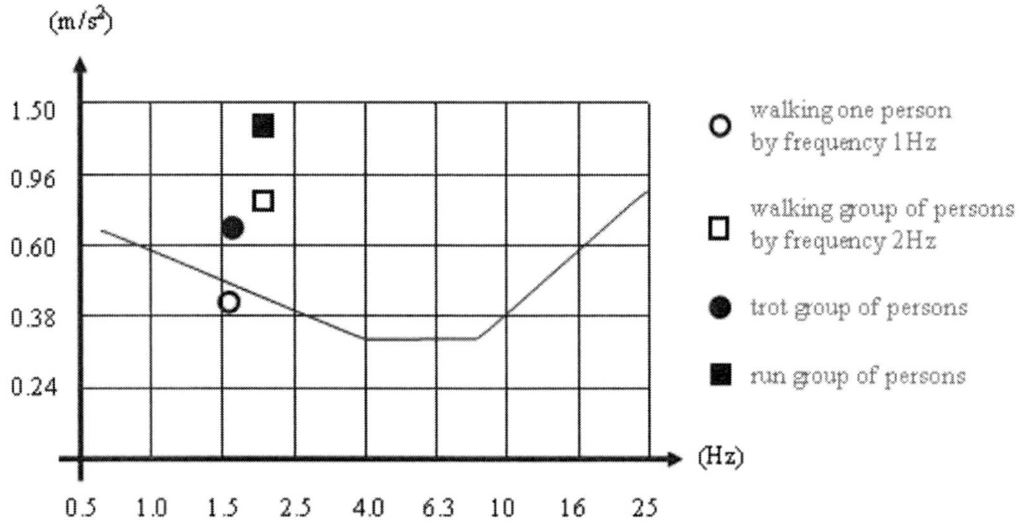

Fig. 6. The results of dynamic tests.

To meet the limit the effective acceleration value shell be below the curve shown in Fig. 6. It is obvious that requirements were not satisfied.

4 Conclusion

The bridge construction put into service must meet the criteria both in terms of load carrying capacity but also in terms of usability perspective, which includes vibration limits on the structure. To determine the models of dynamic characteristic, it is necessary to perform the prescribed dynamic tests. On the basis of dynamic tests, it is possible to propose appropriate measures to correct the behaviour of the structure.

On the observed footbridge, the dynamic load test was conducted. Due the fact that the limits of vibration were exceeded, there was made analysis of effective acceleration value. This analysis also proved the exceeding of the limits. The results proved that is necessary to install vibration dampers.

The project was supported by the Student Grant Competition VSB – TUO. The project registration number is SP2016/136.

References

1. R. Fojtík, A. Lokaj, D. Mareček, Diagnostics of footbridge over the Jizera river in Benešov. Trans Tech Publication Ltd, **752-753**, 124-131 (2015)
2. V. Cecháková, M. Rosmanit, R. Fojtík, FEM modeling and experimental tests of timber bridge structure. Procederia Engineering, **40**, 79-84, (2012), DOI 10.1016/ j.proeng.2012.07.059
3. R. Fojtík, T. Novotný, I. Skotnicová, M. Stolárik, N. Zdražilová, M. Pinkak, Dynamic experimental analysis of steel bridge. Advanced Material Research, **1020**, 215-220, (2014), DOI 10.402/www.scientific.net/ARM.1020.215
4. J. Rumlová, R. Fojtík, *The Timber Tie Beam* : The Analysis of Spatial Framework Joint, Procedia Engineering, **114**, 132-139, (2015), DOI 10.1016/j.proeng.2015.08.051
5. V. Vašková, R. Fojtík, D. Pustka, Monitoring and failures of footbridges made from glued laminated wood. Procedia Engineering, **142**, 87-91, (2016), DOI: 10.1016/ j.proeng.2016.02.017
6. A. Lokaj, J. Vičan, J. Gocál, Theoretical and experimental analysis of timber structure behavior. Building Materials 7, (2010)

A comparison of spectral values obtained in bridge structures via radar interferometry and numerical calculations

Vlastislav Salajka[1,*], *Petr* Hradil[1], *David* Frybort[2] and *Michal* Glockner[3]

[1]Brno University of Technology, Faculty of Civil Engineering, Veveri 331/95, Brno, Czech Republic
[2]CONSULTEST, s.r.o., Sumavska 982/31a, Brno, Czech Republic
[3]Geodezie Ledec nad Sazavou s.r.o., Kozelska 225, Czech Republic

Abstract. Radar interferometry measurements were used to determine the dynamic response of two bridges. The dynamic excitation was performed via trucks crossing the road bumps placed upon each structure. From the relevant records we then determined the spectral values of the bridges. Detailed mathematical models enabled us to compute the natural frequencies and modes (shapes), and these were subsequently compared with their measured counterparts.

1 Introduction

In the monitoring of bridge structures, a basic observed quantity consists in the displacement of structural points caused by the response to the self-weight, traffic, and wind. At present, the static and dynamic behaviour can be also examined via a novel contactless method exploiting radar interferometry; in the wider sense, however, this approach is still being tested. A series of previous experimental measurements showed that the discussed technique can be applied in fast contactless defining of vertical deflections, with the accuracy of up to 0.01 mm in real time. Interestingly, the method is further characterized as suitable for the determination of vibration frequencies in particular objects [1]; as such frequencies may reach as high as 50 Hz, the approach naturally finds application in the monitoring of differently sized bridges, whose dominant vibration frequencies range between 0.5 Hz and 20 Hz.

Before using the method to monitor the static and dynamic motion of a structure, it is advisable to recognize not only the basic static values (deflections of the structure) specified within relevant calculations but also the main natural frequencies and vibration modes, or shapes. Such a set of data enables us to identify suitable positions for the radar reflectors, and it also facilitates the verification of or comparison between the measured values. In many cases, however, appropriate calculations are unavailable or lack the required data; this problem often associates with bridges built during previous decades and

* Corresponding author: salajka.v@fce.vutbr.cz

makes us estimate the positions of the reflectors merely based on common engineering practice.

Radar interferometry remains usable also in situations where the dynamic response is to be determined with respect to evaluating the effects induced by traffic and dynamic wind. The measured values are instrumental for the verification of the initial design and condition of the structure concerned, and they should agree with the results computed using models of the monitored structures.

To demonstrate the applicability of interferometry in measuring the responses of various bridges, we present the outcomes of measurement cycles performed in two markedly different bridges. Both the identification of suitable positions to place the radar reflectors and the selected force excitation procedure (via trucks) exploited design computations simulating the actual load test. We compared the measured and computed deflection values produced by the static testing; the records of the dynamic responses obtained during the excitation of the bridges through the vehicles then enabled us to determine the natural vibration frequencies to be correlated with the computed values.

2 Arch RC bridge

The bridge is a reinforced concrete structure with abutments in the slopes of the embankments, and it embodies a self-anchored arch system having an upper slab deck that rests on four inclined struts and rigid fixing at the crown of the arch. The total length of the structural frame equals 85.9 m, the width is 6.7 m, and the arch span corresponds to 51.3 m, at the rising height of 5.195 m (Fig. 1).

Fig. 1. A view of the discussed bridge with radar and reflectors.

The computational model of the structure to facilitate static and dynamic calculations was set up using the ANSYS system. The full-scale computational model is predominantly based on SOLID187 spatial ten-node finite elements; however, SURF154 and SHELL281 eight-node surface and shell elements, respectively, were also exploited to a large extent. Reinforced concrete is assumed as an isotropic linear elastic material, exhibiting the elastic modulus of E = 33 GPa and the coefficient of lateral contraction equalling ν = 0.2. The specific weight of the material is estimated to be 2 500 kgm^{-3}; the real data are unknown. The specific weight of the reinforced concrete components of the ledges, including the railings, is assumed as 2 600 kgm^{-3}; the corresponding presumed weight of the road then amounts to 2 000 kgm^{-3}. The selected elastic modulus of the ledges is several orders of magnitude lower to prevent the structure from affecting the stiffness of the bridge. The elastic modulus of the road is also selected to be lower, namely 5 GPa. The bearing system is designed as partially pliable.

The static testing comprised two measurement cycles correlated with the positions of two vehicles stationed upon the bridge; their weights, 22.150 Mg and 18.720 Mg, were adopted from relevant weight certificates.

In the actual computation, the two above-indicated values corresponded to the total static load of 400.935 kN, which had been set within the proportional action of the individual wheels.

The measurement records showed the maximum deflection in the central section of the bridge equalling 4.32 mm and 4.20 mm during the first and second test, respectively; the computation-based rate that corresponds to these two values is 4.516 mm. The experimentally obtained deflection at the level of the ledges ranges between 4.41 mm and 4.68 mm, with the computed value being 4.606 mm.

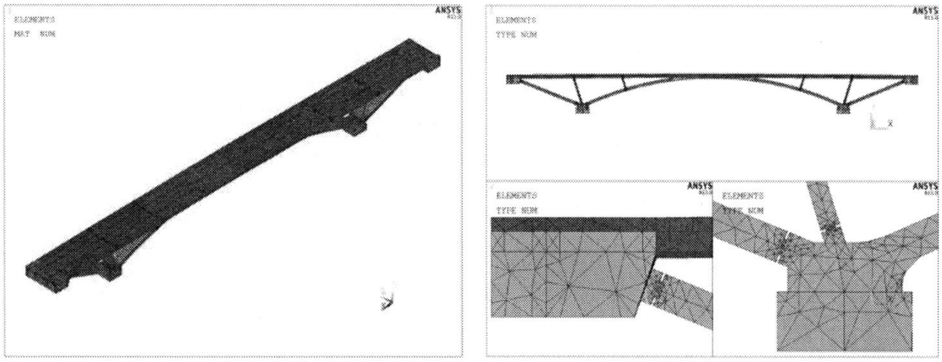

a) The axonometric representation of the model; b) the detailed struts and Freyssinet hinges.
Fig. 2. The mathematical model of the arch bridge.

The dynamic test, involving the heavier vehicle advancing along the structure, produced two measurement sets; we monitored the displacement at both the ledges and the centre of the arch. The excitation was performed via the given vehicle crossing a road bump, whose various locations had been determined from the preliminarily computed natural shapes of the vibrations. In total, twenty four measurement records were acquired; the patterns related to the measured displacement and spectral values are exemplified in Figure 3a) and 3b).

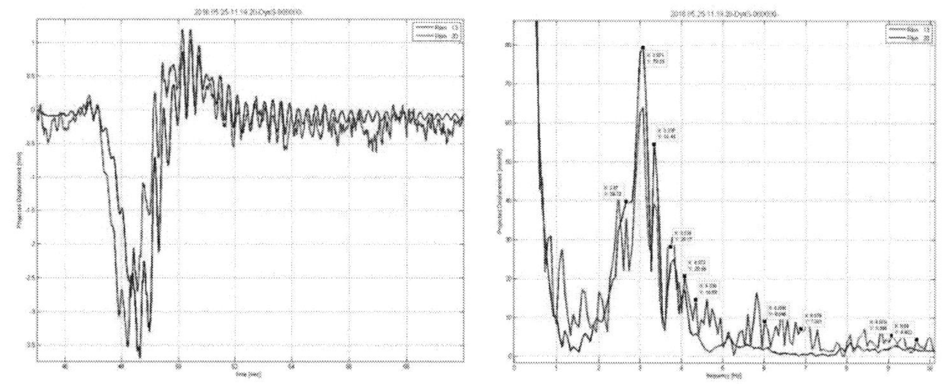

a) The dynamic response patterns; b) the spectral response record.

Fig. 3. The analyzed load case (L5) – a dynamic response pattern – the first set of measurements.

DOI: 10.1051/matecconf/201710700013

The sought frequency values constitute the average of those from the spectral records; the thus obtained values of three frequencies are presented in Table 1.

The same model for the computation of the static response was used to compute the fifteen lowest natural frequencies and modes (shapes). The first five specified natural frequency values complemented with the description of the related vibration shapes are shown in Table 1. For completion, Figs. 4 to 7 display the natural vibration shapes of the first four natural frequencies.

Table 1. The natural frequencies: measurement and computation.

Frequency No.	Computed [Hz]	Measured [Hz]	Vibration shapes
1	**2.675**	**2.460**	1st transverse
2	**3.182**	**3.056**	1st symmetric flexural
3	**3.363**	**3.309**	1st antisymmetric flexural
4	4.158		2nd symmetric flexural
5	5.089		1st torsional

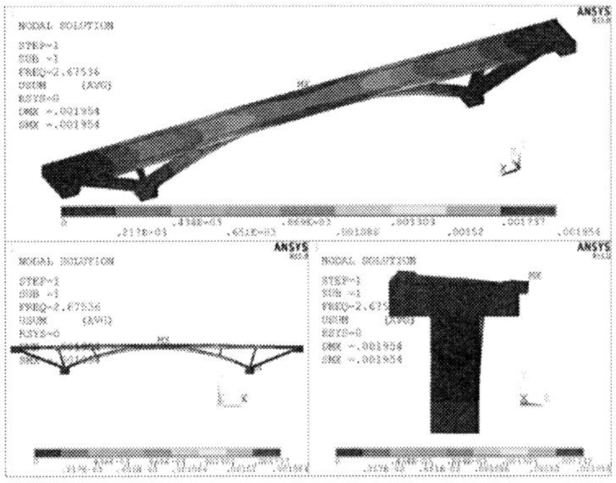

Fig. 4. Natural vibration shape: $f_1 = 2.675$ Hz.

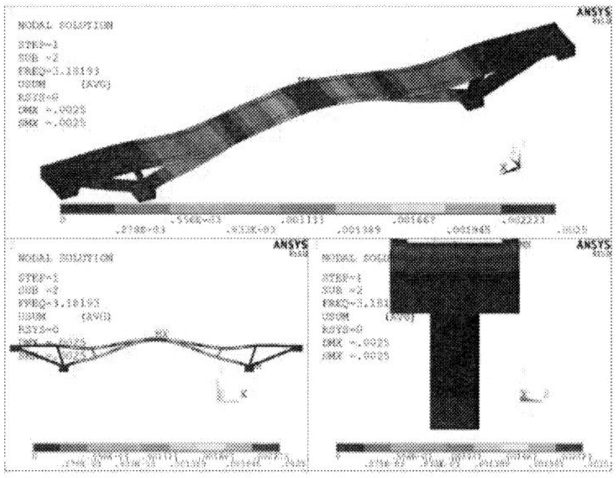

Fig. 5. Natural vibration shape: $f_2 = 3.182$ Hz.

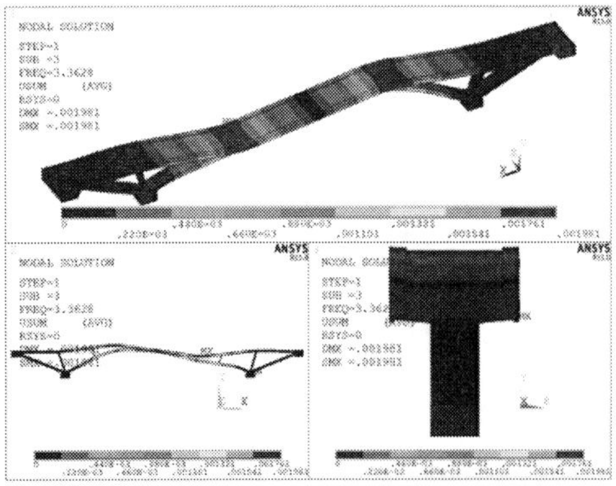

Fig. 6. Natural vibration shape: $f_3 = 3.363$ Hz.

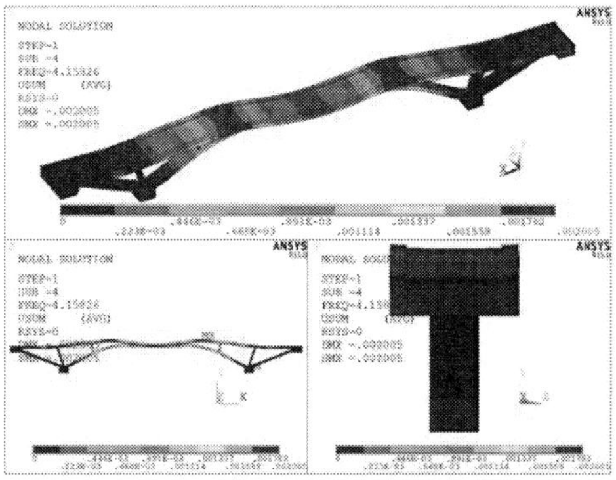

Fig. 7. Natural vibration shape: $f_4 = 4.158$ Hz.

DOI: 10.1051/matecconf/201710700013

3 Continuous multi-span RC slab bridge

Another monitored structure is a bridge whose horizontal portion consists in a continuous prestressed concrete deck having four fields with the span of 13.0 + 17.0 + 17.0 + 13.0 m and obliquely seated on supports. Within the transversal section, the deck is designed in all its fields as a plate of variable thickness. End cross beams are fixed above the abutments. The width of the carrying frame equals 10.05 m, and the frame rests on elastomeric bearings in all the supports. The inner supports are pairs of piers exhibiting different heights.

In this structure too, the model for static and dynamic computations was set up in ANSYS, mainly from SOLID185 spatial eight-node finite elements; these were used to model the piers and the deck. Further, the model exploited SHELL181 four-node shell elements combined with the SOLID185 ones to simulate the distribution of the forces transferred by the modelled bearings, and it also comprised COMBIN14 two-node finite elements to model the stiffnesses of the bearings.

The same material model of reinforced concrete was used; the estimated specific weight of the deck, including the non-carrying frames, amounts to 2 500 kgm^{-3}, and the specific weight of the piers is 2 400 kgm^{-3}.

a) The carrying frames of the bridge;

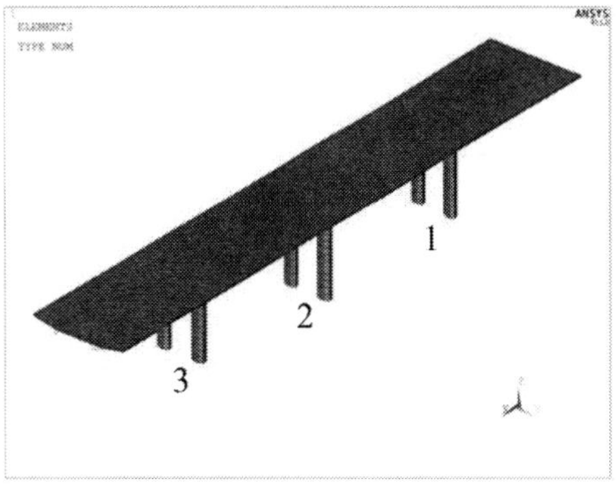

b) the computational model of the bridge, with the piers numbered.

Fig. 8. The axonometric representation of the overall model: subdivision into components.

The static testing comprised two measurement cycles correlated with the positions of two vehicles stationed upon the bridge; these vehicles were positioned between the second and third pairs of pillars. We obtained the values of deflection at the centre of the span of the second field (between pillars 2 and 3), at the ledges, and at the centre of the bottom face of the deck. The weights of the vehicles, 27.580 Mg and 17.900 Mg, were adopted from relevant weight certificates.

In the related computation, the two above-indicated values corresponded to the total static load of 446.159 kN.

The measurement records indicated the maximum deflection in the central section of the bridge being 1.96 mm and 1.98 mm for the first and second vehicles, respectively; the computation-based rate that corresponds to these values is 2.054 mm. The experimentally acquired deflection at the level of the ledges ranges between 1.98 mm and 2.13 mm, with the computed value within the interval from 1.977 mm to 2.271 mm. The larger difference at the ledge on the walkway side can be explained via simplification in the modelling procedure. The stiffness of the sidewalk was not considered in the computation.

In the dynamic test, we measured the deck motion values at locations identical with those observed during the static monitoring. The bridge was excited by the heavier vehicle crossing the applied road threshold at various speeds (30 and 50 km/h). The threshold was positioned to lie either on the axis of the bridge or at its edge, progressively in the second and first fields. The positions had been determined from the preliminarily computed natural vibration shapes. The vehicle was driven on both sides of the bridge, and we obtained 16 measuring records in total.

The first eight specified natural frequency values complemented with the description of the related vibration shapes are shown in Table 2. For completion, Figs. 9 to 14 display the natural vibration shapes of the first five and seventh natural frequencies.

Table 2. The natural frequencies: measurement and computation.

Frequency No.	Computed [Hz]	Measured [Hz]	Vibration shape
1	0.982		Longitudinal
2	3.979		1st transverse
3	**5.386**	**5.518**	1st antisymmetric flexural
4	**7.094**	**7.214**	1st symmetric flexural
5	8.106		1st flexural - support
6	8.115		1st flexural - support
7	**9.360**	**9.445**	2nd antisymmetric + flexural longit.
8	9.398		1st flexural - support

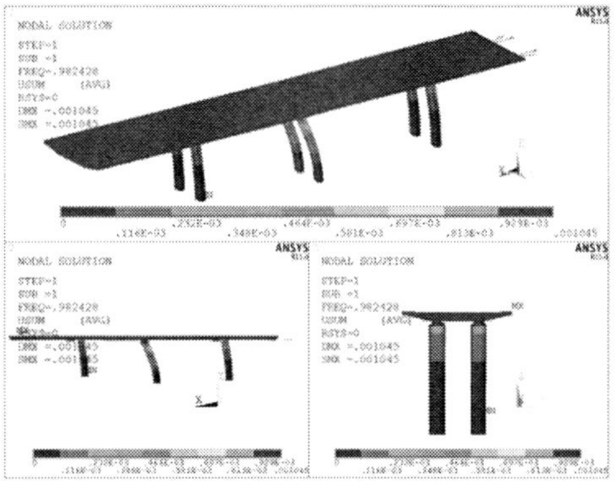

Fig. 9. Natural vibration shape: f_1 = 0.982 Hz.

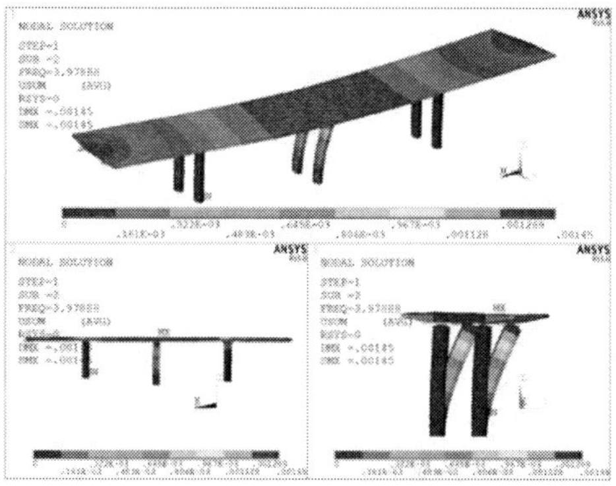

Fig. 10. Natural vibration shape: f_2 = 3.98 Hz.

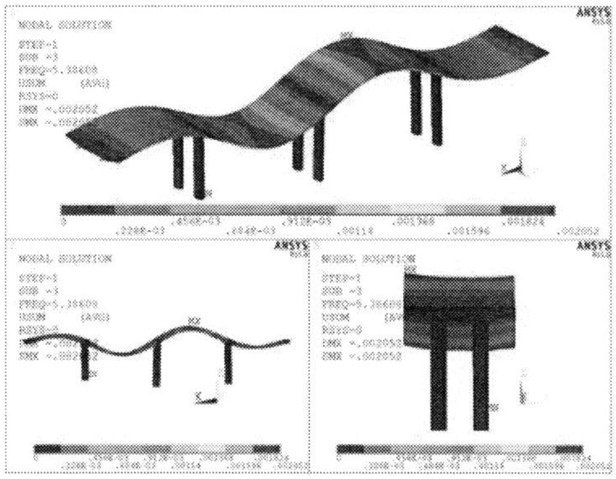

Fig. 11. Natural vibration shape: f_3 = 5.386 Hz.

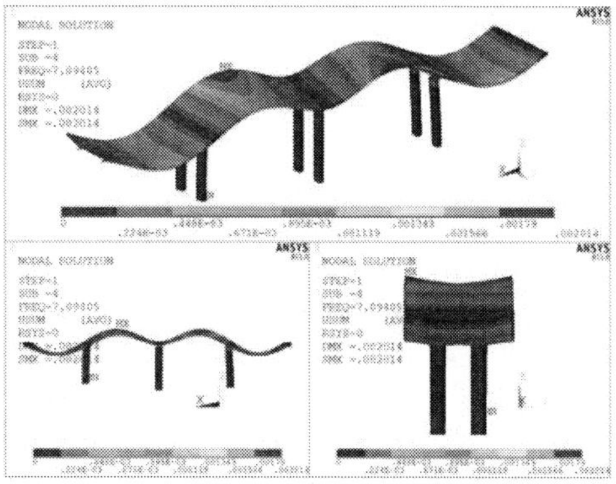

Fig. 12. Natural vibration shape: f_4 = 7.094 Hz.

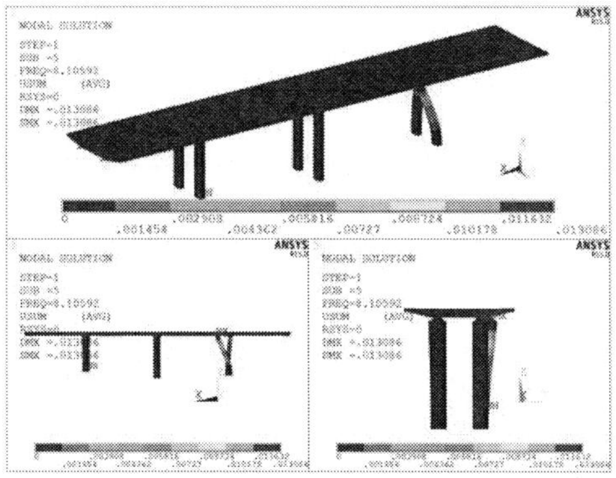

Fig. 13. Natural vibration shape: f_5 = 8.106 Hz.

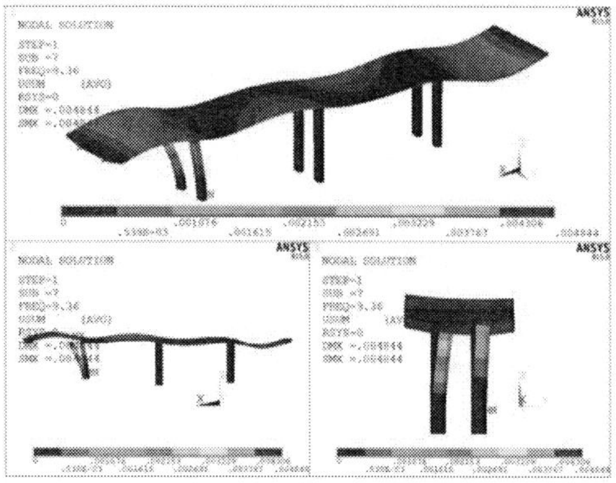

Fig. 14. Natural vibration shape: f_7 = 9.360 Hz.

4 Conclusions

We used the finite element method to set up two similarly designed spatial computational models of two different structures, namely, the reinforced concrete expressway bridges joining the municipalities of Velké Albrechtice and Studénka and urban districts of Polanka nad Odrou and Václavovice (bridge registration codes D1-412.2 and D1-430.2, respectively). All the residential areas are situated in northern Moravia, the latter two being districts within the city of Ostrava.

The static response and natural shapes of vibration were computed in both of the structures. Relevant preliminary calculations enabled us to determine the positions to place the trihedral corner reflectors during the static and dynamic load tests, performed using radar interferometry.

This method was employed to yield the deflection values observed under the load generated by the vehicles used, and it also facilitated the measurement of the dynamic response to a heavy vehicle on the bridge structure. To increase the dynamic effect, the vehicle repeatedly crossed the applied road bump, and the recorded dynamic responses were instrumental for determining the sought frequency spectra.

A comparison of the experimental and computational static response results showed that radar interferometry can be advantageously utilized in static tests and pointed to the high quality of the bridge models set up via the finite element method.

The FEM-based models were exploited in the computation of the natural frequencies and modes (shapes) of the vibrations. For each model, we registered a frequency spectrum of fifteen natural vibration frequencies and shapes; these spectra were then compared with those acquired through measurement. In both cases, the three lowest measured natural frequencies were successfully associated with the calculated ones having corresponding natural vibration shapes.

If we compare the measured and computed values, it becomes obvious that, on condition that the natural frequencies were correctly paired, the results exhibit very good agreement. This type of operation, however, remains an issue hitherto not fully resolved; it would beneficial to create a methodology for processing the time records such that we could clearly determine the natural vibration shapes to which the frequencies found should be assigned.

Significantly, the time records of the dynamic response displacement are normally obtainable and can be employed as the basis for further analysis of the monitored structures. The present paper, comparing the related experiments and computations, proposes that the diagnostics of bridge structures is suitably performable via radar interferometry; this approach, however, could be improved by using multiple radars to suppress the deficiencies arising from the shadowing of parts of the structure.

In terms of the dynamic behaviour, the diagnostics have to be completed with an evaluation methodology to facilitate the assignment of frequencies to corresponding vibration shapes; otherwise, similarly to other relevant methods, it is invariably necessary to rely on computation.

Measurement was performed using interferometric radar IBIS-S, high precision total station Sokkia Net 1AX, videorecorder SONY HDR-CX115 and weather station Conrad D152.

This paper was created with the financial support of project GACzR 17-23578S "Damage assessment identification for reinforced concrete subjected to extreme loading".

References

1. M. Talich, M. Glockner, *A novel approach to defining vertical deflection in railway bridges via terrestrial radar interferometry* (Železniční mosty a tunely, Praha, 2015, in Czech)
2. V. Salajka, *Preparing data for radar interferometry measurements and comparing the measured and computed results* (Technical report, Consultest, Brno, 2016, in Czech)
3. P. Kohnke, ANSYS mechanical APDL Theory Reference, Release 15.0, (ANSYS Inc., 2013)
4. J. Kala, P. Hradil, V. Salajka, V. Juttner, Dynamic response of cable stayed steel bridge with tuned mass damper installed - Numerical and experimental approach, 10th International Conference on Modern Building Materials, Structures and Techniques, Vilnius (2010)
5. V. Salajka, J. Kala, Z. Cada, P. Hradil, Modification of response spectra by probabilistic approach, Proceedings of the European safety and reliability conference, Wroclaw, (2004)

An assessment of the dynamic interaction of the rolling stock and the long-span bridges on high-speed railways

Leonid K. Dyachenko[1,*] and *Andrey* V. Benin[1]

[1]Emperor Alexander I St. Petersburg State Transport University, 190031Moskovsky pr. 9, St. Petersburg, Russia

Abstract. When the high-speed railway traffic is being organized, it becomes necessary to elaborate bridge design standards for high-speed railways. To ensure safety of traffic and reliability of bridges one needs to resolve the dynamic problems of various types of high-speed trains moving along the structures. The article analyzes the magnitude of inertial response as the function of the external stress parameters. A simplified structure assessment technique is offered based on the criterion of passenger comfort while crossing the bridges. The usefulness of the given research arises from the reduction of complexity of the complicated dynamic calculations needed to describe a high-speed train travelling along the artificial structures. Dependences of relative elastic deflection values from the span length are proposed, which allows designing and calculating long-span structures using the simplified dynamic models.

1 Introduction

One of the main peculiarities of bridge design on high-speed railways (HSR) is an obligatory requirement to account for the dynamic interaction of the high-speed rolling stock and the bridge structure. Meanwhile, unlike it is provided by the existing Russian standards, where the dynamic interactions in the 'bridge-train' system are calculated using a system of coefficients (a quasistatic model), at high-speed traffic the dynamic interaction factor should be accounted for by performing a dynamic task [1, 2, 3]. Specificity of the rolling stock power load on the bridge structure is due to the so called speed effect of kinematic excitation, which means that the temporary power action caused by the bridge live load is transferred to the bridge superstructure via the wheel sets of the rolling stock. The most evident power action is posed on the bridge by trains consisting of single-type cars [2, 3].*

While operating a HSR bridge, the speed effect can cause a condition of resonance and lead to dangerous vibrations of the bridge structures. This occurs in case the period of train power action coincides with the main period of the bridge superstructure self-induced vertical vibrations [3]. The dynamic effects not only result in the increase of stress within

* Corresponding author: leonid_dyachenko@mail.ru

the bridge elements, but also they exert an unfavourable influence on the stability of the bridge deck and passenger comfort.

2 Task assignment and design model selection

A dynamic task of the 'bridge-train' system interaction is solved mainly by means of computational simulation. Selection of a design model, thus, depends on the task assigned and – as a consequence – on the necessary specification of the 'bridge-train' system elements. It is possible to use various dynamic models [3]:

I. 'Moving powers on a beam' type. The superstructure is simulated with an elastic beam of finite mass and viscous damping, whereas the train is simulated with a system of moving powers (Fig. 1). Such a model allows to account for the speed effect and to obtain the critical speed values as well as peak values of the superstructure deformations and accelerations.

II. 'Moving masses on a beam' type. The superstructure is simulated similar to the model I, whereas the train is simulated with a system of moving masses with elastic and viscous restraints (Fig. 2). This approach allows to assess the extent of passenger comfort on the basis of acceleration analysis in the car bodies as well as the impact of track and wheel defects on the magnitude of the power of the rolling stock running gear dynamic interaction.

Calculations show that for spans no longer that 50 m the main criterion determining the 'bridge-train' system interaction are the resonance phenomena. The tasks of defining the resonance modes of vibration, critical speeds, dynamic coefficients are evidentially formulated by using Model I. At the same time, when designing long-span bridges according to the Russian and European norms, one should solve a dynamic problem of the 'bridge-train' interaction using Model II [4, 5]. This research comprises an analysis of distinctive problems of the 'bridge-train' system interaction for long-span bridges and is designated to define the main criteria and recommendations to assign the vertical elastic-limit deformations caused by rolling stock as a criterion of providing passenger comfort.

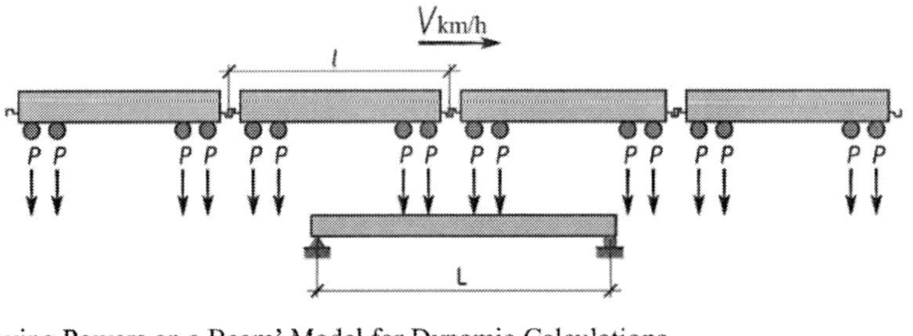

Fig. 1. 'Moving Powers on a Beam' Model for Dynamic Calculations.

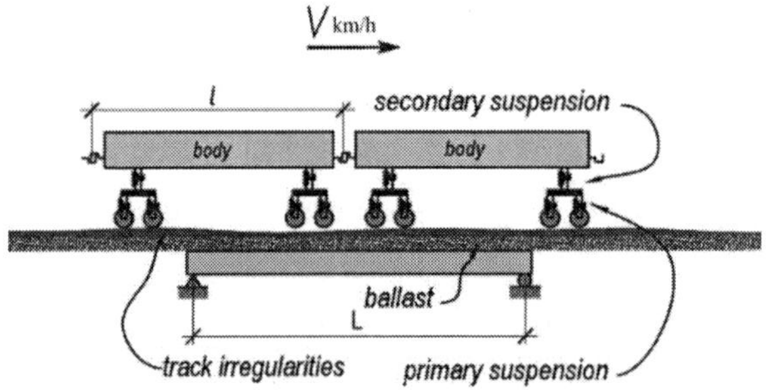

Fig. 2. 'Moving Masses on a Beam' Model for Dynamic Calculations.

The criterion of passenger comfort while travelling along a bridge is ensured by limiting the superstructure elastic deflections, which is a kind of irregularity in the path of motion (kinematic excitation) causing the cars to vibrate. In this case, the vibrations and corresponding car accelerations depend on:

• the car dynamic parameters (rigidity and damping properties of the primary and secondary suspension, distribution of sprung and unsprung masses, car geometry, etc.);

• the magnitude of the superstructure elastic deflection (magnitude of irregularity);

• the magnitude of camber and initial track up-grade.

The main parameter characterizing the dynamic 'car' system is its natural frequency according to the first vertical form (chattering). It is known that the frequency of HSR car vertical vibrations ranges from 0.8 to 1.5 Hz depending on the type of train. Let us introduce a notion of the car resonance speed that is equal to the product of span length multiplied by the car vibration frequency. In other words, the speed at which the kinematic excitation has a frequency equal to the natural car frequency:

$$V_{car}^{crt} = v_{car} \cdot L , \quad (1)$$

where

V_{car}^{crt} – car resonance speed at a bridge;

v_{car} – natural car frequency;

L – span length.

Taking into account the range of possible HSR train types, let us consider the frequencies from 0.8 to 1.5 Hz and define the span lengths which will undergo car resonance vibrations (Table 1).

Table 1. Span lengths corresponding with the car resonance vibrations for various speeds.

V [km/h]	Car Vibration Frequency [Hz]			
	0.8	1	1.215	1.5
200	69.4	55.6	45.7	37.0
250	86.8	69.4	57.2	46.3
300	104.2	83.3	68.6	55.6
350	121.5	97.2	80.0	64.8

The path of train's motion is also of critical importance. Various proportions of span and car lengths cause the path's form to change. Let us suppose that in case the span length L is more than twice as much as the car length, the line of the superstructure elastic deflection will coincide with the sinusoid half-wave and – due to little irregularity of the rolling stock axles load – stay almost unchangeable in time when a train is moving along the structure. At the given proportion of the span and the car lengths resonance phenomena do not arise. Nevertheless, on short and medium spans resonance vibrations of the superstructure might occur when the bridge natural frequency coincides with the frequency of the external action. When resonance vibrations occur, the train path of motion will change according to the superstructure deflection pulsation. It should be noted that this condition not only causes car vibration but also – what is especially important – results in minute local wheel unloading during the contact with rail.

This paper considers more than 50 m long spans and proposes calculations of the 'bridge-train' system dynamic interaction for Velaro RUS ("Sapsan") high-speed train passing along bridge structures. The modal analysis conducted showed that the car vibration frequency according to the vertical form (chattering) is 1.215 Hz.

As a structure model we have selected a chain of superstructures with a single span length, which had been changing in the course of calculation. In the resonance mode of car vibrations the number of spans in a chain plays an important role or – in other words – duration of kinematic action applied to the rolling stock. However, on chains of 10 or more spans vibrations of cars moving along the structure become stable. For structures with shorter spans the adjustment reduction factors are showed in Table 2. It is also interesting how the maximum accelerations in car bodies are distributed according to their position within the train (Fig. 3).

Table 2. Adjustment reduction factors for car body accelerations according to the number of spans.

Adjustment factor	Number of spans					
	1	**2**	**3**	**5**	**7**	**10**
	0.40	0.62	0.74	0.91	0.98	1.00

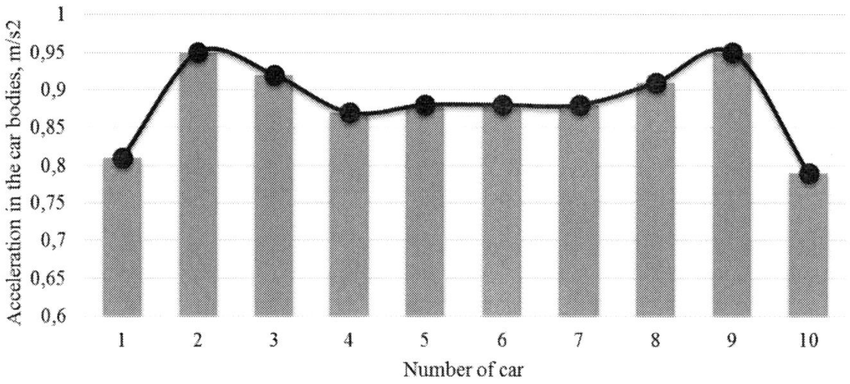

Fig. 3. Distribution of maximum acceleration in the car bodies according to their position within the train moving along the bridge structures.

3 Limitations for vertical elastic deflections

Basing on the models described we have obtained the dependency relations and showed a graph of vertical elastic deflections limitation according to the train speed and span length (Fig. 4). A serious difference from the European norms [5] is the fact that the graph is not limited by the span length of 100 m, which allows to estimate passenger comfort on the long span without making time-consuming dynamic calculations by Model II. Moreover, the graph accounts for the temporary load CK ($K = 8$) adopted in Russia rather that for the European LM71 load, which can be seen as a kind of norms harmonization. Table 3 comprises parameters of the relative deflections (δ) standardization for various traffic speeds.

 DOI: 10.1051/matecconf/201710700014

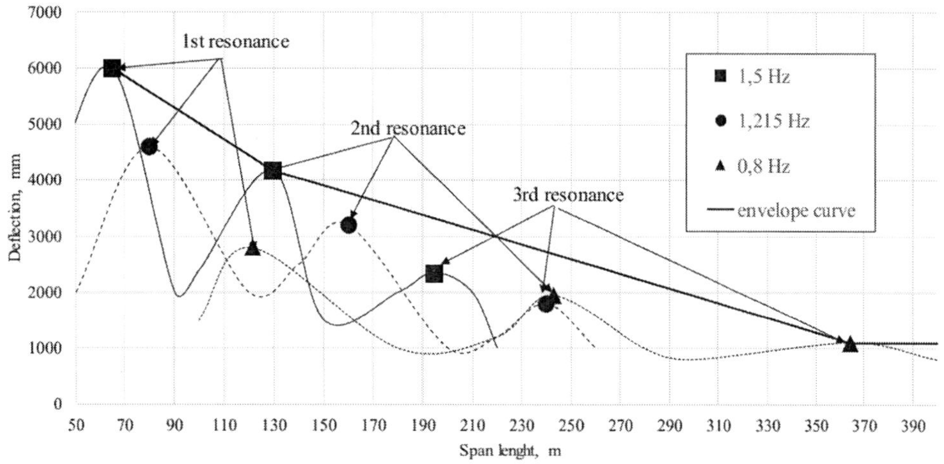

Fig. 4. Limitation of the superstructure vertical deflections aimed to provide maximum accelerations in car bodies of 1.0 m/s² for train speed up to 350 km/h.

Table 3. Parameters of the relative deflections standardization for various traffic speeds

250 [km/h]	**46**	**93**	**140**	**172**	**400**	***L* [m]**
	3920.0	1650.0	1100.0	800.0	800.0	***L/δ***
300 [km/h]	**56**	**110**	**312**	**400**		***L* [m]**
	5100.0	2870.0	800.0	1100		***L/δ***
350 [km/h]	**65**	**130**	**365**	**400**		***L* [m]**
	6000.0	4175.0	1100.0	1100		***L/δ***

As an example of applicability of the presented passenger comfort assessment method we have performed a calculation for a frame-and-beam-type reinforced concrete bridge with 110 + 110 m spans.

According to the norms [4, 5], when defining the admissible elastic deflections, the design length of continuous and frame-type superstructures shall be calculated as a straight average of span lengths multiplied by 1.2 coefficient. Thus, for a frame superstructure the design length is 132 m. At the graph (Fig. 4) one can see that this span corresponds with the second car resonance with 1.5 Hz frequency; that is why it is this train that was taken for calculation. Basing on the same graph we have determined the maximum admissible relative deflection caused by live load C8 [4] and the minimum vertical rigidity of the superstructure. The adjustment reduction coefficient for a two-span structure (equal to 0.62) was also accounted for. Hence, the admissible deflection caused by the load C8 equals 51 mm.

Below one can find the accelerograms of the maximum vertical accelerations (Fig. 5) in the design train cars (natural frequency – 1.5 Hz) moving along a reinforced concrete frame-type superstructure with 110 + 110 m spans at a speed of 350 km/h, which corresponds with the second resonance car speed.

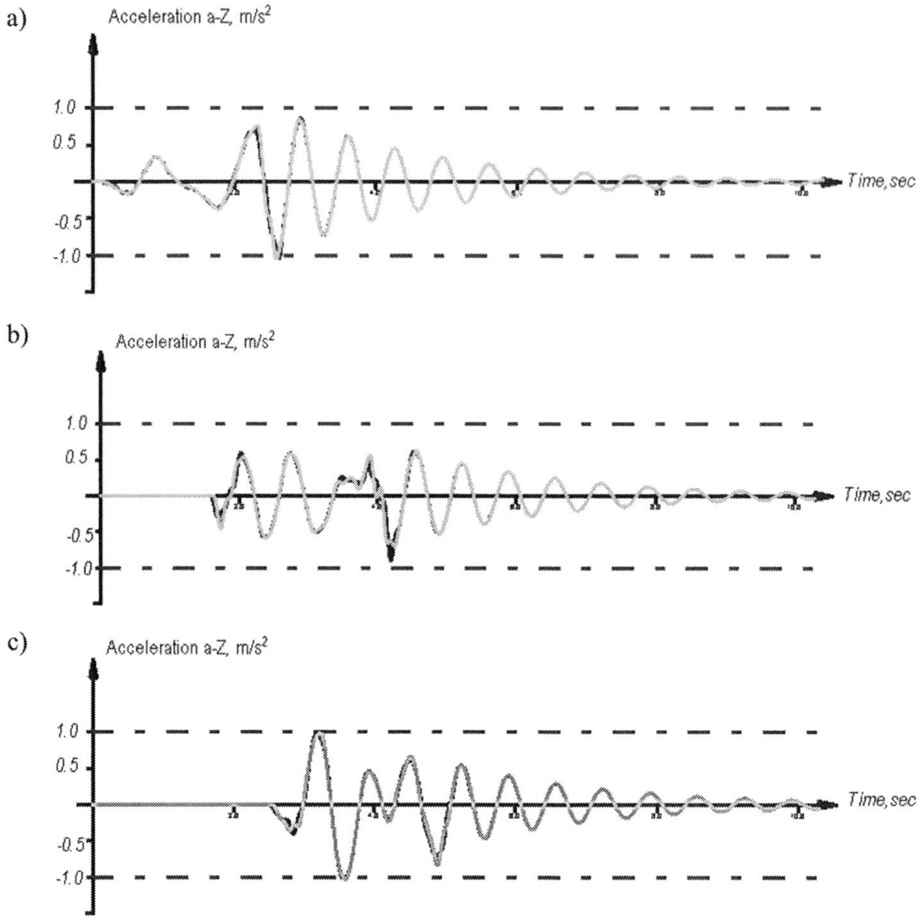

Fig. 5. Vertical accelerations in the design train cars moving along a frame-type superstructure with 110 + 110 m span at a speed of 350 km/h: a – car №2; b – car №6; c – car №9.

4 Conclusion

To ensure safety of high-speed traffic and reliability of bridges during the whole lifetime one needs to resolve the dynamic problems of various types of high-speed trains moving along the structures. The article analyzes the dependences of the 'bridge-train' system interaction from the parameters of external stress. A simplified structure vertical rigidity assessment technique is offered based on the criterion of passenger comfort while crossing the bridges. The usefulness of the given research arises from the reduction of complexity of the complicated dynamic calculations needed to describe a high-speed train travelling along the artificial structures. Dependences of relative elastic deflection values from the span length are proposed, which allows designing and calculating long-span structures using the simplified dynamic models.

References

1. A.V. Benin, L.K. Dyachenko, V.N. Smirnov, Design and construction peculiarities of the 'Moscow–Kazan' high-speed railway mainline, Proceedings of St. Petersburg State Transport University (in Russian), **4,** 45, 15–20 (2015)
2. L.K. Dyachenko, Dynamic calculations of HSR bridge superstructures for passenger trains running at the speed of up to 400 km/h, Proceedings of the International conference New technologies in bridge building from the past to the future (in Russian), St. Petersburg, 91-97 (2015)
3. V.N. Smirnov, A.A. Baranovsky, G.I. Bogdanov, D.E. Vorobyev, L.K. Dyachenko, V.V. Kondratov, *Bridges on high-speed railways* (St. Petersburg, in Russian, 2015)
4. Technical specifications: *Artificial structures on Moscow–Kazan section of Moscow–Kazan–Ekaterinburg high-speed railway mainline. Design and construction norms and requirements*, in Russian (St. Petersburg, 2014)
5. EN 1991-2 (2003): Eurocode 1: Actions on structures. P. 2: Traffic loads on bridges

The verification of a new approach to the experimental estimation of tensile forces in prestressed structural elements by method based on the magnetoelastic principle

Tomáš Klier[1], *Tomáš* Míčka[1], *Tomáš* Plachý[2], *Michal* Polák[2,*], *Tomáš* Smeták[3] and *Miloš* Šimler[3]

[1]Pontex spol. s r. o., Bezová 1658, 147 14 Prague 4, Czech Republic
[2]Czech Technical University in Prague, Faculty of Civil Engineering, Thákurova 7, 166 29 Prague 6, Czech Republic
[3]Freyssinet CS, a. s., Zápy 267, 250 01 Brandýs nad Labem, Czech Republic

Abstract. There are a large number of civil engineering structures where the important structural elements are loaded by large tensile forces. In many practical cases, it is significant to know the current value of tensile force or prestressed stress in these elements for accurate assessment of the reliability of the whole structure. The brief principle and the basic validation results of the new approach to the experimental estimation of tensile forces or stress in prestressed structural elements of engineering structures by the method based on the magnetoelastic principle are described in the paper. The new approach was designed and developed especially for application on existing prestressed concrete structures.

1 Introduction

The basic objectives of the solved problems were research, development, verification and validation of the new approach to experimental estimation of tensile forces in prestressed structural elements produced from ferromagnetic materials, as are prestressed strands, wires and bars, on engineering structures by method based on the magnetoelastic principle.

Structural elements loaded by large tensile forces, as are prestressed reinforcements of prestressed concrete bridges, stays of cable-stayed bridges, suspension cables and vertical suspenders of suspension bridges, suspension cables of suspended roofs, guy wires of radio masts, ground and rock anchors for example, are usually important structural members of a civil engineering construction that influence generally significantly the load bearing capacity of the whole structure. The tensile force in these elements varies in time substantially. The knowledge of its current accurate value is important for exact assessment of the reliability of the entire structure in any of its life cycle, as are its building, putting into operation, monitoring of its construction conditions during its service, preparation of its reconstruction, et cetera.

* Corresponding author: polak@fsv.cvut.cz

The basic intention of the research work described in this paper was a solution of the problem that could be entitled "an accurate experimental estimation of the current value of the tensile force in a particular prestressed structural element on an existing prestressed concrete structure". This task is very common in civil engineering praxis especially in prestressed concrete bridges but is not frequently satisfactorily soluble by any other experimental or theoretical method.

2 Brief description of the magnetoelastic method

2.1 The physical principle

The method is based on an experimental estimation of the magnetic response of the tensile stressed structural element on an external magnetic field. It means, the method is applicable for elements made from ferromagnetic materials only, as is the steel used for production of prestressed reinforcements for example.

The magnetic field intensity H and the magnetic flux density B are ones of the basic physical quantities describing the magnetic field arrangement. The magnetic flux density B compared to the magnetic field intensity H describes a force effect of the magnetism. The hysteresis loop characterizes the relation between B and H which depends on the quantity called the magnetic permeability of material μ that is influenced by particular materials subjected to the effect of the magnetic field and by immediate state of these materials. Some materials or surroundings amplify the force effect of the magnetic field more or less, some others on the contrary fade it.

As was said above, the permeability is given by the kind of material exposed to the effects of the magnetic field, its properties and current conditions. Generally, the value of permeability depends on the level of the affected magnetic field, and not only on its actual value, but also on its history and rate of change, on the actual material temperature and on the current material stress. The ferromagnetic materials are the typical example of ones with this general dependence.

2.2 The principle of the method and the new approach

As was said above, the method described in this paper is based on the magnetoelastic principle. It is known, the magnetic parameters of the ferromagnetic materials are dependent on the level of their mechanical stress among others. The practical consequence is the permeability of steel, what is the function of the tension force or the stress acting on the observed steel structural element and this fact is used by the magnetoelastic method for evaluation of its mechanical stress or tension force.

The principle of the newly proposed arrangement of the magnetoelastic sensor is shown in Fig. 1. The first part of the sensor is the controlled source of the variable homogeneous magnetic field, as is the primary coin supplied by varying current. The second part is the appropriately structured system of Hall sensors for detailed observation of the magnetic field intensity H in the investigated cross section of the studied prestressed structural element. The third one is the secondary coil scanning the magnetic flux density B in the investigated cross section.

The Hall sensor is an electronic component that is used for measuring of the magnetic field intensity H. The modern Hall sensors are small in size and they are sufficiently sensitive for accurate enough observation of the tangential component of the magnetic field close to the surface of the studied prestressed structural element.

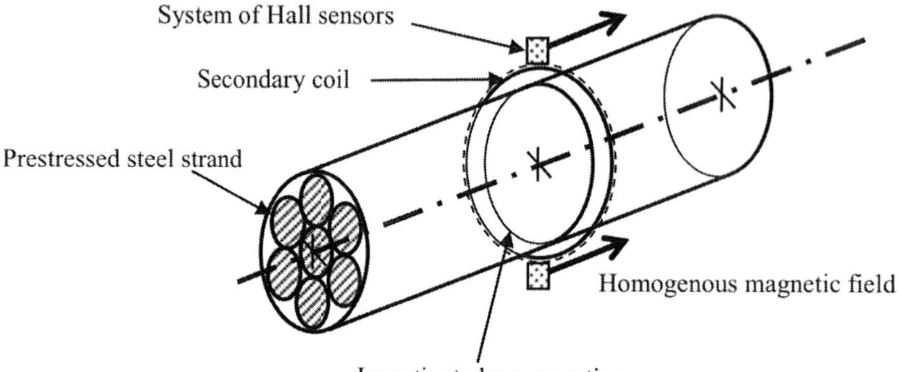

Fig. 1. The principle of the newly proposed arrangement of the magnetoelastic sensor.

3 The verification of the new approach

3.1 Description of the experiments

A number of laboratory experiments was realized during the verification of the new approach to experimental estimation of tensile forces in prestressed structural elements based on the magnetoelastic principle.

Various possibilities for sensor configuration and measurement parameters were examined.

For example, an appropriate placement of the Hall sensors around the observed element for satisfactory characterization of the magnetic field intensity H in the immediate vicinity of the element, an influence of the magnetic surroundings, an effect of the sensor shielding, a suitable time behaviour and time duration of the varying current signal supplying the primary coil and the most suitable dimensionless parameter P used for results evaluation were investigated.

The verification of the new approach was completed by a laboratory experiment that was focused on the systematic investigation of changes in the magnetic behaviour of selected prestressed elements in dependence on its instant temperature and level of the mechanical stress. The experiment was carried out in November 2016 in the experimental centre of the Klokner institute (the Czech Technical University in Prague). The magnetoelastic characteristics of eight prestressed elements were studied during the experiment. Five elements using at present time (two prestressed strands Lp15,7/1860 MPa (below labelled L1 and L2) and three prestressed bars 15/17 made by firms Dywidag and Mikusol (below labelled T1, T2 and T3)) and three elements that were used formerly and were obtained by demolitions of old structures made from prestressed concrete (three patented wires P4,5 of different age (below labelled D1, D2 and D3)), were investigated.

During the experiment, the studied elements were placed in a climatic chamber (see Fig. 2) and loaded in a steel tensile testing machine (see Fig. 2).

The magnetic behaviour of the investigated elements was observed in four temperature steps namely, at -20 °C, at ±0 °C, at +20 °C and at +40 °C. In each temperature step, the elements were loaded in nine or seven force steps selected differently for each particular element according to its design resistance.

For each element and each particular temperature and force step, the hysteresis loop was measured and evaluated. With respect to the quantity and complexity of the obtained experimental results in the form of the hysteresis loops, it was not effective to evaluate measured data in their whole range. At first, several definitions of the different dimensionless parameters *P* were defined. From now on, they represented the shape and character of each measured hysteresis loop.

Secondly, the dependence of the different evaluated parameters P on the temperature and the mechanical stress was statistically evaluated and critically assessed. Base on this rating, the resultant dimensionless parameter P was chosen that is defined as the difference between two values of the quantity B in exactly determined points of the hysteresis loop divided by the greater quantity B. It has been used for finally evaluation of all measured data.

Fig. 2. The exterior view on the climatic chamber and the steel tensile testing machine (on left) and on the magnetoelastic sensor installed on the prestressed strand inside the chamber (on right).

3.2 The conclusions resulted from the verification experiments

The surface appearance of the investigated patent wires D1, D2 and D3 does not allow to differ reliably their kind during a real application of the method on an existing structure though the differences in chemical composition were found between the particular wires (see Table 1). For this reason, the results relation between the stress and parameter *P* (see Fig. 4) was evaluated as the average value for all three studied wires. However, as expected, the evaluated magnetoelastic characteristics for the particular wires have differed slightly each other. But their differences were small and the values of the stress evaluated based on the average relation between the stress and parameter *P* deviated from the real values about ±3% of the largest stress at maximum.

During the verification experiments, the magnetoelastic characteristics of the prestressed strands were studied for one kind of this material only which is used in the civil engineering practice currently. However, several different test samples of this material were used. The stress values evaluated on all samples based on the final relation between the stress and parameter *P* for the prestressed strands deviated from the real values about ±2.0 % of the largest stress at maximum.

The large differences between the magnetoelastic characteristics of the prestressed bars made from different materials were determined. Moreover, it was found, the screw thread of the bar influences negatively the experiment. For obtaining of acceptably accurate results by the magnetoelastic method, the screw thread has to be removed from the bar surface in the neighbourhood of the cross section, where the magnetoelastic sensor would be placed. And it is unrealizable practically on an existing structure in real conditions. The sensitivity of the dimensionless parameter P on the mechanical stress in the investigated prestressed bars is substantially less than these ones obtained for the prestressed strands or wires. This matter of fact is caused partly by the significantly lower design resistance of the bar materials and also by their dissimilar chemical compositions (see Table 1).

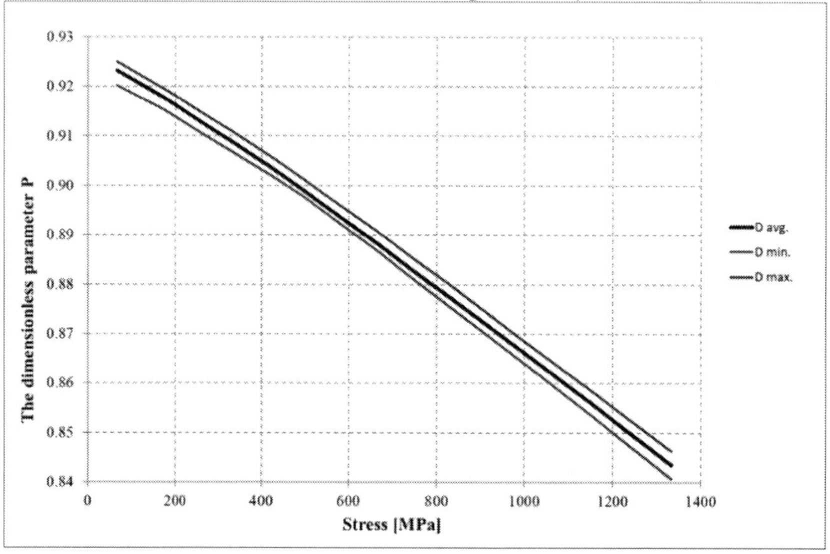

Fig. 3. The results of the statistical evaluation of the verification experiment for the patent wires D1, D2 and D3 and for the temperature step +20°C that describes the relation between the stress and the resultant dimensionless parameter P and its variation.

4 The application in situ on the existing structure

4.1 The experiment description

The usability test of the new approach described in this paper by an application in situ was carried out on a prestessed concrete structure, namely on the existing footbridge across Vltava River in Prague built in 1984.

The load bearing structure of the footbridge is built as the precast segmental construction and its static system is the stress ribbon bridge with three spans 85.5 m, 96.0 m and 67.5 m. The considerable part of the footbridge deck was overflowed by two great floods that happened in August 2002 and in June 2013.

The main reason for realization of the experiment on this particular footbridge were doubts about the real actual value of the prestress in the prestressed cables that was influenced partly by their corrosion and also by partial deterioration of the construction conditions of the footbridge load bearing structure caused by two above mentioned great floods.

The basic information about the used substantial structural elements, about the load bearing structure geometry and about the installation procedure was obtained from the archived footbridge design documentation.

There are used two types of steel cables on the footbridge structure. The cables of type A are the carrying ones of the stress ribbon bridge and the cables of type B are the prestressed ones preventing the dilatation of transversal gaps between the segments.

One magnetoelastic sensor designed for the assembly on the existing structure was installed on the prestressed cable of type B composited from six strands. The position for the sensor installation on the cable was selected not to be significantly affected by corrosion in this place. The cable was studied as the whole because of there was not space between the strands wide enough for the sensor installation on the individual strands.

At first, the hollow (see Fig. 4) was created carefully around the selected cable in which the sensor could be assembled. Secondly, the sensor body was installed on the cable. Thirdly, the secondary coil was reeled on the sensor body by the help of the specially made coiler rotating with the sensor body (see Fig. 4). Fourthly, the primary coil was made in the same way as the secondary one (see Fig. 4). Next, the system of Hall sensors was placed inside the cable in small free hallow between strands and then its position was specified precisely relative to the both coils. At the end, the thorough sensor shielding was installed (see Fig. 4).

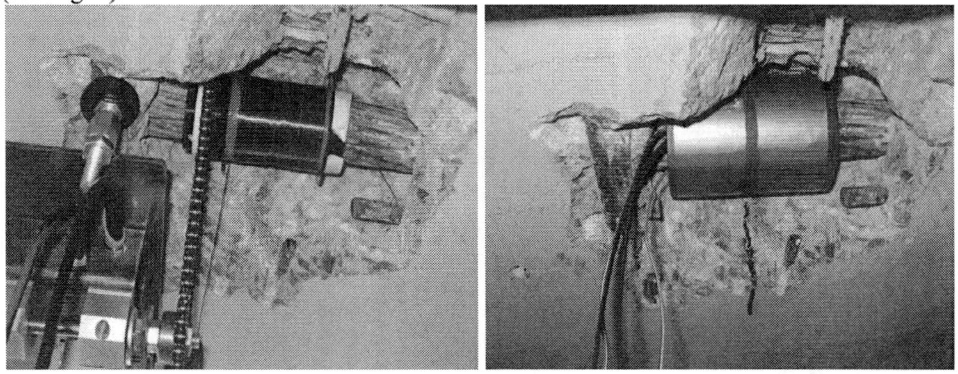

Fig. 4. The view on the created hollow, the studied cable, the coiler, the partially reeled primary coil on the black sensor body (on left) and on the completed magnetoelastic sensor (on right).

4.2 Results of the experiment in situ

The hysteresis loop was obtained as the primary results of the experiment. Next, the dimensionless parameter P was obtained from the measured loop and then the resulting value 825 MPa of the normal stress in the investigated cable was obtained from the evaluation curve similar to the one drawn in Fig. 3.

The assessment of acceptability of the evaluated normal stress in the cable was carried out based on comparison with expected stress in the prestressed cable that was specified from the available information about the footbridge structure. The analysis of the static calculation of the footbridge showed that the normal stress in the cables of type B could be expected in the interval from 680 MPa to 1000 MPa. The wide range of the interval is related to the prestress losses due to successive prestressing of the subtle footbridge structure by fourteen cables of type B. A record about the real procedure of the successive prestressing of the prestressed cables during the footbridge construction enabling a more accurate assessment of the stress in the investigated cable has not been found unfortunately.

The experimentally specified resulting value of the normal stress in the investigated cable is practically located in the middle of the expected interval taken from the static calculation. It means that the experimental resulting value could be characterized as credible.

5 The chemical and microscopic analysis

The chemical analysis of the selected chemical elements for the mass fraction was carried out on the same test samples described above in Chapter 3.1. The results are summarized in the Table 1.

The microscopic analysis was performed on the same test samples too. The steel uniform martensitic and tempering structure was determined for all of them except the prestressed bar T1. The strongly inhomogeneous structure was found in its cross section.

Table 1. The summarization of the results obtained from the chemical analysis of the mass fraction of the chemical elements in the test samples.

Test sample	C [%]	Mn [%]	Si [%]	P [%]	S [%]	Ni [%]	Cr [%]	Co [%]	N [%]	Ba [%]
D1	0.76	0.45	0.21	0.011	0.023	0.023	0.067	0.007	0.005	< 0.01
D2	0.81	0.70	0.16	0.022	0.018	0.030	0.092	0.005	0.006	< 0.01
D3	0.77	0.44	0.19	0.013	0.025	0.022	0.069	0.006	0.005	< 0.01
T3	0.10	0.92	0.39	0.014	0.022	0.170	0.237	0.013	0.008	< 0.01
T2	0.10	1.44	0.31	0.008	0.009	0.071	0.105	0.009	0.012	< 0.01
T1	0.21	1.42	0.53	0.009	0.008	0.108	0.125	0.012	0.011	< 0.01
L1	0.84	0.76	0.27	0.016	0.007	0.057	0.223	0.006	0.005	< 0.01
L2	0.83	0.83	0.22	0.011	0.008	0.035	0.278	0.006	0.005	< 0.01

6 Conclusions

The brief principle and the basic validation results of the new approach to the experimental estimation of tensile forces in prestressed structural elements of engineering structures by the method based on the magnetoelastic principle are described in the paper. The new approach was designed and developed especially for application on existing prestressed concrete structures.

Based on the performed chemical analysis, it could be stated that the chemical elements modifying likely the magnetoelastic behaviour of the studied prestressed elements besides iron (Fe) are also nickel (Ni) and chromium (Cr).

All of the experiments performed in the laboratory and on existing prestressed concrete structures in situ confirmed the applicability of the method that is very suitable for an inspection of the stress in the steel prestressed elements assembled from wires or strands on the existing prestressed concrete structures on which another nondestructive method [1, 2, 3] cannot be used in the most of real cases in the civil engineering praxis.

On the other hand, the approach does not seem appropriate for inspection of the prestressed bars. These ones are usually used on civil engineering structures as temporary mounting prestressed elements and another experimental method for the tensile forces observation, as are a professionally made force transducer or a strain gauge measurement for example, could be easily used in these cases.

According to the author's knowledge, the standard magnetoelastic sensors [4, 5], that have hitherto been used and that have been composed from the primary and secondary coils only, are positioned on an unloaded investigated element before its activation. The sensitivity of the used measuring equipment, which is usually composed from the data logger and the sensors, is specified during tensioning of the element using the information about the actual prestressed force from a tension hydraulic jack. The standard approach is only useable for the observation of the comparative change of the tension force compared with the realized initial reading which has to be carried out in the unloaded state of the element. For the mentioned knowledge, the approach is inappropriate for an objective of the experimental estimation of the current value of the tensile force in a prestressed structural element on an existing concrete structure.

The described new approach also provides an advantage for experiments carried out on newly built prestressed concrete structures compared to experiments performed by standard magnetoelastic sensors [4, 5, 6].

As was found during performed experiments, the advantage of the new approach is a possibility of the significant elimination of risks of an inaccurate evaluation of the current stress in the prestressed elements monitored already in the moment of their activation. These risks are especially attached to the possible change of magnetoelastic characteristic of the standard magnetoelastic sensor [4, 5, 6] caused by a change of sensor magnetic surroundings in the time between its calibration during the prestressed element activation and a performed stress measuring. The variability in the sensor magnetic surroundings is common during building of a civil engineering structure. It could be caused by a concreting of steel-fibre concrete or a removal of a steel massive falsework from the sensor surroundings after concrete hardening for example.

This paper has been supported by the Technology Agency of the Czech Republic project No. TA04030307.

References

1. M. Polák, T. Plachý, Determination of Forces in Roof Cables at Administrative Center Amazon Court, Procedia Engineering **48**, 578-582 (2012)
2. P. Fajman, M. Polák, J. Máca, T. Plachý, The Experimental Observation of the Prestress Forces in the Structural Elements of a Tension Fabric Structure, Appl. Mech. Mater. **486**, 189-194 (2014)
3. M. Polák, T. Plachý, Experimental Evaluation of Tensile Forces in Short Steel Rods, Appl. Mech. Mater. **732**, 333-336 (2015)
4. M. Chandoga, P. Fabo, A. Jaroševič, Measurement of Forces in the Cable Stays of the Apollo Bridge, *Proceedings of the 2nd fib Congress*, Naples, Italy, 5-8 June, 674-675 (2006)
5. A. M. Sarmento, A. Lage, E. Caetano, J. Figueiras, Stress measurement and material defect detection in steel strands by magneto elastic effect. Comparison with other non-destructive measurement techniques, *Proceedings of the 6th International Conference on Bridge Maintenance, Safety and Management IABMAS 2012*, Stresa, Lake Maggiore, Italy, 8 - 12 July, 914-921 (2012)
6. H. J. Wichmann, A. Holst, H. Budelmann, Magnetoelastic stress measurement and material defect detection in prestressed tendons using coil sensors, *Proceedings of 7th International Symposium on Non-Destructive Testing in Civil Engineering NDTCE'09*, Nantes, France, 30 June – 3 July (2009)

Fatigue strength assessment of bridge dilatation shutter in real operation

Miroslav Šulko[1,*] and *Matúš* Margetin[1]

[1]Slovak University of Technology, Faculty of Mechanical Engineering, Department of Applied Mechanics and Mechatronics, Námestie slobody 17, 812 31 Bratislava, Slovakia

Abstract. In general, the bridge dilatation shutters and their components are dynamically loaded structures and so, the fatigue strength for such bridge joints should be analytically evaluated during design process. Due to difficulties arisen in calculation of acting forces on individual components the bridge shutters were assessed in a real operation. In the article, there will be presented results and diagrams obtained from direct measurements of the time depending strain i.e. stress processes in the critical points for that bridge dilatation joint. In addition, the results obtained from measurements and fatigue strength evaluation will be also confronted with testimonials based on applications of standards.

1 Introduction

The bridge dilatation also referred as bridge closures are construction elements of the bridges which are primary used for thermal and humidity compensations. The lifetime and serviceability of the whole bridge can be extended by using such element as a part of such bridge structure. The lamella (scissors) types of those bridge closures are usually used for balancing higher dilatation displacements. If the structure is a relatively simple then during installation a bridge closures there is no necessity adjustment to be performed on a given bridge structure. Principle of the scissor mechanism is obvious and displayed in Fig.1.

The highest loading is transmitted by applied lamellas in the place of the upper aperture for pin as is shown in (Fig.1) for the first lamella. By passing the truck axle above the rail of such dilation closure, the effect is that this place is loaded with combination of the normal force and bending moment. Because the loading of this structure is a cyclically loading, in term of such dynamic character of that bridge operation, there is necessary to assess the bridge not only for expected static loading but also for the fatigue strength of considered structure. For correct assessment of the whole structure, it is very important to identify where it is the most critical point of the structure. For given structure of the bridge, the most critical point of whole structure was the weld of the rail console and stirrup see (Fig.1).

* Corresponding author: miroslav.sulko@stuba.sk

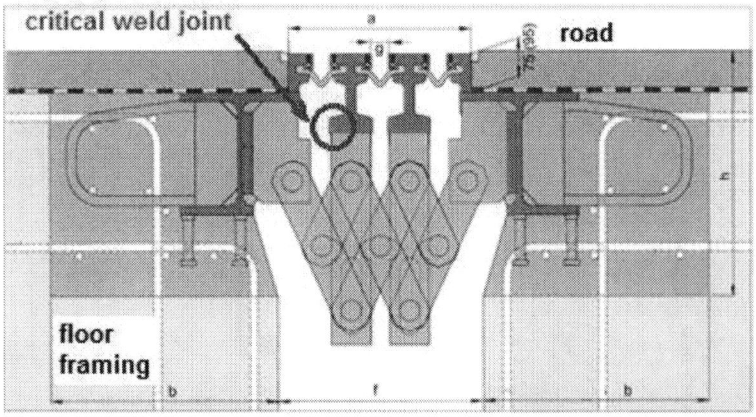

Fig. 1. Principle of usage the bridge closure with specified critical weld.

2 Measuring of operational loading

The assessment of the fatigue strength i.e., the resistance against the fatigue crack of steel structures, describes the standard of Eurocode3 (Design of steel structures – Part 2: Steel bridges), addition E, article. E.3.3 [1], where it is stated that the amplitude of normal stress can be evaluated by

$$\Delta\sigma = \left(\phi_h \Delta\sigma_h + \phi_v \sigma_v\right)\gamma_{Mf} \tag{1}$$

where $\Delta\sigma$ is an overall amplitude of the normal stress,

$\Delta\sigma_h$ is an overall amplitude of the normal stress from horizontal loading (it is about 20% from vertical loading)

$\Delta\sigma_v$ is an overall amplitude of normal stress from vertical loading of the structure

$\phi_h = 2.4$ or $\phi_v = 2$ is a dynamic coefficient of the horizontal or vertical loading

$\gamma_{Mf} = 1.25$ is a safety coefficient

Vertical and horizontal stress amplitudes are designated for maximum prescribed loading of axels 130 kN by Fig. 2.

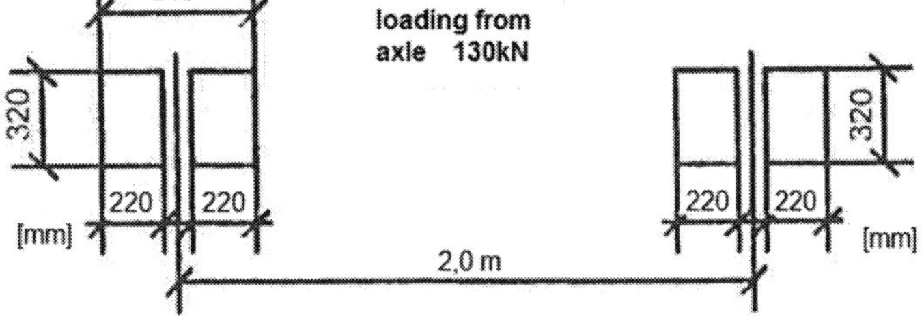

Fig. 2. Vertical loading during fatigue strength assessment.

For the first approximation, the required stresses by (Eq. 1) were obtained using mathematical model created for bridge dilatation MZ 240 N of the length 11.8 m where spacing between individual scissor mechanisms was about 1.4 m overhang of upper "I" beams 180×83 was 0.3 m (Fig. 3).

 DOI: 10.1051/matecconf/201710700016

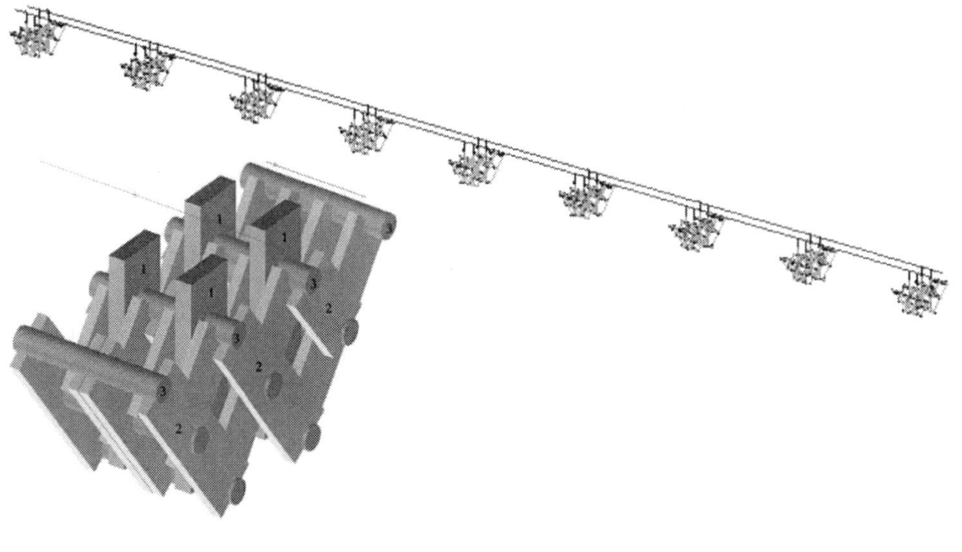

Fig. 3. Mathematical model of bridge dilation MZ240 N with main components: 1 – stirrup, 2 – lamella, 3 – pin.

The calculated values of stresses for the most disadvantageous position the axels transition in three under reviewing places for scissor mechanism are given in Table 1.

The high stress values, as well as requirements on lifetime test realization, led to developing abbreviated bridge dilation with two scissor mechanisms. In the lab, there was closure built-in concrete block and the loading was simulated as a half part of axle transitions with a pressure of 130kN (forced by crank mechanism). Pressure of the half part of an axle was simulated by two hydraulic cylinders. After all lab tests completion that represented about 1.10^6 truck transitions were measured stress values of individual cross-sections, the values are given in Table 1. Significant low stress values comparison with the results from mathematical model led to requirement of the direct measurement of stresses in the real bridge operation with implemented dilatation. The length of the dilatation MZ240 was 17m and two main beams of the shape I 180×83 were supported by 12 scissor mechanisms deployed evenly about 1.4 m. Throughout the dilations installation, there were also installed gauge sensors for sensing strains on the scissor mechanism number 3 a 4 as is shown in Fig. 4a.

The sensor deployment allowed evaluating of individual stress components in lamella (normal stress from vertical force and also from bending moment perpendicular to the direction of drive – Eq.2) and in a stirrup (normal stresses of vertical force and from bending moment in the direction of drive and perpendicularly to the drive direction – Eq.3) and possible shear stresses from torsion loading.

$$\sigma_L = |\sigma_N| + |\sigma_{Mx}| \tag{2}$$

$$\sigma_s = |\sigma_N| + |\sigma_{My}| + |\sigma_{Mx}| \tag{3}$$

DOI: 10.1051/matecconf/201710700016

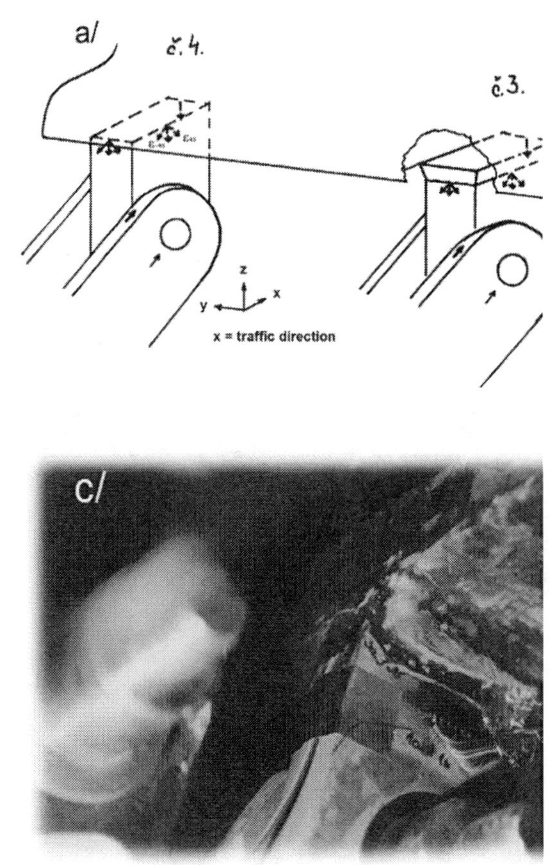

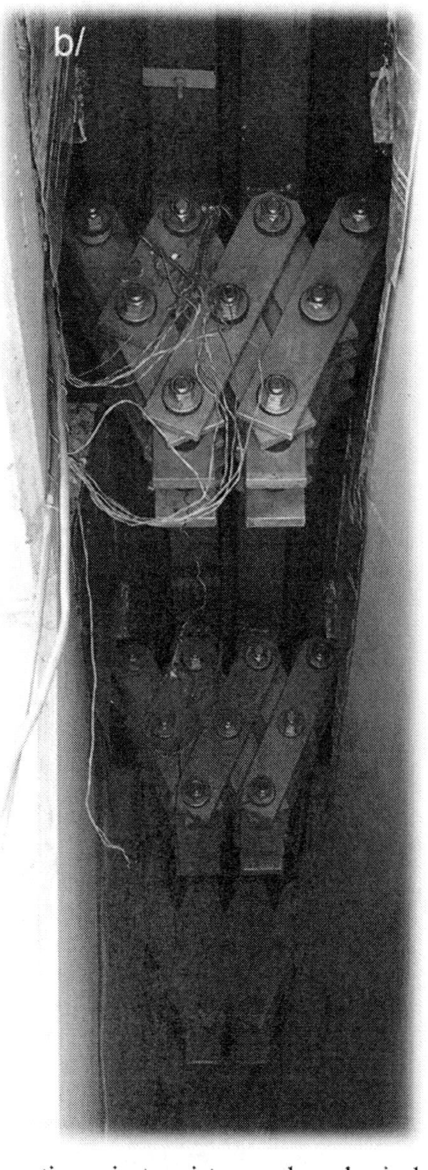

Fig. 4. Gauge sensors applied on bridge dilatation:
a) scheme of deployment,
b) whole view on built-in dilation,
c) detail of sensors applied on stirrup of scissor 3.

The gauge sensors were enclosed with protective mastic against moisture and mechanical damage throughout implementing into the bridge structure. The bridge dilatation was installed on bridge structure of the highway bypass near the city Uherské hradište (Czech Republic) next to support number 29 at the downward part of viaduct. The position of scissor mechanisms with installed sensors is shown in Fig. 5.

Fig. 5. The bridge dilation MZ240 N with installed gauge sensors throughout measurement.

The measurement was performed in the period with expected maximum traffic in Friday. Connection of measuring aperture and setting zero conditions was performed during night shift between 11.00 and 03.00 hour which this adjustment allowed without any problems. The measurement was conducted in time range from 3 a.m. to 2 p.m. Considering on the measured results was shown the continue of measuring is unsupported because the significant stress values form car transitions were registered only during transitions of trucks. During transitions of other vehicles were measured strains almost undetectable (they did not cause any measured loadings on closure).

The diagrams of stresses measured in places were evaluated using the expression of Mohr's circle and eq. 2 and 3. Sampling frequency was up to 500 Hz to precisely capture the peaks of stress during vehicle transitions. Processes of stresses of chosen time segments during observed period (total 11 hours) are displayed in Fig. 6. The displayed segments from the record of measured stresses represent range of stresses in critical place of weld I-beam with stirrup on the scissor mechanism number 4 in length of 210 s. The levels of stresses on the mechanism 3 were significantly lower. The peaks marker as K1 up to K5 are typical stresses for trucks transitions representing the lowest and highest stress levels. The different stress values are the reason of different truck weights (and loading of axels) and also different phasing of stiffness with instant time of transition through the bridge dilatation and different transition velocity. Throughout whole observed time range, values neither higher or lower, displayed in Fig. 6 were not recorded. Unwanted noise of equipment represents total amplitude of 2.5 MPa which is acceptable precision of tensometric measuring devices. Small peaks relatively of huge number slightly overhung this level of noise represent transitions of cars which are by far more numerous as truck transitions.

The process of loading in the aperture of stirrup was considered identical with loading near the weld. The process of loading in lamella of the scissor mechanism number 4 in the place of aperture for pin displayed in the Fig. 7 are significantly lower as in the case displayed in Fig .6.

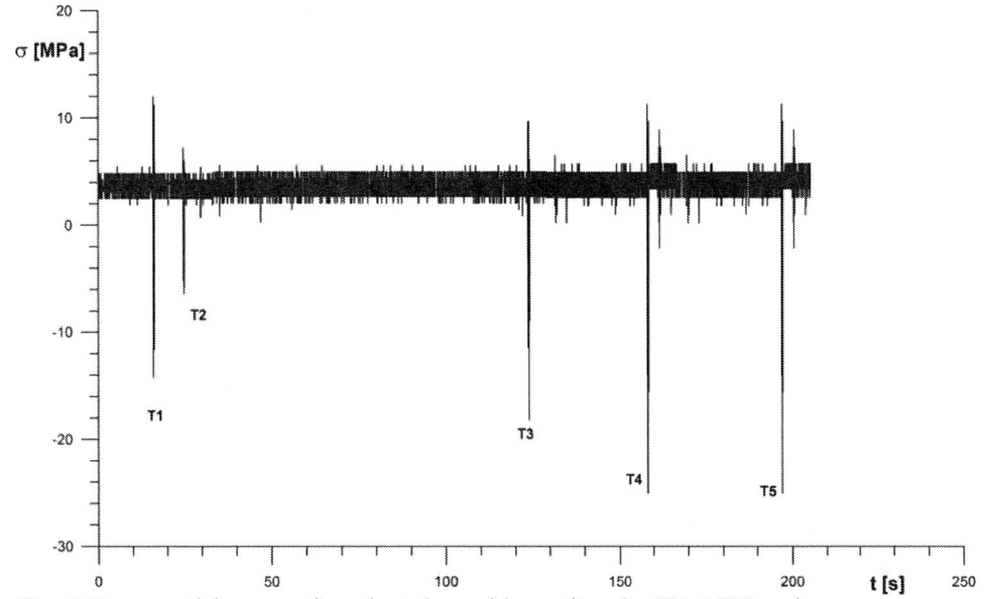

Fig. 6. Process of the stress for selected transitions of trucks (T1až T5) and cars.

3 Fatigue strength assessment

The measuring of loadings directly in operation at the critical part of the bridge dilation was the relevant basis for correct fatigue strength assessment. From the past operation and all gathered experiences, which were collected, was identified that the critical place of structure is the weld of the main I-beam and stirrup of the scissor mechanism. The weld resistance against the fatigue crack (fatigue strength) is possible to consider by various concepts [2-4]. In the term of legislative, the prescription [1] is valid, which for under reviewing the structure node establishes the maximum stress amplitude. On the other hand, in the term of the multiaxiality of a local stress-strain state, there is possible to use several kinds of approaches to assess that state [5, 6]. For determining the corresponding stress amplitude, it is better to use approaches based on equivalent normal stress in the plane of maximum damage [7, 8].

The shear stresses in critical point were measured by pair of inclined gauge sensors ε_{45}, ε_{-45} in rosette (Fig. 4). Their values evaluated by the following term

$$\tau = (\varepsilon_{45} - \varepsilon_{-45})G \tag{3}$$

have not exceeded the level of measuring sensitivity. The acquired results are equal with real loading of the bridge dilation where the shear stress might have been occurred only from inaccuracy of production and assembly. The equivalent amplitude is hence equal to amplitude of normal stress in the place of maximum fibre as it is shown in Fig. 6. Their values together with values obtained by mathematical model as well as allowance values of stress amplitudes for individual judged places are given in Table 1.

Table 1. Calculated and measurement values of stresses.

Place	Normal stress (Eq.1) model [MPa]	Maximum measured amplitude [MPa]	Maximum allowable amplitude [1, 4] [MPa]	Assessment
Stirrup 80×25 weld	93.9	26	36	well
Stirrup aperture for pin	162.5	26	90	well
Tie band aperture for pin	177.9	9	90	well

Conclusion

In the contribution, there is presented the process of obtaining the relevant value of the stress which could allow the correct assessment of the fatigue strength in a critical point of the bridge closure. From three watched places like (the weld of main beam with the stirrup, aperture for pin in stirrup, aperture for pin in tie band) is the most critical point the weld of main beam with the stirrup. The amplitude stress determination using the mathematical model has not led to relevant results because it was difficult to catch the non-linear behaviour of plastic casings in combination with backlashes arisen during production and assembly. The direct measuring of stresses in critical points of dilatation were performed foremost at the miniaturized model and then directly in real operation of highway bypass of the city Uherské Hradište (Czech Republic). During measurements were sensed practically all types of transitions in the term of velocity transitions and loading of axels. The measurable values of stresses caused practically only the transitions of axel trucks (Fig. 6). The magnitudes of stresses in the most loaded fibre of the stirrup near the weld with main beam acquired significantly lower values like they were calculated by mathematical model given in Table 2. Based on the performed measurements, it is possible to conclude that the fatigue strength of the bridge dilatation is sufficient and it satisfies the appropriate prescriptions. This conclusion might be change in the case if the loading condition of the bridge structure are going to change significantly – e.g. by roadway deterioration and getting of the main beam at its surface. Unfortunately, such situation in operation of bridge dilatation could appear and it causes significant increase of the stress component in horizontal direction which even exceeds the value of the stress in vertical direction.

References

1. Eurocode 3. Design of steel structures. Part 1–9: Fatigue, UNI ENV 1993-1-9, 2005.
2. D.F. Socie, G.B. Marquis, *Multiaxial Fatigue* (SAE International, Warrendale, PA, 2000)
3. V. Chmelko, V. Kliman, M. Garan: Procedia Engineering **101**, (2015)
4. V. Chmelko, *Notch effects in operation of machines and structures* (STU 2015)
5. J.A. Bannantine, D.F. Socie, ASTM STP 1122, Philadelphia, 249-275, (1992)
6. I.V. Papadopoulos, Int. Journal of Fatigue **23**, (2001)
7. A. Carpinteri, A. Spagnoli, A.S. Vantadori, Fatigue of Engineering Materials and Structures **26**, (2003)
8. M. Margetin, R. Ďurka, V. Chmelko, Frattura ed Integrita Strutturale **10**, 37 (2016)

Frequency characteristics of a dynamical system at force excitation

Jozef Melcer[1,*] and *Mária* Kúdelčíková[1]

[1]University of Žilina, Faculty of Civil Engineering, Univerzitná 8215/1, 010 26 Žilina, Slovakia

Abstract. Frequency response functions are important characteristics. They define the properties of a dynamic system. Preferably they are used mainly for solving problems in the frequency domain. They are linked to the choice of computational model and to the means of structure excitation. They are complex characteristics, so have their amplitude and phase components. They may be obtained in different ways. Some possibilities of calculating at force excitation are discussed in this paper.

1 Introduction

Dynamic analysis of structures is dependent on the choice of the computational model and the mean of dynamic excitation. In practice the discrete computational model is very often used because the equations of motion have the character of ordinary differential equations. The computational model can be chosen in the spirit of the principles of classical dynamics or in the spirit of the finite element method. Excitation of the structure can be power or kinematic. When choosing a discrete computational model and the force excitation with variable frequency composition it is favourable to use the frequency response functions as characteristics describing the properties of a dynamic system. The paper is devoted to the analysis of such characteristics.

2 Transfer from time to frequency domain

The transfer from time to the frequency domain can be realized by using some integral transformation. The Fourier transform and the Laplace transform are the most commonly used. But also other transformations can be used [1]. For the purpose of this paper the Fourier transform is used. Fourier image of a time function $v(t)$ is denoted as $V(q)$, $V(q) = F\{v(t)\}$, where q is a real number. In our case $q = \omega$ (ω has the meaning of circular frequency in [rad/s]). The complex Fourier transform is defined as

$$V(q) = \int_{-\infty}^{+\infty} v(t) \cdot \mathrm{e}^{-\mathrm{i}q\cdot t} \mathrm{d}t \,. \tag{1}$$

Inverse Fourier transform is defined as

* Corresponding author: jozef.melcer@fstav.uniza.sk

$$v(t) = \frac{1}{2\pi}\int_{-\infty}^{+\infty} V(q)\cdot \mathrm{e}^{\mathrm{i}q\cdot t}\mathrm{d}q\,. \tag{2}$$

The function $v(t)$ and its derivations will be transformed as

$$a\cdot v(t) \quad \rightarrow \quad a\cdot V(q)\,,$$

$$\dot{v}(t) \text{ for } v(\pm\infty)=0 \quad \rightarrow \quad \mathrm{i}q\cdot V(q)\,,$$

$$\ddot{v}(t) \text{ for } v(\pm\infty)=\dot{v}(\pm\infty)=0 \rightarrow -q^2\cdot V(q)\,. \tag{3}$$

3 Definition of frequency response function and its derivation

Dynamical properties of linear system are fully characterized by the response to excitation by simple harmonic functions [2]. Frequency response function $V(p)$ for (p = iω) is defined as the ratio of steady state response to harmonic excitation

$$V(\mathrm{i}\omega) = v_{sst} \,/\, F\mathrm{e}^{\mathrm{i}\omega\cdot t}\,. \tag{4}$$

If the entering value is periodical with unit amplitude

$$F(t) = Ff(t) = 1\mathrm{e}^{\mathrm{i}\omega\cdot t}\quad , \tag{5}$$

the exiting value can be written as

$$v(t) = V(\mathrm{i}\omega)\mathrm{e}^{i\cdot\omega\cdot t}\,. \tag{6}$$

Frequency response $V(\mathrm{i}\omega)$ is the complex function and it can be calculated as vector sum of real Re[$V(\mathrm{i}\omega)$] and imaginary Im[$V(\mathrm{i}\omega)$] parts, Fig. 1.

$$V(\mathrm{i}\omega) = \mathrm{Re}[V(\mathrm{i}\omega)] + \mathrm{Im}[V(\mathrm{i}\omega)] \tag{7}$$

or

$$V(\mathrm{i}\omega) = \left|V(\mathrm{i}\omega)\right|\mathrm{e}^{\mathrm{i}\varphi}\quad , \tag{8}$$

where $\left|V(\mathrm{i}\omega)\right|$ is the absolute value of the frequency response that represents the amplitude of the response $v(t)$ and φ in equation (8) represents the phase of the frequency response.

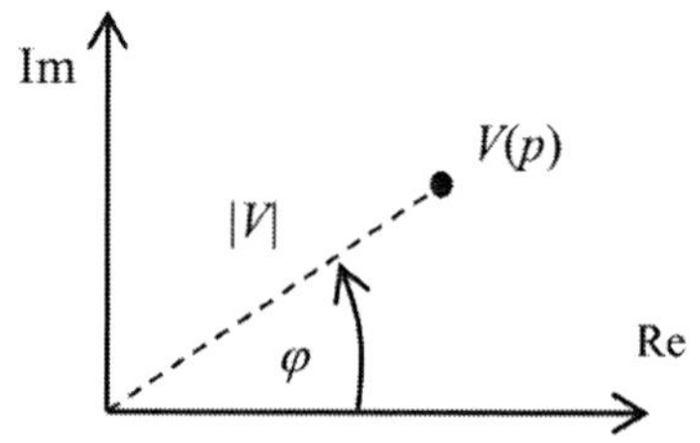

Fig. 1. Frequency response function as a complex function.

When substituting (8) to (6), we obtain

$$v(t) = |V(\mathrm{i}\omega)|\mathrm{e}^{\mathrm{i}\varphi} \cdot \mathrm{e}^{\mathrm{i}\omega t} = |V(\mathrm{i}\omega)|\mathrm{e}^{\mathrm{i}(\omega t+\varphi)}. \tag{9}$$

Then we can write

$$|V(\mathrm{i}\omega)| = \sqrt{\mathrm{Re}^2[V(\mathrm{i}\omega)] + \mathrm{Im}^2[V(\mathrm{i}\omega)]} \tag{10}$$

and

$$\varphi = \varphi(\omega) = \mathrm{arctg}(\mathrm{Im}[V(\mathrm{i}\omega)]/\mathrm{Re}[V(\mathrm{i}\omega)]) \ . \tag{11}$$

Now assume discrete computational model with n degrees of freedom (Fig. 2) excited by discrete forces at the points of lumped masses. The equation of motion describing the forced damped oscillations system can be written in the form

$$[m]_D \cdot \{\ddot{v}(t)\} + 2\omega_b[m]_D \cdot \{\dot{v}(t)\} + [k] \cdot \{v(t)\} = \{F(t)\}, \tag{12}$$

where $[m]_D$ is diagonal mass matrix, $[k]$ is stiffness matrix, ω_b is angular damping frequency, $\{v(t)\}$ is the vector of unknown deflections of mass points and $\{F(t)\}$ is the vector of exciting forces. Derivatives with respect to time are denoted by dot over the symbol [3].

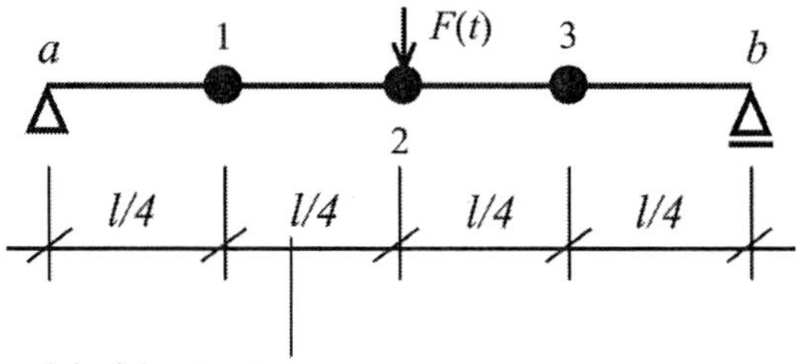

Fig. 2. Computational model of the structure.

Let us apply the Fourier transform to the equation (12). Fourier images of the functions $\{v(t)\}$ and $\{F(t)\}$ are denoted as $\{V(q\}$ and $\{F(q)\}$, where $q = \omega$. Equation (12) is transformed to

$$-q^2[m]_D \cdot \{V(q)\} + iq2\omega_b[m]_D \cdot \{V(q)\} + [k] \cdot \{V(q)\} = \{F(q)\}. \tag{13}$$

Suppose that only k-function of the vector $\{F(q)\}$ is nonzero and all others are zero. It is now possible to define the n^2 frequency responses for $i = 1 \div n$ and $k = 1 \div n$. For the frequency response $\bar{v}_{ik} \equiv \bar{v}_{ik}(q)$ it is valid

$$\bar{v}_{ik} \equiv \bar{v}_{ik}(q) = \frac{V_i(q)}{F_k(q)}. \tag{14}$$

This will provide n systems of simultaneous equations, $k = 1 \div n$, to calculate n frequency responses $\bar{v}_{ik}$ at each step of the solution, for $i = 1 \div n$.

$$-q^2 \cdot [m]_D \cdot \{\bar{v}\} + i \cdot q \cdot 2\omega_b \cdot [m]_D \cdot \{\bar{v}\} + [k] \cdot \{\bar{v}\} = \{F_k\},$$

$$([k] - q^2 \cdot [m]_D) \cdot \{\bar{v}\} + i \cdot q \cdot 2\omega_b \cdot [m]_D \cdot \{\bar{v}\} = \{F_k\},$$

$$(([k]-q^2\cdot[m]_D)+i\cdot q\cdot 2\omega_b\cdot[m]_D)\cdot\{\bar{v}\}=\{F_k\},$$

$$(([k]-\omega^2\cdot[m]_D)+i\cdot\omega\cdot 2\omega_b\cdot[m]_D)\cdot\{\bar{v}\}=\{F_k\}. \qquad (15)$$

Vector $\{F_k\}$ for the kth system of equations contains zero, only in the kth row is number one.

4 Numerical solution

Numerical calculations were applied to discrete computational model of the real bridge construction with one span made from prefabricated elements I73 with the following parameters: span $l = 29.0$ m, elastic modulus $E = 3.85 \cdot 10^{10}$ N/m^2, quadratic moment of the cross-section $I = 2.391711$ m^4, the intensity of mass $\mu = 19.680$ kg/m. Masses of discrete model with 3 degrees of freedom have the following values
$m_1 = m_2 = m_3 = \mu \cdot l/4 = 19680 \cdot 29/4 = 142680$ kg.
Stiffness matrix elements are as follows
$k_{11} = k_{33} = 2.381806419615400 \cdot 10^9$ N/m, $k_{22} = 3.313817627290990 \cdot 10^9$ N/m,
$k_{12} = k_{21} = k_{23} = k_{32} = -2.278249618762560 \cdot 10^9$ N/m,
$k_{13} = k_{31} = 9.320112076755900 \cdot 10^9$ N/m.
Natural frequencies of the model are as follows
$f_{(1)} = 4.0389$ Hz, $f_{(2)} = 16.0432$ Hz, $f_{(1)} = 34.0633$ Hz.
Calculations of frequency response functions have been realized in the frequency range $0 \div 40$ Hz with a step of 0.01 Hz and are displayed in the form of amplitude and phase characteristics. For excitation at point $k = 1$ they are shown in Fig. 3, 4, 5 and for excitation at point $k = 2$ in Fig. 6.

5 Conclusions

Frequency response functions are important characteristics clearly defining the properties of dynamic systems. They are linked to the choice of the computational model and the means of excitation. They can be calculated in various ways, for example by the passage from time to the frequency domain on the basis of the Fourier transform. The usage of derivative procedures can be adopted to solve a wide variety of dynamic tasks [4], [5], [6], [7]. From the above results it can be seen that the values of dominant frequencies are $f_{(1)} = 4.04$ Hz, $f_{(2)} = 16.04$ Hz, $f_{(3)} = 34.06$ Hz. As a result of Maxwell's reciprocity theorem, $\bar{v}_{ik} = \bar{v}_{ki}$. In our specific case, the construction is symmetrical with respect to the axis of symmetry passing through the mass point 2. Due to the symmetry, $\bar{v}_{11} = \bar{v}_{33}$.

This work was supported by the Grant National Agency VEGA of the Slovak Republic, project number 1/0005/16.

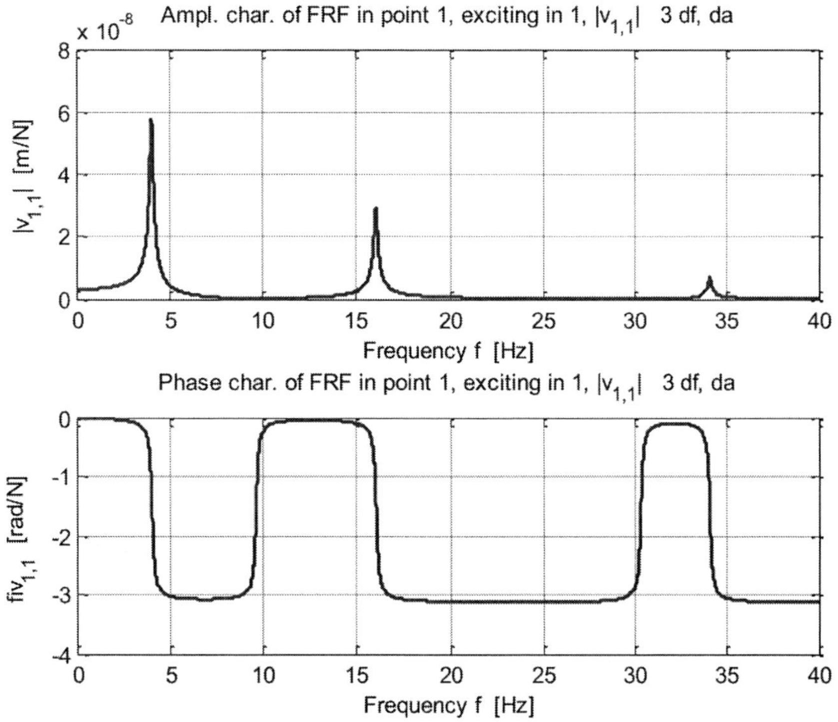

Fig. 3. FRF in point 1, exciting in point 1.

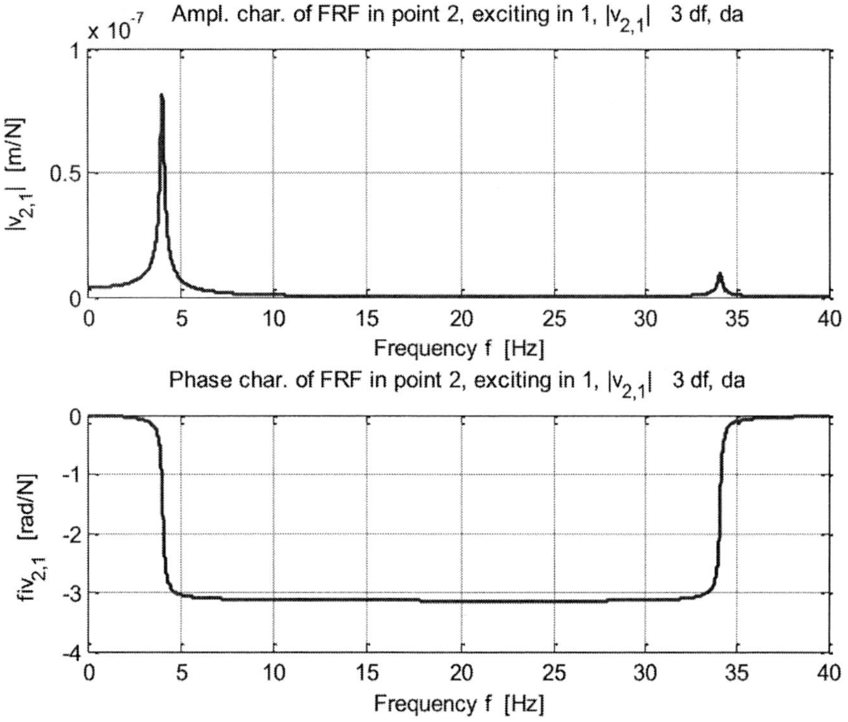

Fig. 4. FRF in point 2, exciting in point 1.

DOI: 10.1051/matecconf/201710700017

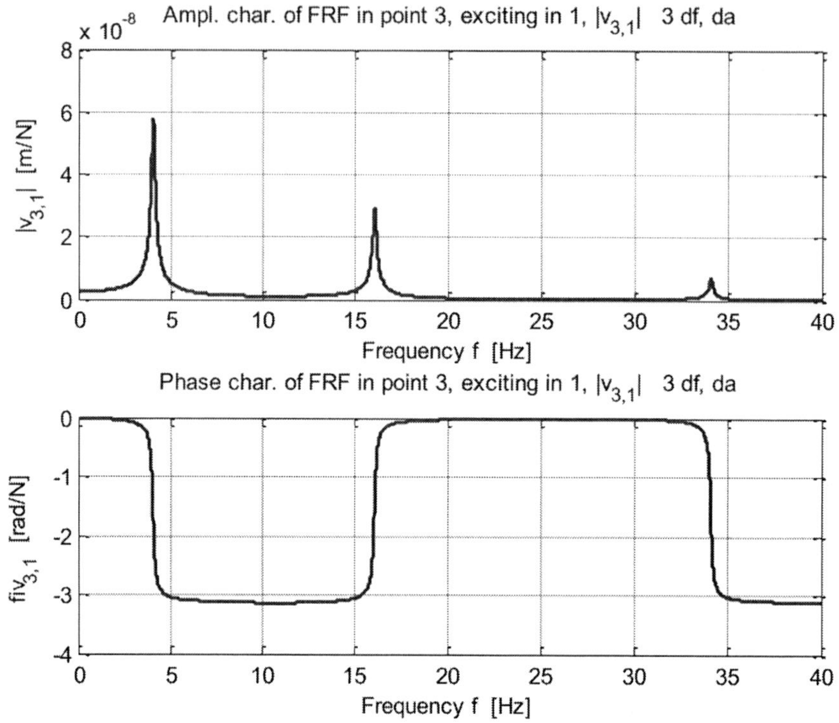

Fig. 5. FRF in point 3, exciting in point 1.

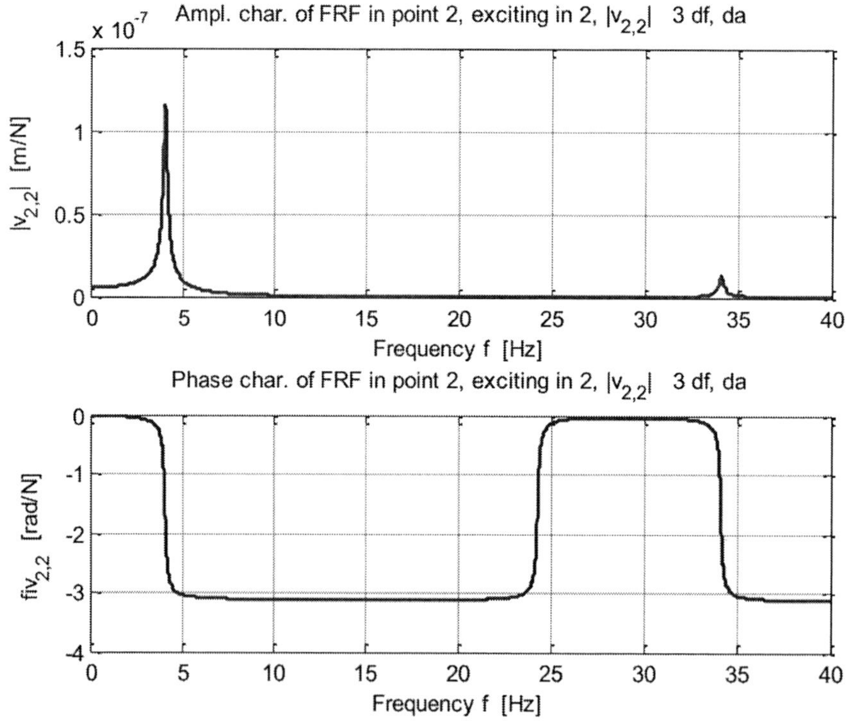

Fig. 6. FRF in point 2, exciting in point 2.

References

1. J. Melcer, et al., *Dynamics of transport structures* (EDIS, Žilina, 2016, in Slovak)
2. M. Baťa, et al., *Dynamics of structures* (SNTL/ALFA, Praha/Bratislava, 1987, in Czech)
3. D. Kuchárová, J. Melcer, *Dynamics of structures* (EDIS, Žilina, 2000, in Slovak)
4. E. Panulinová, Silniční obzor, **62**, 275-279 (2001)
5. K. Kotrasová, E. Kormaníková, International Journal of Mechanics. **10**, 242-252 (2016)
6. M. Mikušová, WIT Trans. Built Env. **134**, 503-514 (2013)
7. I. Olekšáková, O. Ivánková, Adv. Mat. Res. **969**, 148-154 (2014)

Dynamical response of railway switches and crossings

Vlastislav Salajka[1], *Marek* Smolka[2], *Jiří* Kala[1,*] and *Otto* Plášek[3]

[1]Brno University of technology, Dept. of Structural Mechanics, 602 00 Brno, Czech Republic
[2]DT - Vyhybkarna a strojirna, a.s., Dolni 3137/100, Prostejov, Czech Republic
[3]Brno University of technology, Dept. of Railway structures, 602 00 Brno, Czech Republic

Abstract. A procedure for the numerical analysis of the dynamic response during the passage of railway vehicles is described. The solution is based on the finite element method (FEM), which is used for the calculation of track stresses. An FEM model was used with a fine structure that included all components of switches and crossings, including movable parts. The excitation forces are defined on the basis of the assumed interaction between track and vehicle. The track stiffness defined by FEM analyses is used for the calculation of dynamic vertical and lateral wheel load. A special model of a railway vehicle was built with the aim of calculating the forces at points where abrupt stiffness changes occur, as well as geometrical imperfections in the frog structure.

1 Introduction

Switches behaviour optimization from the perspective of reduction of the responses of dynamic effects during vehicles passages brings the need for creation of comprehensive computational geometrically complex models with an appropriate degree of simplification. Numerical solutions using mathematical models created in the finite element method (FEM) comprise a turnout superstructure and substructure including structural layers. The 3D models are demanding in terms of boundary conditions (loads and bonds) modelling and parameters definition [1]. It is necessary to use suitable material models. Results obtained with the aid of these complex models allow a detailed analysis of an individual component of the turnout without the need to introduce considerable simplification [2, 3].
All the details of models require an application of an appropriate program at appropriate work-stations that can compute with enormous data files. The solution is also demanding both for computing time and storage space. A high level of expertise as well as routine knowledge of the applied program is required. Despite the stated demands, obtaining the searched results is virtually practicable.

The issue of finding a feasible design of a rail fastening in the turnout is the subject matter of a turnout with an increased elasticity of track. Searching for the optimal switches arrangement design is based on numerical analyses of the specified J60-1:12-500 turnouts including the analyses of the individual rail fastening of the turnout with corrections related

* Corresponding author: kala.j@fce.vutbr.cz

to the experiment [4]. The FEM are used in which detailed global numerical models of the turnout are created is applied for the calculation. The verified rail fastening using the partial FEM models are put into the global models of the switches on which the dynamic analysis of the vehicle travel effect was carried out. Alternative solutions of responses provided a possibility to determine a suitable solution of the structural arrangement in terms of the strength and the dynamic behaviour with the aim of the turnout life extension.

2 Computation model

The computational turnout model is created by the finite element method in the ANSYS software system. The primary geometry of the turnout model was taken from a 3D CAD model in a SAT format and extended by an access and outgoing part formed by a track length of a plain track from the both sides. The computational model consists of a range of element types. Three-dimensional parts (rails, a crossing, switch tongues, sleepers and bearers, elastic bedding compensating the road bed stiffness and some steel parts of the bedding and rail pads) are modelled by three-dimensional eight-node finite elements. Trough sleepers, base plates and slide plates in the turnout, elements carrying a check rail are modelled by four-node shell elements.

The check rail is modelled in a simplified way by general beam elements. Fastening clamps are modelled in a simplified way by beam elements only carrying axial forces. Additional weight of an electromotor point machine was compensated by concentrated mass on the trough sleeper face modelled by a one-node element. These elements are attached by auxiliary beam elements so that the location of the mass centre of gravity would be maintained. The elastic bedding reflecting road bed subgrade is modelled by four-node surface elements. The final element model is shown in the Figure 1. A great attention was paid to the modelling of the rail fastening and the rail bedding, especially in the switch part. In the rail fastening, with the base plates in the turnout as well as without base plates in the plain track, rail fastening Vossloh types were used [5]. Clamp stiffness characteristics are defined by work diagrams. Clamp modelling is carried out by special elements that have the vertical stiffness. The preload to the clamps is brought by a set value of the relative axial deformation ε_{ax} of the clamp.

Between the rail and the sleeper, there is a layer modelled that varies according to the rail fastening point. The material used for the rail pads is rubber that is characterized by the non-linear behavior. The vertical stiffness of the rail pads in relation to the rate of loading is verified experimentally. The static and dynamic stiffness of the rail pads is distinguished by this [6, 7].

The values of the modulus characterize the stiffness under the dynamic stress of the track and they are selected so that the relative deflection during the travel would correspond to the real values. The two lower layers are coated by elastic plane elements that reflect the stiffness of the surroundings. The stiffness values control the deflection of the railway line model. At an alternative solution the k_W values are changed in relation to the solved alternative. In cross-sections at the ends of the model (a rail and bedding layers section), there are boundary conditions limiting the normal displacement to the given section. There are boundary conditions limiting the horizontal displacement in the lower model level in all the nodes. It should also be mentioned that the tongue ends bear on the rail at the distance of 1.5 m. This contact is modelled by an order of the common displacement of the tongue and rail nodes in the horizontal lateral direction.

DOI: 10.1051/matecconf/201710700018

The rail fastening on the two steel trough sleepers remains without additional elasticity. The modified model of the track with the additional elasticity consists of 407 thousand elements that are located by 522 thousand nodes and it has almost 1.5 mil degrees of freedom, Fig. 1.

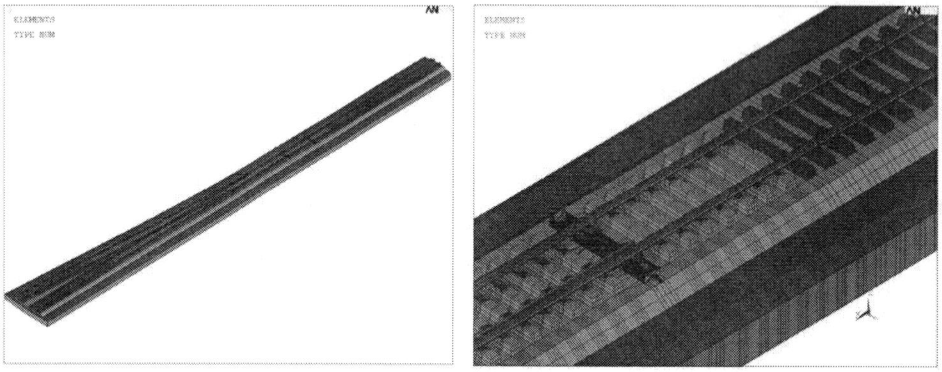

Fig. 1. Computational model.

The stiffness of the turnout running parts varies in relation to the method of their fastening and depending on the stiffness of base courses under the sleepers. Several areas where the rail fastening stiffness varies significantly are considered. The first region that is included in the model is the region in front of the switch. The rails are fastened on the B 91S sleepers here. It is a standard rail fastening W14 type on the sleepers in the ballast bed. The same applies to the region of the model behind the switch. The turnout structure can be divided into the area in front of the switch, a switch part, a middle part, a crossing part and the area behind the switch. Rail bedding additional elasticity is designed in the switch and crossing parts. In the switch part the additional elasticity is at 40 places and in the crossing part it is at 9 places. A slide plate with a pressure spring clip is designed for the switch part. The slide plate structure allows a vertical movement of the sliding surface (a place where the tongue is supported). The model of switch additional elasticity corresponds to the real version. The additional elasticity in the model under the crossing is simply carried out by the pad stiffness modification. The model in the area of the tongue tip is shown in the Fig. 2 where the differences in modelling of the slide plates under the tongues can be seen.

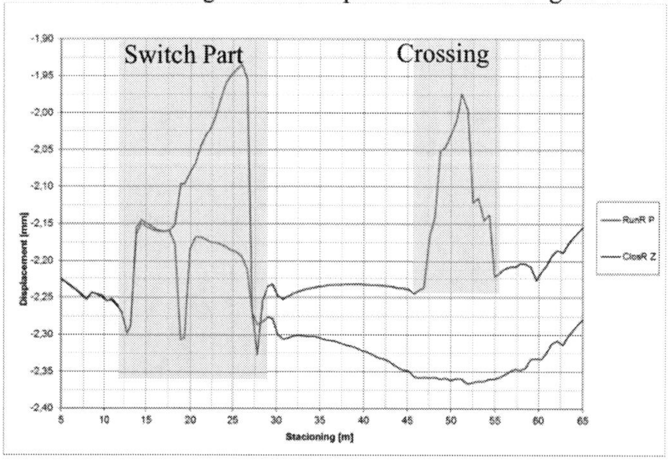

Fig. 2. Deflection of rails with standard rail fastening.

The check of the individual rail fastening in the global model was carried out on partial computational models of the fastening system. Parts of the model were separated from the turnout global model and modified so that they would correspond to the test fastening. Then the elastic layer stiffness in the model was reiteratively determined so that the vertical stiffness of the model fastening is in accordance with the dynamic stiffness obtained from the test.

At the same time, solving the dynamic responses with a model of a vehicle is possible, but at such a great extent of the turnout model virtually impossible. To simplify, the interaction between the vehicle and the running rail of the turnout is substituted by the dynamic force effects that consider discontinuities in the stiffness of the rail track, the impact of the turnout part wear and deficiencies in the rail geological base. The force effects including the static and dynamic parts are represented by a couple of forces, the conducting lateral force Y and the vertical wheel load Q, whose intensity varies in relation to the travelling speed, the rail geometry and the vehicle type.

For the each solution alternative of the vehicle travel through the turnout, the dynamic stiffness of the travel route need to be determined. It is about obtaining stiffness influence lines of the rail head deflection in the discrete points of the FEM model from the static states under the action of the two wheel loads, each of the selected intensity.

3 Dynamic analyses

The static calculations are performed for the monitored alternatives of the structural layers stiffness. After the performance of the loading states calculation with a change of the wheel loads location, a state including the preload influence in the rail fastening has to be subtracted from the each state [8].

Interactions between an eight-wheel vehicle and the track length are determined by a simulative calculation using the influence lines. The calculation of the interaction forces continues on the cooperation with the employees of the University of Pardubice, which deals with vehicle travel simulations. At that workplace, response of the selected vehicle with the set speed and assumed characteristics of the track is determined from the behaviour of dynamic stiffness [9].

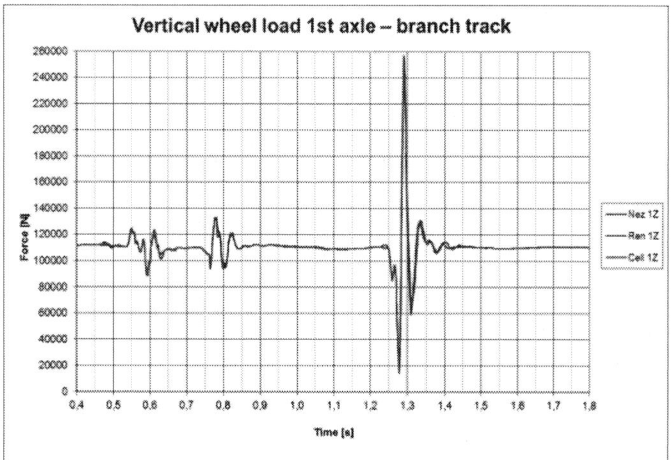

Fig. 3. Vertical wheel load 1st axle – branch track.

The wheel forces obtained like this acting retroactively on the heads of the rails are further used as excitation loading with a variable position corresponding to the travelling vehicle with the specified speed, see the Fig. 3. The dynamic response calculations are carried out by direct integration of equations of motion. The results include a displacement field, deformation and stress in individual instants of time of the solution. Alternations in all the modelled parts of the turnout as well as other design elements can be monitored this way. The static analysis indicates the differences in the turnout stiffness during the passage either in a straight or branch direction, from which suitability or unsuitability of the design of the rail fastening nodes arrangement can be deduced.

The increase of the deflections (see the Fig. 4) and forces in dynamically influenced areas and the endurance strength for the judgement of the endurance of the fastening structure and other derived values, such as speed and acceleration fields, can be read from the dynamic response. This allows you to monitor and to analyse the impact of design changes on the turnout behaviour.

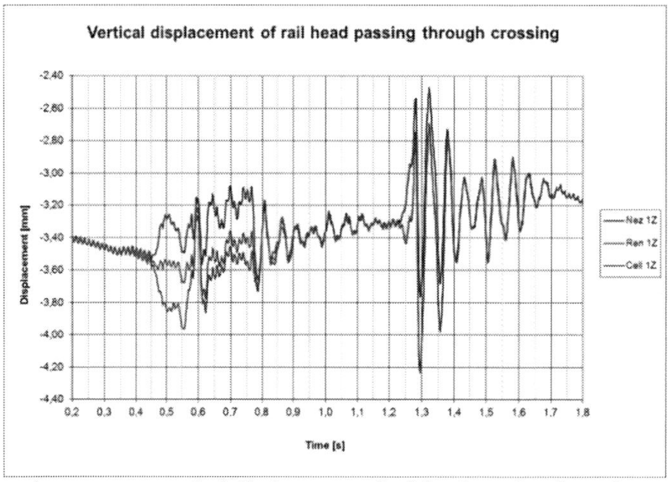

Fig. 4. Dynamic response – vertical displacement of rail head passing through crossing.

4 Conclusions

The calculations of the turnout dynamic response to the passage of the railway eight-wheel vehicle through the turnout in the straight direction against the tongue tip were carried out. There were three alternatives distinguished. The alternative with the original arrangement of the rail bedding, the tongue and the crossing and two alternatives with designed additional elasticity of the track in the switch and in the crossing. The elasticity in the switch part was considered in the alternative of the "suitable" elasticity and in the "soft" alternative. In the crossing part the elasticity is identical for the both elastic alternatives.

It can be concluded that with the aid of the calculations it was not found that the elasticity creates unexpected changes in the displacement and the force action in the turnout structure. On the basis of the performed calculations we can observe that the passage through the turnout with elasticity is smoother and it is more balanced while passing through the switch and the crossing in comparison to the original alternative without elasticity. The extreme stresses in all components of permanent way occur in switches and crossings during a passage of railway vehicles. The additional dynamic loading originates from changes in the track stiffness along the railway turnout or alternatively impact forces cause at points of geometry imperfections. The stresses can be reduced due to the control of track stiffness by using rail fastening with different elasticity of rail pads or special elastic pads in slide plates in switches [10].

This outcome has been achieved with the financial support of project GACR 17-23578S "Damage assessment identification for reinforced concrete subjected to extreme loading" provided by the Czech Science Foundation and under the project No. LO1408 "AdMaS UP - Advanced Materials, Structures and Technologies", supported by Ministry of Education, Youth and Sports under the „National Sustainability Programme I" and with the financial support of the project FAST-S-16-3718 "Advanced numerical methods with complex material models" provided by the Brno University of Technology fund for specific university research.

References

1. P. Král, P. Hradil, J. Kala, F. Hokeš, M. Hušek, Identification of the Parameters of a Concrete Damage Material Model, *Procedia Engineering*, **172**, 578-585 (2017)
2. J. Kala, M. Hušek, High Speed Loading of Concrete Constructions with Transformation of Eroded Mass into the SPH, International Journal of Mechanics, **10**, 145–150 (2016)
3. Z. Kala, J. Valeš Global Sensitivity Analysis of Lateral-torsional Buckling Resistance Based on Finite Element Simulations, Engineering Structures, **134**, 37–47 (2017)
4. V. Salajka, et al.. *Simulation of a vehicle passage over the turnout with an additional realistic elasticity of rail fastening and with the crossing of a new design* (Research Report for DT – Výhybkárna a strojírna, a.s., 45 pp. 2014)
5. C. Esveld, *Modern Railway Track* (2nd Ed. Delft, MRT – Production, 2001)
6. J. Kala, M. Hušek, Improved Element Erosion Function for Concrete-like Materials with the SPH Method, Shock and Vibration, **2016**, 1–13 (2016)
7. F. Hokes, J. Kala, Selecting the objective function during the inverse identification of the parameters of a material model of concrete, Frattura ed Integrita Strutturale, (Fracture and Structural Integrity), **11** (**39**), 7-16, (2017)
8. V. Salajka, V. Kanický, Track Response to Traffic Load Using Non-linear Computing Model. Proceedings Traffic Applied Mechanics'99, University Pardubice,15-20. (1999)
9. P. Král, P. Hradil, J. Kala, Inverse Identification of the Material Parameters of a Nonlinear Concrete Constitutive Model Based on the Triaxial Compression Strength Testing, Frattura ed Integrità Strutturale (Fracture and Structural Integrity), **11** (**39**), 38-46, (2017)
10. W. Stahl, Untersuchung eines für HGV optimierten Schotteroberbaus. Querverschibe-wiederstandmessungen und Schwingungsverhalten. Der Eisenbahningenieur, **5**, 8-16, (2003)

Impact of operation on the geometric parameters of the track in ballastless track transition area

Libor Ižvolt[1], *Michal* Šmalo[1*] and *Janka* Šestáková[1]

[1]University of Žilina, Faculty of Civil Engineering, Department of Railway Engineering and Track Management, Univerzitná 8215/1, 010 26 Žilina, Slovakia

Abstract. The paper deals with the methodology and results of diagnostics in test sections near the tunnel portals of newly built tunnel Turecký vrch on modernized line Nové Mesto nad Váhom – Púchov. Based on the diagnostics results, transition areas are problematic sections of the construction. Transition areas were subjected to detailed analysis and based on trends of the development of geometric quantities are determined trends which predict time of needed maintenance. In order to optimize existing transition areas the design of transition in ANSYS software was modelled and problematic spots of construction were defined.

1 Introduction

The increasing mobility of population and the competition among various transportation systems pose new challenges in the field of railway transport. An essential prerequisite for competitiveness of railway transport is reliable operation of railway tracks, which can be understood as a safe and stable movement of rail vehicles along the track.

The development and verification of various technical solutions of ballastless track demonstrate the efforts to optimize the railway superstructure. An upgraded superstructure would enable long-term quality of track design and track geometry parameters, decrease maintenance costs, extend service life and increase competitiveness and attractiveness of railway tracks. The modernisation of railway infrastructure, which is a highly topical issue in Slovakia, is an ideal opportunity for further application of progressive systems and structural elements of railway tracks, among them also the ballastless track.

2 Diagnostics methodology

The experimental section diagnostics is carried out with a complex continual method by a measuring trolley KRAB^{TM}-Light. An additional method to this measuring is a partial point diagnostics with a manual track gauge. The realization and the partial results of the experimental section diagnostics are described in detail in [1].

* Corresponding author: michal.smalo@fstav.uniza.sk

From the point of view of the diagnostics execution is the monitored section divided into two subsections [2]:

- section 1.1 (line No.1, the southern portal of the Tunnel Turecký vrch) and 2.1 (line No. 2, the southern portal; both sections are 175 m long; 102.360 000 – km 102.535 000):
  - km 102.360 000 – km 102.460 500 the ballasted track,
  - km 102.460 500 – km 102.480 500 the transition area,
  - km 102.480 500 – km 102.535 000 the ballastless track,
- section 1.2 (line No.1, northern portal of the Tunnel Turecký vrch) and 2.2 (line No. 2, northern portal); both sections are 640 m long; km 104.200 000 – km 104.840 000):
  - km 104.200 000 – km 104.720 500 the ballastless track,
  - km 104.720 500 – km 102.740 500 the transition area,
  - km 104.740 500 – km 104.840 000 the ballasted track.

The measuring principle by the manual measuring trolley KRAB^{TM}-Light is contact. The measuring parameters are carried out by the touch of the scanning device with the track skeleton (rail) component of the diagnosed line. The measured results are recorded into the board computer of the measuring trolley. After the transfer to the operating computer is for each part of the experimental section created pursuant to the defined limit values of the track geometry parameters the graph of the measured track geometry parameters, table of the local faults with a list of limit values overruns and the tables of the section evaluation. The section evaluation describes the values of the standard deviations (*SDO*), quality marks (*QM*) and the quality number (*QN*).

The measured data are evaluated by computer in terms of provisions and according to the defined limits of the measured parameters of the track geometry. For each diagnosed subsection is created the graph of the measured parameters of the track geometry, the table of local faults with the list of the limit values overruns of the track alignment of the left and the right rail (*SL*, *SP*), the track gauge (*RK*), the change of gauge (*ZR*), the superelevation (*PK*), the twist fault at three different measuring plants (*ZK*1.8, *ZK*6.0, *ZK*12.0) and the vertical profile of the left and the right rail (*VL*, *VP*). A part of the software evaluation is also the table of the sectional evaluation which expresses the standard deviation values, quality marks (*QM*) of the geometric quantities – the track alignment of the track axis (*SK*), *RK*, *PK* and the vertical profile of the rail (*VK*) – and the quality number (*QN*). The evaluation respects the specifications stated in [3] and [4] for the tracks with the speed from 120 kmph to 160 kmph inclusive (i.e. for the speed zone – *RP*4). The railway infrastructure manager did not incorporate the provisions of the currently valid document [5] into the particular legislation so far. The measured values were evaluated according to the installation requirements for the acceptance of the finished works with the utilization of the new material (MSO), respectively according to the operational and safety limits for *RP*4 (PO1 to PO7).

The diagnostic measurements of the experimental section subsections were executed:

- the initial measurement before the construction commissioning (signed MSO) during the days 10th, 11th July 2012, 2nd, 3rd October 2012,
- the first operational measurement (PO1) during the days 9th, 10th, 22nd April 2013,
- the second operational measurement (PO2) during the days 8th, 9th, 22nd October 2013,
- the third operational measurement (PO3) during the days 25th and 28th May 2014,
- the fourth operational measurement (PO4) on 29th October 2014,
- the fifth operational measurement (PO5) during the days 25th March 2015 and 17th April 2015,
- the sixth operational measurement (PO6) during the days 14th and 15th October 2015,
- the seventh operational measurement (PO7) during the days 22nd and 23rd March 2016,
- the eighth operational measurement (PO8) on 27th November 2016.

3 Diagnostics results and prognosis of the quality development

The monitoring of the experimental section is executed from July 2012 in a half-year period. The initial measurement (MSO) was executed about 2 days before the construction commissioning. Within the initial measurement were recorded 40 local faults, 30 of which in the sections with the conventional permanent way (with the ballasted track), 9 in the section with the ballastless track and 1 local fault was diagnosed in the transition area of the section 2.1. After this measurement the contractor carried out a repair of the track geometry and a microgeometry repair of the rail heads by their grinding.

In the second operational measurement (PO2) the local faults of the vertical profile of the right (*VP*) and the left (*VL*) rail in the transition area of the section 1.1 were for the first time diagnosed. These faults confirm the results of the measurement by the measuring trolley of the Railways of Slovak Republic (ZSR). In the following measurements – PO3 and PO4 – was diagnosed another decrease in quality of the section 1.1, which proves the increased number of the local faults, as well as the decreasing value (increasing) of the general quality number. In November 2014 the railway infrastructure manager executed the maintenance in the transition areas of the sections 1.1 and 2.1. The diagnosed local faults were removed by means of the maintenance. The complex continual diagnostics executed by the measuring trolley KRAB^{TM}-Light and the monitoring executed by the railway infrastructure manager by means of the levelling method from December 2014 to August 2015 in the sections 1.1 and 2.1 prove other settlements of the track level. Their values do not occur as the local faults so far because the operational tolerances were not overrun. However, there is expected the formation and occurrence of the local faults within a short period of time because of the proved quality decrease tendency of the particular parameters of the mentioned sections. The sections 1.2 and 2.2 in the transition areas do not prove any faults of the values. The quality development tendency of the measurements recorded so far does not imply any probability of their occurrence either in the following measurements [6-8].

Within the measurements MSO, PO1, PO2 a PO3 were the quality marks of the alignment of the right (after calculating also the left) rail (QM_{SK}) classified into the interval $2 < QM \leq 3$ (recommendation: to design the repair of the track geometry into the maintenance work plan) in all of the monitored sections. The quality marks in the interval $3 < QM \leq 4$ (recommendation: to perform the repair of the track geometry till the nearest inspection) were reached in the sections 1.1 and 2.1 within the measurement PO4. After the execution of the maintenance before the measurement PO5 were the values of quality marks improved (decreased) into the interval $0 < QM \leq 2$ (the state of the track geometry is satisfactory in the evaluated section) in the section 1.1, respectively $2 < QM \leq 3$ in the section 2.1. By the measurement PO7 was found out a slight impairment QM_{SK} in the section 1.1 to the interval $2 < QM \leq 3$, which was proved also by the measurement PO8.

The quality marks of the gauge deviation QM_{RK} are expresses in all sections and measurements in the interval $0 < QM \leq 2$, which indicates the state of the track geometry as satisfactory. The higher values of the quality marks QM_{RK} of the sections 1.1 and 2.1 result from the higher length allotment of the sections with the ballasted track construction in comparison to the sections 1.2 and 2.2, where considerably exceeds the better quality ballastless track because of the location durability of the rails. The values QM_{RK} acquired within the monitoring executed during the operation (measurements PO1 to PO7) present the gauge as a parameter the quality of which has a permanently satisfactory development.

The quality marks of the superelevation QM_{PK} are in the intervals $0 < QM \leq 2$ in all sections and $2 < QM \leq 3$ in the measurements PO2, PO3, PO4 and PO7 in the section 1.1. The values got into the interval $0 < QM \leq 2$ after the execution of the maintenance (after the measurement PO4). The tendency of the quality marks values to increase in this section recorded within the measurement PO5 and PO6 is, however, proved as stopped by the measurement PO7. The quality mark QM_{PK} is placed into the interval $2 < QM \leq 3$ and in the last measurement (PO8) got the values into the interval $0 < QM \leq 2$.

As the quality mark with the worst development tendency seems to be the quality mark of the vertical profile of the rails *VP*, *VL* QM_{VK} according to the executed measurements.

The faults of the vertical elevation deviation of the rails are the most often occurring failure resulting also from the evaluation of the measurements by the measuring trolley of the ZSR. With the exception of the measuring MSO in the section 2.1 were in all measurements of the monitored sections found the values QM_{VK} $2 < QM \leq 3$ and in the measurements PO1, PO2, PO3 and PO4 of the section 1.1 the values $3 < QM < 4$ indicating the recommendation to perform the repair of the track geometry till the nearest inspection. The values QM_{VK} decreased by the execution of the maintenance carried out in the sections 1.1 and 2.1 before the measurement PO5. The measurement PO5 and PO6 were placed into the interval $2 < QM \leq 3$. In the measurements PO7 and PO8 reached QM_{VK} the values in the interval $3 < QM < 4$ in the section 2.1. The impairment tendency of the quality marks in the section 1.1 predicts the determination of the value QM_{VK} within the interval $2 < QM \leq 3$ by the measurement within a short period of time.

Within the measurements PO1, PO2, PO3 and PO4 was found out a progressive quality decrease of the sections 1.1 and 2.1 presented by the general quality number. The maintenance, executed in November 2014, by means of which the railway infrastructure manager removed the local faults preferentially of the *VK* parameter, considerably decreased the value of the quality numbers of the particular sections from the value 2.36 (measurement PO4) to the value 1.31 (PO5) in the section 1.1 and from the value 1.86 (PO4) to the value 1.59 (PO5) in the section 2.1. The tendency of the results of the measurements PO5, PO6 and PO7 expresses the increase, which means the quality decrease of the sections 1.1 and 2.1. In the sections 1.2 and 2.2 is the quality decrease (increase *QN*) only partial.

The prognosis of the quality development tendency of the particular track geometry parameters is concentrated on the problematical section from the point of view of the track quality. It concerns the transition areas in the subsections 1.1 and 2.1. From the data acquired by the measuring trolley were processed the graphs of the real deviations of the track axis alignment (*SK*), gauge (*RK*), gauge change (*ZR*), superelevation (*PK*), the projected value of which is in the mentioned section 0 mm, the twist fault at three measuring plants (*ZK*1.8, *ZK*6.0, *ZK*12.0) and the vertical profile of the left and the right (*VL*, *VP*) in the section from km 102.454 000 to km 102,454 000. A part of this section are the transition areas in both track lines from km 102.460 500 to km 102.480 500.

The vertical profile of the left and the right rail (*VL*, *VP*) and the development tendencies of this parameter is in regard to the maintenance planning of the transition areas the most critical parameter in the long term.

In the transition area of the section 1.1 detect the tops of the right (after calculation also the left) rail a fault of this parameter on the level of the operational limit overrun defined by the values ±6 mm in PO4 for the right rail, respectively in PO3 and PO4 (the upper limit ±6 mm is overrun) and in PO2, PO3 and PO4 (the lower limit -6 mm) for the left rail. The repair of the track geometry executed in the transition area in November 2014 reduced the deviation values. However, the places with the limiting values of the deviations were remained (the contact of the transition area and ballastless track beginning).

In the transition area of the section 2.1 was not recorded any overrun of the operational limits of the parameters *VL*, *VP* within the PO1 to PO4. However, the railway infrastructure manager carried out the maintenance in the same time at both tracks. The maintenance executed in the transition area of the track No. 2 did not improve the deviation values of the parameter. It came to the translocation of irregularities and places with the limiting values and there was recorded an increase of the deviation *VL* from -1.5 to -3.5 mm and *VP* from -1.0 to -3.0 mm on the contact of the ballasted track with the transition area.

The prognosis of the further quality development of the diagnosed parameters records various tendencies. The planning of the maintenance and repair works depends on the quality of the parameters which values are fastest impairing. In the case of the experimental sections in the area of portals of the Tunnel Turecký vrch it is the vertical profile of the left and the right rail in the transition areas of the subsections 1.1 and 2.1. From the tendency lines created by means of the values *VL* and *VP* measured within PO5, PO6, PO7 and PO8 is possible to predict that in the section 1.1 will be reached the operational limit (±6 mm) of the parameter *VL* probably in September 2017 and the parameter *VP* in May 2018. In the section 2.1 will the *VL* and *VP* reach the value of the operational limit probably in September 2017, respectively in June 2018.The overrun levels of the operational limits should be detected within the second measurement campaign of the measuring trolley of the ZSR (in summer 2017). According to the measurement results is the railway infrastructure manager going to plan the execution of the track geometry repair in the mentioned sections.

4 Numerical modelling of transition areas

The results of the carried out measurements indicate that problematic parts of ballastless track sections are transitions from this structure to the ballasted superstructure. The problem of optimization of the structural solution of transition areas is a highly topical problem for the infrastructure manager, especially with regard to modernisation of corridor tracks where another ballastless track sections are planned. The decreased number of repair interventions in transition areas due to optimization of their structural solution will bring many benefits to the infrastructure manager, specifically in the form of fewer track closures and related maintenance expenses and resulting increased competitiveness of rail transport.

For the sake of optimization of transition areas, dynamic behaviour of existing transition area was analysed using 3-D models in the ANSYS software (Fig. 1 and Fig. 2). The values of vertical displacement (Fig. 3), vertical velocity (Fig. 4) and vertical acceleration (Fig. 5) of these models were then compared. In graphs, there is presented only the most interesting part – the end of ballastless track section and the beginning of transition area.

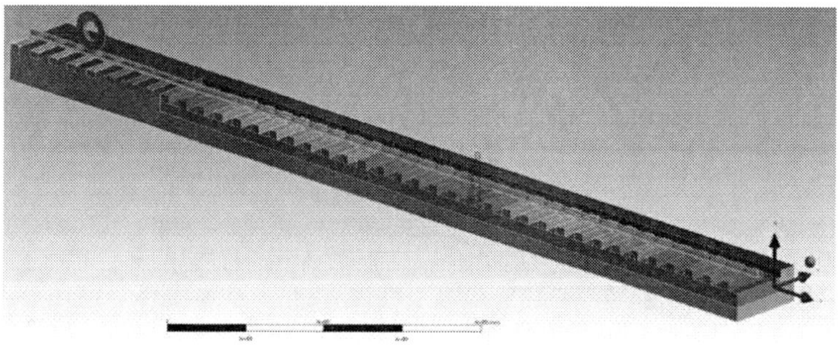

Fig. 1. Contact pairs modelling in *ANSYS* software.

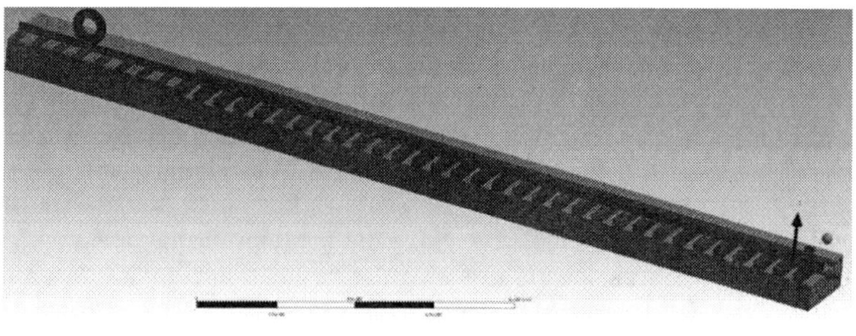

Fig. 2. Transition area model in *ANSYS* software.

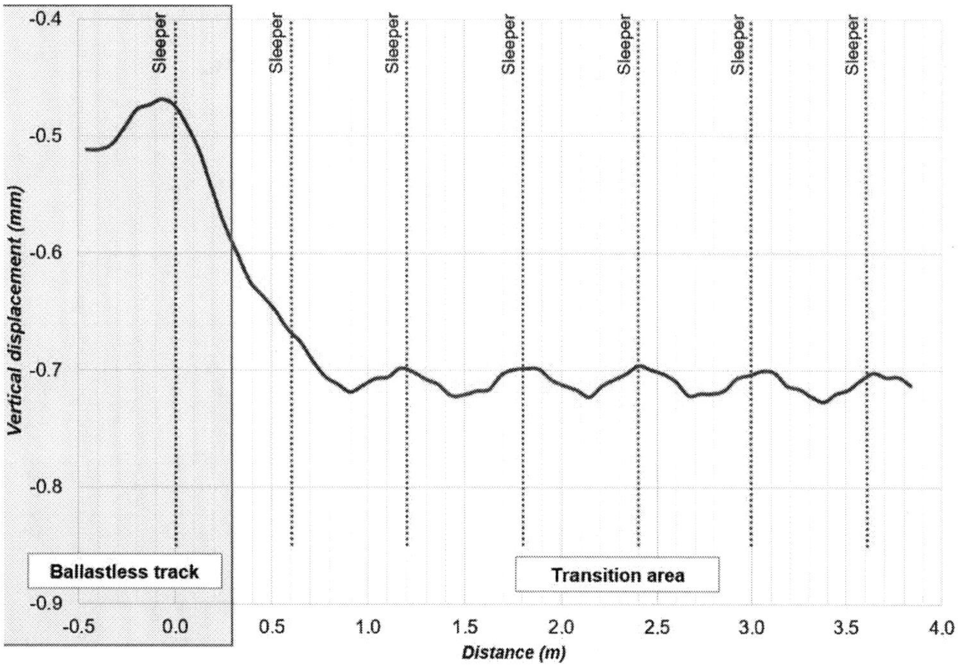

Fig. 3. Vertical displacement in transition area.

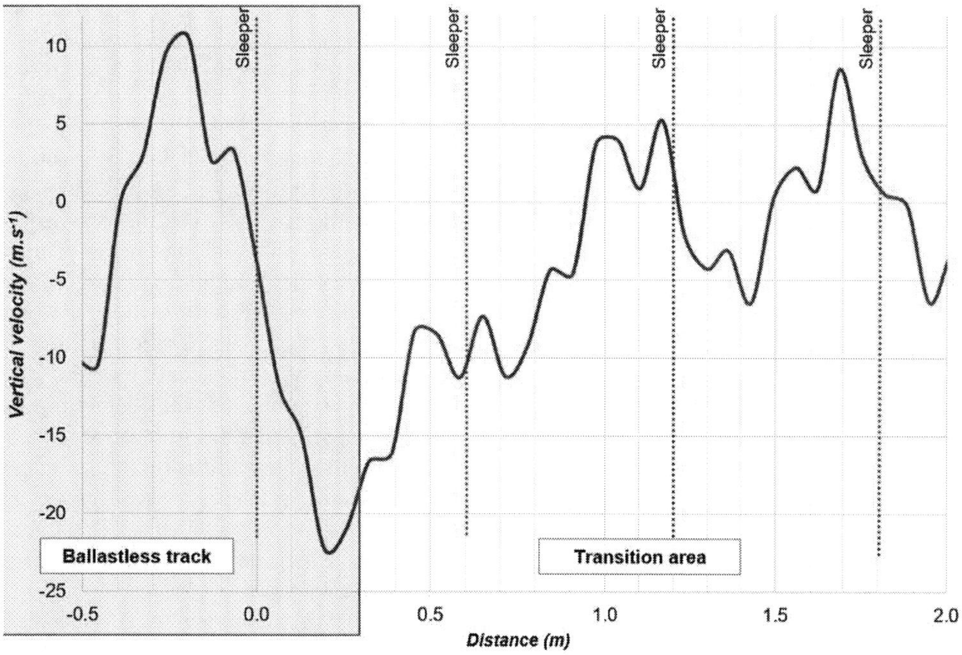

Fig. 4. Vertical velocity in transition area.

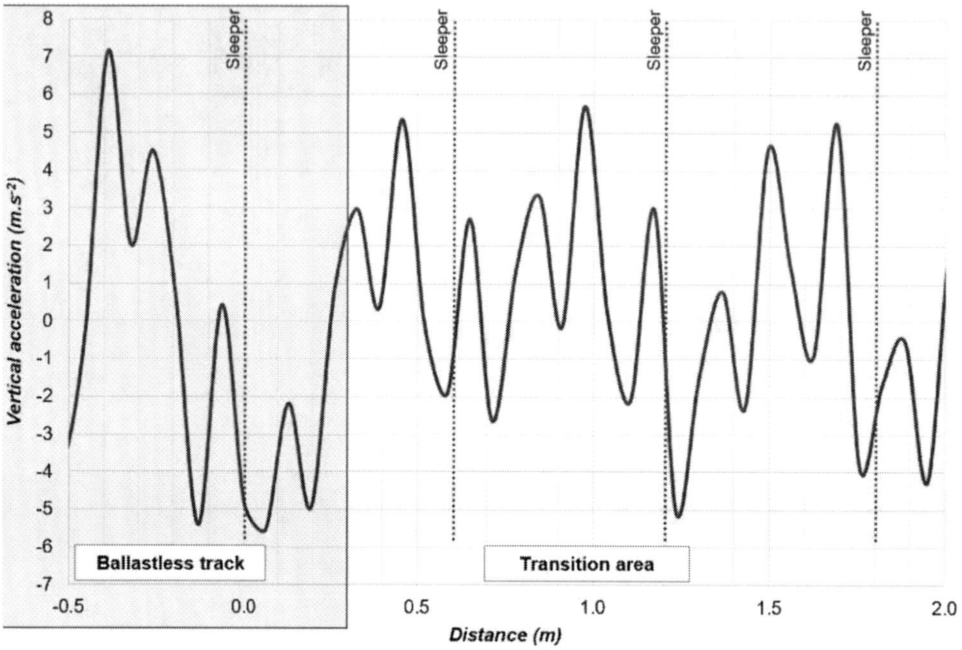

Fig. 5. Vertical acceleration in transition area.

5 Conclusion

The diagnostics of the track design and track geometry parameters was conducted near the portals of the Turecký vrch tunnel for five years. The measurement results were compared to the outputs of the measuring trolley of ZSR in the ballastless track sections and in the transition areas, provided by infrastructure manager. On the basis of conducted measurements and acquired results, we can conclude that the monitored ballastless sections are stable and in the period between measurements do not show any relevant changes of monitored parameters. The sections with ballasted superstructure show certain changes but they can be attributed to standard properties of this structure. The measurements of track design and track geometry by the measuring trolley KRAB^{TM}-Light indicated certain irregularities in the transition areas between the ballasted and ballastless superstructure, especially in the area of the southern tunnel portal, in both track lines. These results were subsequently confirmed by geodetic surveying and by the measuring trolley of the ZSR outputs. The lowerings recorded in these transition areas since the beginning of measurements resulted in exceeding of limit values of quality marks of top level of rails, as well as in occurrence of local errors. Due to this situation, the infrastructure manager conducted the repair intervention. This repair intervention detectably slowed down the trend of error development, but the quality of track design and track geometry of both rails is continuously degrading and further limit exceeding can be expected in the second half of 2017. This situation will certainly require another repair intervention.

The results of the carried out measurements indicate that problematic parts of ballastless track sections are transitions from this structure to the ballasted superstructure. The problem of optimization of the structural solution of transition areas is a highly topical issue for the infrastructure manager, especially with regard to modernisation of corridor tracks where another ballastless track sections are planned. The decreased number of repair interventions in transition areas due to optimization of their structural solution will bring many benefits to the infrastructure manager, specifically in the form of fewer track closures and related maintenance expenses and resulting increased competitiveness of rail transport.

For the sake of optimization of transition areas, dynamic behaviour transition area was monitored using 3-D model in the ANSYS software by the values of vertical displacement, vertical velocity and vertical acceleration.

In future, we are planning to compare existing transition area with optimized solution, e.g. transition area with auxiliary rails or transition area with sleepers using under-sleeper pads. On the basis of numerical modelling of these transition areas, we will try to state which of modelled transition area solution is the most suitable solution for optimization of the existing problematic transition area near the southern portal of the Turecký vrch tunnel and also for future applications of ballastless track system built within ballastless track modernisation in Slovakia.

The paper contains results of the grant VEGA 1/0275/17 Application of numerical methods to define the changes of geometrical track position.

References

1. L. Izvolt, J. Sestakova, M. Smalo, Communications - Scientific Letters of the University of Zilina, **18,** 4, 19-29 (2016)
2. J. Sestakova, Z. Gocalova, Procedia Engineering, **91**, 447-452 (2014), online: http://ac.els-cdn.com/S1877705814030410/1-s2.0-S1877705814030410-main.pdf?_tid=7ac2c2f8-c709-11e4-aaf7-00000aab0f6b&acdnat=1425980605_3c52ef455ea61af04fe1b951e4d82550
3. STN 73 6360 Track Alignment Design and Track Geometry of Normal-gauge Tracks (in Slovak), SUTN Bratislava (1999) and Amendment 1 (2002)
4. ZSR SR 103-7 (S) Measurement and Evaluation of Track Geometry by Measuring Trolley KRAB (in Slovak), ZSR (2008)
5. STN 73 6360-2 Railway Applications. Track. Part 2: Acceptance of Construction Works, Maintenance Works and Assessment of Service Condition Track Gauges 1 435 mm (in Slovak), SUTN Bratislava (2015)
6. J. Sestakova, Civil and Environmental Engineering, **11,** 1, 2-9 (2015), online: http://www.degruyter.com/dg/viewarticle.fullcontentlink:pdfeventlink/$002fj$002fcee.2015.11.issue-1$002fcee-2015-0001$002fcee-2015-0001.pdf?format=INT&t:ac=j$002fcee.2015.11.issue-1$002fcee-2015-0001$002fcee-2015-0001.xml
7. L. Izvolt. et. al. Monitoring of Sections of a Non-conventional Constructions of the Railway Superstructure and the Transition Areas – 7th and 8th Stage. ZSR Modernization of Railway Track Nove Mesto nad Vahom - Puchov, km 100.500 to 159.100, part 24-32-01 Nove Mesto - Trencianske Bohuslavice (in Slovak), University of Zilina (2015)
8. M. Smalo, L. Izvolt, Communications - Scientific Letters of the University of Zilina, **17,** 4, 28-33 (2015)

Dynamic testing of railway metal culvert using geodetic methods

Damian Beben[1,*] and *Wojciech* Anigacz[1]

[1]Faculty of Civil Engineering and Architecture, Opole University of Technology, 45-061 Opole, Poland

Abstract. The objective of the study was to compare and assess suitability of two methods of geodetic measurements (tachymetry i interferometry) used to determine changes of the geometric condition of building and engineering structures. The paper presents the selected results of experimental tests under dynamic loads that were conducted on a railway metal culvert. The dynamic loads were caused by the passages of various trains. The measurements were made for all trains which had been running over the culvert during a 24 hour period. Advantages and disadvantages of both applied methods were characterized. The disadvantage of the tachymetry method is the discreteness of measurements and the lack of the possibility of verifying the results after finishing the field works. The tachymetry measurements were conducted using precise tachymetry manufactured by Leica TC2002. The IBIS microwave coherent radar was used in the interferometry method. Moreover, a special microwave horns IBIS-H23 type with a maximum gain of 23 dBi were used as the interferometer antennas (transmitting and receiving). Conclusions drawn from the tests can be helpful in the measurements of such culverts.

1 Introduction

Non-destructive measurements like GPS, GPR, thermography, photogrammetry, laser scanning, etc. are more and more popular to diagnostic of various engineering structures [1], [2]. The dynamic and static tests of bridges and culverts are conducted in order to confirm structural specifications, or to provide diagnostic surveys for planning maintenance and modernization. Typical experimental tests are usually carried out using accelerometers, strain and inductive (or dial) gauges. Installing most of these sensors require direct contact with the structure. However, examining structures can cause accessibility problems, often requiring using scaffolds. That is why, authors decided to analyse in this article two measurement methods intended for non-contact monitoring of the construction displacements.

The IBIS-S interferometric radar applied in tests, is a precision microwave instrument suitable to the non-contact vibrations and displacements monitoring of engineering structures. The high accuracy and quickness of measurement are among of the main characteristic features of the radar. The comparison of the interferometric radar vs. the

* Corresponding author: d.beben@po.opole.pl

accelerometer used for dynamic monitoring of various large bridges is presented in [3-4]. The differences in displacements of bridges were in the order of one-tenth of millimetres. The interferometry method for non-contact dynamic and static measurement of the vibration of various engineering structures (bridges, towers, dams) was described in [5-8].

The main objective of the study was to compare and assess the suitability of two different geodetic measurement methods (interferometric radar and tachymetry) for determination of displacements of railway culvert. Information on the change in the shape of the bridge structures under the service load are very important for a number of reasons. Knowing the changes in the structure shape, the correctness of the project assumptions can be evaluated. Additionally, after some period of exploitation one can obtain information if the bridge displacements do not exceed the code limit.

2 Short culvert description

The tested railway culvert in the cross section has two spans in shape of closed arch. The effective length of shells is $L1$ = $L2$ = 4.40 m (Fig. 1). The load bearing structure was constructed as two shells assembled from the corrugated metal plate (CMP) sheets. The corrugation depth of 0.05 m, pitch of 0.15 m and plate thickness of 0.003 m were designed. The individual sheets were connected together using high strength bolts 20 mm, and covered with the layers of soil (about 0.20–0.30 m thick) properly compacted to achive I_D = 0.98 (in the Proctor Normal scale). The soil cover over the CMP structures (including ballast, blanket and backfill) equals 2.40 m. The width of the culvert shell at the top is Bt = 16.00 m, whereas at the bottom is Bb = 21.80 mm. The height of culverts is $H1$ = $H2$ = 2.80 mm. In the plan view, the object is situated perpendicularly to the railway line. On the culvert two rail tracks are arranged, first is electrified and the second is designed for non-electrified rolling stocks (Fig. 1).

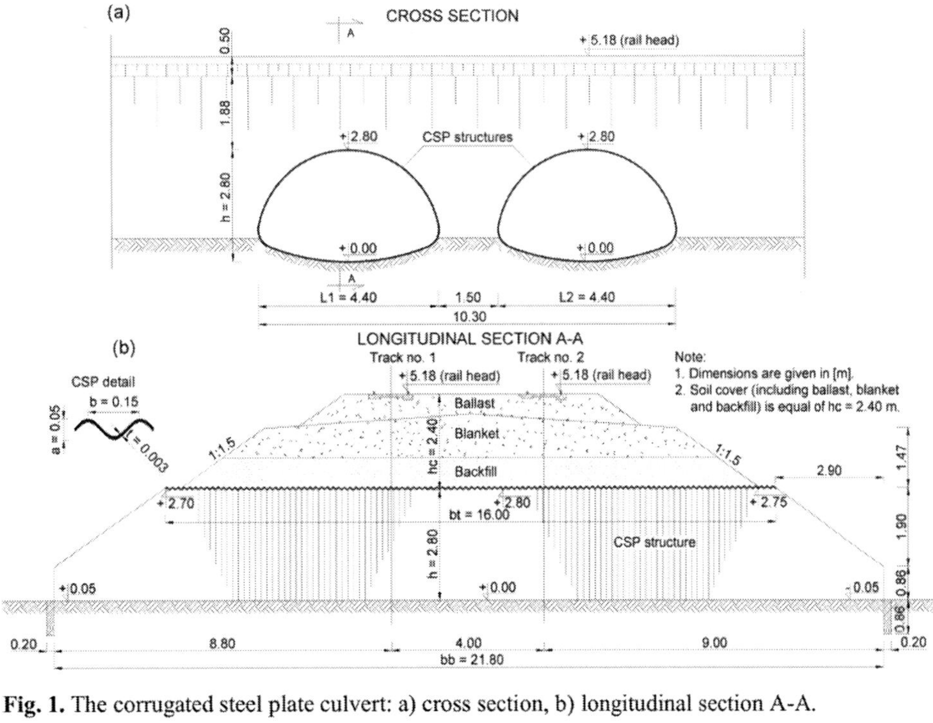

Fig. 1. The corrugated steel plate culvert: a) cross section, b) longitudinal section A-A.

3 Background of applied methods

3.1 General notes

Geodesy has a high range of measurement methods, from which we need to choose the most useful ones for measuring deflection of the CMP culvert. One of the most significant factors influencing the selection of the method is the geometric specificity of the bridge. The culvert span is significant because the accuracy of determination if the location of the measurement point decreases with the distance and is described with universal interrelations between the measured values, according to the formula (1):

$$m_n = \sqrt{m_x^2 + m_y^2 + m_h^2}\ , \tag{1}$$

where: m_n = spatial error, m_x = error of the coordinate $x = d \cdot \sin z \cdot \sin\alpha$, m_y = error of the coordinate $y = d \cdot \sin z \cdot \cos\alpha$, m_h = error of the coordinate $h = d \cdot \cos z$, α = horizontal angle, z = vertical angle (zenithal), d = slant distance.

The experimental tests were carried out after thirteen years of service of the culvert. The culvert displacements were registered during normal service, continuously for 24 hours. On the examined culvert structure, two measurement methods were tested, i.e.: (i) tachymetry, and (ii) interferometric radar. Based on this methods, the displacements of the culvert were determined.

3.2 Radar technique

A coherent microwave beam with a very low power and variable frequencies is emitted by the microwave interferometric radar in direction of the tested object. Next, the reflected signal is received and analysed. Displacements of the reflector are detected and recorded only in the radial direction. The high accuracy of measurements makes possible to carry out tests under conditions in which the displacements are not parallel to the axis of the microwaves beam. It is possible in the case when the change of radial distance is not less than 10 % of the real displacements. It means that the angle between the direction of the microwaves beam axis and the direction of displacements should not exceed 84°.

In these experimental studies, the distance from the radar to the microwave reflector was 14.5 m (bin = 34). The instrument performed recording with a frequency of 94.26 Hz in all sectors up to a distance of 30 m, i.e. for 60 sectors. It was a compromise between the expected range of vibration frequencies and the amount of data recorded. At the stage of

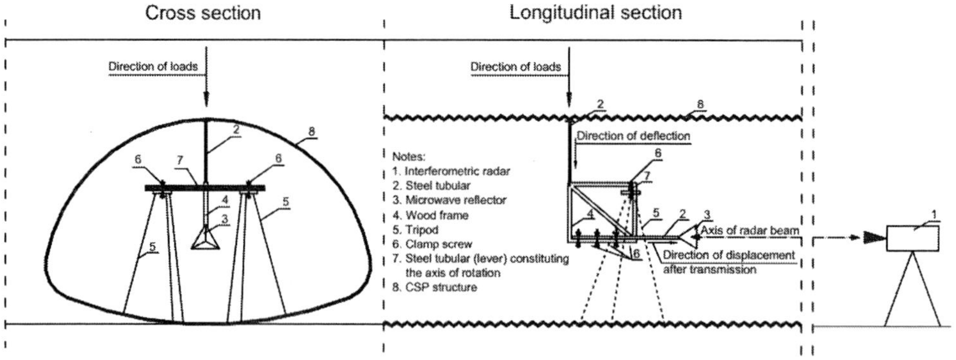

Fig. 2. The "transmission gear" used to experimental tests.

results working out it turned out that frequencies had not exceeded 15 Hz. Typical conditions for the radar measurements could not be fulfilled in the case of this CMP culvert. Thus, a special "transmission gear" must be used. Thanks to this mechanism, the vertical displacements of the culvert were transformed into the horizontal displacements. Such "transmission gear" was specially constructed (Fig. 2). The mechanism operation is based on the assumption that "transmission gear" acts as a lever with one axis of rotation. As a result of preliminary control testing in the lab and field it was found that the specially constructed mechanism was operated in a satisfactory manner.

The tested railway culvert was made from corrugated metal plates (pitch of 150 mm). Such shape of the structure could potentially result in the signal reflections from the multi-surfaces. As a result of this phenomenon, the measurement signals reflected from the reflector might be drowned out. In order to reduce the risk of drowning out the signals, the measurement beam of microwaves was placed horizontally (close to the axis of the culvert). Moreover, a special microwave horns IBIS-H23 type with a maximum gain of 23 dBi were used as the interferometer antennas (transmitting and receiving). During the measurements, it turned out that the signals reflected from the reflector significantly dominated by the noise caused by other sources. The dynamic characteristics of "transmission gear" and the reflector were also determined. It was an extremely important problem, because it was necessary to distinguish the displacements spectra between the "transmission gear" and the metal culvert. For this purpose, the pre-tests in several trials were conducted. The frequencies were equal to 8.8 Hz, and they were damped after about 5 seconds. Therefore, in the main test of displacements caused by passing trains, the frequencies varying from 8.5 to 10.0 Hz were considered as the natural vibration frequencies of the system "transmission gear" – reflector. Details were presented in [9].

3.3 Tachymetry method

The tachymetry method depends on the field determination of the spatial polar coordinates, i.e. distances and vertical and horizontal angles from the set of chosen directions. The accuracy of the determination of the point's position is described by the formula (1). Tachymeters are available theoretically measuring the distances with the accuracy of 1 mm and the directions with the accuracy of 0.5'' (0.15 mgon). The accuracies obtained under the field conditions are usually much lower and oscillate within the range of 1–5 mm for the distances and 3–10'' for the determination of the directions.

The angular resolution of the measurement device has high significance especially for the identification of details, e.g. edges in the case of big distances [10-11]. In the analysed

a) b)

Fig. 3. View on the measurement site using tachymetry method (a) and measurements points installed on the culvert (b).

case, the stations of the tachymeter were placed on both sides of the river in order to lower the target lengths, i.e. distances between the tachymeter station and the measurement point. In the presented case the distances did not exceed 10 m (Fig. 3a).

Despite of such a significant difference between the accuracies obtained in the laboratory conditions and in the field conditions, the tachymetry method is currently the most accurate method of determining the position of the points in space [11]. The characteristic feature of this method is the measurement's discreteness. For the studies, the total station TC2002 manufactured by Leica was used and dozen or so different measurement targets (Fig. 3b). Calculations of the displacements were conducted using the authors' programmes [12].

The results obtained with this method were adopted as the reference data for the remaining methods. The correctness of the obtained results of displacements were verified by a measurement from the second station. In order to identify the measured displacements of the bridge, two-sided telemeter targets were used that enabled measurement of the distance from both sides of the target. The measurement targets were placed in the five points under the railway line (Fig. 3b). The opening in the middle of the target has 5 m diameter. On that basis, it can be concluded that the accuracy of aiming (measurement the direction) was between 0.5–1 mm.

4 Results of measurements and their analysis

The experimental tests were conducted in one cross section (in crown directly under the railway line) of the metal culvert where the maximum displacements and vibration frequencies were expected. The experimental tests were conducted continuously for 24 hours. The forty-one various trains were noticed (eighteen – passenger, eighteen – freight, five – others). Fig. 4a shows the metal railway culvert during tests by using the interferometric radar and tachymetry methods. The total loads of tested culvert amounted to: the passenger trains – 32006 kN, the freight trains – 144211 kN, and others – 4002 kN. Simultaneously, it should be pointed, that the maximum weight of the passenger train amounted to 3236 kN, and the freight train – 16867 kN. The maximum length of the passenger train equalled 178 m, and the freight train amounted to over 520 m. The maximum speeds of passing trains were: 12.5 m/s for the freight trains and 33.3 m/s for the international passenger trains.

In order to evaluate the possibility of geodetic methods application, at the same time, the interferometric radar, tachymetry method and inductive gauge were used. A European

a) b)

Fig. 4. View on : a) the metal culvert under dynamic tests and b) the inductive gauge for deflection measurements. The targets to geodesic measurements and the extensometers to strain measurements can also be seen.

patent no. EP 2116806 B1 for fixing a target plate (mirror) was applied [13] (Fig. 4b).

The inductive gauge and measurement targets were placed in the same cross section of the metal culvert in which the special "transmission gear" was fixed. So, three independent techniques for deflection measurements of the metal culvert were applied.

The vertical displacements of the metal shell culvert were measured during conducted experimental tests using interferometric radar (IR) and tachymetry (T). Results using inductive gauges were presented in [9], [14]. The vibration frequencies of the metal culvert were also recorded using the radar.

Fig. 5 shows example of displacements of the crown of metal culvert. The effect of each axle of the freight train can actually be also observed from the displacement versus time plots. The total weight of passing train was 11846 kN. During train rides, four main phases of displacements were emphasized. The first phase represents impact of locomotive and two heavy wagons. The maximum displacements amounted to f_{IR} = 0.53 mm. The second phase relates to passage of four lighter wagons. In this stage the maximum displacements equalled almost f_{IR} = 0.15 mm. The third phase represents impact of eight heavy wagons. The largest displacements amounted to f_{IR} = 0.22 mm. The fourth phase relates to vibration reduction (damping) after passage of the freight train.

Application of interferometric radar allows to obtain culvert frequencies. Taking into account displacements presented on Fig. 5 (passage of freight train), the frequencies were determined. In the first phase represents impact of locomotive and two heavy wagons the

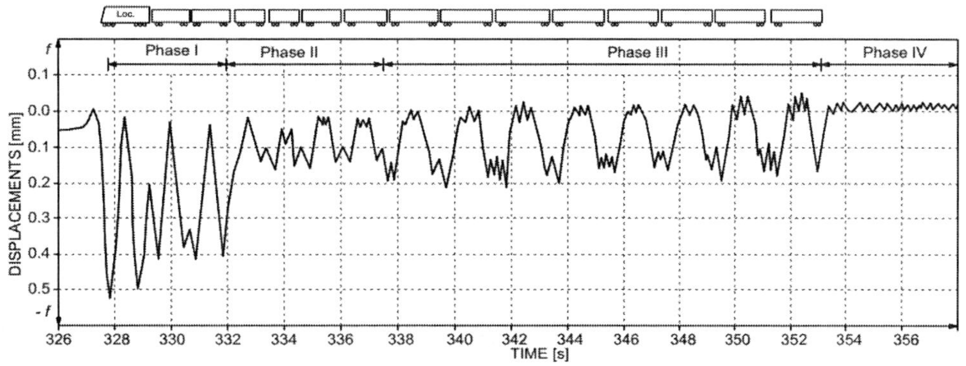

Fig. 5. The courses of vertical displacements measured using the interferometric radar during passage of the freight train.

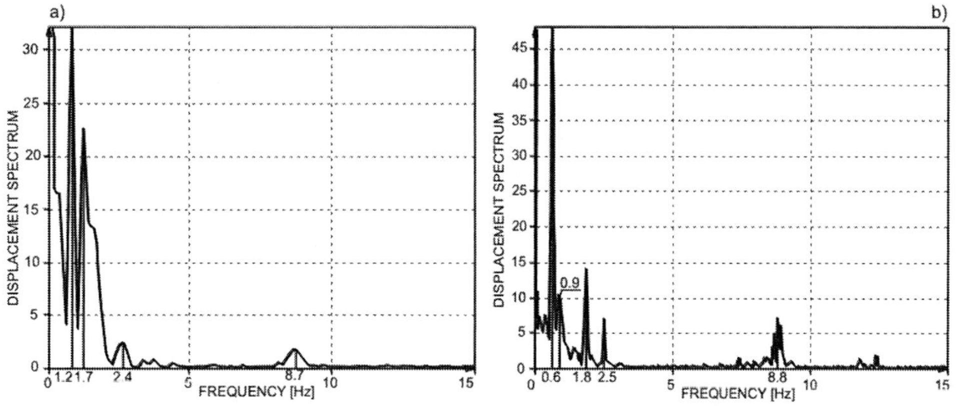

Fig. 6. Spectra of displacements measured using radar at the crown of metal culvert in phase: a) I and b) III, for passage of the freight train.

highest frequencies were 1.2 and 1.7 Hz (Fig. 6a). In the third phase represents impact of eight heavy wagons the largest frequencies were 0.6 Hz (less distinct frequencies were 0.9, 1.8 and 2.5 Hz – Fig. 6b). In the fourth phase (vibration damping after passage of the freight train) the dominate frequencies were in the range of 8.7-8.9 Hz which were identified, according to the preliminary tests, as a natural frequency of the "transmission gear".

Generally, the maximum vertical displacements of the culvert crown do not exceed f_{IR} = 0.65 mm and f_T = 0.7 mm. Displacements received from the interferometric radar were smaller than those obtained from tachymetry method in range of 7–25%. The maximum vertical displacements of the metal culvert crown received from all passing trains using the interferometric radar and the tachymetry method are shown in Fig. 7. Comparing the displacements received from inductive gauge with ones from the interferometry method, they are almost the same. They were obtained during passage of the heavy freight trains. Then, the smallest deflection of the culvert registered during tests was f_{IR} = 0.08 mm. The typical response of the metal railway culvert to passing trains is usually the collection of three frequencies in the range of 0.6 to 3.0 Hz. The highest values of frequencies were caused by passage of the express trains (speed was varying from 70 to 120 km/h).

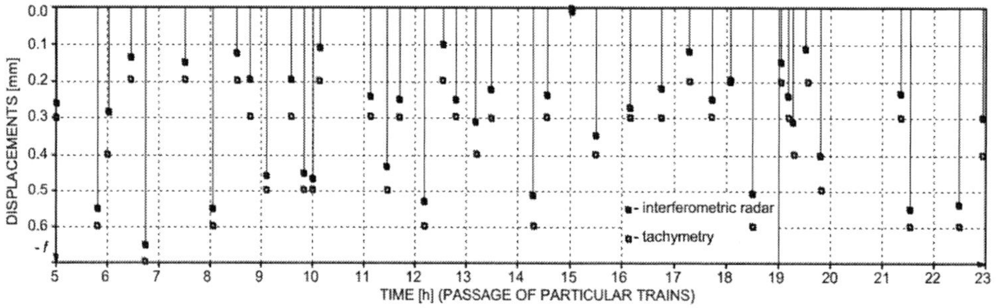

Fig. 7. Maximum displacements on the metal culvert crown using the applied methods.

The metal culvert responds to change of loads very fast. The displacements decrease immediately after passages of trains. The maximum measurement results of the metal culvert using the interferometric radar were quite similar to results received using the tachymetry method (the differences did not exceed 25%). Taking into account the defects of tachymetry method, the results are quite good. Such differences can be caused by manual estimation of the displacement values. The interferometric radar correctly responds to rapid and frequent change of loads. The natural vibrations of shell structure were not observed. It is probably caused by large thickness of the backfill placed around steel shell structure which in this case is a damper of vibration.

5 Conclusions

As a result of the experimental tests, carried out on the metal railway culvert using geodetic methods, the following conclusions can be drawn:

1. The interferometric radar is a suitable measurement instrument to long-term monitoring of small-to-medium culverts or bridges. It allowed to get the displacements and frequencies of this specific type of the metal culverts very fast and with a high accuracy level. The interferometric radar registered even very small values of displacements and frequencies of the metal culvert (with an accuracy in the order of 0.01 mm and 0.1 Hz respectively). The special "transmission gear" need to be used during testing of small culverts and bridges. This necessity results from the possibility of drowning out the

signal by the corrugation surfaces. Generally, the application of the specially constructed "transmission gear" did not influence measurement results significantly.

2. Tachymetry method allows to obtain the higher culvert crown displacements (max. 25 %) than those received using interferometric radar. Such differences can be caused by manual estimation of the displacement values in the tachymetry method. Tachymetry method does not allow to obtain the frequency of the culvert.
3. During experimental tests under service loads any irregularities of behaviour of the metal railway culvert were not found after thirteen years of service. The highest dynamic response of this culvert has been mostly caused by passages of freight trains and passenger express trains. Other types of trains, i.e. slow passenger trains, maintenance-repair trains, locomotives have not a significant dynamic impact on the metal culvert.

References

1. D.M. McCann, M.C. Forde, Review of NDT methods in the assessment of concrete and masonry structures, NDT&E Int. **34**, 71-84 (2001)
2. D. Beben, A. Mordak, W. Anigacz, Ground penetrating radar application to testing of reinforced concrete beams, Procedia Eng. **65**, 242-247 (2013)
3. M. Pieraccini, M. Fratini, F. Parrini, C. Atzeni, G. Partoli, Interferometric radar vs. accelerometer for dynamic monitoring of large structures: An experimental comparison, NDT&E Int. **41**, 258-264 (2008)
4. C. Gentile, G. Bernardini, Output-only model identification of a reinforced concrete bridge from radar-based measurements, NDT&E Int. **41**, 544-553 (2008)
5. D. Dei, M. Pieraccini, M. Fratini, C. Atzeni, G. Bartoli, Detection of vertical bending and torsional movements of a bridge using a coherent radar, NDT&E Int. **42**, 741-747 (2009)
6. C. Gentile, Deflection measurement on vibrating stay cables by non-contact microwave interferometer, NDT&E Int. **43**, 231-240 (2010)
7. M. Fratini, F. Parrini, M. Pieraccini, C. Borri, C. Atzeni, Structural oscillation modes identification by applying controlled loads and using microwave interferometry, NDT&E Int. **42**, 748-752 (2009)
8. M. Pieraccini, F. Parrini, M. Fratini, C. Atzeni, P. Spinelli, M. Micheloni, Static and dynamic testing of bridges through microwave interferometry, NDT&E Int. **40**, 208-214 (2007)
9. D. Beben, Application of the interferometric radar for dynamic tests of corrugated steel plate (CSP) culvert, NDT & E Int. **44**, 405-412 (2011)
10. A.Berenyi, T. Lovas, A. Barsi, L. Dunai, Potential of terrestrial laser scanning in load test measurements of bridges, Period. Polytech. **53**, 25-33 (2009)
11. T. Luhmann, Ch. Müller, *Photogrammetrie Laserscanning Optische 3D-Messtechnik* (Heidelberg: Herbert Wachman Verlag, 2007)
12. W. Anigacz, D. Beben, J. Kwiatkowski, Analysis of suitability of new measurement technologies to examine changes in geometric parameters of a suspension bridge, *Maintenance, Monitoring, Safety, Risk and Resilience of Bridges and Bridge Networks* (Eds. T.N. Bittencourt, D.M. Frangopol, A. Beck, 2204-2211, 2016)
13. W. Anigacz, *A target fixing device*, Patent no. EP 2116806 B1. European Patent Office. Munich (2012)
14. D. Bęben, W. Anigacz, Interferometric radar application for dynamic testing of bridge structures, *Bridge Maintenance, Safety, Management and Life Extension* (Eds. A. Chen, D.M. Frangopol, X. Ruan, 2289-2296, 2014)

Mutual measures of vibration process in differentiating examined construction state

Mariusz Żółtowski[1,*]

[1]UTP University of Science and Technology, Management Department, 85-790 Fordońska 430 Bydgoszcz, Poland

Abstract. In the article author identifies a possible bone of mutual vibration energy measuring application used in the study of standard rail, as part of a dedicated research methodology and detailed methodologies developed in many previous author studies. Theoretical analysis and practical verification of susceptibility testing measurement information complex vibration processes indicates possibilities for their use. Recognizing the need to improve methods of machines and structures testing in order to assess their degradation state - this work presents important results of the research in the verification field of the proposed measures effectiveness of mutual research bench.

1 Introduction

Ensure the safe and trouble-free operation of machines and structures requires a search or improvement of tools, methods and techniques for non-invasive assessment of the emergency state. A common trend in recent years is the scheme and to organize various business areas targeted for specific purposes. Ensuring repeatable perform actions leading to the intended purpose it is possible, using a variety of situations dedicated research developed methodological procedures allowing for action in various areas.

Developed in a number of author's publications [7, 9-11, 13-15] vibration research methodology dedicated to solving technical problems of discipline "building" gives useful procedures endearing research issues:

- acquisition processes in the vibration tests,
- processing processes by vibration. optimization algorithms,
- evaluate the vibration processes relationship in models of decision-making,
- sensitivity vibration parameters,
- statistical inference,
- visualization analysis.

The problem description and study of changes in the components destruction, materials, design and construction are carried out using the measurement of simple and complex vibration diagnostics methods based on modal analysis. An important issue here is

* Corresponding author: mazolto@utp.edu.pl

the description and propagation study of useful vibration energy is already widely used in diagnostic vibration and a little less used in modal analysis methods [16-18].

In this paper the suitability test measures in the monitoring complex process of the rail connections state with the base concrete, which is associated with many problems specific railway construction. The use of composite measurements mapping the interactions measured vibration process is simpler and less subject to error than the complicated procedures simulation based on numerical methods.

2 Mutual measures of vibration process

Research mutual similarities processes vibration through mutual estimators allows to specify:

- interdependencies between processes,
- correlation between sources,
- compliance within one process,
- changes in the road propagation processes in the machine and structure.

Measurement processes mutual indicate a change of source or variation in the transfer passageways signal extortion and answers. For these measures may include [1, 7, 13]:

- cross-correlation function,
- coherence function,
- transfer function,
- FRF function *(Frequency Response Function).*

The process of measurement of the vibration is determined from the measured numerical procedures base signal, which is the value of the measured process or the two measured signals: force and the system response to put force.

The concept of using vibration energy propagation of measurement uses the value of the mean square amplitude of dynamic directly proportional to the energy of the signal and written in the form [7]:

$$U = \int_0^T x(t)^2 \mathrm{d}t \tag{1}$$

A measure of the propagation efficiency of vibration energy H binding force signal $x(t)$ of forcing $F(t)$ takes the form:

$$H = \frac{\int_0^T x(t)^2 \mathrm{d}t}{\int_0^T F(t)^2 \mathrm{d}t} \tag{2}$$

Due to the large range of amplitudes of vibration signals is a deliberate to use logarithmic scale for comparing the numerical values of these measures:

$$H_{[\log]} = \log(H) \tag{3}$$

Large potential research opportunities create the analysis results described in the form of a matrix using measures built at different intervals integration or pre-processed waveforms.

The concept measures the propagation of vibration energy refer to the Parseval's theorem [7], which indicates the identity of the signal energy presented in the time and frequency domains:

$$\int_{-\infty}^{\infty} x(t)^2 \mathrm{d}t = \int_{-\infty}^{\infty} \left|X(f)^2\right| \mathrm{d}f \tag{4}$$

It is thus possible to define the measure of vibration energy propagation also in the frequency domain, and then the equation (2) becomes:

$$H(f) = \frac{\int_{-\infty}^{\infty} \left|X(f)^2\right| \mathrm{d}f}{\int_{-\infty}^{\infty} \left|F(f)^2\right| \mathrm{d}f} \tag{5}$$

In this relationship $X(f)$ is the power spectral density of the signal force, $F(f)$ - the power spectral density of the signal force, and f is the frequency in Hertz.

The essence of the measurement of input and response system for e designate of measures of mutual shown in Fig. 1.

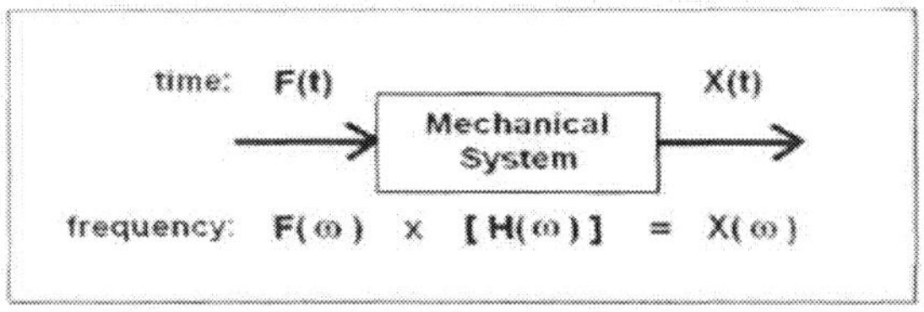

Fig. 1. Model components present measurement metrics mutual.

The opportunities created by the presented approach allows to solve many technical problems during testing of real objects, which measure defined below each other are widely use [3, 5, 7, 13].

2.1 Correlation function of the vibration signal

Estimator function is a signal that can be obtained by averaging in the time domain and examining the mutual relationship between the two segments of the delayed time interval T. It has the following form [16, 18, 19]:

$$R_{uu}(\tau) = \frac{1}{T} \int_{-T}^{T} u(t,\theta)\, u(t+\tau,\theta) \mathrm{d}t \tag{6}$$

In practical applications observation time T is chosen sufficiently high so that a correlation estimator properties are independent of the integration time T. A prerequisite for obtaining this property is the selection of T, that the time was much greater than the correlation time of the process τ_k. Defined correlation function may be in the range (-1, 1). A very important and convenient property of this estimate is its relationship with the power spectral density (for operational modal analysis) by Fourier transformation, because:

$$R_{uu}(\tau) = \int_{-\infty}^{\infty} G_{uu}(f) \cdot \exp(2\pi \mathrm{j} f \tau) \mathrm{d}f$$

$$G_{uu}(\tau) = \int_{-\infty}^{\infty} R_{uu}(f) \cdot \exp(-2\pi \mathrm{j} f \tau) \mathrm{d}\tau \tag{7}$$

This shows that the correlation function and power spectral density are the same as if the estimators of the test process, but the area they determine is different. Therefore, some of the properties of the process can be better pinpoint the time-domain retardation τ pressure than the frequency f.

2.2 Cross-correlation function

For the purpose of examining the similarity of processes at two points of the material and check if there is only one source of vibration in Fig. 2 shows the flow pattern signal $u(t)$ of such a case.

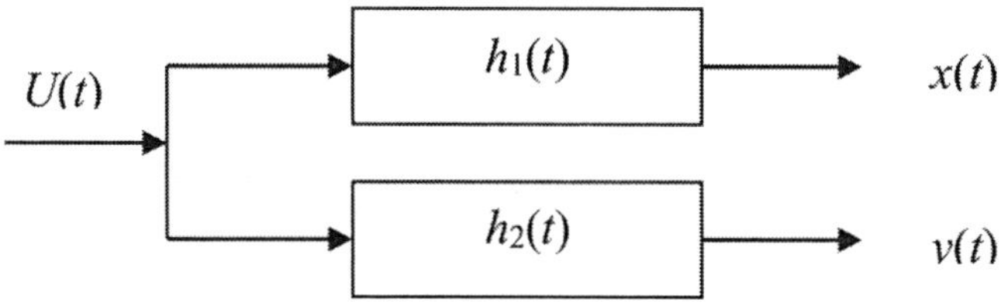

Fig. 2. Signal flow model for assessing the similarity of processes $x(t)$ and $y(t)$.

The initial process of the vibration that two different ways by the dynamic systems of the pulse transfer function $h_1(t)$, $h_2(t)$ enters the measuring points X and Y to form new processes vibrating designated $x(t)$ and $y(t)$, [6, 16, 14].

The cross-correlation $R_{xy}(\tau)$, which is in the form of unor-plated should be equal to one if the signals $x(t)$ and $y(t)$ are the same, it is defined as:

$$R_{xy}(\tau) = \frac{1}{2T} \int_{-T}^{T} x(t) \cdot y(t+\tau) \mathrm{d}t \tag{8}$$

wherein: $R_{xy}(\tau) = R_{yx}(-\tau)$, and $Q_{xy}(0) = 1$ if $x(t) \to y(t)$.

In terms of diagnostic correlation function makes it possible to detect changes in signal power ratio determined for the noise interference. It is therefore a measure of the functional arrangement of these moments of the process tested. In diagnostic applications it is sometimes used to measure the delay time determining signal path and signal detection in noise and playing them.

2.3 The coherence function

A measure of the consistency of the two vibration processes, $x(t)$ and $y(t)$ is the coherence function defined as follows [12, 14, 15]:

$$\gamma_{xy}^2(f) \equiv \frac{|G_{xy}(f)|^2}{G_{xx}(f) \cdot G_{yy}(f)} \tag{9}$$

Calculating the spectral density of the source of the process $x(t)$ and the known transfer function $H_1(f)$, $H_2(f)$, we have:

$$\begin{aligned} G_{xy}(f) &= H_1(f) \cdot H_2(f)\, G_{uu}(f) \\ G_{xx}(f) &= [H_1(f)]^2\, G_{uu}(f) \\ G_{yy}(f) &= [H_2(f)]^2\, G_{uu}(f) \end{aligned} \tag{10}$$

Hence, we obtain:

$$\gamma_{xy}^2(f) = \frac{|H_1(f) \cdot H_2(f)|^2 G_{uu}^2(f)}{|H_1(f)|^2 \cdot G_{uu}(f) \cdot |H_2(f)|^2 \cdot G_{uu}(f)} = 1 \tag{11}$$

Regardless of the nature of transmittance passageways coherence function always assumes the value 1 if only the signals $x(t)$ and $y(t)$ come from the same source. For more than one source coherence function which is a non-negative it is always less than unity ⟨0-1⟩. The coherence function as a measure of the consistency of the two signals has good diagnostic properties, because the appearance of a new signal failure violates the consistency of the previous ones, so that coherence function decreases.

It is the local measure of similarity between signals from different sources of vibration. In practical action, measured at different times living facility, detects the appearance of defects and changes in transmittance. The function, which of the two points defines is, provides more information about the technical condition object.

2.4 Transmittance

The dynamic properties of linear with constant parameters can be described using a *transfer function* defined in the time domain or with a *transmittance* defined in the frequency domain. These measures fully characterize the dynamic behavior of the object (linear systems, for which it is satisfied superposition principle) and is defined as the ratio of output to input [6, 7].

The transformation of input to output in the time domain can be written as:

$$y(t) = \int_0^\infty h(\tau)x(t-\tau)\mathrm{d}\tau \tag{12}$$

where: $y(t)$, $x(t)$, - input and output signals; $h(t)$ - pulse transition function.

Pulse transition function is defined as a response to the force in the form of a Dirac's pulse:

$$y(t) = \int_0^\infty h(\tau)\delta(t-\tau)\mathrm{d}\tau = h(t) \tag{13}$$

The transmittance of the frequency response is defined as:

$$H(\mathrm{j}\overline{\omega}) = \frac{Y(\mathrm{j}\overline{\omega})}{X(\mathrm{j}\overline{\omega})} \tag{14}$$

where: $Y(\mathrm{j}\overline{\omega})$ - Fourier Transform output signal; $X(\mathrm{j}\overline{\omega})$ - Transform of the input signal.

Characteristic of the frequency (transmittance) is the size of the complex, m and the real part and the imaginary part:

$$H(\mathrm{j}\overline{\omega}) = \mathrm{Re}\{H(\mathrm{j}\overline{\omega})\} + \mathrm{Im}\{H(\mathrm{j}\overline{\omega})\} \tag{15}$$

The transfer function $H(\mathrm{j}\omega)$ and the impulse response $h(t)$ are related to a clear relationship:

$$H(\mathrm{j}\overline{\omega}) = \int_0^\infty h(\tau)e^{-\mathrm{j}\overline{\omega}\tau}\mathrm{d}\tau \tag{16}$$

Transmittance carries a lot of information about the behavior of the system under load. During the test masonry using transmittance are assessed changes in stiffness (real part) and damping (the imaginary part) reflecting the change in the destruction of the materials tested under varying load.

MATEC Web of Conferences **107**, 00021 (2017) DOI: 10.1051/matecconf/201710700021
DYN-WIND'2017

2.5 FRF – Frequency response function

It is the function describing the frequency relation between the strength force $F(t)$ and the vibration acceleration $a(t)$ as the response signal measured at two points in the system under study. Spectral function of transition *(FRF - Frequency Response Function)* can be described as a quotient of the Fourier transform vibration exciting force $F(\omega)$, the Fourier transform of the response signal $A(\omega)$

$$FRF = \frac{\sum F(\omega)}{\sum A(\omega)} \tag{17}$$

The *FRF* as the main measure of experimental modal analysis is an inversion defined previously transmittance $H(f)$ and is referred to as real and imaginary parts. The method of its determination on the actual objects is shown in Fig. 3.

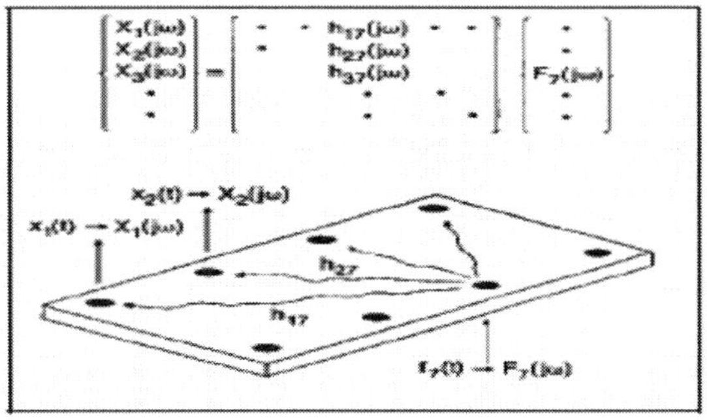

Fig. 3. Research *FRF* facility.

The *FRF* is a complex function of the frequency, which the module shows the characteristics of amplitude - the frequency, and the argument is the angle of the phase shift between the power of forcing a response [7, 13, 17]. The FRF is an inversion of the transmittance, which is an important observation resulting from the analysis and testing of these measures global propagation of energy.

Due to the possibility of the use of force or strength kinematic introduced different characteristics depending on the nature of the physical processes the input and output, highlighting: the vulnerability of dynamic mobility, acceleration, dynamic stiffness, impedance and dynamic mass [4, 8, 10].

3 Research methodology

Verification of the indicated measures of mutual propagation of vibration energy was carried out at a laboratory in the EAFIT University of Colombia. Being investigated size vibration energy flow between the points of attachment rails to the sleepers of the track has been switched on Fig. 4.

Point 3 :
all bold O.K.

Point 4 :
bad bold: [b1]

Point 5 :
bad bold : [b1, b2]

Point 6 :
bad bolds: [b1, b2, b3]

Point 7 :
bad bolds: [b2 b4]

Point 8:
bad bold: [b4]

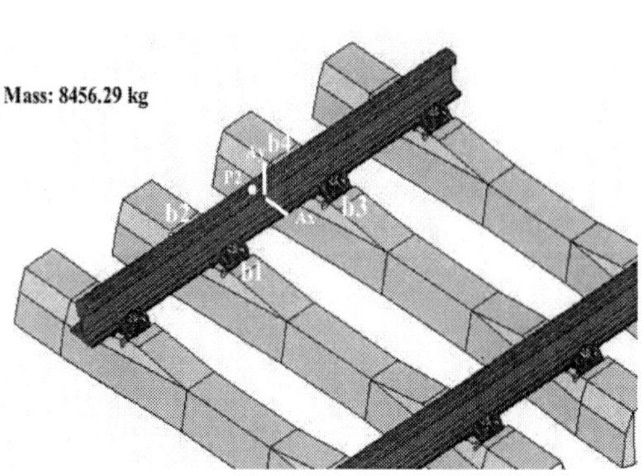

Fig. 4. Diagram of the test with the defined measuring points.

Using the studies measured vibration signals extortion and response software was set appropriate measure of each other, where the sample waveforms are shown in Fig. 5-7.

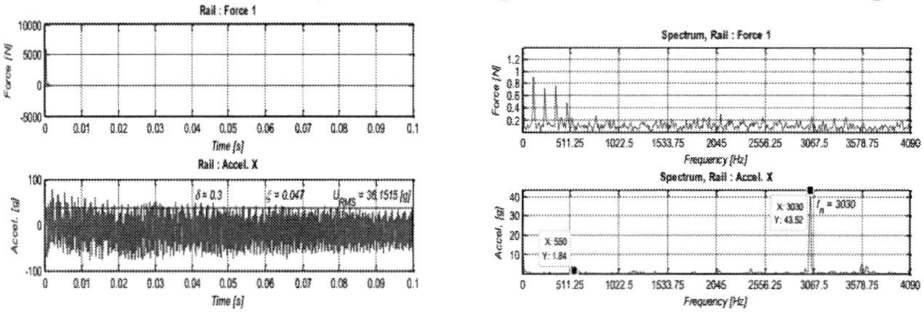

Fig. 5. Examples of the results of the exciting force and response in the time domain and frequency.

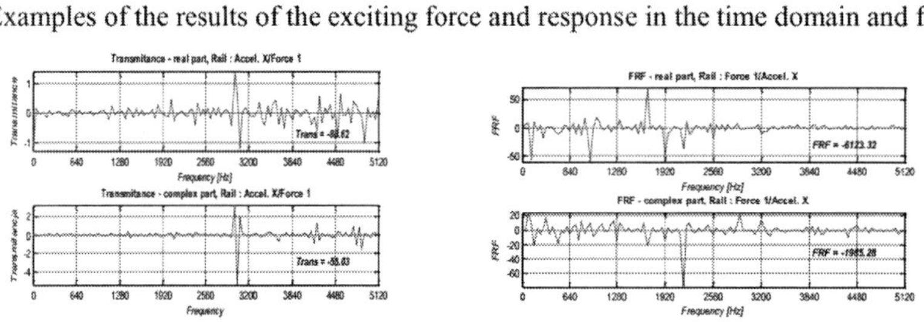

Fig. 6. The response signal and determined its coherence function.

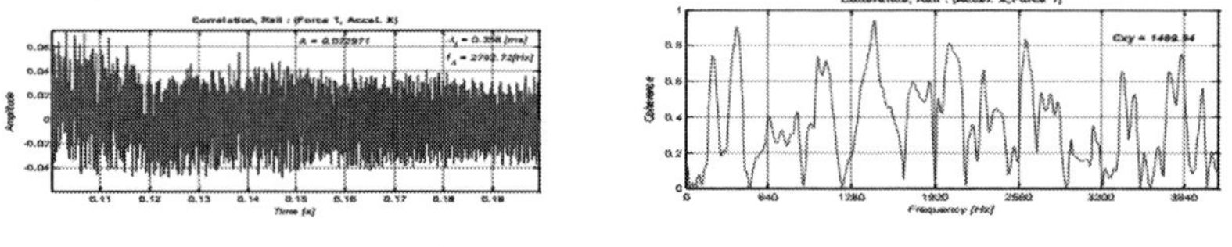

Fig. 7. Examples of waveforms transmittance and FRF.

The completed experiment studied a la the various states of attachment (pressure) between the indicated attachment points give a lot of diff s values (Table 1).

Table 1. Results of bench studies indicated measures of mutual.

Reception points	FRFr	FRFu	H(f)r	H(f)u	Cohere
1 - 2	288.71	-15.05	137.47	4.91	197.03
2 - 1	327.21	248.69	5.67	4.32	112.65
3 - 4	163.73	352.91	-3.42	-0.54	89.41
4 - 3	380.32	-29.54	108,18	11.01	195.27
1 - 3	83,97	892.43	-9.88	-6.97	109.06
1 - 4	587.91	-15.35	55.93	1.92	192.37
2 - 3	456.41	506.42	7.73	18.02	139.29
4 - 2	639.01	-118.72	58.17	10.72	192.04
2 - 4	1,213.83	15.03	17,91	17.99	125.36
3 - 1	679.94	157.66	16,34	11.73	139.08
3 - 2	985.87	-64.41	39.13	6.75	183.6
4 - 1	799.05	38.18	4.55	24.81	106.61

Correct inference requires redundancy of information and statistical analysis to obtain a model of bonding between the assumed technical condition and the value of mutual measures. To implement this project proposed in the work of the author's methodology dedicated research facilities in the "construction" [1,.2,.9] uses redundancy procedures, OPTIMUM, SVD and modeling cause - effect. The essence and the algorithm good measure of changes in the method of OPTIMUM (Fig. 8), and in the method of SVD (rys.8B) is shown in Fig. 8.

A B

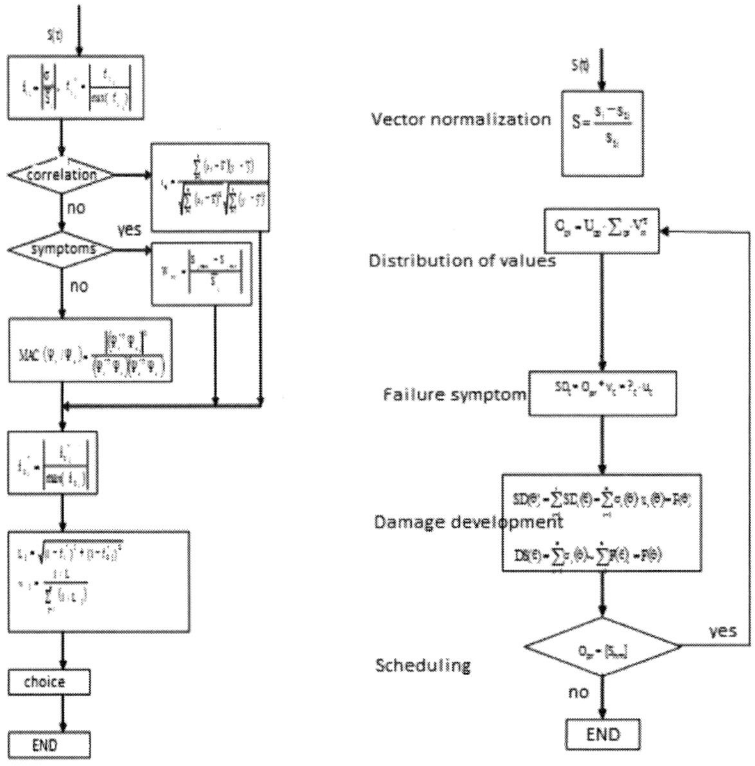

Fig. 8. Optimum algorithms for carrying out the method (A) and SVD (B).

Examples of the results of the proposed optimization methods usefulness of measures of mutual are shown in Fig. 9.

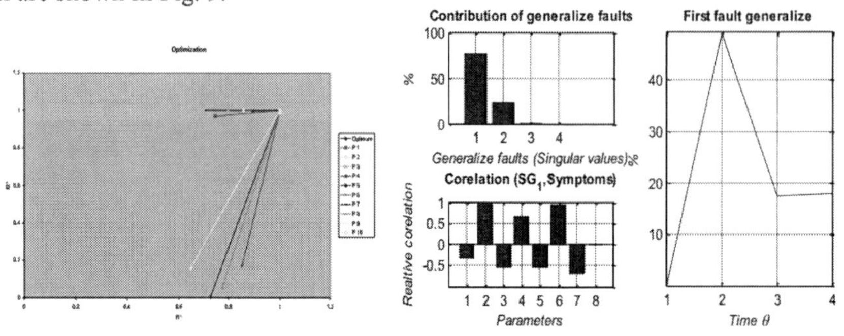

Fig. 9. The results of the OPTIMUM (A) and SVD (B) method.

Using the information contained in Table 1 processed according to the proposed methodology can confirm the good properties of measurement destruction of the cases examined the destruction of joints in order: *FRF$_r$, FRF$_u$, H(f)$_y$, H(f)$_u$, cohere.*

Taking into account all conducted a sensitivity analysis of the proposed measures of mutual propagation of vibration energy, using different statistical methods used in this work can continue to propose the following as the main measure of the state of destruction: *FRF$_y$, H(f)$_R$, cohere.*

For the indicated reduction measures of destruction was determined regression model binding shown in Fig. 10.

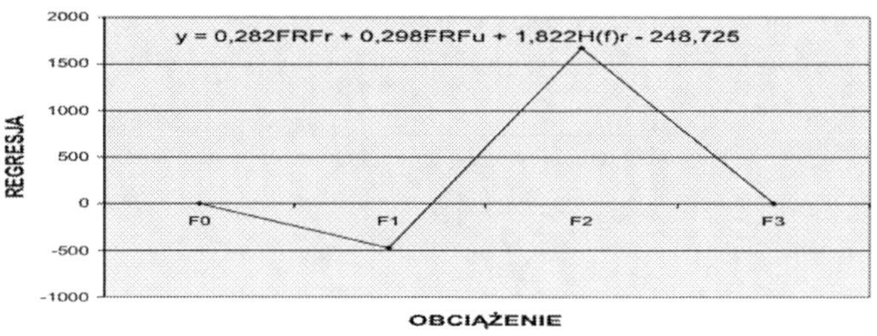

Fig. 10. Binding measures model of mutual technical connection condition.

Presented procedure is part of the proposed test method of cause - effect using various tools recommended by science literature[10, 13, 16] shown in Fig. 11.

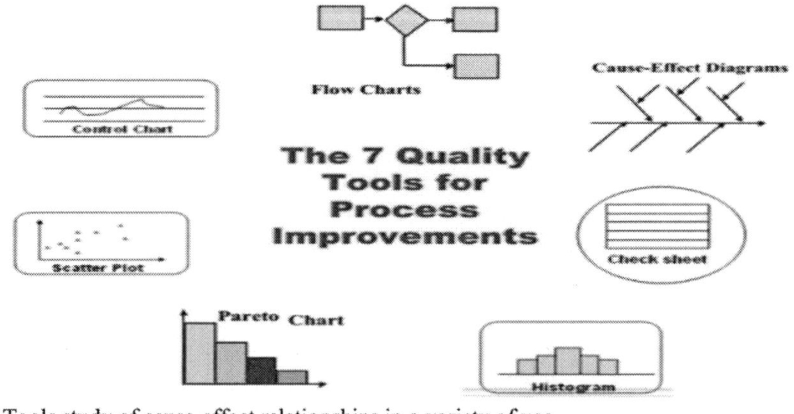

Fig. 11. Tools study of cause-effect relationships in a variety of use.

The presented methodology of the study allows to simplify the painstaking research conducted in the field of traditional airworthy calls on the raceways, replacing the work of many service employees to maintain fitness for work programmed utilizing available information technology proposed in this study (Fig.12).

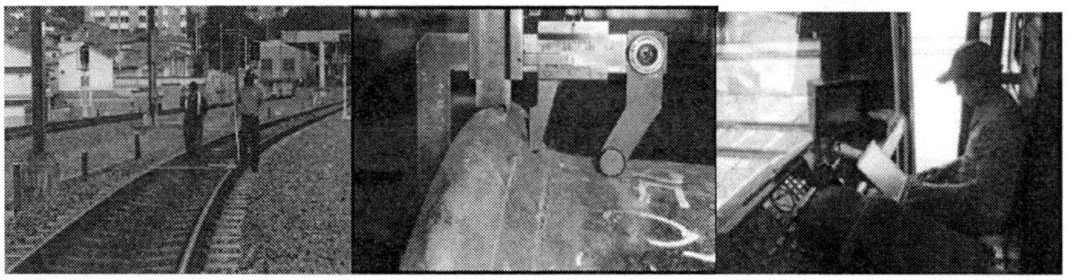

Fig. 12. The modernization work of maintenance services [4].

4 Summary

The substance of the development includes the results shows usefulness of mutual propagation measures of vibration energy than the various forms of applications in civil engineering. The main assumption was realized by exploratory research to present the possibilities of practical use of complex estimators vibration - as a tool to solve many fundamental problems with the condition of degradation and building construction elements dynamics.

These issues, it studies exemplify dedicated to the degradation state of the construction methodology procedure for determining the physical aspects and practical applications of propagation of vibration energy measures. The proposed methodology proceedings indicate that the important role and manner of acquisition and use of vibration composite measurements in the construction industry, revealing at the same time the need to take this issue in the area of research the destruction of existing structures. Served briefly elements of vibration theory focused on the physical aspects of the description and use of vibration estimators in the construction industry bring the issues of modeling the dynamic state of different structures and rules for their use.

Changes identification in the building structures degradation state becomes a method of assessing the structure operational status and modal analysis methods and related vibration estimates reflects well the elements and segments degradation.

New ways of obtaining and processing information contained in the processes of vibration often leads to improvement of various operating systems constructions, often leads to creative change existing practical solutions.

References

1. C. Cempel, *Vibroacoustical Condition Monitoring* (Ellis Ltd.New York, 1991)
2. H. Tylicki, B. Żółtowski, *Rozpoznawanie stanu maszyn* (ITE - PIB Radom, 2010)
3. T. Uhl, *Computer-aided identification of mechanical structure models* (Scientific Technical Publishers, 1997)
4. B. Żółtowski, L. *Castaneda, Structural diagnosis of rail vehicles and method for redesign.* Diagnostyka, **15** (2014)
5. B. Żółtowski, M. Łukasiewicz, *Diagnostyka drganiowa maszy* (ITE-PIB, 2012)
6. B. Żółtowski, M. Łukasiewicz, T. Kałaczyński, *Techniki informatyczne w badaniach stanu maszyn* (UTP Bydgoszcz, 2012)

7. B. Żółtowski, M. Żółtowski, *Vibration signals in mechanical engineering and construction* (ITE-PIB, 2015)
8. B. Żółtowski, C. Cempel, *Engineering of diagnostics machines* (PTDT, ITE – PIB, 2004)
9. M. Żółtowski, *Operacyjna analiza modalna w badaniach konstrukcji budowlanych* (WU UTP Bydgoszcz, 2012)
10. M. Żółtowski, *Informatyczne systemy zarządzania w inżynierii produkcji* (ITE-PIB 2011)
11. M. Żółtowski, *Investigations of harbour brick structures by using operational modal analysis.* Polish Maritime Research, **21**, 1, (2014)
12. M. Żółtowski, B. Żółtowski, *Vibrations signal to the description of structural damage of dynamic the technical systems*, XIII International Technical Systems Degradation Conference (2015)
13. M. Żółtowski, *Opis drganiowy konstrukcji budowlanych.* Logistyka, 6 (2014)
14. M. Żółtowski, *Investigations of harbour brick structures by using operational modal analysis.* Polish Maritime Research, **21**, 1 (2014)
15. M. Żółtowski, M. Liss, *The use of modal analysis in the evaluation of welded steel structures.* Studies and Proceedings of Polish Association for Knowledge Management **79** (2016)
16. M. Żółtowski, M. Liss, *Zastosowanie eksperymentalnej analizy modalnej w ocenie zmian sztywności prostego elementu konstrukcyjnego.* Studies and Proceedings of Polish Association for Knowledge Management, **80** (2016)
17. M. Żółtowski, R.M. Martinod, *Technical Condition Assessment of Masonry Structural Components using Frequency Response Function (FRF).* Masonry International Journal of the International Masonry Society, **29**, 1 (2016)
18. M. Żółtowski, R.M. Martinod, *Quality identification methodology applied to wall-elements based on modal analysis.* Civil Engineering the Athens Institute for Education and Research, Emerald, Athens (2015)

MATEC Web of Conferences **107**, 00022 (2017) DOI: 10.1051/matecconf/201710700022
DYN-WIND'2017

Influence of tire inflating pressure at dynamic forces acting on articulated trolleybus

Pavel Polach[1,*]

[1]Výzkumný a zkušební ústav Plzeň s.r.o., 301 00 Plzeň, Tylova 1581/46, Czech Republic

Abstract. ŠKODA VÝZKUM s.r.o. cooperated on the development of the NEOPLAN DMA low-floor articulated trolleybus intended for the Boston city. Multibody models and finite element models of the trolleybus were utilized in the stage of the vehicle design. At the end of the stage of computer modelling and testing of the trolleybus prototype a decision to reduce tire inflating of vehicles in serial production (together with the change the type of shock absorbers used in trolleybus suspension) was made. The impact of this change on forces acting in the trolleybus suspension elements (i.e. in air springs and shock absorbers) and radius rods on the trolleybus chassis when running on an uneven test track was investigated using multibody simulations. Time histories of the forces calculated utilising multibody models were used as input data of the trolleybus finite element models.

1 Introduction

Optimum dynamic properties of a vehicle intended for public transport can usually be achieved by the proper chassis structural design including the used spring-damper elements (i.e. springs, shock absorbers, bushings and finally tires). The design must be the compromise of requirements for the vehicle behaviour during driving manoeuvres, for riding comfort, for the vehicle body and the chassis parts lifetime when driving on an uneven road surface and for the passengers' safety (e.g. [1]).

In 2003 Neoplan USA Corporation, the American producer of buses and trolleybuses, started to develop the NEOPLAN DMA low-floor articulated trolleybus intended for the city of Boston (see Fig. 1). ŠKODA VÝZKUM s.r.o. (now Výzkumný a zkušební ústav Plzeň s.r.o.) cooperated on the development of the trolleybus. Multibody models and finite element (FE) models of the trolleybus were utilized in the stage of the trolleybus design. At the end of computer modelling and testing the trolleybus prototype a decision to reduce the tire inflating of vehicles in serial production (together with the change of the type of shock absorbers used in trolleybus suspension) was made because high stresses were calculated and measured in the trolleybus chassis in the place of the centre axle mounting during test drives on an uneven road surface. Lower tires inflating (100 psi instead of original 120 psi) results in their less steep radial force-deformation characteristics and their lower radial damping coefficient. On the one hand it leads to smaller stiffness of tires and on the other

* Corresponding author: polach@vzuplzen.cz

hand to larger deflections of tires. The impact of this change on forces acting in the trolleybus suspension elements (i.e. in air springs and shock absorbers) and radius rods on the trolleybus chassis when running on an uneven test track was investigated using the multibody simulations. Time histories of the forces calculated using multibody models were then used as input data of the trolleybus FE models. Utilizing the trolleybus FE models created in the COSMOS/M software [2] the critical places of the trolleybus body structure from the point of view of high stresses were determined [3]. At the measurement on the real trolleybus prototype in these places there were measured stresses using strain gauges. The experimental measurement results were utilized for the confirmation of the stresses reducing.

Fig. 1. The NEOPLAN DMA low-floor articulated trolleybus.

The influence of the already mentioned using shock absorbers of another type that have less steep force-velocity characteristics in rebound stage on forces acting in the trolleybus suspension elements is presented in [4, 5] – utilisation of another type of shock absorbers turned out to be suitable because maximum values of forces acting in the trolleybus suspension elements and radius rods decreased.

Generally, the assessment of vehicle's dynamic properties is related to three essential motions: longitudinal motion (driving and braking), lateral motion (guidance and steering) and vertical motion (suspension and damping) [6]. Driving on the uneven road surface can reveal many facts about the vehicle vertical dynamic properties and about the suitability of the used suspension elements of axles. Especially time histories of relative deflections of springs, relative velocities in the shock absorbers, stress acting in the axles' radius rods and/or radius arms and acceleration in various points in the vehicle interior are the monitored quantities [7]. On the basis of relative deflections of springs, relative velocities of shock absorbers and stresses acting in radius rods and/or radius arms it is possible to determine time histories and extreme values of the forces acting in the suspension elements of axles which can be utilized in connection with the suitable computational methods for the stress analysis of structures, for the prediction of the fatigue life of the body and the chassis parts of the tested vehicle (e.g. [8]).

Many professional books and papers deal with the investigation and approaches to the evaluation of vehicle dynamics using computer simulations. A smaller part of them is focused on the vehicles of public transport but despite this fact they are numerous. That is why a relevant literature is mentioned only in brief. The problematics of modelling vehicles in order to investigate their driving properties is globally solved e.g. in [1, 6, 8, 9]. Global investigation of vehicle dynamics using computer simulations is dealt with e.g. in [10], where numerous quantities informing on the vehicle driving properties are monitored.

A good survey of literature dealing with vertical dynamics of public transport and heavy vehicles is given e.g. in [11] and the literature dealing with the optimization of suspension elements of those types of vehicles in [12]. E.g. Margolis and Edeal [13] created a 2 degree-of-freedom, small vertical motion bus frame model to which the engine, cab, and load were added in addition to suspension forces. Michelberger et al. [14] identified a discrete transfer function for a free-free beam model of a two-axis bus, and then estimated modal characteristics based on the measured data. The principal aim of paper [15] is to evaluate the feasibility of using gradient-based approximation methods for the optimization of the spring and damper characteristics of an off-road vehicle, for both ride comfort and handling. Eriksson and Friberg [16] optimized the linear spring and damper characteristics of the engine mounting system on a city bus for the ride comfort. Andersson and Eriksson [17] optimized the non-linear spring and damper characteristics of a full city bus vehicle model that was validated against test data. Kowarska et al. [18, 19] was engaged in virtual designing of the bus suspension elements.

2 Multibody models of the trolleybus

In order to obtain a tool for dynamic analysis (e.g. [9, 20]) multibody models of an empty (21 tons weight) and a fully loaded (31 tons weight) NEOPLAN DMA low-floor articulated trolleybus were created [21] (see Fig. 2). The multibody models were created in the *alaska 2.3* simulation tool [22].

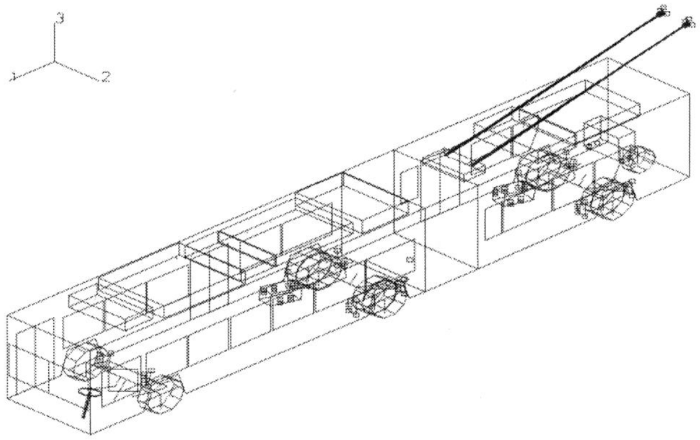

Fig. 2. Simple visualization of the NEOPLAN DMA trolleybus multibody model in the *alaska 2.3* simulation tool.

The multibody models of the NEOPLAN DMA trolleybus are formed by 54 rigid bodies mutually coupled by 63 kinematic joints. The number of degrees of freedom of multibody models in kinematic joints is 164. Kinematic scheme of the multibody models and sources of data used at multibody models creation are given in [21].

The rigid bodies correspond to the trolleybus individual structural parts and are defined by mass, centre of gravity coordinates and mass moments of inertia. Air springs and hydraulic shock absorbers in axles' suspension and bushings in the places of mounting certain trolleybus structural parts are modelled by connecting the corresponding bodies by nonlinear spring-damper elements. When simulating driving on an uneven road surface the contact point model of tires is used; radial stiffness and radial damping properties of tires are modelled by nonlinear spring-damper elements considering the possibility of the tire bounce from the road surface.

Generally, the aim of the simulations with the multibody models of the NEOPLAN DMA low-floor articulated trolleybus is the determination of time histories or FFT results of time histories of monitored kinematic and dynamic quantities in the course of the chosen operational situation.

3 Simulations

As it has been already stated, the impact of changing the tire inflating on forces acting in the (empty) NEOPLAN DMA low-floor articulated trolleybus suspension elements and radius rods were investigated during the simulations of running on an uneven test track.

An uneven track test was carried out in the standard way according to the ŠKODA VÝZKUM road vehicles testing methodology (so called bump test; e.g. [5]). An artificial test track was created on a common bitumen road with a set of four portable standard bumps (according to the ČSN 30 0560 Czech Standard Obstacle II – see Fig. 3).

Vertical coordinates $z(x)$ of the standardized artificial obstacle are given by the formula

$$z(x)=\sqrt{R^2-\left(x-\frac{d}{2}\right)^2}-(R-h), \tag{1}$$

where R = 551 mm is the obstacle radius, h = 60 mm is the obstacle height, d = 500 mm is the obstacle length and x is the obstacle coordinate in the vehicle driving direction [mm].

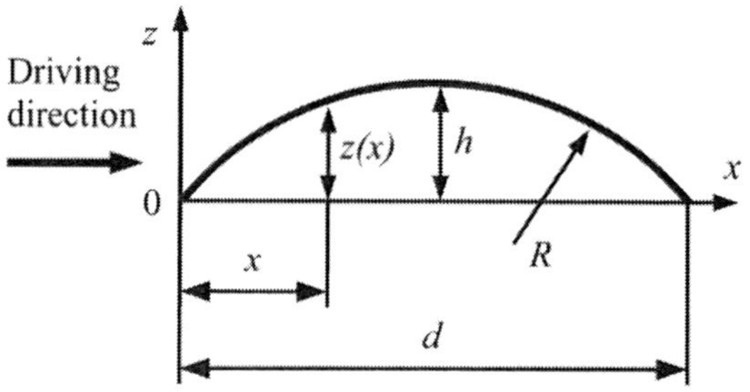

Fig. 3. The standardized artificial obstacle.

Artificial obstacles are spaced out on the smooth road surface 20 meters apart of a horizontal road. The first obstacle is run over only with right wheels, the second one with both and the third one only with left wheels (Fig. 4) at trolleybus speed 40 km/h. During the real NEOPLAN DMA trolleybus test the relative displacements between the axles and the chassis frame were measured (among others) and were compared with the results of the simulations with the multibody model [4].

The results of the simulations with the multibody model, i.e. the forces acting in the trolleybus suspension elements and radius rods on the trolleybus chassis, were used for dynamic calculation of the body structure stresses applying the finite element analysis (FEA).

Another reason for performing the bump test was the FEA model verification by measuring the stress-time histories at selected structural joints. The test results and/or its FEA simulation were also used for the preliminary evaluation of the critical joints fatigue life using design spectra. They represent the symmetrical loading cycles' histogram, the main parameters of which (maximum stress amplitude and predominant frequency of mechanical vibration) can be obtained from the bump test and the other ones can be derived from expected operational conditions. On the basis of empirical knowledge it is assumed that the maximum stress amplitudes at heavy damaged roads achieve from 80% to 100% of the bump test values. In the vehicle operation phase the design spectra can be compared with the real spectra and can be further modified.

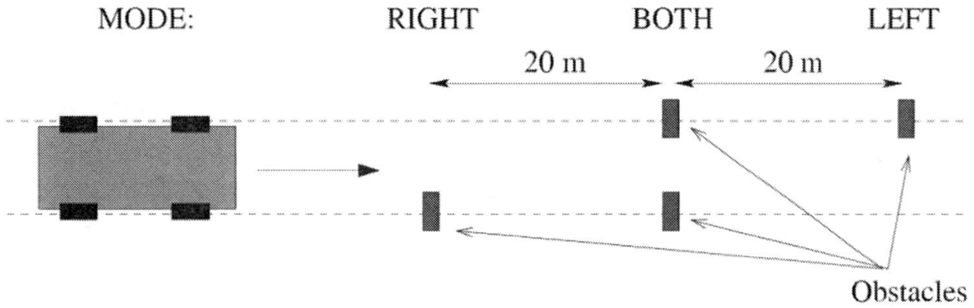

Fig. 4. Scheme of the track according to the ŠKODA VÝZKUM road vehicles testing methodology.

4 Results

Forces acting in axles' suspension elements and axial forces acting in axles' radius rods of the NEOPLAN DMA low-floor articulated trolleybus were the monitored quantities at simulations with the trolleybus multibody model.

Only the monitored quantities on the right side of the trolleybus and at running over the standardized obstacle with the right wheels are presented in this paper (just as at examining the influence on the change of the type of shock absorbers in [5]). It is given especially by the paper extent and also by the fact that time histories of the monitored quantities on the right side of the trolleybus and at running over the standardized obstacle with all the wheels and with only left wheels do not bring essential new pieces of knowledge for the evaluation of the impact of the change in the shock absorbers on the vehicle dynamic behaviour. Time histories of the dynamic forces acting in the centre air spring, time histories of the forces acting in the centre shock absorbers and time histories of the forces acting in the centre radius rods when simulating drives are given in Fig. 5 and 6. Extreme values taken from time histories of the monitored quantities are given in Table 1. Complete results are given in [4].

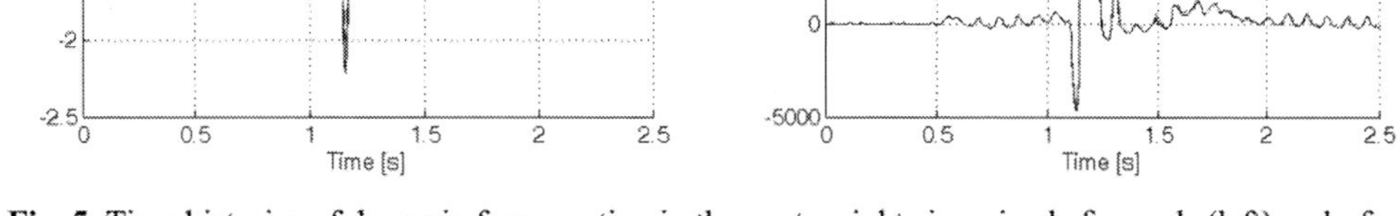

Fig. 5. Time histories of dynamic forces acting in the centre right air spring before axle (left) and of forces acting in the centre right shock absorber before axle (right).

Signs of the monitored quantities in air springs and shock absorbers and of axial forces acting in radius rods of the articulated trolleybus axles (Table 1 and graphs in Fig. 5 and 6) indicate acting of adjoining constructional parts (i.e. front body chassis, rear body chassis and axles) on those suspensions elements (e.g. minus sign at relative deflection of air spring corresponds to its compression, plus sign at the force acting in the shock absorber corresponds to its rebound, etc.).

Table 1. Extreme values of the monitored quantities at running over the standardized obstacle with the right wheels.

Monitored quantity	Placing of the element of the axle suspension	Value	Pressure 120 psi	Pressure 100 psi
Dynamic force in the air spring [kN]	Front right	min	-19.2	-17.3
		max	3.6	3.4
	Centre right before axle	min	-22.1	-20.3
		max	1.6	1.4
	Centre right behind axle	min	-19.3	-17.6
		max	1.6	1.4
	Rear right before axle	min	-22.2	-20.2
		max	2.2	2.0
	Rear right behind axle	min	-18.9	-17.4
		max	2.2	2.0
Force in the shock absorber [kN]	Front right	min	-3.1	-2.9
		max	14.9	14.6
	Centre right before axle	min	-4.7	-4.4
		max	15.6	15.6
	Centre right behind axle	min	-4.6	-4.3
		max	15.6	15.6
	Rear right before axle	min	-4.4	-4.1
		max	16.2	16.2
	Rear right behind axle	min	-4.3	-4.0
		max	16.3	16.1
Axial force in the radius rod [kN]	Front upper right	min	-19.3	-17.3
		max	14.5	13.3
	Front lower right	min	-27.3	-27.6
		max	48.1	45.0
	Centre upper right	min	-25.2	-23.1
		max	30.0	27.4
	Centre lower right	min	-35.0	-36.4
		max	38.3	38.1
	Rear upper right	min	-22.1	-20.9
		max	24.9	23.5
	Rear lower right	min	-46.4	-47.7
		max	50.1	48.9

When comparing extreme values of time histories of the monitored quantities taking into account the lower tires inflating, the consequence of their smaller radial stiffness and damping is evident [4]. It manifests itself in smaller dynamic forces and relative deflections in the air springs (see Fig. 5) and in the axial forces acting in the radius rods of the axles (see Fig. 6). Impact of the lower tires inflating does not manifest itself in the forces acting in the shock absorbers (see Fig. 5). Generally, at lower tires inflating lower extreme forces act in the trolleybus chassis and thereby there are lower extreme values of stresses in the trolleybus chassis.

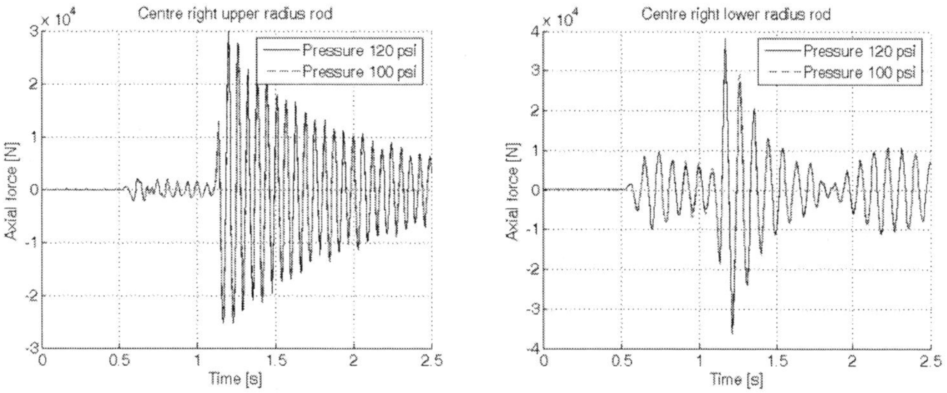

Fig. 6. Time histories of axial forces acting in the centre right upper radius rod (left) and in the centre right lower radius rod.

5 Conclusions

Influence of the lower tires inflating of the NEOPLAN DMA low-floor articulated trolleybus on forces acting in the trolleybus suspension elements and radius rods on the trolleybus chassis and axles is introduced in this paper. The impact of this change on these forces when running on an uneven test track was investigated using the multibody simulations. Multibody models of the trolleybus and simulations of its running were performed using the *alaska* simulation tool.

On the basis of the simulations results it was possible to state that the lower tires inflating is suitable because the maximum values of forces acting in the trolleybus suspension elements and radius rods decreased.

It would be possible to extend the simulations with the NEOPLAN DMA trolleybus multibody model by the simulations of other operational situations, the results of which would be interesting for the trolleybus producer. But due to the Neoplan USA Corporation having gone bankrupt further improvement of the NEOPLAN DMA trolleybus multibody model can be found only in the sphere of wishes.

Acknowledgement: The paper has originated in the framework of institutional support for the long-time conception development of the research institution provided by the Ministry of Industry and Trade of the Czech Republic.

References

1. G. Genta, L. Morello, *The Automotive Chassis. Volume 2: System Design* (Springer Science+Business Media B.V. 2009)
2. *COSMOS/M, Finite Element Analysis System, User Guide, Version 2.5* (SRAC, Los Angeles, 1999)
3. M. Smola, J. Jankovec, M. Hejman, Static and Dynamic FEM Calculations of the NEOPLAN DMA Low-floor Articulated Trolleybus, *Research Report VYZ 0693/2004* (ŠKODA VÝZKUM s.r.o. Plzeň, 2004)
4. P. Polach, Verified Multibody Model of the Empty NEOPLAN DMA Low-floor Articulated Trolleybus and Results of Running over the Large Road Unevenness Simulations, *Research Report VYZ 0726/2004* (ŠKODA VÝZKUM s.r.o. Plzeň, 2004)
5. P. Polach, Manufacturing Technology, **15**, 1, 77–86 (2015)

6. K. Popp, W. Schiehlen, *Ground Vehicle Dynamics* (Springer-Verlag, Berlin Heidelberg, 2010)
7. T. D. Gillespie, S. M. Karamihas, Int. J. Heavy Veh. Syst. **7,** 1, 52–63 (2000)
8. F. Vlk, *Dynamics of motor vehicles* (VLK Publishing House, Brno, in Czech, 2000)
9. M. Blundell, D. Harty, *The Multibody Systems Approach to Vehicle Dynamics* (Elsevier Butterworth-Heinemann, Oxford, 2004)
10. S. Hegazy, H. Rahnejat, K. Hussain, P. I. Mech. Eng. K-J. Mul. **213,** 1, 19–31 (1999)
11. D. G. Rideout, Modern Mechanical Engineering, **2,** 4, 176–189 (2012)
12. M.J. Thoresson, P.E. Uys, P.S. Els, J.A. Snyman, Math. Comput. Model. **50,** 9–10, 1421–1436 (2009)
13. D. Margolis, D. Edeal, SAE Technical Paper 892488 (1989)
14. P. Michelberger, A. Keresztes, J. Bokor, P. Variaki, SAE Technical Paper 845120 (1984)
15. P.S. Els, P.E. Uys, J.A. Snyman, M.J. Thoresson, Math. Comput. Model. **43,** 7–8, 787–801 (2006)
16. P. Eriksson, O. Friberg, Struct. Multidiscip. O. **20,** 1, 67–75 (2000)
17. D. Andersson, P. Eriksson, Vehicle Syst. Dyn., Suppl. **41**, 547–556 (2004)
18. I. Kowarska, J. Skarbek, K. Kuczek, T. Uhl, Journal of KONES Powertrain and Transport, **18,** 4, 205–212 (2011)
19. I. Kowarska, J. Korta, K. Kuczek, T. Uhl, Shock Vib. **2014**, ID 201952 (2014)
20. S. Hegazy, H. Rahnejat, K. Hussain, Vehicle Syst. Dyn. **34,** 1, 1–24 (2000)
21. P. Polach, Manufacturing Technology, **13,** 4, 515–520 (2013)
22. P. Maißer, C.-D. Wolf, A. Keil, K. Hendel, U. Jungnickel, H. Hermsdorf, P. A. Tuan, G. Kielau, O. Enge, U. Parsche, T. Härtel, H. Freudenberg, *alaska, User Manual, Version 2.3* (Institute of Mechatronics, Chemnitz, 1998)

Psychomotor conditions of bus drivers subjected to noise and vibration in the working environment

Anna Romańska-Zapała[1], *Alicja* Kowalska-Koczwara[2,*], *Aleksander* Korchut[3] and *Krzysztof* Stypuła[2]

[1] Cracow University of Technology, Malopolska Laboratory of Efficient Building, 31-155 Kraków, Warszawska 24 St., Poland

[2] Cracow University of Technology, Institute of Structural Mechanics, 31-155 Kraków, Warszawska 24 St., Poland

[3] University of Social Sciences and Humanities, 40-326 Katowice, Technikow 9 St., Poland

Abstract. The influence of vibration on human body is an issue that from several decades is a subject of many research and experiments. But most of these research concern on negative influences on human body. The paper presents methodology and results of research conducted on bus drivers of National Urban Transport in Tychy. The study involved fourteen drivers. First stage of the experiment included biomechanical and medical tests. After that psychological tests were made determining such factors like personality characteristics and stress resistance. Second stage included monitoring of psychomotor condition of drivers during night and day shifts, before and after work.

1 Introduction

Bus drivers of the public transport have various physical factors in their workplace. The most significant influence on the driver's body have noise and vibrations, both hand-arm and whole body vibration [1]. These factors cause functional and physiological effects [2] [3] [4] [5]. Drivers are exposed to deterioration of their health and occurrence of occupational diseases [6].

Working as a bus driver is considered a high-risk group for the appearance of work-related stress [7], this classifies it into professions that require special psychomotor skills. Drivers are required, on the one hand, to set specific ethical standards or skills for social functioning and, on the other hand, irreproachable mental health and specific psychological and cognitive traits.

* Corresponding author: akowalska@pk.edu.pl

So, it is true that the driver largely determines the proper functioning of communication in both urban agglomerations and off-shore centers as well as has an influence on the safety in public transport [8]. Evidence of the importance of the occupation of the bus driver in public transport is the number of publications on this subject. At the same time, in Poland as well as in many European countries, a new field has been formed from applied psychology dealing exclusively with the driver's profession. This field is the transportation psychology. Its competencies include periodic driver examinations of response parameters, visual coordination, perceptual sensitivity and resistance to stress-generating situations.

Among the many factors of the drivers' working environment, that adversely influence driver's health and psychophysical efficiency, stress has a significant role [3]. Driving a bus under heavy traffic conditions (currently the norm) is a challenge at the present time [9]. The scope of driver's duties is differential. It forces the driver to perform several tasks simultaneously - observing the road and the inside of the vehicle, reacting to the changing road situation, maintaining the time regime. Simultaneous contact with the passengers and controlling the emotions accompanying with these activities lead in consequence to a state of permanent stress [10].

Many studies conducted on drivers (including bus drivers) describe the most common physiological changes accompanying with work. These are: high blood pressure and pulse rate, accelerated breathing with congestion, excessive sweating and muscle fatigue. The consequences of these changes are:

- diseases of the circulatory system including increased mortality
- bone diseases, mainly degeneration of the spine
- psychosomatic diseases (peptic ulcer disease)
- obesity

In addition to the costs associated with physical health, constant fatigue and forced position lead to disturbances in the nervous system and its functional representation [11]. This mainly concerns visual and motor coordination and cognitive functions, including attention deficit disorder. Constant tension of skeletal muscles (especially upper limbs) causes movements to become inaccurate. The effectiveness of motor activities decreases.

The nervous system often reacts in a defensive manner to the whole body. Nervous system previously fatigued than the rest of the system protects the entire body from exhaustion. It makes possible to maintain a functional reserve.

Additional disadvantages are external factors. They are excessive noise, vehicle vibration, exhaust emissions, poor visibility and unfavorable meteorological conditions.

It should be added, however, that bus drivers are a professional group subjected to periodic selection [12]. Workers with special psychophysical features are admitted to this type of work, characterized by easiness of adaptation to difficult working conditions.

This article presents the process and results of studies of bus drivers of the National Urban Transport in Tychy (Państwowa Komunikacja Miejska w Tychach).

2 Purpose and methodology of research

The aim of the study was to determine the influence of the working environment of city bus drivers working in a two-shift system on their cognitive and motor skills.

The study involved fourteen bus drivers employed in the National Urban Transport in Tychy. The criterion of inclusion in the experiment was the admission of the driver to work on the basis of valid periodic medical examinations and psychological tests as well as the dispatcher agreement. The tested drivers are only men aged 24 to 49 years. These studies were conducted in two ways.

The first stage of research was related to determining the working conditions of the bus driver. For this purpose, the driver's seat was subjected to vibration and measurements of whole body vibrations were made in accordance with existing regulations [13]. For the measurement, a four-channel sound and vibration SVAN 948 sensor, which is mounted on the driver's seat was used. The measurements were carried out during driving with passengers on urban and suburban roads. Research was performed during ride on a full bus route. In the conducted experiment, the measurements included drivers working in two shifts system on seven buses.

The second stage of the research concerned the cognitive and motor skills of the examined drivers. A computer-aided set of psychological tests of cognitive and motor skills was used for the research. The set consisted of a computer, keyboard and exposition monitor. It is a non-invasive method i.e. it does not directly affect the person's body. The study consisted of performing the commands and research tasks displayed on the monitor. The results of the research are two parameters: time of performance and the level of accuracy. The device identified, among others, time of motor response to task stimulus and number of correct and incorrect responses. On the basis of the obtained data it was possible to determine the current level of changes in the organism of a subject under the influence of a specific factor (stimulus). In the research following tests were used:

- Line Test

Efficiency test of perceptual vision (focus and selectivity of the image) and attention. The task of the investigated person was to enter on the keyboard the correct number of horizontal lines displayed on the monitor.

- Summary Test

Studying the speed and efficiency of memory-based operations, attention (dividing and concentrating), and perceptiveness.
The task is to execute in memory the operation of adding numbers displayed on the monitor and then typing the result on the keyboard.

- Memory Test

Research on the function of attention and memory task (operational memory).
The task is to execute in memory the operation of adding numbers displayed on the monitor and then, after some time, typing the result on the keyboard.

- Test of letters substitution

Visual-motor coordination and concentration on the task. In this task, the investigated person searched in the letter collection the assigned numbers from which he "constructed" the specified word.

Each of the tests consisted of three parallel test matrices based on difficult-to-remember simple task formulas. This limited the ability to learn text tasks in the next study. All measurements were performed under repetitive conditions. These measurements represent a pilot study.

3 Results and discussions

The results of vibration tests are compiled and summarized in Table 1. The type of bus, its mileage and age are given. The vibrations that influence the driver through the chair for each of the buses are shown in the last column of Table 1.

Table 1. Summary of results for individual buses.

	Bus type	Mileage [km]	Age [months]	RMS [m/s^2]
1	articulated	26916	4	0.322
2	not articulated	456031	51	0.372
3	articulated	729284	198	0.412
4	articulated	4835	1	0.349
5	not articulated	517855	107	0.355
6	not articulated	415805	51	0.323
7	not articulated	263022	51	0.383

The obtained results show the highest weighted value of vibration acceleration in the Z direction, which is the main vertical axis of the driver (along the spine). With regard to exposure to vibration at the workplace this limit level for 8 is equal 0.5 m/s^2 [13], exceeding the values stated in the regulation, the employer is obliged to take actions reducing the occupational risk.

There is a low dependence of the observed vibration acceleration on the mileage of the vehicle. In all vehicles, the driver's seat was characterized by high vibration damping ability. In the case of cognitive and motor efficiency tests, the examined drivers used a computer-aided set of psychological tests examining cognitive and motor skills. The study consisted of performing the commands and research tasks displayed on the monitor. The results of the study were two parameters: time of performance and the level of correctness.

The device identified, among others, time of motor response to task stimulus, number of correct and incorrect responses. On the basis of the obtained data it was possible to determine the current level of changes in the organism of the subject under the influence of the stimulus. Measurements were made three times:

- First measurement on the second day of the experiment (morning group 2R and afternoon group 2P),
- Second measurement in the fourth day of the experiment (morning group 4R and an afternoon group 4P),
- Third measurement in the sixth day of the experiment (morning group 6R and an afternoon group 6P).

Only the execution times of individual test tasks were analyzed. Line Test results are presented in Figure 1, and the t-test for dependent samples are listed in Table 2.

Table 2. Summary results for the Line Test.

Correlation between measurements	Average time to complete the task	Standard deviation	Number of drivers	*T* student value	Degree of freedom	*p* - level of significance
1 and 2	23.53214	3.99235	14	4.07975	13	0.001302
1 and 3	22.56429	6.31341	14	5.37759	13	0.000126
2 and 3	22.56429	6.31341	14	0.77774	13	0.450648

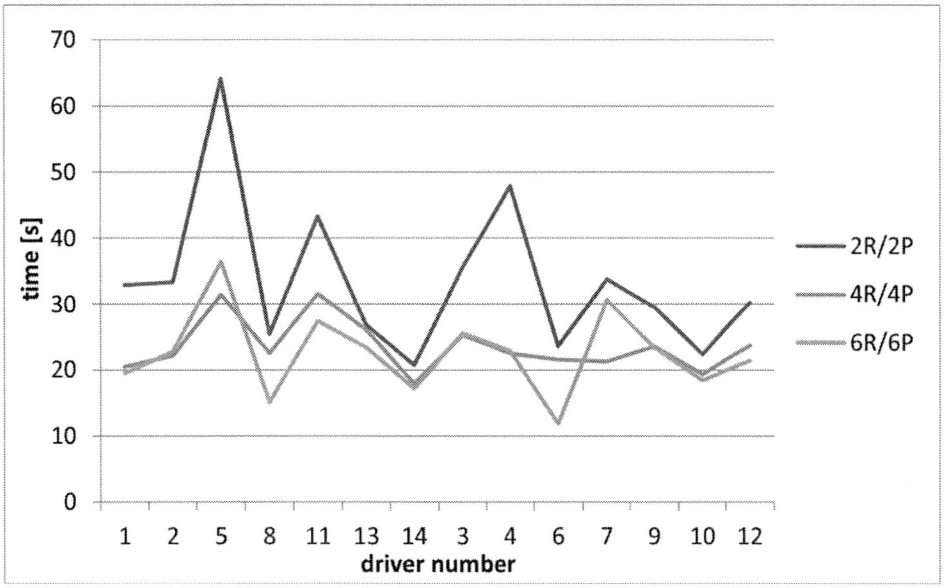

Fig. 1. Line Test results on the second (2R/2P), fourth (4R/4P) and sixth (6R/6P) day of the morning and afternoon group experiment.

As can be seen from the above table, the duration of the test in the Line Test significantly decreases in the second study (compared to the first one) $p < 0.01$, whereas in the third study the change compared to the second test is not statistically significant.

Summary Test

The results of the Summary Test are presented in Figure 2 and the test values for the dependent samples are listed in Table 3.

Table 3. Summary results for the Summary Test.

Correlation between measurements	Average time to complete the task	Standard deviation	Number of drivers	*T* student value	Degree of freedom	*p* - level of significance
1 and 2	74.45071	15.05633	14	4.33991	13	0.000802
1 and 3	64.09929	12.77247	14	7.59144	13	0.000004
2 and 3	64.09929	12.77247	14	3.93807	13	0.001699

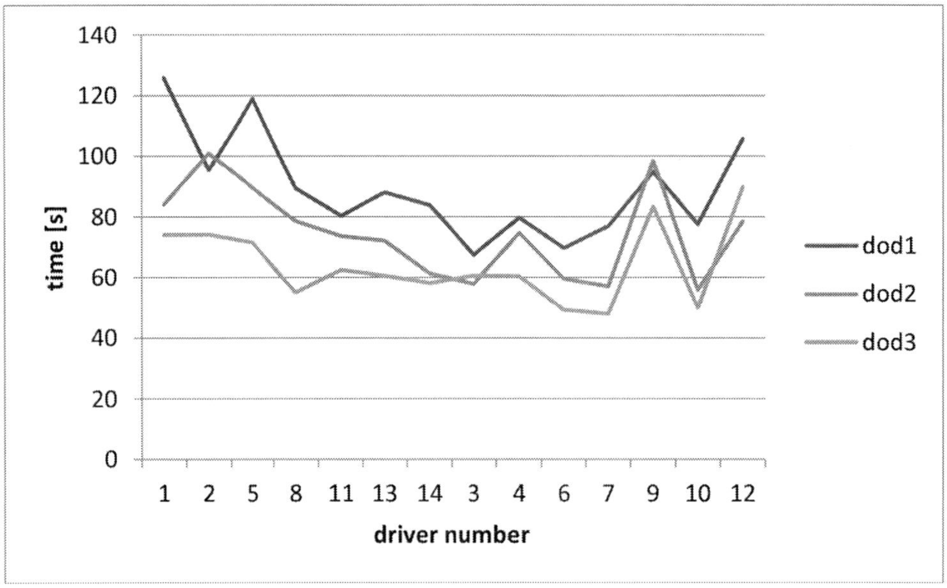

Fig. 2. Summary Test results on the second (dod1), fourth (dod2) and sixth (dod3) day of the morning and afternoon group experiment.

In Summary Tests from sample to sample the performance time is improved, all differences statistically significant at least $p < 0.001$.

Memory Test

The results of the Memory Test are presented in Figure 3 and the test values for the dependent samples are listed in Table 4.

Table 4. Summary results for the Memory Test

Correlation between measurements	Average time to complete the task	Standard deviation	Number of drivers	T student value	Degree of freedom	p - level of significance
1 and 2	58.37000	12.02796	14	4.02088	13	0.001454
1 and 3	49.65857	10.35056	14	4.97452	13	0.000254
2 and 3	49.65857	10.35056	14	3.58296	13	0.003339

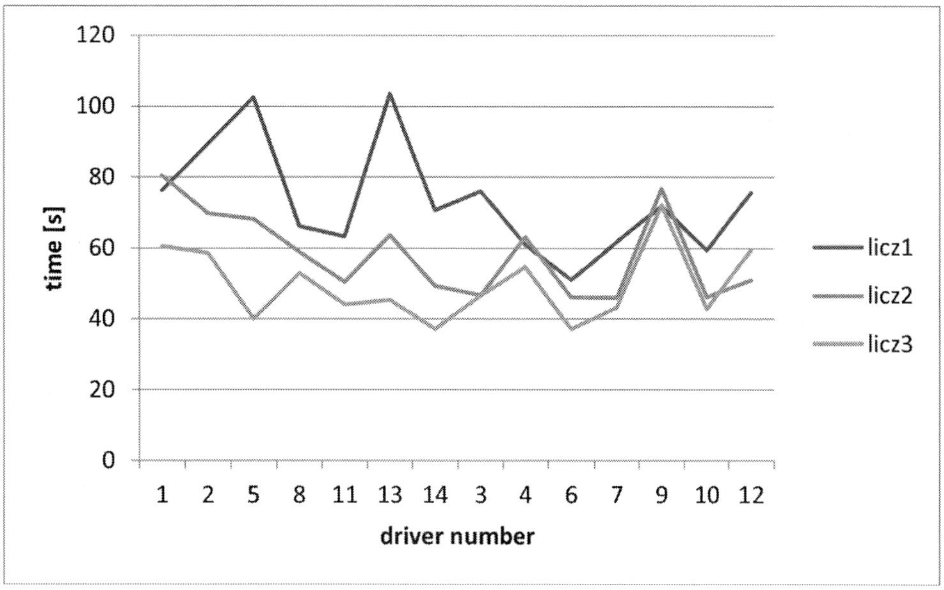

Fig. 3. Memory Test results on the second (licz1), fourth (licz2) and sixth (licz3) day of the morning and afternoon group experiment.

In the Memory Test, clear effect of the acquisition of practice can be seen. Any subsequent attempt brings significantly lower (better) execution times of test tasks.

Test of letters substitution

The results of the Test of letters substitution are presented in Figure 4 and the test values for the dependent samples are listed in Table 5.

In the Test of letters substitution, the differences in execution times between the various tests are statistically significant.

Table 5. Summary of results for the Test of letters substitution

Correlation between measurements	Average time to complete the task	Standard deviation	Number of drivers	*T* student value	Degree of freedom	*p* - level of significance
1 and 2	84.06000	14.50680	14	4.05902	13	0.001353
1 and 3	75.16000	10.61157	14	7.63643	13	0.000004
2 and 3	75.16000	10.61157	14	3.42972	13	0.004479

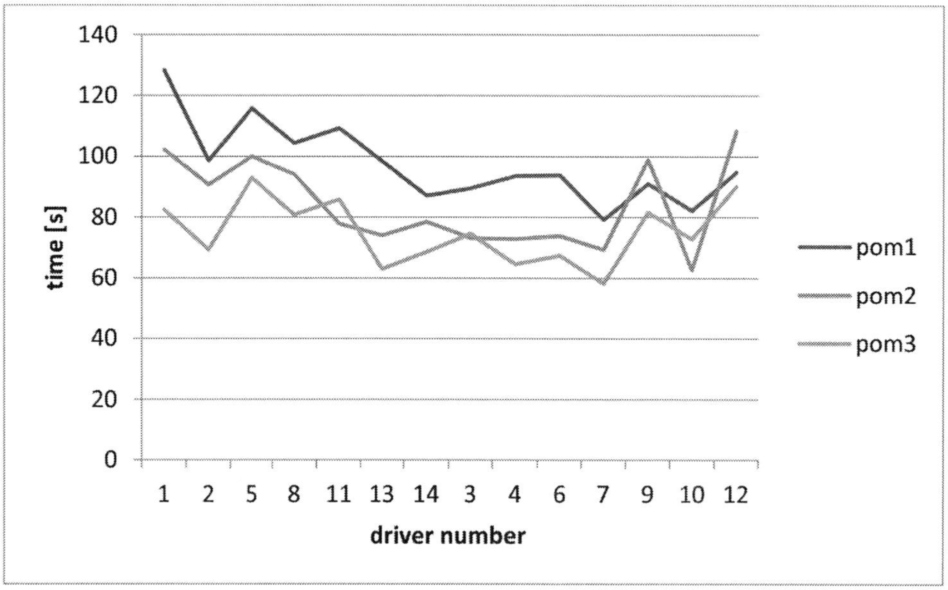

Fig. 4. Test of letters substitution results on the second (pom1), fourth (pom2) and sixth (pom3) day of the morning and afternoon group experiment.

4 Conclusions

The results from the individual measurements clearly show the ability to adapt to the task. The selection of professional predispositions enables the selection of persons of exceptional psychomotor efficiency, in a sense of protection the selected group of drivers against the burden of workplace specificity.

It should be remembered that the measurement of psychomotor efficiency is one of the components of a wider project exploring the influence of the external environment (including noise and noise) on the driver of the bus. Features such as personality can also play a significant role in the profession of bus driver [14].

While the health implications of the driver have a negative dimension, then in the cognitive dimension stimulation of the nervous system can be positive. At the same time, in all four tests there is an improvement in the "afternoon" group (the first seven drivers are in the morning group, the next seven are in the afternoon group). Although the differences between the two groups did not reach statistical significance but the shape of the curve suggests the influence of the time of day on the results, which requires verification.

References

1. P.S. Bruno, Q. R. Marcos, C. Amanda, Z. H. T. Paulo, Annoyance evaluation and the effect of noise on the health of bus drivers, Noise & Health, September-October 2013, **15**, 66, 301-6 (2013)
2. M. J. Griffin, Handbook of human vibration, Academic Press, London (1994)
3. T. M. Kiihlmann, Coping with occupational stress among urban bus and tram drivers. Journal of Occupational Psychology **63**, 89-96, (1990)
4. E. Wągrowska-Koski, redakcja, Zagrożenia zdrowia kierowców pojazdów silnikowych związane ze szkodliwymi i uciążliwymi warunkami środowiska pracy, Instytut Medycyny Pracy im. prof. J. Nofera, Łódź (2007)
5. G. Aronsson, A. Rissler, Psychophysiological Stress Reactions in Female and Male Urban Bus Drivers Journal of Occupational Health Psychology **3**, 2, 122-129, (1998)
6. pod red. D. Koradeckiej, Bezpieczeństwo pracy i ergonomia, tom 1 i 2, CIOP, Warszawa (1997)
7. H. Biglari, M. Hossein Ebrahimi, M. Salehi, M. Poursadeghiyan, I. Ahmadnezhad and M. Abbasi, Relationship between occupational stress and cardiovascular diseases risk factors in drivers. International Journal of Occupational Medicine and Environmental Health **29(6)**, 895–901, (2016)
8. E. Razmpa, K. Sadegh Niat, B. Saedi, Urban Bus Drivers' Sleep Problems and Crash Accidents, Indian J Otolaryngol Head Neck Surg **63(3)**, 269–273, (2011)
9. S. Hoseinabadi, S. Porabdeyan, M. Zare, S. Amiri, M. Ghasemi and A. Mansori, Does Traffic Stress Affect Distance Estimation and Recognition Accuracy in Urban Bus Drivers? Archives of Environmental & Occupational Health **70**, 214–217, (2013)
10. M. A. Machin, P. N. Hoare, The role of workload and driver coping styles in predicting bus drivers' need for recovery, positive and negative affect, and physical symptoms. Anxiety, Stress, & Coping, October 2008; **21(4)**, 359-375, (2008)
11. U. Maheswari Krishnaswamy, M.S. Chhabria, A. Rao, Excessive sleepiness, sleep hygiene, and coping strategies among night bus drivers: A cross-sectional study, Indian Journal of Occupational and Environmental Medicine, (2017)
12. J. Boada-Grau, A. J. Prizmic-Kuzmica, M.D. González-Fernández and A. Vigil-Colet, Spanish version of Bus Drivers' Job Demands Scale (BDJD-24), Psicothema, **25**, 2, 258-265, (2013)
13. Minister of Economy and Labor Ordinance of 5 August 2005 on occupational safety and health in work involving noise or mechanical vibration exposure. (in Polish - Dz. U. 2005, 157, pos. 1318)
14. S. Zivković, V. Nikolić and M. Markič, Influence of professional drivers' personality traits on road traffic safety: case study, International Journal of Injury Control and Safety, Promotion, **22**, 2, 100–110, (2013)

Crash avoidance systems and collision safety devices for vehicle occupants

Miroslava Mikusova[1*]

[1]University of Žilina, Faculty of Operation and Economics of Transport and Communications, Univerzitná 8215/1, 010 26 Žilina, Slovakia

Abstract. Numerous efforts have been made to improve safety on our roads, including the vehicle safety, from passive elements to advanced active systems that are affecting vehicle stability and influencing vehicle dynamics. This paper presents a basic classification of vehicle safety systems, devices and futures. Presented information was compiled from different sources, and is based on the function of these systems over time and their cooperation characteristics. The purpose of this text is to establish a general structure for classification of these systems that make possible the providing of their overview in a comprehensive way.

1 Introduction

Road crashes are uncompromising affairs that can have very negative impact on vehicles and vehicle occupants involved in them. Strong forces that are actuating on human body during a car accident can cause serious injuries and the damage could be very significant, even fatal.

There are many different factors that affect severity of the injury, e.g. kind of crash (rollover, side-on crash head-on crash, etc.), vehicle speed, use/not use of seat belts, vehicle airbags, seat position, kind of collided object (another vehicle, solid object, etc.), and so on. In each case injuries are broadly related to the amount of kinetic energy applied to the human frame. Majority of kinetic forces released in road traffic crashes cannot be tolerated by the human body. For above mentioned reasons it is necessary to take a variety of measures to prevent the crash occurrence and to increase passenger safety.

In vehicles´ development and design it is given increasing emphasis to comply with the requirements of vehicle safety. These requirements are set by the applicable regulatory provisions and also by the requirements of the customers. Precisely because of the customer requirements, vehicle safety becomes one of the cornerstones of a competitiveness in contemporary market. Vehicle manufacturers dedicate to these issues more and more attention in the development of new vehicles and they are working also on the development of a variety of security features that could be installed in vehicles additionally. Modern car is usually designed with a particular emphasis on ensuring a high level of safety of the crew and other road users (pedestrians, cyclists, other vehicles, etc.). To ensure the highest levels of safety the modern vehicles are equipped with systems for driver support that are in

* Corresponding author: miroslava.mikusova@fpedas.uniza.sk

certain cases even capable replace the driver (e.g. automatic maneuvering system that moves a vehicle from a traffic lane into a parking spot, automatic car accident detection and notification system that identifies the vehicle's position and calls for help, and so on.). All these elements comprise integrated vehicle safety. Integrated vehicle safety systems can be into two basic categories: active safety and passive safety.

2 Active safety

The *active safety* features, devices and systems help maintain vehicle under control and prevent an accident occurrence. Active safety is a result of harmonic structure of the vehicle chassis, which takes into account the wheel guidance, suspension, steering feedback and stability of the brakes. It guaranties the optimum vehicle dynamic behaviour also during the process of evading obstacles.

As a part of the active safety we can recognize:

1. Autonomous active safety systems

• permanently supporting driver,
• actuating in a moment of expected impact;

2. Active safety systems cooperating

• with other vehicles,
• with infrastructure.

2.1 Autonomous active safety systems

The best way to reduce road accidents is prevent them. For this purpose were designed active safety systems. These systems can be classified in terms of their function over time to continuously operating systems and systems that affect vehicle activity at the moment of potential danger (in risky situations that could conclude in traffic accident).

2.1.1 Autonomous active safety systems permanently supporting the driver

Active safety systems that provide permanent support for the driver are also called driver assistance systems and are known as ADAS (from Eng. Advanced Driver Assistance Systems) and DSS systems (from Eng. Driver Support Systems). Autonomous permanently active safety systems support the driver adaptation to the current situation on the road, anticipate the needs and, if necessary, take the initiative [1].

The most known representatives of these systems are Power steering, Electronic Stability Program – ESP, Anti-lock Braking System – ABS, Electronic Brake-force Distribution – EBD, Anti-slip Regulation – ASR, Electronic differential lock – EDS, Break Assist System – BAS, Advanced Emergency Braking System – AEBS, Multi-collision brake – MKB, Regulator of engine torque – MSR, Downhill Assist Control – DAC, Active Body Control – ABC, Speed Alert, Adaptive Cruise Control – ACC, Intelligent Speed Adaptation – ISA, Seat Belts Reminder – SBR, Signalization of an improperly closed doors, Lane Departure Warning System – LW, Blind Spot Monitoring – BLIS, City Safety, Collision Avoidance System – CAS, Pre-crash Sensing, Parking Assistance System – PDC, Driver Monitoring System – DMS, Lighting Xenonheadlight, Adaptive Headlights – AHL, Vision Enhancement System, *Alcohol* Interlocs, Route Guidance and Navigation Systems, Tire-pressure Monitoring Systems (Intelligent Tyre System – ITS, Tyre Pressure Monitoring System – TPMS, Deflation Detection System - DDS), Run Flat tyre – RFT, Heads-Up Display – HUD and Automatic Dimming Mirror – ADM [2-4].

The systems for permanent support of driver actions include the core vehicle technologies that are affecting vehicle stability and dynamics, systems that maintain distance, warning systems and systems for road accident prevention, driver drowsiness detection systems, vision improving systems, pedestrian detection systems, alcohol control systems, navigation systems and other. Systems of vehicle dynamics can be marked as a basic active safety systems.

They ensure traction and stabilization control of vehicle in different types of traffic situations and under various driving conditions. Many times there are critical road conditions (e.g. icy road) or critical road incidents (that requires e.g. emergency braking) or cases when correction of driver's error is needed (for example when cornering or during fast changes of vehicle direction). Some of these systems are often used when braking. Traction control is particularly acute during starting and acceleration of the vehicle when the traction systems' mission is to improve the transmission of torque from the drive axle to the ground. Other systems, if necessary, are pumping without breaking the wheels. Almost all of these systems utilize a substantial part of the ABS system.

The purpose of the maintaining the distance control systems are assuring a safe distance from the vehicle ahead. A device used for this purpose is a part of the most automatic driving control systems.

Warning systems and systems to prevent collision detect obstacles (vehicles or static objects) with digital CCD cameras, millimetre-wave radars, laser radars, piezo ceramic devices (Polaroid), or are detecting distance on all vehicle sides (Delco). This group includes systems to warn against frontal collisions, systems to warn against a side collision, systems for night vision and systems to warn against lane departure. In some cases, these systems are connected with the automatic braking of the vehicle. The main difference between the standoff and preventive action against the collision occurrence is a lack of ability of ACC systems to adequately respond to solid objects. In recent years there is a strong trend towards the integration of the functions of all the systems of this group [5].

Driver drowsiness detection systems are operating by analysing the driver's steering bavoir or selected driver's physiological parameters. They are based on an analysis of the stability of steering wheel (measured in relation to keep vehicle in the lane) or on the condition of the driver by measured by specified physiological parameters. Whenever the driver's performance, measured by either method, exceeds a certain limit the driver is considered drowsy, respectively sleepy and specific predefined countermeasures are activated.

Another group of active safety futures are vision improving systems. There are two basic groups of these systems.

The first group - *active VES* - use additional sensors, information sources such as digital maps and special facilities that seek headlights of the vehicle on the part of road space ahead of the vehicle, which should be subject of justified interest of the driver. Intelligent vehicle systems from this group can additionally adapt the light distribution according to the vehicle speed.

Systems of the second group - *passive VES* – are using invisible light sources on the road that light over a greater distance or a greater width comparing with the basic vehicle headlights. Reflected radiance is captured by special sensors that transform it into visible light. Then additional image of a road space ahead of the vehicle is reproduced to the driver.

Pedestrian detection systems are designed for the purpose of detecting pedestrians (or single barriers) located in the closeness of the vehicle or in its roadway. Various development workplaces are using different techniques, different sensor technologies and different ranges of space where the system is looking for obstacles or pedestrians. For example the trend of Japanese vehicle producers is the realization of the idea of detecting pedestrians at night and address the problem of the blind spot.

Among the active safety systems continuously supporting the activity of the driver we can also include the usual elements of the vehicle, such as vehicle lights, wipers, heaters, heated rear glasses and mirrors and so on. The reason why these systems are considered as an active safety systems is that they maintain comfort of the driver and thereby facilitate maintaining his vigilance and attention while driving [6].

2.1.2 Autonomous active safety systems actuating in a moment of expected impact

With a purpose of increasing active vehicle safety some car producers are assembled systems that are activated in a moment of the expected impact in the higher classes of their vehicles. These so called Intelligent Vehicles, automatically effectuate series of steps in the case of imminent danger and expectance of the accident, including:
- mechanical or visual driver warning (light vibration of the steering wheel or seat);
- activation of the braking system (adjustment of the brake elements into working position);
- tensioning seat belts, to shorten the time of a stretch in a case of possible impact;
- anchoring electrically operated seats to the optimum position in terms of safety;
- activate the automatic vehicle braking system if the collision could not be stave off.

2.2 Cooperating active safety systems

Safe and intelligent vehicles are characterized not only by the ability to communicate with each other but they also have the ability to interact with intelligent infrastructure. Overview of the active safety systems able to cooperate with other elements of the transport system is provided in Table 1. These systems are currently in development and testing phase.

Table 1. Cooperating active safety systems.

Systems'categorization		Name of the systems
Active safety systems	cooperating with other vehicles	Car2Car systems (communication between vehicles themselves)
	cooperating with infrastructure	systems Car2infrastructure (communication between vehicle and any stationary communication equipment, including portable traffic signs and warning devices)

The implementation of intelligent vehicles that are based on cooperating active safety systems into ordinary traffic flow could bring numerous benefits such as e.g. [7]:
- road safety improvement;
- reduced fuel consumption;
- improved traffic flow associated with communication between vehicles;
- reduced congestion in urban areas.

2.2.1 Active safety systems cooperating with other vehicles

In addition to the vehicle active safety systems that communicate with each other, there are exist also cooperating systems that make possible communication between vehicles. They are called Car2Car systems.

Car2Car system consists of vehicles that are in a range of each other and can create an ad hoc network in which the vehicles know their location, speed and direction between them [8]. Car2Car system allows providing and exchange of warnings and information between vehicles.

2.2.2 Active safety systems cooperating with infrastructure

Active safety systems cooperating with infrastructure are known as Car2Infrastructure Systems. They incorporate all the communication technology between the vehicles and any stationary communication equipment (stationary in this case refers to static at a given time). This group includes also vehicle communication with portable traffic signs or warning devices.

3 Passive safety

According to statistics the most often cause of traffic accidents is reduced alertness of the driver when driving [9]. This occurs mainly due to driver´s fatigue or because the driver is devoted to activities which conduct its attention away while driving, such e.g. turning on the radio or mobile use and communication with other passengers. In this case, even the use of the widest range of active safety features cannot prevent the accidents. In this situation are applied elements of passive safety.

Passive safety include all the vehicle facilities and measures that minimize the impact and damage during an accident. It is especially important if the driver can not actively intervene in the events in road traffic. Elements of passive safety should minimize the consequences of an accident that is already in progress, protect the vehicle occupants prior to the occurrence of injuries or reduce their extent and minimize resulting damages.

The passive safety systems can be divided into:

1. Safety systems activated at the moment of impact;

2. Safety systems activated after the impact

- autonomous,
- cooperating.

Table 2. Passive safety systems.

Systems'categorization		Name of the system
Passive safety systems	activated at the moment of impact	belt tensioners child seats airbags system for protection of vehicle occupants active headrests active front seat rollover protection vehicles system airbag system for pedestrian protection active front bonnet active spoiler active bumper
	activated after the impact	fuel cut-off unlock the vehicle disconnection of the vehicle battery launch warning light

3.1 Passive safety systems activated at the moment of impact

Passive safety systems active at the moment of impact includes features as:

- belt tensioners,
- child seats,
- airbags systems for protection of vehicle occupants (driver airbag, passenger airbag, side airbags for the front seats, front and side airbags for the rear seats, roof airbags),
- active headrests,
- active front seat,
- rollover protection vehicle system [10].

Other passive safety features, activated at the moment of the impact, that actuate as a protection of vulnerable road users (particularly pedestrians) are [11]:

- airbags system for pedestrian protection,
- active front bonnet,
- active spoiler.

3.2 Passive safety systems activated after the moment of impact

Passive safety systems activated after the impact in road accident can be divided into two groups: autonomous systems and cooperating systems.

Autonomous passive safety systems are designed to protect the vehicle occupants against any possible dangers after the occurrence of the accident.

In this case the most commonly applied autonomous vehicle operations are:

- fuel cut-off,
- unlock the vehicle,
- battery disconnection,
- turning on the warning lights.

Safety cooperating systems activated after the crash includes automatic reporting system for emergency situations known as an E-Call.

E-Call is the name of pan-European system for vehicle, which in case of an accident automatically call the emergency unit of the integrated rescue system. This emergency system is based on use of satellites and classic SIM card (like in mobile phones). In the event of a critical situation or accident, E-Call can significantly reduce the response time of emergency services. This emergency call system can be activated manually or automatically by the vehicle passengers if the sensors register the car collision. E-Call then contact the closest coordination centre 112 and transmits important information such as location of the accident, the time of the accident and the type of vehicle. After that the operator is able to send rescue services to the place of accident without any co-operation with other people or with a crew that may not be conscious. The system does not send only location information of the vehicle at the time of the accident, but also data from 60 seconds before the accident, and 15 seconds after the accident, containing for example, information on overload of the vehicle in the moment of impact. Based on all this information the operator can assess accident severity and deduce if it is not only a false alarm. This system can be also very helpful in determining the driver at fault.

3.3 Passive safety outside the vehicle

Passive safety is not just about vehicles. As elements of passive safety can be considered the highlight border traffic signs by fluorescence method, different traffic lights, their different sizes or colour. Furthermore, it can also be construction of road signs, which allows them to break away from surface in the case of impact and avoid that vehicle get "wrapped" around them. For passive safety guardrails we can also be seen, for example, or even a fence along the highway, which should prevent the entry of unwanted wildlife and pedestrians on the roads. The cities are the different fences that prevent free movement of pedestrians.

4 Conclusion

Currently, vehicles tend to be mostly equipped with smart technologies in the field of safety, which would avoid the risks associated with driving or minimize the consequences in the event of an accident. These systems provide an improved control of the vehicle in different driving situations, increase the safety and comfort of passengers, and facilitate communication and control of the situation on the road. Vehicles are equipped with a large number of sensors, transducers, actuators and control units whose task is to coordinate cooperation of all electronic and mechatronic components of the vehicle and to improve communication between other road users. Deploying these systems into vehicles is partly optional (a competitive advantage in terms of safety) and partly mandatory due to the legislation for vehicle type approval to operate in traffic. Actual developments in vehicle safety focus on linking existing systems to increase their efficiency and lower their prices, allowing these systems can be applied to the lower categories of vehicles [12]. This massive extension of safety systems contributes to an overall increase of all types of vehicle safety (operational, perceptual, etc.). The question is how this share of the development will be reflected in real reduction of the number of road accidents, as drivers supported by a quantity of systems and technologies thanks to their sense of security often apply riskier vehicle driven manner.

References

1. S.C. Mongones, P. Fischbeck, P. Jaramillo, Safety science, **91**, 122-131 (2017)
2. A. Steinfeld, D. Duggins, J. Gowdy, J. Kozar, R. MacLachlan, C. Mertz, A. Suppe, C. Thorpe, C. Wang, IEEE ITSC, 343-348 (2004)
3. A.Torok, Transport, **7**, 1-6 (2015) DOI: 10.3846/16484142.2015.1062798
4. K. Jamroz, W. Kustra, M. Budzynski, J. Zukowska, Transportation Research Procedia **14**, 3905-3914 (2016) DOI: 10.1016/j.trpro.2016.05.479
5. J. B. Cicchino, D.S. Zuby, Traffic Injury Prevention, 1-7 (2017) 10.1080/15389588.2016.1247446
6. A. Vahidi, A. Eskandarian, IEEE 4, 143– 153 (2003) http://dx.doi.org/10.1109/TITS.2003.821292
7. M. M. Waldrop, Nature **518**, 20-23 (2015) http://dx.doi.org/10.1038/518020a
8. E. Uhlemann, IEEE Vehicular Technology Magazine, **10**, 4, 20-23 (2015)
9. F. Alonso, C. Esteban, J.Sanmartin, S.A. Useche, Science Journal of Public Health, **4**, 6, 482-488 (2016)
10. T. Gandhi, M. M. Trivedi, IEEE **8**, 3, 413-430 (2007)
11. H. Hamande, T. Serre, C. Masson, R. Anderson, Accident Analysis and Prevention, **82**, 53-60 (2015)
12. E. Ledezma-Zavala, R.A. Ramrey-Mendoza, International Journal on Interactive Design and Manufacturing, 1-9 (2016)

Modelling of moving load effect on concrete pavements

Daniela Kuchárová[1,*] and *Gabriela* Lajčáková[1]

[1]University of Žilina, Faculty of Civil Engineering, Univerzitná 8215/1, 010 26 Žilina, Slovakia

Abstract. Concrete roads may have different layout arrangement. The structures made of isolated prefabricated slabs are used for the construction of temporary roadways. The paper deals with numerical simulation of moving load effect on isolated slabs on elastic foundation. The computing model of a lorry and computing model of a slab are introduced. The deflections at the middle of the slab and tire forces of vehicle are modeled under various conditions. The influence of speed of vehicle motion and influence of initial conditions are evaluated. The results are presented in time domain in graphical and numerical manner.

1 Introduction

Concrete slabs are widely used in road building. Isolated plates are usually used for the temporary pavement structures. Knowledge of stress and strain is required for correct design of the slabs. Numerical methods enable to simulate moving load effect in dynamic regime with satisfactory precision. The computing models of vehicles can be created on various levels. For the purpose of this task the plane computing model is adopted. The pavement is modelled as the thin slab on elastic foundation. The assumption about deflection plate of the slab is introduced to reduce the partial differential equation of motion on ordinary differential equation. The problem is solved by numerical way. The results depend on the speed of the vehicle and on its initial conditions. The deflections in the middle of the slab and tire forces are calculated. The results are presented in numerical and graphical manner.

2 Computing model of vehicle

A plane computing model of the Tatra 815 lorry is used for the solution of this problem [1]. It is the multibody computing model with 8 degrees of freedom, Fig. 1. The model has three mass bodies with five degrees of freedom and three massless degrees of freedom. Functions $r_i(t)$, $(i = 1 - 5)$ describe the vibration of mass bodies. The model can simulate the heave and pitch effects of mass bodies. The model is excited kinematically by running along the road. Contact forces $F_j(t)$, $(j = 3, 4, 5)$ correspond to the massless degrees of freedom. The

* Corresponding author: daniela.kucharova@fstav.uniza.sk

equations of motion for the calculation of functions $r_i(t)$ have the form of ordinary differential equations.

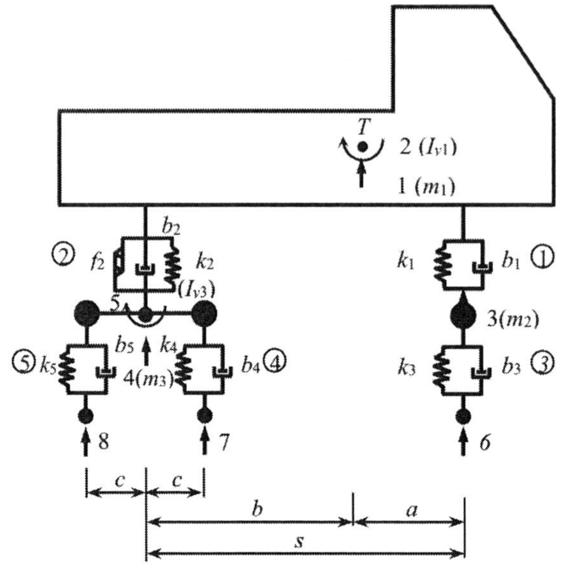

Fig. 1. Plane computing model of the Tatra 815 vehicle.

$$\ddot{r}_1(t) = -\{+k_1 \cdot d_1(t) + b_1 \cdot \dot{d}_1(t) + k_2 \cdot d_2(t) + b_2 \cdot \dot{d}_2(t) + f_2 \cdot \dot{d}_2(t)/\dot{d}_c\} / m_1,$$

$$\ddot{r}_2(t) = -\{-a \cdot k_1 \cdot d_1(t) - a \cdot b_1 \cdot \dot{d}_1(t) + b \cdot k_2 \cdot d_2(t) + b \cdot b_2 \cdot \dot{d}_2(t) + f_2 \cdot \dot{d}_2(t)/\dot{d}_c\} / I_{y1},$$

$$\ddot{r}_3(t) = -\{-k_1 \cdot d_1(t) - b_1 \cdot \dot{d}_1(t) + k_3 \cdot d_3(t) + b_3 \cdot \dot{d}_3(t)\} / m_2,$$

$$\ddot{r}_4(t) = -\{-k_2 \cdot d_2(t) - b_2 \cdot \dot{d}_2(t) - f_2 \cdot \dot{d}_2(t)/\dot{d}_c + k_4 \cdot d_4(t) + b_4 \cdot \dot{d}_4(t) + k_5 \cdot d_5(t) + b_5 \cdot \dot{d}_5(t)\} / m_3,$$

$$\ddot{r}_5(t) = -\{-c \cdot k_4 \cdot d_4(t) - c \cdot b_4 \cdot \dot{d}_4(t) + c \cdot k_5 \cdot d_5(t) + c \cdot b_5 \cdot \dot{d}_5(t)\} / I_{y3}. \quad (1)$$

The terms for calculation of tire forces are

$$F_3(t) = -G_3 + k_3 \cdot d_3(t) + b_3 \cdot \dot{d}_3(t) = -g \cdot \left(m_1 \cdot \frac{b}{s} + m_2\right) + k_3 \cdot d_3(t) + b_3 \cdot \dot{d}_3(t),$$

$$F_4(t) = -G_4 + k_4 \cdot d_4(t) + b_4 \cdot \dot{d}_4(t) = -\frac{1}{2} \cdot g \cdot \left(m_1 \cdot \frac{a}{s} + m_3\right) + k_4 \cdot d_4(t) + b_4 \cdot \dot{d}_4(t),$$

$$F_5(t) = -G_5 + k_5 \cdot d_5(t) + b_5 \cdot \dot{d}_5(t) = -\frac{1}{2} \cdot g \cdot \left(m_1 \cdot \frac{a}{s} + m_3\right) + k_5 \cdot d_5(t) + b_5 \cdot \dot{d}_5(t). \quad (2)$$

Values $d_i(t)$, (i = 1 – 4) represent the deformation of connected members of the model in time t. The derivations with respect to time are denoted by the dot above the symbol of dependent variable. G_j (j = 3, 4, 5) represents the gravity forces acting in the contact points.

Symbols k, b, f describe the stiffness, damping and friction properties of connecting members, m, I_y represent the mass and mass moment of inertia of mass bodies.

3 Computing model of slab

The slab computing model is created in the sense of the Kirchhoff theory of thin slabs on elastic foundation [2]. The equation of motion describing the slab vibration has the form

$$D\left(\frac{\partial^4 w}{\partial x^4}+2\frac{\partial^4 w}{\partial x^2 \partial y^2}+\frac{\partial^4 w}{\partial y^4}\right)+K\cdot w+\mu\frac{\partial^2 w}{\partial t^2}+2\mu\omega_b\frac{\partial w}{\partial t}=p(x,y,t). \tag{3}$$

It is a partial differential equation, which will be solved using the Fourier method. The assumption about the shape of deflection plate of slab $w_0(x,y)$ due to the load is adopted and wanted function $w(x,y,t)$ describing the shape of deflection plate of the slab in time t is expressed as

$$w(x,y,t)=q(t)w_0(x,y), \tag{4}$$

where

$$w_0(x,y)=\sin\frac{\pi x}{l_x}\sin\frac{\pi y}{l_y}. \tag{5}$$

Function $w_0(x,y)$ is the known function and it depends on coordinates x, y only. Function $q(t)$ is the unknown function and it dependents on time t. Function $q(t)$ has the meaning of the generalized Lagrange coordinate. It represents the proportionality coefficient between dynamic and static deflection in the time t. The meaning of other symbols is as follows: D represents the slab stiffness [Nm^2/m], K is the modulus of compressibility of elastic foundation [N/m^3], μ is the mass intensity of the slab [kg/m^2], ω_b is damped angular frequency [rad/s], l_x, l_y are the dimensions of the slab [m]. Expression $p(x,y,t)$ on the right hand side of equation (3) is the intensity of continuous dynamic load. By substituting assumption (4) into (3) we obtain

$$\ddot{q}(t)\left\{\mu\sin\frac{\pi x}{l_x}\sin\frac{\pi y}{l_y}\right\}+\dot{q}(t)\left\{2\mu\omega_b\sin\frac{\pi x}{l_x}\sin\frac{\pi y}{l_y}\right\}+$$

$$+q(t)\left\{\sin\frac{\pi x}{l_x}\sin\frac{\pi y}{l_y}D\left[\left(\frac{\pi}{l_x}\right)^4+2\left(\frac{\pi}{l_x}\right)^2\left(\frac{\pi}{l_y}\right)^2+\left(\frac{\pi}{l_y}\right)^4+\frac{K}{D}\right]\right\}=p(x,y,t). \tag{6}$$

In case of discrete moving load contact force $F_j(t)$ must be transformed on continuous load $p(x,y,t)$. It may be done by the advance proposed by Dirac [2]

$$p(x,y,t)=\sum_j \varepsilon_j F_j(t)\delta(x-x_j)\delta(y-y_j), \tag{7}$$

$$p(x,y,t)=\sum_j \varepsilon_j \sum_{m=1}^{\infty}\sum_{n=1}^{\infty} p_{mn,j}(t)\sin\frac{m\pi x}{l_x}\sin\frac{n\pi y}{l_y}, \tag{8}$$

where

$$p_{mn,j}(t) = \frac{2}{l_x}\frac{2}{l_y}\int_0^{l_x}\int_0^{l_y} F_j(t)\delta(x-x_j)\delta(y-y_j)\sin\frac{m\pi x}{l_x}\sin\frac{n\pi y}{l_y}\,dxdy = F_j(t)\frac{4}{l_x l_y}\sin\frac{m\pi x_j}{l_x}\sin\frac{n\pi y_j}{l_y} \qquad (9)$$

$\varepsilon_j = 1$ when the force F_j acts on the slab, otherwise $\varepsilon_j = 0$. Then

$$p(x,y,t) = \sum_j \varepsilon_j \sum_{m=1}^{\infty}\sum_{n=1}^{\infty} F_j(t)\frac{4}{l_x l_y}\sin\frac{m\pi x_j}{l_x}\sin\frac{n\pi y_j}{l_y}\sin\frac{m\pi x}{l_x}\sin\frac{n\pi y}{l_y}. \qquad (10)$$

The convergence of the series in equation (10) is possible without major detriment to accuracy, taking into account only the first member of the series. Then the expression (10) is simplified

$$p(x,y,t) = \sum_j \varepsilon_j \cdot F_j(t)\cdot\frac{4}{l_x l_y}\cdot\sin\frac{\pi x_j}{l_x}\cdot\sin\frac{\pi y_j}{l_y}\cdot\sin\frac{\pi x}{l_x}\cdot\sin\frac{\pi y}{l_y}. \qquad (11)$$

Substituting $F_j(t)$ in equation (11) by equation (2) and replacing $p(x,y,t)$ in equation (6) by (11) we obtain the final shape of the slab equation of motion. It can be written in a simplified form as

$$\ddot{q}(t) = -[\ \dot{q}(t)\cdot a_{d2} + q(t)\cdot a_{d3} + r_3(t)\cdot a_{d4} + \dot{r}_3(t)\cdot a_{d5} + r_4(t)\cdot a_{d6} + \dot{r}_4(t)\cdot a_{d7} +$$
$$+ r_5(t)\cdot a_{d8} + \dot{r}_5(t)\cdot a_{d9} + a_{d10} + a_{d11} + a_{d12}\ ]/a_{d1}. \qquad (12)$$

4 Vehicle and slab parameters and numerical solution

The vehicle computing model has the following mass parameters: $m_1 = 11475$ kg, $m_2 = 455$ kg, $m_3 = 1070$ kg, I_{y1}=31149 kgm^2, I_{y3}=466 kgm^2. The gravity forces acting in contact points are: $G_3 = (m_1 \cdot b/s + m_2)g = 33208$ N, $G_4 = G_5 = 0.5(m_1 \cdot a/s + m_3)g = 47161$ N.

The model of the slab on layered foundation showed in Fig. 2 has following parameters: slab depth $h = 0.24$ m, slab proportions in longitudinal and transverse direction $l_x = 6.0$ m; $l_y = 3.75$ m, concrete modulus of elasticity $E = 37500$ MPa, Poisson ratio $\nu = 0.20$. The layers of foundation 2 – 5 are included in the computing model as Winkler elastic foundation. The modulus of foundation compressibility $K = 171.8$ MN/m^3 was calculated by the use of computer program LAYMED [3]. Mass intensity of slab $\mu = \rho \cdot h = 2500 \cdot 0.24 = 600$ kg/m^2, damping angular frequency $\omega_b = 0.1$ rad/s.

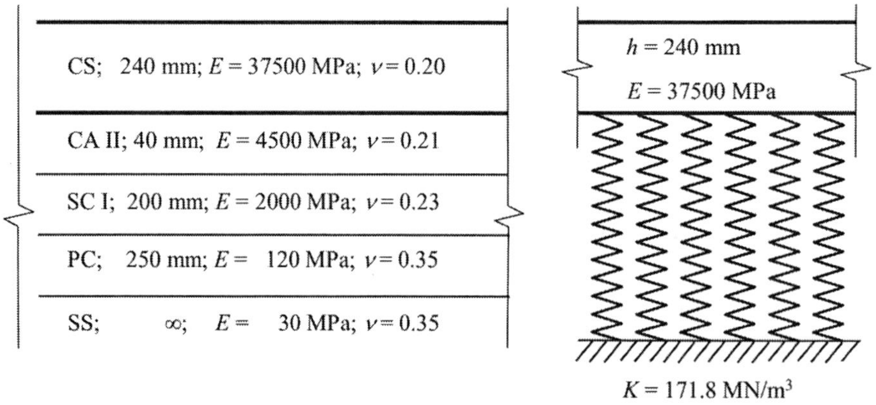

Fig. 2. Composition of the pavement with concrete slab.

Synchronous vibration of vehicle – slab in our case is described by a system of the 2nd order ordinary differential equations. Equations (1) describe the vibration of vehicle and equation (12) describes the slab vibration. The computer program in MATLAB was created for the numerical solution of equations of motion and for displaying of the obtained results. The 4th order Runge-Kutta step-by-step integration method was used for the numerical solution [4].

5 Results of numerical solution

Vehicle always enters the slab vibrant. The results of solution are influenced by vehicle initial conditions at the moment of vehicle entry on the slab. Two variants were considered during numerical solution, Fig. 3.

Variant a – vehicle springs are aggravated, mass body m_1 amplitude is down oriented from the equilibrium position. Initial conditions are as follows: $r_1(0) = -0.01$ m, $r_2(0) = +0.003$ m, $r_3(0) - r_5(0) = 0.0$ m, $\dot{r}_1(0) - \dot{r}_5(0) = 0.0$ m/s.

Variant b – vehicle springs are lightened, mass body m_1 amplitude is up oriented from the equilibrium position. Initial conditions are as follows: $r_1(0) = +0.01$ m, $r_2(0) = -0.003$ m, $r_3(0) - r_5(0) = 0.0$ m, $\dot{r}_1(0) - \dot{r}_5(0) = 0.0$ m/s.

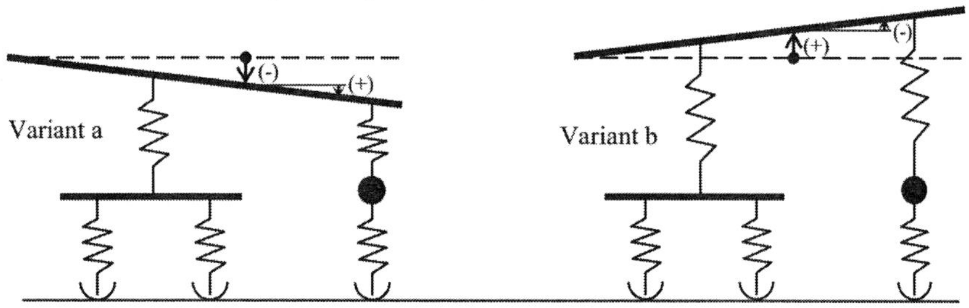

Fig. 3. Initial conditions of vehicle.

Numerical solution was realized in dependence on the speed of the vehicle motion in the interval of speeds 5 – 130 km/h with the step of 5 km/h. For every speed of the vehicle motion the time course of the middle slab vertical displacements and the time course of contact forces were calculated. Maxima of middle slab vertical displacements and dispersion of contact forces are plotted versus speed of the vehicle motion. Demonstration of the obtained results for Variant a is presented in Figs. 4, 5, 6, 7 and for Variant b in Figs. 8, 9, 10, 11.

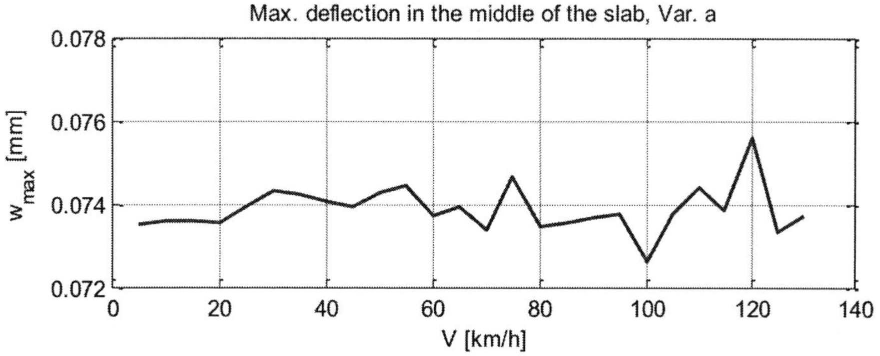

Fig. 4. Maximal deflection in the middle of the slab, Variant a.

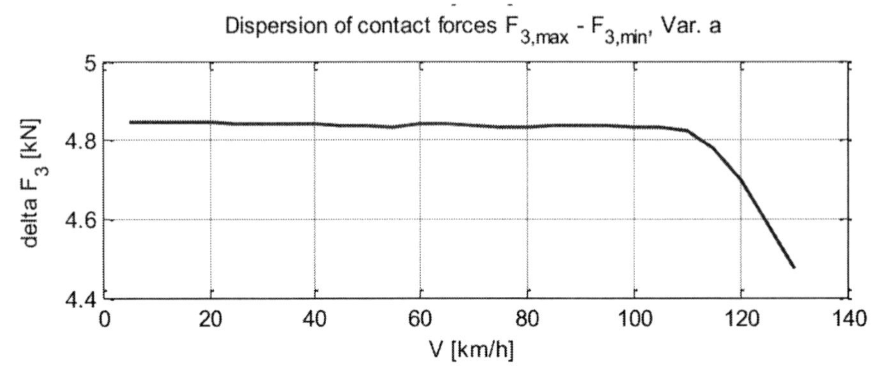

Fig. 5. Dispersion of contact force F_3, Variant a.

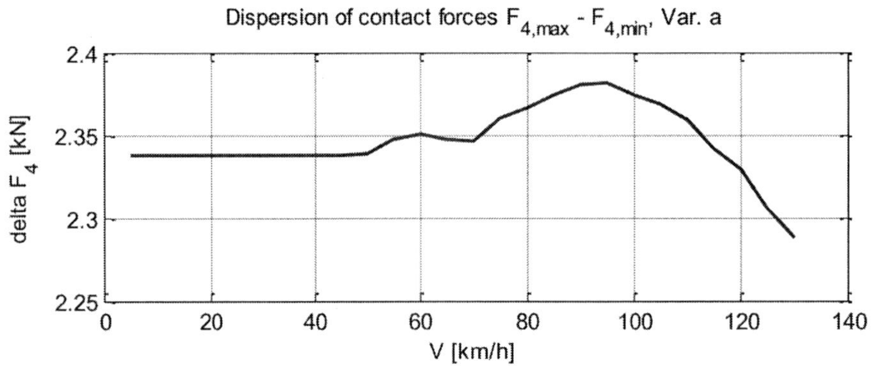

Fig. 6. Dispersion of contact force F_4, Variant a.

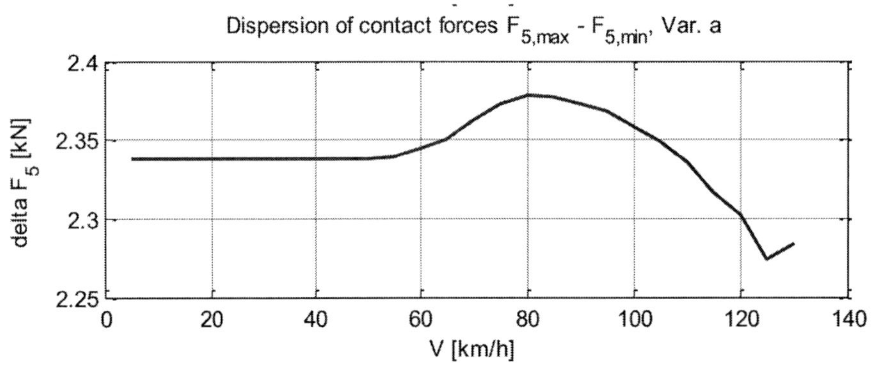

Fig. 7. Dispersion of contact force F_5, Variant a.

 DOI: 10.1051/matecconf/201710700025

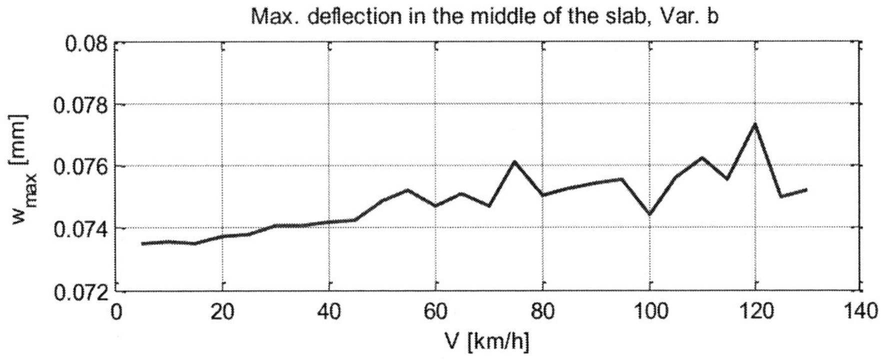

Fig. 8. Maximal deflection in the middle of the slab, Variant b.

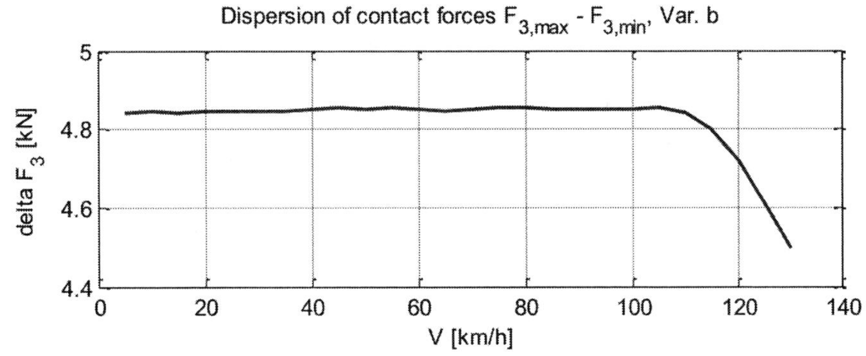

Fig. 9. Dispersion of contact force F_3, Variant b.

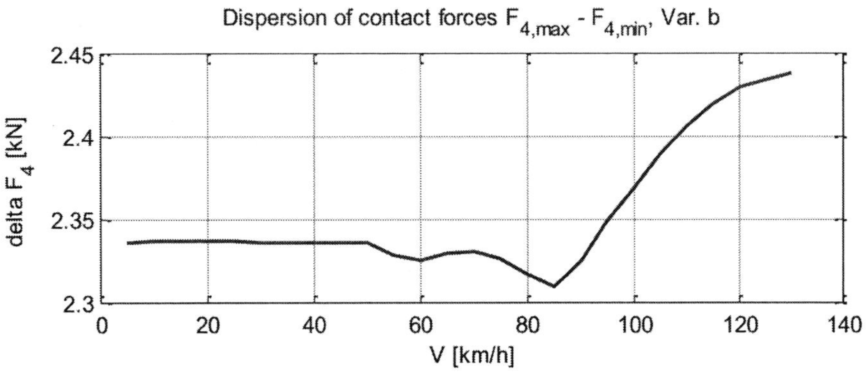

Fig. 10. Dispersion of contact force F_4, Variant b.

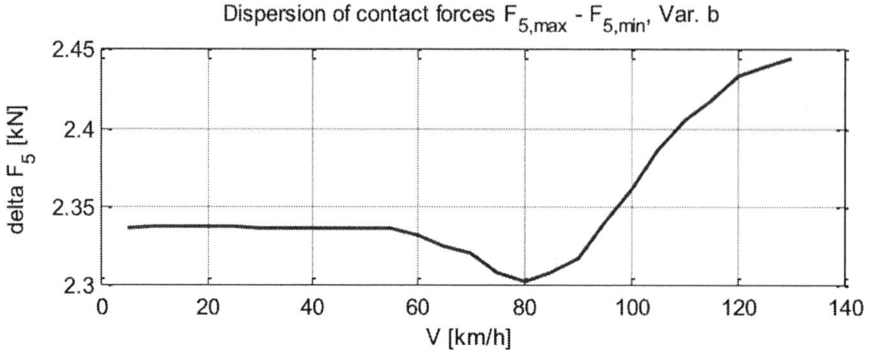

Fig. 11. Dispersion of contact force F_5, Variant b.

6 Conclusion

The movement of vehicles along isolated concrete slabs on the elastic foundation is a real engineering task and can be solved in a numerical way. The slab dynamic deflections, internal forces and the vehicle tire forces can be observed. The task was analyzed under two variants of the vehicle initial conditions. The vehicle enters the slab with aggravated springs (Variant a) and the vehicle enters the slab with lightened springs (Variant b).

The absolute maxima of the middle slab vertical dynamic deflections in the interval of speed of vehicle motion 5 – 130 km/h are practically the same for both variants (Variant a, w_{max} = 0.07562 mm, V = 120 km/h, Variant b, w_{max} = 0.07728 mm, V = 120 km/h). The absolute minima of the middle slab vertical dynamic deflections in the interval of speed of vehicle motion 5 – 130 km/h are different for individual variants (Variant a, w_{min} = 0.07261 mm, V = 100 km/h, Variant b, w_{min} = 0.07349 mm, V = 5 km/h).

The dispersion of contact forces F_3 is practically the same in both cases (Variant a, $\Delta F_{3,min}$ = 4.4727 kN, V = 130 km/h, $\Delta F_{3,max}$ = 4.8439 kN, V = 15 km/h, Variant b, $\Delta F_{3,min}$ = 4.4967 kN, V = 130 km/h, $\Delta F_{3,max}$ = 4.8548 kN, V = 55 km/h). The dispersion of contact forces F_4 and F_5 is practically the same but for individual variants different (Variant a, $\Delta F_{4,min}$ = 2.2876 kN, V = 130 km/h, $\Delta F_{4,max}$ = 2.3812 kN, V = 95 km/h, $\Delta F_{5,min}$ = 2.2739 kN, V = 125 km/h, $\Delta F_{5,max}$ = 2.3781 kN, V = 80 km/h, Variant b, $\Delta F_{4,min}$ = 2.3099 kN, V = 85 km/h, $\Delta F_{4,max}$ = 2.4384 kN, V = 130 km/h, $\Delta F_{5,min}$ = 2.3028 kN, V = 80 km/h, $\Delta F_{5,max}$ = 2.4454 kN, V = 130 km/h).

This work was supported by the Grant National Agency VEGA of the Slovak Republic, project number 1/0005/16.

References

1. J. Melcer, et al., *Dynamics of transport structures* (EDIS, Žilina, 2016, in Slovak)
2. L. Frýba, *Vibration of solids and structures under moving load* (ACADEMIA, Praha, Noordhoff International Publishing, Groningen, 1972)
3. B. Novotný, A. Hanuška, *Theory of layered half-space* (VEDA, SAV, Bratislava, 1983, in Slovak).
4. J. Melcer, G. Lajčáková, *Application of program system MATLAB for the solution of the structural dynamics problems* (EDIS, Žilina, 2011, in Slovak)

The comparisons of computational models of pavement due to the dynamic load act

Veronika Valašková[1,*]

[1] University of Žilina, Faculty of Civil Engineering, Univerzitná 8215/1, 01026 Žilina, Slovakia

Abstract. Finite element method (FEM) is frequently used to solve current problems of the structural mechanics in engineering practice. The dynamic interaction of the two systems is one of the topics that can be solved using FEM. To simplify the solution of the interaction, it is possible to divide the system in two independent subsystems, vehicle and pavement. Computing software ADINA, based on the principles of the finite element modelling, was adopted to solve the given problem. The results from the first computational model can be used as the inputs for the second model. To analyse stress state of the pavement with different composition, three specific types of pavements were selected – concrete pavement, semi-rigid pavement and flexible pavement. The pavements represent the typical pavement types used for the regular road structures. This article is focused on the analysis of the stress state and the displacements of the pavement induced by the moving load from the traffic.

1 Introduction

Pavements are structures that are exposed to direct dynamic effects of the moving load. Unevenness of the pavement surface is the main source of the kinematic excitation for the vehicle. One of the most important kinematic excitation for the vehicle is also road obstacle. The obstacle has significant influence on the magnitude of the contact forces between the pavement surface and the vehicle tires. The actual load amplitude is variable of the time and the frequency domain. There are two basic approaches to obtain the required data – computational or experimental methods [1, 2].

There are many possibilities how to create the computational model of the pavement and the vehicle. 2D half-part model of the Tatra 815 vehicle and the ground represented by 3 types of the pavements with different composition were created. Results of the stress-strain analysis of the numerical models are presented in this paper. The first step of the numerical simulation involves the determination of the contact forces caused by the moving vehicle passing the obstacle. The results from the first step are used as the input for the second calculation step, evaluation of stresses in the pavement. Three models of the pavements were prepared, one for each pavement type. Pavement types were selected for typical road pavements – a concrete pavement, a semi-rigid pavement and a flexible pavement.

* Corresponding author: veronika.valaskova@fstav.uniza.sk

The theoretical solution is based on the Finite Element Method (FEM), which is adopted in numerous calculation softwares.

2 Dynamic subsystem vehicle

One of the most important parts of numerical simulation is the selection of the appropriate model of the vehicle. The corresponding computational model represents the implementation of the theoretical mathematical formulations. Simplification of the studied vehicle to the half-part model is often used in 2D analyses. The vehicle model is the combination of the mass, spring and the beam elements. The discrete model of the vehicle with finite degrees of freedom simplifies the solution from the mathematical point of view. This assumption converts partial differential equations to the ordinary differential equations. The half-part model of the Tatra 815 lorry was used for the simulations presented in this article (Fig. 1).

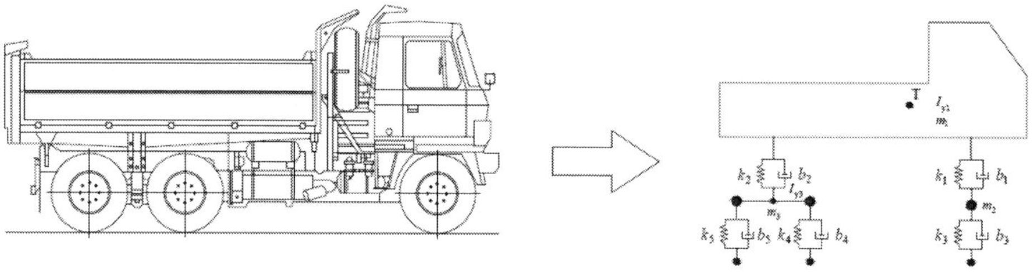

Fig. 1. Half-part model of the Tatra 815 lorry.

The half-part model of the lorry is defined by three diagonal matrices. They are mass **m**, stiffness **k** and damping **b** matrices, which contain experimentally measured values [2].

$\mathbf{m} = \{m_1, I_{y1}, m_2, m_3, I_{y3}\}_D = \{11475; 31149; 455; 1070; 466\}_D$ [kg, kg·m^2]
$\mathbf{k} = \{k_1, k_2, k_3, k_4, k_5\}_D = \{143716.5; 761256; 1275300; 2511360; 2511360\}_D$ [N·m^{-1}]
$\mathbf{b} = \{b_1, b_2, b_3, b_4, b_5\}_D = \{19228, 260197, 2746, 5494, 5494\}_D$ [kg·s^{-1}]

Natural frequencies determined as a result of the experiment measurements are consistent with the calculated natural frequencies.
$\mathbf{f} = \{f_{(1)}; f_{(2)}; f_{(3)}; f_{(4)}; f_{(5)}\} = \{1.13; 1.45; 8.89; 10.91; 11.71\}$ [Hz]

3 Dynamic subsystem pavement

Three different types of road pavements were used for the analyses – the concrete pavement, the semi-rigid pavement and the flexible pavement. Material properties of the particular construction layers have been determined in accordance with the road pavement catalogues. Axisymmetric numerical model of the pavement to simulate three-dimensional stress state in the pavement structure was adopted. Each layer was modelled separately with consideration of the contact using the interface element.

DOI: 10.1051/matecconf/201710700026

3.1 Composition of the pavement layers

Material characteristics of the particular pavement layers are in Tables 1 to 3. Since the dynamic loads inflict very small deformations to the ground, it is necessary to use "dynamic" values of modulus of elasticity E. The "dynamic" values are then higher than the "static" values.

Table 1. Composition of the layers for the concrete pavement.

Number of layer	Material	Designation of the material	Thickness (mm)	Modulus of elasticity E (MPa)	Poisson's ratio
1	Cement- concrete	CC	260	35000	0.2
2	Cement bound granular mixture	CBGM	310	4500	0.22
3	Hydraulic road binder bound granular mixture	HBBM	150	2000	0.25
4	Subgrade	S	∞	80	0.35

Table 2. Composition of the layers for the semi-rigid pavement.

Number of layer	Material	Designation of the material	Thickness (mm)	Modulus of elasticity E (MPa)	Poisson's ratio
1	Stone mastix asphalt	SMA 11	40	5500	0.33
2	Asphalt concrete	AC 22	50	6000	0.3
3	Asphalt concrete - L	AC 22-L	50	3050	0.33
4	Cement bound granular mixtures	CBGM C3/4	150	2000	0.22
5	Graded aggregate	GA	150	400	0.3
6	Subgrade	S	∞	60	0.35

Table 3. Composition of the layers of the flexible pavement.

Number of layer	Material	Designation of the material	Thickness (mm)	Modulus of elasticity E (MPa)	Poisson's ratio
1	Stone mastix asphalt	SMA 11	40	5500	0.33
2	Asphalt concrete	AC 22	80	6000	0.3
3	Asphalt concrete - L	AC 22-L	100	3000	0.33
4	Unbound mixture	UM	300	600	0.25
5	Crushed gravel	CG	200	350	0.3
6	Subgrade	S	∞	90	0.35

4 Model description

4.1 Model of vehicle passing over obstacle

The numerical model of the vehicle passing over the obstacle on the pavement was compiled in ADINA FEM computational software. The pavement and the obstacle are modelled as rigid surfaces consisting of the beam elements. The obstacle is defined as a part of the pavement surface. The pavement consists of the part for acceleration (length 15 m), the obstacle (length 0.94 m) and the part for deceleration (length 15 m). The shape of the obstacle is similar to the shapes that are recommended for the dynamic tests of the bridges. This shape is also identical with retarders on the pavement [3].

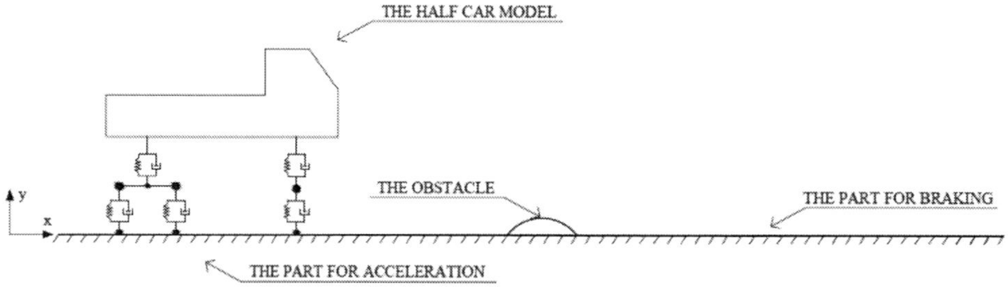

Fig. 2. The numerical model of the vehicle with the obstacle.

The obstacle was modelled as function h(x), where $h_0 = 0.06$ m and $2l_0 = 0.94$ m (Fig. 3).

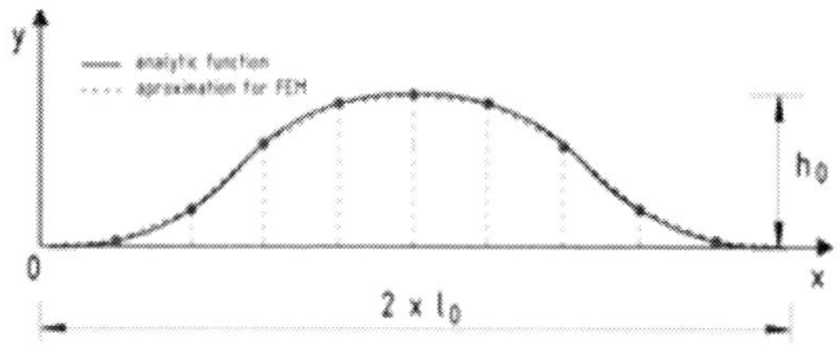

$$h(x) = h_0 . \frac{1 - \cos\left(\frac{2\pi x}{2l_0}\right)}{2} \quad (1)$$

Fig. 3. The shape of the obstacle.

The vehicle passes over the obstacle at the constant velocity of 40 km/h for all simulations. The basic time step for the simulation was selected 1/20 s. The time step when the vehicle passes the obstacle was then changed to 1/100 s to obtain more accurate results. Calculated force for front tire reached 83 200 N. The contact force was then distributed over idealized squared tire contact area 250 × 250 mm. The tire pressure was applied as a uniform load in the second step analysis with intensity 1 020 408 Pa. The load was then applied on the road pavement surface to investigate the stress-strain state of the ground.

4.2 Pavement models

The representative distributing area for loading was selected for a 2D numerical model. The selection of the distribute area was based on the experimental measurements. The length of the tire load is 250 mm. The whole model was modelled in ADINA software as an axisymmetric computational model. The distribute area was then modelled with length 125 mm. The scheme of the numerical model is shown in Fig. 4.

DOI: 10.1051/matecconf/201710700026

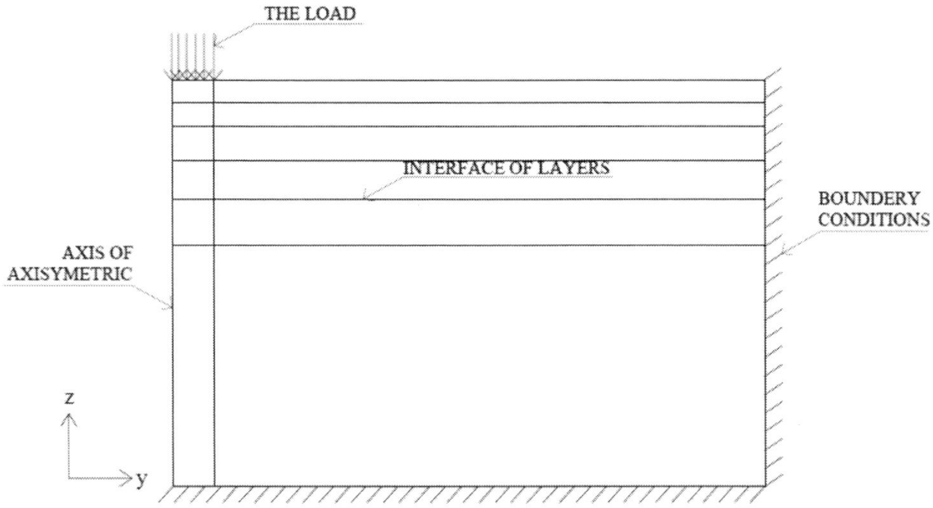

Fig. 4. Scheme of the numerical model of the pavement.

5 Description of numerical solver

ADINA is a commercial engineering simulation software, which is used in industry and academia to solve structural, fluid and heat transfer, and electromagnetic problems. The ADINA software can also be used to solve multiphysics problems, including fluid-structure interactions and thermo-mechanical problems.
ADINA is the acronym for Automatic Dynamic Incremental Nonlinear Analysis. The program consists of four core moduli:

- ADINA Structures for linear and nonlinear analysis of solids and structures,
- ADINA Thermal for analysis of heat transfer in solids and field problems,
- ADINA CFD for analysis of compressible and incompressible flows,
- ADINA EM for analysis of electromagnetic phenomena.

These moduli can be used fully coupled together to solve multiphysics problems where the response of the system is affected by the interaction of several distinct physical fields (e.g. fluid-structure interaction, thermo-mechanical analysis, piezoelectric coupling, Joule heating, fluid flow-mass transfer coupling, electromagnetic forces on fluids and structures, etc.) [4].

An effective implicit time integration scheme was proposed for the finite element solution of nonlinear dynamics problems. Numerical time integration is widely used in the finite element solutions of transient wave propagation problems. However, accurate solutions are difficult to obtain because of numerical dispersion and dissipation, resulting from period elongations and amplitude decays. In the solution of transient wave propagations, the errors from the spatial and temporal discretization appear together and affect each other. Analyses of these errors have led to the use of linear combinations of consistent and lumped mass matrices. Another approach is to reduce the solution errors by evaluating the mass and stiffness matrices using modified spatial integration rules. Such schemes can improve the accuracy of solutions of multi-dimensional wave propagation problems when using certain elements and meshes.

The dispersion errors resulting from the spatial discretization coupled with the temporal discretization can be analysed. The Bathe method and the Newmark trapezoidal rule are considered. For this purpose, it is analysed the solution obtained for the scalar wave governed by:

$$\frac{\partial^2 u}{\partial t^2} - c_0^2 \nabla^2 u = 0 \tag{2}$$

where u is the field variable and c_0 is the wave velocity. Here, body forces are not considered since we focus on the dispersion associated with the propagations of disturbances due to initial conditions. The associated finite element discretization gives:

$$\mathbf{M}\ddot{\mathbf{U}} + c_0^2 \mathbf{K}\mathbf{U} = \mathbf{0} \tag{3}$$

where **K** and **M** are the stiffness and mass matrices, and for element *(m)* with volume $V^{(m)}$

$$\mathbf{M}^{(m)} = \int_{V^{(m)}} \mathbf{H}^{(m)^T} dV^{(m)} \tag{4}$$

$$\mathbf{K}^{(m)} = \int_{V^{(m)}} \left(\nabla \mathbf{H}^{(m)}\right)^T \left(\nabla \mathbf{H}^{(m)}\right) dV^{(m)} \tag{5}$$

Here, $\mathbf{H}^{(m)}$ and **U** are the element interpolation matrix and the nodal values of the solution, respectively. Matrices **M** and **K** in Eq. (3) are obtained by the usual summation process. In the Bathe method, the following relations are employed:

$$^{t+\Delta t/2}\dot{\mathbf{U}} = {}^{t}\dot{\mathbf{U}} + \left[\frac{\Delta t}{4}\right]\left({}^{t}\ddot{\mathbf{U}} + {}^{t+\Delta t/2}\ddot{\mathbf{U}}\right) \tag{6}$$

$$^{t+\Delta t/2}\mathbf{U} = {}^{t}\mathbf{U} + \left[\frac{\Delta t}{4}\right]\left({}^{t}\dot{\mathbf{U}} + {}^{t+\Delta t/2}\dot{\mathbf{U}}\right) \tag{7}$$

$$^{t+\Delta t}\dot{\mathbf{U}} = \frac{1}{\Delta t}{}^{t}\mathbf{U} - \left[\frac{4}{\Delta t}\right]{}^{t+\Delta t/2}\mathbf{U} + \frac{3}{\Delta t}{}^{t+\Delta t}\mathbf{U} \tag{8}$$

$$^{t+\Delta t}\ddot{\mathbf{U}} = \frac{1}{\Delta t}{}^{t}\dot{\mathbf{U}} - \left[\frac{4}{\Delta t}\right]{}^{t+\Delta t/2}\dot{\mathbf{U}} + \frac{3}{\Delta t}{}^{t+\Delta t}\dot{\mathbf{U}} \tag{9}$$

Using Eq. (3) at times t, $t + \Delta t/2$ and $t + \Delta t$, where t denotes the current time and Δt the time step, with Eqs. (6) – (9), it is obtained a linear multistep form of the Bathe method:

$$\left(72\mathbf{M} + 8c_0^2\Delta t^2\mathbf{K}\right)^{t+\Delta t}\mathbf{U} + \left(-144\mathbf{M} + 5c_0^2\Delta t^2\mathbf{K}\right)^{t+\frac{\Delta t}{2}}\mathbf{U} + \left(72\mathbf{M} + 5c_0^2\Delta t^k\mathbf{K}\right)^{t}\mathbf{U} = \mathbf{0} \tag{10}$$

Using the definition of the *CFL* number, $CFL = \frac{c_0 \Delta t}{h}$, where h is the "characteristic length" of a finite element (or fundamental length used), Eq. (10) becomes:

$$\left(72\mathbf{M} + 8\gamma K\right)^{t+\Delta t}\mathbf{U} + \left(-144\mathbf{M} + 5\gamma\mathbf{K}\right)^{t+\frac{\Delta t}{2}}\mathbf{U} + \left(72\mathbf{M} + 5\gamma\mathbf{K}\right)^{t}\mathbf{U} = \mathbf{0} \tag{11}$$

where $\gamma = CFL^2h^2$ [5].

6 Results

Figures 5 to 7 represent the outputs of the numerical simulations for the horizontal stress, vertical stress and the vertical displacement in the layers of the road pavement. Results have been obtained in axisymmetric 2D simulations in XZ plane. Overall, the results show minor differences between values of the stresses and the displacement for the pavements with different composition and thickness of the layers.

6.1 Concrete pavement

The results of the stress-strain analysis from the numerical simulation are plotted in Fig. 5. The horizontal stress reached the maximum value 720 kPa, which is located at the interface of the cement-concrete layer and cement bound granular mixture layer. The maximum value of the vertical stress is 133 kPa and is located in the first cement-concrete layer. The maximum vertical displacement in Z-direction reached 0.14 mm.

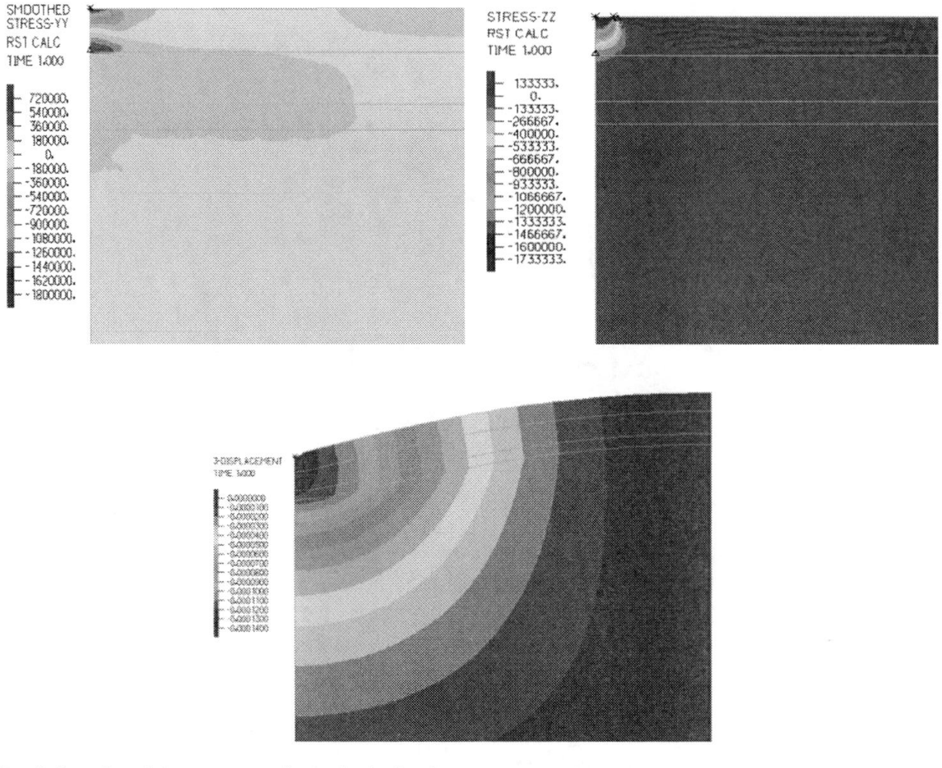

Fig. 5. Results of the stress-strain analysis for the concrete pavement.

6.2 Semi-rigid pavement

The results of the stress-strain analysis from the numerical simulation are plotted in Fig. 6. The horizontal stress reached the maximum value 675 kPa, which is located at the interface of graded aggregate layer and the subgrade. The maximum value of the vertical stress is 133 kPa and is located in the first stone mastix asphalt layer. The maximum vertical displacement in Z-direction reached 0.42 mm.

DOI: 10.1051/matecconf/201710700026

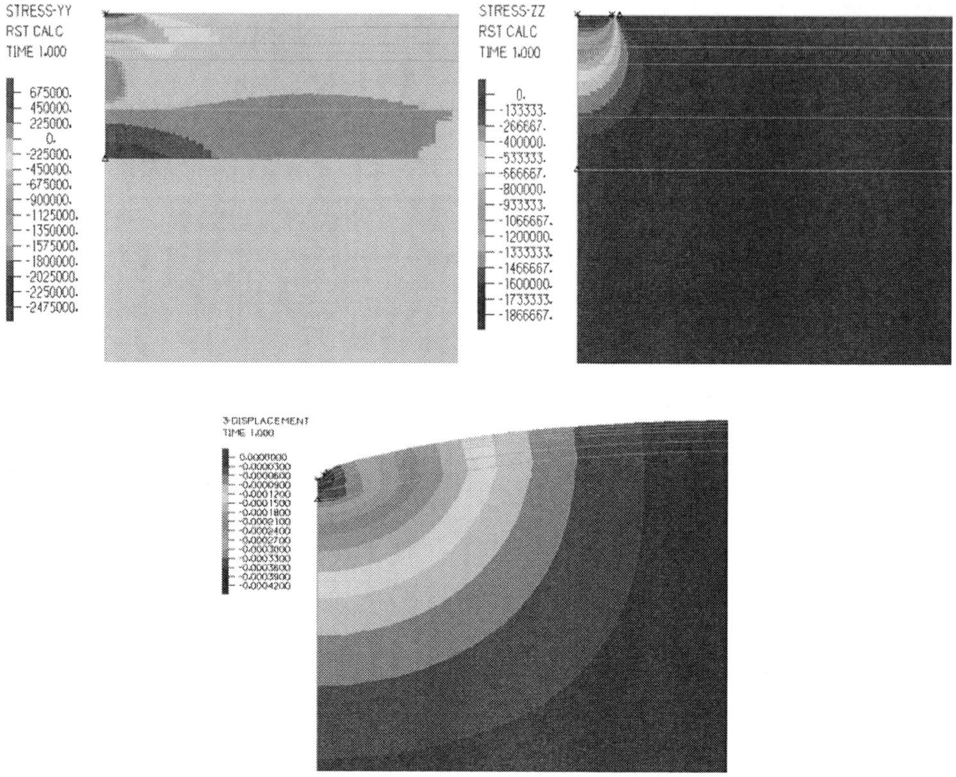

Fig. 6. Results of the stress-strain analysis for the semi-rigid pavement.

6.3 Flexible pavement

The results of the stress-strain analysis from the numerical simulation are plotted in Fig. 7. The horizontal stress reached the maximum value 675 kPa, which is located at the interface of the asphalt concrete layer and the unbound mixture layer. The maximum value of the vertical stress is 150 kPa and is located in first stone mastix asphalt layer. The maximum vertical displacement in Z-direction reached 0.315 mm.

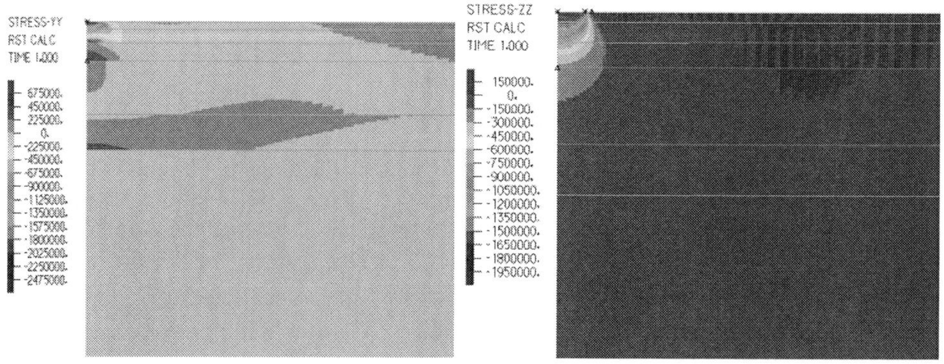

 DOI: 10.1051/matecconf/201710700026

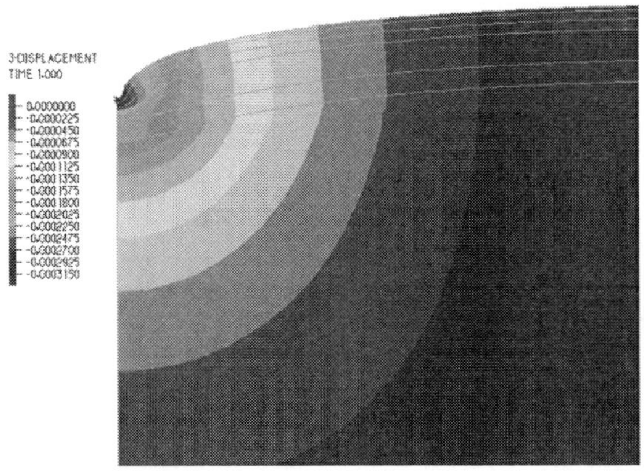

Fig. 7. Results of the stress-strain analysis for the flexible pavement.

7 Conclusion

The presented results show that the utilization of the FEM methods can be useful for solving the phenomenon of the vehicle-ground interaction. The stress-strain analysis show minor differences in the stress distribution in the pavement structure for the selected pavement types in terms of the maximum reached values but vertical displacements differ for asphalt concrete and cement concrete pavements. This variance is caused by the higher stiffness of the cement concrete slab of the concrete pavement. Further investigation will involve fully three dimensional analyses of the stress-strain state of the ground under dynamic effects of the moving load.

This paper was supported by the Grant National Agency VEGA of the Slovak Republic (Grant No. 1/0005/16).

References

[1] L. Fryba, *Vibration of Solids and Structures under Moving Loads* (ACADEMIA, Praha, Nordhoff International Publishing, Groningen, 1972)

[2] I. Martinicka, Pozemné komunikácie a dráhy, **6,** 41-50 (2010)

[3] Ľ. Daniel, V. Valašková, J. Kortiš, Civil and Environmental Engineering, **10,** 108-112 (2014)

[4] ADINA R & D. *Theory and Modeling Guide. Volume I* (ADINA. ADINA R & D, Inc., Watertown, MA, USA, 2012)

[5] N. Gunwoo, H. Seounghyun, K. J. Bathe, Comput. Struct. **123**, 93-105 (2013)

Experience with the application of the spectral analysis of surface waves method on a falling weight deflectometer time history

Jozef Komacka[1,*]

[1]University of Žilina, Faculty of Civil Engineering, Univerzitná 8215/1, 010 26 Žilina, Slovakia

Abstract. Falling Weight Deflectometers are commonly used in pavement diagnostics and the outputs of measurements are a basis for pavement performance evaluation. Usually, a surface modulus of a pavement is determined or the modulus of a pavement layer is back-calculated using maximum values of a load force and measured deflections. A Falling Weight Deflectometer time history contains data describing a total response of a pavement to a load in time domain. The theory and principles of the Spectral Analysis of Surface Waves method were applied to these measurement outputs to determine Rayleigh wave velocity and a compression-tension modulus of a pavement. The problems with the determination of travel time of a wave were identified. The calculated compression-tension modulus differed noticeably depending on a used distance between receivers. Analysis showed the depth of the propagation of a wave or the bulk density of an investigated area is probably not a reason of the difference in the values of the compression-tension modulus calculated for the various distances of receivers.

1 Introduction

Falling Weight Deflectometrs (FWD) are commonly used in pavement bearing capacity diagnostic and the outputs of measurements are a basis for performance evaluation of a pavement structure. The outputs of diagnostics by a FWD contain a lot of data regarding the change of a load force and deflections values in time domain. This time history describes total pavement response to a load. Common practice in pavement performance evaluation is to use only the maximum values of a load force and deflections measured at various spacing. A surface modulus is calculated from the peak value of a load and deflection at the axis of a load plate using the linear elastic half space theory. It characterizes total stiffness of a pavement and its subgrade without possibility to determine which part of the system is weak when a pavement has low performance. This disadvantage can be eliminated via back-calculation of modulus of pavement layers and subgrade based on the multilayered linear elastic system theory or finite element methods. Both mentioned approaches use the static interpretation of a load although it is dynamic in reality. Modeling of a pavement response to a dynamic load in a pavement model loaded by an impulse with

* Corresponding author: komacka@fstav.utc.sk

a shape comparable to a FWD load impulse [1] indicated problems with the boundary conditions of the calculation model and longtime of calculations.

The application of the Spectral Analysis of Surface Waves (SASW) method is another possibility to respect dynamic nature of the FWD load. The SASW method consists of generation, measurement and processing of dispersive elastic waves in a layered system. Dispersion of one kind of surface waves, named Rayleigh`s waves, is used to assess the stiffness of an investigated system. The application of the SASW method consists of field measurements of surface wave velocity at numerous wavelengths. The typical configuration to perform field measurements using the SASW method consists of a source, one [2], two or more vertical acceleration transducers located on a surface at the equidistance from the source and a data acquisition system. A vertical load is applied to generate surface waves and the propagation of these waves along a surface is recorded by receivers. A pavement bearing capacity diagnostics procedure by a FWD is basically the same as investigation by the SASW method. A vertical load is generated by a falling mass and pavement response is recorded by sensors spaced in different distances from a load axis. Number of the sensors obviously varies from 5 to 8 and they are positioned at various distances up to 2.1 m from the axis of a load. Due to similarity of the principles of both methods the theory used for evaluation of SASW measurements was applied to FWD outputs.

2 Theoretical backgrounds

Rayleigh`s waves are two dimensional surface waves generated by a vertical impulse load. They consist of longitudinal and transversal oscillations that propagate in a plane that is predetermined by the direction of a wave. The velocity of Rayleigh`s wave can be determined using the velocity of a shear wave, the correlation or spectral analysis methods or direct determination of Rayleigh`s wave velocity. The last one is very convenient approach from practical point of view. The velocity of Rayleigh`s wave c_R is determined from the formula

$$c_R = L/t \tag{1}$$

where L = the distance of the receivers used for determination of a wave velocity; and t = travel time of a wave between the receivers.

Since the distance of the receivers is always known, only the response of the receivers in time domain has to be recorded. Moreover, the distance between various receivers can be used to determine the wave velocity.

The known velocity of the Rayleigh`s wave is a base to calculate a shear modulus G using the formula

$$G = \rho \cdot (c_R/r_v)^2 \tag{2}$$

where ρ = the bulk density of investigated area; and r_v = the velocity ratio of Rayleigh`s and shear wave.

Subsequently, compression-tension modulus E can be calculated by

$$E = 2 \cdot G\,(1+\mu) \tag{3}$$

where μ = Poisson`s ratio of investigated area.

As for surface waves (i.e. also for Rayleigh`s waves) it is important that energy of a wave reduces along a depth. The depth of propagation of a Rayleigh`s wave depends on its wavelength λ_R that is a function of the velocity c_R of the Rayleigh`s wave and the frequency of a load impulse f_R

$$\lambda_R = c_R / f_R. \tag{4}$$

Waves with short wavelength propagate near ground surface whereas longer wavelengths travel deeper. Experience indicates the depth of the propagation of Rayleigh`s waves is approximately equal to its wavelength.

3 Experimental

The compression-tension modulus of a pavement structure was determined based on the theory of the SASW method. The essential input, the velocity of the Rayleigh`s wave, was calculated using the FWD time history data. There are data relating to time change of a load impulse and surface deflections at known distances from the axis of a load plate (Fig. 1).

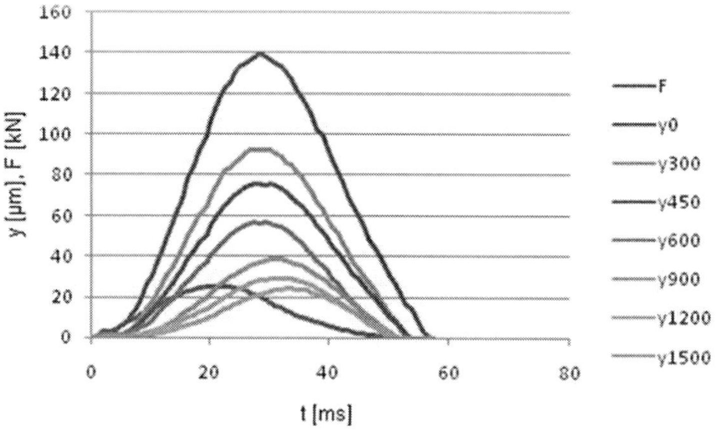

Fig. 1. Example of FWD time history.

The travel time of a Rayleigh`s wave between receivers can be determined using the front of a wave or the maximum value of a surface deflection (Fig. 2). The analysis of the time history data showed the use of the maximal value of a surface deflection is a better approach because it was not possible to exactly determine the arrival time of a wave at the positions of the receivers. The probably reason was that the small changes of the deflections at beginning were not properly registered due to accuracy (sensitivity) of the FWD receivers. Next problem was identified when the maximal values of the deflections were intended to use. There were the same values at the consecutive time records. Therefore, the approximation of the course of the values close to a maximum value was performed (Fig. 3). The maximum deflection value and relating time was determined using the approximation curve. When time of the maximum values determined by the approximation was plotted against the distance of the FWD receivers it was found in some cases that time not increased with the distance as it was expected. Therefore, the regression curve was derived to determine final time of the maximum values of the deflections at the FWD receivers (Fig. 4). These values were used to calculate the velocity of Rayleigh`s wave using known spacing of the FWD receivers.

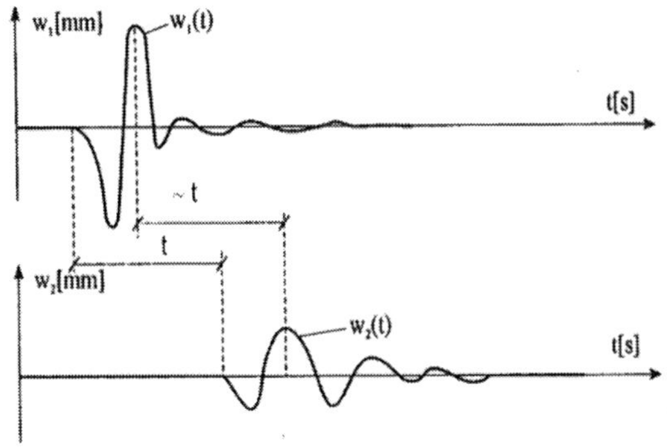

Fig. 2. Determination of travel time of wave.

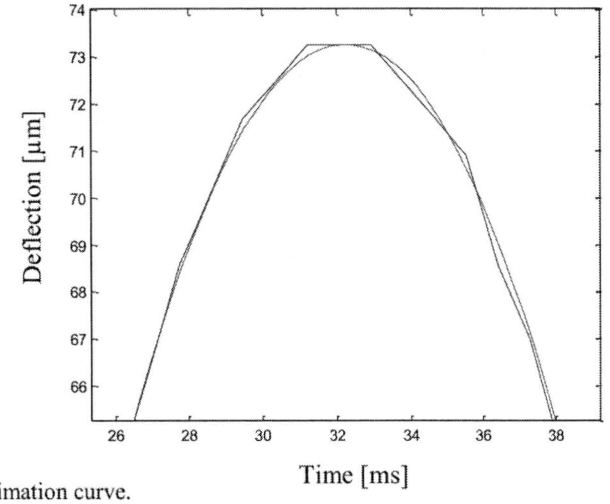

Fig. 3. The approximation curve.

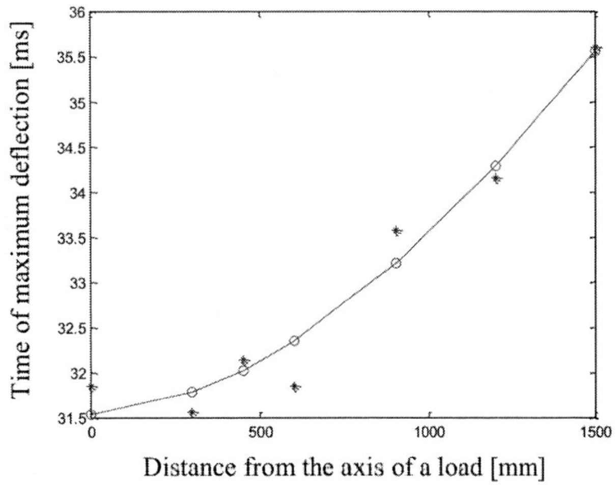

Fig. 4. Time of the maximum deflections at the FWD receivers.

The shear modulus and afterwards compression-tension modulus was calculated for the distance between the receiver positioned at the axis of the FWD load plate and others receivers that were placed in a line on a beam in the distance of 300 mm, 450 mm, 600 mm, 900 mm, 1200 mm and 1500 mm from the axis of the FWD load plate. The compression-tension modulus calculated for six different distances of the FWD receivers from the axis of a load is plotted in Fig. 5.

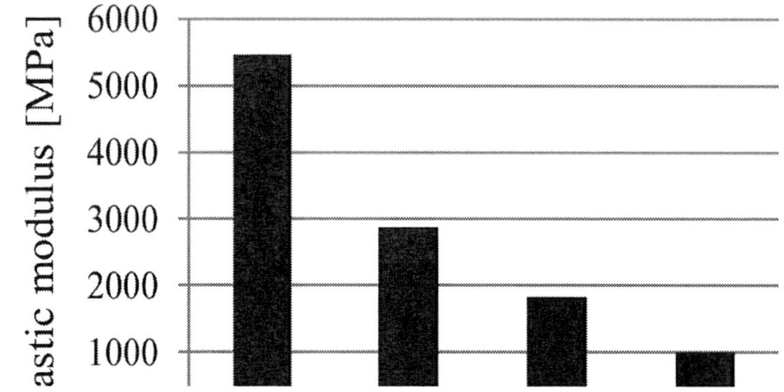

Fig. 5. Compression-tension modulus calculated for the different distances of a receiver to the axis of a load.

It is evident the compression-tension modulus values are strong different and decrease when the distance increases. It means the calculated velocity of a Rayleigh`s wave was different when the various spacing of the receivers was used. Assuming the problem could be connected with the use of the receiver in the axis of a load the receiver at the distance of 300 mm was taken into account as a base and new distances to the more distant receivers were used for the calculation of Rayleigh`s wave velocity and both modulus. Nevertheless, the results were basically the same (Fig. 6). This finding was also verified for next numerous distances calculated from others possible combinations of the receivers.

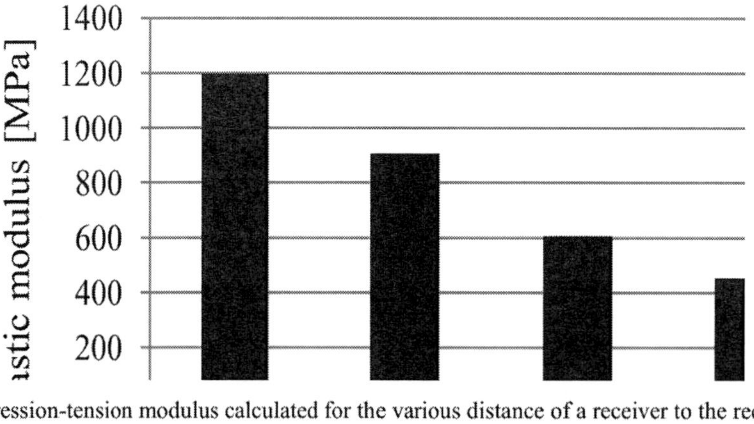

Fig. 6. Compression-tension modulus calculated for the various distance of a receiver to the receiver at 300 mm from the axis of a load.

Keeping in mind the frequency of a load impulse can influence the wavelength of Rayleigh`s wave and the depth of its propagation the load impulse of used FWD was analysed. Data in time domain (see Fig. 1, the blue line labelled as "F") was transferred into frequency domain using the discrete Fourier transformation (DFT). The result confirmed an expectation and only one dominant frequency was identified in the FWD load impulse. When the same value of a frequency is used in the formula (4), the different velocity of a wave leads to a different wavelength. As a wavelength relate to the depth of the propagation of a wave it could be assumed the differences in the depth of the propagation of a wave is a cause of the difference in the values of compression-tension modulus. However, the wavelength (and subsequently the depth of the propagation of a wave) is directly proportional to the velocity of a wave. It means a higher velocity leads to a higher wavelength and deeper propagation. Based on this, the differences in the values of compression-tension modulus should relate to material differences in great depth (more than 15 m in this case). As it does not seem real attention was drawn to the calculation of a shear modulus according to the formula (2).

In addition to the speed of a wave the bulk density of area is another input parameter. The influence of a bulk density to the velocity of a wave could be found in [3]. In this case the same value of the bulk density was used for all distances of the receivers. It resulted to the different values of shear modulus due to the dissimilar velocities of the wave. If the values of the shear modulus should be the same for any distance of the receivers (assuming the homogeneity of investigated area) the values of the bulk density has to be different. This assumes a wave propagates to different depths and a bulk density changes along the depth. It is only possible in a layered system with the different bulk density of layers. This is a real situation in the case of asphalt pavements. Therefore, the bulk densities were calculated to have the same value of a shear modulus regardless the distance of the receivers. The values of the shear modulus were chosen from the range of the values calculated in the previous evaluation. A part of outputs is shown in Table1.

Table 1. Calculated bulk density for variable values of constituents.

	Distance of receivers in mm					
	300	450	600	900	1200	1500
Shear modulus	2025 MPa					
Wave velocity in m/s	985	714	570	421	345	299
Bulk density in g/cm^3	2.087	3.972	6.233	11.425	17.013	22.651
Shear modulus	186 MPa					
Wave velocity in m/s	985	714	570	421	345	299
Bulk density in g/cm^3	0.192	0.365	0.572	1.049	1.563	2.081

The calculated bulk densities are very different and mostly unrealistic (very high or very low) when compared to the real bulk density (approx. 2.084 g·cm^{-3}). It cannot be concluded the variability of the shear modulus calculated for the various distances of the receiver relate to the differences in the bulk density of investigated area.

4 Conclusions

The application of the SASW method on the FWD time history data indicated some problems. It was not possible to exactly determine the arrival time of a wave when the front of a wave was chosen. The better solution was to use the maximum value of a pavement surface deflection. Nevertheless, an approximation was necessary to determine final time of the maximum values of the deflections at the receivers.

The compression-tension modulus calculated from the various distances of the receivers differed noticeably. The higher values were determined for the shorter distances of the receivers. The analysis of the load impulse in frequency domain proved the existence only one dominant frequency. Therefore, the possible influence of the depth of the propagation of a wave was assessed. Taking into account the calculated wavelength the depth of the propagation of a wave was estimated. It was found the material differences in the great depth should be a reason of difference in the values of compression-tension modulus.

As this is probably not real the influence of the bulk density of investigated area was analysed. It was concluded the calculated values are very different and mostly unrealistic and therefore they could not be the cause of the variability of the shear modulus calculated for the various distances of receivers.

The author is grateful for the financial support from the research project VEGA 1/005/16 Modelling of selected dynamic problems in time and frequency domain

References

1. N. Rapanová, J. Kortiš, Transport and telecommunication **14 (3)**, 230-236 (2013)
2. N. Ryden et al, J Geotech Geoenviron **130 (6)**, 636-645 (2004)
3. J. Komačka, N. Rapanová, Road horizon (Silniční obzor) **73 (7-8)**, 205-209 (2012)

Properties of road unevenness inducing the kinematical excitation of vehicles

Mária Kúdelčíková[1,*] and *Jozef* Melcer[1]

[1] University of Žilina, Faculty of Civil Engineering, Univerzitná 8215/1, 010 26 Žilina, Slovakia

Abstract. The submitted paper is devoted to the mapping of the surface road profile and to the mathematical description of unevenness in one vehicle track. Its statistical parameters are analyzed and the classification of the road into a category based on power spectral density of unevenness is made.

1 Introduction

Modeling of dynamic effect of moving load on the road is a serious engineering problem. Analysis of the problem implies: creation of computational models of vehicles, creation of computational models of roads, modeling of runway surface unevenness, method of solving equation of motion, processing and presentation of results. Surface unevenness represents an important role in this process, because it is the source of kinematic excitation of the vehicle and therefore the source of dynamic components of the contact forces [1, 2]. The submitted paper is devoted to the mapping of the surface road profile, mathematical description of unevenness in one vehicle track, analyzing its statistical parameters and the classification of the road into a category based on power spectral density of unevenness.

2 Unevenness mapping

Unevenness mapping of the roadway was carried out by the prototype of new developed devices like vibrograph. The result of mapping was road profile in one vehicle track, Fig. 1.

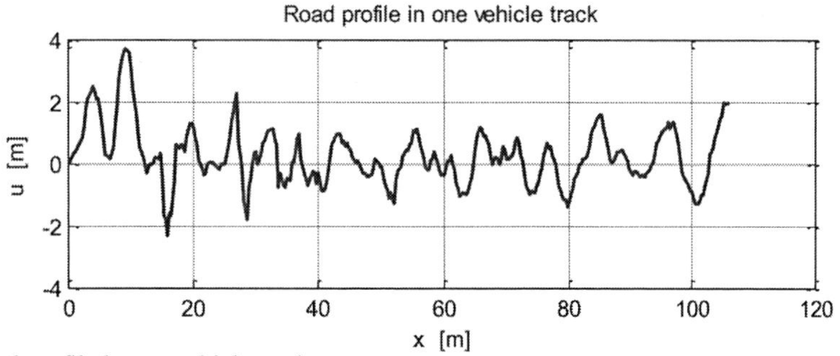

Fig. 1. Road profile in one vehicle track.

* Corresponding author: maria.kudelcikova@fstav.uniza.sk

It was mapped the section 105.9 m long in increments of 0.3 m.

3 Amplitude characteristics of unevenness

The original record had 354 samples. The record was re-sampled. New record has $N = 2^{13} = 8192$ samples. All other analyzes were performed with such re-sampled record. The following characteristics were calculated:

Mean value

$$\bar{u} = \frac{1}{N}\sum_{i=1}^{N} u_i = 0.232256\,\text{mm}. \tag{1}$$

Arithmetic mean deviation R_a

$$R_a = \frac{1}{N}\sum_{i=1}^{N} |u_i - \bar{u}| = 0.690666\,\text{mm}. \tag{2}$$

Root mean square average deviation R_q

$$R_q = \sqrt{\frac{1}{N}\sum_{i=1}^{N} (u_i - \bar{u})^2} = 0.931226\,\text{mm}. \tag{3}$$

Dispersion σ^2

$$\sigma^2 = \frac{1}{N}\sum_{i=1}^{N} (u_i - \bar{u})^2 = 0.867182\,\text{mm}^2. \tag{4}$$

Effective values - Root mean square value *RMS*

$$\bar{u} = \sqrt{\frac{1}{N}\sum_{i=1}^{N} u_i^2} = 0.959753\,\text{mm}. \tag{5}$$

Asymmetry coefficient R_{sk}

$$R_{sk} = \frac{\sum_{i=1}^{N} (u_i - \bar{u})^3}{N \cdot \sigma^3} = 0.839111. \tag{6}$$

Kurtosis R_{ku}

$$R_{ku} = \frac{\sum_{i=1}^{N} (u_i - \bar{u})^4}{N \cdot \sigma^4} = 4.788243. \tag{7}$$

Greatest depth of unevenness

$$R_v = \min(u_i) = \text{-}2.327134\,\text{mm}. \tag{8}$$

The largest height of unevenness

$$R_p = \max(u_i) = 3.709063 \text{ mm}. \tag{9}$$

Overall height of the profile

$$R_t = |R_v| + R_p = 2.327134 + 3.709063 = 6.036197 \text{ mm}. \tag{10}$$

The histogram of amplitude distribution is shown in Fig. 2.

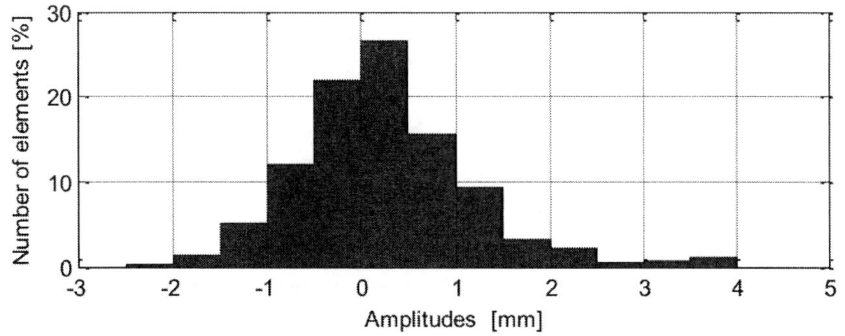

Fig. 2. Histogram of amplitude distribution.

4 Spectral characteristics of unevenness

Passing from length to frequency domain can be carried out by Fast Fourier Transform. To each real number u_k it is assigned a complex number U_k by the relation

$$U_k = \sum_{n=1}^{N} u_n \cdot e^{-2\pi \cdot i \cdot (k-1)(n-1)/N}, \text{ for } 1 \le k \le N. \tag{11}$$

N is the number of samples of the record. The amplitude and phase angle are then calculated for each complex number, so the amplitude and phase spectra can be calculated for the record. There are a few possibilities how to present the amplitude spectrum:

Peak Amplitude *PA*

$$PA_I = \sqrt{\mathrm{Re}[U_I] \cdot \mathrm{Re}[U_I] + \mathrm{Im}[U_I] \cdot \mathrm{Im}[U_I]} / N, \text{ for } I = 0, \tag{12a}$$

$$PA_I = 2 \cdot \sqrt{\mathrm{Re}[U_I] \cdot \mathrm{Re}[U_I] + \mathrm{Im}[U_I] \cdot \mathrm{Im}[U_I]} / N, \quad I = 1 \div (N/2) - 1 \tag{12b}$$

Auto Spectrum *AS*

$$AS_I = PA_I \cdot PA_I, \quad \text{for } I = 0 \div (N/2) - 1. \tag{13}$$

Root Mean Square Amplitude *RMSA*

$$RMSA_I = PA_I, \qquad \text{for } I = 0, \tag{14a}$$

$$RMSA_I = PA_I / \sqrt{2}, \quad \text{for } I = 1 \div (N/2) - 1. \tag{14b}$$

Power Spectrum *PS*

$$PS_I = AS_I\,, \qquad \text{for } I = 0, \tag{15a}$$

$$PS_I = AS_I\,/\,2\,, \quad \text{for } I = 1 \div (N/2) - 1. \tag{15b}$$

The spectrum is usually shown as a function of wavenumber $\Omega = 2\pi / L$ in [rad/m] or as a function of wavelength L in [m]. The PSD of unevenness u as a function of Ω is shown in Fig. 3 and peak amplitude of unevenness u as a function of L is shown in Fig. 4.

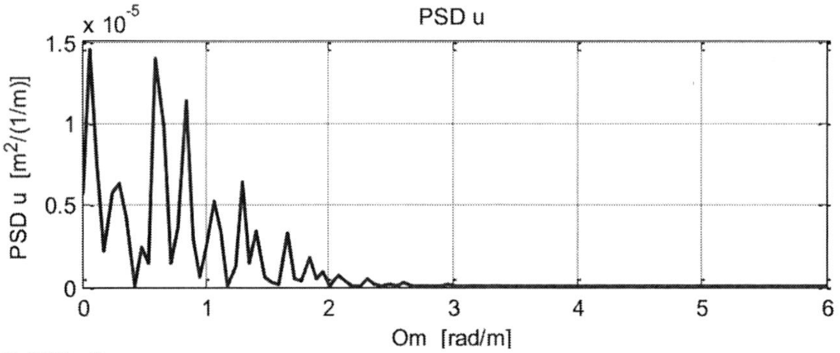

Fig. 3. PSD of unevenness *u*.

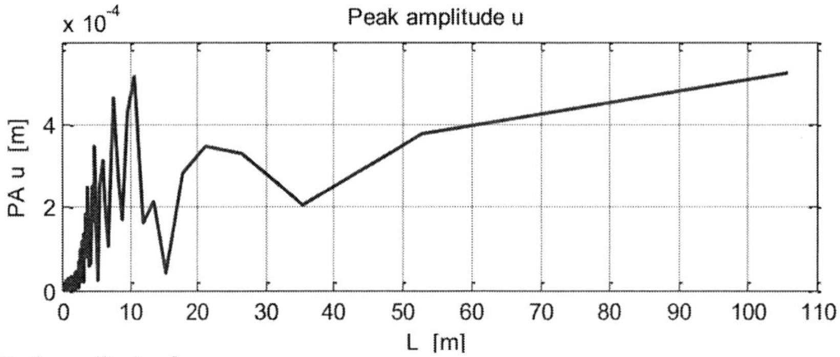

Fig. 4. Peak amplitude of unevenness *u*.

The standard ISO 8608 [3] classified the road under the PSD into 8 categories A – H. Our road profile can be classified in the category B, Fig. 5.

5 Conclusion

Road unevenness represents the source of kinematic excitation of vehicle. It directly affects the vibration of the vehicle and the value of the dynamic components of the contact forces. The knowledge of real road profile and its amplitude and spectral characteristics is needed for classification of the road and for the numerical simulation of vehicle motion along the road. Our profile is 105.9 m long. Overall height of the profile is 6.03 mm. The greatest depth and the largest height are -2.32 and 3.71 mm, respectively. The profile contains the waves of wavelength L in the interval 0.97 – 105.9 m. By the Standard ISO 8608 the profile can be classified as category B. It can be used as input value for numerical simulation of motion of plane vehicle computing model along the road.

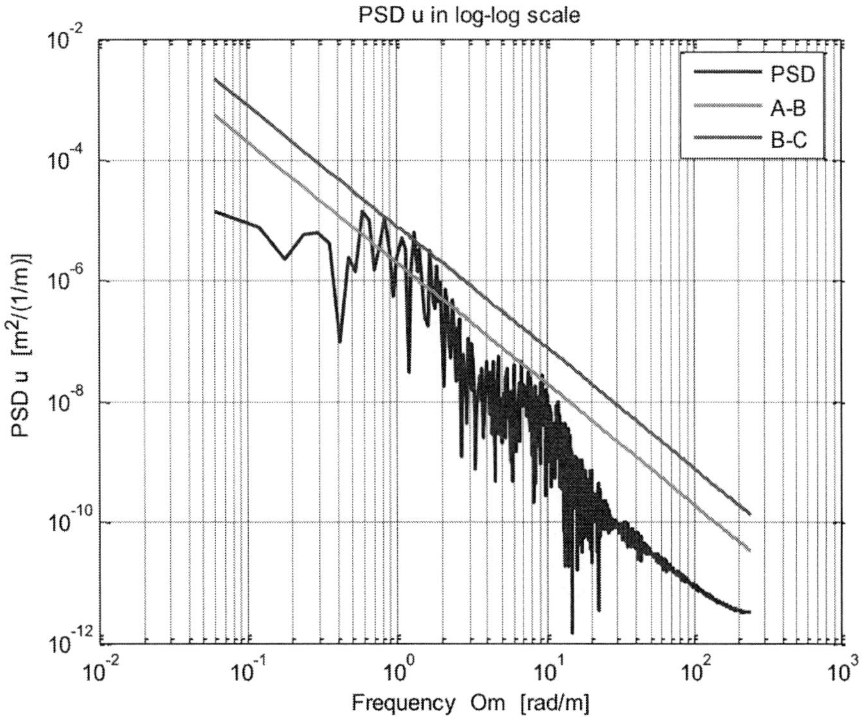

Fig. 5. Classification of unevenness *u*.

This work was supported by the Grant National Agency VEGA of the Slovak Republic, project number 1/0005/16.

References

1. G. Rill, *Road vehicle dynamics, fundamentals and modeling* (CRC Press, Taylor & Francis Group, London, New York, 2012)
2. M. Decký, M. Kováč, *Longitudinal unevenness of roads* (EDIS, Žilina, 2014, in Slovak)
3. ISO 8608, Mechanical vibration – road surface profiles – reporting of measures data, International standard, (1995)

Trajectory of a road vehicle during road maintenance

Darina Stachová[1,*]

[1] Department of Technical Sciences and Informatics, Faculty of Security Engineering, University of Žilina, Univerzitná 8215/1, 010 26 Žilina, Slovakia

Abstract. Consider a vehicle moving on a road whose usage over time creates an uneven surface on the road. Road unevenness that we encounter on surface communications often arises as a consequence of dynamical effects of moving vehicles, of weather changes, and due to road construction works. This article concerns with mathematical modeling of the trajectory of a road vehicle moving on such a surface during the course of road maintenance.

1 Introduction

Consider a vehicle moving on a road whose usage over time creates an uneven surface on the road. The problem we want to focus on here is mathematical modeling of the trajectory that this vehicle traverses while moving on such a road. For simplicity, we identify the motion of the vehicle with the motion of the center axis of a moving wheel [1]. Unevenness that we encounter on surface communications (besides intentionally positioned speed bumps) are typically a consequence of dynamical effects of moving vehicles, due to changes in weather, and due to reconstruction of the road surface. A vehicle moving on a road must often adjust its speed based on the state and shape of the road. This fact is a reason to focus on modeling of the vehicle's trajectory on a road with damaged surface [2]. In particular, driving on such a road is safe and economical, if the trajectory constitutes a smooth curve, i.e. a curve without singular points. This, of course, assumes that the road

2 Computational model of a road

For the purposes of this contribution, let us consider a road to be a surface communication with hard and uneven surface. Mathematical description of a computational model of such a road then reduces to a mathematical model of unevenness of the trajectory of a vehicle [3].

In the course of a road reconstruction, the following three situations can occur:

1. vehicle driving through a right-angle asphalt cutout after the removal of the asphalt layer from the road surface,

* Corresponding author: Darina.Stachova@fbi.uniza.sk

2. vehicle driving through a trapezoidal asphalt cutout after the removal of the asphalt layer from the road surface,
3. vehicle driving through a step over a steel board positioned to cover the terrain during excavation works.

We shall assume that the wheels of the vehicle undergo no deformation during the motion and maintain their circular shape throughout. This is a reasonable assumption to make, since in practice the changes of wheel shape are negligible (thanks to modern technologies) when compared to the changes in road surface that we consider here. We shall therefore consider the problems that arise when rolling a circular wheel on a road with damaged surface [4].

2.1 Road with a right-angle cutout

Assume that the surface of the road has the shape of a right-angle cutout as illustrated in Fig. 1, where we denote by l the length of the cutout, by h the depth of the cutout, and by r the radius of the moving wheel.

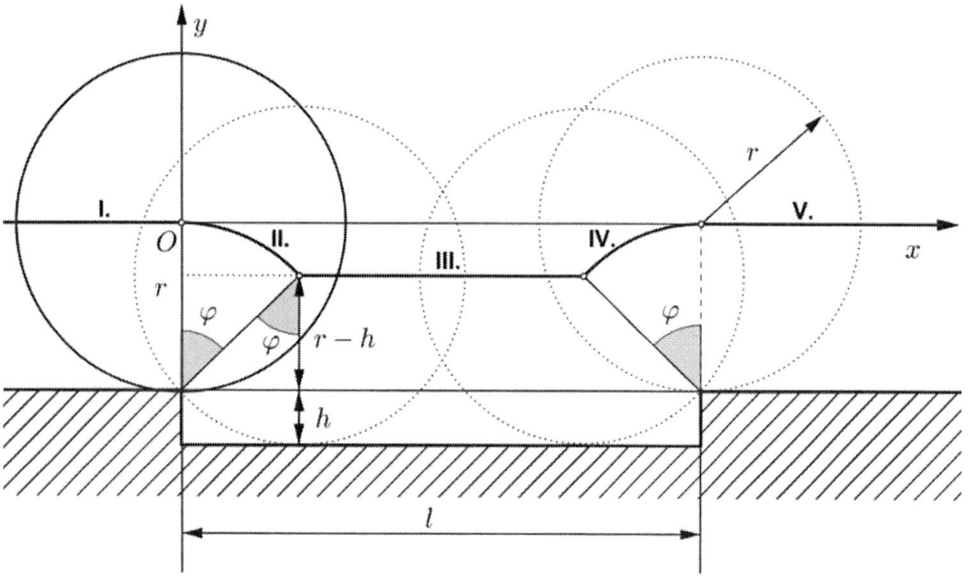

Fig. 1. Road with a right-angle cutout.

The shape of the road surface is described by a function that is not differentiable everywhere. The jumps in the function value at the beginning and the end of the cutout constitute from the mathematical perspective point singularities, which are relevant also for answering non-mathematical questions. The (axis of the) wheel when driving through the cutout does not precisely copy the shape of the cutout, since the motion of the wheel axis in the vertical direction is caused by kinematic excitation of the suspended mass of the vehicle. The axis traverses a trajectory consisting of line segments and circular arcs denoted in Fig. 1 as I., II., III., IV., and V. When moving across the two ends of the cutout, the axis moves along the circular arcs II. and IV., which can be represented in the (O, x, x) coordinate system in different ways, for instance, in the explicit form. Since the angle φ is small compared to the depth h, we can use polynomial approximations (this choice is also meaningful due to the technical and practical use of such approximations) [5]. Since the circular arcs are short, we can sufficiently accurately approximate them with section of parabolas $y(x) = \frac{-x^2}{2p}$, resp. $y(x) = \frac{-(x-l)^2}{2p}$, where p is the parameter to be determined.

The complete description of the trajectory is then as follows.

I. $y(x) = 0$ $\quad x \in (-\infty; 0)$ (1)

II. $y(x) = \frac{-x^2}{2r-h}$ $\quad x \in \left(0; \sqrt{2rh - h^2}\right)$ (2)

III. $y(x) = -h$ $\quad x \in \left(\sqrt{2rh - h^2}; l - \sqrt{2rh - h^2}\right)$ (3)

IV. $y(x) = \frac{-(x-l)^2}{2r-h}$ $\quad x \in \left(l - \sqrt{2rh - h^2}; l\right)$ (4)

V. $y(x) = 0$ $\quad x \in (l; \infty)$ (5)

Critical points of this trajectory occur when transitioning from segment I. to II, and from IV. to V. In these points the vehicle has a two-point contact with the surface, which creates a singularity in the trajectory.

2.2 Road with a trapezoidal cutout

Assume that the road surface has a shape of a trapezoidal cutout. This is illustrated in Fig. 2. The size of the depth angle φ determines the shape and the components of the wheel's trajectory.

2.2.1 Road with a trapezoidal cutout of type A

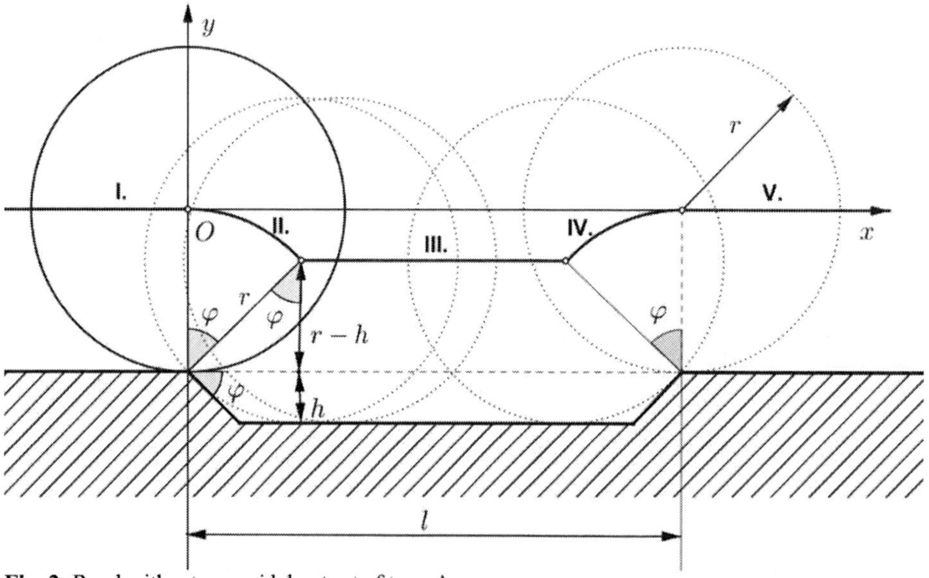

Fig. 2. Road with a trapezoidal cutout of type A.

If the angle φ satisfies: $\arccos\frac{r-h}{r} \le \varphi \le \frac{\pi}{2}$, then the two-point contact of the wheel with the surface only occurs at the beginning and the end of the cutout. The mathematical description of the trajectory is identical to Case 1, i.e. it is again composed of five segments described by the equations (1) – (5) listed above.

2.2.2 Road with a trapezoidal cutout of type B

If the angle φ satisfies: $\varphi < \arccos\frac{r-h}{r}$, as illustrated in Fig. 3, then again a two-point contact of the wheel with the surface occurs in two places, however, this time not at the beginning and the end of the cutout. The mathematical description of the trajectory now consists of seven segments and can be expressed as follows.

I. $y(x) = 0 \qquad x \in (-\infty; 0)$ (6)

II. $y(x) = \frac{-1}{r(1+\cos\varphi)}x^2 \qquad x \in (0; r\cdot\cos\varphi)$ (7)

III. $y(x) = (-\mathrm{tg}\varphi)x + r\left(\frac{1}{\cos\varphi} - 1\right) \qquad x \in \left(r\cdot\sin\varphi; \frac{r}{\sin\varphi} + \frac{h-r}{\mathrm{tg}\varphi}\right)$ (8)

IV. $y(x) = -h \qquad x \in \left(\frac{r}{\sin\varphi} + \frac{h-r}{tg\varphi}; l - \frac{r}{\sin\varphi} - \frac{h-r}{tg\varphi}\right)$ (9)

V. $y(x) = (\mathrm{tg}\varphi)\cdot(x-l) + r\left(\frac{1}{\cos\varphi} - 1\right) \qquad x \in \left(l - \frac{r}{\sin\varphi} - \frac{h-r}{\mathrm{tg}\varphi}; l - r\cdot\sin\varphi\right)$ (10)

VI. $y(x) = \frac{-1}{r(1+\cos\varphi)}(x-l)^2 \qquad x \in (l - r\cdot\sin\varphi; l)$ (11)

VII. $y(x) = 0 \qquad x \in (l; \infty)$ (12)

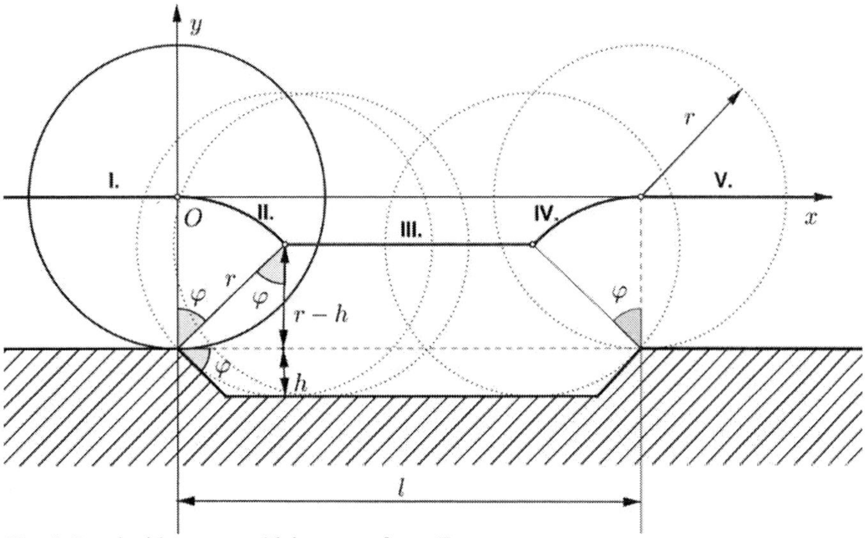

Fig. 3. Road with a trapezoidal cutout of type B.

2.3 Rectangular step

In the last case, we look at the road surface in the shape of a rectangular terrain step, as illustrated in Fig. 4.

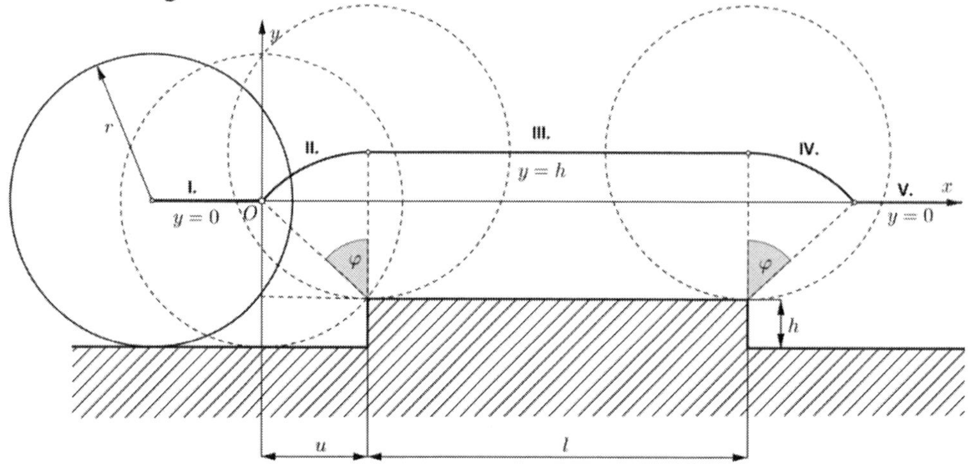

Fig. 4. Road surface with a terrain step.

During excavation works in the course of construction on the road, the resulting trenches are often covered by steel plates to allow vehicles to continue using the road. The surface of this road can be then modeled using a rectangular step in the coordinate system (O, x, y), as shown in Fig. 4, where we denote by l the length of the step, by h the height of the step, by u the distance of the step from the y axis, and by r the radius of the moving wheel. The axis of the wheel traverses a trajectory comprised of five parts. The singular points are this time found before and after the step. The jumps in the function value at the beginning and the end of the step constitute in the trajectory of the wheel the motion along circular arcs II. and IV. by turning the wheel by angle φ. Therefore $u = r \cdot \sin\varphi$ and $|r-h| = r \cdot \cos\varphi$. The two arcs can be expressed in the Cartesian coordinates explicitly as follows.

II. $$(x-u)^2 + (y-h+r)^2 - r^2 = 0 \qquad \text{for} \quad x \in \langle 0; u \rangle \qquad (14)$$

IV. $$(x-l-u)^2 + (y-h+r)^2 - r^2 = 0 \qquad \text{for} \quad x \in \langle l+u; l+2u \rangle \qquad (15)$$

Since the arcs are short (because h is relatively small compared to r), we approximate the arcs using parabolas of second order as follows.

II. $$y = h - \frac{h}{u^2}(x-u)^2 \qquad x \in \langle 0; u \rangle \qquad (16)$$

IV. $$y = h - \frac{h}{u^2}(x-l-u)^2 \qquad x \in \langle l+u; l+2u \rangle \qquad (17)$$

3 Conclusions

The vehicle–road interaction is a current problem that is a focus of attention of engineers around the world. The best way to solve these types of problems is a combination of numerical and experimental methods. Numerical methods today provide an effective tool for solving these problems using the state-of-the-art computers, which allow solving these problems practically in real time.

The search for and removal of singular points on the road is monitored for ensuring safe ride on damaged roads. The existence of these points causes the shape of the trajectory of the vehicle to not agree with the shape of the road surface in places where the road is damaged. In general, solving some of the problems is not simple. In particular, the problems become difficult in cases when the wheel loses contact with the road surface. This is, on the one hand, due to the fact that at the points of the loss of contact the character of vibrations changes from forced to free, and on the other hand, the points when the contact is restored cause impact between the wheel and the surface. In particular, to explain the conditions at the moment of the impact requires a separate complex analysis.

The results of numerical and experimental analyses are applied in the design of road vehicles as well as in the road design, and to improve comfort of passenger. Acquired results are also used when designing optimal parameters of the road when it comes to its reliability, longevity, and maintenance costs.

This work was produced as part of the grant project VEGA SR 1/0005/16.

References

1. J. Melcer, D. Stachová, Proceedings of the 2nd International Conference Dynamics of Civil Engineering and Transport Structures and Wind Engineering, DYN-WIND 2017, 178-181 (2003)
2. J. Melcer, Maintenance and its impact on roads and highways, Seminar with international participation, Žilina, Slovakia, 55-58 (1999)
3. J. Melcer, *Dynamic calculations of highway bridges*, (EDIS, Žilina, 1997, in Slovak)
4. D. Stachová, Proceedings of the Symposium on Computer Geometry SCG '2006, **15**, 130-135 (2006)
5. D. Stachová, Proceedings of the Symposium on Computer Geometry SCG '2008, **17**, 87-90 (2008)

Consideration of impact assessment of vibrations for strength and compression in silty soils

Janusz Ukleja[1,*]

[1]Faculty of Civil Engineering and Architecture, Opole University of Technology, Katowicka 48, 45-061 Opole, Poland

Abstract. Strength and deformation of soil are determined in laboratory tests through shear strength and compression tests. However they do not take into account the impact of dynamic vibration which can cause thixotropy or fluidization of soil. To allow for that, a method which modifies the course of laboratory tests has been proposed. It includes dynamic vibration into shear strength and compression tests of soil. It makes it possible to measure and classify the tested ground with respect to susceptibility to thixotropy caused by dynamic vibrations which lower strength parameters of water saturated silty soils. Earlier research of soil strength parameters tended to underestimate the impact of dynamic vibrations which sometimes constitute initiatory impulse for numerous landslides in rockmass, especially in flysch.

1. Preface

1.1 Introduction

Two physical and mechanical parameters decide about strength and deformation of soil: the first one being shear strength, the second one - soil compression. Shear strength is resistance of soil to static stress in the considered fragment. In order to determine shear strength, laboratory tests are held with intermediate shearing test apparatus or triaxial compression test apparatus, whereas soil compression is determined in oedometric tests. In the procedure of both the test types, a soil sample is made to reach its natural structure. The triaxial method reflects shearing conditions in the isotropic layer of high thickness, whereas the immediate shearing method additionally allows to reflect conditions found in narrow contact zones of soil layers of different characteristics, where landslides can be initiated. Results of the above mentioned tests are obtained with assumption of consistency or insignificant changes in the soil structure within the whole period of testing in a laboratory. They, above all, do not consider a possibility of changes in ground framework structure at

* Corresponding author: ukleja2012@gmail.com

the time of testing caused by dynamic impact. In site conditions the external dynamic impact can result from:

a) vehicles travelling on soil surcharge,
b) vibrations of machines and equipment,
c) vicinity of belt conveyor flights (for instance in open-pit mines),
d) shooting or rockbust (for instance in underground mining),
e) natural factors such as earthquakes or thunders and others.

The above mentioned factors caused by dynamic impact can significantly change stress conditions and form strength parameters of soil. It mainly considers soils of considerable content of silt fraction (silty soils), especially when they are found in water saturated environment. Dynamic vibrations can have multiple impact on changes in soil strength:

– increase in strength caused by self-thickening of soil can be observed in case of unsaturated coarse soils,
– in case of water saturated cohesive silty soils a phenomenon called thixotropy can occur . It causes a significant decrease in particle friction resulting in lessening or loosing contact strength, which is caused by pore pressure and results in transformation into liquid state,
– in case of water saturated non-cohesive soils with no drainage, liquefaction of soil can occur as a result of dynamic vibrations of cyclic load resulting in increase in pore water pressure, and in extreme cases to the extent when effective stress reaches zero.

The above discussed issues have partially been reflected in numerous publications, for example [1-7]. The test method described by Taslagy, Chan and Morgenstern [4 and 5] is equally worth consideration. They tested the effect of sand vibration [4]. The analysis was based on innovatory method of laboratory tests published in [5] in 2015. The idea of the solution has been referred to and used in this article.

1.2 Characteristics of influence of dynamic impact on soil

Soil vibrations caused by dynamic impact can initiate thixotropy in cohesive (silty) soils and liquefaction of non-cohesive soils. In non-cohesive soils liquefaction can be the result of vibrations in high water saturation conditions and lack of drainage. It is manifested by reduction of particle friction causing their mutual relocation. In this case we speak about quicksand occurring in sands and silty sands. It can be stopped by lowering water pressure in ground. In the case of cohesive soils, the framework of which is made of small particles with dominance of silt, long lasting structure disturbances can occur leading to changing its state to liquid. However, to make it happen, not only water saturation is needed, but also an impulse. This role can be taken by all sorts of vibrations. The changes are hardly reversible, sometimes impossible to reverse, despite ceasing of their initial reason.

Thixotropy consists in solation (into a liquid body of sol) of colloid saturated soils (small particles) from gelatinised state (gelatinous body) under the influence of quakes and vibrations or increase of pore water pressure (Fig. 1). Soils containing small particles of clay of colloidal sizes (< 0.0002 mm) are characterised by thixotropy. It also occurs when the ground framework is partially composed of particles significantly bigger than colloids, for instance silt or fine sand. [7].

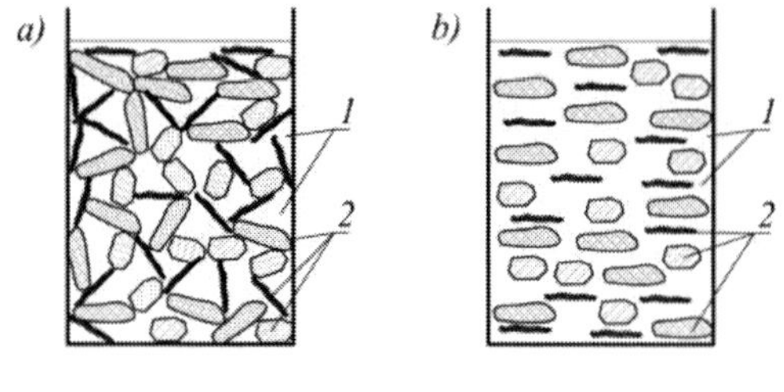

Fig. 1. Various forms of thixotropic structure.

Gelatinisation of soil is possible under the influence of mechanical stimuli when mineral particles in the solution get freedom of motion and become able to rotate. This phenomenon is partially reversible and the time of getting a stable state depends on the type of soil.

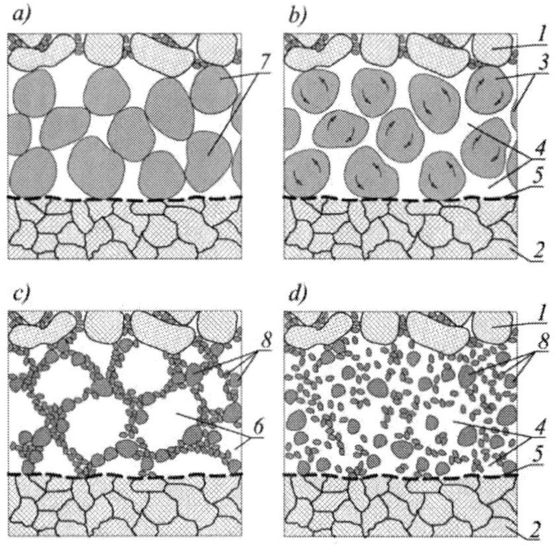

Fig. 2. Characteristic forms of soil sensitive to thixotropy and fluidization.

Second mutual clinging of neighbouring particles with their sharp edges is caused by forces of molecular and electrostatic attraction between them (Fig. 1a). Special susceptibility to thixotropy causing transition from firm to soft or even very soft state is observed mainly in silt, silty clay, sandy silt, sandy clay, silty sand, unwashed rock-mantle of all types of rock.

1.3 Influence of thixotropy and fluidization on rock mass stability

The phenomenon of thixotropy has major influence of ground parameters and, as a result, stability of rock mass. Decrease in ground strength can result in landslides initiated by vibrations, lasting longer than the vibrations themselves. It causes hazard of the rock mass stability loss in slopes and scarps or in ground under buildings [6]. However, to let that happen, the following favourable circumstances are necessary:

a) full water saturation of the potential slide layer of ground as a result of precipitation or rise of ground water level,
b) Simultaneous presence of dynamic stimuli, especially harmonic vibrations of given frequency and amplitude and duration, which make thixotropy or fluidization possible in the slide layer of ground,
c) Possible additional external load, for example with trees, wind, vehicles, etc.

Fig. 2 shows characteristic forms of soil structure sensitive to vibrations lessening or causing lack of friction among particles in the ground framework. Both the coarse soils (Fig. 2a and 2b) and the fine ones (Fig. 2c and 2d) can, in such conditions, become fluidised or undergo thixotropy caused by vibrations.

Sandy alluvia composed of smoothed grains, especially with addition of silt and clay are predestined to decrease in internal friction in their ground framework. This is caused by possibility of mutual rotation of individual grains (Fig. 2b). Silty soils of cellular structure (Fig. 2c) lose their apparent coherence to reach liquid state as a result of quakes Fig. 2d. Therefore it seems advisable to undertake laboratory tests of soils of similar structure to determine changes in their strength under the impact of dynamic vibrations.

Until now it was common to neglect these issues in laboratory tests despite their significance. Observations and analyses prove that the phenomenon appears to have considerable influence on stability and safety of numerous structures, such as scarps and slopes in open pit mines.

2. Methods applied to determine soil strength

2.1 Immediate shearing test apparatus

Coulomb interaction constitutes basis for determination of soil shear strength [7], however it does not take into consideration the impact of dynamic stimuli. As it results from this interaction, the amount of static stress constituting soil shear strength depends on friction resistance and cohesion:

$$\tau'_f = \sigma' \cdot \mathrm{tg}\phi' + c', \tag{1}$$

where: σ' – effective normal stress to the shear surface $(\sigma' = \sigma - u)$,
σ – normal stress to the shear surface,
u – water pressure in soil pores,
τ'_f – resistance of the ground against static stress (when Δu=0),
ϕ' – proper (effective) internal friction angle (when Δu=0),
c' – effective cohesion (when Δu=0).

2.2 Oedometer

To determine the rate of ground subsidence the module of ground compression M is used, determined from the following interaction:

$$M_o, M = \frac{\Delta\sigma}{\varepsilon} = \frac{\Delta\sigma \cdot h}{\Delta h} \tag{2}$$

where: M_o, M - oedometric module of primary compression, secondary,
$\Delta\sigma_i$ – increase in stress in the ground framework
$\varepsilon = \Delta h / h$ – relative deformation of ground sample,
h – sample height before tests,
Δh – subsidence of the sample as a result of stress change by $\Delta\sigma_i$.

3. Proposed modified method of testing

Modified testing equipment can ensure consideration of vibration impact on soil strength parameters. It should take into account transition of vibrations from the equipment to the tested soil sample. Such equipment was described in [4] and tested with success [5] with respect to sand, proving impact of vibrations on its strength. The range of amplitudes and frequency of vibrations applied at tests should reflect their parameters determined by measurements or expected for natural conditions. The same approach could be used to consider increase in subsidence caused by vibrations.

3.1 Modified tests with immediate shearing test apparatus

When there is a possibility for dynamic vibration to occur, one can determine their influence upon soil strength with use of methodology described in 3 with the following modifications to interaction (1):

$$\tau'_t = \sigma' \cdot \mathrm{tg}\phi_t + c_t \tag{3}$$

where: τ'_t – resistance of the soil against static stress under dynamic vibrations
σ' – effective normal stress,
ϕ'_t – effective internal friction angle with consideration of the effect of dynamic vibrations,
c'_t – effective cohesion with consideration of the effect of dynamic vibrations,.

It is therefore possible to determine minimal values ϕ'^{min}_t and c'^{min}_t for a number of tests with forced vibrations of various amplitudes and frequencies. When the obtained results are compared to soil strength parameters ϕ' and c', determined for soil not subjected to vibrations, one can reach a coefficient to correct the obtained results in a typical way (with no vibrations used). A coefficient of vibration impact υ can be a measure of potential influence of these vibrations on ground and its strength, and it is determined from the following interaction:

$$\upsilon = \frac{\sigma'^{max} \cdot \mathrm{tg}\phi'^{min}_t + c'^{min}_t}{\sigma'^{max} \cdot \mathrm{tg}\phi' + c'} \leq 1 \tag{4}$$

where: ϕ'^{min}_t, c'^{min}_t– minimum values of effective friction angle and cohesion, from all tests,
ϕ', c'– effective internal friction angle and cohesion characteristic of soil not subjected to dynamic vibrations,
σ'^{max} – maximum value of normal effective stress used in test.

The therefrom determined coefficient υ, used for the shear strength determined in tests, allows to consider its decrease as a result of vibrations initiating thixotropy (ϕ' and c' are multiplied by coefficient υ). It can be assumed that, as the author suggests, soils with relation to vibrations remain insensitive (υ =1), slightly sensitive ($1 > \upsilon \geq 0.8$), sensitive ($0.8 > \upsilon \geq 0.5$) and very sensitive ($\upsilon < 0.5$).

3.2 Modified oedometric test of soil compression

As a result of vibrations in the vicinity of a structure, additional deformations related to soil thickening in response to them can be observed. It is possible to take the phenomenon into account in laboratory tests. To do so one needs to apply an apparatus modelling close to natural dynamic vibrations in soil compression oedometer. The following interaction is then obtained:

$$\xi = \frac{M^{min}_t}{M} = \left(\frac{\Delta\sigma \cdot h}{\Delta h^{max}_t}\right) : \left(\frac{\Delta\sigma \cdot h}{\Delta h}\right) = \frac{\Delta h}{\Delta h^{max}_t} \leq 1 \tag{5}$$

where: M – oedometric compression module determined with no vibrations used,

M_t^{min} – minimal oedometric compression module for various vibration types,
Δh – change to sample height caused by increase in stress by $\Delta\sigma$ without vibrations,
Δh_t^{max} – maximum change of sample height as a result of change in stress by $\Delta\sigma$ and for dynamic vibrations of various parameters,
ξ – coefficient of dynamic vibrations impact on soil subsidence .

The range of tested values $\Delta\sigma$ and related Δh need to be determined for adequate sections of primary and secondary cohesion curve. The same $\Delta\sigma$ intersections should be applied for test results with and without dynamic vibrations impact.

Coefficient ξ should be used to determine subsidence in soils susceptible to dynamic vibrations. The calculated subsidence values for soil not subjected to dynamic vibrations should be divided by the coefficient ξ.

4. Conclusions

Vibrations occurring in water saturated soils have two sorts of unfavourable effects; they cause decrease in shear strength and increase in subsidence. This happens because of restructuring of ground structure, especially in soils susceptible to thixotropy or fluidization. A method modifying course of laboratory tests has been proposed to take the above phenomena into account. It considers using forced vibrations to test shear strength and compression of soil which is especially useful for bedded grounds. The proposed method allows to measure and classify given soil with regard to its susceptibility to fluidization and thixotropy as well as self -thickening , caused by dynamic vibrations. The hereby described coefficients allow to take into account weakening of ground as a result of possible dynamic vibrations in soils susceptible to thixoctropy, fluidization or self-thickening, at the time its stability is being calculated. The method allows to determine the indirect values of ground strength parameters with consideration of measured or potential dynamic vibrations. It also makes it possible to determine the values of strength parameters by means of analogies with use of coefficients taking into account the least favourable impact of dynamic vibrations.

References

1. B. Fatahi, S. Tabatabaiefar S, *Effects of Soil Plasticity on Seismic Performance of Mid-Rise Building Frames Resting on Soft Soils*, Advan. in Struct. Eng. **17**, 1387 (2014)
2. D. Park, B. Kutter, J. DeJong, *Effects of Thixotropy and Cement Content on the Sensitivity of Soft Remolded Clay*, Journ. of Geotech. and Geoenv. Eng. **141**, 2, DOI: 10.1061/(ASCE)GT.1943-5606.0001221, (2015)
3. D. Park, B. Kutter, *Static and seismic stability of sensitive clay slopes*, Soil Dynam. Earthqu. Engin. **79**, 118 (2015)
4. A. Taslagyan, D. Chan, N. Morgenstern, *A Direct Shear Apparatus With Vibrational Loading, Geotechnical Testing Journal* **38**, 1 (2015)
5. A. Taslagyan, D. Chan, N. Morgenstern, *Effect of vibration on the critical state of dry granular soils*, Granular Matter, Springer **17**, 687 (2015)
6. J. Ukleja, *Geotechniczne aspekty stabilizacji obszarów zagrożonych osuwiskami*, (Ofic. Wyd. Pol. Opol. , Studia i Monografie, **369**, Opole, 2013)
7. Z. Wiłun: *Zarys geotechniki*, (Wyd. Kom. i Łączn., Warsaw 1982)

Assessment of slope stability

Slavka Harabinova[1,*] and *Eva* Panulinova[1]

[1]TU Košice, Faculty of Civil Engineering, Vysokoškoľská 4, 042 00 Košice, Slovakia

Abstract. The paper deals with the assessment of slope stability on the road II/595 near the village Zlatno before and after the landslide caused by floods in 2010. The calculation of factor of safety was made using GEO 5 software. For proposal comprehensive assessment and possible remedial work is necessary to know the geological conditions and choose the appropriate method for assessing slope stability. The critical factors of safety have been determined by Petterson, Bishop and Sarma Methods. After finding the most unfavorable slip surface a proposal of remediation measures has been made. The paper presents various solutions to ensure slope stability of the road. The most appropriate remedial action is Variant III - reinforced slope due to its greater factor of safety.

1 Introduction

Currently, frequent flood events are taking place in our area. Floods are associated with rapid erosion processes, which involve the destruction of land resources and consequently may lead to landslides of drenched soil. In 2010, severe floods hit most of the territory of the Slovak Republic. It was the biggest flooding in the last 50 years.

Climatic factors combined with the erosion activity watercourses and groundwater are major causes of slope deformations. In 2010, a result of exceptional rainfall and flood situation increased the number of slope deformations. Continuous rain caused significant elevations in almost all watercourses and floods. Slope that becomes saturated with water might be unstable and prone to landslide.

2 Landslide on the road II/595

The penetration of rainwater on the right embankment slope of the road body was the consequence of a landslide on the road II/595 near the village Zlatno. The main cause of the landslide was primarily long-term rainfall in the months of May and June, when daily rainfall was 25-50 mm throughout Slovakia, but in the north and east of Slovakia was up to 80 mm. Slope saturation by water is a primary cause of landslides.

The area of interest is located in the village Zlatno. Road section is guided in a slope. In 2010 a landslide occurred in this area (from 24.886 to 24.932 km). Asphalt layer of road was broken and slide down after the landslide activity (Fig. 1). Landslide narrowed the width of the roadway, thereby reducing the security of vehicle passages in this section [1].

* Corresponding author: slavka.harabinova@tuke.sk

According to the geomorphological division of Slovakia the territory is located in the Western Carpathians, in the Slovak Ore Mountains. The territory belongs to the Banská Bystrica region of the district Poltár. In hydrological terms, this territory belongs to the basin of the Ipeľ River and is drained by a tributary of the Poltarica River. Climatically the area is evaluated as moderately warm and moderately wet [2].

Fig. 1. Situation and photo documentation of landslide.

From a geological point of view, subsoil is formed by Paleozoic rocks – granitoids, which are unevenly cover deluvial sediments. They are sandy clays and clayey sands with fragments of rocks. Two boreholes JZ-1 and JZ-2 have been used for geotechnical investigation [2]. Geological profiles are shown in Fig. 2.

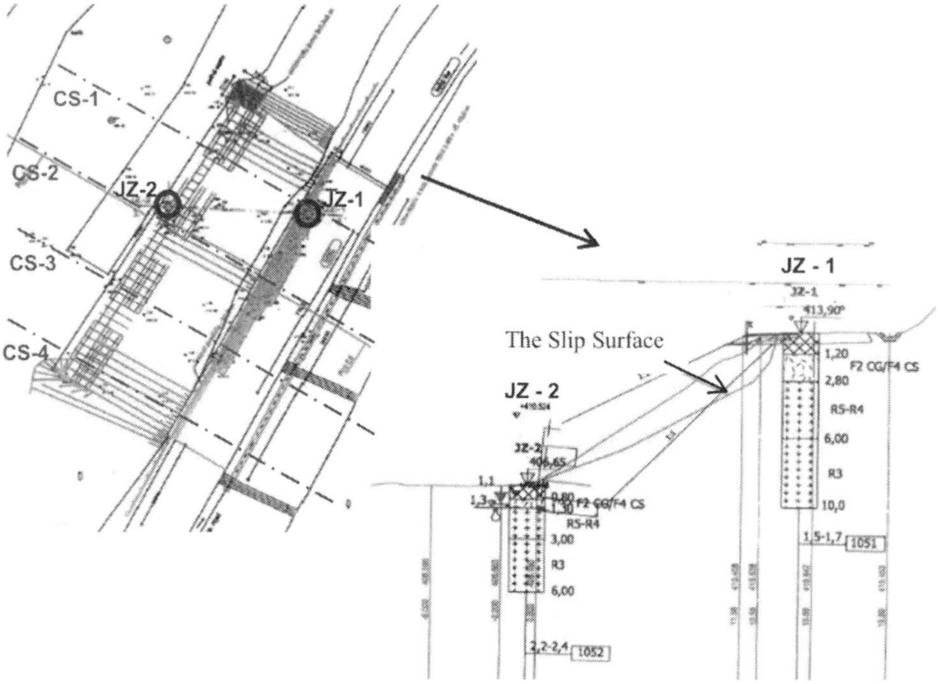

Fig. 2. The location of boreholes and geological profile of boreholes JZ-1 and JZ-2.

Soil classification was made according to STN 72 1001 [8] and in accordance with EN 1997-1 [7]. The values of the geotechnical characteristics are given in the Table 1 and 2 [2].

Table 1. The geotechnical parameters of soils.

Properties	Group F2 – CG (Gravelly Clay) Group F4 – CS (Sandy Clay)	Group S5 – SC (Clayey Sand)
Poisson's ratio υ [-]	0.35	0.35
Unit weight γ [kN.m^{-3}]	19.0	18.5
Deformation modulus E_{def} [MPa]	6	9
Total stress parameters – cohesion c_u [kPa]	55	-
Total stress parameters – angle of friction φ_u [°]	0	-
Effective stress parameters – cohesion c_{ef} [kPa]	13	8
Effective stress parameters – angle of friction φ_{ef} [°]	25	27

Table 2. The geotechnical parameters of rocks.

Properties	Weathered rock layers Group R5-R4	Weathered rock layers Group R3
Poisson's ratio υ [-]	0.35	0.35
Deformation modulus E_{def} [MPa]	6	9

3 Assessment of slope stability on the road II/595

Currently, there are several methods of calculating slope stability [3-5]. Most often there are the methods based on the assumption that failure occurs towards slip surface. The shape of slip surface depends mainly on the physical and mechanical properties of soils or their arrangement in the profile. The stability analysis in question takes into account two basic principles. The first is an assumption that the slip surface developed will be circular (Petterson and Bishop Methods) and the second principle is that the slip surface will be polygonal (Sarma Method) [6, 10, 11].

The slopes of the road are made up of fine-grained soils and rocks. That is the reason why Petterson, Bishop and Sarma Methods were selected for calculation and assessment of slope stability of the road. Calculation and assessment of slope stability on the road II/595 near the village Zlatno was carried out by using program "Slope stability", which is a sub-program of GEO 5 by company FINE Ltd [9].

The slope stability analysis was carried out in compliance with STN EN 1997-1, Design Approach 3 [7]. According to EN 1997-1 [7] was assessed stability of slopes according to the "limit states theory". A factor of safety was calculated as the ratio of the forces resisting movement (thus ensuring the slope stability) to those driving movement (thus threatening the slope stability), i.e. the ratio between the passive and active forces. In general, if the factor of safety of a slope is within the interval between 0 and 1.0, the slope is actively unstable. The value over 1.0 indicates that the slope is considered stable. The calculated factor of safety has been compared to limit value of stability degree.

Assessment of slope stability on the road II/595 near the village Zlatno was made on the landslide place, in four cross-sections CS-1, CS-2, CS-3, and CS-4 (from 24.886 to 24.932 km). The load on the construction of the road was 16.8 kN/m^2 and the axle load was 115 kN. Calculation of slope stability was realized in several variants.

Variant I - The original condition, the dry state (i.e. before the landslide). In this variant, the groundwater level was considered as in geological profile (JZ-1 a JZ-2).

Variant II – Condition after landslide, the saturated state (i.e. after the flood). In this variant with maximum groundwater level was considered, i.e. situation that caused the landslide.

Whereas the landslide has large dimensions, it is possible as a remedial measure to propose anchoring of reinforcing elements into the body of the slope. As a first method of remediation measures to slope stability of the road reinforcement of slope with geosynthetics materials was proposed.

Variant III - Reinforced slope, i.e. the slope reinforcement with geogrid. For remediation of slope, geogrid with tensile strength 173 kN/m was used. By static calculation in the GEO 5 program, the required length and number of geogrids has been determined.

For comparison as another method of remediation of slope stability of roads a **Variant IV** – Reinforced slope with ground anchors was chosen. Ground anchors can be used to improve the stability of a slope for road applications. Ground anchors can be used also as a remedial measure for landslide slope. The proposal was made by calculating in the GEO 5 program. Length of ground anchor is 10 m, bond length of ground anchor is 1.5 m and anchorage strength is 150 kN.

The calculated factors of safety (F_s) for these variants are shown in Table 3.

Table 3. Calculated factors of safety for Variant I - Variant IV.

The cross-section	**Variant I** Factor of safety Fs and Evaluation			**Variant II** Factor of safety Fs and Evaluation		
	Petterson Methods	Bishop Methods	Sarma Method	Petterson Methods	Bishop Methods	Sarma Method
CS-1	0.88	0.97	1.04	0.60	0.77	0.81
CS-2	0.97	1.06	1.24	0.65	0.79	0.89
CS-3	0.89	0.99	1.12	0.61	0.77	0.85
CS-4	0.86	0.96	1.04	0.58	0.74	0.80
The cross-section	**Variant III** Factor of safety Fs and Evaluation			**Variant IV** Factor of safety Fs and Evaluation		
	Petterson Methods	Bishop Methods	Sarma Method	Petterson Methods	Bishop Methods	Sarma Method
CS-1	1.42	1.98	2.14	1.12	1.69	1.83
CS-2	1.46	1.99	2.18	1.09	1.54	1.63
CS-3	1.37	1.86	1.97	1.17	1.69	1.78
CS-4	1.53	2.01	2.14	1.06	1.56	1.67

Based on the calculations and the results listed in Table 3 it can be seen that the stability of slopes did not satisfy the assessment of slope stability before the flood situation (Variant I – Petterson Method).

Sarma Method is more appropriate for this calculation, because the slope is formed of rock. The calculated factors of safety by Bishop and Sarma Methods point to the fact that at any overrun load is an increased risk of landslide (Variant I).

Due to rain and infiltration of rainwater into the slope (Variant II), there was a landslide, as confirmed by the calculated factors of safety for all cross-sections.

Both the proposed remedial measures (Variant III and Variant IV) improve the stability of the slope. Variant III is more appropriate than Variant IV, because it has a greater factor of safety. In terms of the factor of safety it is preferable Variant III.

 DOI: 10.1051/matecconf/201710700031

In Fig. 3 you can see the arrangement of geogrid in the slope (Variant III). Variant IV is shown in Fig. 4.

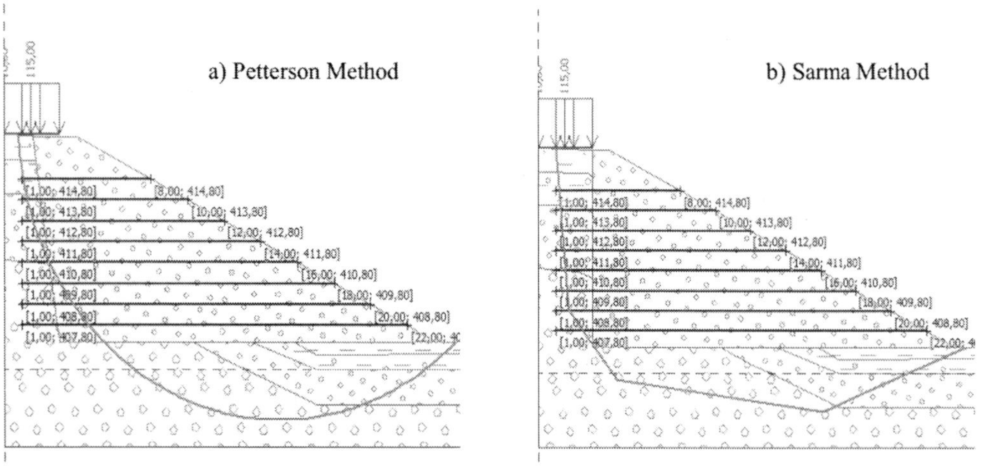

Fig. 3. Variant III - Reinforced slope.

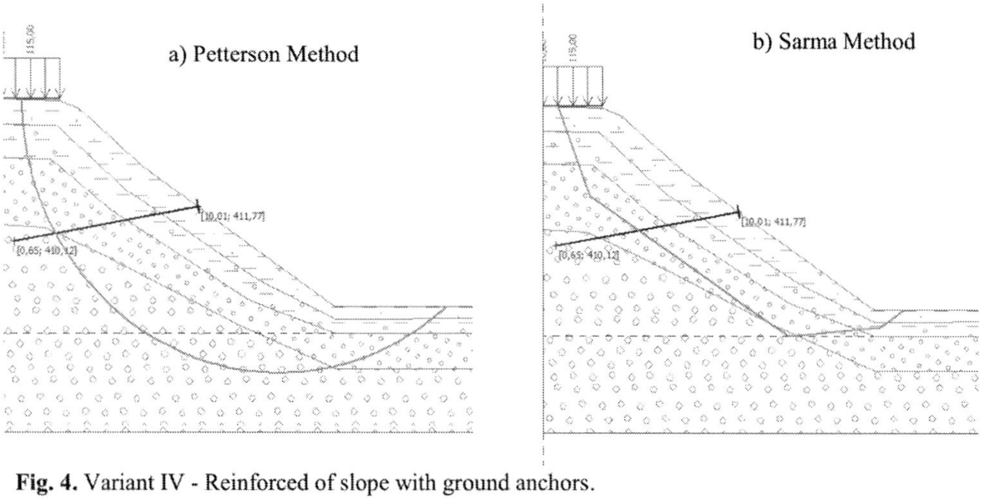

Fig. 4. Variant IV - Reinforced of slope with ground anchors.

4 Conclusion

Calculation of slope stability of road body was realized in four variants, in which we analysed possibilities to solve and improve the stability slope after the flood. Assessment of slope stability on the road II / 595 has been done before and after the landslide, caused by the flood (Variant I - The dry state and Variant II - The saturated state). In view of the fact that the body of the road is formed of rocks groups R3 to R5, it is more appropriate for calculation of factor of safety to use Sarma Method.

The calculated factors of safety by Sarma Method (Variant I) confirmed that any overrun load is an increased risk of landslide. As confirmed by the calculated factors of safety for Variant II, due to rain and infiltration of rainwater into the slope there was a landslide.

The proposal of remediation measures (Variant III and Variant IV) have been made after finding the most unfavorable slip surface. Variant III was proposed as a reinforced of slope with geogrid and Variant IV was proposed as a reinforced of slope with ground anchors. By comparison of the factors of safety, Variant III is the best one.

Roads are considered as significant structures and therefore it is necessary to pay high attention to the design and assessment of these constructions. Because of their importance, security and reliability throughout their lifetime remains the top priority.

This work was supported by the Scientific Grant Agency of the Ministry of Education of Slovak Republic and the Slovak Academy of Sciences under Project VEGA 1/0477/15.

References

1. P. Maretta, J. Chrensc, Road II/595 Veľká Ves – Kokava nad Rimavicou, km 24.886 00 – 24.932 00 - Remediation of landslide (2011)
2. INGEO-ighp, s.r.o, Remediation of landslides on the road (2011)
3. G. Bulatov, Y. Ibraeva, P. Tarasevskii, X. Rakova, D. Zejak, MATEC Web of Conferences **73** (2016)
4. N. Vatin, N. Lavrov, N. Korzhavin, Applied Mechanics and Materials, 641–642 (2014)
5. G. Bulatov, N. Vatin, D. Nemova, Y. Ibraeva, P. Tarasevskii, Applied Mechanics and Materials, 725–726 (2014)
6. E. Panulinova, S. Harabinova, Advanced Material Research, **969,** 245-248 (2014)
7. STN EN 1997-1 Eurocode 7. Geotechnical design, Part 1: General rules (2005)
8. STN 72 1001 Classification of soils and rocks (2010)
9. GEO 5 sowtver © Fine Ltd.
10. D. H. Cornforth, *Landslides in Practice, Investigation, Analysis, and Remedial Preventive Options in Soils* (John Wiley & Sons, 2005)
11. M. Cala, D. Flum, A. Roduner, R. Ruegger, S. Wartmann, *TECCO® Slope Stabilization System and RUVOLUM® Dimensioning Method* (Romanshorn, Switzerland, 2012)

MATEC Web of Conferences **107**, 00032 (2017) DOI: 10.1051/matecconf/201710700032
DYN-WIND'2017

Application of inclined injection piles as a means of seismic protection of slope

Evgenii Kolosov [1], *Kamil* Agishev[2] and *Natalia* Kolosova [1,*]

[1] Peter the Great Saint-Petersburg Polytechnic University, Institute of Civil Engineering, 195220 St. Petersburg, Russia
[2] TÜV International, 190000 St. Petersburg, Russia

Abstract. The using of inclined injection piles in an area with high seismic activity is described in this work with the purpose of their application in the construction of various facilities in seismic active areas. In this paper the authors define the possibility of their using for the stability of the slopes and slope construction in the earthquake conditions. The authors analyze sustainability slope both with the use of inclined injection piles and without them. They offer a methodology for determining the embedment of inclined injection piles into the thickness of slope.

1 History of a pre-calculation of CFA piles in seismically active regions

CFA piles - a kind of bored piles, the main distinguishing features are:

- small diameter - from 0.13 to 0.35 m;
- greater flexibility (with a length of piles over 10 m);
- used material (cement-sand mixture);
- process of molding of the concrete mix after filling the wells with the 0.2-0.4 MPa pressure.

In the domestic construction industry CFA piles have appeared in the middle of 1980s. They are mostly used to strengthen the foundations of buildings [1]. Currently, the range of use has significantly expanded; these piles use to protect against landslides, strengthening embankments, device enclosing walls in the ground, in the construction in dense urban areas and for other constructions [2-4].

The ability of the piles to resist both vertical and horizontal loads determined their application in seismically active areas [1, 5].

However, it should be noted that in the design of structures using CFA piles mainly used calculation methods for bored piles used in bridge construction. These piles compared to bored piles considerably differed by large diameter and their calculation methodology was quite complex and time-consuming [5, 6].

As shown by field tests of CFA piles the value of the bearing capacity of soil is much higher in various geotechnical conditions on the combined effect of vertical and horizontal

* Corresponding author: nb.kolosova@yandex.ru

forces [1]. Preliminary bearing capacity of piles at the construction site exceeded the bearing capacity of the pile calculated according to the procedures of the building regulations (excluding manufacturing technology piles), approximately 1.7 times. There were also significant differences in the determination of the incorporation of root anchor into the body of the slope [1].

The use of CFA piles as a stabilizing structure studied insufficiently in terms of earthquake. In this paper, the authors first conducted a comparative performance evaluation of CFA piles in normal conditions and under seismic activity.

As a design tool the authors use the software package PLAXIS 2D, which has demonstrated quite good results in the solution of various problems of soil mechanics [6,7].

2 Calculation of inclined CFA piles with application PLAXIS 2D complex program

2.1 Calculation of slope stability in seismic conditions.

The calculation of slope stability was carried out for two combinations of loads (including seismic component and without it). The cross section of the CFA pile is shown in Fig.1. Here is one of typical size roof bolt 40/18 mm and the CFA pile diameter equal to 150 mm.

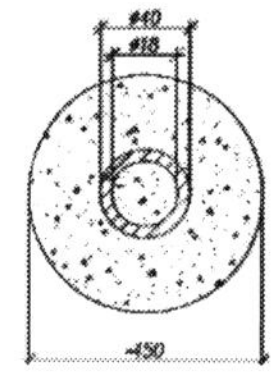

Fig. 1. The cross section of the CFA pile.

Mohr-Coulomb model was used as the estimated model. The stability calculation was performed based on the seismic forces that cause additional horizontal and vertical strains in the ground massif. And the role of vertical component of the seismic force was relatively small and was not taken into account when assessing sustainability.

Seismic force Q_c in accordance with [8] is defined as the ratio of the weight of the soil compartment, which is undergoing a seismic impact:

$$Q = \mu P \qquad (1)$$

μ - coefficient of dynamic seismicity (equal to 0.1 for areas with 9 points seismicity).

The direction of seismic force coincides with the direction of shear force in the corresponding compartment. As settlement scheme were used a real soil slope in one of the southern regions of the country exposed to seismic effects (Fig.2). The estimated seismic intensity was 9 points.

In the upper part of the slope was the retaining wall. The slope surface was represented by a broken line. The upper part value of the slope (CFA piles from №1 to №15) was 43° in relation to the horizon. The lower part value of the slope (from №15 to №19) was 28° in relation to the horizon. The height mark of the earth's surface was changing from +610.000 (pile №1) to +586.600 (pile №19). The total horizontal distance of the slope was 50.0 meters.

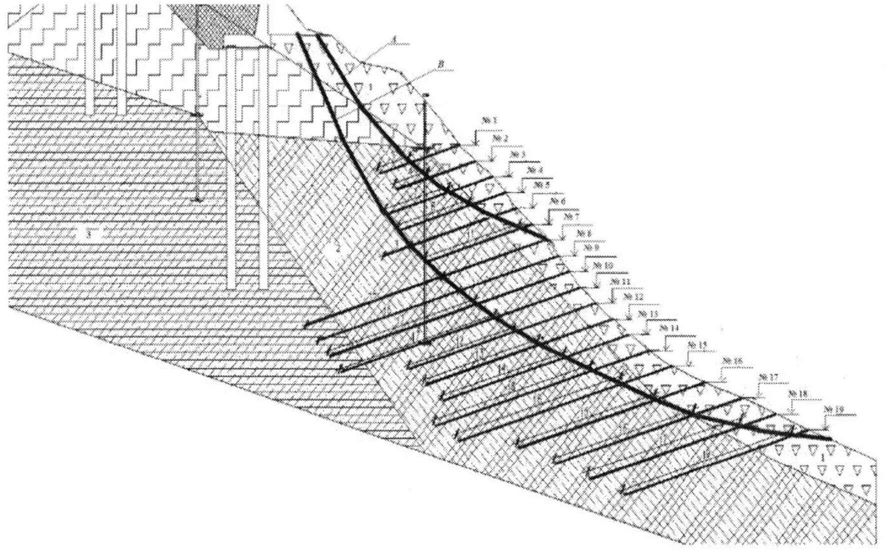

Fig. 2. The estimated scheme of the slope. *A* – collapse curve without seismic load. *B* – collapse curve with taking into account seismic loads.

Composing the thickness of the slope soils represents three types: 1 – clayey medium fine gravel from 1.5 m to 4.0 m; 2 – clayey medium fine gravel fragments with low strength; 3 – claystone with very low strength. The main physical and mechanical characteristics are presented in Table 1.

Table 1. The main physical and mechanical characteristics of soil slope.

№	Name of soil	Density	Porosity coefficient	Modulus of deformation	The angle of internal friction	Cohesion	The number of plasticity	Fluidity value
		ρ	e	E	φ_1	c_1	I_p	I_l
		[kg/м³]		[MPa]	[°]	[kPa]		
1	Clayey medium fine gravel	2070	0.609	29.5	23.9	12.9	0.123	0.173
2	Clayey medium fine gravel fragments with low strength	2060	0.538	31.7	18.3	17	0.08	0.12
3	Claystone with very low strength	2110	0.5	-	-	-	0.09	< 0

Calculation of slope stability is determined in accordance with p.p.5.9 [8]. Estimated value of the soil mass of generalized resistance to the effects of shear should be determined based on the condition that the ratio between the normal σ_n and tangential τ_n stresses across the sliding surface of the corresponding limit state sliding wedge meets the following condition

$$\tau_n = \sigma_n \text{tg}\varphi_1 + c_1 \quad (2)$$

φ_1 and c_1 – values of internal friction angle and specific cohesion of the soil in which occurs the shift of the soil.

The calculation was performed in the PLAXSIS 2D software package by reducing the strength characteristics of the soil. This process is called Phi-c reduction and present as a separate type of calculation. If use algorithm Phi-c reduction the strength parameters of the soil $\text{tg}\varphi$ and c were subsequently reduced until destruction of the structure.

The coefficient of slope's stability was determined as the ratio of the real resistance of the soil shear to the minimum shear strength required maintaining the balance.

According to the [8] for the structures of the normal level of responsibility the normalized slope stability factor is:

K_S = 1.15 - at the basic load combinations,

K_S = 1.05 - in special combinations of loads.

In our case, the basic mix includes a static state of the slope, taking into account the load on the road, and water saturation of the soil, and a special combination - seismic loads.

At first was determined a coefficient of slope stability for static condition without use of CFA piles. Its value amounted to K_S = 0.95, which is less than the permissible values. This meant that the slope is not stable.

In the case of using CFA piles, the factor of stability for the basic load combinations amounted to K_S = 1.29, and in special combinations - K_S = 1.1. Realized calculation objectively demonstrated the feasibility of using CFA piles as strengthening of the structure during operation of the slope in seismic conditions.

2.2 Calculation of bearing capacity and the depth of embedment of the injection anchor into the ground

In addition to the calculation of slope stability was carried out the calculation for the determination of the anchor bearing capacity on the ground and the required value of the root sealing into the soil with the influence of seismic forces.

The calculation was performed for two load cases:

– without considering seismic forces (upper curve slide),

– with the seismic component (lower curve slide).

The bearing capacity of injection anchors F_d is determined by the known formula:

$$F_d = \pi D_k l_k (1 + \sin\varphi_1) \cdot (\sigma_s \text{tg}\varphi_1 + c_1) K_p \gamma_c , \quad (3)$$

D_k - the diameter of the anchor root sealing, l_k - the length of the anchor root, φ_1 - the angle of internal friction of the soil, c_1 - specific adhesion of the soil, σ_s - the stress of the ground by the side surface, K_p - coefficient depend on the ratio of borehole diameter to the diameter of the sealing, γ_c - the coefficient of working conditions.

The stress of the ground at the side surfaces σ_s is determined depending on the specific weight of the soil subject to the weighing action of water γ_1, depth of the center of the seal anchor from the surface of soil coefficient of lateral earth pressure ξ_0 and the angle of the anchor to the horizontal α .

As initial values taken the following values: $\gamma_1 = 23.52$ kN/m^3; $c_1 = 17.0$ kPa; $\varphi_1 = 18.3$; $\xi_0 = 0.55$; diameter of the root sealing $D_k = 0.12$m, the anchors angle to the horizon equal to $\alpha = 15^o$ and $K_p = 1.0$; $\gamma_c = 0{,}9$ – the adopted coefficient of working conditions.

The value of embedment root of the injection anchor in the body of the slope was determined from the formula (3). The main results obtained in the calculation shown in Table 2.

Table 2. Bearing capacity and drilling depth of injection anchor.

Row number	Typical size roof bolt	Distance between anchors	Anchor's actual length outside sliding surface	Calculated embedment length (normal load)	Calculated embedment length (seismic load)	Bearing capacity in ground (seismic conditions)
	d	Δl	l_{kf}	l_{k1}	l_{k2}	F_d
	mm/mm	m	m	m	m	kN
1	40/18	2.0	3.00	2.00	8.10	83.3
2	40/18	2.0	4.00	1.00	8.10	110.7
3	40/18	2.0	7.00	4.00	8.70	203.8
4	40/18	2.0	8.00	6.00	8.80	235.2
5	40/18	2.0	12.00	5.00	8.80	352.8
6	40/18	2.0	11.00	4.00	14.70	488.0
7	40/18	1.5	15.00	14.00	6.70	361.6
8	40/18	1.5	16.00	5.00	19.53	906.5
9	40/18	1.5	17.00	5.00	20.00	982.9
10	40/18	1.5	17.00	5.00	21.00	1026.1
11	40/18	2.0	12.00	4.00	20.40	706.6
12	40/19	2.0	13.00	4.00	16.20	623.3
13	40/20	2.0	14.00	4.00	17.80	732.1
14	40/21	2.0	18.00	4.00	15.70	844.8
15	40/22	2.0	18.00	4.00	15.70	844.8
16	40/23	2.0	15.00	4.00	14.60	662.5
17	40/24	2.0	16.00	5.00	11.00	560.6
18	40/25	2.0	17.00	3.00	8.40	483.1
19	40/26	2.0	19.00	2.00	8.70	554.7

The table shows data about actual embedment length for each anchor pile and calculated length under normal load and seismic component.

3 Conclusion

The first: calculation of slope stability using the software PLAXIS 2D showed that:

- The normative factor of slope stability was not provided below the retaining wall during the combined loads and seismic influence.
- To ensure the slope stability below the retaining wall proposed fixation of the slope by CFA piles with length from 9 to 27 meters with step 2×2 and 1.5×2 meters.

The second: comparing calculating data of the injection anchor sealing depth in the ground with the actual values of the embedment depth determined that:

- The actual length of each anchor embedment much higher (by 1.5 - 9.5 times) compared to the estimated for normal operating slope conditions.
- The value of anchor embedment into the ground is insufficient for piles № 1 - №4; №6; №8 – №13 in seismic conditions.
- The actual value of anchor embedment into the ground exceeds the calculated value for piles №5. №7, and №14 – №19 when working in seismic conditions. Exceeds 2 or more times for piles №7, №18 and №19.

In the conclusion should be noted the effective use of CFA piles as increasing slope stability elements with various combinations of load.

According to the author's opinion, CFA piles are possible use not only for strengthening the slope. Taking into account given calculation results, they can be used in other building structures.

References

1. V. N. Paramonov, *Bearing capacity of piles in geotechnical conditions of Saint-Petersburg*. Proceedings of the International seminar on soil mechanics, 250-252, (2000)
2. A.S. Scalat, *Evaluation of existing building in Israel for seismic hazard*, Wiley Inter Science. Earthquake Engineering Dynamics **6**, 3, 317- 325 (2007)
3. M. Pils, S. Parolai, M. Stupazzini, R. Paolucci & J. Zschau, *Modelling basin effects on earthquake ground motion in the Santiago de Chile basin by a spectral element code*, Geophysical Journal International **187**, 929-945 (2011)
4. G. Ya. Bulatov, N. B. Kolosova, E. S. Kolosov, A. B. Teplov, *Dependence the bearing capacity of the pile on the shapes of cross-section*, Construction of unique buildings and structures **10, (15),**15-24 (2013)
5. V. V. Konyushkov, V. M. Ulitsky, *Estimation of the bearing capacity of CFA piles*, Bulletin of civil engineers **2**, 52-57 (2007)
6. A. K. Bugrov, E. S. Kolosov, *Practical methods of taking into account the nonlinear deformation of soils*, IV International scientific-practical conference, COLL Actual problems of modern Foundation with energy saving technologies Penza, 14-21 (2014)
7. PLAXIS Software. https://www.plaxis.com/
8. SNiP 22-02-2003, Engineering protection of territories, buildings and structures against dangerous geological processes. Basic provisions (2003)

Modelling the pile load test

Ľubomír Prekop[1,*]

[1]Slovak University of Technology, Faculty of Civil Engineering, Department of Structural Mechanics, Radlinskeho 11, 810 05 Bratislava, Slovakia

Abstract. This paper deals with the modelling of the load test of horizontal resistance of reinforced concrete piles. The pile belongs to group of piles with reinforced concrete heads. The head is pressed with steel arches of a bridge on motorway D1 Jablonov - Studenec. Pile model was created in ANSYS with several models of foundation having properties found out from geotechnical survey. Finally some crucial results obtained from computer models are presented and compared with these obtained from experiment.

1 Introduction

This paper deals with modelling the horizontal load test resistance of reinforced concrete piles. The pile is a part of a group of piles with reinforced concrete heads. The head is pressed with steel arches bridge construction on motorway bridge D1 Jablonov – Studenec.

1.1 Description of the object

Bridge structure consists of three parts (Fig. 1), its total span is 668.6 meters:

- 3 pre-stressed reinforced concrete clearways of spans 26.0 m; 32.0 m and 26.3 m; total span of 84.3 meters;
- 7 major steel arch bridge fields with spans of 2×60 m, 2×70 m and 3×80 m; comprehensive length of 500 m;
- 3 pre-stressed reinforced concrete clearways of spans 26.0 m; 32.0 m and 26.3 m; total span of 84.3 meters.

Fig. 1. General view of the bridge structure.

* Corresponding author: lubomir.prekop@stuba.sk

The assessed pile is located in 7th pillar supporting steel arches of the bridge (Fig. 2, Fig. 3). In the preliminary stage, followed by a detailed engineering geological survey, 10 exploring boreholes as a basis of 7 subsoil layers models were realized. From them the calculation model of the subsoil at the site of exploration was used according to drilled borehole 205-7/V 7. During the phase of control survey overall 8 boreholes were completed.

Fig. 2. View of steel bridge arch.

Fig. 3. Detail view of steel bridge arch and bridge support.

2 Load test of a pile

Different load tests on horizontal direction of piles were conducted. The purpose of the tests was to create a basis for any design modifications of the pile foundation of the bridge and verify the bearing capacity of piles for horizontal loads and assess the deflection - load relationship in the given geological conditions as well. A detailed description can be found in [1]. In order to verify the bearing capacity of piles loaded by horizontal forces, the test of lateral resistance was performed on a pair of piles. The test was carried out by mutual mashing of piles by loading device. Load assembly was formed with beams, jacks and power measuring box.

2.1 Foundation conditions

Geological profile at the site of tested pile is as follows:

- 0.0 – 8.5 m proluvial sandy clay with intermediate plasticity (F4/CS),
- 8.5 – 10.5 m paleogene - heavily weathered and decomposed claystone (R5, F6/CI),

DOI: 10.1051/matecconf/201710700033

- 10.5 – 13.5 m paleogene - heavily weathered and decomposed claystone (R5-R6, F6/CI),
- 13.5 – 14.7 m paleogene - replacing of 0.1-0.3 m thick layers of heavily weathered sandstone (R5, R4).

Ground water encountered in cracked claystone was found at a depth of 13.6 meters.

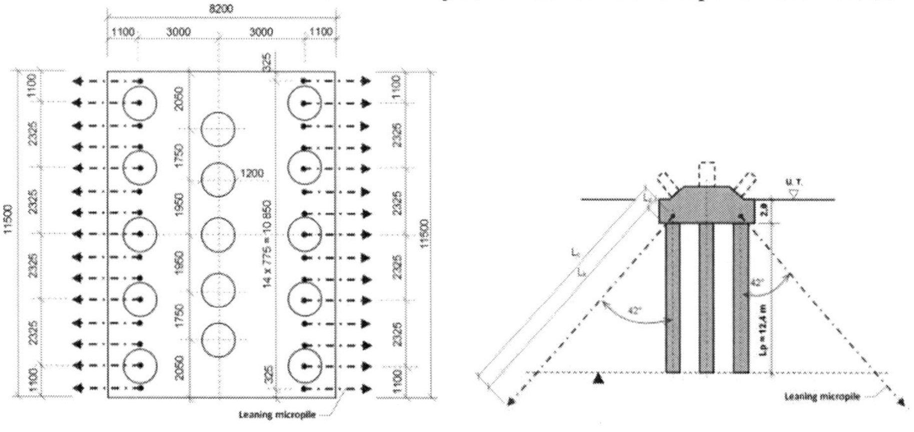

Fig. 4. Pillar Nr.7 – suggested arrangement of piles and piles of oblique.

3 Numerical model of structure

A numerical model of pile was created as 3D model using software ANSYS (Fig. 5). The nominal diameter of the pile was 1.2 m and the length of 12.4 m. Two stages of loads were selected: 280 kN and 420 kN.

Reinforced concrete piles were modelled using SOLID65 elements and surrounding ground mass by SOLID45. These elements were assigned by properties of the individual layers of soil. Contact of piles with geological environment was modelled using TARGE170 and CONTA173 elements using the contact surface - surface [3].

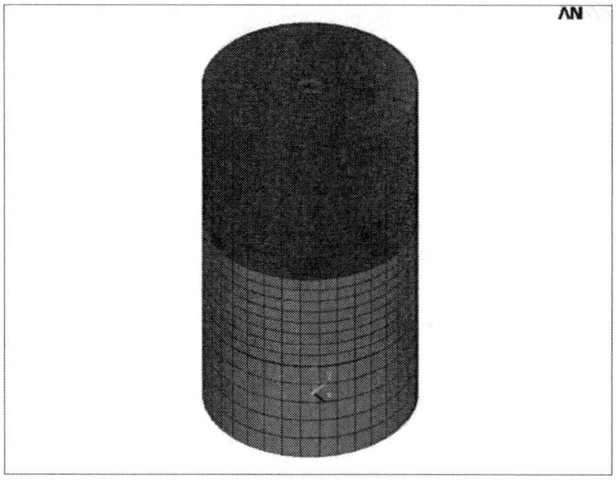

Fig. 5. FEM model.

4 Results

The results obtained for the horizontal deformation of the piles are shown in the graphs in Fig. 6. From the deformations it can be seen that the coincidence of the measurement results and the numerical model is very good.

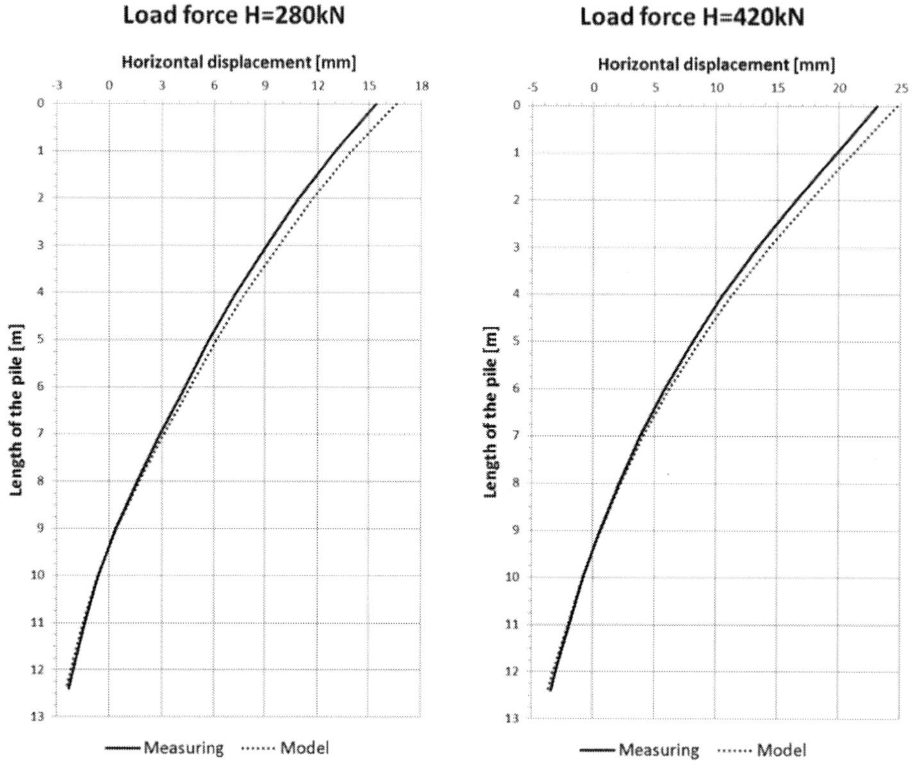

Fig. 6. Comparison of horizontal displacement of the pile.

5 Conclusion

As it is clear from final diagram, the results obtained by numerical modelling correspond very well with the measured values from the experiment. Higher values of horizontal deformation of the numerical model can be justified because of rather complicated modelling of contact piles – multi layered soil.

This paper was supported by Grant Agency VEGA, project No. 1/0544/15.

References

1. Report of load tests of piles, Project: Motorway D1 Jablonov - Studenec, building SO 205 Bridge on D1 at km 1 650 - pillar P7, VUIS - Foundation engineering, Ltd., Bratislava, August, In Slovak (2009)
2. STN EN 1997-730091 Eurocode 7: Geotechnical design Part 1: General rules. In Slovak
3. ANSYS ® User's Manual for Revision 11, Swanson Analysis Systems, Inc.

MATEC Web of Conferences 107, 00034 (2017) DOI: 10.1051/matecconf/201710700034
DYN-WIND'2017

Effect of steel fibres dosage in alkali-activated slag mortar on acoustic emission obtained during three-point bending tests

Libor Topolář[1,*], *Hana* Šimonová[1], *Pavel* Schmid[1] and *Pavel* Rovnaník[1]

[1]Brno University of Technology, Faculty of Civil Engineering, Veveří 331/95, Brno 602 00, Czech Republic

Abstract. The acoustic emission phenomenon is directly associated with nucleation of cracks in structural materials during loading. This paper analyses acoustic emission signals captured during three-point bending fracture tests of alkali-activated slag mortar specimens with different amount of steel fibres. Typical parameters of acoustic emission signals were identified for different mixtures to further describe the under-the-stress behaviour and failure development. The acoustic emission signals from crack growth were continuously monitored using acoustic emission sensors mounted on the specimen surface. Acoustic emission results are accompanied by selected mechanical fracture parameters determined via evaluation of load versus displacement diagrams recorded during three-point bending tests.

1 Introduction

Alkali-activated binders are relatively new binders, alternative to the conventional Portland cement, formed through the alkaline activation of different industrial by-products such as blast furnace slag and fly ash. In last two decades, alkali-activated cement and concrete have attracted attention all over the world due to their benefit of low energy cost and good durability compared to Portland types of cement [1, 2].

The composites based on alkali-activated slag are quasi-brittle materials with low tensile strength properties. This negative property can be changed by the inclusion of short discrete fibres into the matrix which prevent or control initiation, propagation, or coalescence of microcracks [3, 4]. Steel, glass, carbon, wooden, synthetic and natural fibres are used for this functionality. Addition of fibres to the brittle concrete or mortar can improve many of its technical features (e.g. tensile strength, flexural strength, fracture toughness, resistance to fatigue, impact, wear and thermal shock) [4, 5].

Addition of fibres is also an efficient method to improve shrinkage behaviour of alkali-activated slag based composites [6, 7]. The most significant effect of fibres is a prevention of microcracks from the propagation in composites. The extension and propagation of microcracks that occur due to the internal stresses in concrete are forestalled by stress transfer capability of randomly distributed fibres. According to their shape and quantity,

* Corresponding author: libor.topolar@vutbr.cz

fibres carry some of the stress that occurs in the matrix itself and transfer the other part of the stress to the stable cement matrix [8–10].

2 Experimental methods

2.1 Materials and specimen preparation

Five different mixtures of alkali-activated slag mortars were prepared (Table 1). The first one was reference without addition of fibres. The others contain the steel fibres in amount 5, 10, 15 and 20% from the weight of the slag. The mean length of steel fibres was 6 mm and mean diameter was 0.17 mm. The mixtures were cast into 40 × 40 × 160 mm prismatic moulds and left to set. The hardened specimens were immersed in water for 27 days, then pulled out of water and allowed to dry spontaneously under ambient conditions for 24 hours prior to three-point bending test.

Table 1. Mixture proportions of alkali-activated slag composites.

Mixture component	REF	D_05	D_10	D_15	D_20
Slag [g]	450	450	450	450	450
Sodium silicate [g]	90	90	90	90	90
Aggregate [g]	1350	1350	1350	1350	1350
Water [ml]	185	190	190	195	195
Steel fibres [g]	–	22.5	45	67.5	90

2.2 Testing methods

2.2.1 Acoustic emission method

Acoustic emission (AE) method is a non-destructive technique that is used to monitor damage mechanism in composites [11]. It is known that increasing cracks give rise to AE activity which can be recorded using sensors mounted on the surface of the composites under investigation. The degree of damage and the characteristics of received signals are strongly dependent on the damage process and enable the monitoring of composite behaviour during the fracture test [12]. The universal measurement and diagnostic system DAKEL-XEDO© (made by ZD Rpety-Dakel company) was used for the AE measurements. The guard sensors eliminate noise that is generated from outside the area of interest. In this work, four acoustic MIDI emission sensors (type IDK-09) were used, all having identical frequency range and being attached to the surface by beeswax (Figure 1).

2.2.2 Fracture test procedure

Three-point bending tests were carried out on a LabTest 6-1000.1.10 multi-purpose mechanical testing machine with the measuring range 0–1000 kN at a laboratory centre AdMaS, Faculty of Civil Engineering, Brno University of Technology. The beam specimens with central notch were tested in three-point bending configuration. The initial central notch was cut before testing by diamond blade saw, depth of notch was approximately 1/3 of the depth of specimen (Fig. 1). A load–displacement (deflection in the middle of span length) F–d diagram was recorded during testing and consequently used for the determination of elasticity modulus E from the first (almost linear) part of the F–d diagram, and for the calculation of effective fracture toughness K_{Ic} and specific fracture energy G_F* values. The effective fracture toughness value was determined using the

Effective Crack Model [13], which combines the linear elastic fracture mechanics and crack length approaches. An estimate of fracture energy was obtained from the *F–d* diagram according to the RILEM method (using the work-of-fracture value) [14]. Informative compressive strength f_c values were also determined for all specimens on the fragments remaining after the fracture experiments had been performed.

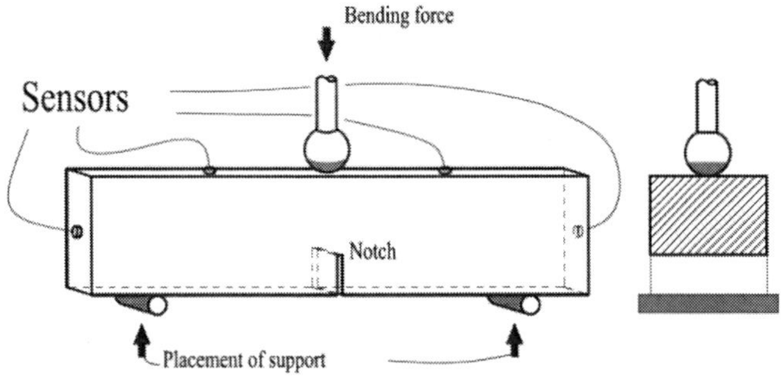

Fig. 1. The arrangement of experiment with placement of acoustic emission sensors on specimen during the three-point bending test.

3 Results

3.1 Acoustic emission method

To describe AE signals which are formed during the three-point bending test in the specimens, we focused our attention on the number of AE events and amplitude of these signals. The number of times the signal rises and crosses the threshold is the number of the AE events. The second observed parameter is amplitude of AE signals, which is the greatest voltage measured in a waveform. This is a very important parameter in AE inspection because it determines the detectability of the signal. Signals with amplitudes below the operator-defined minimum threshold are not recorded. Results of analysis of AE signals captured during three-point bending fracture tests are introduced in Fig. 2 in the form of the arithmetic mean (obtained from 3 independent measurements) and standard deviations as error bars.

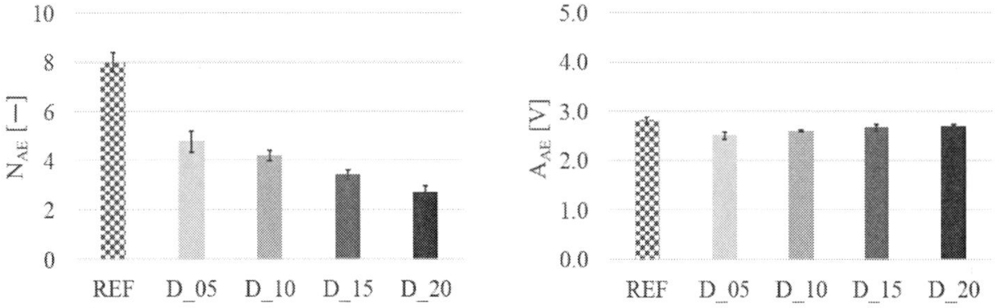

Fig. 2. Determined values (mean ± standard deviation) of number of AE events (left) and amplitude of AE signals (right) record to ultimate tensile capacity.

Fig. 2 and Fig. 3 present the dependence of a number of AE events and amplitude of AE signals on the amount of steel fibres in specimens. The number of AE events and thus the number of micro-cracks created is decreased with the increasing amount of steel fibres, before maximum load. The diagram on the right (Fig. 2) shows that although the number of

AE events decreased the size of created microcracks is quite uniform. It is probably caused by the properties of the matrix.

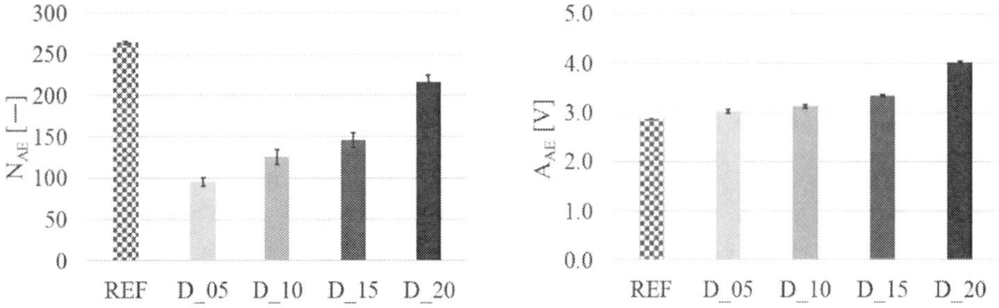

Fig. 3. Determined values (mean ± standard deviation) of number of AE events (left) and amplitude of AE signals (right) record during the whole measurement.

From the record of the whole measurement it is evident that the number of AE events increased with amount of steel fibres which is, in this case, caused by the pulling the steel fibres out of the matrix. REF specimens have the highest number of AE events which is caused by the total destruction of specimens during three-point bending test. This destruction did not occur at specimens with steel fibres. As can be observed, the amplitude of AE signals increases as the amount of steel fibres in the specimen becomes larger.

3.2 Mechanical fracture parameters

The acoustic emission results are accompanied by selected mechanical fracture parameters determined via evaluation of *F–d* diagrams recorded during three-point bending tests. Mean values and coefficients of variation are introduced in Table 2.

Table 2. Mean values of selected mechanical fracture parameters (coefficient of variation in %).

Parameter [unit]	REF	D_05	D_10	D_15	D_20
E [GPa]	15. 1 (4.1)	21.1 (10.6)	20.3 (13.2)	20.0 (9.7)	18.4 (22.6)
K_{Ic} [MPa·m$^{1/2}$]	0.843 (8.1)	0.821 (16.0)	0.971 (10.8)	1.170 (10.0)	1.304 (25.6)
G_F* [J/m^2]	113.1 (17.7)	155.4 (2.3)	249.0 (6.6)	299.7 (10.0)	327.4 (18.5)
f_c [MPa]	64.9 (1.7)	77.1 (10.7)	91.2 (2.8)	92.9 (2.9)	92.1 (2.7)

The modulus of elasticity value increased with the addition of fibres. The highest value was in case of 5% addition of. Then, values gradually decreased with higher dosage of fibres but the coefficient of variation increased more than twice. This might have been caused by relatively high volume of fibres in matrix, the distribution of which is not uniform. The effective fracture toughness increased with higher dosage of steel fibres, too. The value for D_20 mixture was by 55% higher compared to the reference one. The most significant increase was observed for specific fracture energy. This parameter increased almost three times in case of the highest fibre dosage. The informative value of compressive strength increased with addition of steel fibres by approximately 40% if at least 10% of fibres was added.

4 Conclusions

In this investigation, several experimental tests were carried out on alkali-activated slag with steel fibres specimens loaded up to failure are analysed. Specimens with central notch were subjected to the three-point bending test. During the experiments, the AE technique was used to monitor the progress of the specimens' failure. It is obvious, that different types of damage generate different AE signals. These differences can be related to the amount of fibres in the structure. A small number of cracks generate a small number of events before the failure occurs. This superior performance may be attributed to higher bond properties between fibre and alkali-activated slag mortar. The properties of the AE signals can be linked to the overall fracture behaviour of the materials. The value of all selected mechanical fracture parameters was increased with the addition of fibres to the alkali-activated slag matrix. The addition of fibres improves resistance to crack propagation through the alkali-activated matrix.

This outcome has been achieved with the financial support of the Czech Science Foundation, project No. GA 16-00567S, and the Ministry of Education, Youth and Sports of the Czech Republic under the "National Sustainability Programme I", project No. LO1408 (AdMaS UP).

References

1. C. Shi, P.V. Krivenko, D. Roy, *Alkali-activated cements and concretes* (CRC Press, Taylor & Francis, 2006)
2. F. Puertas, T. Amat, A. Fernandez-Jimenez, T. Vazquez, Cement Concrete Res, **33** (2003)
3. D.J. Hannant, *Fiber cements and fiber concretes* (John Wiley & Sons Ltd, 1978)
4. M. Şahmaran, I.Ö. Yaman, Constr Build Mater, **21** (2007)
5. F. Bencardino, L. Rizzuti, G. Spadea, R.N. Swamy, Compos Part B-Eng, **24** (2010)
6. S. Bernal, R.D. Gutierrez, S. Delvasto, E. Rodriguez, Constr Build Mater, **24** (2010)
7. V. Corinaldesi, G. Moriconi, Cement Concrete Res, **34** (2004)
8. Ş. Yazıcı, I.G, V. Tabak, Constr Build Mater, **21** (2007)
9. A. Bentur, Mat Sci Series, **1** (1989)
10. A. Bentur, S. Mindness, *Fibre reinforced cementitious composites* (E & FN Spon, 1990)
11. Ch.U. Grosse, M. Ohtsu, *Acoustic Emission Testing* (Springer, 2008)
12. M. Ohtsu, Mater Sci Res Int, **4** (1998)
13. B.L. Karihaloo, *Fracture Mechanics and Structural Concrete* (Longman Pub Group, 1995)
14. RILEM Committee FMT 50, Materials and Structures, **18** (1985)

MATEC Web of Conferences **107**, 00035 (2017) DOI: 10.1051/matecconf/201710700035
DYN-WIND'2017

Impact of graphite admixture on electrical properties of alkali-activated slag mortars

Ivo Kusak[1,*], *Miroslav* Lunak[1] and *Pavel* Rovnanik[2]

[1]Institute of Physics, Faculty of Civil Engineering, Brno University of Technology, Veveří 95, 602 00 Brno, Czech Republic
[2]Institute of Chemistry, Faculty of Civil Engineering, Brno University of Technology, Veveří 95, 602 00 Brno, Czech Republic

Abstract One of the objectives of the applied research's striving consists in providing the users with new slag-mortar-based materials. Basic research, in its turn, aims at examining the newly created materials from the viewpoint of all possible testing methods. Slag mortar specimens were subjected to electrical analysis carried out by means of an ZNC vector analyser and an SPEAG-made DAK-12 coaxial probe within the frequency range from 100 MHz to 3GHz and, furthermore, a dedicated automatically measuring device within the frequency range from 40 Hz to 1 MHz. The frequency spectra of interest were measured on various copolymer specimens differing from each other by the content of the carbon powder. Higher graphite powder content increases the electrical conductivity of cement/slag-based building materials, which thus become easier to measure by means of electromagnetic measuring methods. Carbon admixture may also improve the material's antistatic properties.

1 Introduction

The development of alkali-activated slag (AAS) cements has been object of intensive research recently [1]. The AAS cements are manufactured by mixing fine-ground glass slag with strong alkaline solutions, such as water glass NaOH and Na_2CO_3. As far as their mechanical properties are concerned, the AAS cements are comparable with the PC (Portland cement), particularly in the case where water glass is used as activator. They also feature better endurance, better resistance against aggressive environment, better frost resistance and lower amount of hydration heat released. Among their most important drawbacks, there are the high autogenous as well as dry-up-related shrinkage, which is particularly characteristic for activators in the form of water glass. If NaOH and Na_2CO_3 are used, they show similar shrinkage as the PC. Among other drawbacks there are their propensity to form efflorescences and, in some cases, to too fast setting [2, 3].

To enhance the applicability of electrical measuring methods, graphite powder appears to be a convenient admixture, increasing the electrical conductivity of the bodies under test. Thus obtained increased electrical conductivity will enhance the measurability and provide for better utilization of the impedance spectroscopy method. The amount of the graphite

* Corresponding author: kusak.i@fce.vutbr.cz

powder admixture is of the order of several per cent of the specimen total weight. Attainment of the so-called percolation threshold could be detected by the measurement [4, 5].

2 Experiment setup

Following brands of input raw materials were used for the specimen preparation: SMŠ 380 granulated blast-furnace slag, SUSIL MP 2.0 dried water glass, PG1-3 sand, COND 8 96 powdered graphite, Triton X-100 nonionic detergent, Lukosan S.defoamer.

First, the suspension of COND powdered graphite with Triton X100 detergent and about 100 ml of water were added to the water glass to be stirred in a mixer for 1 minute. Subsequently, SMŠ 380 slag and 3 gradings of sand (PG1-3) were added. Finally, Lukosan S defoamer was added. The moulding being completed, the specimens were immersed into water. After 28 days, they were taken out to be kept in the open air for 7 days in order to stabilize their water content. Subsequently, the specimens were dried up at a temperature of 105 °C until a constant mass was reached.

Three test bodies of dimensions 40 × 40 × 160 mm from each of the fifteen mixtures were manufactured. The results obtained from each test body are compared with those of the reference specimen.

Table 1. Chemical composition of granulated blast-furnace slag (%).

SiO_2	Al_2O_3	Fe_2O_3	CaO	MgO	K_2O	Na_2O	MnO	SO_3
39.75	6.61	0.46	39.03	10.45	0.63	0.38	0.37	0.71

Table 2. Mixture composition.

Mixture	**Slag** [g]	**Water glass** [g]	**Sand** [g]	**Graphite** [g]	**Triton X-100** [ml]	**Lukosan S** [ml]	**Water** [ml]
G0				0	0	0	185
G1				4.5	30	5	150
G2				9	30	5	155
G3				13.5	30	5	160
G4				18	30	5	165
G5				22.5	60	10	135
G6				27	60	10	140
G7	450	90	1350	31.5	90	15	110
G8				36	90	15	115
G9				40.5	120	20	85
G10				45	120	20	90
G15				67.5	180	30	75
G20				90	240	40	70
G25				112.5	300	40	70
G30				135	390	40	70

The bodies to be tested (Table 2) were characterized by means of the impedance spectroscopy method [6]. The specimens were measured in the frequency range from 40 Hz to 1 MHz. An R&S ZNC vector analyser with DAK12 coaxial probe (made by Speag) was used to measure at higher frequencies (from 100 MHz to 3 GHz). The electrical conductivity σ was measured within this frequency range.

Micro-images of alkali-activated slag mortars were taken at TESCAN MIRA3 XMU by means of a scanning electron microscope in the SEM mode. The experiments were carried out on dry specimens, the surface of which was coated with dust gold using an accelerating voltage of 20 kV.

3 Results

Electrical properties have been measured in two frequency bands of the high-frequency electric field.

Fig. 1 shows the specimen electrical resistance vs. frequency plot within the frequency range from 40 Hz to 1 MHz with graphite powder content as a parameter. The electric resistivity decreases when the graphite amount and the electric field frequency increase. However, there is quite a distinct peak, which is located on the reference mixture curve (G0) at a frequency of about 1 kHz. A tiny peak can be observed at this frequency for the mixture, which was enriched by adding 1% of graphite. Only a monotonous decrease of the electric resistivity without any pronounced anomalies is observed when higher graphite amounts are added. The resistivity of the composites, which contain higher amounts of the powdered graphite, reaches its minimum value at relatively low frequencies below 10 kHz. The electric resistivity of all tested specimens decreased to a similar level above 100 kHz. To analyse the influence of the added graphite amount, a reference frequency of 1 kHz was chosen (at this frequency, highest differences in the resistivity are observed).

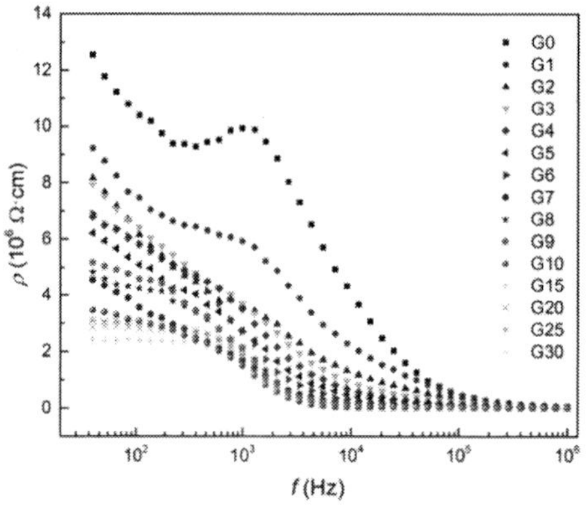

Fig. 1. Electric resistivity of AAS composites with 0 to 30% of powdered graphite in the frequency range from 40 Hz to 1 MHz (the frequency is plotted in log scale).

Fig. 2a shows that even a small amount of added graphite reduces the electric resistivity substantially. The electric resistivity decreases when the amount of the filling substance (graphite) is growing. Its most effective value in terms of the graphite amount to the electric resistance ratio is reached at 10% of graphite. Such an amount of graphite may be supposed to be sufficiently high for ensuring contact between adjacent particles and, therefore, appears to be the best suited from the electric resistance (conductivity) viewpoint. It may therefore be stated that a higher graphite content than 10% influences the electric resistance of AAS composites to a lesser extent in the low and medium frequency range.

Fig. 2 illustrates the electrical conductivity vs. graphite content plot for six different frequencies of the electric field, namely, 0.5 GHz, 1 GHz, 1.5 GHz, 2 GHz, 2.5 GHz and 3 GHz. The mixture containing 10% of graphite features a substantial growth of the electric conductivity for all selected frequencies of the electric field. An electric conductivity limit is evident to occur at lower frequencies (0.5 GHz, 1 GHz and 1.5 GHz) for at least 25% of graphite added. At other frequencies, a steady growth of the electrical conductivity is observed irrespective of the graphite content (corresponding, for example, to a G30 mixture with 30% of graphite). Generally, the steepest growth of the electrical conductivity is observed below and up to 10% of added graphite. The growth is slower for higher graphite percentage. Electric conductivity versus frequency relationship is clearly evident here. The most noticeable differences can be observed at a frequency of 3 GHz and the powdered graphite content of 30%. The higher the electric field frequency the higher value of the electric conductivity is reached for 30% graphite specimens.

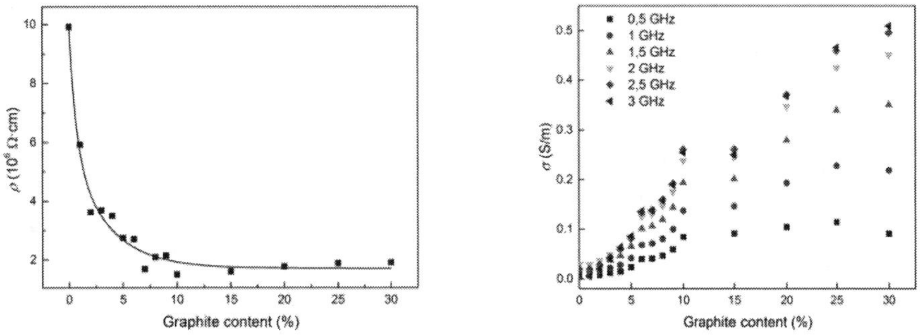

Fig. 2. a) Change in electric resistivity of AAS composites versus the graphite content at 1 kHz; b) electrical conductivity of AAS composites with 0–30% of graphite powder at different frequencies.

Fig. 3a, 3b and 4a show the morphology of the fracture surface of AAS composites with conducting filling substance. Graphite particles in the form of relatively thin slabs with lamellar structure are clearly visible (Fig. 4b). In view of the fact that in this case these particles are sparsely dispersed (the AAS matrix contains only 5% of graphite), only few graphite particles can be seen here. Amorphous structure, which is prevailingly seen in the microscopic image, corresponds to the alkali-activated slag that is predominating here (Fig. 3a). Gradual increase of the graphite amount results in the particle denser packing, which in turn enhances the electric conductivity. After 3% of the powdered graphite is added, the graphite will occupy about 50% of the matrix volume and the particles will be in close contact with each other (Fig. 4a). Such an amount enables the contact conductivity of the AAS matrix, the graphite internal resistance becoming the limiting factor. We would like to thank Dr Patrik Bayer for microscopic measurements (SEM – Scanning Electron Microscope).

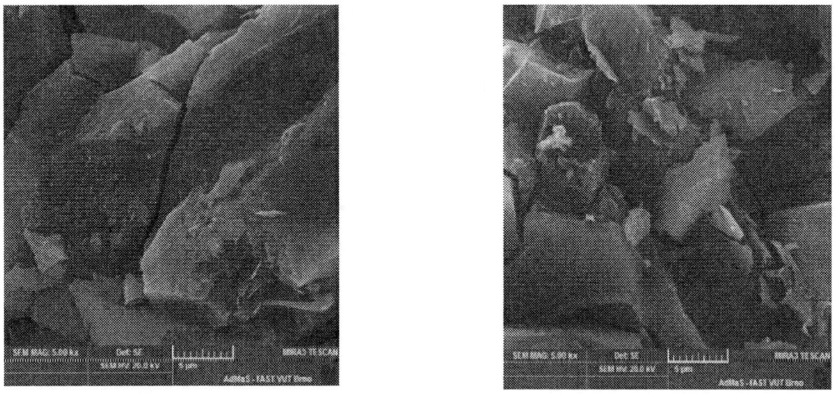

Fig. 3. a) Microstructure of AAS composites with 5 % of graphite; b) 10 % of graphite.

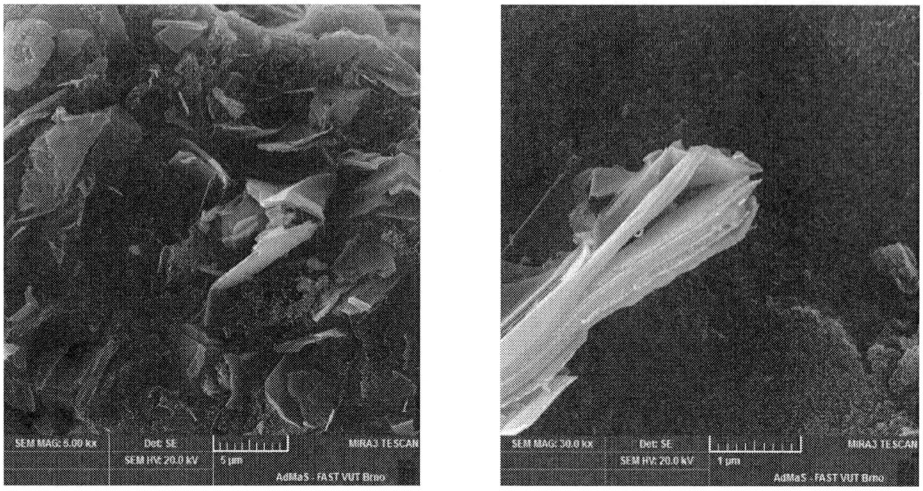

Fig. 4. a) Microstructure of AAS composites with 30 % of graphite; b) lamellar structure of graphite particles.

4 Conclusion

The present paper deals with the change of electric parameters of alkali-activated slag mortars to which powdered graphite has been added. This graphite admixture improves the electric conductivity of the materials in question, making them easier to measure by electromagnetic-principle-based methods. An ideal graphite admixture is described from the viewpoint of both its amount and the most effective increase of the electrical conductivity. Another quantity to study was the electric resistivity and its frequency dependence. The micro-structure of the AAS composites with conducting filling substance has also been analysed by means of a scanning electron microscope.

This paper has been worked out under the project GAČR No. 16-02261S and under the project No.S-16-2967 supported by Faculty of Civil Engineering BUT.

References

1. R. Zhao, J. G. Sanjayan, Mag. Concrete Res. **63**, 163 (2011)
2. F. Puertas et al., J. Eur Ceram. Soc. **31**, 2043 (2011)
3. A. Fernandez-Jimenez: Cement Concrete Res. **29**, 1313 (1999)
4. S. Matsutani, Y. Shimosako, Y. Wang, Physica A. **391**, 5802 (2012)
5. K. Y. Kim, T. S. Zun, K. P. Park, Cement Concrete Res. **50**, 34 (2013)
6. I. Kusak; M. Lunak, P. Schauer, Appl. Mech. Mater. **248**, 370 (2013)

Development of shrinkage and fracture parameters in selected fine-grained cement-based composites

Barbara Kucharczyková[1,*], *Hana* Šimonová[1], *Petr* Misák[1] and *Zbyněk* Keršner[1]

[1]Brno University of Technology, Faculty of Civil Engineering, Veveří 331/95, 602 00 Brno, Czech Republic

Abstract. The paper summarizes results of a pilot study aimed at the evaluation of an experimental investigation focused on determination of the material characteristics development of selected fine-grained cement-based composites during their ageing. The composition of composites being investigated differed only in a water to cement (*w/c*) ratio and in amount of superplasticizer. Quite extensive experiments were performed with the aim to determine shrinkage, dynamic a static modulus of elasticity and fracture properties on test specimens exposed to free drying during the whole time of its ageing (including the early stage of setting and hardening). The article presents especially results (including their statistical evaluation) of shrinkage and fracture parameters development within 90 days of composites' ageing. Experimental results show the dependence of the investigated characteristics on the value of *w/c* ratio. The most visible effect was observed in the case of shrinkage development. The curing conditions were reflected especially in high variability of the test results.

1 Introduction

Cement-based composites are one of the most frequently used materials in building practice. In order to achieve even better physical and mechanical characteristics, the composition of the cement-based composites is continuously developing. The main aim of this effort is to obtain a building material with a high quality and high durability.

One of the most commonly monitored characteristics is their resistance to the formation and propagation of cracks during whole time of ageing. Generally speaking, the first source of cracking are volume changes which take place in the early stage of setting and hardening. Shortly after the pouring of fresh mixture into the moulds plastic shrinkage occurs. Its magnitude strongly depends on the ambient conditions and appropriate curing conditions have to be maintained to eliminate the cracking. Other substantial source of early cracking, mainly in composites with a low water to cement (*w/c*) ratio, can be also the occurrence of autogenous shrinkage especially when it is restrained [1, 2]. However, the major causes of shrinkage cracking are probably deformations due to drying which start after the removal the final structural element from the mould and appropriate curing conditions are not

* Corresponding author: barbara.kucharczykova@vutbr.cz

maintained [3, 4]. The above mentioned sources of cracking may be causes of decrease in strength as well as durability of structural members [5, 6].

To quantify degree of material deterioration, an appropriate approach has to be applied. Commonly used tests for determination of materials' strength parameters are not sufficient to describe the actual degree of their inner structure deterioration. In this case, more suitable approaches are those based on the quantification of the fracture parameters which are strongly dependent on the damage of macro- and micro-bonds in the inner structure of the tested material [7].

2 Experimental part

2.1 Materials

The test specimens used in the experiment were manufactured from three different fine-grained cement-based mixtures (A, B and C). The composition of the investigated composites differed only in a *w/c* ratio and in an amount of superplasticizer, rest of components was always the same. A design of mix proportions originates from the standard ČSN EN 196 - 1 [8]. The fresh composites were made using the standardized quartzite sand (*s*) with maximum nominal grain size of 2 mm, CEM I 42.5 R Portland cement (*c*) and water (*w*) at a proportions mentioned in Table. 1. "C" mixture contains also polycarboxylate ether-based superplasticizer SIKA SVC 4035 in an amount of 1% by cement mass. A mixing device with controllable mixing speed was used to prepare the fresh mixtures. Basic information about mix proportions and parameters of fresh-state composites are shown in Table 1. The workability was determined according to ČSN EN 1015-3 [9] using the flow table with a calibrated scale designed for the testing of fresh mortars. Bulk density of the fresh composite was determined using a calibrated vessel with the volume of 1 dm^3 in accordance with ČSN EN 1015-6 [10].

Table 1. Mix proportion and properties of fresh concrete.

	Units	Composite A	Composite B	Composite C
s:c:w	–	3:1:0.5	3:1:0.47	3:1:0.35
Plasticizer	% by cement mass	–	–	1.0
w/c ratio	–	0.5	0.47	0.35
Mixing speed	revolutions/min	20	30	40
Workability	mm	145	120	130
Bulk density	kg/m^3	2200	2230	2270

Two groups of test specimens were prepared from each mixture. The first group contained test specimens with dimensions of 100×60×1000 mm intended especially for measurement of shrinkage development (6 specimen from each mixture). The second group contained test specimens with dimensions of 40×40×160 mm which were subjected, among others, to fracture tests (12 specimen from each mixture).

2.2 Test procedures

2.2.1 Shrinkage measurement

The test procedure intended for determination of shrinkage development was designed to measure relative length changes along the central axis of the test specimens placed in the moulds with dimensions of 100×60×1000 mm. The measurement was performed using an inductive sensor leaning against the movable head of the mould whereas the head on the other side was fixed. The measurement started approx. 1 hour after the fresh composite was placed into the moulds. In this way the early relative length changes (up to ca. 72 hours of specimens' ageing) were recorded. Once early measurement was finished, the specimens were extracted from the moulds, placed on the table, and left to dry freely. The long-term measurement was performed using a strain gauge which was fixed onto the markers embedded into the upper surface of the test specimens during their manufacturing. The spacing of gauging points was 200 mm. Refer to [11] for more details about the test procedure.

All measurements were performed in a climate control chamber at a stable temperature of 21 ± 2 °C and relative humidity of 60 ± 10%. The measurements were finished approx. at the specimens' age of 90 days.

2.2.2 Fracture parameters evaluation

The mechanical fracture parameters were determined from the records captured during the fracture test of beam specimens with a central edge notch in three-point bending. All specimens were subjected to the fracture test using the test machine with controllable speed of displacement increment which was, for the purpose of experiment, set to value of 0.02 mm/min. The record in form of load versus displacement (deflection in the middle of span length) diagrams were evaluated using the effective crack extension method [12] and work-of-fracture method [13]. The first (almost linear) part of the load versus displacement (*L–d*) diagrams were used for the determination of the elasticity modulus value. Because of stability loss during loading, it was not possible to reconstruct the descending part of *L–d* diagrams. Therefore, the work of fracture value W_F* is determined as area under *L–d* diagrams before stability loss occurred. Refer to [14] for details about determination of above mentioned mechanical fracture parameters.

3 Results and discussion

The results of performed tests are introduced in following figures in form of boxplots. Six specimens for shrinkage and twelve specimens for the rest of investigated parameters were used in the experiment. Boxplots provide information about the distribution of obtained results. Boxplot shows median, 25th and 75th percentiles, and the range of data. The monitored materials' parameters were following: bulk density D, shrinkage ε, modulus of elasticity E_c, effective fracture toughness K_{Ice} and specific fracture energy G_F* (determined using already mentioned work of fracture W_F* value) and informative compressive strength f_c determined on the fragments of test beams after fracture test were finished.

Fig. 1 shows increase in values of bulk density of "C" composite which corresponds to the reduced value of a *w/c* ratio (see Table 1). This reduction was reflected in the development of shrinkage (see Fig. 2) where the increased value of early shrinkage for "C" composite was recorded.

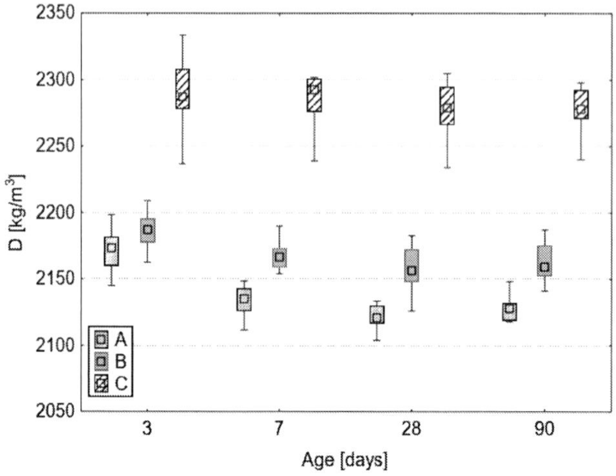

Fig. 1. Bulk density at 3, 7, 28 and 90 days of composites' ageing.

This rapid increase affected also the final steady-state value of total shrinkage which was more than 2 times higher compared to "A" and "B" composites. Initial expansion of "A" and B" composites was recorded within first 3 days of ageing. This expansion is connected with a rather high *w/c* ratio. "A" and "B" composites show very similar shrinkage process despite the reduced *w/c* ratio for "B" composite. This reduction was reflected slightly in the values of bulk density (Fig. 1).

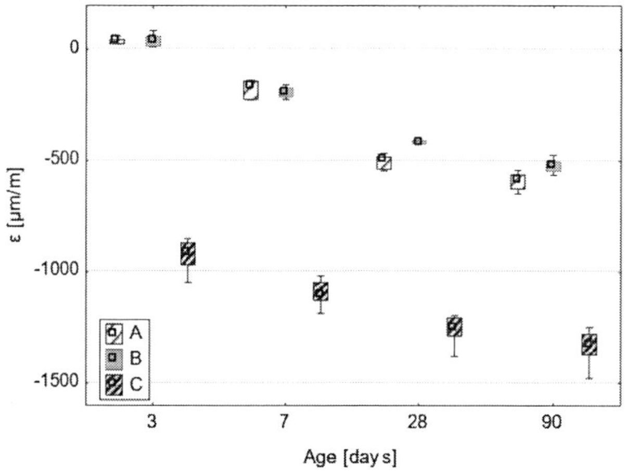

Fig. 2. Shrinkage values at 3, 7, 28 and 90 days of composites' ageing.

Concerning the mechanical fracture parameters, it can be suggested that these increased values of shrinkage (for "C" composite) did not cause a sharp drop in any of monitored parameters (see Figs. 4–6). Note, that relatively high variability was recorded for all results of performed tests (including the values of informative compressive strength – see Fig. 3). The variability can be connected with the curing conditions. Note also that no special wet or water curing conditions were maintained during the ageing. All tested specimens (from Set "A", "B" and "C") were approx. 3 days in moulds while the upper surface was exposed to free drying. After removing from the moulds, the specimens were stored in a laboratory and left to dry freely during the entire time of ageing. This fact may be reflected in magnitudes of mechanical parameters [6], [15] and also in variability of the tests results. Following aspects must be taken into account in relation to the results interpretation. First, the water needed to full cement hydration evaporated too early – a decrease in mechanical parameters can be expected. Second, there was a non-uniform distribution of water in volume of the particular test specimens during their ageing, especially within first 3 days the upper surface dried much more. Third, these "un-cured" composites are also more sensitive to micro-cracking. All above mentioned aspects are not always the cause of the decrease in absolute value of investigated parameters, but are reflected in the variability of results [16].

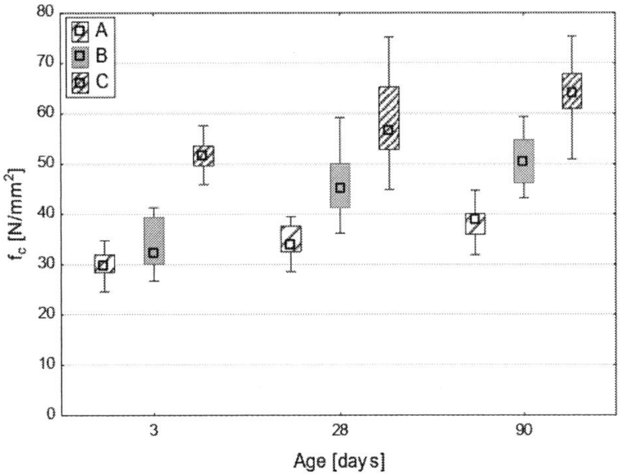

Fig. 3. Compressive strength at 3, 28 and 90 days of composites' ageing.

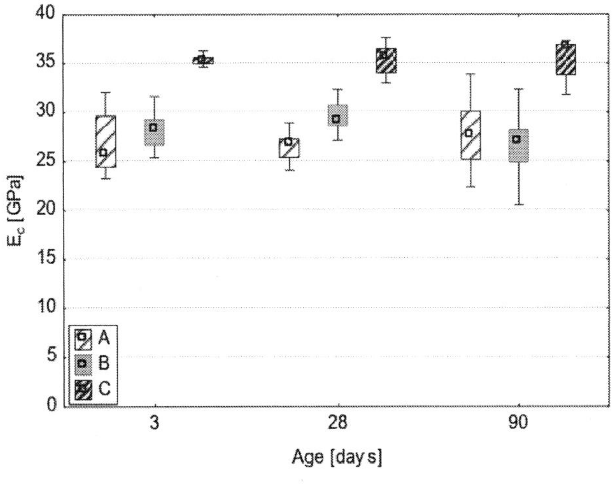

Fig. 4. Modulus of elasticity values at 3, 28 and 90 days of composites' ageing.

The high variability of modulus of elasticity values is rather remarkable, especially in the case of "A" and "B" composites. This phenomenon may originate in composition and workability of fresh mixtures. In the case of "A" composite, bleeding was observed after pouring the moulds during the manufacturing of test specimens. In the case of "B" composite, the fresh mixture exhibited poor workability. Both these technological aspects can essentially influence overall development of physical and mechanical parameters of hardened composite. Evaluation of the modulus of elasticity from (*L–d*) diagram recorded during the fracture tests can be more sensitive to these technological aspects.

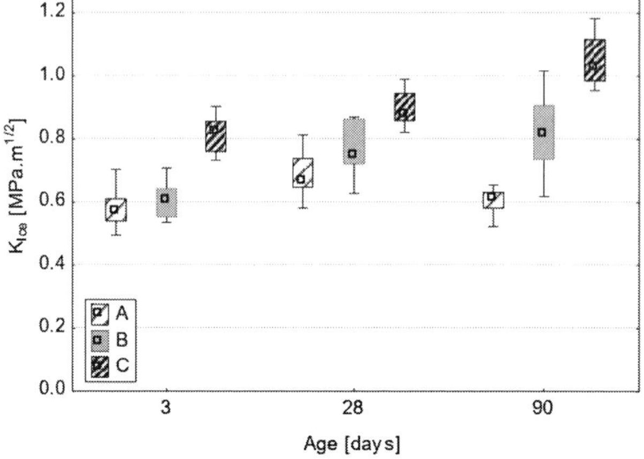

Fig. 5. Effective fracture toughness values at 3, 28 and 90 days of composites' ageing.

Fig. 5 and Fig. 6 show development of effective fracture toughness and specific fracture energy values. These two parameters should be most sensitive to materials' inhomogeneity and micro-cracks. A decrease of both fracture parameters was recorded at the age of 90 days for "A" composite (the one with the highest *w/c*). This decrease was not recorded in any other monitored mechanical parameters.

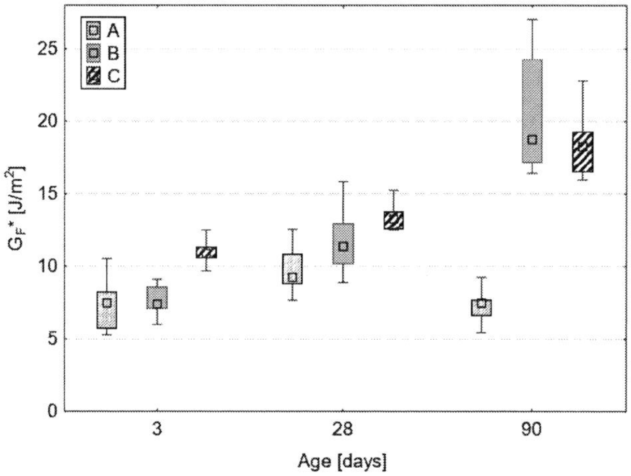

Fig. 6. Specific fracture energy values at 3, 28 and 90 days of composites' ageing.

4 Conclusion

The main aim of the performed experiment was to determine the development of shrinkage and fracture parameters for three cement-based composites which differed in the value of a *w/c* ratio and presence of plasticizer. To observe the influence of insufficient curing on the development of investigated parameters, all of the test specimens were exposed to free desiccation (drying) during whole time of their ageing, including the early stage of setting and hardening. The experimental results show that the actual value of a *w/c* ratio substantially influences the magnitude and development of all investigated parameters. The most visible effect was observed in shrinkage development which was reflected in increased values of shrinkage during the whole time of ageing. The high shrinkage recorded at the age of 3 days (approx. 900 μm/m) for mixture "C" (the lowest *w/c* with addition of plasticizer) may lead to crack formation if it is restrained. The poor curing conditions were reflected especially in high variability of the test results and in the decrease of fracture parameters.

This outcome has been achieved with the financial support of the Czech Science Foundation under project No. 17-14302S „Experimental analysis of the early-age volume changes in cement-based composites ".

References

1. P. Lura, O. M. Jensen, J. Weiss. Mater Struct **42,** 8 (2009)
2. L. Østergaard, PhD Thesis, Denmark, supervisor: L. Damkilde, H. Stang (2003)
3. L.Liu, X. Wang, H. Chen, Ch. Wan, M. Zhang, Computers & Structures **178** (2017)
4. P. Havlásek, PhD Thesis, Prague, supervisor: M. Jirásek (2014)
5. M. N. Hassoun, A. A. Al-Manaseer, *Structural concrete: theory and design*, 6th ed. (Wiley, USA, 2015)
6. P.-C. Aïtcin, *36th Conference on OUR WORLD IN CONCRETE & STRUCTURES,* http://cipremier.com/100036002 [online], (Singapore, CI-Premier PTE LTD, 2011)
7. Z.P. Bažant, *5th International RILEM Symposium on Creep and Shrinkage of Concrete (ConCreep 5)*, [online], (E & FN Spon, London, 1993)
8. ČSN EN 196-1, *Determination of strength*, (Prague: ČNI, 2005)
9. ČSN EN 1015-3, *Determination of consistence of fresh mortar*, (Prague: ČNI, 2000)
10. ČSN EN 1015-6, *Determination of bulk density of fresh mortar*, (Prague: ČNI, 1999)
11. B. Kucharczyková, L. Topolář, P. Daněk, D. Kocáb, P. Misák, Adv. Mater. Sci. Eng. (2017)
12. B. L. Karihaloo, *Fracture mechanics and structural concrete* (Wiley, New York, 1995)
13. RILEM TC-50 FCM Recommendation, Mater. Struct. **18** (1985)
14. P. Rovnaník, H. Šimonová, L. Topolář, P. Bayer, P. Schmid, Z. Keršner, Constr. Build. Mater. **199** (2016)
15. D. Kocáb, M. Králíková, P. Cikrle, P. Misák, B. Kucharczyková, Experimental Analysis of the Influence of Concrete Curing on the Development of its Elastic Modulus over Time, Mater. Technol., (MIT, Slovenia, to be published 2017)
16. P. C. Aïtcin, *High-Performance Concrete*, Czech version (IC ČKAIT, Prague, 2005)

The use of the acoustic emission method for the monitoring of changes in the internal structure of polymer-cement mortars when testing the static compressive modulus of elasticity

Dalibor Kocáb[1,*], *Libor* Topolář[1], *Barbara* Kucharczyková[1], *Michaela* Hoduláková[1] and *Petr* Daněk[1]

[1]Brno University of Technology, Faculty of Civil Engineering, Veveří 331/95, 602 00 Brno, Czech Republic

Abstract. The paper focuses on the experimental determination of the dynamic and static compressive modulus of elasticity of selected polymer-cement mortars and their comparison. The dynamic modulus of elasticity was determined by means of the ultrasonic pulse velocity test and the resonance method. The static compressive modulus of elasticity test was accompanied by a measurement of the material's behaviour using the acoustic emission method, the principle of which is the detection of mechanical dilation waves produced by changes and microscopic faults in the structure of the material. The advantage of the acoustic emission method is the possibility of detecting a crack before it can be found e.g. using a microscope. The outcome of the experiment is a detailed evaluation of the behaviour of polymer-cement mortars made using the values of the elastic modulus and acoustic response of the material during loading.

1 Introduction

Polymer-cement mortars (PCM) have been used in civil engineering mainly for the purposes of rehabilitation and protection of concrete structures [1]. These composites combine a cement matrix with an organic polymer matrix [2]. The composition of PCMs improves some properties, such as workability, flexural strength, adhesion or chemical resistance [3, 4]. Cement mortars can be modified by a variety of polymers, e.g. latexes, redispersible polymer powders, liquid resins, monomers or acrylic and epoxy emulsions. Polymers can be considered an important component in the manufacturing of modern, sustainable building materials [5].

The suitability of PCMs and the technology of their application are usually given by the technical description of each individual composite. The variety purposes, and therewith connected risk of improper application, can be detrimental to the development of physical and mechanical properties of PCMs, which can in turn endanger the functionality of the rehabilitation system [6]. In order to correctly understand

* Corresponding author: dalibor.kocab@vutbr.cz

the behaviour of a PCM, it is necessary to know its modulus of elasticity as well as its response to loading.

2 Experiment

2.1 Material and specimens

The specimens were made from two fine-grained polymer-modified mortars based on Portland cement. Tests were performed as part of a pilot experiment, which is why the paper does not include the product designation of the mortars used. For the purposes of evaluation, the composites are identified as "V" and "VII". Specimens of the "V" set were made from a two-component mix – the liquid component is an aqueous copolymer dispersion and the powder component consists of a mixture of Portland cements and mineral fillers. Specimens of set "VII" were made from a single-component powder mix, which comprises silica sand, microfibre and Portland cement, including plasticisers, polymers and the prescribed amount of water.

The fresh PCMs were mixed using a handheld mixer and in keeping with the producer's instructions. Once the components have been mixed, each PCM was made into 6 cylindrical specimens with the nominal diameter of 40 mm and nominal length of 160 mm (apart from specimens made for the observation of shrinkage, cf. [7]). The specimens were made for the determination of the dynamic and static modulus of elasticity. For the first three days the specimens were left in their moulds and stored in laboratory conditions with a temperature of (21 ± 2) °C and relative humidity of (60 ± 10) %, where they were also stored after demoulding. The specimens were not cured during ageing and their surface was intentionally left to dry freely.

2.2 Testing procedures

When the specimens reached 28 days of age, all 6 of them were tested for the dynamic modulus of elasticity using two electroacoustic methods – the ultrasonic pulse velocity test and the resonance method. A Pundit PL-200 ultrasonic testing instrument with sensors operating at a frequency of 150 kHz was used to determine the time it takes for an ultrasonic wave to transit along the length of a specimen, see Fig. 1a. Next, the modulus of elasticity E_{cu} was calculated according to the method described in ČSN 73 1371 [8]. The test cylinders were then tested for the natural frequency of longitudinal, flexural and torsional vibration using a Handyscope HS4 oscilloscope with an acoustic emission (AE) sensor, see Fig. 1b. Dynamic moduli E_{crL} and E_{crf} were calculated in accordance with ČSN 73 1372 [9]. For more details about the dynamic methods used, see e.g. [10].

The static compressive modulus of elasticity E_c was determined by means of cyclic loading according to ČSN ISO 1920-10 [11] in a testing press FORM+TEST 200 kN. The longitudinal deformations of the specimens were measured by 100 mm long mechanical strain gauges (Hollan's gauge) with dial indicators (Fig. 2a). The static modulus of elasticity test was completed by gradual increase of compressive stress until specimen failure (Fig. 2b).

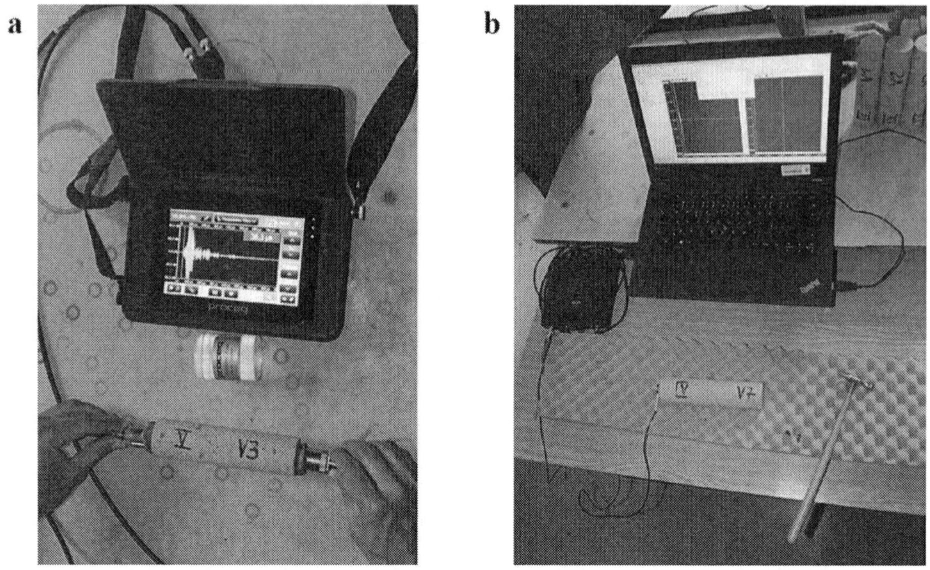

Fig. 1. a) Determination of the time necessary for an ultrasonic pulse to travel through a specimen, b) determination of the natural frequency of a specimen's longitudinal vibration.

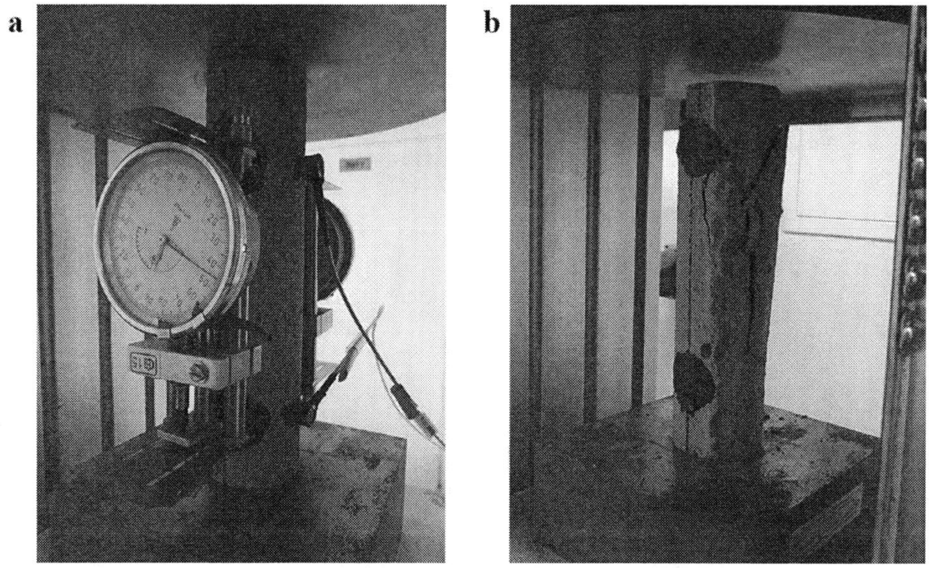

Fig. 2. a) A specimen mounted in a press during the static modulus of elasticity test, b) specimen at the end of the static modulus of elasticity test completed by the determination of compressive strength.

The formation and subsequent propagation of microcracks inside the composite during the cyclic loading test of E_c was monitored by the acoustic emission method – a specimen fitted with AE sensors can be seen in Fig. 2a. Unlike other non-destructive techniques, the AE method is a passive inspection method capable of examining the entire volume of a structure, structural element or a specimen. An advantage of AE over other defectoscopic methods is the continuous monitoring of an object and lower time requirements compared to the gradual testing made by other methods. However, the AE method is only able to detect active damage, which releases energy inside the body being examined in response to stimulation by internal or external stress, as seen in Fig. 3. An AE event is caused by irreversible dislocation and degradation affecting the material's microstructure. The released energy is transformed into a mechanical pulse travelling through the material as a longitudinal or flexural elastic wave. When the wave is detected by an AE sensor it is converted to an electric signal and is known as an AE signal [12, 13, 14]. The AE method detects and characterises the development of the damage. It aims to "listen" for acoustic activity emitted by processes taking place inside the material, such as plastic deformation, formation of microstructure, crack initiation and propagation, etc. The AE method is able to detect and evaluate (even localise) the activity of faults and defects as they develop (and only when they develop) [15].

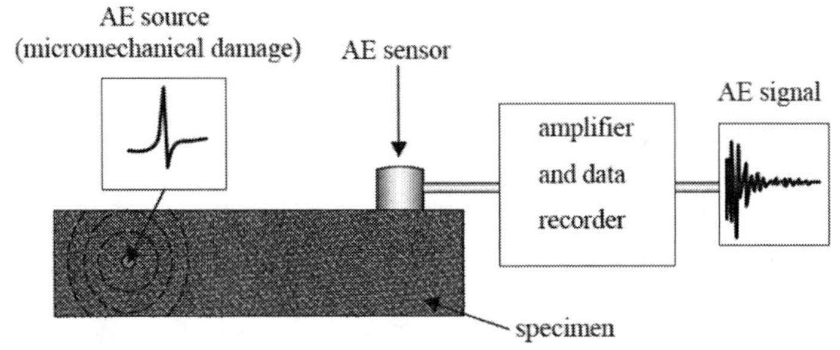

Fig. 3. Principle of acoustic emission measurement [12].

3 Results and discussion

3.1 Elastic modulus

Table 1 shows the test results for the dynamic moduli of elasticity, the static modulus of elasticity and compressive strength of the two PCMs. The results show that composite "VII" has better strength and elastic properties at the age of examination. This is most clearly visible in the static tests, where value of the static modulus of elasticity of composite "VII" is approx. 5 GPa higher and compressive strength approx. 15 MPa higher than those of composite "V".

3.2 Acoustic emission

The chart (Fig. 4) shows the number of AE events that occurred during the test. For clarity, the values are plotted on a logarithmic scale. The chart shows mortar "VII" generated far more AE events during the test, which suggests that a higher number of microcracks had formed than in PCM identified as "V". What is quite interesting is the different trend in the number of events in the two composites during each loading cycle. While mortar "VII" exhibited the Kaiser effect (a sort of "material memory") and the number of AE events occurring during cyclic loading slowly decreased, mortar "V" saw more AE events during the third cycle than during the second one. Also during loading until failure the two PCMs behaved differently. While the number of events recorded for mortar "VII" during loading until failure was similar to the number recorded during the preloading cycle, mortar "V" produced much more AE events during loading until failure than during previous loading cycles.

Table 1. Bulk density, elastic moduli and compressive strength of the PCM including average values and sample standard deviations (S. s. d) at the age of 28 days.

Specimen	Bulk density	Elastic modulus [GPa]				Compressive strength
	D [mm]	E_{cu}	E_{crL}	E_{crf}	E_c	f_c [MPa]
V-1	1 804	20.07	17.56	17.84	13.66	32.88
V-2	1 728	17.85	15.56	15.46	12.27	31.51
V-3	1 840	20.90	18.45	18.75	13.90	33.68
V-4	1 740	18.71	16.51	16.91	14.32	29.90
V-5	1 776	19.10	16.81	16.77	12.95	32.85
V-6	1 773	19.52	17.20	17.15	13.86	34.21
Average:	**1 780**	**19.4**	**17.0**	**17.1**	**13.5**	**32.5**
S. s. d.:	**41**	**1.07**	**0.98**	**1.1**	**0.75**	**1.57**
VII-1	1858	23.03	21.62	21.56	18.83	50.17
VII-2	1923	23.36	22.12	22.94	18.64	50.72
VII-3	1796	20.73	19.24	19.08	17.19	46.32
VII-4	1803	20.91	19.61	20.38	17.98	46.21
VII-5	1897	22.90	21.71	21.54	18.21	47.49
VII-6	1823	21.12	19.22	18.90	17.00	46.89
Average:	**1 850**	**22.0**	**20.6**	**20.7**	**18.0**	**48.0**
S. s. d.:	**52**	**1.21**	**1.36**	**1.58**	**0.75**	**1.98**

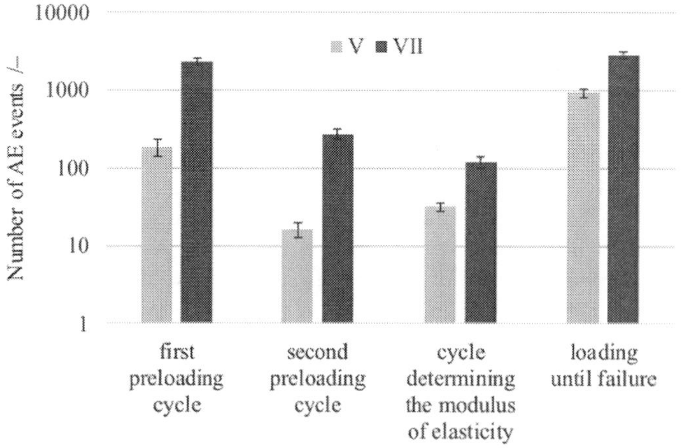

Fig. 4. Number of AE events during the static modulus of elasticity test.

Given the average amplitude measured during the loading cycles (Fig. 5) as well as the progress of loading until failure, it can be said that microcracks that formed in specimens made from mortar "VII" were of roughly the same size throughout the entire test. On the other hand, the cracks formed in mortar "V" slowly increased in size during cyclic loading.

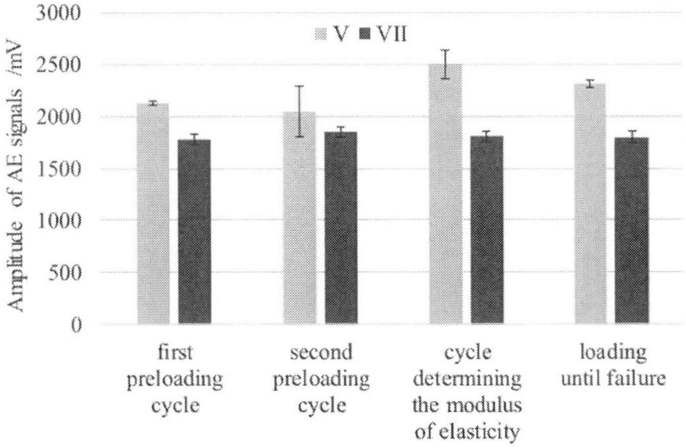

Fig. 5. Amplitude of AE signals during the static modulus of elasticity test.

The energy released by AE signals (Fig. 6) is again plotted on a logarithmic scale. It is clear that the energy released during AE signals was greater in specimens of mortar "VII" than in mortar "V".

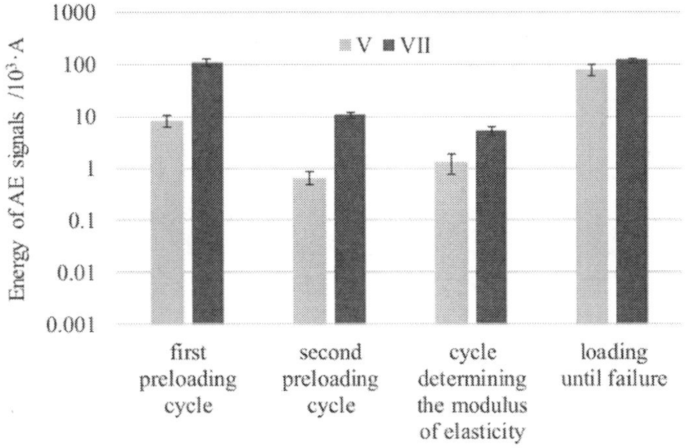

Fig. 6. Energy of AE signals during the static modulus of elasticity test.

4 Conclusion

The experiment results show that the PCMs tested herein (designed primarily for concrete rehabilitation) reached lower values of dynamic and static moduli of elasticity (average E_c value is only 13.5 GPa in PCM "V" and 18.0 GPa in PCM "VII") than cement-based mortars. PCM are used because of properties other than elastic parameters [1-6]; however, it is still very useful to know the values of their elastic moduli. Polymers, by which the cement mortar is modified, generally have low values of the elastic modulus, which decreases the elastic modulus of the whole composite. The use of PCMs in combination with e.g. concrete with a high modulus of elasticity can result in faults caused by the very different values of the elastic modulus.

The presented acoustic emission measurement results show that PCM "VII" suffers from a high number of cracks during loading, however, they are likely small in size. Contrarily, during the loading of PCM "V" a smaller number of larger cracks appears. This material is currently subject to further research.

This paper has been written as a part of project No. 17-14302S „Experimental analysis of the early-age volume changes in cement-based composites", supported by the GAČR - Czech Science Foundation.

References

1. W. Malorny, M. Plath, *Advanced Materials Research* (Trans Tech Publications Ltd, Switzerland, **668**, 2013)
2. Y. Ohama, *Handbook of polymer-modified concrete and mortars properties and process technology* (Noyes Publications, Park Ridge, 1995)
3. P. Łukowski, *Materials* (MDPI AG, Switzerland, **9**, 2016)
4. L. Aggarwal, P. Thapliyal, S. Karade, Construction and Building Materials (Elsevier B.V. **21**, 2007)
5. D. Van Gemert, L. Czarnecki, M. Maultzsch, H. Schorn, A. Beeldens, P. Łukowski, E. Knapen, Cement and Concrete Composites (Elsevier B.V. **27**, 2005)
6. L. Courard, J. Lenaers, F. Michel, A. Garbacz, *Construction and Building Materials* (Elsevier B.V. **25**, 2011)
7. B. Kucharczyková, H. Šimonová, D. Kocáb, Z. Keršner, *Modelování v mechanice* (VŠB – TUO, Ostrava, 2017)
8. ČSN 73 1371 *Nedestruktivní zkoušení betonu - Ultrazvuková impulzová metoda zkoušení betonu* (2011)
9. ČSN 73 1372 *Nedestruktivní zkoušení betonu - Rezonanční metoda zkoušení betonu* (2012)
10. D. Kocáb, M. Králíková, P. Cikrle, P. Misák, B. Kucharczyková, Mater. Technol. (Slovenia, to be published 2017)
11. ČSN ISO 1920-10 *Zkoušení betonu - Část 10: Stanovení statického modulu pružnosi v tlaku* (2016)
12. M. Kreidl, R. Šmíd, *Technická diagnostika: senzory, metody, analýza signálu* (BEN - technická literatura, Praha, 2006)
13. V. Malhorta, N. Carino, *Handbook on nondestructive testing of concrete* (CRC Press, Boca Raton, 2004)
14. M. Matysík, L. Topolář, P. Daněk, H. Šimonová, T. Vymazal, I. Plšková, *Advanced Materials Research* (Trans Tech Publications Ltd, Switzerland, **1000**, 2014)
15. B. Kopec, *Nedestruktivní zkoušení materiálů a konstrukcí* (CERM, Brno, 2008)

Experimental assessment of the influence of multiple cyclic loading on the dynamic modulus of elasticity of concrete

Petr Misák[1,*], *Dalibor* Kocáb[1], *Petr* Daněk[1], *Barbara* Kucharczyková[1] and *Libor* Topolář[1]

[1]Brno University of Technology, Faculty of Civil Engineering, Veveří 95, 602 00 Brno, Czech Republic

Abstract. The goal of the experiment was to determine the influence that cyclic loading has on concrete dynamic modulus of elasticity and compressive strength. Each test specimen was subjected to cyclic loading with the loading force equivalent to one third of the expected compressive strength value similarly to the static modulus of elasticity test with the difference that the total number of loading cycles was 3000. The dynamic modulus of elasticity was determined on every specimen before and after loading by means of the ultrasonic pulse velocity test and resonance method. The result of the experiment is the degree to which the dynamic modulus of elasticity decreased after multiple cyclic loading. A decrease was observed in all specimens.

1 Introduction

The elastic properties of hardened structural concrete have been seeing increased attention lately. The elastic modulus is undoubtedly one of the most important elastic properties of hardened concrete [1, 2]. This property affects primarily deformation under load, the magnitude of deformation and shrinkage. Generally speaking, the higher the value of the modulus of elasticity, the less a structure deforms. During its service in a structure, concrete enters alternating states of loading and unloading, or states of greater and lesser loading. If situation in which a structure is at one point loaded and at another this load is reduced or removed can be called a loading cycle, then a structure can be considered to be cyclically loaded during its use. These cycles are irregular and can vary depending on the structure's purpose and manner of use.

This paper discusses a part of an experiment, which focused on the cyclic compressive loading of hardened concrete. Test specimens were loaded by 3000 loading cycles to elicit elastic deformation. The goal of the experiment was to determine the influence of such multiple cyclic loading on the dynamic compressive and tensile modulus of elasticity, the dynamic shear modulus of elasticity, the dynamic value of Poisson's ratio and compressive strength. The basic purpose of the experiment was the determination of how the observed properties develop depending on the number of loading cycles. This experiment is only a minor part (a pilot experiment) of a larger effort, the goal of which is to observe the change

* Corresponding author: petr.misak@vutbr.cz

in the static modulus of elasticity during cyclic loading alongside with the development of acoustic emission characteristics.

2 Experiment

The tests were performed with ordinary concrete of the C 35/45 strength class. Table 1 shows its composition and basic properties.

Table 1. Composition and basic properties of fresh and hardened concrete.

CEM I 42,5 R [kg]	338
Sand Bratčice 0-4 mm [kg]	905
Aggregate Olbramovice 4-8 mm [kg]	183
Aggregate Olbramovice 8-16 mm [kg]	667
Water [kg]	176
Plasticiser Sika ViscoCrete 4035 [kg]	1.77
w/c [-]	0.48
Flow table test (fresh concrete) [mm]	340
Slump test (fresh concrete) [mm]	70
Air content (fresh concrete) [%]	4.5
Density (fresh concrete) [kg/m^3]	2 270
Compressive strength after 28 days (cube, hardened concrete) [N/mm^2]	52.2
Flexural strength after 28 days (hardened concrete) [N/mm^2]	5.6

The tests were performed with 12 cylindrical specimens with the diameter of 150 mm and height of 300 mm. First, all the specimens were subjected to two tests. The ultrasonic pulse velocity test was used to determine the time taken by pulse to get through the concrete and the resonance method was applied to find out the actual frequency of longitudinal, flexural and torsional vibration. Fig. 2 shows the results of these tests. These parameters were then used in the calculation of the dynamic compressive/flexural and shear modulus of elasticity as well as Poisson's ratio. For more details about the dynamic methods, see e.g. [3-6]. Next, three specimens were tested for compressive strength. The remaining 9 specimens were cyclically loaded by up to one third of the determined value of compressive strength, similarly to the static elastic modulus test according to [7], except the number of loading cycles was 3000. After the cyclic loading test, all the specimens were tested for dynamic parameters and in the end also compressive strength. Because of the time needed for testing one specimen, which often took up to 3 days, it was necessary to make sure the testing time did not affect the results. This is why all tests were made at the age of 2 years, at which the properties of concrete are far more constant than at the age of e.g. 28 days.

The loading was performed in accordance with method B described in [7]. The testing procedure was configured in a way that allowed performing as many loading cycles as

 DOI: 10.1051/matecconf/201710700038

possible within several dozens of hours, while still performing a static loading test. The goal of the experiment was to observe the behaviour of concrete during loading in the elastic region of the stress-strain curve; for this reason, the upper loading stress was set to $\sigma_a = f_c/3$, where f_c is compressive strength measured on reference specimens of the same shape and dimensions as the test specimens. According to [7], the basic loading force should be within $0.1 \cdot f_c \leq \sigma_b \leq 0.15 \cdot f_c$. This experiment used a lower loading stress of the highest possible value, i.e. $\sigma_b = 0.15 \cdot f_c$. The reason for this was to save time when transferring between the loading stresses. For the same reason, the highest allowed loading rate was chosen to be 0.80 MPa/s with a holding time of 3 s. The cyclic loading was performed using a mechanical testing press LaborTech with a loading force range of 0–1000 kN; the loading rate was set by a compressive force increment.

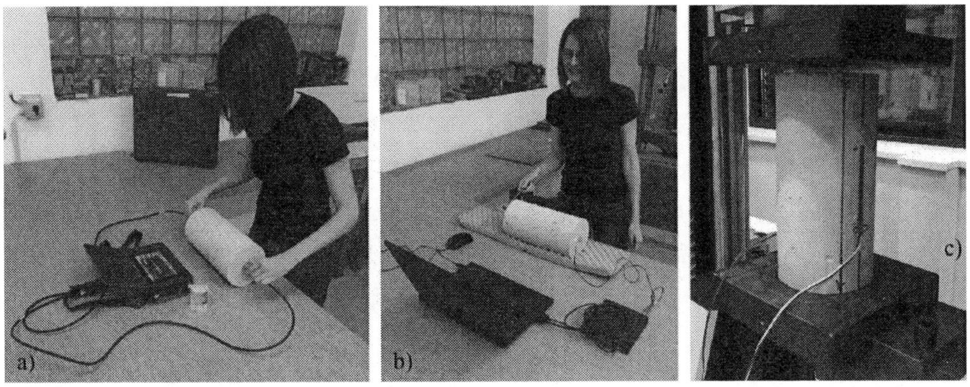

Fig. 1. a) Ultrasonic pulse velocity test. b) Determination of the natural frequency by the resonance method; c) Testing specimen.

3 Results and discussion

The result of the experiment described here is an evaluation of the drop in the dynamic modulus of elasticity. The following properties were determined:

- E_{cu} – dynamic flexural and compressive modulus of elasticity determined by the ultrasonic pulse velocity test;
- E_{crL} – dynamic flexural and compressive modulus of elasticity determined from the natural frequency of longitudinal vibration – resonance method;
- E_{crf} – dynamic flexural and compressive modulus of elasticity determined from the natural frequency of flexural vibration – resonance method;
- G_{cr} – dynamic shear modulus of elasticity determined by the resonance method;
- ν_{cr} – dynamic Poisson's ratio determined by the resonance method;
- f_c – compressive strength.

All the non-destructively determinable properties (i.e. all except compressive strength) were determined before and after the cyclic loading using the same method and the same specimens. Only compressive strength was measured on three unloaded specimens and nine loaded ones. The results for all the properties are shown as standard boxplots in Fig. 2.

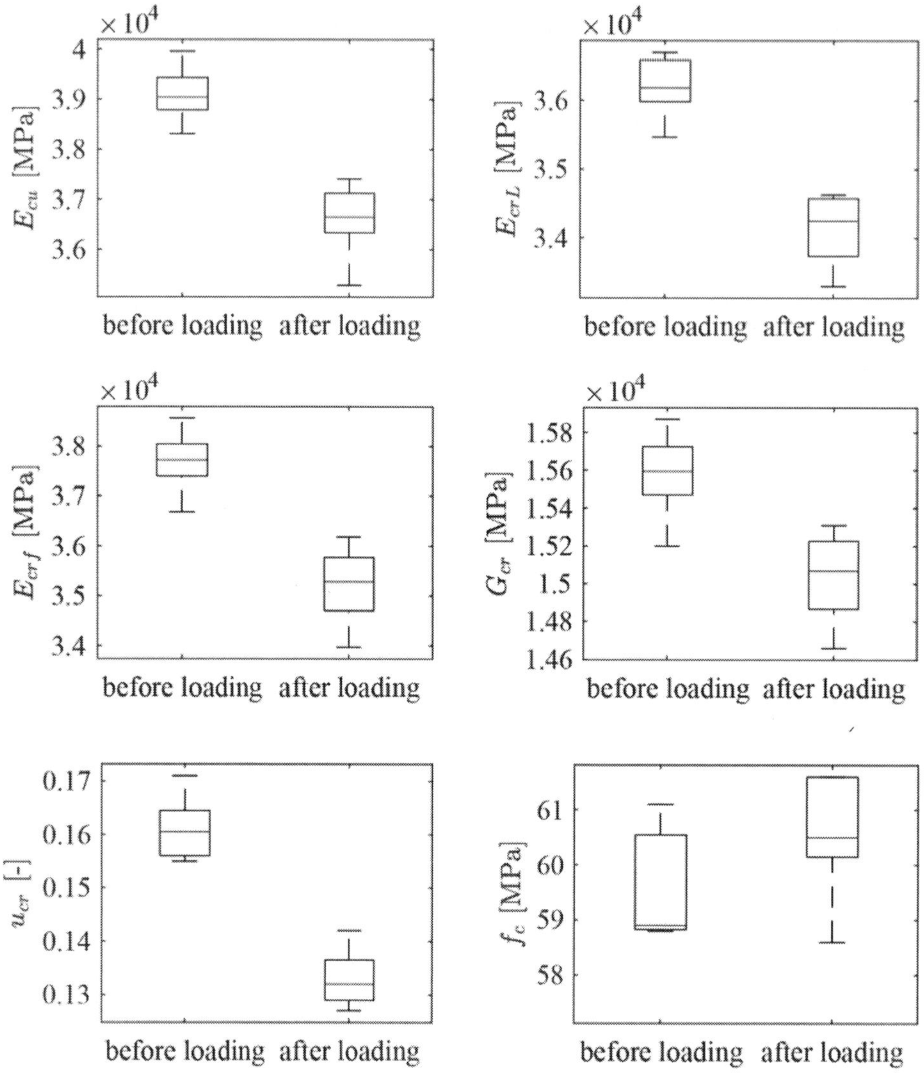

Fig. 2. Boxplots of the results; red line – median; blue box – 25% and 75% quantile; black lines – minimum and maximum value.

The test results show that the multiple cyclic loading brought a decrease in all the observed properties with the exception of compressive strength, where no reduction was found. However, the compressive strength values can be influenced by the number of specimens used – only three specimens were tested for compressive strength before the cyclic loading. The drop in dynamic properties was measured by means of statistical tests. After the verification of data normality, the mean value equality hypothesis was refuted by the ANOVA test. Next, a one-tailed t-test was made, which returned the same result. The decrease in the dynamic properties caused by multiple loading can thus be considered statistically significant.

The authors of [8] attribute the drop in the elastic modulus to the formation of microcracks and creep. This explanation would correspond to the results presented herein, as microcracks and microfailures in the internal structure of concrete have a major influence on the dynamic modulus of elasticity without having any serious effect on compressive strength [1, 9].

4 Conclusion

The experiment results invite the following conclusions:

- All the observed properties with the exception of compressive strength registered a decrease after multiple cyclic loading. No decrease in compressive strength after multiple loading cycles has been recorded.
- The decrease in the values of the observed properties corresponds with previously published conclusions [8], which, however, largely concern changes in the values of the static modulus.
- An evaluation of the dependence of the static modulus of elasticity on number of cycles, which was determined during every cycle of multiple cyclic loading. These findings are going to be published in near future.

This paper has been written as a part of project No. 17-14302S „Experimental analysis of the early-age volume changes in cement-based composites ", supported by the GAČR - Czech Science Foundation and project FAST-S-17-4693 supported by Brno University of Technology.

References

1. A. Neville, *Properties of Concrete* (Prentice Hall, New Jersey, 2012)
2. H. S. Abdelgader, A. S. Elbaden, A. Diouri, N. Khachani, M. Alami Talbi. MATEC Web of Conferences, 11 (**2014**)
3. ČSN 73 1371 *Nedestruktivní zkoušení betonu - Ultrazvuková impulzová metoda zkoušení betonu* (2011)
4. ČSN 73 1372 *Nedestruktivní zkoušení betonu - Rezonanční metoda zkoušení betonu* (2012)
5. V. M. Malhotra, N. J. Carino, *Handbook on non-destructive testing of concrete* (CRC Press LLC, West Conshohocken, 2004)
6. D. Kocáb, M. Králíková, P. Cikrle, P. Misák, B. Kucharczyková, Mat. and Tech. (to be published)
7. ČSN EN 12390-13 *Zkoušení ztvrdlého betonu - Část 13: Stanovení sečnového modulu pružnosti v tlaku* (2014)
8. A. Neville, J. J. Brooks, *Concrete technology* (Harlow, Prentice Hall, 2010)
9. ASTM C 666/C, *Standard Test Method for Resistance of Concrete to Rapid Freezing and Thawing* (2003)

Fracture parameters of concrete after exposure to high temperatures: pilot tests

Hana Šimonová[1,*], *Petr* Halfar[1], *Iva* Rozsypalová[1], *Petr* Daněk[1] and *Zbyněk* Keršner[1]

[1]Brno University of Technology, Faculty of Civil Engineering, Veveří 331/95, 602 00 Brno, Czech Republic

Abstract. Experimental concrete panels were after an appropriate aging time loaded by high temperatures (550, 600, 800 and 1000 °C) in a furnace intended for fire tests of building materials. These panels were heated according to the standard temperature-time curve according to EN 1991-1-2. One of panels was a reference without temperature load. Test specimens were obtained as cores drilled out from panels after performing fire tests. The cylindrical specimens were provided with a central chevron type notch and subsequently tested in three-point bending fracture test. The load versus displacement (deflection in the middle of span) diagrams were recorded during testing and basic fracture parameters were subsequently evaluated.

1 Introduction

Concrete belongs to the plentifully used building materials for a wide range of applications. Structures and construction components utilizing this composite were commonly designed assuming normal service temperatures. However, high temperatures acting on concrete – for example in connection with certain technological procedures, fire etc. – cause wide range of physical and chemical processes, which result in changes in the microstructure of composites and thus affects the mechanical properties of concrete [1].

The currently valid standard for the assessment of existing structures ISO 13822:2010 [2] states that assessing of the real condition of building structures have to reflect the actual properties of the materials, material parameters characterized in the initial proposal of construction, in standards or in regulations. It is necessary considered the degradation and the potential influence of the load during the structure life to determine real material parameters. The standard highlights the need to consider extraordinary loads, among which also belongs the fire. However, currently there are no guidelines for effective determination of the material characteristics needed for the evaluation of the static state of structures damaged by fire.

This paper is focused to determination basic fracture parameters via evaluation of pilot fracture tests of cylindrical specimens which were obtained as cores drilled out from the concrete panels after their exposure to fire.

* Corresponding author: simonova.h@vutbr.cz

2 Specimens

The normal concrete with ordinary material characteristics in terms of fire resistance was used for the production of experimental panels. It was a C30/37 strength class concrete (selected as widely used for building structures), exposure class XC3 and quartz aggregate (16 mm maximal size) was used. In total, the seven concrete panels were made of nominal dimensions 2300 × 1300 × 150 mm intended for subsequent temperature loading.

Experimental panels (P1, P2, P3, and P7) were after an appropriate aging time loaded by high temperatures (maximal values were 550, 600, 800, and 1000 °C) in a furnace intended for fire tests of building materials, see Fig. 1; the furnace equipment belongs to the AdMaS (Advanced Materials, Structures and Technologies) science centre, part of the Faculty of Civil Engineering at Brno University of Technology [3]. Panels were heated according to the standard temperature-time curve according to EN 1991-1-2 [4].

Fig. 1. Concrete panel P7 covered furnace of AdMaS science centre [3] and loaded by high temperature – maximal value was approximately 1000 °C in this case.

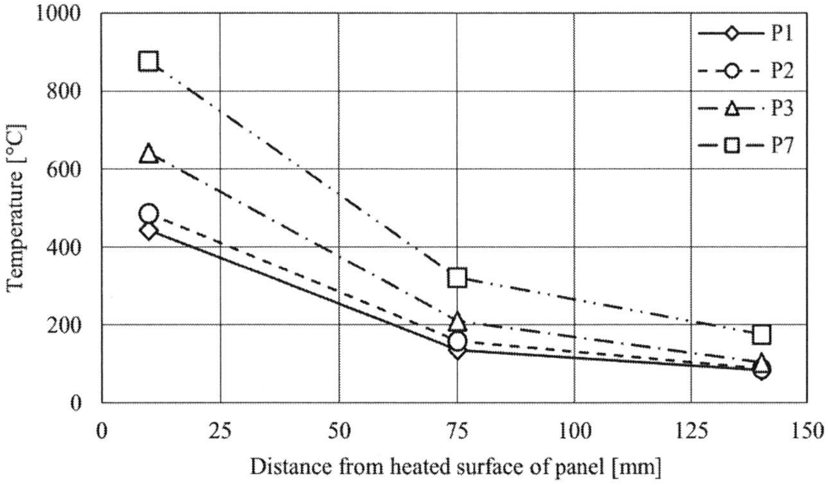

Fig. 2. Measured temperature versus distance from heated surface of panels P1, P2, P3, and P7.

Test specimens were obtained as cores drilled out from concrete panels after performing fire tests. Nominal diameter of these cylinders was 75 mm. Three specimens from each panel were used for subsequent fracture tests. Temperature sensors were placed in three locations in longitudinal axis of panels (as can be seen in Fig. 1) and in each of these places in three (or four) locations along the thickness of the panel. Temperature introduced in Fig. 2 represents mean value determined from sensor in middle of panel and from the sensor placed nearest the place from which the specimens were drilled.

One of panels was a reference without temperature load (P4). Five specimens from this panel were used for fracture tests.

3 Fracture tests

The cylindrical specimens (drilled cores) were provided with an initial stress concentrator in the form of chevron notch and subsequently tested in three-point bending fracture test, see Fig. 3 – scheme and selected realization of the test.

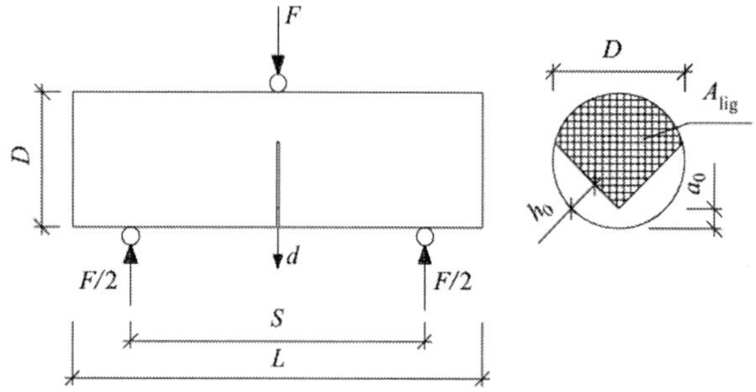

Fig. 3. Scheme of three-point bending fracture test of specimen with chevron notch (left) and illustration of realization of the test.

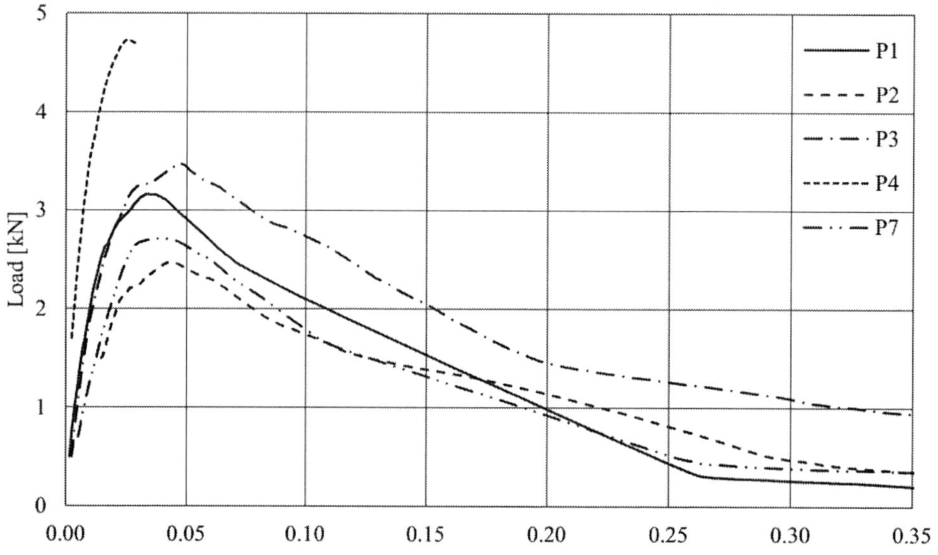

Fig. 4. Selected load–deflection diagrams.

Specimens were loaded under displacement control, therefore the load versus displacement (deflection in the middle of span) diagrams (*F–d*) were recorded during testing from which basic fracture parameters [5, 6] were subsequently evaluated. Selected *F–d* diagrams are depicted in Fig. 4. Nominal thickness of panels – and thus length of the specimens – was 150 mm, for that reason span length was only 127 mm. The fracture tests were carried out using a Heckert FPZ 100/1 testing machine with measuring range of 0–10 kN. All specimens after fracture tests are captured on photo in Fig. 5.

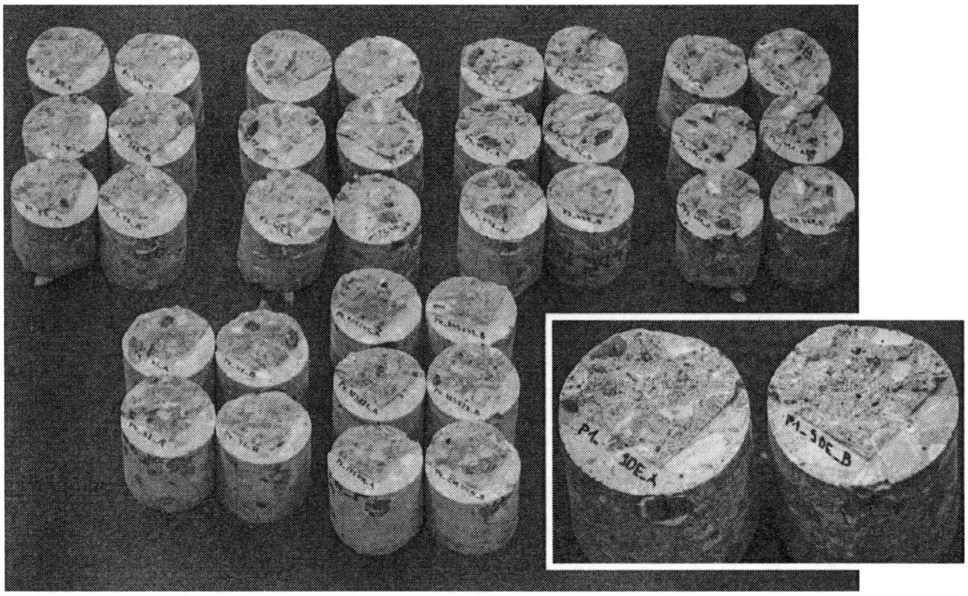

Fig. 5. All tested specimens (and selected detail) after fracture tests.

4 Methods

The fracture parameters of concrete were calculated from corrected *F–d* diagrams. Maximal load F_{max} was used for fracture toughness assessment with help of geometrical function A_{min} [7] and span S:

$$K_{Ic} = A_{min} \cdot \frac{F_{max}}{D^{1.5}}, \text{ where } A_{min} = \frac{S}{D} \cdot \left[1.835 + 7.15 \cdot \frac{a_0}{D} + 9.85 \cdot \left(\frac{a_0}{D} \right)^2 \right], \tag{1}$$

where D is specimen diameter and a_0 is notch tip depth, see Fig. 3.

The work of fracture value W_F, and then the specific fracture energy G_F value, was assessed from the whole *F–d* diagram [8]:

$$W_F = \int F(d)\,\mathrm{d}d \ , \tag{2}$$

$$G_F = \frac{W_F}{A_{lig}} \ , \tag{3}$$

where A_{lig} is area of ligament, see Fig. 3.

5 Results

Basic fracture characteristics evaluated using Eq. (1) and (3) – fracture toughness and specific fracture energy – are summarized in Figs. 6 and 7 in form of mean values and standard deviations.

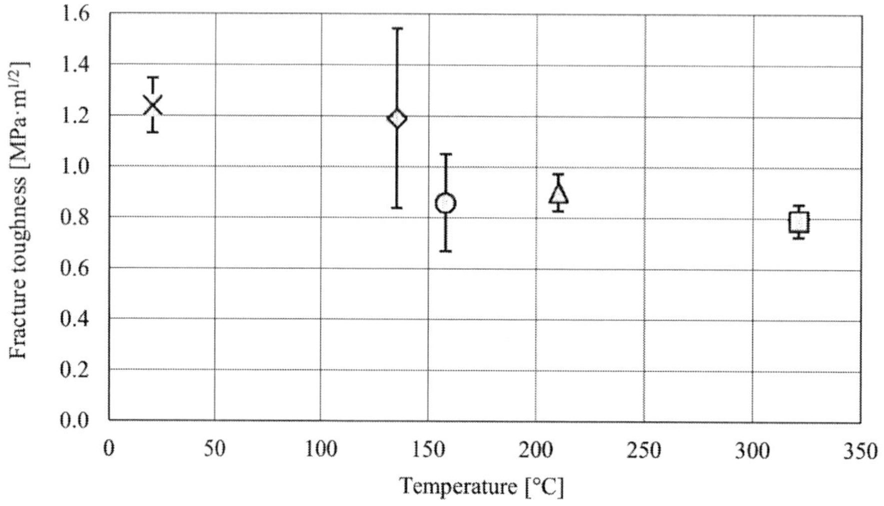

Fig. 6. Fracture toughness versus temperature: mean values and standard deviations.

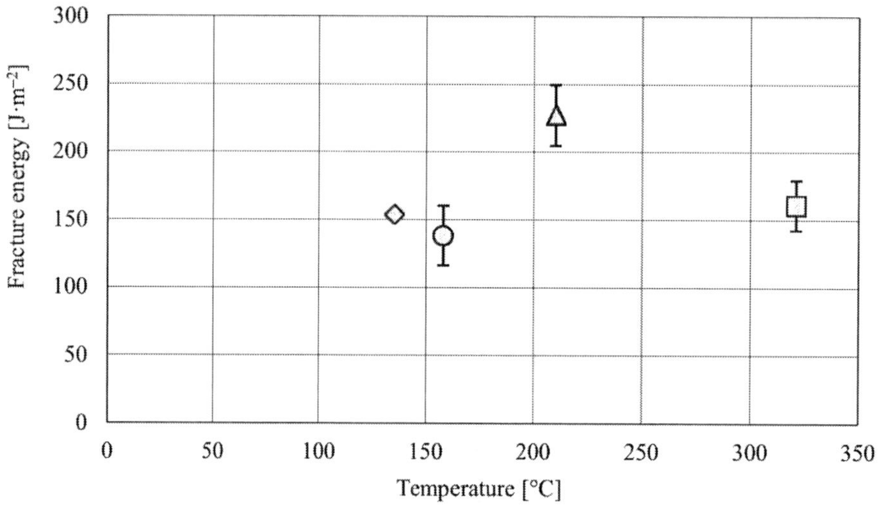

Fig. 7. Fracture energy versus temperature: mean values and standard deviations.

With increased temperature, a gradual decrease in fracture toughness values can be seen. Compared to the reference concrete specimen, the maximum decrease in fracture toughness was about 40%. The results showed variability typical for concrete: the coefficient of variation being below 10%, with the exemption of the P1 and P2 panel specimens (the maximum temperature was 550 °C/600 °C on the heated surface and about 135 °C/158 °C in the middle of the panel/specimen), where the result variability reached 20–30%.

Fracture energy results proved to be complicated to evaluate, seeing as they fluctuated significantly with temperature increase. The *F–d* diagrams in post peak part were often disrupted by sections where stability of loading was lost. Most importantly, this was the case for the reference specimens, where no values could be evaluated whatsoever (see Fig. 4).

6 Conclusions

As results in this paper showed, there are limitations to highly heated specimen fracture parameter assessment. The results were probably most influenced by the following factors: (i) the specimen diameter was too small in relation to the maximal aggregate size of used concrete, (ii) the specimen length was not sufficient for standard three-point bending fracture test evaluation, and (iii) the specimen relatively small size did not allow comfortable measurement device connection. Based on these pilot findings, fracture parameter assessment is planned to be conducted on beam specimens cut out of the panels.

This outcome has been achieved with the financial support of the Czech Science Foundation under project No. 16-18702S "AMIRI – Aggregate-Matrix-Interface Related Issues in Silicate-based Composites".

References

1. I. Hager, Bull Pol Acad Sci-Te **61**, 1 (2013)
2. ISO 13822:2010, *Bases for design of structures – Assessment of existing structures*, (ISO, Switzerland, 2010)
3. http://www.admas.eu/en/, March 10, 2017
4. EN 1991-1-2:2002/AC:2009, *Eurocode 1: Actions on structures – Part 1-2: General actions – Actions on structures exposed to fire*, (CEN, Brussels, 2002)
5. B. L. Karihaloo, *Fracture Mechanics and Structural Concrete*, (Longman Scientific & Technical, New York, 1995)
6. V. Veselý, P. Frantík, Adv Eng Softw **72,** 66–76 (2014)
7. F. Ouchterlony, Int J Rock Mech Min **25,** 71–96 (1988)
8. RILEM TC-50 FMC Recommendation, Mater Struct **18,** 285–290 (1985)

Flexure response of thermal loaded concrete specimens by acoustic emission method

Lubos Pazdera[1,*], *Libor* Topolar[1], *Karel* Mikulasek[1], *Jaroslav* Smutny[1], *Michaela* Hodulakova[1], *Zdenek* Chobola[1] and *Herbert* Seelman[2]

[1] Brno University of Technology, Faculty of Civil Engineering, Veveri 331/95 Brno, Czech Republic
[2] Rail Data Services Austria GmbH & Co KG, Steinergasse 11/5, A-1230 Wien, Austria

Abstract. The response to fire of concrete structural members depends on the thermal, mechanical, and deformation properties of concrete. These properties vary significantly with temperature depending on the composition and characteristics of the concrete batch mix as well as heating rate and other environmental conditions. The paper presents the effects of a high temperature on selected physical properties of concrete. The main aim of the article is the evaluation of monitoring concrete properties loaded in a few thermal steps up to 1200 °C. Therefore, the concrete specimens were heated in a programmable laboratory furnace at a heating rate of 5 °C/min. The specimens were loaded at six temperatures, 200 °C, 400 °C, 600 °C, 800 °C, 1000 °C, and 1200 °C maintained for 60 minutes. The acoustic emission activity and some material characteristics were evaluated in a three-point bending test.

1 Introduction

Many cases of fires occurring in buildings, tunnels and other building structures stimulate interest in the behaviour of concrete at high temperatures. In a fire, the temperature may rise up to 1300 °C, leading to damage in a concrete structure. [1]. Concrete is a poor heat conductor.

It is efficient to use acoustic emission technology to study crack damage location of concrete [2]. Non-destructive evaluation techniques are used to assess the condition of concrete structures, predict future performance, and monitor repair systems. One of the best-known non-destructive evaluation techniques is the acoustic emission method of detecting active defects in the tested materials [3]. Acoustic emission is a technique used widely for structural health monitoring purposes of several types of structures [4]. Piezoelectric sensors are mounted on the surface of the material or structure to record the motion of the surface under the elastic excitation of the cracking sources [5]. The rate and the characteristics of the received signals are strongly dependent on the damage process and enable the monitoring of fracture, corrosion and healing in cementitious materials [6]. There are two different ways to analyze cracks. Events are either described by acoustic emission parameters or analyzed, for example, by frequency spectra. Cracking in concrete

* Corresponding author: pazdera.l@fce.vutbr.cz

is a major problem in the design and durability of concrete structures. Various experimental methods have already been employed to detect the fracture process [7].

As a non-destructive method, acoustic emission can often be used to detect a failure at a very early stage of damage, long before a structure completely fails [8]. The article presents an experiment analysing the acoustic emission signals from the three-point bend testing of thermal loaded concrete specimens. Quantitative acoustic emission techniques were used to measure the micro fracture properties [9].

In engineering applications, active defects are a major hidden danger of structure security. The origin of micro-cracks in specimens is detected by using acoustic emission techniques with data samples collected, and, subsequently, the typical acoustic emission data acquired through experiment analyzed by using frequency spectra and other techniques. [10].

The frequency spectra of the recorded waveforms were obtained by Fast Fourier Transform [11]. The signals represented in the time domain can be investigated in other domains such as the frequency domain or the time–frequency domain. The Fourier transform clearly shows the natural frequency of the signals [12].

There are changes expected if concrete is heated up to [13]

200 °C	slow capillary water loss and reduction in cohesive forces as water expand, ettringite dehydration, C-S-H gel dehydration, gypsum decomposition ($CaSO_4\ 2H_2O$), physically bound water loss
400 °C	break up of some siliceous aggregates (flint),
600 °C	portlandite decomposition $Ca(OH)_2 \rightarrow CaO + H2O$, quartz phase change $\beta-\alpha$ in aggregates and sands
800 °C	second phase of the C-S-H decomposition, formation of β-C_2S
1000 °C	dolomite decomposition calcite decomposition $CaCO_3 \rightarrow CaO + CO_2$, carbon dioxide release, ceramic binding initiation which replaces hydraulic bonds
1200 °C	basalt melting
over 1200 °C	total decomposition of concrete, melting.

2 Experimental set-up

The three-point bending tests were performed on notched beams of cross section 100 mm × 100 mm and length 400 mm (notch depth of 33 mm was placed in centre) as shown in Fig. 1. The beams were made from cement CEM I 42.5R of 345 kg/m^3, sand 0/4 of 848 kg/m^3, coarse aggregate 8/16 of 980 kg/m^3, water of 160 kg/m^3 and super-plasticizer of 3 kg/m^3. The specimens of common temperature were heated at a heating rate of 5 °C/min up to a defined temperature maintained for 60 minutes and then placed in a room with a normal temperature. The defined temperatures were 200 °C, 400 °C, 600 °C, 800 °C, 1000 °C, and 1200 °C. The beams were loaded at a constant rate of extension until destruction [14]. In the sequel, the groups of specimens are marked by the corresponding temperatures, i.e. 200 °C is marked 200 and so on.

The fracture tests were carried out using a Heckert FPZ 100/1 testing machine within a range from 0 to 10 kN at a laboratory of the Institute of Building Testing, Faculty of Civil Engineering, Brno University of Technology [15]. Two identical rigid metallic cylindrical supporting rollers and one loading ball were used for performing the three-point bending test. The distance of the supporting roller–specimen contact point from the centre of the beam was 300 mm. The loading was applied to a ball at the top-middle zone of the specimen in order to achieve bending. A load-deflection diagram, which was used to calculate the elasticity modulus of the first part of the diagram, and of the effective fracture toughness, was recorded using an HBM SPIDER 8 device [16].

An acoustic emission measuring system XEDO made by DAKEL (Czech Republic) was used for the acoustic emission measurements. This system consisted of eight channels. The guard sensor eliminated mechanical and electrical noise. In this study, four acoustic type MIDI emission sensors (made by DAKEL) were used all having identical frequency ranges attached to the surface by beeswax [17]. Each sensor has a diameter of 6 mm, height of 6.3 mm its advantage being small sensor & exciter, compact full metal case, low noise, higher wear hardness and high frequency range. The sensors were placed around the expected location of the fracture process zone (Fig. 1).

Fig. 1. Experimental set up.

A specimen was held in pressure and the acoustic emission sensors were placed in three different places as shown in Fig. 1. The acoustic emission measurement started after preloading the specimen. The "continuous" acoustic emission parameters such as acoustic emission energy, count etc. and some events were recorded. The signals (events) recorded were analyzed after finishing the experiment (off-line) by fast Fourier transform.

3 Results

Fig. 2 shows the shapes of stress and strength curves depending on the deformation and elongation respectively for each sample group. For the 20, 200, 400, and 600 sample groups, the curve shapes are similar showing a diminishing maximum strength with growing temperature. The shift of the maximum strength to higher elongations means a higher modulus of elasticity. For the remaining samples, that is, those exposed to temperatures higher that 600 °C, the curve shapes are substantially different. Thus, from this temperature on, the changes of the concrete structure are very marked.

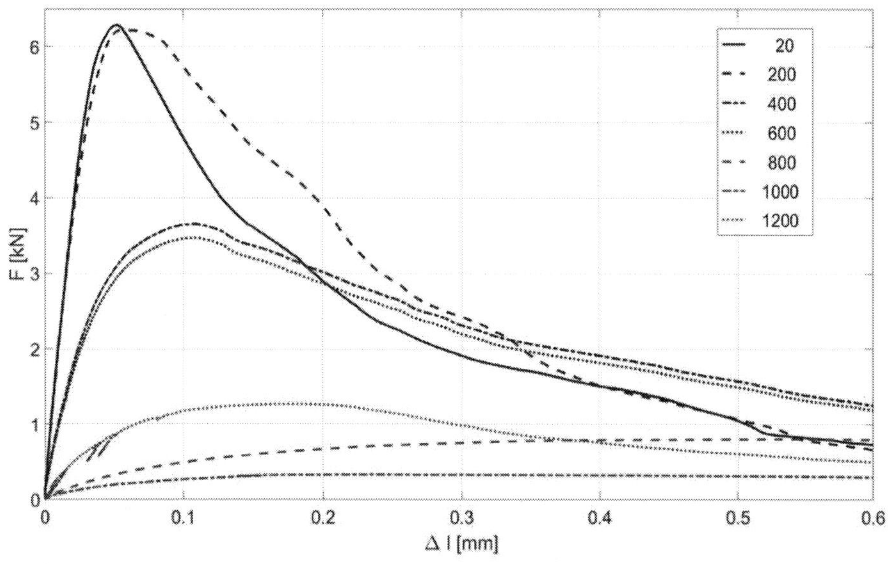

Fig. 2. The graph dependence of force on displacement of the three-point bending test of seven types of specimens.

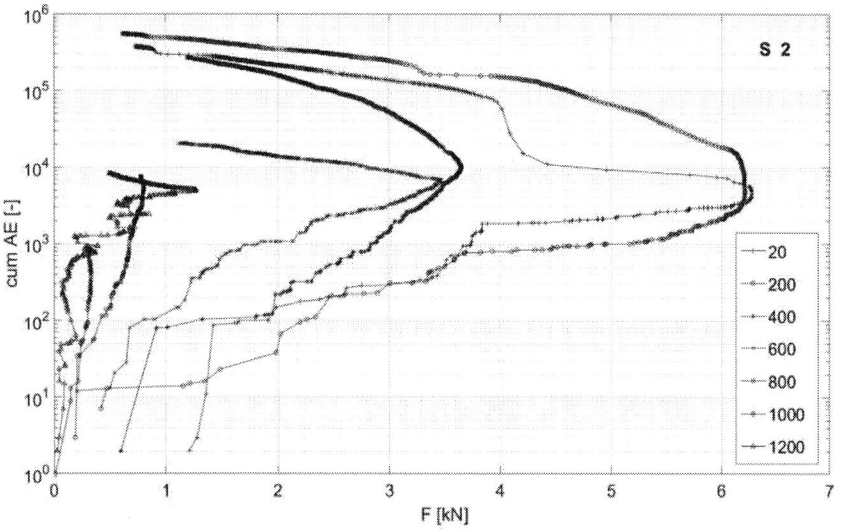

Fig. 3. The graph dependence of cumulative acoustic emission activity on loading of the three-point bending test of seven types of specimens degraded different temperature (in legend).

Fig. 3 shows the acoustic emission activity curve depending on the loading force for each sample group. In samples exposed to temperatures of up to 800 °C, the acoustic emission activity curves are of a very similar nature. Samples 20 and 200 show the possibility of a standard use for building purposes, particularly, with the curve shapes being much the same with growing strength. As can be seen in Fig. 3, the acoustic emission activity sharply increases at approximately the half of the maximum strength.

 DOI: 10.1051/matecconf/201710700040

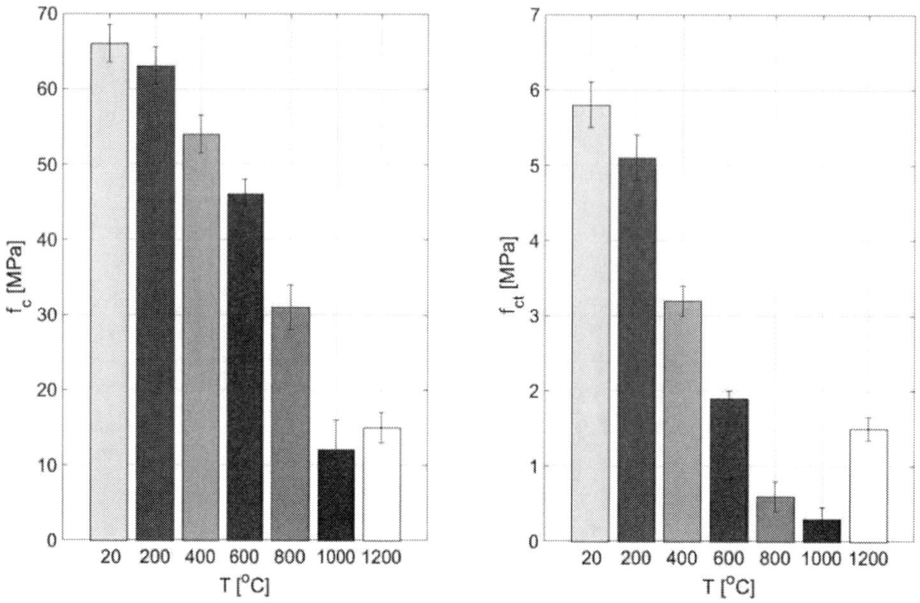

Fig. 4. Compressive strength f_c and flexural bending strength f_{ct} of specimens degraded different temperature T.

Fig. 4 shows the compressive strength and flexural bending strength depending on degradation temperature. The decreasing values of both strengths up to 1000 °C is obvious.

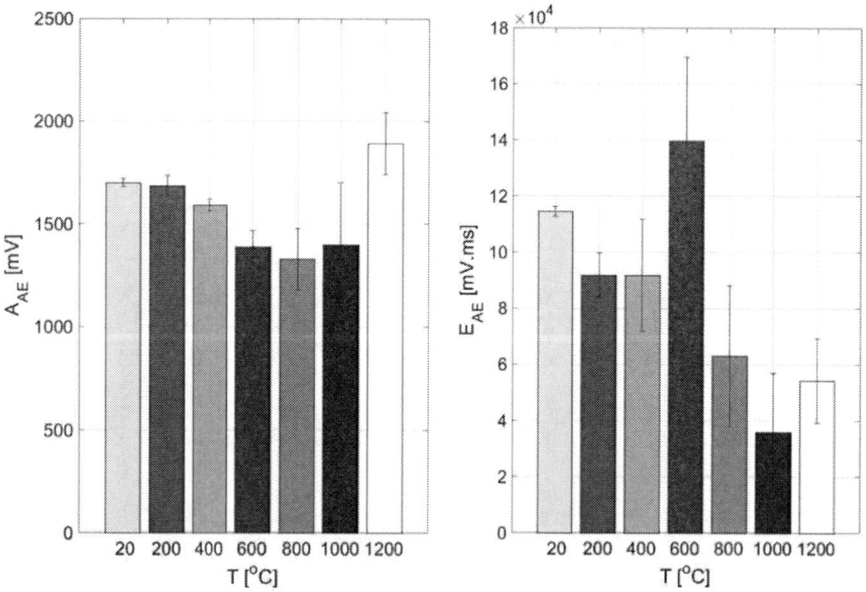

Fig. 5. Acoustic emission amplitude A_{AE} (left) and acoustic emission energy E_{AE} of specimens degraded different temperature T.

Table 1. Acoustic emission events of temperature degraded specimens

Specimens Degraded temperature [°C]	20	200	400	600	800	1000	1200
Number of events Average	140	114	180	68	67	20	86
Number of events deviation	5	6	15	4	7	6	5

The evaluation of the acoustic emission activities of the monitored specimens being rather sophisticated, the graph as shown in Figure 3 can actually only be correctly interpreted by acoustic emission experts with long previous experience. This is the reason why the graphs in Figure 5 have been created for clearer presentation. The average acoustic emission amplitude is decreasing for specimens degraded in the temperature range of 600 °C to 1000 °C, but higher at a degraded temperature of 1200 °C. The material changes are different from a temperature of 800 °C. Acoustic emission energy in Fig. 5 has a shape similar to the compressive strength and flexural bending strength in Fig. 4. A similar result of the number of acoustic emission events is in Table 1, where the progress is similar to the common parameters in Fig. 4.

4 Conclusion

The acoustic emission method provides a detailed description of the behaviour of a sample under load. Monitoring its activity with as the loading strength grows makes it possible to determine by Figure 3 the limit amount of the acoustic emission energy released after which structure deformations can be expected – 1000 °C in our case. Its great advantage is that the amount of the energy released can be determined during the loading.

The behaviour of a sample in a loading temperature over 600 °C is already very difficult to follow, as the sample structure is no longer compact. Note that the samples were measured at a temperature of 20 °C being previously heated to the desired temperature and then cooled down as described earlier.

The experiment describes the sample state under an extreme load such as a fire. After subsequent measurements, the behaviour of new mixes under an extreme temperature load can be defined in terms of the internal energy of the structure, which can be detected by the acoustic emission method very well. The data acquired can be employed to advantage in mathematical models to describe the behaviour of structures.

This paper has been written under the project GAČR No.16-02261S supported by Czech Science Foundation and the project No. LO1408 "AdMaS UP - Advanced Materials, Structures and Technologies", supported by Ministry of Education, Youth and Sports under the „National Sustainability Programme I" and under the projects No. FAST-S-16-2967 and No. FAST-J-17-4554 and No. FAST-S-16-2967 both supported by the Brno University of Technology.

References

1. X.H. Wang, J.J. Xiang, H.W. Hu, W. Xie, X.B. Li, *Ultrasonics*, **60** (2015)
2. K. Ohno, K. Uji, A. Ueno, M. Ohtsu, *Construction and Building Materials*, **67**, SI (2014)

3. Y. Jiang, C.H. Lee, J.P. Wu, *Insight*, **54**, 3 (2012)
4. Y. Xu, D. Xu, J. Qu, X. Cheng, H. Jiao, S. Huang, *Advances in Acoustic Emission Technology Springer Proceedings in Physics*, **158** (2015)
5. Ch. Stergiopoulos, I. Stavrakas, D. Triantis, F. Vallianatos, J. Stonham, *Physica A: Statistical Mechanics and its Applications*, **419** (2015)
6. S. Muralidhara, B.K.R. Prasad, H. Eskandari, B.L. Karihaloo, *Constr Build Mater*, **24**, 4 (2010)
7. J. Blom, M. El Kadi, J. Wastiels, D.G. Aggelis, *Construction and Building Materials*, **70** (2014)
8. J. Smutny, L. Pazdera, *Insight*, **46**, 10 (2004)
9. L. Topolar, L. Pazdera, P. Cikrle, *Applied Mechanics and Materials*, **486** (2014)
10. L. Pazdera, L. Topolar, *Russian Journal of Nondestructive Testing*, **50**, 2 (2014)
11. J. Smutny, *DT & E International*, **37**, 2 (2004)
12. F. Sha, D.Y. Xu, S.F. Huang, X. Cheng, *Applied Mechanics and Materials* (2013),
13. J. Piasta, *Materials and Structures*, **17**, 6 (1989)
14. H. Simonova, I. Havlikova, P. Danek, Z. Kersner, T. Vymazal, *Materiali in Tehnologije*, **49**, 3 (2015)
15. T. Vymazal, N. Zizkova, P. Misak, C*eramics-Silikaty*, **53**, 3 (2009)
16. I. Plskova, Z. Chobola, M. Matysik, *Ceramics-Silikaty*, **55**, 2 (2011)
17. B. Kucharczykova, P. Misak, T. Vymazal, *Russian Journal of Nondestructive Testing*, **46**, 3 (2010)

Strength characteristics of concrete exposed to the elevated temperatures according to the temperature-time curve ISO 834

Iva Rozsypalová[1,*], *Petr* Daněk[1], *Hana* Šimonová[1] and *Zbyněk* Keršner[1]

[1]Brno University of Technology, Faculty of Civil Engineering, Veveří 331/95, 602 00 Brno, Czech Republic

Abstract. While exposed to high temperatures caused by fire, the concrete undergoes a sequence of physical and chemical structure changes causing a mechanical degradation. This paper concerns an experimental strength determination of a temperature stressed concrete. The concrete was temperature loaded according to temperature-time curve ISO 834 and left on a top temperature level for 60 minutes afterwards. This temperature heating rise is in accordance with a common fire expansion in a structure. The concrete panels sized 150 × 1300 × 2300 mm were temperature loaded up to 550, 600, 800 a 1000 °C in a horizontal position in gas furnace for fire testing of structure elements in research Centre AdMaS. The temperatures of the gas in the furnace and the panel was measured during the whole experiment by using the thermocouples. After the fire test, the specimens were drilled out using a 100 mm diameter core drill. The compressive strength and splitting tensile strength tests were made and the results were compared to the reference specimen's test results. The objective results of a commonly used strength class concrete loaded by elevated temperature corresponding to the real fire exposure differ common linear heat exposure test results and are considered being very valuable.

1 Introduction

The fire means a serious danger for the structure safety. Thanks to the fire safety precautions, the significant number of fire accidents is prevented and the consequences concerning loss of lives and property are minimized.

According to EN 1990:2002 [1] the fire related stress is classified as an extraordinary load condition. While designing the structure the fire load and a usual mechanical load is concerned [2].

Despite all the precautions it is still very difficult to prevent many of the fire occurrences and it is necessary to assess the amount of damage and the real condition of a fire-hit structure. Many research teams have been interesting in the fire-stressed concrete

* Corresponding author: iva.rozsypalova@vutbr.cz

behaviour and the general opinion considers both compressive and splitting tensile strength decrease after the fire exposure [3, 4].

The residual strength characteristics after the high temperature exposure depend on several factors. Using a siliceous aggregate generally leads to a lower residual strength compared to a calcareous aggregate usage. However, the statement is valid only up to approximately 700 °C. Above this temperature, the residual strength differences are minimized. The main differences were observed depending on the cooling process. The fast water based temperature drop caused up to 40% lower residual strength compared to the slow and spontaneous cooling [3].

The experiments evaluating the concrete characteristics during or after high temperature exposure are based on the gentle temperature increase in general, after that the maximum temperature is maintained for a specified time. The heating is usually performed in an electronic furnace. The common heating rate for concrete characteristics testing is 0.5 °C/min [4] to 10 °C/min [5]. This slow heating is chosen to restrain the thermal gradients. Specimens are dried slowly, therefore the pore pressure (caused by water evaporation) does not rise up to the high values and thus the thermal damages in the concrete are minimized. Avery high heating rate can cause cracking in the concrete due to the thermal stress [5].

Real fires occurring in existing structures take place very differently from these considered models. The heating rate during a fire in buildings has been studied e.g. the research [6], refers to an experiments in Ostrava, where a compartment in the three-storey steel frame building was exposed to fire before demolition. The wooden cribs were placed into eight piles and created the fire load density per unite area of 1039 $MJ^{-1}m^{-2}$ (which represents the fire load of 58.5 kgm^{-2}). Gas temperature conditions within the fire test are shown in in Fig. 1. Primarily there are shown the records of thermocouples TG1 to TG4, they were located 0.3 m below the ceiling to measure gas temperature in fire compartment in Fig. 1. Thermocouples TG5 and TG6 were placed in front of the fire compartment (0.5 and 1 m from the front wall). Maximum temperature 1050 °C was reached after 62 minutes from an ignition of the wood fuel. The research pointed out to the rapid growth in the early stage of fire when the maximum gas temperature rises up to 550 °C in approximately 8 minutes [6].

An equivalent experiment was conducted in a compartment on the 8-storey steel-concrete composite frame building at the Cardington laboratory during the Building Research Establishment and large-scale fire test. Measured temperatures are depicted in Fig. 2. There is compared temperatures meassured in the comparement with the predicted temperature according to the parametric curve presented in Eurocode 1, Annex A. Despite the more moderate temperature increase after triggering a fire, the maximum temperature of gas reached 1108 °C and was achieved already in 54 minutes [7].

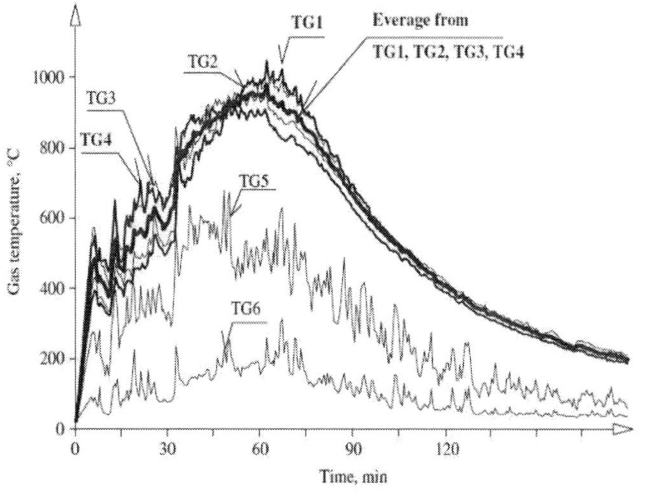

Fig. 1. Measured gas temperatures in the experimental fire compartment (Ostrava) [6].

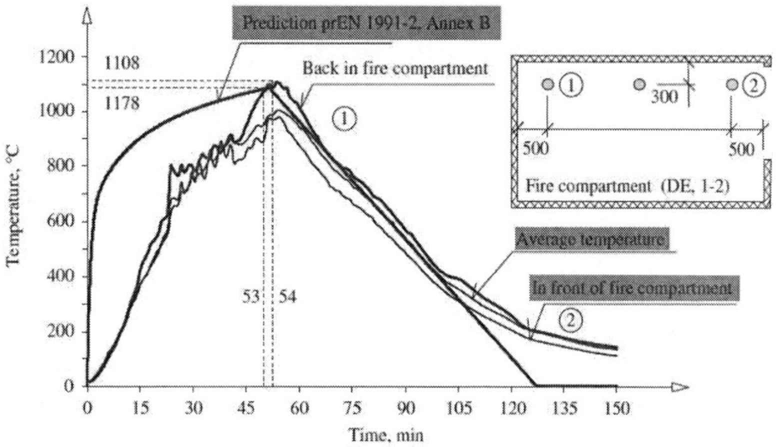

Fig. 2. Measured gas temperature in the experimental fire compartment (Cardington) [7].

According to the previously mentioned experimental results the most realistic heating rate was chosen for realization of this experiment. The fire tests were controlled in accordance with standard temperature-time curve known as ISO 834, followed by heating to the maximum temperature level in the furnace for 60 minutes. ISO 834 curve is established by EN 1991-1-2:2002/AC:2009 as a result of many experiments determining the development of temperatures at ordinary fires conditions in the structures [2].

2 Experiment

The panels of nominal dimensions 150 × 1300 × 2300 mm prepared from commonly used structural concrete of strength class C30/37 were tested to determine the properties of concrete after the experimental simulation of real fires.

There were casted 5 panels in total for fire tests at 550, 600, 800 and 1000 °C and one reference panel without heating. The panels were reinforced by reinforcing mesh on the bottom face. Total of 10 thermocouples (type K) was placed into each plate for temperature measurement during thermal tests. Thermocouples were positioned along the height of the three sections of the panels so that can be recorded temperatures on the heated and the unheated surface of the plate and in the middle of the panel thickness. Moreover, the midspan cross-section temperatures were controlled also at the one quarter of the height of the panel thickness from the bottom (heated) side of panel.

The panels were thermally loaded by using gas furnace for studying of building materials and construction elements at high temperatures in AdMaS Research Centre, see the photo in Fig. 3. The furnace chamber is heated by two massive gas burners placed diagonally on the shorter sides of the furnace. The aim of research was to simulate the load acting on the concrete elements in the conventional building construction, therefore, the material was heated according to the standard temperature-time curve ISO 834 up to the required nominal temperatures 550, 600, 800 and 1000 °C, after achieving the nominal temperature were maintained for 60 minutes. Then the panels were allowed to cool to ambient temperature. The real history of gas temperature in the furnace during experiment is illustrated by the temperature-time curves shown in Fig. 4. Concretes were loaded at 5 months of age, which meets the requirements for minimum age specified in Recommendation of RILEM TC 129-MHT [8].

Fig. 3. Gas furnace for fire testing of the structural elements at AdMaS Centre during heating of panel P7 (maximal temperature 1000 °C).

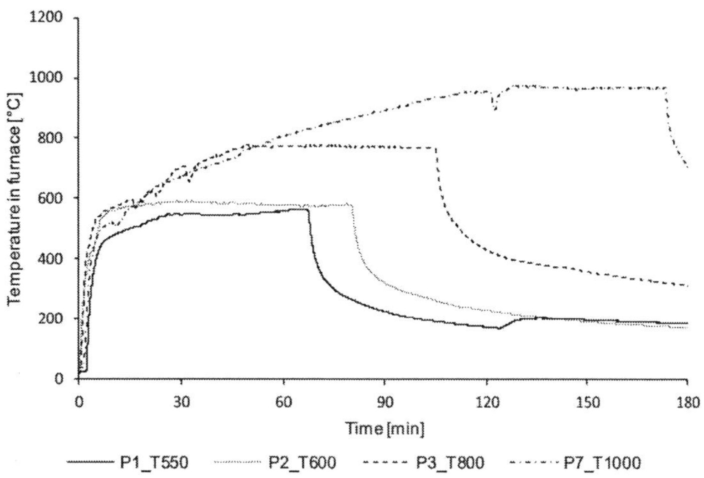

Fig. 4. Selected temperature-time curves measured in furnace during experiments carried out in AdMaS Centre.

Samples for testing were taken out by core drilling after cooling. Six cores (cylinders) of 100 mm diameters were got from each panel. Thereafter the cores were shortened to the 100 mm length (the larger thickness of material was removed from the side without thermal loading to determine the most unfavourable result). The specimens were tested for the compressive strength and the splitting tensile strength in accordance with the procedures specified in EN 12390-3:2009 [9] and EN 12390-6:2009 [10]. Furthermore, for these panels exposed to high temperatures were determined basic fracture parameters by means of evaluation of pilot fracture tests of cylindrical specimens with an initial stress concentrator in the form of a chevron notch and subsequently tested in three-point bending test, see more in [11].

3 Results

The values of compressive strength and splitting tensile strength of concrete exposed to the elevated temperatures according to the temperature curve of ISO 834 prolonged by the maximum temperature maintaining for one hour were determined by evaluating the experiments.

Reference specimens without temperature exposition reached a mean compressive strength 54.0 MPa with a coefficient of variation 4.4%. After thermal subjection was noticed a reduction of concrete strength characteristics, the results are shown below. The heating up to 1000 °C reduced the strength to 66%. Especially, a significant increase in the variation of the compressive strength of the average values for specimens, the values exceed the 13 % after heating to 600 and 800 °C. Results of compressive strength tests are shown in Fig. 5 and in Table 1.

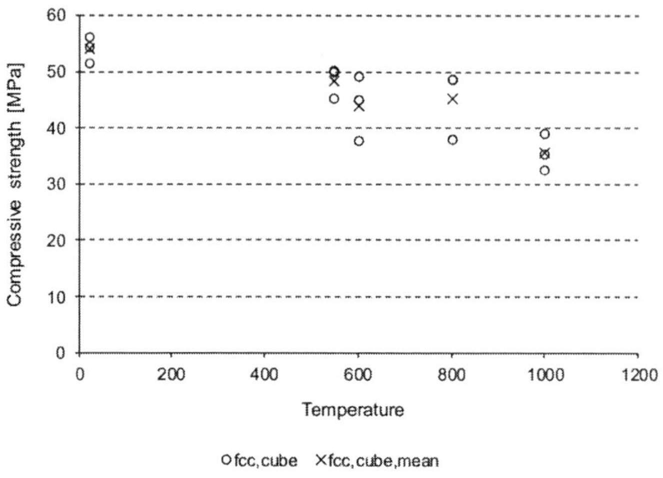

Fig. 5. Results of compressive strength tests of concrete exposed to elevated temperatures according to temperature curve ISO 834.

Table 1. Compressive strength (and bulk density) of concrete exposed to elevated temperatures.

specimen	gas temperature [°C]	ρ [kg.m^{-3}]	fcc,cube [MPa]	fcc,cube,mean [MPa]	coefficient of variation [%]
P4_C5	20	2263	56.1	54.0	4.4
P4_D4		2261	54.4		
P4_E3		2254	51.5		
P1_3C	550	2260	45.3	48.4	5.6
P1_5C		2226	49.8		
P1_4F		2293	50.1		
P2_9F	600	2255	37.8	44.0	13.1
P2_7CD		2226	49.2		
P2_9D		2253	45.0		
P3_8D	800	2211	38.0	45.1	13.6
P3_7DE		2230	48.7		
P3_7FE		2247	48.7		
P7_87ED	1000	2229	39.0	35.7	9.1
P7_87E		2206	32.6		
P7_9E		2197	35.5		

The splitting tensile strength results of unheated specimens reached a mean value 3.3 MPa with a coefficient of variation about 7%. Specimens with temperature loading up to 1000 °C decrease the strength to 65%. There were remarkable changes in the variation coefficient, it is about 12–23% for exposed material. Results of splitting tensile strength tests are shown in Fig. 6 and in Table 2.

 DOI: 10.1051/matecconf/201710700041

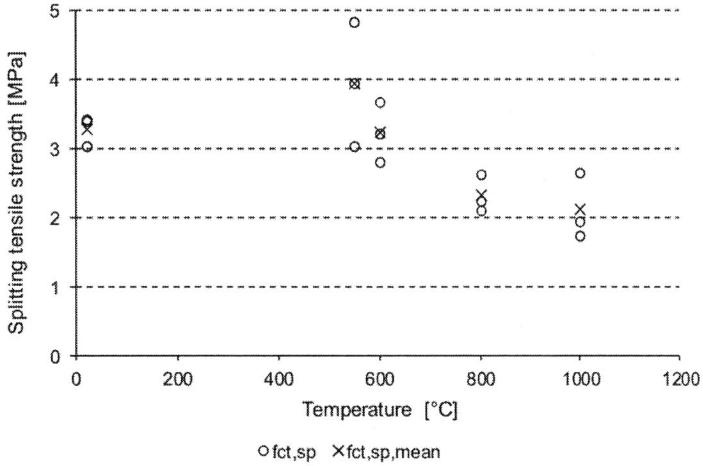

Fig. 6. Results of splitting tensile strength tests of concrete exposed to elevated temperatures according to temperature-time curve ISO 834.

Table 2. Tensile splitting strength (and bulk density) of concrete exposed to elevated temperatures.

specimen	gas temperature [°C]	ρ [kg.m^{-3}]	fct,sp [MPa]	fct,sp,mean [MPa]	coefficient of variation [%]
P4_3B	20	2305	3.41	3.28	6.8
P4_6B		2472	3.40		
P4_6D		2265	3.02		
P1_5D	550	2263	3.02	3.93	23.1
P1_3D		2272	3.93		
P1_5E		2292	4.83		
P2_7DE	600	2248	2.80	3.22	13.4
P2_9C		2246	3.66		
P2_7FE		2122	3.21		
P3_7D	800	2222	2.09	2.31	11.7
P3_7CD		2192	2.61		
P3_9D		2211	2.23		
P7_8D	1000	2226	2.64	2.10	22.7
P7_87CD		2232	1.94		
P7_10D		2260	1.73		

4 Conclusion

The outcomes of this article are compressive and splitting tensile strengths of concrete after high temperature exposition according to standard temperature-time curve ISO 834. It corresponds to the heating rate in real fires in structures. Rapid temperature increment causes development of cracks in concrete. The values of strength characteristics were affected by this phenomenon: compared to the reference concrete specimens, the maximum decrease was about 35%. Increase of variation coefficient was observed in both cases of strengths, significantly in values of splitting tensile strength. Knowledge of compressive and splitting tensile strength values determined above method are suitable for use in practice to assessment of concrete structures damaged by fire.

This outcome has been achieved with the support of the project No. LO1408 "AdMaS UP – Advanced Materials, Structures and Technologies" supported by Ministry of Education, Youth and Sports under the „National Sustainability Programme I".

References

1. EN 1990:2002, *Eurocode: Basis of structural design* (CEN, Brussels, 2002)
2. EN 1991-1-2:2002/AC:2009, *Eurocode 1: Actions on structures – Part 1-2: General actions – Actions on structures exposed to fire* (CEN, Brussels, 2002)
3. I. Hager, Bull Pol Acad Sci-Te **61**, 1 (2013)
4. K. Krzemień, I. Hager, Constr Build Mater **96,** 155–163 (2015)
5. R. Felicetti, F. Lo Monte, P. Pimisenta, F. Meftah, *MATEC web conf.* 6 (2013)
6. F. Wald, J. Chlouba, A. Uhlíř, P. Kallerová, M. Štujbetová, Fire Safety J. **44** (2009)
7. F. Wald, L. Simoñes Da Silva, D. B. Moore, T. Lennon, M. Chladná, A. Santiago, M. Beneš, L. Borges, Fire Safety J. **41** (2006)
8. RILEM TC 129-MHT, Mater Struct, **33,** 224-228 (1997)
9. EN 12390-3:2009, *Testing hardened concrete – Part 3: Compressive strength of test specimens* (CEN, Brussels, 2009)
10. EN 12390- 6:2009, *Testing hardened concrete – Part 6: Tensile splitting strength of test specimens* (CEN, Brussels, 2009)
11. H. Šimonová, P. Halfar, I. Rozsypalová, P. Daněk, Z. Keršner, *MATEC web conf.* (to be published)

Parameter identification for coupled elasto-plasto-damage model for overheated concrete

Andrey Benin[1,*], *Sergey* Semenov[1], *Artem* Semenov[2] and *Galina* Bogdanova[1]

[1]Emperor Alexander I St. Petersburg State Transport University, N.A. Beleloubski Laboratory of Mechanics, 190031, Moskovsky pr. 9, St. Petersburg, Russia.
[2]Peter the Great St. Petersburg Polytechnic University, Mechanics and control processes, 195251, Polytechnicheskaya st. 29, St. Petersburg, Russia.

Abstract. The methodology of experimental acquisition and subsequent processing of data for identification of nonlinear elasto-plasto-damage concrete model parameters are considered. Typical measurement equipment and data acquisition process are described. The experiments were carried out on standard specimens of cubic and prismatic form under compression and dog-bone-shaped specimens under tension. Elasticity modulus in tension and compression, ultimate strengths in compression and tension, damage evolutions during deformations process and other mechanical characteristics were obtained. The results of parameter identification and validation for typical structural concrete including regular and overheated concrete are presented and discussed.

1 Introduction

The elasto-plastic models with account of damage [1-3] provides effective tools for the modelling nonlinear concrete behavior with taking into account irreversible deformation, softening effect, inelastic volumetric expansion and stiffness degradation. There are many implementations of such type models, for example, in commercial software ABAQUS [4]. One of the main difficulties is an identification of the model parameters for the computation of real concrete and reinforced concrete structures [5-8]. There are few resent researches aimed to obtain all parameters for elasto-plasto-damage model [9, 10]. Calibration of concrete parameters based on digital image correlation was proposed in [11]. Aims of the current research are:

- design of test setup for the correct control test and obtaining stress-strain curves;
- developing of identification methodology for model parameter obtaining (excepting biaxial loading) for finite-element computations;
- carrying out experiments using recommended standard specimens [12], identifying model constants;
- validation of identified parameters using finite-element computations.

* Corresponding author: benin.andrey@mail.ru

2 Test setup

The test setup (Fig. 1a) is based on Shimadzu AGX300 electromechanical test machine. The fixtures (Fig. 1b) for tensile tests and deformation sensor frame (Fig.1c) were designed and produced.

a)　　b)　　c)

Fig. 1. Test machine (a), tensile test fixtures (b) and deformation sensor frame (c).

There are two Heidenhain ST1278 length gauge sensors used for strain measurement. The sensors were interfaced with test machine controller and result strain was calculated as average of two sensors measurements for neglecting influence of bending effect. The strain sensor frame was installed directly on the specimen. This approach allows obtaining the direct strain measurement during testing and as result to measure correctly elastic modulus (see, for example, comparison of modules measured by different methods in table 2 [10]). Such design allows to perform experiments in the temperature chamber and using quartz glass tubes for precisely strain measurement.

3 Specimens

Specimens were prepared according to the recommendation of Russian standard [12]. There are three types of specimens:

- cubes 100×100×100 mm for concrete class (cube strength) definition,
- prisms 100×100×400 mm for compression experiments,
- dog-bone specimens for tension experiments.

Concrete mixture recipe is 400 kg/m^3 of CEM I 42.5R, 700 kg/m^3 of 0/4 gabbro sand, 1125 kg/m^3 of 5/10 gabbro coarse aggregate, 120 kg/m3 of fly-ash, 230 kg/m^3 of water and 35 kg/m^3 of plasticizers and modifiers. Specimens were casted in standard steel forms, extracted from forms after 3 days and cured in the 95% humidity and 20°C temperature environment during 28 days. The standard cube strength class of prepared specimens is B50.

A part of specimens have a heating treatment up to the maximum temperature 600°C during 1 hour. The heating rate was 10°C/h.

4 Experimental results and parameter identification

Fig. 2a and b represent typical cyclic stress-strain diagrams containing the post-peak behaviour under compression and tension for the specimens from the overheated concrete. These curves are basis for the parameter determination of the evolution law for the plastic strain and damage.

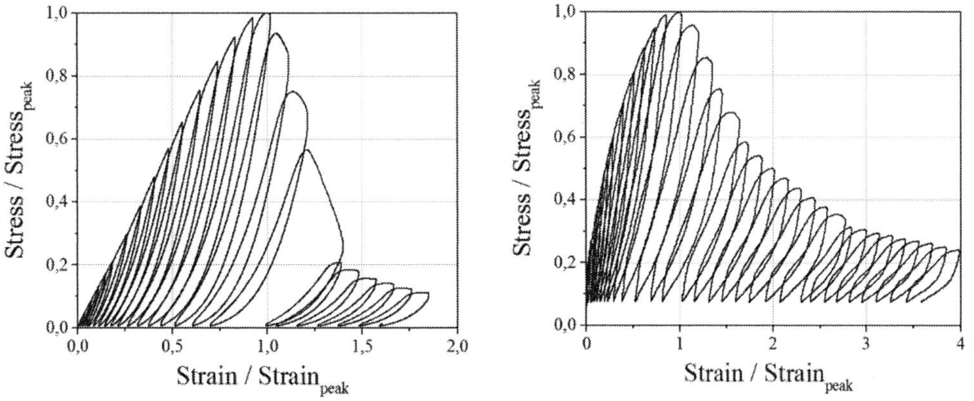

a) b)

Fig. 2. Typical compression (a) and tension (b) cyclic diagrams for overheated concrete.

The progressive elastic stiffness degradation from cycle to cycle provides information for the damage calculation. The damage (in compression) is considered as a scalar variable D, which is equal to 0 for virgin material and equal to 1 at the failure. The damage variable at k-th cycle can be evaluated as:

$$D_k = 1 - E_k / E_0, \tag{1}$$

where E_0 is the value of the undamaged modulus and E_k is the values of the damaged moduli at the at k-th cycle (see Fig. 3).

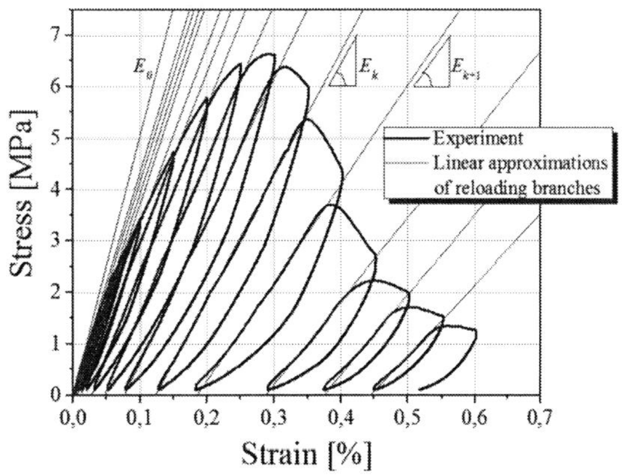

Fig. 3. Illustration of the determination of the elastic reloading moduli of the damaged material.

Due to non-linear character of both unloading and reloading branches of cyclic stress-strain curve it is non-trivial problem to calculate appropriate elastic moduli of the damaged material. The loading and reloading branches, tangential, averaged and secant slopes can be used. The results of multivariant computations show that the elastic reloading moduli of the damaged material determined as the slopes of linear best-fit approximations of the ascending parts of the experimental reloading branches (see Fig. 3) provide the best variant.

For the further analysis it is rational to recalculate the damage as function of inelastic strain instead of function of number of cycles. The axial inelastic strain is determined as

$$\varepsilon^{in} = \varepsilon - \sigma / E_0, \quad (2)$$

The results of damaged elastic moduli and damage calculation as function of inelastic strain is given in Fig. 4 for regular and overheated concrete.

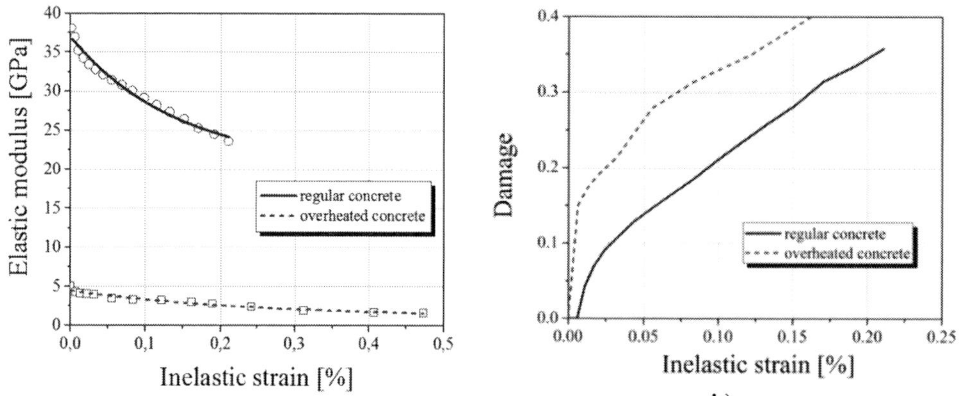

a) b)

Fig. 4. Elastic modulus degradation (a) and damage evolution (b) with increasing of inelastic strain for regular and overheated concretes.

5 Constitutive equations

The constitutive equation of elastic-plastic material with scalar isotropic damage for three dimensional general case takes the following form:

$$\boldsymbol{\sigma} = (1-D)\cdot {}^4\mathbf{C}_0^e \cdot\cdot (\boldsymbol{\varepsilon} - \boldsymbol{\varepsilon}^p) = {}^4\mathbf{C}^e \cdot\cdot (\boldsymbol{\varepsilon} - \boldsymbol{\varepsilon}^p), \quad (3)$$

where $\boldsymbol{\sigma}$ is Cauchy stress tensor, $\boldsymbol{\varepsilon}$ is the strain tensor, $\boldsymbol{\varepsilon}^p$ is the plastic strain tensor, ${}^4\mathbf{C}_0^e$ the initial (undamaged) elastic stiffness of the material, while ${}^4\mathbf{C}^e = (1-D)\cdot {}^4\mathbf{C}_0^e$ is the degraded elastic stiffness tensor. The effective stress tensor is defined by the relation $\overline{\boldsymbol{\sigma}} = {}^4\mathbf{C}_0^{el} \cdot\cdot (\boldsymbol{\varepsilon} - \boldsymbol{\varepsilon}^p) = (1-D)^{-1}\boldsymbol{\sigma}$.

The evolution of the scalar degradation variable is defined by the function $D = D(\overline{\boldsymbol{\sigma}}, \tilde{\varepsilon}^{in})$, governed by a set of the effective stress tensor $\overline{\boldsymbol{\sigma}}$ and hardening (softening) variables $\tilde{\varepsilon}^{in}$. In the used in further Lubliner model [2], the stiffness degradation is initially isotropic and defined by degradation variable D_c in a compression zone and variable D_t in a tension zone.

Plastic flow is governed by a flow potential function $G(\overline{\boldsymbol{\sigma}})$ according to non-associative flow rule:

$$\frac{d\boldsymbol{\varepsilon}^p}{dt} = \lambda \frac{\partial G(\overline{\boldsymbol{\sigma}})}{\partial \overline{\boldsymbol{\sigma}}} . \tag{4}$$

The flow potential G is accepted in the form:

$$G = \sqrt{(f_c - m \cdot f_t \cdot \tan\beta)^2 + 3\bar{J}_2} + \tfrac{1}{3}\bar{I}_1 \cdot \tan\beta , \tag{5}$$

where f_t and f_c are the uniaxial tensile and compressive strengths of concrete, respectively, β is the dilation angle measured in the $\frac{1}{3}\bar{I}_1 - \sqrt{3\bar{J}_2}$ plane at high confining pressure, while m is an eccentricity of the plastic potential surface, $\bar{I}_1 = \mathbf{1} \cdot\cdot \overline{\boldsymbol{\sigma}}$ and $\bar{J}_2 = \frac{1}{2} dev\overline{\boldsymbol{\sigma}} \cdot\cdot dev\overline{\boldsymbol{\sigma}}$.

6 Validation

The results of comparison of simulation results with experimental data under monotonic compression for regular and for overheated concrete are given in Fig. 5. A good agreement is observed in both cases. The model parameters are identified with help of the relations (1) and (2). The constitutive equations (3)-(5) are used in simulations. Note that the considered approach allows to describe strain hardening, softening and post-peak behaviour under compression and tension

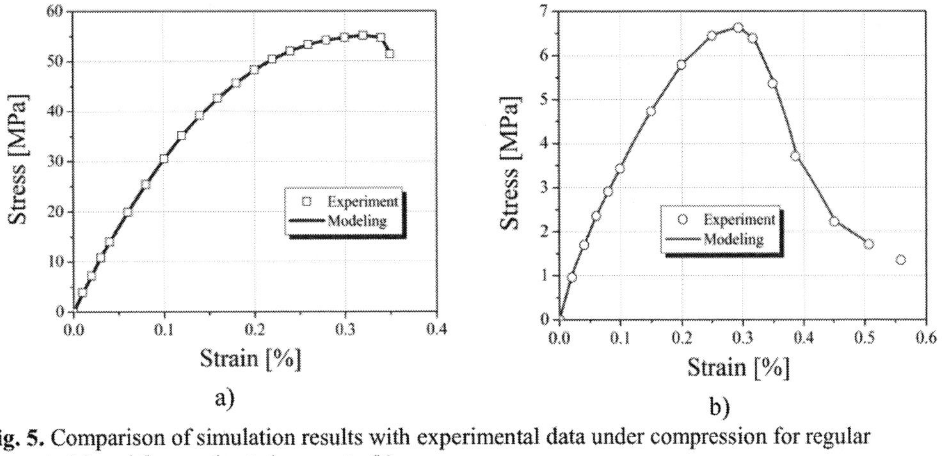

Fig. 5. Comparison of simulation results with experimental data under compression for regular concrete (a) and for overheated concrete (b).

The validation results of proposed identification procedure under cyclic loading for regular and for overheated concrete are given in Fig. 6. Prediction accuracy is lower in comparison with monotonic case, but a satisfactory agreement results of experiment and simulations is observed in this case too.

The considered model is limited by the possibility to predict only the linear unloading. This reduces the accuracy of the computations. One of the possible ways to further improve the material model is to consider structural or multisurface models.

Examples of further application of the considered elasto-plasto-damage model in solving real problems of practice are given in [5-8].

DOI: 10.1051/matecconf/201710700042

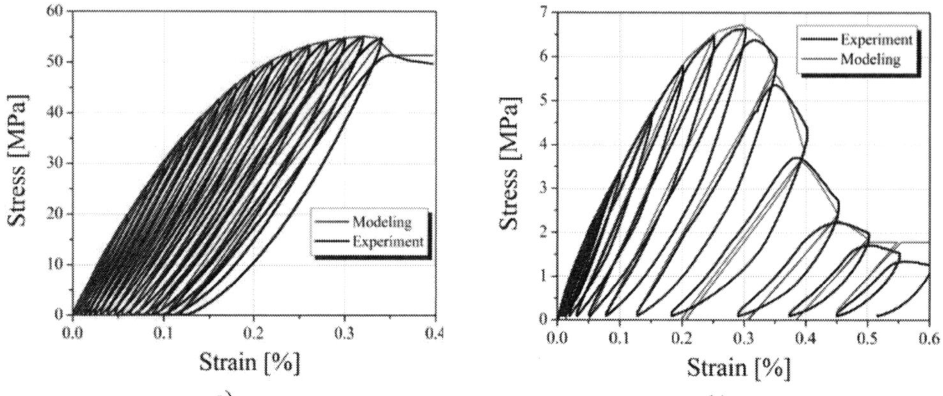

Fig. 6. Comparison of simulation results with experimental data under cyclic compression for regular concrete (a) and for overheated concrete (b).

7 Conclusion

The parameter identification procedure of nonlinear elasto-plasto-damage concrete model is proposed and validated. The results of multivariant calculations of damage on the base of experimental cyclic stress-strain curves under compression show that the elastic reloading moduli of the damaged material determined as the slopes of linear best-fit approximations of the ascending parts of the experimental reloading branches provide the best variant. Comparison of simulation results with experimental data under monotonic and cyclic compression demonstrates a good agreement for regular and for overheated concrete.

This research was partially supported by the Russian Science Foundation (grant No. 15-19-00091).

References

1 J. Mazars, G. Pyaudier-Cabot, J. Eng. Mech. **115**, 345-365 (1989)
2 J. Lubliner, J. Oliver, S. Oller, E. Oñate, A, Int. J. Solids Struct. **25**, 299–326 (1989)
3 J. Lee, G.L. Fenves, J. Eng. Mech. **124**, 892–900 (1998)
4 ABAQUS User's manual, ver. 6.12 Abaqus Inc. (2012)
5 A. Benin, A. Semenov, S. Semenov. Advances in Civil Engineering and Building Materials Proc. 2nd Int. Conf. on Civil Eng. and Building Materials, 233-237 (2012)
6 A.V. Benin, A.S. Semenov, S.G. Semenov, B.E. Melnikov. Magazine of Civil Eng. **40**, 86-99 (2013)
7 A.V. Benin, A.S. Semenov, S.G. Semenov, B.E. Melnikov. Magazine of Civil Eng. **45**, 23-40 (2014)
8 A.V. Benin, A.S. Semenov, S.G. Semenov. Advanced Materials Research. **831**, 364-369 (2014)
9 T. Jankowiak, T. Lodygowski, Found. Civ. Environ. Eng. **6**, 53-69 (2005)
10 M. Neuenschwander, M. Knobloch, M. Fontana, Cem. Concr. Res. **79**, 57–75 (2016)
11 GOST 10180-12 Concretes. Methods for strength determination using reference specimens

Comparison of creep of the cement pastes included fly ash

Pavel Padevět[1,*] and *Petr* Bittnar[1]

[1]CTU in Prague, Faculty of Civil Engineering, Thakurova 7, Prague 6, 166 29, Czech Republic

Abstract. The paper is devoted to comparison of creep of cement pastes containing fly ash admixture. The size of creep in time depends on the amount of components of the cement paste. Attention is paid to the content of classical fly ash in cement paste and its impact on the size of creep. The moisture of cement pastes is distinguished because it significantly affects the rheological properties of the material.

1 Introduction

The size of creep of cement pastes among the properties of materials used in the design of building structures. Cement pastes are the basic parts of a composite called concrete. Of course, the concrete also contains ingredients sand and gravel. But the basis is the same, a mixture of cement and water.

The properties of the cement paste can be influenced by additives and admixtures [1]. This paper deals with a mixture containing fly ash. The fly ash in cement paste causes secondary hydration. Lowers the bulk density can vary the mechanical properties and also can affect the size and development of creep.

Production of ash from coal-fired power plants in the Czech Republic is high against production in other countries. The fly ash is formed in the combustion of coal. The fly ash is not considered as a waste today but as raw material which is use for example in the construction industry. Type of fly ash affects the properties of the concrete, as well. The fly ash which is generated in low-temperature combustion of coal is not in our angle of interest because its SO_3 content adversely affects the material properties of the concrete over time. Conversely, the ash which is formed during high temperature combustion of coal is suitable for use in concrete. Such fly ash is called classical. Therefore, this work is dedicated to measuring of creep of the cement pastes containing classical fly ash.

The addition of fly ash to cement or concrete is one of the possible uses. Amount of fly ash affects the rate of increase of the concrete strength and the amount of generated heat of hydration in concrete.

The concrete containing fly ash has a slower evolution of hydration heat [2]. This beneficial effect is preferably applied in massive structures. Evolution of compression strength of concrete with fly ash is slower. Structures which do not require high strength at

* Corresponding author: pavel.padevet@fsv.cvut.cz

the beginning it is possible to suitably realize from concrete containing fly ash. How affect the content of fly ash in cement paste the size of creep, attempts to answer this paper.

2 Measurement of creep

Measurements are carried out on small specimens measuring 70 mm in length and 10 mm diameter. Bodies had the shape of cylinder. Small specimens are suitable for measurement of creep and shrinkage in the test apparatus. Measurements are taken in lever mechanisms, which are always equipped with three optoelectronic gauges of deformation.
The article presents a comparison of these mixtures:

- 50:50
- 60:40
- 70:30
- 100:0

Where, the first number represents the content of cement, and the second one expresses content of fly ash in the mixture. The ratio expresses the weight representation of both parts. Water ratio of mixture was 0.4. It is defined as the ratio between weight of water and cement. Fly ash is not considered in water-cement ratio is considered as filler. The mixtures were prepared from Portland cement CEM I 42.5 R.

All experiments were carries out in one year age after production. All sets were consisted of six specimens. Sets of bodies with a content of fly ash contained 4 samples of dried, and 2 samples water saturated. Set of specimens without content of fly ash contained 3 samples of dried and 3 samples saturated.

The drying process had length 24 hours at the temperature 105°C. Water saturated specimens were stored in water basin for 24 hours, too. The length measurement was usually between 25-30 days.

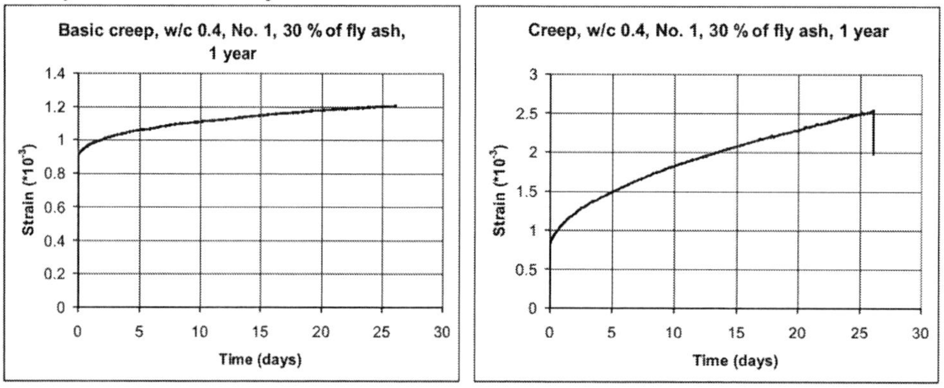

Fig. 1. Graphs of creep cement paste with 30% fly ash content. On the left the basic creep of the dried paste. On the right is the creep of water-saturated paste.

Basic creep is calculated as the difference of creep of the dried paste and shrinkage of the dried paste. On the other hand, data of creep of water saturated paste are used from measurement on the specimen. Creep of saturated pastes contains the influence of their shrinkage.

The values in the Table 1 are related to the 25 day of measurement. The size of creep is expressed in dimensionless units -milistrains. The basic unit is the meter. The values in graphs are related to units of millimeters. The value of basic creep in the table is the average of measurements on the two specimens and without influence of the shrinkage. Relative values are calculated to the length of bodies of specimens. Creep values are related to measurements of two specimens, but with influence of the shrinkage. Positive values in the table reflect the shortening of bodies.

Table 1. Size of creep after 25 days.

Cement : fly ash	**Basic creep (mm/mm ×10^{-3})**	**Creep (mm/mm ×10^{-3})**
60:40	0.210	1.827
70:30	0.256	0.803
50:50	0.339	2.050
100:0	0.279	1.103

3 Results of measurements

Fig. 2 and 3 clearly shows the results of creep, which are listed in Table 1. The result of basic creep of cement pastes containing fly ash is very interesting. The range between the highest and lowest creep is 0.129 milistrain. This is a relatively small margin. The value of creep of pure cement paste is between the values of creep. Cement pastes with fly ash in dry condition achieves low values of creep. Nevertheless, it is evident that with increasing ash content in the cement paste increases the size of creep.

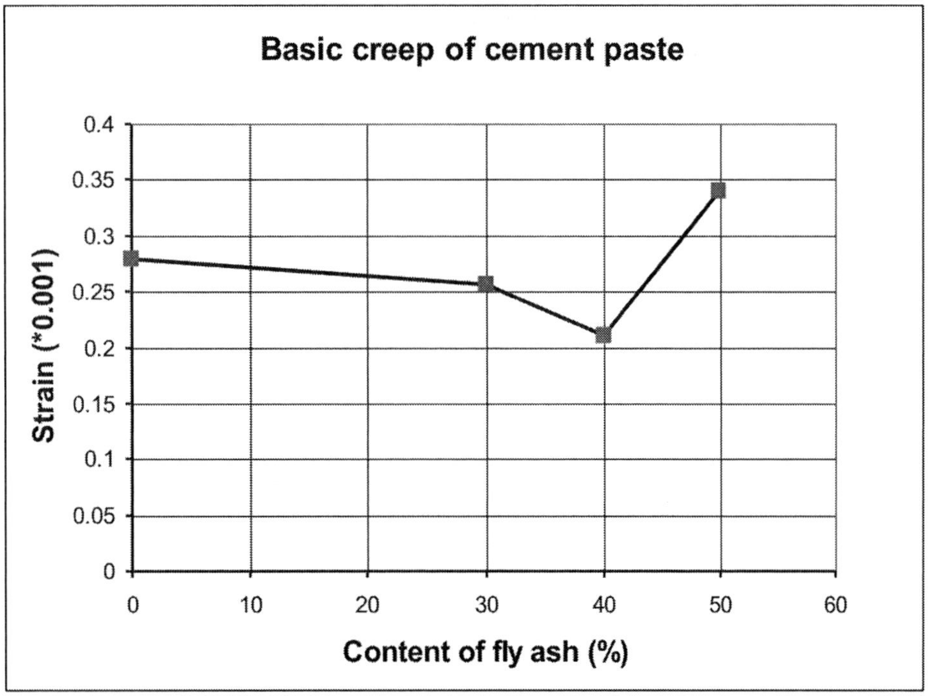

Fig. 2. Results of basic creep of the cement pastes with content of fly ash.

Water saturated cement paste with fly ash reached higher values of creep than the creep of dried pastes. The growth of strain is related to the growth of fly ash content in cement paste.

Increasing the fly ash content of 20% has led to an increase of strain by more than 100%. Creep of pure cement paste corresponds to the creep pastes containing 32% of fly ash.

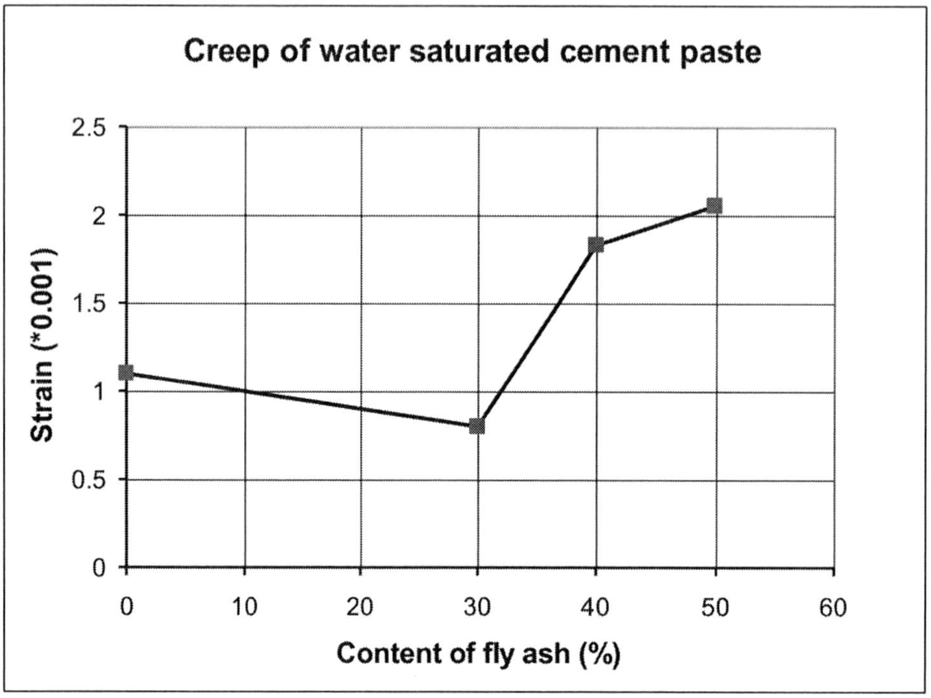

Fig. 3. Results of creep of the water saturated cement pastes with content of fly ash.

4 Summary

Enlarging the size of the creep occurs together with increasing fly ash content in the cement paste. Amount of free water has a significant impact on the size of creep. Dried cement paste with fly ash has a small scattering of creep sizes for different amount fly ash in pastes. In contrast, the size of creep of water-saturated cement paste is higher with increasing ash. The most suitable amount of classical ash in the saturated and dried cement paste is 30%.

This work has been supported by project SGS project No.16/201/OHK1/3T/11 "Advanced from design of materials to the constructions".

References

1. P. Padevět, T. Otcovská, O, Zobal, Variation of Material Properties of Cement Pastes with Various types of Fly Ash During Maturation, WSEAS Transactions on Applied and Theoretical Mechanics **9**, 88-96 (2014)
2. D. P. Bentz, E. Schlangen, and E. J. Garboczi, *Materials Science of Concrete IV*, (edited by J.P. Skalny and S. Mindess, American Ceramic Society, Westerville, OH, pp. 155-200,1994)

MATEC Web of Conferences **107**, 00044 (2017) DOI: 10.1051/matecconf/201710700044
DYN-WIND'2017

Substantiation of concrete core rational parameters for bending composite structures

Glib Vatulia[1], *Svetlana* Berestianskaya[1], *Elena* Opanasenko[1] and *Anastasiya* Berestianskaya[1,*]

[1] Ukrainian State University of Railway Transport, Structural Mechanics and Hydraulics Department, Feyerbakh sq. 7, 61050 Kharkiv, Ukraine

Abstract. In order to provide bending structures rationalization for reducing the materials consumption, labor and power inputs, construction or renovation terms, the authors considered the possibility of utilizing the structures with external steel sheet reinforcement and concrete layer made from fibers of different types. Experimental researches of various authors, both domestic and overseas, have been analyzed during the preliminary investigations. As a result, the steel and basalt fibers were selected for further inquiry, proved their rational sizes, percentage to concrete mass in structures worked under thermal and force impacts. It was developed the algorithm and software, helps to determine the stress-strain state and carrying capacity of composite floor slabs with different end and load conditions. It was concluded the necessity of physical-mechanical and thermal physic properties clarification of heated fibrous concrete. The experiment planning was performed to obtain the temperature dependences of strength and modulus of deformation, thermal conductivity and specific heat capacity of fibrous concrete mix.

1 Introduction

Since the concrete has been ever more extensively used in civil engineering, and concrete structures have been operated in more severe conditions, concrete quality should constantly improve to achieve better strength, crack resistance, fire resistance, etc. Today, in addition to using traditional steel concrete, structures with modified concrete are used more widely with introduction of various additives, including fibers that allow enhancing of the performance of the construction material. Fibers make the material more durable by binding the internal structure. Adding fibers to concrete and their disperse location in the material volume allowed to formulate the concept of the concrete (cement) matrix-based composite material. Qualitative characteristics of fiber-reinforced concrete are several times better than those of the conventional concrete [1].

The scope of application of fibrous concrete is very wide. Each field of application imposes specific requirements to fibrous concrete constructions concerning both mechanical and rheological properties. For instance, one of the important characteristics of flexural elements is bending rigidity. Fibrous concrete enables to increase significantly

* Corresponding author: anastasiia.berestianska@gmail.com

strength against this kind of deformation [2]. In [3] experimental studies of different authors were analyzed and a conclusion on expediency of use of metal and basalt fiber for thin slabs was drawn.

2 Types of fiber reinforcement

2.1 Steel fiber

Steel fiber is a durable and generally recognized material for quality improvement of the concrete [4]. Due to some specific properties, steel fiber reinforced concrete outperform ordinary concrete and has significant share in the total global production (12-16%) of concrete. Metal fiber, unlike the polypropylene and glass ones, improve mechanical characteristics of concrete after strength gain, i.e. have forcing functions.

Steel fiber reinforced concrete has increased tear strength, practically never shows shrinkage or cracks in use. A number of prerequisites should be met to obtain high-performance steel fiber reinforced concrete: fibers should have equal properties and nominal size, have good adhesive bonding to matrix and concrete, be evenly distributed in concrete matrix, and their material should prevent occurrence and progress of corrosion and chemical reaction with the matrix material. Concrete reinforcement with metal fiber promotes improvement of its strength properties: ultimate tensile, compressive and bending strength increase, as well as its impact resistance [5]. Deformability, durability, wear resistance, frost resistance, heat resistance, water tightness, reliability and corrosion resistance of concrete structures increases significantly. Improvement of physical and mechanical properties of steel fibrous concrete enables to reduce weight of concrete construction.

Fiber diameter and its length influence the properties of steel fiber concrete. The cross-section shape of steel fiber can vary, so equivalent diameter is often used. The efficiency of fiber reinforcement increases with decreasing diameter, however, it also makes the mixture preparing and product manufacturing technologies more complicated. The length-to-diameter ratio, shape, as well as the condition of the surface are important characteristics of the fiber which affect its physical mechanical properties and stress-strain behavior. The greater the saturation of the mixture with fiber is, the more contact zones there are and the better the strength and deformation characteristics of the material is. For better adhesive binding of fiber to concrete, it is desirable that it has a periodic section, bent ends or a wavy shape. Based on the analysis made in [6, 7], it was concluded that the use of Chelyabinka fiber is efficient. This fiber manufactured according to TU 1276–001–70832021–2005 of rolled steel (band, sheet) is a steel strip with anchors on the ends in the shape of segments of circle with radius conjugation to the straight sections of the strip. The ends of the strip are turned to each other at an arbitrary angle (Fig. 1).

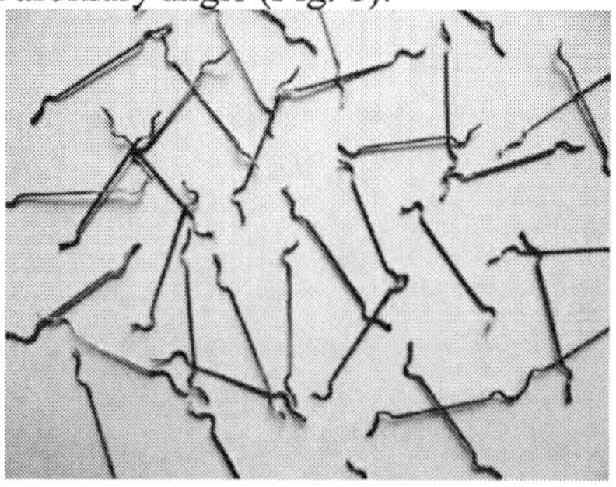

Fig. 1 Chelyabinka fiber.

Chelyabinka has higher physical and mechanical properties compared to other types of fibers, as evidenced by the results of tests of CNIIS OJSC (Tab.1). According to Volvek Plus, Chelyabinka fiber has better strength characteristics in comparison with other fibers [8].

Table 1. The test results of fiber concrete specimens.

Test type	Chelyabinka R, [MPa]	HAREX R, [MPa]	DRAMIX R, [MPa]	MagFiber-stroy R, [MPa]
Compression	49.4	50.0	47.2	49.2
Stretch	3.34	2.82	2.72	2.64
Bend	7.89	6.57	7.41	5.57

According to [9], minimum value of the fibrous reinforcement factor can be determined using the formula:

$$\mu_{\min} = \frac{1.5 \cdot C \cdot R_{bt}}{R_f \cdot k_{or}^2 \left(1 - \frac{30}{R_f} - \frac{l_{f,an}}{l_f}\right)} \tag{1}$$

where C - taken on the basis of [9];

l_f - fiber length;

R_f - reference resistance to fiber reinforcement strain;

R_{bt} reference concrete resistance to axial tension according to Group I of limit states;

k_{or} - factor of fiber orientation relative to direction of main tension stresses;

$l_{f,an}$ - minimum length of embedding of steel fiber into concrete, corresponds to its rupture when picked from matrix concrete.

2.2 Basalt fiber

Adding basalt fibers increase the tensile strength of concrete. They have a number of benefits, since they are one of the hardest mineral fibers. Basalt fibers are segments of complex basalt fiber as loose monofilaments (Fig. 2). Fiber length can be 3, 6, 13, 15, 18, 25, 27, 30 mm, probably 40, 50 mm, diameter of a single fiber is 13-20 mm.

The studies demonstrate that the concrete quality improves substantially even if a little of basalt fibers is added. Resistance to stress, durability, and resistance to cracking increase, while deformation probability reduces. It is also critical that the use of this material enables to reduce the total weight of the constructions without any to the detriment of strength [10]. Basalt fiber is unique since it is manufactured from the natural stone and has perfect chemical resistance properties. Basalt fiber has a number of significant benefits, ranging from the price and environmental safety, physical technical properties to the difficulty to falsify [11].

Fig. 2. Basalt fiber.

Adding basalt fiber increases ultimate bending, compressive and axial tensile strength, frost resistance, water resistance, fire resistance, surface resistance to abrasion and to impact and dynamic loads of the concrete construction, while the probability of shrinkage cracking reduces.

In [6] the results of experimental studies of different authors were analyzed and a conclusion on rational parameters and percentage of basalt fiber was drawn. According to N.G. Vasilovskaya, I.G. Yendzhiyevskaya and I.G. Kalugin [12], basalt fiber with the length of 12 mm with the content of 0.2% gives the highest ultimate compressive strength – 88.6 MPa (Fig. 3, a) and one of the highest ultimate bending strength – 30.7 MPa (Fig. 3, b).

3 Calculation method of steel concrete slabs at thermal power influences

In the work [13] it was developed a mathematical apparatus for calculating reinforced concrete rectangular slab at thermal power influences with hinged support, and also provided a fire protection of the structure. Methodology of assessing the fire resistance of the main tenets of the theory contains steel-concrete slabs, which takes into account the addition of power, temperature effects, and is a development study [14, 15, 16, 17]. The equilibrium conditions of the element reinforced concrete slabs, obtained in [15] are being used:

$$\frac{\partial^2}{\partial x^2}(M_T - M_x) + \frac{\partial^2}{\partial y^2}(M_T - M_y) - 2\frac{\partial^2 M_{xy}}{\partial x \partial y} = q(x,y) \tag{2}$$

a) b)

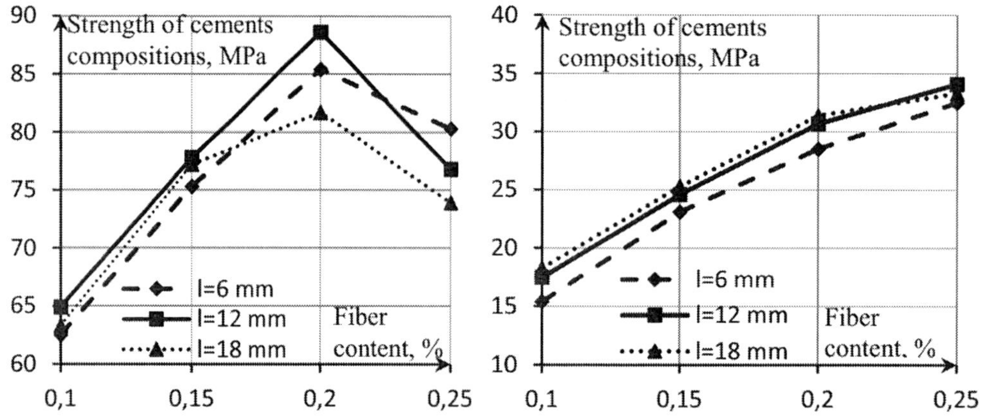

Fig. 3. Strength of cement compositions a) compressive, b) bending.

Temperature bending moments in the concrete and steel sheet are defined as follows:

$$M_T = M_T^c + M_T^s$$

$$M_T^c = -\int_{x_{ti}}^{x_i} \frac{\alpha_c E_c (T - T_0)}{3(1-\nu_c)} x dx \; ; \; M_T^s = -\int_{h_c}^{h_c+\delta} \frac{\alpha_s E_s (T - T_0)}{3(1-\nu_s)} x dx \tag{3}$$

where T_0 is the initial temperature;

α_c, α_s is the thermal expansion coefficients of concrete and steel.

Moments M_x, M_y, M_{xy} are associated with stiffness coefficients and curvature dependences [15];

$$\begin{vmatrix} M_x \\ M_y \\ M_{xy} \end{vmatrix} = \begin{vmatrix} D_{11} & D_{12} & D_{13} \\ D_{21} & D_{22} & D_{23} \\ D_{31} & D_{32} & D_{33} \end{vmatrix} \begin{Bmatrix} K_x \\ K_y \\ 2K_{xy} \end{Bmatrix}, \tag{4}$$

where $D_{i,j}$ – bending rigidity;
K_x, K_y, K_{xy} – curves.

In [3, 17] temperature action patterns and methods of evaluation of the stress-strained state of construction at thermal power influence with regard to different support conditions are considered.

Modeling of deformation process reinforced concrete slabs under load was carried out in steps of 10 kN/m^2. The action temperature was taken into account to the equation of the standard fire and simulated as a supplement to the load at each point of the finite-difference grid. The temperature fields were determined at an interval of time until the moisture evaporates – 0.67 min., after evaporation – 1.67 min. The finite difference method (FDM) was used as a numerical method for solving.

Fire resistance design was characterized by its ability to resist the effects of temperature and determines the time t, when the slab loses load bearing capacity. In turn, the bearing capacity was characterized by the following factors taking place in any point of the finite-difference grid: the strength of concrete steel sheet, the strength of contact [16].

We suggest calculating a fiber-reinforced steel concrete slab at thermal power influence using the above method. It applies physical-mechanical and thermal physic properties of fibrous concrete. Therefore, experimental studies are planned which would allow to find modulus of elasticity, Poisson ratio and limit strength of fiber-reinforced concretes with different kinds of fibers at different temperatures, as well as thermal conductivity. The experiment was planned before conducting so that the necessary number of samples could be determined.

Conduction of the study can be presented in diagram form as a "black box" (Fig. 4). The object is influenced by factors x - types of reinforcement; $y = f(x)$ – response of the system for the impact of such factors – ultimate strength, i.e. the result of the experiment. Function $f(x)$ is a mathematical model of object behavior to be determined. This function is called regression.

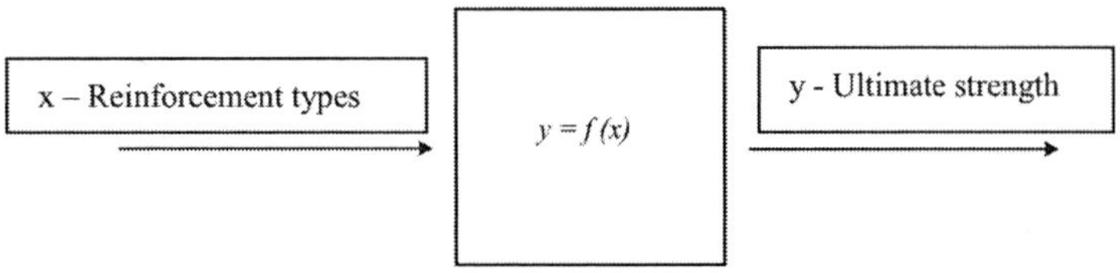

Fig. 4. Study diagram.

A fixed set of factor levels defines one of the possible states of the "black box". At the same time, this is the condition for conduction of one of the possible experiments. If we go through all possible sets of states, the total number of different experiments N will be obtained [18, 19].

Strength (for example, cubic) of fibrous concrete depending on the material heating temperature and the type of fiber is to be determined. The percentage and length of the fiber is taken in accordance with the data of the reference sources and is not subject to study. Thus, the factors are the temperature and the type of fiber; the response of the system is the ultimate strength of fibrous concrete.

We consider three types of reinforcement (steel fiber, basalt fiber and a control concrete sample without fiber). The dependence of the strength on temperature will be determined for the following temperatures: 20°C, 60°C, 90°C, 120°C, 200°C, 400°C, 600°C, 800°C.

The number of possible different experiments N is determined by expression

$$N = p^k \times n = 72, \tag{5}$$

where $p = 3$ – number of levels;
$k = 2$ – number of factors;
$n = 8$ – number of temperature values.

4 Conclusion

As a result of the analysis of the experimental studies by various authors, steel and basalt fibers were selected for further investigation. Their rational dimensions, percentage ratio to the mass of concrete in structures are justified. We came to the conclusion is that it is necessary to refine the physic mechanical and thermophysical properties of heated fiber concrete. For this purpose the experiment was planned, that made it possible to determine the required number of samples to obtain the temperature dependences of the strength and modulus of deformation, the thermal conductivity, and the specific heat of the fiber concrete mixture.

References

1. F.N. Rabinovich, *Kompoziti na osnove dispersno-armirovannykh betonov,* Izdatel'stvo Assotsiatsii stroitel'nikh vuzov 559, (2004)
2. E.V. Opanaseko, A.A. Berestyanskaya, Zbirnik naukovykh prats Natsionalnogo universytetu vodnogo gospodarstva ta pryrodokorystuvannya **30**, 57 (2015)
3. G.L. Vatulya, S.Yu. Berestyanskaya, E.V. Opanasenko, A.A. Berestyanskaya, Journal of Civil Engineering and Construction **5**, **2**, 98 (2016)
4. V. Bhikshma, K. Manipal, Asian Journal of Civil Engineering **13** **(2)**, 155 (2012)
5. K.V. Talantova, *Stroitel'nye konstruktsie s zadannymi svoystvami na osnove stalefibrobetona*, avtoreferat DSc in Technical Sciences, 30 (2010)
6. M.A. Verevicheva, A.A. Berestyanskaya, S.V. Derizemlya, Sbornik nauchnih trudov Dnepropetrovsk PGASA **82**, 60, (2015)
7. A.A. Berestyanskaya, VII Mezdunarodnyy molodezhnyy forum, eResurs, 215 (2015)
8. Volvek-Plus [ofic. sayt]. URL: http://wolwekplus.ru
9. *Design of concrete structures - Part 1-2: General rules - Structural fire design*, EN 1992-1-2. Eurocode 2, 99 (2004)
10. I.V. Borovskikh, *Vysokoprochnyy tonkozernistyy bazal'tofibrobeton*, avtoreferat PhD in Technical Sciences 24, (2009)
11. S.E. Artemenko, *Polymer composite materials made from carbon, basalt, and glass fibers, structure and properties, fiber chemistry* **35** **(3)**, 226 (2003)
12. N.G. Vasilovskaya, I.G. Yendzhievskaya, I.G. Kalugin, Vestnik Tomskogo gosudarstvennogo arkhitekturno-stroitelnogo universiteta **3**, 153 (2011)
13. S.Yu. Berestyanskaya, *Napryazhenno-deformirovannoe sostoyanie stalebetonnykh plit pri silovikh i temperaturnikh vozdeistviyakh*, avtoreferat PhD in Technical Sciences, (2003)
14. E.D. Chikhladze, A.I. Zhakin., M.A. Verevicheva, Izv. VUZov. Stroitelstvo **1**, 111 (1998)

15. E.D. Chikhladze, A.I. Zhakin, M.A. Verevicheva, Sbornik nauchnikh trudov KSART, **40**, 97 (2000)
16. E.D. Chikhladze, A.D. Arslankhanov, Stroitelnaya mekhanika **2**, 22 (1990)
17. E.D. Chikhladze, G.L. Vatulya, Proceedings of the 5th International Conference on Dynamics of Civil Engineering and Transport Structures and Wind Engineering - Zilina, 181 (2011)
18. E.A. Lyubchenko, O.A. Chudnova, *Planirovanie i organizatsyya eksperimenta*, Izd. TGEU (2010)
19. K.M. Khamkhanov, *Osnovy planirovanie eksperimenta*, Metodicheskoe posobie (2001)

Analysis of bond behaviour in strengthened reinforced concrete beam with carbon fiber reinforced polymer lamella

Milan Žmindák[1,*], *Pavol* Novák[1] and *Vladimír* Dekýš[1]

[1]University of Žilina, Faculty of Mechanical Engineering, Department of Applied Mechanics, Univerzitná 1, 010 26 Žilina, Slovakia

Abstract. Carbon fiber reinforced polymer (CFRP) can be used to strengthen the reinforced concrete (RC) beams. The debonding is the main failure model and strengthening is limited. The paper deals with the finite element analysis of influence of CFRP strip on stress distribution in a steel reinforced concrete beam loaded by four-point bending on RC beam with and without CFRP lamella. Simulation of the delamination is modelled by a cohesion zone material model using ABAQUS and ANSYS software.

1 Introduction

In the past decades, the use of composites for strengthening and repairing concrete structures has gained significant importance in civil engineering [1-3]. There are a large number of new, innovative building materials on the market. In order to avoid a drastic impact of building materials now or in the future, these materials as well as their manufacture and use must comply with the ecological standards. The trend of developing new building materials is not characterised by inventing something new, but rather by improving something long-existing. The aim of this paper is to create a model to simulate a four-point bending loading experiment on a reinforced concrete beam with and without carbon-fibre reinforced polymer.

2 FEM simulation

Experimental measurements of four-point bending of a reinforced concrete beam were carried out in our laboratory. Fig. 1 shows that the contact between the load and the beam, as well as the contact between the supports and the beam. The surface representing half profile of the T-beam is divided into smaller areas, borderlines of which will be used later to create a mesh of link/beam elements forming the steel reinforcement. The element Solid65 was used for discretization the steel reinforcement. The dimensions of the full-size beams are 3600 mm × 350 mm × 400 mm and the span between the supports was 3310 mm. Moreover, the lower part of the beam includes a milled groove in the covering layer. The groove is 4 mm wide, 20 mm deep, and 3 000 mm long, positioned symmetrically in relation to the transverse plane of the beam symmetry (Fig. 2).

* Corresponding author: milan.zmindak@fstroj.uniza.sk

Fig. 1. Experimental stand for four-point bending test.

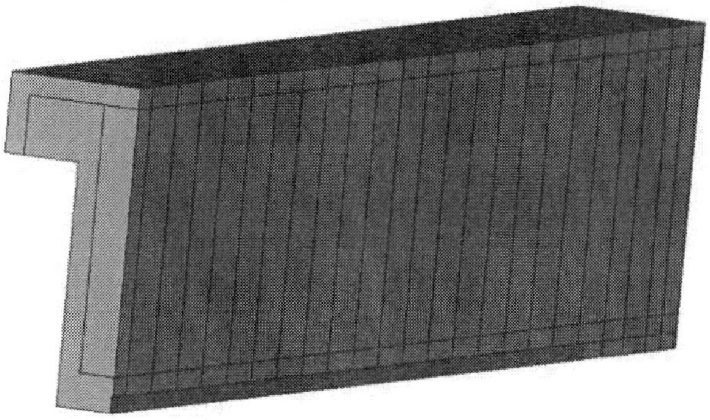

Fig. 2. Geometric model of the reinforced concrete beam with a gap or the composite lamella.

Bonding of the lamella with the concrete is modelled using the conta173 contact elements, considering the contact with adhesive bonding (bonded always contact), and using a cohesive material model (CZM – Cohesive Zone Material). We used the data provided by the manufacturer of the glue. We assumed a dominant shear mode of breaking the bonded joint. This material model requires entering τ_{max} = 12 MPa (maximum equivalent contact shear stress), $\delta_t = 1$ mm (tangential displacement after debonding). This value is not stated by the manufacturer and must be obtained through an experiment [4].

In the post-processing, for the contact defined in the above manner, we can evaluate the generally known data, such as contact status, contact gap, contact pressure, as well as the following variables: DPARAM - debonding parameter - this parameter takes a value from 0 to 1, where the value 1 corresponds with the breaking of the bonded joint, and the value 0 corresponds with the initial unbroken joint, DTSTARI – the time step at which debonding occurred.

2.1 Boundary conditions

The solution is significantly influenced also by the manner used for the beam positioning modelling. In order to prevent the formation of stress concentrators, we modelled a plane support with a width corresponding to the experiment, and a joint in the support width centre. The above was done using shell elements – shell181. The nodes in the support width centre were removed of all degrees of freedom for displacement, and we retained all degrees of freedom for rotation. Contact with the beam was done using the conta173 and target170 contact elements. A support modelled in this manner has no longer caused the formation of stress concentrators in its vicinity; therefore no cracks occurred in this area [5, 6].

The load acting on the beam in its upper part was created as a quarter of the total force F = 110 kN acting on the small area formed by the experiment requirements. The small area is 150 mm wide and its distance from the beam centre is 500 mm [7].

2.2 Material properties

RC structures strengthened by FRP are based on combination of concrete, steel and FRP. This section summarizes the material properties of all components:

Concrete: the concrete class C35/45-C10, 04-D_{max}16-S3 according to STN EN 206-1 was used. It properties are: Young's modulus E = 34 GPa, ductility stated by the manufacturer $\varepsilon_{cu} = 0.035$, Poisson's rate $\mu = 0.2$, the concrete density ρ = 2 400 $kg \cdot m^{-3}$.

Steel reinforcement: the reinforcing steel B500 B according to STN EN 10027 was used: ultimate tensile strength f_y = 500 MPa, ultimate tensile strength f_t = 550 MPa, Young's modulus E = 200 GPa, ductility $\varepsilon = 0.005$, Poisson's ratio $\mu = 0.3$.

Composite lamella: the simulated composite lamella is made by the BASF company. Its trade name is MBRACE CFK 150/2 000. The composite lamella dimensions are 20 × 1.4 × 3000 mm. For this lamella type, the manufacturer states the Young's modulus larger than 165 GPa, Poisson's number $\mu = 0.2$ and the recommended design tensile load is 51.8 kN, which means that the uniaxial tension stress is 1 850 MPa at the lamella cross-section of 28 mm^2.

High-strength epoxy glue: to affix the lamella into the groove in the beam it was used high-strength epoxy adhesive with a high glass transition temperature for the MBRACE lamella system by BASF, with the trade name MBRACE Laminate Adhesive HT. Its bond strength is >14 MPa, compressive strength is 73 MPa, bending elastic modulus is 4263 MPa, glass transition temperature is ≥ 40°, shrinkage/ expansion is ≤ 0.1 % [5].

3 FEM solution

In the calculation we considered the geometric linearity of the task because the achieved bending is sufficiently small against the beam dimensions. We spread the calculation over several loading states. We did so, among other things, also due to a good convergence of the beam contact with the lower articulated support. Another reason was the concrete deterioration at higher loading values, and it was appropriate to reduce the loading increments for a faster task solving. The concrete failure model caused that it was impossible to work with standard values for the convergence criteria (these were too strict). The value of force imbalance, which was considered as the convergence criterion, was set to 200 N.

4 Analysis of results

Comparisons of the load-tensile displacement plots from the FE analyses with the experimental data for the reinforced concrete beam at midspan are shown in Fig. 3. This figure shows that the displacements from FE analyses correlate well with those from experimental data in the range loading force 0 to 40 kN. Higher displacement value in the experimental beam is possibly due to a local effect caused by the major cracks, which take place close to the midspan. Fig. 4 shows the maximum deflection in the middle of the reinforced concrete beam with lamella, and it reaches the value w_{max} = 7.57 mm. This value is slightly reduced compared with the model with CFRP lamella [8]. Due to bending, the free end of the beam behind the support is displaced upwards by 0.89 mm. Fig. 5 shows the first principal stress in the concrete. This value indicates the sites that are critical for the concrete failure due to a high tensile stress, which concrete naturally "does not like". High values are present in the beam lower side. The maximum is reached due to concrete failure consideration, and it appears near the steel reinforcement.

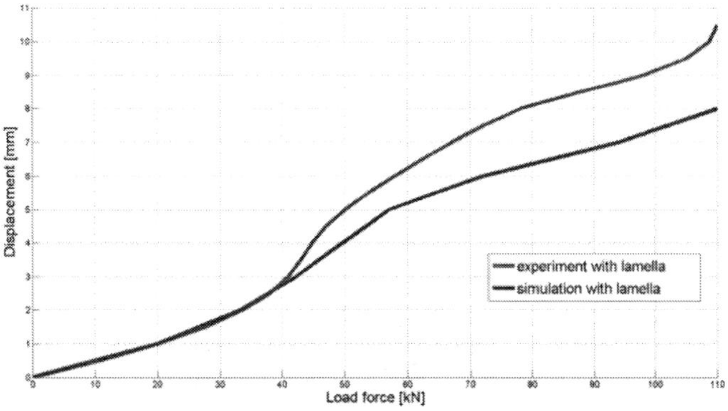

Fig. 3. The plot of displacement versus loading force.

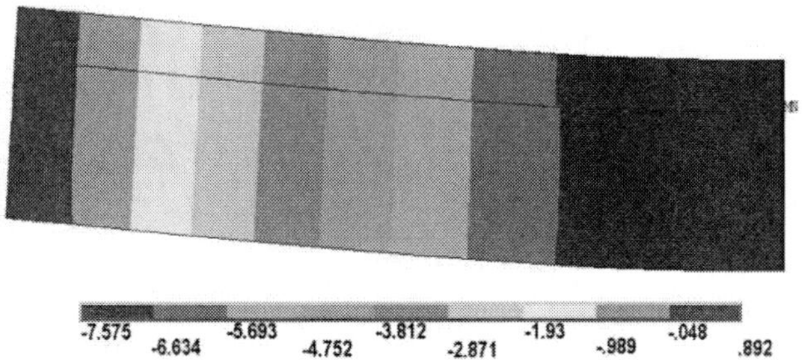

Fig. 4. Displacement of the reinforced concrete beam with a lamella.

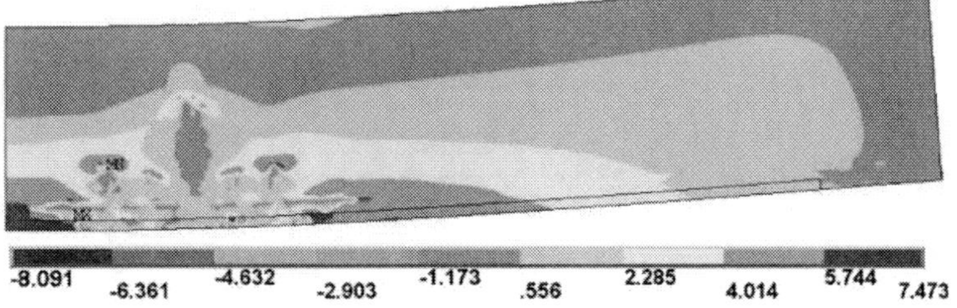

Fig. 5. First principal stress in the concrete.

Due to the lamella groove, and the lamella with a high elasticity modulus, concentration of stresses occurs near the original beam/groove interface. The maximum value dropped and moved to this area from the steel reinforcement site.

Fig. 5 shows the third principal stress in the concrete. This value shows the areas where concrete crushing might occur in case of exceeding the concrete compressive strength. The maximum reached value is -28.1 MPa is below the used concrete compressive strength, therefore there should be no crushing. The stress value was higher compared to the previous case and, like in the first principal stress, the value moved from the steel reinforcement to the site of the lamella groove [8].

The course of axial stresses in the reinforcement is shown in Fig. 7. As we can see, compressive stress occurs in the upper part of the reinforcement. The maximum compressive stress emerged near the action of the loading pressure, and the compressive stress value in the reinforcement reached the value $p_{max,steel}$ = 164.32 MPa. The maximum tensile stress emerged in the bottom part of the steel reinforcement, at the site under the loading force, and its value reached $\sigma_{tmax,steel}$ = 424.63 MPa. This value is lowered compared with the model without the composite reinforcement, but it is still near the value of the yield strength R_e.

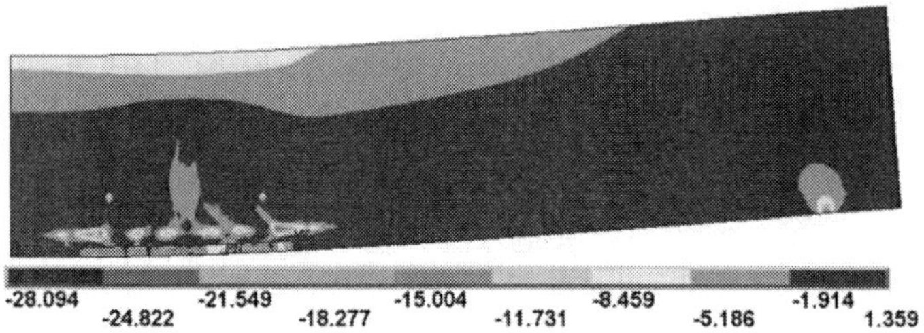

Fig. 6. Third principal stress in the concrete.

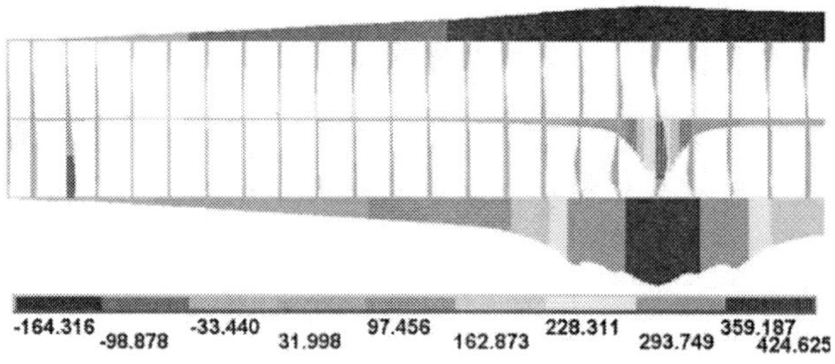

Fig. 7. Axial stresses in the reinforcement.

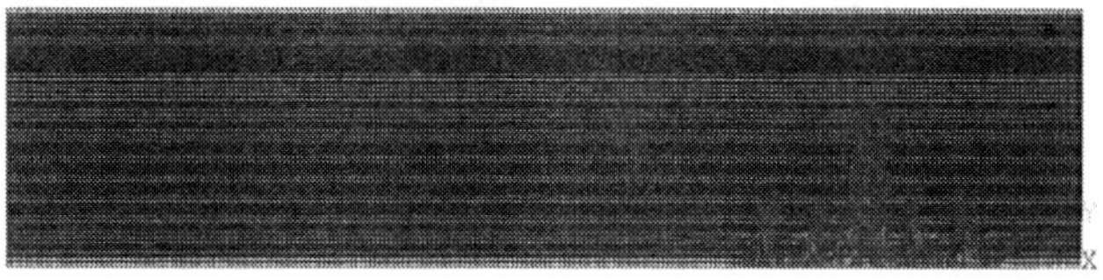

Fig. 8. Depiction of cracks in the concrete.

Fig. 8 shows the pink-coloured cracks formed in the concrete as a result of exceeding the concrete tensile strength. These cracks occur around the T-shaped steel stirrups.

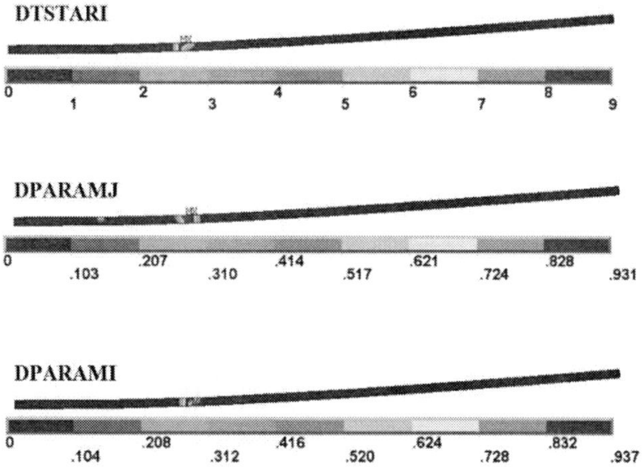

Fig. 9. The probability of adhesion loss with loading steps.

The shape and length of cracks has changed significantly compared with the case without a lamella. Although there is still one significantly long crack at the site under the acting load, other cracks are of approximately the same length

From Fig. 9 we can see that the DPARAMJ and DPARAMI values indicates the probability of adhesion loss of the glue and the lamella. The maximum value is approximately 0.94, which represents a 94% probability of adhesion loss. According to the DTSTARI value, this should occur in the last, 9th loading step, at the site filled with red.

5 Conclusions

The results obtained show that the use of the above-mentioned composite lamella changes the stress distribution in the concrete matrix as well as in the steel reinforcement. This is manifested by the reduction in the maximum stress values in the tensile domain for both the concrete and the reinforcement, and in case of the concrete also by increased maximum compression stresses, which is favourable for concrete. The maximum bending (deflection) values are very similar in the two cases. It also results from the data obtained that there will be no adhesion compromise between the concrete and the lamella (the maximum value is 94% of the bond ultimate shear strength), but there will occur delamination in the concrete. The shape and size of the cracks in the concrete have changed as well.

The authors gratefully acknowledge for support the Slovak and Technology Assistance Agency register under number APVV-0736-12 and Slovak Grant Agency VEGA 1/0983/15.

References

1. E. J. Barbero, *Finite Element Analysis of Composite Materials Using ANSYS* (CRC Press 2014)
2. L. Dahmani, A. Khennane, S. Kaci, Strength Mat. **42** (2010)
3. M. Halapi, Z. Murcinkova, In: Int. Conf. on Manufacturing Engineering and Materials, (2016)

4. M. Zmindak, P. Novak, V. Dekys, Advanced Materials Research **897** (2014)
5. E. Kormanikova, K. Kotrasova, Dynamic of Civil Engineering and Transport Structures and Wind Engineering, Book Series: Applied Mechanics and Materials **617**, (2014)
6. E. Kormanikova, K. Kotrasova, Key Eng. Mat. **635**, 85-88 (2015)
7. A. Cavojcova, M. Moravcik, F. Bahleda, J. Jost, Procedia Engng. **92** (2014)
8. M. Žmindák, P. Novák, P. Pastorek, Proceedings of Mechanics of Composite Materials and Structures, Sušice (2016)

Using DOProC method in reliability assessment of steel elements exposed to fatigue

Martin Krejsa[1,*], *Jiri* Brozovsky[1], *David* Mikolasek[1], *Petr* Lehner[1] and *Premysl* Parenica[2]

[1]VŠB-Technical University of Ostrava, Faculty of Civil Engineering, Department of Structural Mechanics, Ludvika Podeste 1875/17, 708 33 Ostrava – Poruba, Czech Republic
[2] VŠB-Technical University of Ostrava, Faculty of Civil Engineering, Department of Building Structures, Ludvika Podeste 1875/17, 708 33 Ostrava – Poruba, Czech Republic

Abstract. Fatigue crack damage depends on a number of stress range cycles. This is a time factor in the course of reliability for the entire designed service life. Three sizes are important for the characteristics of the propagation of fatigue cracks - initial size, detectable size and acceptable size. The theoretical model of fatigue crack progression can be based on a linear fracture mechanic. Depending on location of an initial crack, the crack may propagate in structural element e.g. from the edge or from the surface. When determining the required degree of reliability, it is possible to specify the time of the first inspection of the construction which will focus on the fatigue damage. Using a conditional probability and Bayesian approach, times for subsequent inspections can be determined. For probabilistic modelling of fatigue crack progression was used the original and new probabilistic method - the Direct Optimized Probabilistic Calculation ("DOProC"), which uses a purely numerical approach without any simulation techniques or approximation approach based on optimized numerical integration.

1 Introduction to probabilistic computation

In process of designing a structure under static or dynamic loads [1], various reliability criteria should be met in accordance with the relevant standards. Common studies of the misalignment of failure probability P_f (or the reliability index β) over time tend to focus only on structural details, however, a comprehensive probabilistic methodology generally applicable to bridge structures is currently missing. Numerous stochastic methods, which enable the determination of failure probability, respectively the reliability index, have been developed [2]. A substantial part of these methods are based on the crude Monte Carlo simulation method (MC), whose disadvantage is poor efficiency due to the need of a high number of simulation steps. Advanced and stratified simulation methods, for e.g. Markov chain Monte Carlo simulations (MCMC), also applied in fatigue damage prognosis, strive to increase the efficiency of these computational methods. An alternative solution is the evaluation of reliability using the reliability index, which can be efficiently determined using the Latin Hypercube Sampling method (LHS). The LHS method is capable of

* Corresponding author: martin.krejsa@vsb.cz

evaluating the reliability index from a small number of simulation steps (hundreds to thousands).

Another category of stochastic approaches for the quantification of the reliability of structures are represented by methods, which determine the failure probability on the basis of numerical integration, for e.g. the Direct Optimized Probabilistic Calculation method – DOProC, which was comprehensively published, e.g. in [3, 4]. It appears that this method is very effective for many probabilistic problems. It is also distinguished by higher accuracy than that of simulation methods, resp. approximation ones. This calculation method was applied in the solution of some engineering problems, among others, the assessment of reliability of steel bridge structures loaded by fatigue [5]. Probabilistic modelling of fatigue crack progression leads to designing a system of regular inspections of structures [6].

2 Fatigue crack propagation using linear fracture propagation

When investigating into the propagation, the fatigue crack that deteriorates a certain area of the structure components is described with one dimension only a [7]. In order to describe the propagation of the crack, the linear elastic fracture mechanics is typically used. This method defines the limit of propagation rate of the crack (da/dN) and range of the stress rate coefficient in the face of the crack using the Paris-Erdogan law [8]:

$$\frac{\mathrm{d}a}{\mathrm{d}N} = C \cdot \Delta K^m \tag{1}$$

where C, m are material constants, that are determined experimentally [9], a is the crack size, N is the number of loading cycles and ΔK is the stress rate coefficient. The fatigue crack will propagate in a stable way only if the initial crack a_0 exists in the place where the stress is concentrated. This place is located, e.g., at the edge or on the surface of the element.

Having modified (1) according [5], the following formula will be achieved:

$$\int_{a_1}^{a_2} \frac{\mathrm{d}a}{\left(\sqrt{\pi \cdot a} \cdot F_{(a)}\right)^m} = \int_{N_1}^{N_2} C \cdot \Delta\sigma^m \cdot \mathrm{d}N \tag{2}$$

where number of cycles from N_1 to N_2 is needed to increase the crack from the a_1 to a_2, $\Delta\sigma$ stress range, and $F_{(a)}$ is the calibration function which represents the course of propagation of the crack. Depending on location of an initial crack, the crack may propagate from the edge or from the surface [10].

3 Probabilistic reliability assessment

Three sizes are important for the characteristics of the propagation of fatigue cracks. The first size is the initiation size of the crack a_0 that corresponds to a random failure in an element subject to random loads. Existence of the initiation cracks during the propagation should be revealed, along the detectable length of the crack a_d, during inspections. The crack propagates in a stable way until it reaches the third important size - acceptable length of the crack a_{ac}, which is a limit for the required reliability.

The primary assumption is that the primary design should take into account the effects of the extreme loading and the fatigue resistance should be assessed then. This means, the safety margin is defined e.g. by:

$$g(R,E) = G = R - E, \tag{3}$$

where R is the random resistance of the element and E represents random variable effects of the extreme load.

The probabilistic methods should be used for the investigation into the propagation rate of the fatigue crack until the acceptable size is reached because the input variables include uncertainties and reliability should be taken into account. If the length of the crack length a_1 in (2) equals to the initial length a_0 and if a_2 in (2) equals to the acceptable crack length a_{ac}, the left-hand side of (2) can be regarded as the resistance of the structure - R:

$$R_{(a_{ac})} = \int_{a_0}^{a_{ac}} \frac{\mathrm{d}a}{\left(\sqrt{\pi \cdot a} \cdot F_{(a)}\right)^m} . \tag{4}$$

Similarly, it is possible to define the cumulated effect of loads that is equal to the right side (randomly variable effects of the extreme load) of (2):

$$E_{(N)} = \int_{N_0}^{N} C \cdot \Delta\sigma^m \cdot \mathrm{d}N = C \cdot \Delta\sigma^m \cdot (N - N_0) , \tag{5}$$

where N is the total number of oscillations of stress peaks ($\Delta\sigma$) for the change of the length from a_0 to a_{ac}, and N_0 is the number of oscillations in the time of initialisation of the fatigue crack (typically, the number of oscillations is zero).

It is possible to define a reliability function G_{fail} :

$$G_{fail(\mathbf{X})} = R_{(a_{ac})} - E_{(N)}, \tag{6}$$

where **X** is a vector of random physical properties such as mechanical properties, geometry of the structure, load effects and dimensions of the fatigue crack.

The analysis of the reliability function (6) gives a failure probability P_f, which equals to:

$$P_f = P\left(G_{fail(\mathbf{X})} < 0\right) = P\left(R_{(a_{ac})} - E_{(N)} < 0\right) . \tag{7}$$

4 Design of structural inspection system

Because it is not certain in the probabilistic calculation whether the initiation crack exists and what the initiation crack size is and because other inaccuracies influence the calculation of the crack propagation, a specialised inspection is necessary to check the size of the measureable crack in a specific period of time. The acceptable crack size influences the time of the inspection. If no fatigue cracks are found, the analyses of inspection results give conditional probability during occurrence.

While the fatigue crack is propagating, it is possible to define following random phenomena that are related to the growth of the fatigue crack and may occur in any time, t, during the service life of the structure:

$U_{(t)}$ **phenomenon:** No fatigue crack failure has not been revealed within the t–time and the fatigue crack size $a_{(t)}$ has not reached the detectable crack size, a_d . This means:

$$a_{(t)} < a_d , \tag{8}$$

$D_{(t)}$ **phenomenon:** A fatigue crack failure has been revealed within the t–time and the fatigue crack size $a_{(t)}$ is still below the acceptable crack size a_{ac}. This means:

$$a_d \leq a_{(t)} < a_{ac} \text{ ,} \tag{9}$$

***F*$_{(t)}$ phenomenon:** A failure has been revealed within the t–time and the fatigue crack size $a_{(t)}$ has reached the acceptable crack size a_{ac}. This means:

$$a_{(t)} \geq a_{ac} \text{ .} \tag{10}$$

If the crack is not revealed within the t-time, this may mean that there is not any fatigue crack in the structural element. This might be also an initiative phase of nucleation of the fatigue crack (when a crack appears in the material) and this phenomenon is not taken into account in the fracture mechanics. Even if the fatigue crack is not revealed it is likely that it exists there but the fatigue crack size is so small that it cannot be detected under existing conditions.

Using the phenomena above, it is possible to define following probabilities:

The probability that the failure is not detected within the t-time, this means the probability that the fatigue crack size $a_{(t)}$ is below the measurable crack size a_d :

$$P(U_{(t)}) = P(a_{(t)} < a_d) \text{ .} \tag{11}$$

The probability that the failure detected within the t-time has the crack size $a_{(t)}$ that is less than the acceptable size a_{ac}:

$$P(D_{(t)}) = P(a_d \leq a_{(t)} < a_{ac}) \text{ .} \tag{12}$$

The probability that the failure occurs within the t-time, this means the probability that the fatigue crack size $a_{(t)}$ reaches the acceptable size a_{ac}:

$$P(F_{(t)}) = P(a_{ac} \leq a_{(t)}) \text{ .} \tag{13}$$

Those three phenomena cover the complete spectrum of phenomena that might occur in the t-time. This means:

$$P(U_{(t)}) + P(D_{(t)}) + P(F_{(t)}) = 1 \text{ .} \tag{14}$$

The probable course of the growth of the fatigue crack is shown in Fig. 1. The probabilities (11) - (13) can be determined in any period of time, t, using the probabilistic methods. The probabilistic calculation is carried out in time steps where one step typically equals to one year of the service life of the structure. When the probability of failure $P(F_{(t)})$ reaches the designed failure probability P_d [11], an inspection should be carried out in order to find out fatigue cracks, if any, in the structural element. The inspection provides information about conditions of the structure. Such conditions can be taken into account when carrying out further probabilistic calculations. The inspection in the t time may result in any of the three mentioned phenomena.

Using the inspection results for the t time, it is possible to define the probability of the mentioned phenomena in times $T > t_I$. For that purpose, the conditional probability should be taken into consideration. If the event of interest is A and the event B is known or assumed to have occurred, probability of A under the condition B is:

$$P(A|B) = \frac{P(A \cap B)}{P(B)} \text{ , where } P(B) > 0 \text{.} \tag{15}$$

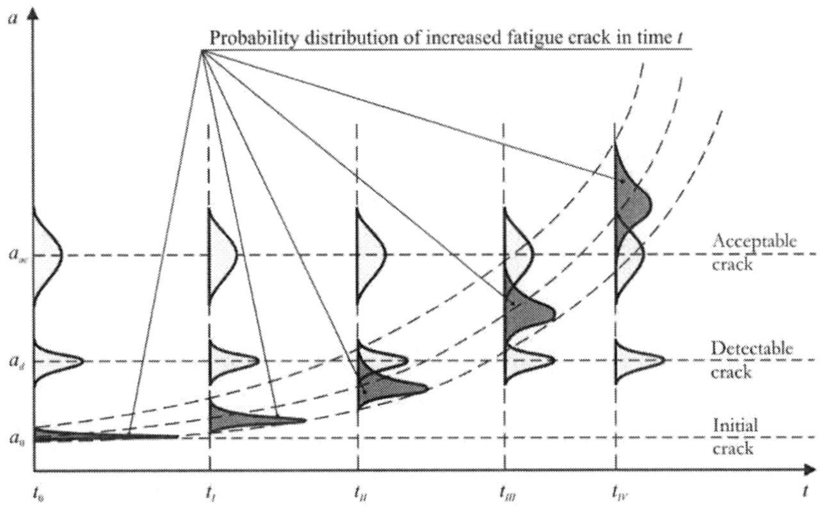

Fig. 1. Probabilistic growth of the fatigue crack in the course of time *t*.

In order to specify the time for the next inspection, it is necessary to determine the conditioned probabilities which can be expressed using the full probability rule:

$$P(A)=P(H_1)\cdot P(A\mid H_1)+P(H_2)\cdot P(A\mid H_2)+\ldots+P(H_n)\cdot P(A\mid H_n)\,, \tag{16}$$

where H_1, H_2, … H_n is a system of mutually incompatible phenomena, and A is the random event, which can occur simultaneously only with any of the phenomena H_i. For the probabilistic design of structural inspection system the phenomena H_1, H_2, … H_n represent events U, D and F. Bayes' theorem describes the probability of an event, based on prior knowledge of conditions that might be related to the selected event:

$$P(A\mid B)=\frac{P(B\mid A)\cdot P(A)}{P(B)}\,,\ \text{where}\ P(B)>0\,. \tag{17}$$

In (17) A and B are random events under analysis, $P(A)$ and $P(B)$ are the probabilities of observing A and B without regard to each other, $P(A|B)$ is a conditional probability according (15), which define the probability of observing event A given that B is true, and $P(B|A)$ is the probability of observing event B given that A is true.

In the Bayesian (or epistemological) interpretation, probability measures a "degree of belief." Bayes' theorem then links the degree of belief in a proposition before and after accounting for evidence. For proposition A and evidence B, $P(A)$ - the "prior", is the initial degree of belief in A, $P(A|B)$ - the "posterior", is the degree of belief having accounted for B and the quotient $P(B|A)/P(B)$ represents the support B provides for A.

With these findings it can be defined the probability of event F in time T with respect to the results of structural inspection in time t_I:

$$P\left(F_{(T)}\middle|U_{(t_I)}\right)=\frac{P\left(F_{(T)}\right)-P\left(F_{(t_I)}\right)-P\left(D_{(t_I)}\right)P\left(F_{(T)}\middle|D_{(t_I)}\right)}{P\left(U_{(t_I)}\right)} \tag{18}$$

and

$$P\left(F_{(T)} \middle| D_{(t_I)}\right) = \frac{P\left(F_{(T)}\right) - P\left(F_{(t_I)}\right) - P\left(U_{(t_I)}\right) P\left(F_{(T)} \middle| U_{(t_I)}\right)}{P\left(D_{(t_I)}\right)} \ . \tag{19}$$

If re-distribution of stress from a point that is weakened by the crack is not taken into account, the crack propagation crack is usually rather high in the practical range of detectable values. If a fatigue crack is found during the inspection, it is necessary to monitor the safe growth of the crack or to take actions that will slow down or stop further propagation of the fatigue crack. In order to time the inspections well, (18) is most important. It defines the failure probability in $T > t_I$ provided that no fatigue cracks have been revealed during the last inspection. It is clear from the equation that the results of the failure probability are influenced by mutual relations between the three crack sizes - the initial crack size a_0, measurable crack size a_d and acceptable crack size a_{ac}.

The probabilities in (18) can be calculated in any $T > t_I$ time using probabilistic methods. When the failure probability (18) reaches the designed failure probability P_d, an inspection should be carried out in order to reveal fatigue cracks, if any, in the construction component. The inspection may result in one of the three mentioned phenomena with corresponding probabilities. The entire calculation can be repeated in order to ensure well-timed inspections in the future.

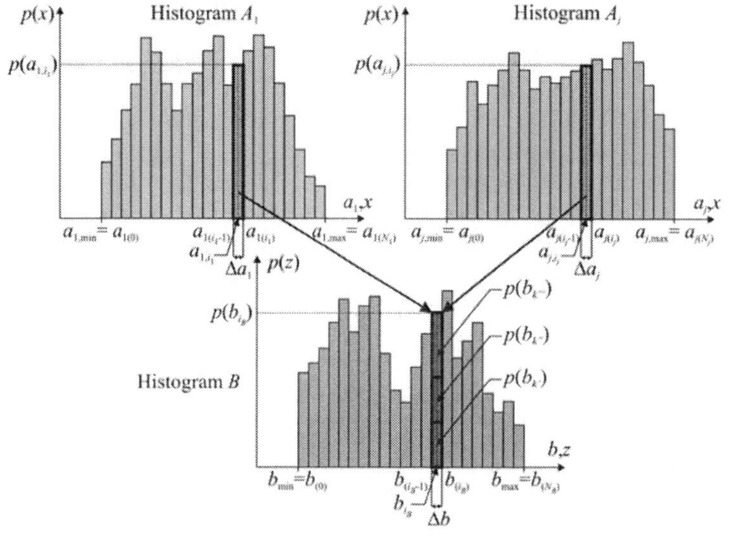

Fig. 2. Principle of the performing the numerical operations with the histograms of two random variables, which is similar to computation of structural resistance R (4) or load effect E (5).

5 Probabilistic computation using DOProC method

The basic computation algorithm of DOProC is based on general terms and procedures used in probabilistic theories. Let the histogram B be an arbitrary function f of histograms A_j where j ranges from 1 to n. Then:

$$B = f\left(A_1, A_2, A_3, \ldots, A_j, \ldots A_n\right) \ . \tag{20}$$

Each histogram A_j consists of i_j interval, where i_j ranges from 1 to N_j. Performing these numerical operations of the probabilistic calculation with two random variables A_1 and A_j expressed by histograms is shown in scheme in the Fig. 2.

The number of intervals i_j in each histogram of the random variables A_j can be similar as the number of i_B intervals in the histogram of the resulting random variable B. The number of intervals is very important for the total number of needed numerical operations and required computing time. On top of this, the accuracy of the calculation depends considerably on the number of intervals. If there are too many random quantities, the tasks require too much time even if advanced computational facilities are available. Therefore, efforts have been made to optimize calculations in order to reduce the number of operations, while retaining reliable calculation results (for details see [3]).

Using this probabilistic approach is possible to determine the probabilities of random events U, D and F according (11) - (13), and, of course, using conditional probability (18) and the time of the designed failure probability P_d the time of the inspections focused on the structural elements susceptible to fatigue damage (see Fig. 3).

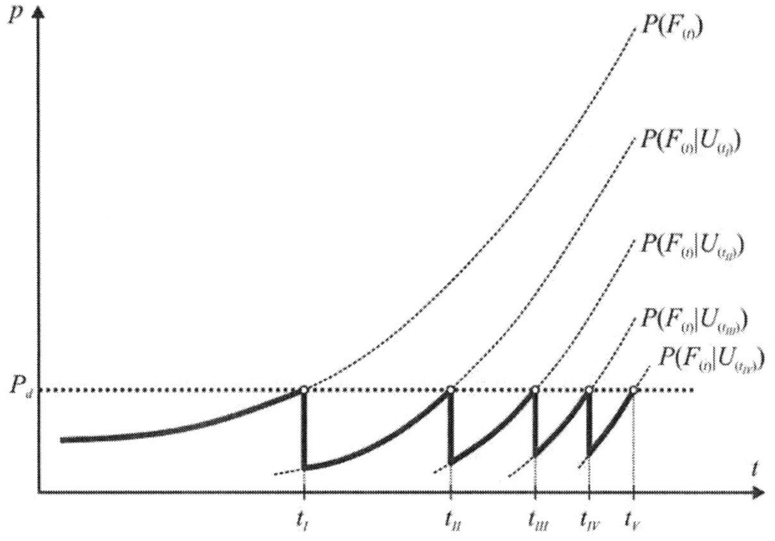

Fig. 3. Probabilities of failure Pf calculated according (7) and (18) with the times of 5 inspections.

6 Conclusion

Propagation of the fatigue cracks and possible prediction of such propagation in time since the start of variable loading effects is the case when probabilistic methods must be used because too many uncertainties influence the determination of the input values.

The computation uses the newly developed Direct Optimized Probabilistic Calculation ("DOProC"), which appears to be a very efficient tool to make probabilistic assessment of the structural reliability on the basis of the exact definition of the acceptable size of the fatigue crack.

The DOProC method and its application in probabilistic prediction of fatigue crack damage can considerably improve estimation of maintenance costs for the structures and bridges subject to cyclic loads. If this methodology is developed further, the goal of investigations seems to be, in particular, application of Bayesian networks in the computational model [12, 13], which describes propagation of fatigue cracks in the system.

This contribution has been developed as a part of the research project GACR 17-01589S "Advanced computational and probabilistic modelling of steel structures taking account fatigue damage" supported by the Czech Grant Agency and also has been completed thanks to the financial support provided to VSB-Technical University of Ostrava by the Czech Ministry of Education, Youth and Sports from the budget for conceptual development of science, research and innovations for the 2017 year.

References

1. K. Kotrasova, I. Grajciar, E. Kormanikova, A case study on the seismic behavior of tanks considering soil-structure-fluid interaction, Journal of Vibrational Engineering and Technologies **3**, 3, 315-330 (2015)
2. J. Kralik, J. Kralik jr., Failure probability of NPP communication bridge under the extreme loads, Applied Mechanics and Materials **617,** 81-85 (2014) DOI: 10.4028/www.scientific.net/AMM.617.81.
3. P. Janas, M. Krejsa, V. Krejsa, *Přímý Optimalizovaný Pravděpodobnostní Výpočet – Direct Opzimized Probabilistic Calculation*, 1st ed. (VSB-Technical University of Ostrava, 2015, in Czech)
4. M. Krejsa, P. Janas, V. Krejsa, Structural Reliability Analysis Using DOProC Method, Procedia Engineering **142,** 34-41 (2016). DOI: 10.1016/j.proeng.2016.02.010
5. M. Krejsa, L. Koubova, J. Flodr, J. Protivinsky, Q.T. Nguyen, Probabilistic prediction of fatigue damage based on linear fracture mechanics, Frattura ed Integrita Strutturale **11**, 39, 143-159 (2017). DOI: 10.3221/IGF-ESIS.39.15
6. M. Krejsa, Z. Kala, S. Seitl, Inspection based probabilistic modeling of fatigue crack progression, Procedia Engineering **142,** 145-152, (2016). DOI: 10.1016/j.proeng.2016.02.025
7. R. Halama, M. Fusek, Z. Poruba, Influence of mean stress and stress amplitude on uniaxial and biaxial ratcheting of ST52 steel and its prediction by the AbdelKarim–Ohno model, International Journal of Fatigue **91,** 313-321, (2016). DOI: 10.1016/j.ijfatigue.2016.04.033
8. S. Seitl, T. Thienpont, W. De Corte, Fatigue crack behaviour: comparing three-point bend test and wedge splitting test data on vibrated concrete using Paris' law, Frattura ed Integrita Strutturale **39,** 110-117, (2017). DOI: 10.3221/IGF-ESIS.39.12
9. I. Major, M. Major, Traveling waves in a thin layer composed of nonlinear hyperelastic zahorski's material, Journal of Theoretical and Applied Mechanics **47**, 1, 109-126, (2009)
10. M. Krejsa, V. Tomica, Determination of Inspections of Structures Subject to Fatigue, Transactions of the VSB – Technical University of Ostrava, Civil Engineering Series **11**, 1, 1-9, (2011). DOI: 10.2478/v10160-011-0007-x
11. J. Vican, J. Gocal, J. Odrobinak, M. Moravcik, P. Kotes, Determination of railway bridges loading capacity, Procedia Engineering **111,** 839-844, (2015). DOI: 10.1016/j.proeng.2015.07.155
12. S. Mahadevan, R. Zhang, N. Smith, Bayesian networks for system reliability reassessment, Struct. Saf. **23,** 231-251, (2001). DOI: 10.1016/S0167-4730(01)00017-0
13. J. Luque, D. Straub, Reliability analysis and updating of deteriorating systems with dynamic Bayesian networks, Struct. Saf. **62,** 34-46, (2016). DOI: 10.1016/j.strusafe.2016.03.004

Numerical thermal analysis of the vertical external partition made as the frame thin-walled steel structure

Maciej Major[1,*] and *Mariusz* Kosiń[1]

[1] Technical Sciences, Department of Steel Constructions and Building Materials, Faculty of Civil Engineering, Częstochowa University of Technology, 42-200 Częstochowa, Poland
[2] Technical Sciences, Department of Organization and Building Technology, Faculty of Civil Engineering, Częstochowa University of Technology, 42-200 Częstochowa, Poland

Abstract. The article presents numerical thermal analysis of the vertical external partitions made in the lightweight steel framing technology. Steel posts that perform the structural role lead to the formation of linear thermal bridges and have a negative effect on the level of thermal transmittance *U*. Therefore, optimal solutions are being explored for such technologies. One of the solutions is to use perforated Thermo sections. The effect of perforated Thermo sections on energy loss was verified through comparison to the wall made of solid sections. Furthermore, the calculations analysed the effect of linear thermal bridges that are formed on wall connections in the corner. Computer simulation was employed to emphasize the significant differences in the temperature distribution in both analysed wall structures that resulted from constructional solutions.

1 Introduction

The problems concerning energy loss for the external walls in buildings continue to be particularly important as they affect maintenance costs in the heating season. The use of modern technologies based on framing structures made of cold-rolled steel sections that have been widely used in building structures (see [2]) substantially reduces the time to complete the investment (construction of a building), but the analysis of this structure from the standpoint of building's physics is needed. In order to evaluate heat flux density and temperature distribution in such walls, it is justified to use numerical software based on finite element method (MES) [3-5].

The aim of the study was to develop and compare characteristics of heat protection offered by the external wall, temperature distribution and value of heat loss through heat transfer from the heated space (residential room) to the environment through the wall made using the technology of a light steel framing. The corners of the vertical wall made of two thin-wall sections were analysed for three values of insulation thickness. Based on [6], the internal temperature was adopted as Θ_i = +20°C whereas based on [7], the external

* Corresponding author: mmajor@bud.pcz.czest.pl

temperature was adopted as Θ_e = -20°C. For these assumptions, two variants of the geometrical solution for the wall corners were compared.

2 Characterization of walls adopted for thermal analysis

Six-layer walls were adopted for the analysis, with their load-bearing components showing different structures (Fig. 1). The C-shaped section was adopted in all the cases, with the cross dimensions: h = 140 mm, b = 38 mm, c = 18 mm, t = 1.5 mm and r = 3 mm.

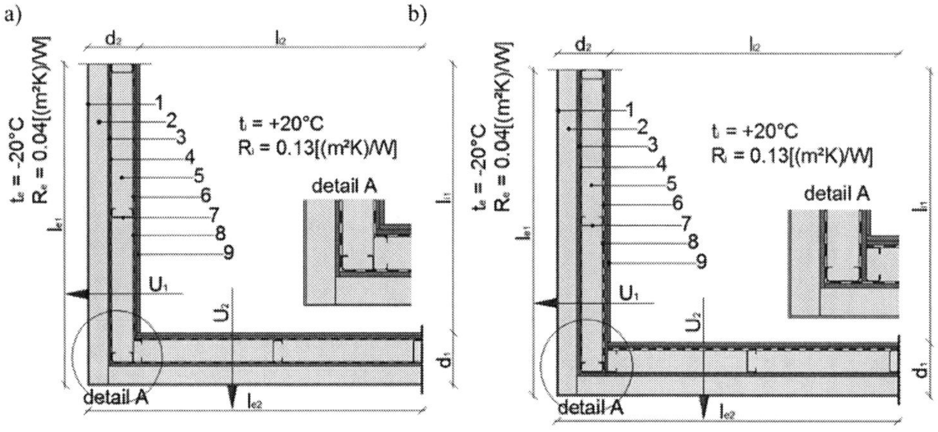

Fig. 1. Computational model for the analysed wall made using the technology of a lightweight steel framing; a) Variant I; b) Variant II; for the corner's solution. 1 – thin-layer plaster, thk. 0.5 cm ; 2 – mineral wool, thk. 16 cm/ 8 cm/ 5cm; 3 – cement-bonded particleboard, thk. 1.2 cm ; 4 – vapor-permeable membrane; 5 – mineral wool, 14 cm ; 6 – steel profile C140, thk. 0.15 cm ; 7 – vapor barrier; 8 – cement-bonded particleboard, thk. 1.2 cm ; 9 – gypsum plasterboard, thk. 1.25 cm.

3 Heat parameters of the analysed walls

The material characterization for the analysed solutions of the corners of the external wall is presented in Table 1. Using the algorithm below through computer simulation, the thermal transmittance U was evaluated for the analysed solutions of the walls [8]:
- calculation of the value of the mean horizontal component of heat flux density for the part of joint q_1 and q_2,
- calculation of the thermal transmittance based on:

$$U_1 = q_1/(\Theta_i - \Theta_e) \quad [\mathrm{W/(m^2 K)}] \tag{1}$$

$$U_2 = q_2/(\Theta_i - \Theta_e) \quad [\mathrm{W/(m^2 K)}] \tag{2}$$

An analytical method was used to verify the values of thermal transmittance. The greatest difference in the analytically and numerically calculated thermal transmittance U was 6%. It was assumed for the linear thermal transmittance Ψ that the height of the corner is 1 m. Therefore, after calculation of the linear thermal transmittance, its value can be multiplied by the bridge's length [9].

Heat flux Φ for the heat flowing through the joint for each variant was calculated using numerical methods, see Table 2. Table 3 presents heat parameters of the analysed walls.

Numerical calculations showed that thermal transmittance for Thermo sections was higher than thermal transmittance for solid sections. The difference between U values reaches up to 15%. In light of the holes with small surface in a perforated section and the simplified grid of finite components in the numerical model, it can be presumed that building a more complex grid would reduce this difference.

Table 1. Material data for analyzed partitions.

Ordinal	Type of material	Thickness d [m]	Thermal conductivity λ [W/(m·K)]
1	Thin-layer plaster	0.0045	0.7
2	Mineral wool	0.16; 0.08; 0.05	0.035
3	Cement-bonded particleboard	0.012	0.215
4	Vapor-permeable membrane	-	-
5	Steel profile cold formed C140	0.14	50
6	Mineral wool	0.09	0.035
7	Vapor barrier	-	-
8	Cement-bonded particleboard	0.012	0.215
9	Gypsum plasterboard	0.0125	0.25

Furthermore, quite different situation is observed for the heat flux Φ where energy loss through the joint in the form of the wall corner is smaller for the Thermo sections. The same pattern is observed for the linear thermal transmittance Ψ. In certain variants, the value of linear thermal transmittance obtained with respect to external dimensions Ψ_e showed an increase, see Table 3.

Table 2. The values of the total heat flux Φ for the analyzed variants of wall corners.

Total heat flux Φ flowing through the connector		Value [W]		
		5 cm	8 cm	16 cm
1	Variant I P*	28.369	22.903	15.485
2	Variant I T*	27.537	22.378	15.236
3	Variant II P*	28.626	23.104	15.547
4	Variant II T*	27.904	22.627	15.328

P* - full profile, T* - THERMO profile

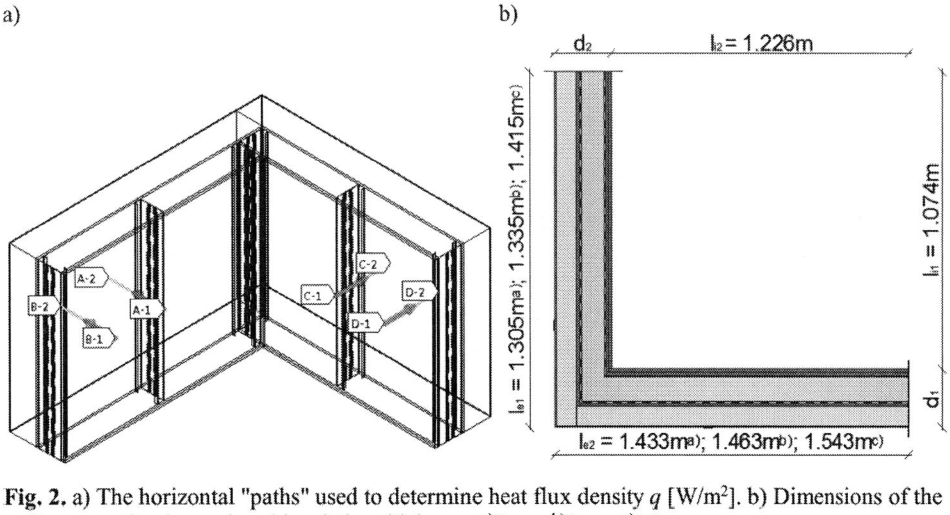

Fig. 2. a) The horizontal "paths" used to determine heat flux density q [W/m^2]. b) Dimensions of the wall corners for the analysed insulation thickness: [a]5 cm; [b]8 cm, [c]16 cm.

The difference between individual values may also result from the horizontal "paths" (see Fig. 2a) in order to calculate heat flux density q and based on equations (1) and (2). Different situation is observed for the heat flux Φ where the value was calculated for the whole wall.

Table 3. Thermal parameters for analyzed exterior partitions.

Ordinal	The characteristic thermal parameters corner of the exterior wall			Value – Insulation thickness 5 cm	8 cm	16 cm
1	The heat transfer coefficient U [W/(m²·K)] of each part of the connector	Variant I P*	U_1	0.231	0.203	0.137
			U_2	0.231	0.203	0.137
		Variant I T*	U_1	0.237	0.211	0.144
			U_2	0.237	0.211	0.144
		Variant II P*	U_1	0.227	0.200	0.130
			U_2	0.227	0.200	0.130
		Variant II T*	U_1	0.232	0.212	0.144
			U_2	0.232	0.212	0.144
2	Linear coupling coefficient of thermal L^{2D} [W/(m·K)] $L^{2D}=\frac{\Phi}{l\cdot(t_i-t_e)}$	Variant I P*	L^{2D}	0.709	0.573	0.387
		Variant I T*	L^{2D}	0.688	0.559	0.381
		Variant II P*	L^{2D}	0.716	0.578	0.389
		Variant II T*	L^{2D}	0.698	0.566	0.383
3	Linear heat transfer coefficient (the internal dimensions) Ψ_i [W/(m·K)] $\Psi_i = L^{2D} - (l_{i1}\cdot U_1 + l_{i2}\cdot U_2)$	Variant I P*	Ψ_i	0.183	0.129	0.082
		Variant I T*	Ψ_i	0.161	0.072	0.035
		Variant II P*	Ψ_i	0.211	0.126	0.094
		Variant II T*	Ψ_i	0.172	0.043	0.042
4	Linear heat transfer coefficient (the external dimensions) Ψ_e [W/(m·K)] $\Psi_e = L^{2D} - (l_{e1}\cdot U_1 + l_{e2}\cdot U_2)$	Variant I P*	Ψ_e	0.082	0.034	-0.006
		Variant I T*	Ψ_e	0.045	-0.034	-0.065
		Variant II P*	Ψ_e	0.115	0.028	0.01
		Variant II T*	Ψ_e	0.072	-0.071	-0.055

P* - full profile, T* - THERMO profile

4 Heat loss through heat transfer

Heat loss in the building's corner was dimensioned according to the external dimension l_e and external dimensions l_i (Fig. 2b). In order to calculate heat loss through heat transfer, the calculation of the value of thermal transmittance for the wall with heat bridges U_{kc} was based on the simplified method [10]. It consists in adoption of the corrected value of thermal transmittance:

$$U_{kc} = U_k + \Delta U_{tb} \quad [\mathrm{W/(m^2 \cdot K)}] \tag{3}$$

where:
U_{kc} – corrected thermal transmittance of the building component (k) including linear thermal bridges [W/(m^2·K)];
U_k – thermal transmittance for building component (k), [W/(m^2·K)];
ΔU_{tb} – correction coefficient depending on the type of building component, [W/(m^2·K)].

For the calculated corrected thermal transmittance U_{kc} (Table 3) and adopted wall surface area A=20 m^2 the heat loss coefficient H_{ti} was calculated using external dimensions.

$$H_{ti} = A_k \cdot U_k \quad [\mathrm{W/K}] \tag{4}$$

where:
A_k – cross-section of the analysed wall,
U_k – corrected thermal transmittance of the building component (k) including linear thermal bridges, [W/(m^2·K)].

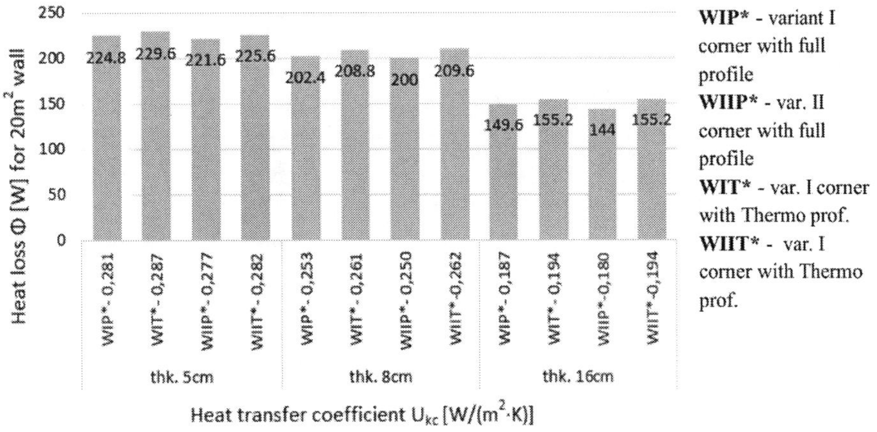

Fig. 3. Heat loss through 20 m^2 wall surface in the lightweight framing structure.

Heat loss Φ_{te} from the heated space through heat transfer through the wall was calculated according to (5).

$$\Phi_{t,e} = H_{t,i} \cdot (\Theta_i - \Theta_e) \quad [\mathrm{W}] \tag{5}$$

where:
$H_{t,i}$ – heat loss coefficient, [W/K];
Θ_i – adopted internal temperature 293.15 [K];
Θ_e – adopted external temperature 253.15 [K].

The results of heat loss for the analysed walls were presented in Fig. 4.

5 The results of numerical analysis

The numerical thermal analysis was conducted using ANSYS software in the Steady - State Thermal module [11,12]. Tables 4 and 5 show temperature distribution and vector

DOI: 10.1051/matecconf/201710700047

distribution of heat flux density for the variants of wall corners. In the walls with 5 cm and 8 cm insulation, the heat flux density is transferred to the outside. Different situation is with the insulation with thickness of 16 cm. Furthermore, the temperature distribution at the internal side of the wall corner reveals lower temperature for insulation thickness of 5 and 8 cm. This situation occurs for all the variants of wall corners.

Table 4. The results of computer simulation for the solid section (variants I and II) of the corner solution: temperature distribution and vector distribution of heat flux density.

		Temperature distribution	**Vector heat flux density distribution**
Full profile -Variant I	**Insulation thickness 5 cm**		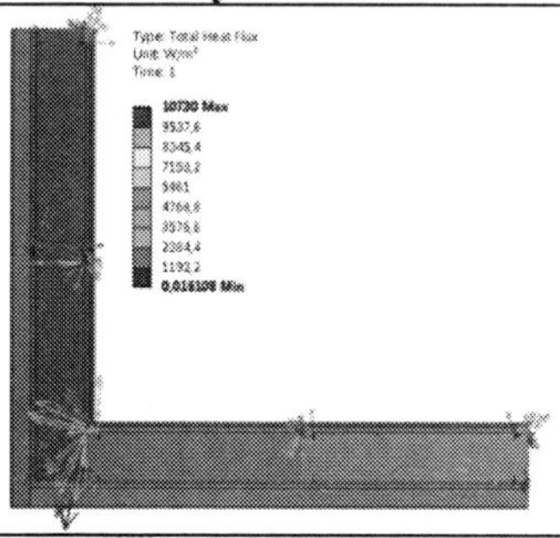
	Insulation thickness 16 cm	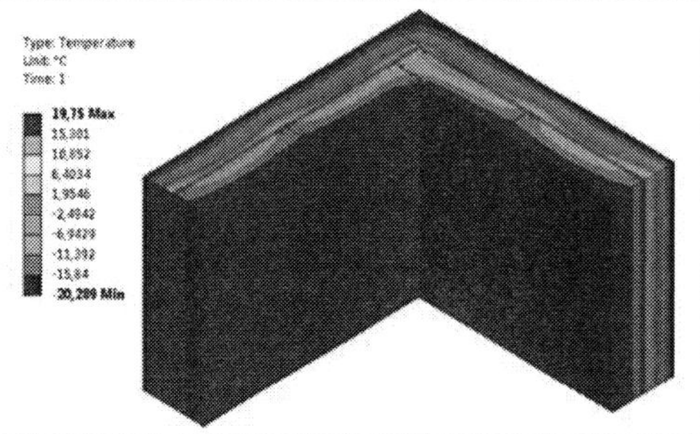	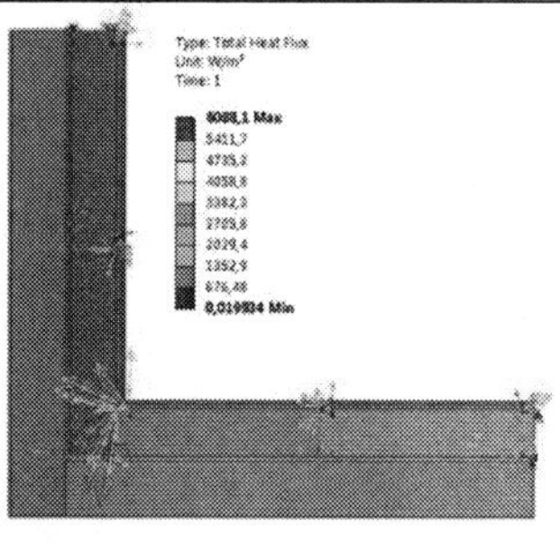
Full profile -Variant II	**Insulation thickness 5 cm**	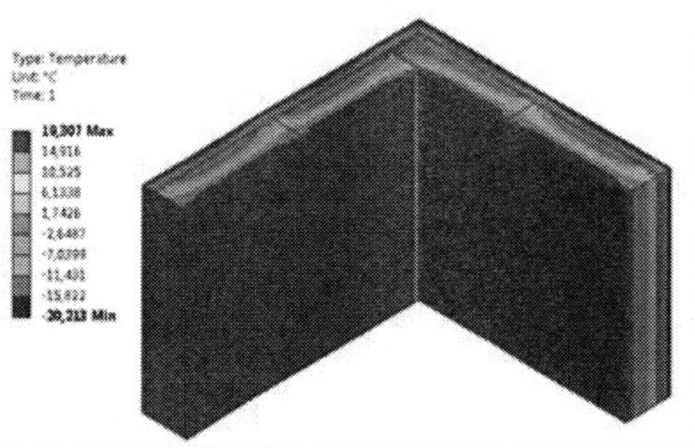	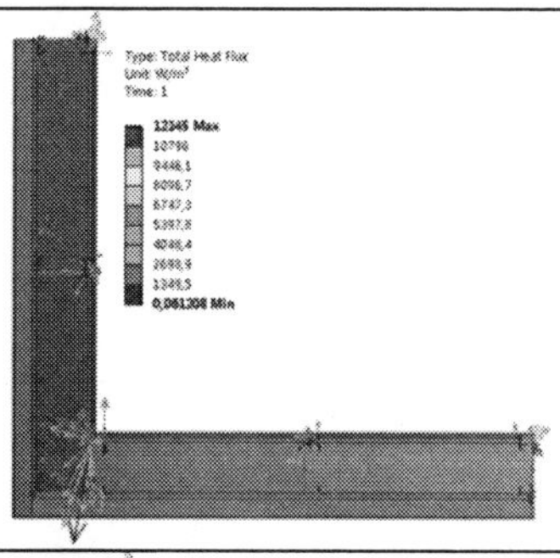
	Insulation thickness 16 cm		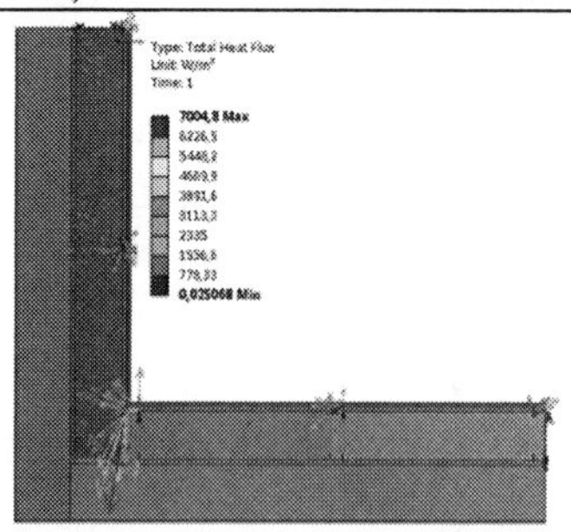

Table 5. The results of computer simulation for the THERMO section (variants I and II) of the corner solution: temperature distribution and vector distribution of heat flux density.

		Temperature distribution	**Vector heat flux density distribution**
THERMO profile -Variant I	**Insulation thickness 5 cm**		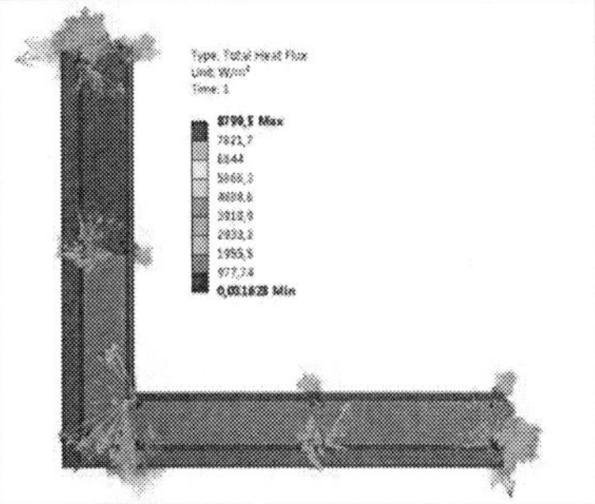
	Insulation thickness 16 cm	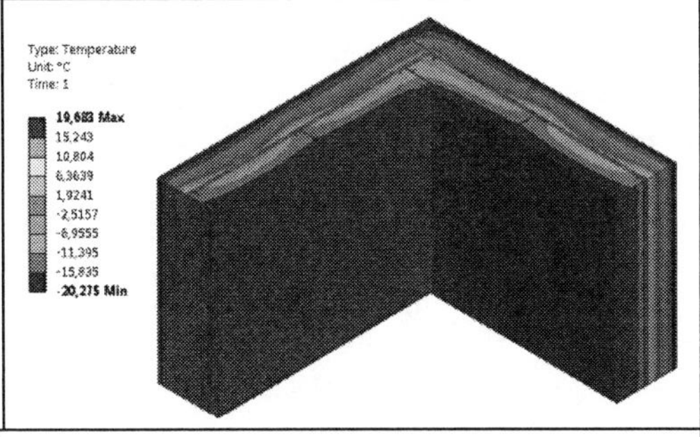	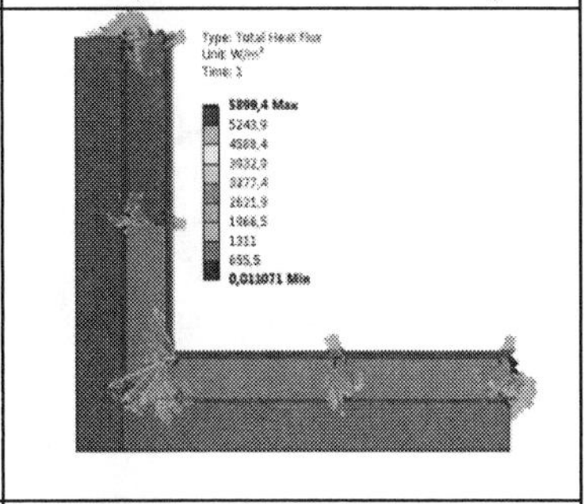
THERMO profile -Variant II	**Insulation thickness 5 cm**	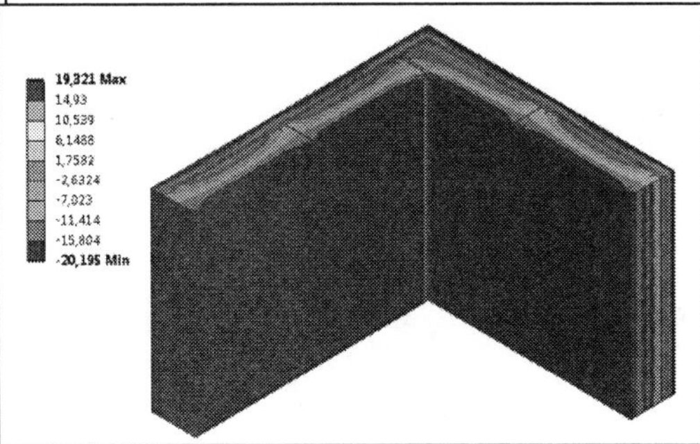	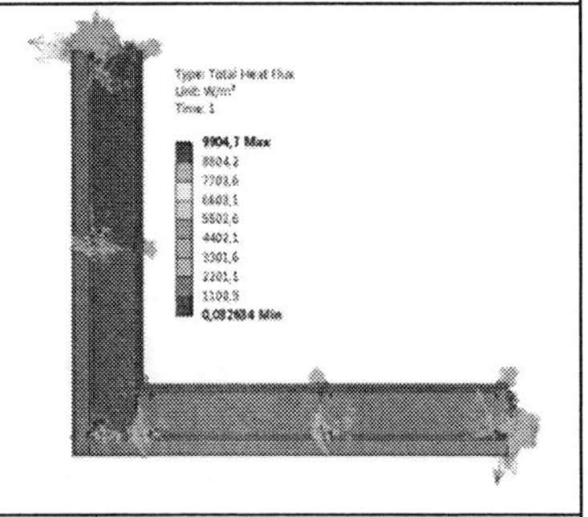
	Insulation thickness 16 cm	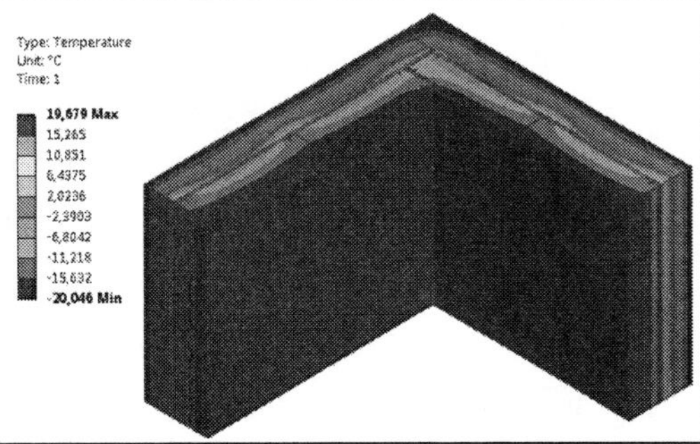	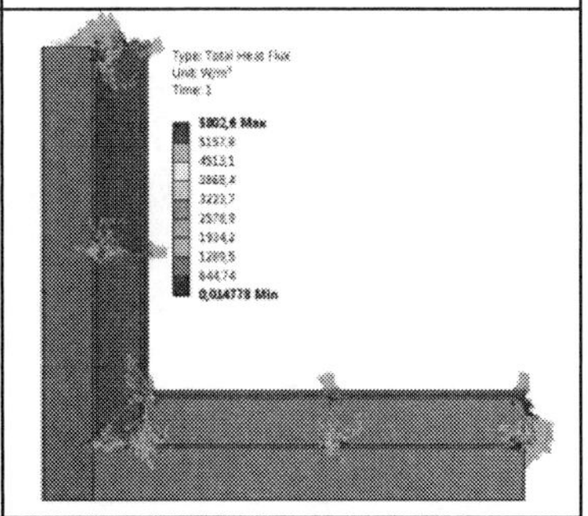

The tables abowe show that the lowest temperature at the internal side is 19.267°C, calculated for the Thermo section (Variant I). Furthermore, the highest temperature of 19.750°C was found for the solid section (Variant I). The biggest difference in the internal temperature between the wall corner variants is around 0.5%.

6 Summary

The study evaluated temperature distribution and energy loss in walls with steel framing structure made of cold-rolled sections. Two types of sections and two variants of geometrical solution were analysed for the wall corner. Tables 2 and 3 contain thermal parameters of the analysed walls. The values of total heat flux Φ for Thermo sections for the variants analysed in the study is by ca. 3% smaller compared to solid sections. Furthermore, the difference in the heat loss $\Phi_{t,e}$ for the analysed wall is by ca. 5% greater in solid sections. However, with the simplified method of calculations and adoption of the simplified grid of finite elements, this difference would change, with higher values found for Thermo sections.

The analysis for two different sections did not show significant changes in terms of temperature distribution. The calculations did not reveal a significant effect on the improvement in the thermal insulation properties of the wall through by using Thermo sections.

Numerical computation software based on finite element method (FEM) is useful in analysis of homogeneous structures, such as the framing structure of external walls analysed in this study. With numerical computations, analysis of technological and construction solutions can be made at the design stage, which has a substantial effect on the functional properties of the building. Finding weak points of the wall that are characterized by the most substantial heat loss offers opportunities for the improvement of adopted technological and construction solutions. However, the focus should be on discretization of the analysed geometry.

References

1. I. Major, *Jakościowe i ekologiczne aspekty w technologiach budowlanych*, 86-102, In polish (2013)
2. J. Flodr, M. Krejsa, D. Mikolášek, O. Sucharda, L. Žídek, Appl. Mech. Mat. **617**, 171-174 (2014)
3. B. Balogh, J. Lógó, Transactions of the VŠB – Technical University of Ostrava, Civil Engineering Series, **15**, 2 (2016)
4. I. Major, Civil and environmental engineering, Scientific - Technical Journal, **10**, 1, 42–48 (2014)
5. M. Major, I. Major, Advanced Materials Research, **1020**, 165-170 (2014)
6. *Rozporządzenie Ministra Infrastruktury z dnia 12 kwietnia 2002 r. w sprawie warunków technicznych, jakim powinny odpowiadać budynki i ich usytuowanie*, Dz.U. z 2002 r., Nr 75, poz. 690, In polish (2002)
7. PN-EN 12831:2006. *Instalacje ogrzewcze w budynkach – Metoda obliczania projektowego obciążenia cieplnego*, in polish (2006)
8. A. Gołaś, M. Ryś, *Modelowanie Inżynierskie*, In polish (2011)
9. BuildDesk, Straty ciepła przez przewodzenie a sposób wymiarowania. PN - EN ISO 10211-1; Materiały edukacyjne dla doradców i audytorów energetycznych, In polish
10. PN-EN ISO 14683:2008, Mostki cieplne w budynkach. Liniowy współczynnik przenikania ciepła. Metody uproszczone i wartości orientacyjne, In polish
11. *ANSYS Mechanical APDL Thermal Analysis Guide* (2013)
12. E. Madenci, I. Guven, *The Finite Element Method and Applications in Engineering Using ANSYS®* (Springer, 2007)

Impact analysis of input parameter for damage detection code

Michal Venglar[1,*] and *Milan* Sokol[1]

[1]Slovak University of Technology in Bratislava, Faculty of Civil Engineering, Department of Structural Mechanics, Radlinskeho 11, 81005, Bratislava, Slovakia

Abstract. The primary aim of the article is to analyse impact of appropriate values of input parameter for an effective solution of the self-developed code used for damage detection. The code was prepared to determine the change of bending stiffness in Microsoft Office Excel Visual Basic for Applications. The code used non-destructive vibration based method, i.e. the FE model updating method. A steel bar was assumed for numerical calculation. Time consumption of calculation, precision of identification and degree of possible damage detection were investigated. The values of computation time depend on the input values, the desired limit of the accepted error. Then, data from a laboratory experiment was analysed. The damage detection was done in accordance with the suitable input data from a parametric study of the steel bar.

1 Introduction

Nowadays, damage detection gets ahead throughout the civil, mechanical and aerospace engineering communities [1-3]. For civil engineers, it is caused by increasing demands on safety and reduction of maintenance cost [4-5]. Another reason is that many bridges are obsolete and need for their rehabilitation [6]. In accordance with [7], the average age of bridges in Slovakia is over 43 years. In many cases, lack of long-lasting maintenance or periodic inspection may result in necessity of later expensive complete reconstruction or construction of a new bridge structure which is often even more expansive. The circumstance from the USA confirms this fact. The 94-year-old Greenfield Bridge in Pittsburgh had to be demolished in 2005 and a new structure have replaced the old one. Structural health monitoring can help us to avoid above-mentioned situations.

Many methods are used for monitoring of structures e.g. visual controls, destructive methods, non-destructive methods. The Finite Element Model Updating (FEMU) method is one of vibration based non-destructive methods. Several approaches of FEMU method were summarized in [8-9]. The FEMU method is also applied to damage detection with neglecting damping of structure in a simple self-developed code. The used version of the code has been customized in comparison to its published version in [10]. The paper is devoted to analysis of input data and their impact on accuracy of damage assessment and analysis of minimum

* Corresponding author: michal.venglar@stuba.sk

degree of possible damage that can be detect on a numerical model of a steel bar. The achieved date will be used for system identification of an experimental laboratory specimen.

2 Study of input parameter LIMIT

The theory and description of the code was described in [10]. However, it has been modified a little bit for this study. The self-developed code in Excel VBA uses more input parameters, which affect the accuracy of the results and the time consumption of calculations. The modal analysis of the steel beam was done in the FEM software and the results were used as the input data for calculations. Specified changes of stiffness were applied for the 35 elements in case of the modal analysis and for damage detection. The length of the used element was 100 mm, so the length of the investigated beam was 3500 mm. The second calculations were prepared in a way that the only one element (1200 mm from the right side of the beam) had changed the height of the cross-section (Fig. 1). Damaged scenarios for the one element and appropriate eigen frequencies for the beam are shown in the Fig. 2.

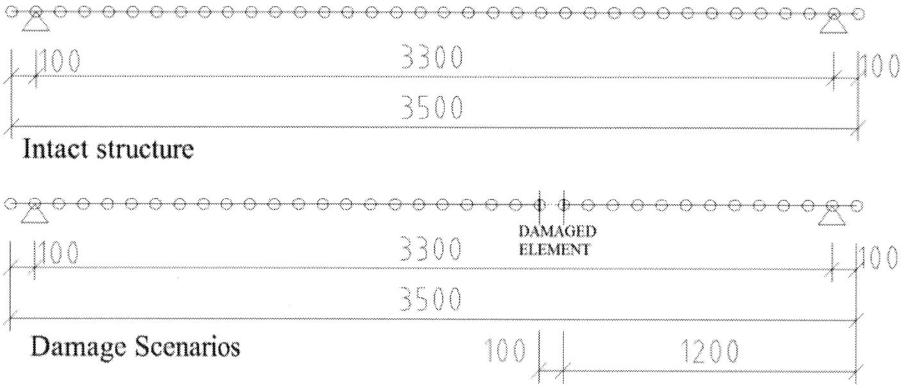

Fig. 1. The investigated steel beam.

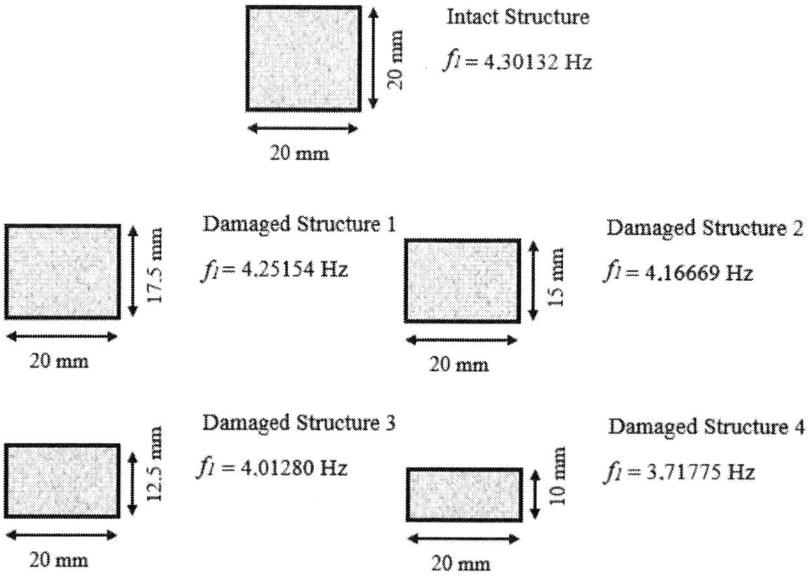

Fig. 2. Damage scenarios and corresponding eigen frequencies of structure.

The PC used for the calculations consisted of 8-core processor and it worked with 3.25 GB of RAM. The code ran only in the 32-bit operating system.

2.1 Accuracy and time consumption

The limit value of auxiliary error (indicated by LIMIT in accordance with [10]) was monitored during this parametric study for the Damage Scenario 4 (Fig. 2). The auxiliary error FAULT equivalents the sum of squares of differences between the desired and calculated displacements at nodes, thus:

$$FAULT = \frac{\sum (Y_i - Y\_Y_i)^2}{\sum Y_i^{\,2}} \, 100\% \tag{1}$$

where Y_Y_i is the bending mode-shape and Y_i *is* the desired state of the appropriate bending mode-shape (Fig. 3).

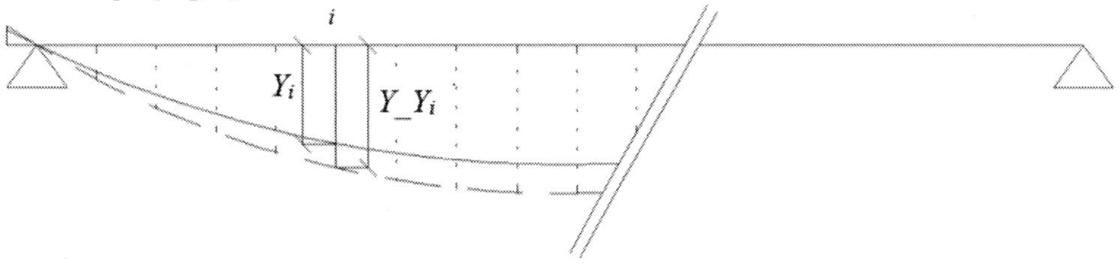

Fig. 3. Description of the code (desired displacement - Y_i, calculated displacement - Y_Y_i).

The whole calculation runs in a while-loop cycle. The while-loop cycle ends when the FAULT value is smaller than the LIMIT value. This is the only one condition which completes the calculation of damage. During the parametric study seven cases of LIMIT value has been investigated. The LIMIT value was considered in logarithmic scale from 0.5% to 0.0000005%.

The parametric study gives logical results and it is possible to indicate that time consumption of the calculation is the highest in the case of the LIMIT value – the smallest auxiliary error. Analyses with value of LIMIT from 0.5% to 0.0005% are below the needed accuracy and the future application is not recommended. Other calculations exceeded time consumption of 6 minutes. On the other hand, the difference between real and identified stiffness has dropped below difference of 20%. This value can be acceptable. Finally, the difference for the identified stiffness achieved zero when the exact LIMIT value 0.0000005% was assumed. The calculated difference is negligible in this case. However, the computation time has increased approximately up to 2 hours.

Based on the above-mentioned, it can be concluded that the optimum ratio of time consumption and the minimal difference has been reached for the LIMIT value which is about 0.000005%. This parameter was also used in the next chapters of the paper.

2.2 Degree of damage detection

Damage of structure can assume various proportions. Big damages could be very dangerous and it is possible to identify them. On the other hand, smaller damages should also be identified. Four damage scenarios were considered. In accordance with the Fig. 1, the height of damaged element has been gradually changing. The height was assumed between 10 and 20 mm.

The appropriate LIMIT value was used for damage detection in the mentioned scenarios. For calculating the difference in the damage detection, the following formula was used:

$$Difference = \frac{(A-B)}{max(A,B)} 100\% \quad (2)$$

where A is the identified bending stiffness and B the desired (real) bending stiffness. The Table 1 shows that various damages could be detect by the developed code. The small change in height (Damaged structure 1) can also be detected. The difference in detection is approximately 19%.

Table 1. Comparison of damage detection for more scenarios.

Damage scenario	Identified stiffness *EI* [Nm²]	Real stiffness *EI* [Nm²]	Difference [%]	Time consumption
Intact	2791.98	2793.00	-0	1 min 24 s
Damaged 1	2328.30	1875.78	+19	4 min 25 s
Damaged 2	1179.12	1459.25	+19	13 min 51 s
Damaged 3	733.60	683.59	+7	22 min 9 s
Damaged 4	365.14	351.00	+4	27 min 42 s

3 System identification of experimental laboratory model

The application of above-mentioned system identification was conducted in different case than in study of input parameter. The investigated experimental laboratory model was made from wooden and plaster boards. Three wooden boards were used for the main beam and the deck was made of the 3 plasterboard layers with height of 12.5 mm. The deck was 300 mm wide and 37.5 mm high in total. The composite beam was 4 metres long and it was supported as the simple supported beam with a single span of 3.9 metres.

Firstly, data for solid model were acquired. Secondly, a stiffener on the main beam was added for simpler modelling of the stiffness change (damage). The stiffener was also located 1200 mm from the right side of the specimen but its length was 200 mm. The real bending stiffness *EI* with the stiffener was 375 kNm^2. Acceleration was measured at 18 points along the experimental specimen, so the stiffener represented one half of the distance between placed accelerometers. Analyses of the measurement data (in accordance with [11]) were used for identifying of the first vertical mode-shape.

Using optimized parameters, the more satisfactory degree of the accuracy of the damage detection was achieved in comparison with unoptimized input parameter. The difference in the identified stiffness was reduced (Table 2). The optimized input value of the parameter LIMIT gave more accurate results, but the damage assessment was not good enough. On the other hand, the difference about 19% is still acceptable.

Table 2. Identification of stiffener in self-developed code.

Method	Identified stiffness *EI* [kNm²]	Difference [%]
FEMU (unoptimized)	247	-34
FEMU (optimized)	314	-19

4 Conclusions

The parametric study verified the time consumption of calculations, as well as the desired accuracy of the damage assessment. The difference between calculated and desired displacement was minimized and the results were accurately determined. The identification of the stiffness was successful with acceptable differences in above-mentioned parameter by using the sufficiently small LIMIT value. The price for the high accuracy of the assessment was enormous time-consumption. It appears to be reasonable (the ratio between accuracy and time-consumption) to use the optimized value of accuracy.

The various damage scenarios were investigated. The code could find quite small damage like the reduced height of the cross-section. The smallest investigated reduction was about 10% of the total height.

The LIMIT value which had been found in the parametric study was applied for the damage detection (in this case the addition of the stiffener), which was carried out on the experimental laboratory model. The achieved results of the damage detection were better than with unoptimized parameter. The location of the damage has been properly identified, as well as the value of the identified stiffness which was closer to the exact value.

This paper has been supported by the Slovak Research and Development Agency (SRDA), i.e. a grant from research program No. APVV-0236-12.

References

1. G. Pandey, E.T. Thostenson, D. Heider, Progr. Electromagn. Res. **137**, 551 (2013)
2. M. Radzienski, Ł. Dolinski, M. Krawczuk, M. Palacz, Mech. Syst. Sig. Process. **39**, 388 (2013)
3. J. Niu, Z. Zong, F. Chu, Sci. Chin. Technol. Sci. **58**, 701 (2015)
4. H. Wenzel, D. Pichler, *Ambient Vibration Monitoring* (John Wiley & Sons, Ltd, Chichester, 2005)
5. H. Wenzel, *Health Monitoring of Bridges* (John Wiley & Sons, Ltd, Chichester, 2009)
6. P. Franchetti, M. Frizzarin, A. Leonardi, F. Zeni, MATEC Web of Conferences **24**, 09008 (2015)
7. P. Paulik, *Bridges in Slovakia* (Jaga, Bratislava, 2014)
8. E.P. Carden, P. Fanning, Struct. Health Monit. **3**, 355 (2004)
9. T. Marwala, *Finite-Element-Model Updating Using Computational Intelligence Techniques: Application to Structural Dynamics* (Springer, London, 2010)
10. M. Sokol, R. Aroch, M. Venglar, M. Fabry, T. Zivner, Appl. Mech. and Mat. **769**, 192 (2015)
11. M. Sokol, R. Flesch, J. Bridge Eng. **10**, 77 (2010)

Delamination of laminate plate under tearing load mode

Eva Kormaníková[1,*] and *Kamila* Kotrasová[1]

[1]Technical University of Košice, Faculty of Civil Engineering, Department of Structural Mechanics, Vysokoškolská 4, 04200 Košice, Slovakia

Abstract. The paper presents the mixed mode delamination of laminate plate made of two sublaminates. To this purpose the load mode III of delamination is proposed as failure orthotropic model. Sublaminates are modeled by using shear deformable elements in program ANSYS. The response mixed-mode components of vector of energy release rates along the delamination front are calculated.

1 Introduction

The analysis of damage mechanisms in laminated composites has received a lot of interest for many years. Because of the low tensile strength in the transverse directions, interlaminar fracture represents a very important design consideration. As a matter of fact, because of their low interlaminar toughness, composite laminates often exhibit delamination when subjected to stress concentrations near manufacturing defects or to low velocity impacts. Delamination may arise at the edge of interlaminar defects in laminates loaded in compression, due to a global or local buckling of layers [1]. Moreover, delamination may occur as a failure mechanism in rehabilitated structures originating from interfacial defects between the repair composite laminate, made of fibre-reinforced plastic, and the existing concrete system [2].

A more efficient alternative for delamination analysis is to use plate theory. Classical delamination models [3], [4], [5] which consider the laminate as an assemblage of two plates in the cracked zone and of a unique plate in the undelaminated one, may give an accurate estimate of the total energy release rate G. Unfortunately, they are unable to accurately predict the individual energy release rates due to an inaccurate description of the local crack tip strain state and to take into account for shear deformations. As a consequence, the classical plate-based delamination models have been improved according to different methods. Delamination was modeled by adopting sublaminates governed by transverse shear deformable laminate theory, thus obtaining a reasonable approximation to the mode separation solution [6].

* Corresponding author: eva.kormanikova@tuke.sk

2 Mechanics of delaminated composite structure

The mechanisms that lead to failure in composite materials are not yet fully understood, especially for matrix or fiber compression. Strength-based failure criteria are commonly used with the FEM to predict failure events in composite structures and different continuum-based criteria have been derived to relate the internal stresses and experimental measures of material strength to the onset of failure. In Fig. 1 a laminate contains a single in-plane delamination crack of area Ω_D with a smooth front $\partial\Omega_D$. The laminate thickness is denoted by h_0. The *x-y* plane is taken to be the mid-plane of the laminate, and the z-axis is taken positive downwards from the mid-plane. For sublaminate modeling plate finite elements could be developed from shell elements. In this case each sublaminate is represented by an assemblage of the first order shear deformable (FSDT) plate elements bonded by zero-thickness interfaces in the transverse direction as shown in Fig. 1 [7]. Accordingly, the displacements in the *z*-th plate element, in terms of a global reference system located at the laminate mid-surface, are expressed [8]:

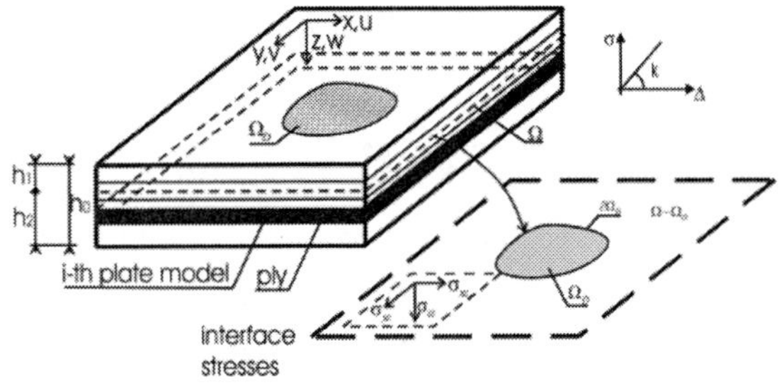

Fig. 1. Mechanics of the delaminated composite structure.

$$
\begin{aligned}
U_i(x,y,z) &= u_i(x,y) + (z - z_i)\psi_{xi}(x,y) \\
V_i(x,y,z) &= v_i(x,y) + (z - z_i)\psi_{yi}(x,y) \\
W_i(x,y,z) &= w_i(x,y)
\end{aligned}
\tag{1}
$$

where U_i, V_i refer to the in-plane displacements, and W_i to the transverse displacements through the thickness of the *i*-th plate element, u_i, v_i, w_i are the displacements at the mid-surface of the *i*-th plate element, respectively, and $\psi_{xi}(x,y), \psi_{yi}(x,y)$ denote rotations of transverse normals about *y* and *x* axis, respectively.

At the reference surfaces, the membrane strain vector ε_i, the curvature κ_i, and transverse shear strain γ_i respectively are defined as

$$
\begin{Bmatrix} \varepsilon_{xi} \\ \varepsilon_{yi} \\ \gamma_{xyi} \end{Bmatrix} = \begin{Bmatrix} \dfrac{\partial u_i}{\partial x} \\ \dfrac{\partial v_i}{\partial y} \\ \dfrac{\partial u_i}{\partial y} + \dfrac{\partial v_i}{\partial x} \end{Bmatrix}, \quad
\begin{Bmatrix} \kappa_{xi} \\ \kappa_{yi} \\ \kappa_{xyi} \end{Bmatrix} = \begin{Bmatrix} \dfrac{\partial \psi_{xi}}{\partial x} \\ \dfrac{\partial \psi_{yi}}{\partial y} \\ \dfrac{\partial \psi_{xi}}{\partial y} + \dfrac{\partial \psi_{yi}}{\partial x} \end{Bmatrix}, \quad
\begin{Bmatrix} \gamma_{yzi} \\ \gamma_{xzi} \end{Bmatrix} = \begin{Bmatrix} \psi_{xi} + \dfrac{\partial w_i}{\partial y} \\ \psi_{xi} + \dfrac{\partial w_i}{\partial x} \end{Bmatrix}
\tag{2}
$$

By considering each plate as composed by one or several physical fiber-reinforced plies with their material axes arbitrarily oriented, the constitutive relations between stress resultants and corresponding strains are:

$$\begin{Bmatrix} \boldsymbol{N}_i \\ \boldsymbol{M}_i \end{Bmatrix} = \begin{bmatrix} \boldsymbol{A}_i & \boldsymbol{B}_i \\ \boldsymbol{B}_i & \boldsymbol{D}_i \end{bmatrix} \begin{Bmatrix} \boldsymbol{\varepsilon}_i \\ \boldsymbol{\kappa}_i \end{Bmatrix}, \quad \boldsymbol{V}_i = \boldsymbol{H}_i \boldsymbol{\gamma}_i \tag{3}$$

where $\boldsymbol{N}_i = \{N_{xi}, N_{yi}, N_{xyi}\}^T$ is the membrane force resultant vector, $\boldsymbol{M}_i = \{M_{xi}, M_{yi}, M_{xyi}\}^T$ the moment resultant vector and $\boldsymbol{V}_i = \{V_{yzi}, V_{xzi}\}^T$ the transverse shear force resultant vector. In addition, $\boldsymbol{A}_i$, $\boldsymbol{D}_i$, $\boldsymbol{B}_i$ denote the classical extensional stiffness matrix, bending stiffness matrix and bending-extensional coupling stiffness matrix, respectively, whereas $\boldsymbol{H}_i$ is the shear stiffness matrix [8].

3 Global and local approaches

The stress singularities, captured by using the local approach, and stress resultant discontinuities, obtained in the context of the global approach, may be both responsible for energy release rates $G(s)$. Consequently they must represent the same mechanism of energy dissipation during delamination growth. The equivalence between local (interface) and global (plate) approaches can be proved by considering the regularized version of the modified total potential energy Π_r and extracting discontinuities directly from the interface strain energy.

Using the interface constitutive equation $\{\sigma\} = [K]\{\Delta\}$ to compute interlaminar stresses, leads to the following total $G(s)$ expression

$$G(s) = \lim_{k \to \infty} \frac{1}{2} \left[k_z \Delta w^2(s) + k_{xy} \Delta u^2(s) + k_{yx} \Delta v^2(s) \right] = G_I(s) + G_{II}(s) + G_{III}(s), \quad \Delta w(s) \geq 0 \tag{4}$$

where s is a curvilinear coordinate along the delamination front. In (4) $G(s)$ is the local ERR function along the delamination front $\partial\Omega_D$, defined by

$$-\int_{\partial\Omega_D} G(s) \dot{\Omega}_D(s) ds = \dot{\Pi}_r \tag{5}$$

The mathematical correspondence between the local and global approaches can be thought as a direct consequence of the equivalence between the Lagrangian and the penalty methods to impose the adhesion constraint. These results show that two alternative methodologies leading to the same results can be used to compute the total energy release rate and its mode components: the one provided by the "strong" interface model", which imposes perfect adhesion ahead the delamination front in a limit process as the interface stiffness parameters approach infinity according to a penalty formulation; the other one based on the "collapsed" interface model in which adhesion is enforced directly by means of a Lagrangian multipliers formulation. However, the "strong" interface model leads to very simple formulas for both the total energy release rate and its individual mode components based on the limit values of interface strain energy densities which are very efficient when used in conjunction with the numerical finite element formulation. On the other hand, the "collapsed" interface model leads to more complex analytical formulas involving jumps in plate strains and stresses, which are necessary since it is not possible to retrieve interlaminar stresses and relative displacements between the surfaces of the delamination directly from the interface model. As a consequence, in the numerical calculations the "strong" interface model will be adopted.

4 Delaminated orthotropic plate

A square plate comprising two subplates with thicknesses h_1 = 0.5 mm, h_2 =1 mm is investigated. The subplates are uniform loaded in tearing mode (Mode III) (Fig. 2, right). The plate geometry is shown in Fig. 2, left. The mechanical characteristics of the upper sublaminate (AS4D/9310) are: E_x = 134 GPa, $E_y = E_z$ = 7.7 GPa, G_{yz} = 2.76 GPa, $G_{xy} = G_{xz}$ = 4.3 GPa, $\nu_{xy} = \nu_{xz}$ = 0.3, ν_{yz} = 0.4. The mechanical characteristics of the lower subplate (T300/5208) are: E_x = 136 GPa, $E_y = E_z$ = 9.8 GPa, G_{yz} =5.2 GPa, $G_{xy} = G_{xz}$ = 4.7 GPa, $\nu_{xy} = \nu_{xz}$ = 0.28, ν_{yz} = 0.15. The response mode components of energy release rate along the delamination front are calculated.

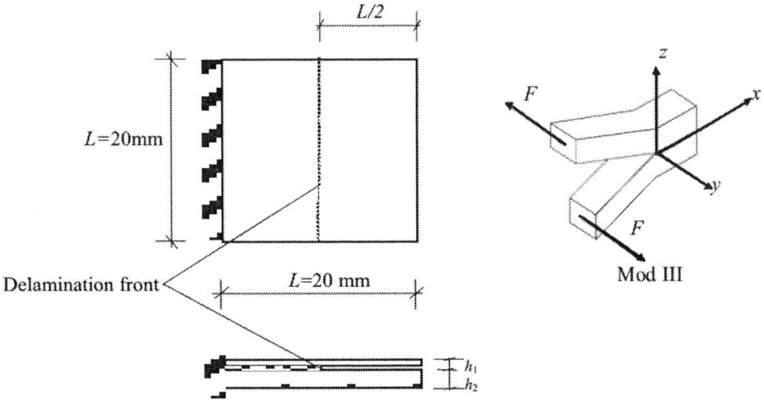

Fig. 2. A square plate geometry (left), tearing mode of delamination (right).

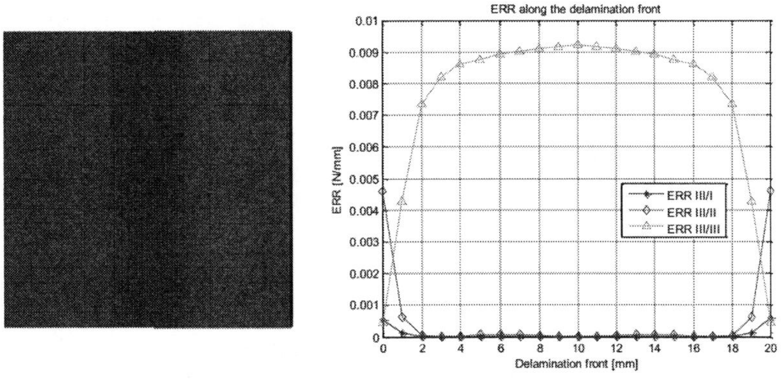

Fig. 3. Finite element model (left), results of individual components of ERR III (right).

Two plates have been used to model the delaminated plate in the thickness direction by using ANSYS. Each is modeled with SHELL181 elements. The projection on the x-y plane of the finite element mesh, assumed equal for the plate and the interface models. The mesh is refined in a zone of 5×20 mm^2 centered with respect to the delamination front. In this zone, the length of the plate and interface elements in the x-direction is 0.125 mm, whereas along the y-direction it is 1 mm (Fig. 3, left). A force of 1 N/mm is imposed in the y-direction (tearing load mode III (Fig. 2)). Two subplates have been used to model the delaminated plate in the thickness direction. Interface elements, necessary to connect the two sublaminates, are implemented by using a combination of rigid links MPC 184 defined by two nodes and three degrees of freedom at each node. The delamination front was created by spring elements COMBIN14, in each offset node of the delamination front by three elements. The offset nodes can be generated by rigid links. The interface between subplates was modeled by constrain equations. When CE are used in place of rigid links, for each mid-plane node of the upper plate model, three coincident nodes located on the lower surface of the plate are created. Similarly, for each mid-plane node of the lower plate model, three coincident nodes located on the upper surface of the plate are created. Three COMBIN14 elements connected to the three pairs of coincident nodes placed at the delamination plane are introduced, each one acting in differ translation direction. For mixed-mode response conditions, a complete modeling of interface elements has been implemented. The stiffness of the spring elements binding the subplates is chosen as 10^8 N/mm^3, that is appropriate for the normal and tangential stiffness.

5 Conclusion

A two-subplate orthotropic model [9] loaded by mode III was proposed as failure model. The material characteristics of the sublaminates were calculated by homogenization [10]. Subplates were modeled by using shear deformable elements of FEM [11] in program ANSYS. The individual components of ERR for loading mode III and response modes I, II and III, respectively along the delamination front are plotted in Fig. 3, right. Response mode III of ERR has maximum values in the middle of the model, therefore it is expected that the delamination proceeds from this place. Results have shown that response mode III is predominant and the response modes I and II are negligible, except of free edges. Based on these results a designer must pay attention at the free edges of delamination front [12-15].

This work was supported by the Scientific Grant Agency of the Ministry of Education of Slovak Republic and the Slovak Academy of Sciences under Projects VEGA 1/0477/15 and 1/0078/16.

References

1. D. Bruno, F. Greco, Int. J. Solids Struct. **37**, 6239–6276 (2000)
2. Y. M. Lim, V. C. Li, Cement Concr. Compos. **19,** 373–385 (1997)
3. B. Cochelin, M. Potier-Ferry, Comput. Methods, Appl. Mech. **89**, 361–380 (1991)
4. J. - K. Kim, Comput. Structures **26,** 6, 975–983 (1997)
5. D. Bruno, F. Greco, Int. J. Solids Struct. **38/50–51**, 9149–9177 (2001)
6. Z. Zou, S. R. Reid, P. D. Soden, S. Li, J. Solids Structures **38**, 2597–2613 (2001)
7. M. Zmindak, V. Dekys, P. Novak, Adv. Mat. Res. **969**, 176-181, (2014)
8. E. J. Barbero, *Finite Element Analysis of Composite Materials* (CRC Press, Boca Raton, 2008)
9. E. J. Barbero, *Introduction to Composite Materials Design* (Taylor and Francis, NY., 1999)
10. M. Sejnoha, L. Kucikova, V. Kralik, J. Sejnoha, Civil-Comp Proceedings **108**, (2015)
11. V. Michalcova, L. Lausova, Civil-Comp Proceedings, **108**, (2015)|
12. M. Krejsa, L. Koubova, J. Flodr, J. Protivinsky, Q. T. Nguyen, Frattura ed Integrita Strutturale **11(39)**, 143-159, (2017)
13. M. Mihalikova, M. Nemet, M. Vojtko, Key Eng. Mat. **635**, 118-121, (2015)
14. J. Melcer, G. Lajcakova, Adv. Mat. Res. **969**, 85-88, (2014)
15. E. Kormanikova, K. Kotrasova, Key Eng. Mat. **635**, 85-88, (2015)

Determination of the critical load of eccentrically compressed columns by the second group of limit states

Evgeniy Galagurya[1*], *Maksym* Kovalov[1], *Larisa* Kravtsiv [1] and *Igor* Bychenok [1]

[1] Ukrainian State University of Railway Transport, Structural Mechanics and Hydraulics Department, Feuerbach sq., 7, Kharkiv, 61050, Ukraine

Abstract. Determination of critical load values of eccentrically loaded columns was subject of studies by such authors as Vatulia G. L., Galagurya E. I., Glazunov Y. V., Zhakin I. A., Opanasenko E. V., Storozhenko L. I., Timoshenko S. P., Trull B. A., Hoff N. D., Chikhladze E. D. and other researchers. However, those works have not studied properly work of statically determinable and statically indeterminable columns, compressed eccentrically at one end and being under the axial compression at the other one. The article deals with the problem of determining the critical load for the following design patterns: a column that is loaded at the ends by longitudinal forces by equal and single-direction eccentricities; a column that is rigidly clamped at the bottom, hinged at the top and loaded by longitudinal force at the upper end eccentrically; column rigidly clamped at the bottom, loaded at the free end of the longitudinal force eccentrically; a hinged column, loaded at the upper end of the longitudinal force eccentrically. Critical load of excentrically loaded columns with different types of bearing and load patterns can be determined using the obtained equations. Allowable deflection value according to the regulatory documents and geometric characteristics of the considered columns should be specified to solve these equations.

1 Introduction

Works by Chikhladze E. [1, 2, 3], Vatulia G. [3], Glazunov Y, Zhakin I., Opanasenko E., Storozhenko L. deal with the experimental and theoretical study of stress-strain state of the centrically and eccentrically compressed columns and other supporting structures. Those works have contributed to solving the difficult problem of assessing the strength and stiffness of columns. However, not all problems have been solved. In particular, action of statically determinate and statically indeterminate columns compressed eccentrically at one end and being under axial compressions at the other has not been studied thoroughly. In this regard, this article provides solutions that allow determining critical load of the following design patterns (Fig. 1).

* Corresponding author: galagurya@rambler.ru

2 Problem definition

First the relation between load and deflection which occur in columns under various load patterns should be obtained to determine the critical load of eccentrically compressed columns.

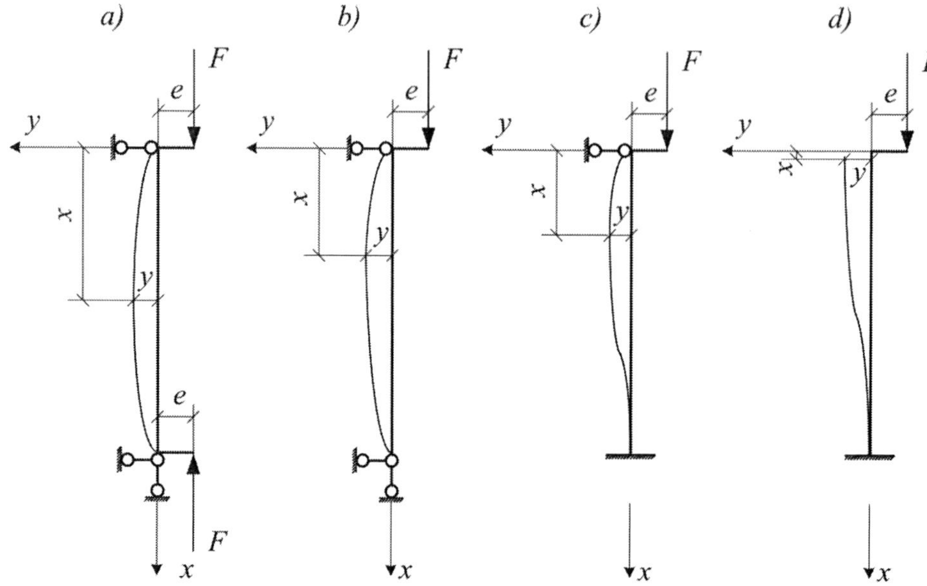

Fig. 1. Design patterns of the columns.

2.1 Column eccentrically loaded by two compressive forces

Consider a column loaded by two compressive forces applied with eccentricity *e* (Fig. 1a). The differential equation of the deflected axis has the form:

$$y'' + k^2 \cdot y = -k^2 \cdot e \tag{1}$$

where:

$$k^2 = \frac{F}{EI} \tag{2}$$

solution of equation (1):

$$y = e\left(\cos kx + \frac{1-\cos kl}{\sin kl}\sin kx - 1\right) \tag{3}$$

The distance between the cross-section with maximum deflection and the origin of coordinates is $x = 0.5l$. The maximum deflection is defined by the formula:

$$y = e\left(\cos 0.5kl + \frac{1-\cos kl}{\sin kl}\sin 0.5kl - 1\right) \tag{4}$$

After we insert the equation (2) into the equation (4), we get the following

$$y = e\left(\cos\left(0.5l\sqrt{\frac{F}{EI}}\right) + \frac{1-\cos\left(l\sqrt{\frac{F}{EI}}\right)}{\sin\left(l\sqrt{\frac{F}{EI}}\right)}\sin\left(0.5l\sqrt{\frac{F}{EI}}\right) - 1\right) \quad (5)$$

We can express the value of critical force from the equation (5) using the software

$$F = \frac{4\text{arctg}\left(\frac{e\cdot\sqrt{2ey+y^2}}{y+e}\right)^2 EI}{l^2} \quad (6)$$

Substitute the value of the allowable deflection of the column according to [4, 5, 6] to find the critical force.

2.2 Hinged column loaded by force F applied eccentrically at the upper end

Consider a hinged column loaded by eccentrically applied force F at the upper end (Fig. 1b). The differential equation of the curved axis is:

$$y'' + k^2 \cdot y = \frac{k^2 \cdot e \cdot x}{l} - k^2 \cdot e \quad (7)$$

solution of the equation (7):

$$y = e\left(\cos kx - \text{ctg}kl \cdot \sin kx - 1 + \frac{x}{l}\right) \quad (8)$$

The section which has the maximum deflection will be critical. The point of extreme (8) and the value of x should be found to determine the critical section. Using the software we can find the cross section with the maximum deflection:

$$x = \frac{2\text{arctg}\left(\frac{kl\cdot\text{tg}kl - \sqrt{\frac{k^2l^2-1}{\cos(kl)^2}+1}}{kl+\text{tg}kl}\right)}{k} \quad (9)$$

After we insert the equation (2) and the equation (9) into the equation (8), we get the equation (10). Since the function in the equation (10) is indeterminate, we cannot calculate the value of the critical force analytically, but knowing the geometric characteristics of the column and the allowable deflection, we can get the value of the critical force.

$$y = e\cdot\cos\left(2\operatorname{arctg}\left(\frac{\sqrt{\frac{F}{EI}}l\cdot\operatorname{tg}\sqrt{\frac{F}{EI}}l-\sqrt{\frac{\frac{F}{EI}l^2-1}{\cos\left(\sqrt{\frac{F}{EI}}l\right)^2}+1}}{\sqrt{\frac{F}{EI}}l+\operatorname{tg}\sqrt{\frac{F}{EI}}l}\right)\right)-$$

$$-e\cdot\operatorname{ctg}\sqrt{\frac{F}{EI}}l\cdot\sin\left(2\operatorname{arctg}\left(\frac{\sqrt{\frac{F}{EI}}l\cdot\operatorname{tg}\sqrt{\frac{F}{EI}}l-\sqrt{\frac{\frac{F}{EI}l^2-1}{\cos\left(\sqrt{\frac{F}{EI}}l\right)^2}+1}}{\sqrt{\frac{F}{EI}}l+\operatorname{tg}\sqrt{\frac{F}{EI}}l}\right)\right)-$$

$$-e+e\frac{2\operatorname{arctg}\left(\frac{\sqrt{\frac{F}{EI}}l\cdot\operatorname{tg}\sqrt{\frac{F}{EI}}l-\sqrt{\frac{\frac{F}{EI}l^2-1}{\cos\left(\sqrt{\frac{F}{EI}}l\right)^2}+1}}{\sqrt{\frac{F}{EI}}l+\operatorname{tg}\sqrt{\frac{F}{EI}}l}\right)}{\sqrt{\frac{F}{EI}}l} \tag{10}$$

2.3 Column fixed at the bottom, hinged on top and loaded eccentrically by force applied at the upper end

Consider a column which is rigidly fixed at the bottom, hinged on top and loaded eccentrically by longitudinal force applied at the upper end (Fig. 1c). The differential equation of the deflected axis will have the following form:

$$y''+k^2\cdot y=\frac{k^2\cdot e\cdot x}{2\cdot l}-k^2 \tag{11}$$

Solution of the equation (11):

$$y=e\left(\cos kx-\left(\frac{\cos kl+0.5}{\sin kl}\right)\cdot\sin kx-1+\frac{x}{2\cdot l}\right) \tag{12}$$

The point of extreme (12) should be found to determine the critical section. As a result, we will get the distance x to the section with the maximum deflection:

$$x=\frac{1}{k}\cdot 2\mathrm{arctg}\left(\frac{-4kl\mathrm{tg}(0.5kl)}{kl\mathrm{tg}(0.5kl)^2-3kl-6\mathrm{tg}(0.5kl)}\right)+$$
$$+\frac{\sqrt{k^2l^2\mathrm{tg}(0.5kl)^4+10k^2l^2\mathrm{tg}(0.5kl)^2+9k^2l^2-36\mathrm{tg}(0.5kl)}}{kl\mathrm{tg}(0.5kl)^2-3kl-6\mathrm{tg}(0.5kl)} \tag{13}$$

After we insert the equation (2) and the equation (13) into the equation (12), we get the equation (14), in which the function is indeterminate and we cannot calculate the value of the critical force analytically, but knowing the geometric characteristics of the column and the allowable deflection, we can get the value of the critical force.

$$y=e\cos\left(2\mathrm{arctg}\left(\frac{l\sqrt{\frac{F}{EI}}\mathrm{tg}(l\sqrt{\frac{F}{EI}})-\sqrt{\frac{Fl^2}{EI\cos(l\sqrt{\frac{F}{EI}})^2}+1-\frac{1}{\cos(l\sqrt{\frac{F}{EI}})^2}}}{l\sqrt{\frac{F}{EI}}+\mathrm{tg}(l\sqrt{\frac{F}{EI}})}\right)\right)-$$
$$-e\mathrm{ctg}(l\sqrt{\frac{F}{EI}})\sin\left(2\mathrm{arctg}\left(\frac{l\sqrt{\frac{F}{EI}}\mathrm{tg}(l\sqrt{\frac{F}{EI}})-\sqrt{\frac{Fl^2}{EI\cos(l\sqrt{\frac{F}{EI}})^2}+1-\frac{1}{\cos(l\sqrt{\frac{F}{EI}})^2}}}{l\sqrt{\frac{F}{EI}}+\mathrm{tg}(l\sqrt{\frac{F}{EI}})}\right)\right)-$$
$$-e+e\frac{2\mathrm{arctg}\left(\frac{l\sqrt{\frac{F}{EI}}\mathrm{tg}(l\sqrt{\frac{F}{EI}})-\sqrt{\frac{Fl^2}{EI\cos(l\sqrt{\frac{F}{EI}})^2}+1-\frac{1}{\cos(l\sqrt{\frac{F}{EI}})^2}}}{l\sqrt{\frac{F}{EI}}+\mathrm{tg}(l\sqrt{\frac{F}{EI}})}\right)}{l\sqrt{\frac{F}{EI}}} \tag{14}$$

2.4 Column rigidly fixed at the bottom, loaded at the free end by eccentric force

Consider a column which is rigidly fixed at the bottom, loaded at the free end by eccentric force (Fig. 1d). The differential equation of the deflected axis will have the following form:

$$y''+k^2\cdot y=-k^2\cdot e \tag{15}$$

Solution of the equation (15):

$$y=e\left(\cos kx+\frac{1-\cos kl}{\sin kl}\sin kx-1\right) \tag{16}$$

After we set the first-order derivative of the function (16) to 0, we will determine the value of x where the function tends to the maximum value (i.e. find the cross-section with the maximum deflection). We use the software for this purpose.

$$x = -\frac{\operatorname{arctg}\left(\dfrac{\cos(0.5kl)-1}{\sin(0.5kl)}\right)}{k} \quad (17)$$

After we insert the equation (2) and the equation (17) into the equation (16), we get the following

$$y = e\left(\begin{aligned}&\cos\left(-\operatorname{arctg}\left(\frac{\cos(0.5l\sqrt{\frac{F}{EI}})-1}{\sin(0.5l\sqrt{\frac{F}{EI}})}\right)\right)+\\&+\frac{1-\cos l\sqrt{\frac{F}{EI}}}{\sin l\sqrt{\frac{F}{EI}}}\sin\left(-\operatorname{arctg}\left(\frac{\cos(0.5l\sqrt{\frac{F}{EI}})-1}{\sin(0.5l\sqrt{\frac{F}{EI}})}\right)\right)-1\end{aligned}\right) \quad (18)$$

We can express the value of critical force from the equation (18) using the software

$$F = \frac{4\operatorname{arctg}\left(\dfrac{2e\cdot\sqrt{2ey+y^2}}{(y+e)^2}\right)^2 EI}{l^2} \quad (19)$$

The value of the allowable deflection of the column according to [4, 5, 6] should be used to calculate the critical force.

3 Experimental studies

Experimental studies have been carried out using the example of a column loaded by eccentrically applied force F at one end (Fig. 1b) to verify the obtained equations. A steel square electric-welded cold-deformed pipe (Fig. 2) with a length of $l = 900$ *mm* has been used as a column.

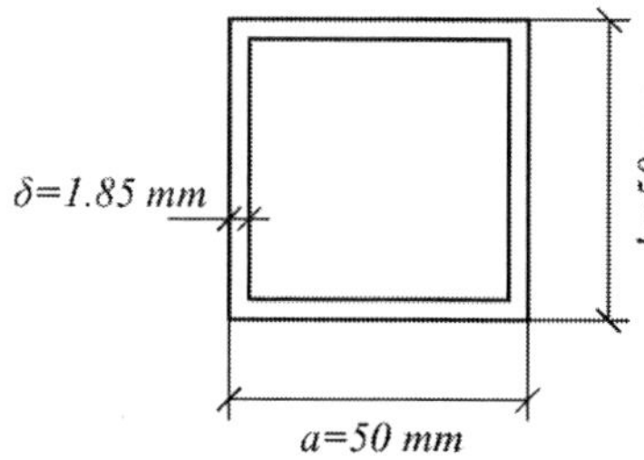

Fig. 2. Cross-sectional diagram of the prototype model.

In order to obtain the data on the physicomechanical characteristics of steel, steel strips have been tested according to [7]. After testing of control samples, the following characteristics have been obtained: yield strength of steel σ_a = 245 MPa, modulus of elasticity of steel $E_a = 1.27\cdot10^5$ MPa.

Steel columns have been tested using the hydraulic press PMM125. Bearing assemblies provided a hinged pattern for sample fastening. The load was applied with an eccentricity of 25 mm. During the tests, displacement of the sample was measured. Dial gauge indicators IMIG with the division value of 0.01 mm were used for this purpose. The instruments were installed in the critical section estimated according to formula (9).

Table 1 and Fig. 3 show the values of deflections in critical sections of experimental samples and their comparison with theoretical values, while Table 2 summarizes the values of critical loads of prototypes and their comparison with the theoretical values.

Table 1. Comparison of deflections.

F [kN]	Sample 1			Sample 2			Sample 3		
	δ_{exp} [mm]	δ_{theor} [mm]	Δ [%]	δ_{exp} [mm]	δ_{theor} [mm]	Δ [%]	δ_{exp} [mm]	δ_{theor} [mm]	Δ [%]
0	0	0		0	0		0	0	
10	0.09	0.08	11	0.098	0.08	22	0.07	0.08	12
20	0.205	0.168	22	0.172	0.168	2.3	0.17	0.168	1.1
30	0.289	0.266	8.6	0.243	0.266	8.6	0.253	0.266	4.8
40	0.4	0.375	6.6	0.393	0.375	4.8	0.372	0.375	0.8
50	0.5	0.499	0.2	0.497	0.499	0.4	0.498	0.499	0.2

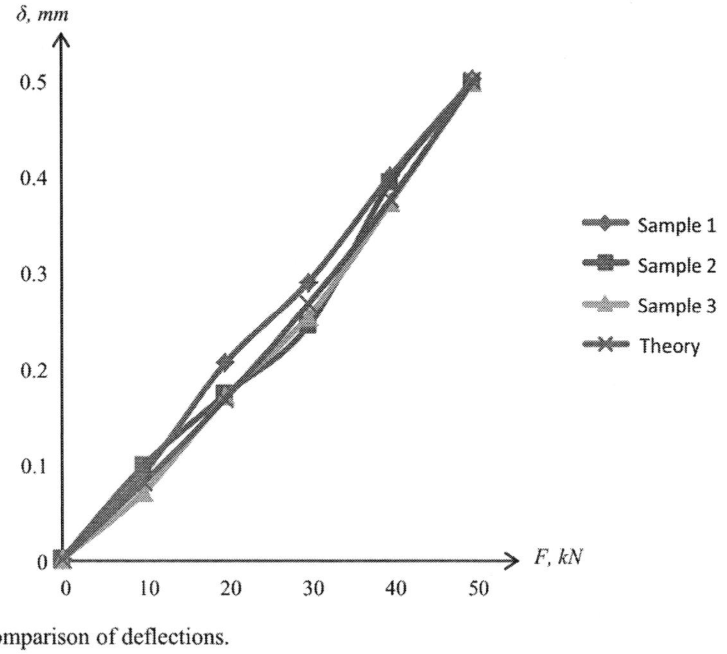

Fig. 3. Comparison of deflections.

Table 2. Comparison of loads.

Sample No.	F_{exp} [kN]	F_{theor} [kN]	Δ [%]
1	48	51.91	7.5
2	50	51.91	3.6
3	49	51.91	5.6

4 Conclusions

Equations to determine the critical load of eccentrically compressed columns have been obtained. Allowable deflection value according to the [4, 5, 6] and geometric characteristics of the considered columns should be specified to solve these equations. These formulas have been confirmed experimentally for a hinged column. As a result of the conducted experimental studies, data on movements in the critical section of experimental columns and their carrying capacity value were obtained.

References

1. E.D. Chikhladze, Proceedings of Institute of KhIRTE. No. 21 (1993)
2. E.D. Chikhladze, A.D.Arslankhanov, Proceedings of the universities. Building. No. 1 (1992)
3. E.D. Chikhladze, G.L.Vatulya, Kiev, Transport of Ukraine, (2006)
4. DSTU B. V.1.2-3: 2006. Deflections and moving. Requirements engineering. State Standard of Ukraine (2007)
5. DSTU B. V.2.6-75: 2008. Construction of houses and buildings. Construction metal building. General specifications. State Standard of Ukraine (2008)
6. DBN V 2.6-163: 2010. Steel constructions. The rules of designing, manufacturing and installation. State building codes Ukraine. K. Minregionbud Ukraine (2010)
7. DSTU 7305:2013. Metals. Methods of tension test. State Standard of Ukraine (2013)

Regression equations for circular CFST columns carrying capacity evaluation

Glib Vatulia[1*], *Maryna* Rezunenko[1], *Yevhen* Orel[1] and *Dmytro* Petrenko[1]

[1] Ukrainian State University of Railway Transport, Structural Mechanics and Hydraulics Department, Feuerbach sq. 7, 61050 Kharkiv, Ukraine

Abstract. Within the last decades, a considerable amount of experimental studies have been carried out by numerous researchers across the world with the purpose to study the carrying capacity of concrete-filled steel tubular (CFST) columns and evaluation of their stressed-strained state. The array of the obtained results have allowed designing a mathematical model to determine the maximum carrying capacity value of such constructions using the methods of mathematical statistics. The authors obtained three types of regression equations for short and long circular CFST columns with different geometrical and physical properties under axial compression. Statistical quality of the obtained models was verified by both regression equation quality in general and statistical significance of the equation parameters. The comparison of the obtained carrying capacity values with the results calculated by Eurocode 4 and AIJ methodologies allows making a conclusion on the sufficient calculation accuracy of the designed mathematical models.

1 Introduction

Since the steel concrete columns have been commonly used in the civil engineering, the issues of research of their carrying capacity and evaluation of the stressed-strained state still remain very important today. The extensive experimental data set on carrying capacity of the columns under various loading conditions assembled within the latest decades by the researchers across the world has enabled to study the behaviour of the columns using the techniques of mathematical statistics [1-6].

The authors of this research have built mathematical models to evaluate the maximum carrying capacity of circular CFST columns under axial compression. One of the most widely used statistical techniques was used for this purpose, namely, regression analysis [7].

Initially, two models were supposed to be built for circular CFST short ($L/D < 4$) and long ($L/D \geq 4$) columns, where L, D – length and outer diameter of the column, respectively.

* Corresponding author: glebvatulya@gmail.com

However, analysis of the obtained results showed that in case of the long columns it would be preferable to consider separately the samples with the casing thickness $t < 2.5$ mm and those with the thickness $t \geq 2.5$ mm.

2 Regression equation

2.1 CFST long thin-walled columns

In the first case, thin-wall steel concrete circular columns with $t < 2.5$ mm, ratio $L/D \geq 4$, column diameter varying within $76 \leq D \leq 210$ mm, length $500 \leq L \leq 2750$ mm, yield strength of steel $186 \leq \sigma_y \leq 490$ MPa, modulus of elasticity of steel $1.68 \cdot 10^5 \leq E_s \leq 2.17 \cdot 10^5$ MPa, prism strength of concrete $10 \leq f_{c,prizm} \leq 50$ MPa, modulus of elasticity of concrete $2.0 \cdot 10^4 \leq E_c \leq 3.6 \cdot 10^4$ MPa were considered. In the series of $n = 40$ observations, a factor regression model was obtained as follows:

$$\overline{N} = a_0 + a_1 \frac{D^2 E_s}{\sigma_y} + a_2 \frac{DL}{t} + a_3 D f_{c,prizm} + a_4 \left(\frac{D}{t}\right)^2 \frac{1}{E_c} =$$

$$= 163.3128 + 0.0309 \frac{D^2 E_s}{\sigma_y} - 0.0021 \frac{DL}{t} + 0.0526 D f_{c,prizm} + 0.381 \left(\frac{D}{t}\right)^2 \frac{1}{E_c}. \quad (1)$$

2.2 CFST long thick-walled columns

In the second case, circular CFST columns with the casing thickness at least 2.5 mm, ratio $L/D \geq 4$, column diameter varying within $88 \leq D \leq 267$ mm, length $485 \leq L \leq 4956$ mm, yield strength of steel $200 \leq \sigma_y \leq 490$ MPa, modulus of elasticity of steel $1.68 \cdot 10^5 \leq E_s \leq 2.17 \cdot 10^5$ MPa, prism strength of concrete $18.2 \leq f_{c,prizm} \leq 50$ MPa, modulus of elasticity of concrete $2.0 \cdot 10^4 \leq E_c \leq 3.6 \cdot 10^4$ MPa were considered. A regression model built on the basis of $n = 224$ observations is written as:

$$\overline{N} = a_0 + a_1 D E_s + a_2 (D - 2t) L \sigma_y + a_3 D^2 f_{c,prizm} + a_4 \frac{(D-2t)^2 t}{E_c} + a_5 \frac{(D-2t)\, t}{\sigma_y} =$$

$$= 213.5814 + 0.0291 D E_s - 4.5 \cdot 10^{-6} (D - 2t) L \sigma_y + 0.0007 D^2 f_{c,prizm} + \quad (2)$$

$$+0.2322 \frac{(D-2t)^2 t}{E_c} - 1518.73 \frac{(D-2t)\, t}{\sigma_y}.$$

2.3 CFST short columns

In the third case, circular CFST short columns with ratio $L/D < 4$, column diameter varying within $86 \leq D \leq 820$ mm, length $200 \leq L \leq 2460$ mm, yield strength of steel $226 \leq \sigma_y \leq 490$ MPa, modulus of elasticity of steel $1.68 \cdot 10^5 \leq E_s \leq 2.17 \cdot 10^5$ MPa, prism strength of concrete $18 \leq f_{c,prizm} \leq 50$ MPa, modulus of elasticity of concrete $2.0 \cdot 10^4 \leq E_c \leq 3.6 \cdot 10^4$ MPa were considered. Based on the study of the results of 156 experiments, the regression equation is written as:

$$\overline{N} = a_0 + a_1 D t + a_2 D^2 f_{c,prizm} + a_3 L \sigma_y^2 + a_4 t^2 \sigma_y L = -81{,}3206 +$$

$$+0.9297 D t + 7.53 \cdot 10^{-4} D^2 f_{c,prizm} + 7.75 \cdot 10^{-6} L \sigma_y^2 + 8.94 \cdot 10^{-8} t^2 \sigma_y L. \quad (3)$$

DOI: 10.1051/matecconf/201710700051

3 Confidence intervals and comparative analysis

Confidence intervals for regression equation parameters with probability $P = 0.95$ are shown in Table 1.

Table 1. Confidence intervals.

Model 1	Model 2	Model 3
$108.13 \le a_0 \le 218.5$; $0.029 \le a_1 \le 0.033$; $-0.0025 \le a_2 \le -0.0017$; $0.039 \le a_3 \le 0.067$; $0.113 \le a_4 \le 0.649$	$119.173 \le a_0 \le 307.99$; $0.0219 \le a_1 \le 0.0364$; $-4.8\times10^{-6} \le a_2 \le -4.2\cdot10^{-6}$; $6.5\times10^{-4} \le a_3 \le 7.8\cdot10^{-4}$; $0.231 \le a_4 \le 0.2506$; $-1722.28 \le a_5 \le -1315.17$.	$-134.305 \le a_0 \le -28.3359$; $-0.848 \le a_1 \le -1.012$; $7.39\cdot10^{-4} \le a_2 \le 7.67\cdot10^{-4}$; $6.75\cdot10^{-6} \le a_3 \le 8.74\cdot10^{-6}$; $6.66\cdot10^{-8} \le a_4 \le 1.12\cdot10^{-7}$

Comparative analysis of experimental and theoretical values of the carrying capacity of circular CFST columns for all the above regression models are shown in Fig. 1-3.

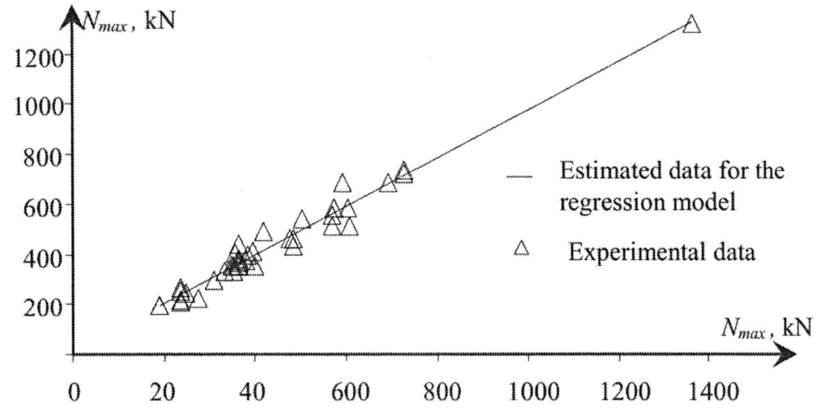

Fig. 1. Carrying capacity of concrete-filled steel tubular columns (t < 2.5 mm, $L/D \ge 4$).

DOI: 10.1051/matecconf/201710700051

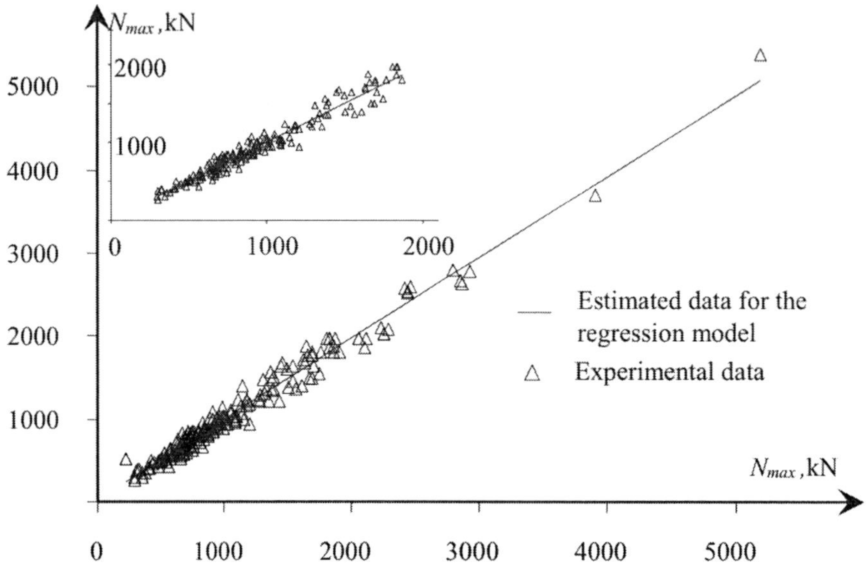

Fig. 2. Carrying capacity of concrete-filled steel tubular columns (t ≥ 2.5 mm, L/D ≥ 4).

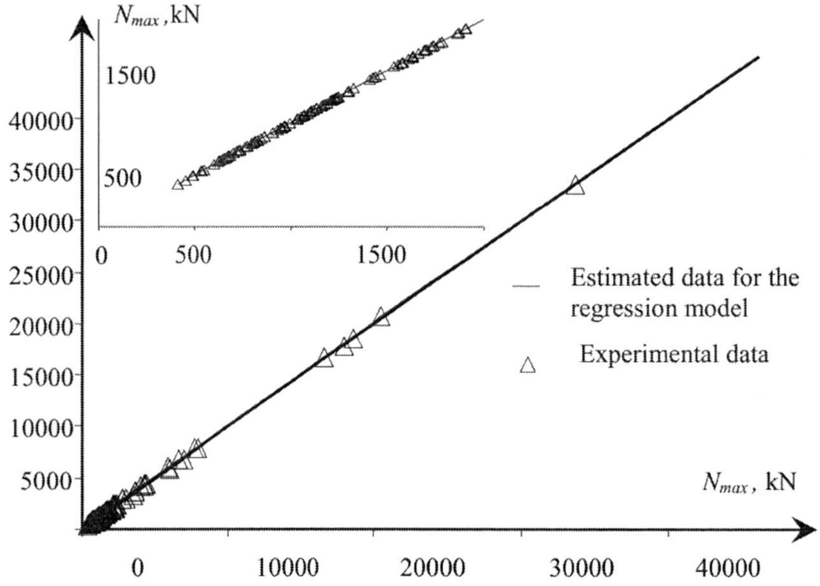

Fig. 3. Carrying capacity of concrete-filled steel tubular columns (L/D < 4).

Dependences of absolute error $\Delta N_{max} = \left|\frac{N_{theor}}{N_{exp}}\right|$ on geometrical parameters of the columns and correlation thereof are shown in Fig. 4÷6 for models (1) ÷ (3), respectively.

 DOI: 10.1051/matecconf/201710700051

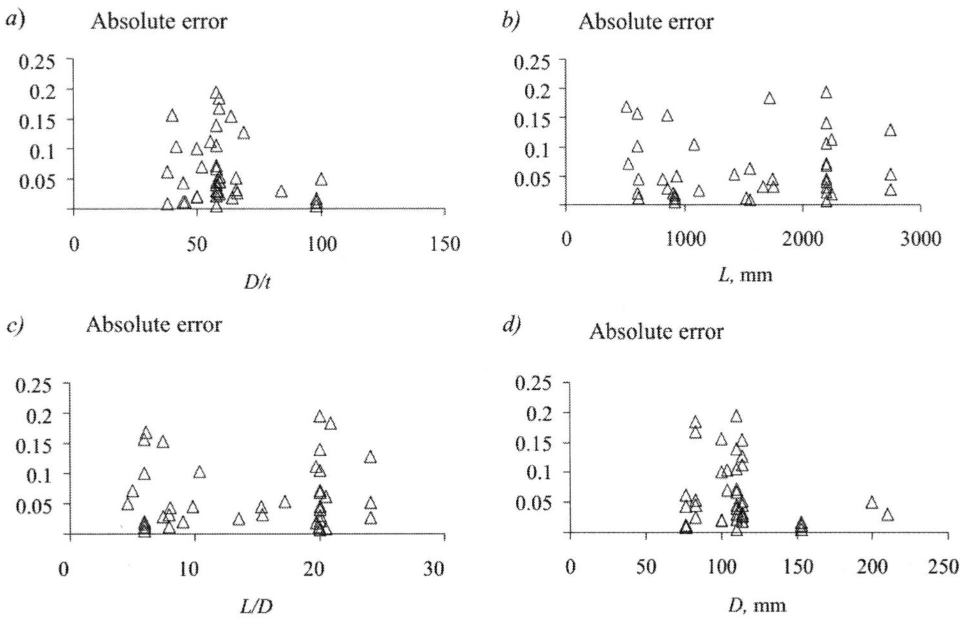

Fig. 4. Dependences of absolute error of the regression model 1:
a – on *D*/*t* value; *b* – on the length of the column *L*; *c* – on *L*/*D* value,
d – on external diameter *D*.

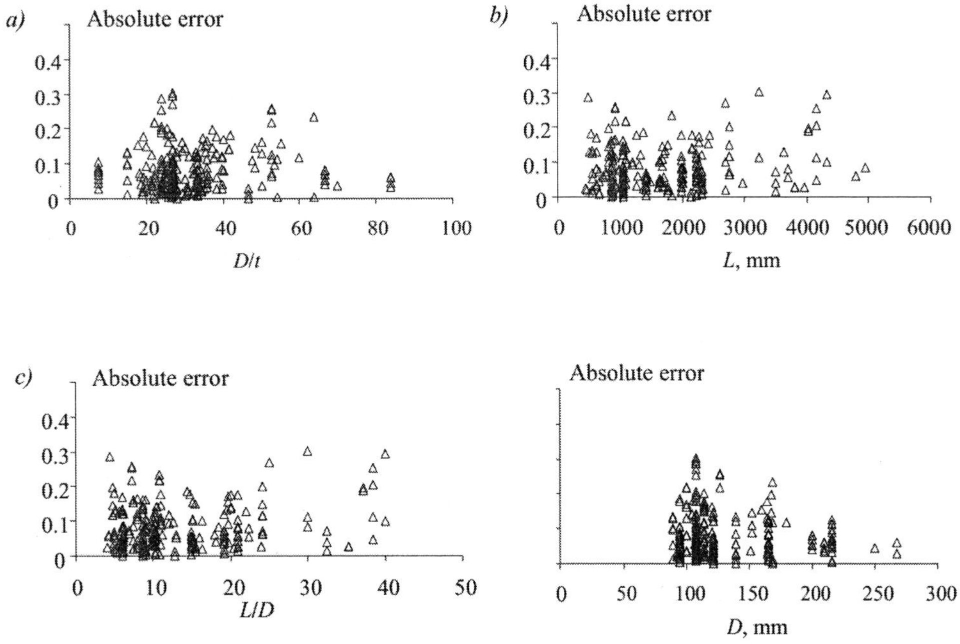

Fig. 5. Dependences of absolute error of the regression model 2:
a – on *D*/*t* value; *b* – on the length of the column *L*; *c* – on *L*/*D* value
d – on external diameter *D*.

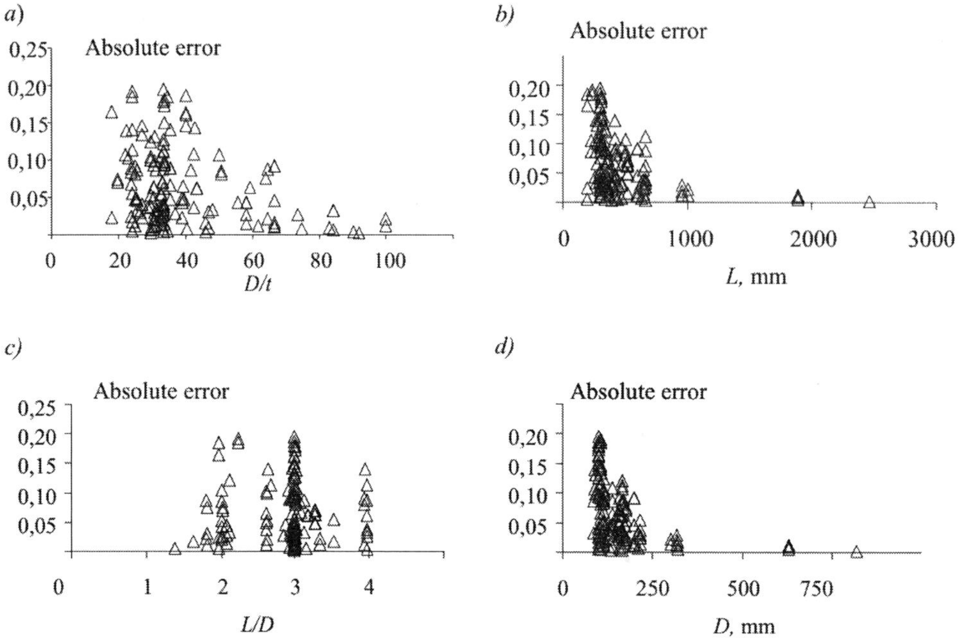

Fig. 6. Dependences of absolute error of the regression model 3:
a – on *D*/*t* value; *b* – on the length of the column *L*; *c* – on *L*/*D* value
d – on external diameter *D*.

In Table 2, quality parameters of the built models, such as coefficient of determination R^2 and mean approximation error are given

$$\Delta = \frac{1}{n}\sum_{i=1}^{n}\frac{|N_{exp}-N_{theor}|}{N_{exp}}\,100\%.$$

Table 2. Quality parameters of the models.

	Model 1	Model 2	Model 3
Coefficient of determination R^2	0.989	0.988	0.99
Mean approximation error	6.30%	7.68%	6.41%

Study of the remainder $(N_{exp} - N_{theor})$ for models (1) ÷ (3) shows that they are normally distributed with zero probabilistic expectation and lie within the interval [-3*S*; 3*S*], where *S* - corrected mean quadratic deviation which is 47.94 ($t < 2.5$ mm, $L/D \geq 4$), 97.55 ($t \geq 2.5$ mm, $L/D \geq 4$), 104.74 ($L/D < 4$). Moreover, in all cases, about 71% remainder are in the interval [-*S*; *S*], 98%, 94% and 96%, respectively – in the interval [-2*S*; 2*S*].

In general, the statistical significance of the regression equations is verified with the Fisher criterion, while the statistical significance of the equation parameters is verified with the Student criterion.

The performed mathematical analysis of the obtained models, as well as comparison of the results with the carrying capacity value of the circular CFST columns calculated using the methods of Eurocode 4 [8] and Architecture Institute of Japan (AIJ) [9] allow to make a conclusion that the degree of confidence of the regression equations (1) ÷(3) is sufficiently high.

References

1. E.D. Chikhladze, G.L. Vatulia, Y.P. Kitov, *Basis for calculation and design of composite and steel-concrete constructions*, Transport of Ukraine, 104 (2006)
2. G.L. Vatulia, *Rozrakhunok ta proektuvannya stalebetonnykh ta kombinovanykh konstruktsiy*. Kharkiv, 409 (2015)
3. T. Yamamoto, J. Struct. Constr. Eng., Transactions of AIJ **561**, 237-244 (2002)
4. A.L. Krishan, A.S. Melnitchuk, News of KSUAE, Kaza, **3 (29)**, 6-50 (2014)
5. L.I. Storozhenko, D.A. Ermolenko, O.I. Lapenko, *Tube confined concrete*, ASMI, Poltava, 306 (2010)
6. I. Nishiyama, S. Morino, K. Sakino, H. Nakahara, *Summary of Research on Concrete-Filled Structural Steel Tube Column System Carried Out Under The US-JAPAN Cooperative Research Program on Composite and Hybrid Structures*, Tokyo, 176 (2002)
7. I. Vuchkov, *Prikladnoy lineynyiy regressionnyiy analiz*. Finance and Statistics, 239 (1987)
8. *Design of composite steel and concrete structures – Part 1-1: General Rules and Rules for Bridges. EN 1994-1: 2001. Eurocode 4*, Brussels, 96 (2006)
9. *Recommendations for design and construction of concrete filled steel tubular structures*, Architectural Institute of Japan (AIJ), Tokyo, 333 (1997)

Stability analysis of an open shallow cylindrical shell with imperfection under external pressure

Martin Psotny[1,*] and *Jozef* Havran[1]

[1]Slovak University of Technology, Faculty of Civil Engineering, Department of Structural Mechanics, Radlinského 11, 810 05 Bratislava, Slovakia

Abstract. Elastic shallow generalized cylindrical shells of an open cross-section subjected to the various forms of external pressure are analysed in the paper numerically using the finite element method. Load - displacement paths are calculated for the perfect and imperfect geometry, respectively. Special attention is paid to the influence of initial geometric imperfection on the limit load level of fundamental equilibrium path of nonlinear analysis. ANSYS system was used for analysis, arc-length method was chosen for obtaining fundamental load - displacement path of solution.

1 Introduction

Cylindrical shells with open cross-section are structural elements widely used in many branches of engineering. Such shells occur as parts of aircraft and marine structures in mechanical engineering, create covers of large span structures in civil engineering. Their middle surface is generally created by sliding generator along any curve (directrix). Presented paper considers only right cylinders and the circle, parabola, vertical and horizontal ellipses as a directrix. The effort was to assume comparable value of the average curvature of the considered curves.

These shells subjected to the external pressure are liable to the buckling due to dominant compression membrane forces within the shell. It is the reason, why the stability problem has been analysed since the beginning of the twenty century. It was then when the first very slender structures of barrel shells appeared. Among the basic books dealing with this problematics are included works by Bushnell [1] and by Yamaki [2]. The shell stability handbook [3] treats various important shell buckling problems of practical interest.

Solving stability of the thin shell, it is often insufficient to determine the elastic critical load from eigenvalue buckling analysis, i.e. the load, when perfect shell starts buckling. Nonlinear analysis is necessary, resulting in a full load-displacement response. It is also necessary to include initial imperfections of real shell into the solution and determine limit load level more accurately. Significant works dealing with imperfections in real structures are [4-5]. Experience shows that small deviations from the prescribed geometry of a perfect thin-walled cylinder may cause a strong reduction of the buckling load and of the load-carrying capacity of shell. When the perfect geometry is broken ability to use membrane action is heavily reduced.

* Corresponding author: martin.psotny@stuba.sk

2 FEM in nonlinear theory

The geometrically nonlinear theory represents a basis for the reliable description of the postbuckling behaviour of the imperfect shell. Applying the finite element method to the problem of geometrical nonlinearity, problem leads to a set of nonlinear algebraic equations. This system can be written as:

$$\boldsymbol{K}_{inc}\,\Delta\boldsymbol{a} + \boldsymbol{F}_{int} - \boldsymbol{F}_{ext} - \Delta\boldsymbol{F}_{ext} = \boldsymbol{0}, \tag{1}$$

where $\boldsymbol{K}_{inc}$ is the incremental stiffness matrix of shell, $\boldsymbol{F}_{int}$ is the internal force vector of shell, $\boldsymbol{F}_{ext}$ is the external load of shell and $\Delta\boldsymbol{F}_{ext}$ is the increment of the external load of shell.

The solution of presented system cannot be achieved directly, and is based on a step by step incremental process that causes a deviation from the equilibrium nonlinear path. To correct this deviation, an iterative technique such as the Newton-Raphson method is used. Murray and Wilson [6] first presented idea of combining incremental (Euler) and iterative (Newton-Raphson) methods for solving nonlinear problems. Early works involving critical points and snap-through effect were written by Sharifi and Popov [7], and Sabir and Lock [8]. Using arc-length method to pass limit points on load-displacement paths introduced Riks in [9]. Getting through this problem using displacement control procedure presented Batoz and Dhatt [10]. Detection of critical points using arc-length method was introduced by Wriggers and Simo [11]. Works of Bathe [12] dominate in application of FEM to geometric nonlinear problems. Crisfield [13] gave the most important modification of the arc-length method suggesting the fixation of incremental length during load increment (cylindrical ALM). Fafard and Massicotte [14] and Teng and Luo [15] submitted further significant modifications of the method.

3 Numerical analysis, conclusions

Illustrative example of isotropic shallow shell loaded by the external pressure (Fig. 1) is presented. Eigenvalue buckling analysis is calculated first. These results offer an image about location of critical points of nonlinear solution, help with settings in the management of nonlinear calculation process. Results of fully nonlinear analysis follow (ideal shell and subsequently structure with initial imperfection).

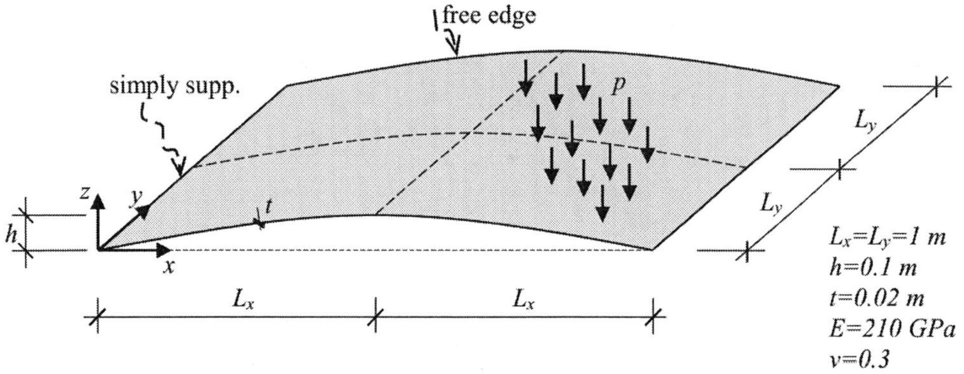

Fig. 1. Shallow cylindrical shell. Geometry and material properties.

 DOI: 10.1051/matecconf/201710700052

Presented results were obtained by division into 40×40 elements. Boundary conditions are considered as simply supported along boundary generators (UX, UY and UZ applied on lines), and free boundary directrices. Uniform external pressure (in normal direction) was assumed first, later uniform pressure in vertical direction was applied. System ANSYS [16] was used for analysis, FEM model was created using element type SHELL181 (4 nodes, 6 DOF at each node). The arc-length method was chosen for analysis, the reference arc-length radius is calculated from the load increment. Only fundamental path of nonlinear solution has been presented.

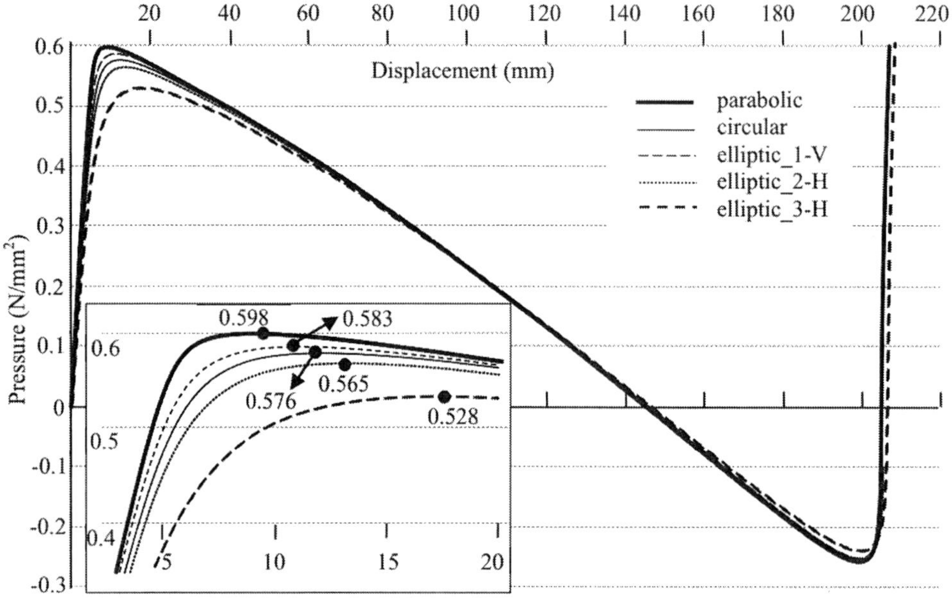

Fig. 2. Fundamental paths of perfect geometry for different shapes of directrix.

In Fig. 2, the fundamental path of the analysis of an ideal cylindrical shell having parameters from Figure 1 can be seen. Applied load acts in the normal direction. Parabolic, circular and elliptic (vertical and horizontal ellipse) shapes of arch were considered. Due to considerable shallowness of the shell ($h/2L_x$=1/20) shapes of the middle surface are very similar. However, in Fig. 2 (bottom left detail) one can see non-negligible differences in values of load in limit points of particular paths. Obtained results are analyzed and compared also in Table 1. Load in limit point of the circular arch was chosen as the comparative value. The results for the load acting in the vertical direction have been presented in the table, too. In this case, also the component of the load acting in tangential direction affects the shell; value of load acting in normal direction has been changing along the arch. Due to this, the value of load in limit point is lower (since the shell is very shallow, differences are small).

Let us now analyze the shell with geometrical imperfection. The initial displacements were assumed as the out of plane displacements only [17] as a product of selected buckling mode (dimensionless) and appropriately selected constant α_0.

Table 1. Comparison of the loads in limit points for different shapes of directrix.

	External pressure [N/mm²] (normal direction)			Uniform pressure [N/mm²] (vertical direction)		
directrix	load [N/mm²]	Δ [%]	displ. [mm]	load [N/mm²]	Δ [%]	displ. [mm]
parabola	0.598	+3.82	9.16	0.585	+2.99	10.1
ellipse_1 (vertical)	0.583	+1.22	11.40			
circle	0.576	100	12.49	0.568	100	15.0
ellipse_2 (horizontal)	0.565	-1.91	13.29	0.559	-1.58	15.7
ellipse_3 (horizontal)	0.528	-8.33	17.66	0.525	-7.57	17.5

Solution in Fig. 3 presents, in addition to load - displacement paths of an ideal shell (symmetrical and non-symmetrical path), also paths which represent solution of the shell with geometric imperfection coincided with first buckling mode and multiplied by parameter $\alpha_0 = 0.5$ mm and 1.5 mm, respectively. The decrease of load in the limit point of load - displacement path of an imperfect shell can be seen in the detail in the top right, corresponding to the increase of the magnitude of an initial imperfection.

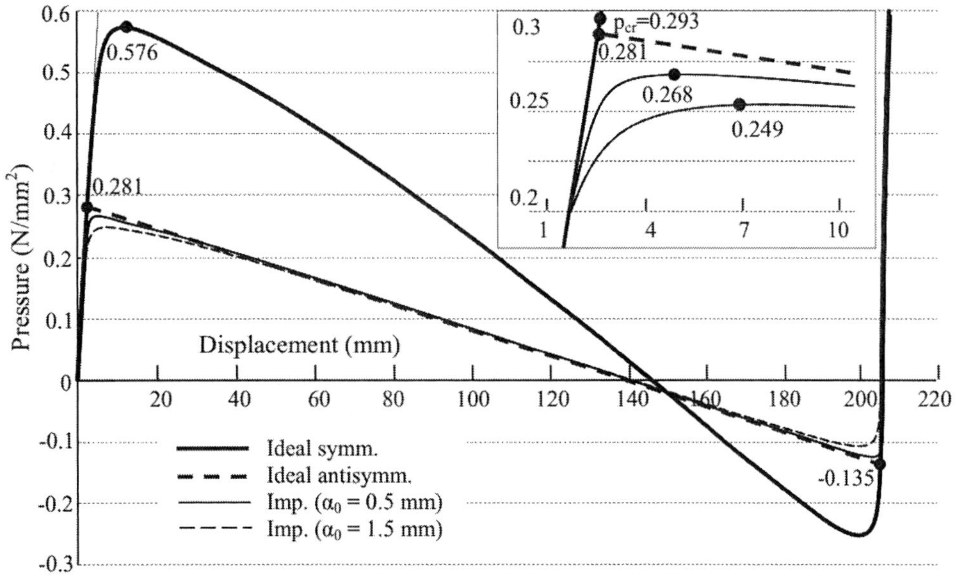

Fig. 3. Shallow circular cylindrical shell. Equilibrium paths for perfect shell and shell with geometric imperfection.

Obtained results are compared also in Table 2. Here the results from nonlinear analysis (NA) - ideal shell and shell with initial imperfections (for different values of α_0) are compared with results from eigenvalue buckling analysis (EBA) – critical load coresponded to first eigenmode.

Table 2. Circular cylindrical shell, external pressure – results comparison.

analysis	load [N/mm^2]	Δ [%]
EBA p_{cr}	0.293	100
NA perfect shell	0.281	-4.10
NA α_0 = 0.5 mm	0.268	-8.53
NA α_0 = 1.5 mm	0.249	-15.02

In the first part of the paper the effect of the cross-section of a shallow shell upon its load-carrying capacity has been investigated. It has been proven that even in the case of a very shallow shell it is important how the curvature changes along the arch. The reduction of curvature from the top towards the supports is preferable.

In the second part of the paper, the effects of the initial geometric imperfection upon the load-carrying capacity of a shell has been analysed. The reduction of the load-carrying capacity in dependence on the magnitude of assumed geometric imperfection has been presented.

Presented results have been arranged due to the research supported by the Slovak Scientific Grant Agency, project No. 1/0272/15.
The authors would like to thank for financial contribution from the STU Grant scheme for Support of Young Researchers.

References

1. D. Bushnell, *Computerized Buckling Analysis of Shells* (Martinus Nijhoff, Dordrecht, 1985)
2. N. Yamaki, *Elastic Stability of Circular Cylindrical Shells* (North-Holland, Amsterdam, 1984)
3. L. A. Samuelson, S. Eggwertz, *Shell Stability Handbook* (Elsevier, London, 1992)
4. J. G. Teng, Buckling of thin shells: Recent advances and trends. Appl. Mech. Rev. **49,** 4, 263–274 (1996)
5. J. Arbocz, J. H. Starnes, Future directions and challenges in shell stability analysis. Thin. Wall. Struct. **40,** 9, 729–754 (2002)
6. D. W. Murray, E. L. Wilson, Finite element postbuckling analysis of thin elastic plates, AIAA J. 7, 1915–1920 (1969)
7. P. Sharifi, E. P. Popov, Nonlinear buckling analysis of sandwich arches, J. Eng. Mech. Div. **97**, 1397–1412 (1971)
8. A. B. Sabir, A. C. Lock, The application of finite elements to the large-deflection geometrically non-linear behavior of cylindrical shells in variational methods in engineering, *Proceedings on an international conference held at the University of Southampton*, Session VII, Southampton University Press, Southampton, GB, 67–76 (1972)

9. E. Riks, The application of Newton's method to the problem of elastic stability, J. Appl. Mech. **39**, 1060–1066 (1972)
10. J. L. Batoz, G. Dhatt, Incremental displacement algorithms for non-linear problems. Int. J. Numer. Meth. Eng. **14**, 1262–1266 (1979)
11. P. Wriggers, J. C. Simo, A general procedure for the direct computation of turning and bifurcation points, Int. J. Numer. Meth. Eng. **30**, 155–176 (1990)
12. K. J. Bathe, *Finite element procedures in engineering analysis* (Prentice-Hall, Englewood Cliffs, NJ., 1982)
13. M. A. Crisfield, A fast incremental/iterative solution procedure that handles snap-through, Comput. Struct. **13,** 55–62 (1981)
14. M. Fafard, B. Massicotte, Geometrical interpretation of the arc-length method. Comput. Struct. **46,** 4, 603–615 (1993)
15. J. G. Teng, Y. F. Luo, A user-controlled arc-length method for convergence to predefined deformation states, Commun. Numer. Meth. En. **14**, 51–58 (1998)
16. ANSYS User's Manual 13.0. Swanson Analysis Systems, Inc. (2010)
17. J. Ravinger, Vibration of Imperfect Thin-Walled Panel. Part 1: Theory and Illustrative Examples. Part 2: Numerical Results and Experiment. Thin. Wall. Struct. **19,** 1, 1–36 (1994)

Axial force and bending stiffness

Ľuboš Šnirc[1]*, *Alžbeta* Grmanová[1] and *Ján* Ravinger[1]

[1] Slovak University of Technology, Faculty of Civil Engineering, Department of Structural Mechanics, 810 05 Bratislava, Radlinského 11, Slovakia

Abstract. In the solution of effects of axial force on the bending stiffness, geometrical nonlinear analysis must be used. Tensile axial force generates an increase of the bending stiffness. In case of low bending stiffness (e.g. cable) the task is simplified and it is possible to determine the tensile force needed to achieve required bending stiffness. When considering the compression force, many problems occur. This force cannot exceed the critical force. This principle is not valid, if there is an initial imperfection, which is perpendicular to the transversal load. Knowing the frequency of beam can help to identify the size of axial force. In case of compression force one gets a combination of stability and vibration.

1 Introduction

Generally, the axial force does not affect the distribution of bending moments in the beam. However, it does not mean that the axial force cannot affect the bending stiffness of a beam. Nowadays many interesting constructional elements have been used in modern structural engineering, e.g. glass plates supported by tension ropes. To analyse this type of problems, the geometric nonlinear analysis must be used [1-2].

2 Geometric nonlinear analysis of beam

In the general solution of a straight beam one needs to consider an elongation of beam, which is given by the relation between the deformed element and its original length. In a further analysis the linear elastic material will be considered.

$$\sigma = E(\varepsilon - \varepsilon_0) \tag{1}$$

where E means Young modulus and the index "0" marks an initial strain.

The analysis starts with calculation of the total potential energy and in further calculation the Ritz variation method will be used. This type of problem generally leads to the system of cubic algebraic equations, if the square is considered in the elongation. Cubic terms are mutually eliminated in the case of a straight beam

* Corresponding author: lubos.snirc@stuba.sk

3 Beam loaded with axial force and uniform load

In Fig. 1 there are shown two cases of beam loading, which will be analyzed.

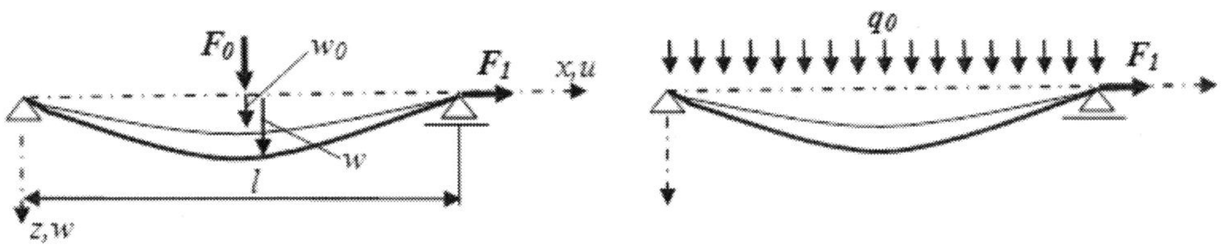

Fig. 1. Beam loaded with axial force and uniform load.

The beam with an initial imperfection will be considered. Two models of transversal loading - concentrated force in the middle and uniform load along the full length of the beam have been used.

As it was already mentioned, Ritz variation method with consideration of following general variation functions will be used.

$$w = \alpha_1 \sin\frac{\pi x}{l},\; w_0 = \alpha_0 \sin\frac{\pi x}{l},\; u = \alpha_2\, x + \alpha_3 \sin\frac{2\pi x}{l} \tag{2}$$

More detailed studies can be found in papers [3-5]. After all mathematical operations, problem leads to the equation:

$$F_E(\alpha_1 - \alpha_0) + F_1\alpha_1 = F_0\frac{2l}{\pi^2} \qquad \left\langle F_E(\alpha_1 - \alpha_0) + F_1\alpha_1 = \frac{4l^2}{\pi^3}q_0 \right\rangle \tag{3}$$

where $F_E = \frac{\pi^2 EI}{l^2}$ is Euler critical force, EI – bending stiffness, l – span of beam.

Equation in brackets $\langle\ \rangle$ refers to the example of system with uniform loading. In the case of a straight beam without imperfection, a check can be done considering $(\alpha_0 = 0)$ with zero axial force. Then the deflection in the middle is calculated.

$$\alpha_1 = F_0\frac{2l^3}{\pi^4 EI} = \frac{1}{48.7045}\frac{F_0 l^3}{EI}\; exact\_solution \frac{1}{48}\frac{F_0 l^3}{EI}$$
$$\left\langle \alpha_1 = q_0\frac{4l^4}{\pi^5 EI} = \frac{5}{382.52}\frac{q_0 l^4}{EI}\; exact\_solution \frac{5}{385}\frac{q_0 l^4}{EI} \right\rangle \tag{4}$$

It is obvious that the obtained results are not identical. This is the consequence of our consideration, that deflection function has a sinus form. However, the difference in forces is only 1.5% and in uniform loads it is 0.64%.

Another simple check can be done for the case of acting just compression force and without transversal loading. Subsequently, well known result for buckling of the beam with an initial imperfection is obtained.

$$\frac{F}{F_E} = \left(1 - \frac{\alpha_0}{\alpha_1}\right) \tag{5}$$

where $F = -F_1$

4 Beam with low bending stiffness loaded with tensile axial force

As a special example can be shown a solution of beam loaded with tensile axial force while having extreme low bending stiffness. A good example for this is a steel rope. In this case the Euler critical force can be considered to be equal to zero ($F_E = 0$). Equation (3) is simplified.

$$F_1\alpha_1 = F_0 \frac{2l}{\pi^2} \quad \left\langle F_1\alpha_1 = \frac{4l^2}{\pi^3} q_0 \right\rangle \tag{6}$$

Let us solve the example with transversal load represented by concentrated force first. The value of axial force in rope is determined:

$$F_1 = F_0 \frac{2l}{\pi^2 \alpha_1} \tag{7}$$

Stiffness in deflection is limited with a ratio $n = \dfrac{l}{\alpha_1}$, where values for $n = 250 - 800$ are considered. With the substitution of equation (7) the following equation is obtained.

$$F_1 = F_0 \frac{2n}{\pi^2} \tag{8}$$

In our example it can be supposed $F_0 = 1000$ N, $n = 500$. Then the force in rope is $F_1 = 101321$ N.

The most interesting fact is that the span of beam does not need to be mentioned in this equation. Taking the transverse uniform load in account, analogically the equation is obtained.

$$F_1 = q_0 \frac{4l \cdot n}{\pi^3} \tag{9}$$

In this case the values used in the example are $l = 5.0$ m, $q_0 = 200$ N/m, $n = 500$. The force in rope is $F_1 = 64503$ N.

5 Stability and vibration of beam

In the case of compressive axial force one gets the stability problems. Another interesting problem is the combination of stability and vibration of the beam [6-7]. Using linear dynamics, problem is geometrically linear and the free vibration can be obtained from the following equation:

$$\left| \mathbf{K}_L - \omega^2 \mathbf{K}_M \right|_{\det} = 0 \tag{10}$$

where $\mathbf{K}_L$ is stiffness matrix, ω – eigen circular frequency , $\mathbf{K}_M$ – mass matrix.

Usually in solution of the stability problem using the theory of linear stability is sufficient, where it is supposed that axial forces in the beam or membrane forces in the thin wall structures are not dependent on the transversal deformation. From this type of problems, one is able to determine critical forces or critical stresses and modes of buckling. Supposing the linear stability also in the problem of vibration of a given structure, equation (10) will change:

$$\left|\mathbf{K}_{\mathrm{L}} - \lambda \mathbf{K}_{G} - \omega^2 \mathbf{K}_{\mathrm{M}}\right|_{\det} = 0 \tag{11}$$

where $\mathbf{K}_G$ is geometric matrix, λ – multiplier of critical load.

Incremental stiffness matrix contains also initial imperfections, level of loading and also deformations of the system. Detailed explanation of theoretical and also numerical background of the given problem was analyzed in works referred in the end [8]. At this point we will introduce only some of interesting results.

5.1 Vibration and buckling of beam

Basic example of the combination of stability and vibration is the simply supported beam loaded with the axial force at its ends. Even such a simple example shows the complexity of a given problem. In the point of the acting force, using of the moveable support is needed. With the same approach in analysis of the case of vibration, in the level of critical force the zero frequency is obtained. However, this result does not match reality. For example if the mine foreman checks timbering by knocking, the overload of supporting columns is accompanied with sound of high frequencies. In order to get the result that matching the reality, the non-moveable support must be used in the moment of vibration. Results are graphically presented in Fig. 2.

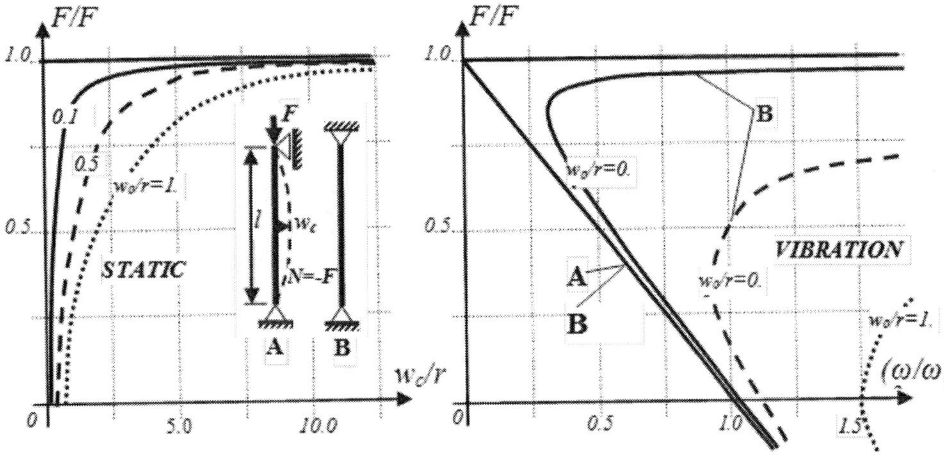

Fig. 2. Stability and vibration of a beam with initial imperfection.

5.2 Effects of the initial deformation on vibration of a beam

In the case of beam the mode of vibration is identical with the mode of buckling. That is the reason why the dependence between the load and the square of eigen circular frequency is linear. In static problems the most unfavourable mode of initial imperfection is the one identical with the mode of buckling corresponding to the lowest critical force („mode 1"). The vibration of beam is shown in Fig. 3. Although the mode of initial imperfection of the beam is close to the "second mode of buckling" it will vibrate in the first mode.

 DOI: 10.1051/matecconf/201710700053

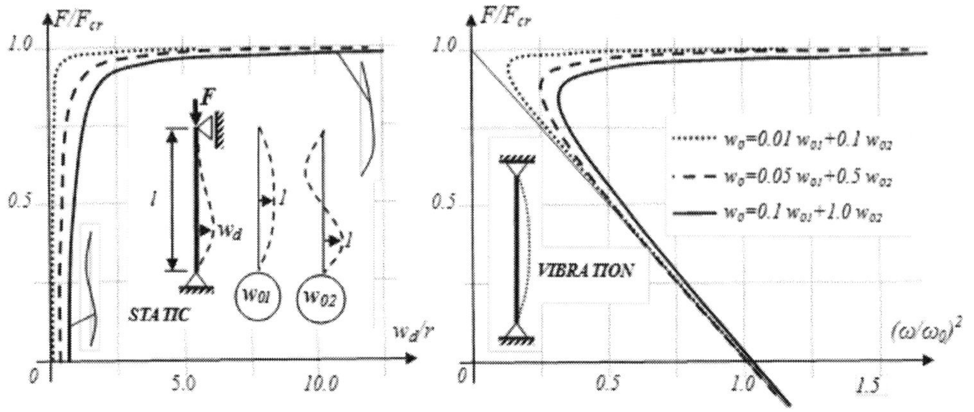

Fig. 3. Stability and vibration of a beam with initial imperfection in second mode of buckling.

5.3 Experimental measurement

As is well known precise research of the structural theory is a combination of theoretical, numerical and experimental investigations. Authors had designed special equipment for experimental laboratory measurements in order to meet this requirement. The scheme of this equipment is shown in Fig. 4. Many measurements have been made using this equipment.

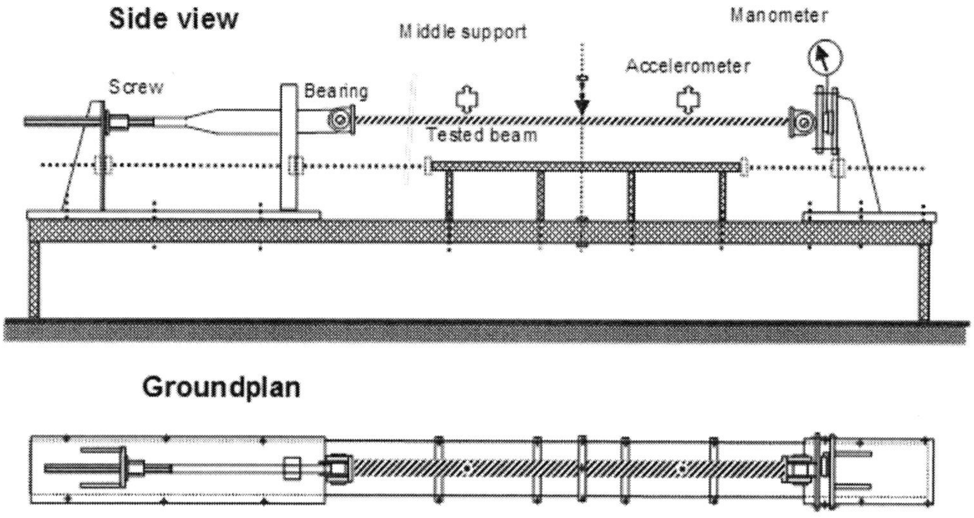

Fig. 4. Scheme of the test set-up.

6 Conclusion

The paper presents some problems of structural stability. For proper analysis of stability problems, geometric nonlinear approach is required. Just the change of compression force to tension force and the use of geometric nonlinear analysis is enough to obtain an interesting result in the determination of effects of axial force on the bending stiffness of a beam. Due to the combination of stability and vibration it is possible to determine the frequency of system considering the level of load as well as the effects of magnitude and mode of initial deformation. Comparison of theoretically and numerically obtained values of frequencies with the frequency measured on the real construction is the basis for the identification of properties or quality of structure within the scope of non-destructive methods.

This work has been undertaken as a part of a project No. 1/0272/15 funded by the Slovak Scientific Grant Agency VEGA.

References

1. Z. P. Bažant, L. Cedolin, *Stability of Structures: Elastic, Inelastic, Fracture and Damage Theories* (Oxford University Press, New York, Oxford, 1991)
2. V. V. Bolotin, *Dinamičeskaja ustojčivosť uprugich system* (CITL, Moskva, 1956, English translation: Holden Day. San Francisco, 1994)
3. J. Ravinger, *Vibration of an Imperfect Thin-Walled Panel.* Part 1: Theory and Illustrative Examples. Part 2: Numerical Results and Experiment. Thin-Walled Structures **19,** 1-36 (1994)
4. J. Ravinger, P. Kleiman, *Natural Vibration of Imperfect Columns and Frames*. Building Research Journal **50**, 1, 49-68 (2002)
5. J. Ravinger, *Stability & Vibration* (STU Bratislava, 2012)
6. K. Kolakowski, K. Kowal-Michalska, *Statics, Dynamics and Stability of Structures*, Vol. 2. Statics, Dynamics and Stability of Structural Elements and Systems. (Quick-Druck, Lodz, 2012)
7. M. Krolak, R. J. Mania, *Statics, Dynamics and Stability of Structures*, Vol. 1. Stability of Thin-Walled Plate Structures (Quick-Druck, Lodz, 2011)
8. D. Burgreen, *Free Vibration of Pined Column with Constant Distance Between Pin-ends*. J. Appl. Mechan. **18,** 135-139 (1951)

MATEC Web of Conferences **107**, 00054 (2017) DOI: 10.1051/matecconf/201710700054
DYN-WIND'2017

Comparison of the fans´ behaviour and the induced grandstand vibrations on two different football stadia

Tomáš Plachý[1,*], *Michal* Polák[1] and *Martin* Verner[1]

[1]Czech Technical University in Prague, Faculty of Civil Engineering, Department of Mechanics, Thákurova 7, 166 29 Praha 6, Czech Republic

Abstract. The paper presents comparison of the fans´ behaviour on two football stadia in Prague, stadium of AC Sparta Praha and stadium of SK Slavia Praha. The behaviour of fans was monitored by camera and classified to different categories. The response (acceleration) of the grandstands was measured on both stadia at the most exposed positions. The comparison of the response of two grandstands with different bearing systems (a steel cantilevered structure and a reinforced concrete frame structure) induced by fans is presented and discussed in the conclusion.

1 Introduction

It is well known that the synchronized movement of a large group of spectators at the sport stadium can cause significant dynamic load on grandstands and subsequently cause their vibration. In some extreme cases, the induced vibrations are for the spectators uncomfortable and even visually observable. Excessive vibration can initiate panic of crowd and stampede. This can lead to injury and even death of some people in the crowd [1] and even cause the collapse of the grandstands [1]. It is difficult to predict the dynamic load caused by a moving group of people because its size and frequency composition depending on the type of physical activity of people are changing significantly.

A series of experiments focused on fans' behaviour and induced grandstand vibrations during football matches were carried out at the home stadium of the SK Slavia Praha football club from the Czech Republic which followed similar experiments performed on a steel cantilever grandstand structure at the AC Sparta football stadium in Prague [2, 3, 4]

The paper presents the main results of experimental analyses carried out during two matches SK Slavia Praha versus FK Teplice and AC Sparta Praha versus FK Teplice. The analyses were focused on the behaviour of spectators and on monitoring of the induced vibrations in the chosen grandstand sectors of the football stadia, where the most active fans were concentrated during a football matches. The used types of cheering and changes in fans´ behaviour in various situations during the matches were investigated.

* Corresponding author: plachy@fsv.cvut.cz

2 Description of the stadia

The paper presents comparison of the fans′ behaviour on two football stadia in Prague, in the Czech Republic, stadium of AC Sparta Praha and stadium of SK Slavia Praha. The load bearing structures of the grandstands in both observed stadium sectors are different. The grandstand on stadium of AC Sparta is built as a steel cantilevered structure (see Fig. 1 - left) and the one on SK Slavia Praha stadium is constructed as a reinforced concrete frame structure (see Fig. 2 - left).

2.1 Stadium of AC Sparta Praha

The football stadium of the club AC Sparta Praha has been used for a long time since 1917. In the year 1934 it was decided to rebuild it for a capacity of 45 000 spectators, but in the same year the main grandstand was burned down. So in the year 1937, new reinforced concrete one was opened. Between years 1967 and 1969 all other grandstands were replaced by new ones made from iron and concrete with the steel cantilever main beams. The stadium was completely modernized in 1994, only seated places remain and the stadium meets all international standards. The stadium has a heated lawn football pitch since the year 2002.

Fig. 1. The investigated grandstand at AC Sparta Praha stadium with positions of the accelerometers.

The dynamic response of the grandstand was observed at the ends of grandstand main beams where the maximal levels of induced vibrations were expected (Fig. 1). The part of the grandstand selected for an experiment was chosen based on the position where the most active fans usually gathered. The piezoelectric transducers Brüel&Kjær of type 4507 B005 were attached to the lower beam flanges by using magnets. The measurement station Front-end 3050-B-040 Brüel&Kjær and software Pulse were used for data acquisition. The obtained vibration records were modified by use of the Butterworth low-pass filter with cut off frequency 20 Hz.

2.2 Stadium of SK Slavia Praha

The home stadium of SK Slavia Prague football club is called Eden Arena. It was put into operation in May 2008 and it is the most modern and with capacity of 20 617 spectators also the largest football stadium in the Czech Republic. All places to sit are covered by a roof so they provide comfortable conditions for spectators during bad weather. On the other hand, the cantilever steel roof is constructed that the views from the grandstands are not restricted by any column (see Fig. 2 - right).

The investigated part of the grandstand is the reinforced concrete structure assembled from prefabricated elements. The basic structural members of the grandstand in the observed area are noticeable in Fig. 2 (left). The L-shaped reinforced concrete horizontal beams that create the walking and sitting area of the grandstand are simply supported on the inclined main beams and these ones are simply supported too but on the vertical columns. The joints between the inclined main girders and columns are recognizable in Fig. 2 (left).

Fig. 2. The investigated grandstand at SK Slavia Praha stadium with positions of the accelerometers.

The dynamic response of the grandstand was observed in the middle of the L-shaped reinforced concrete horizontal beam (Fig. 2 – left) and on the inclined main beams under the support of the L-shaped beam (Fig. 2 – left). The position was chosen based on assumption where the highest concentration of the most active fans usually is. The same devices and measurement system were used as in the above mentioned case of AC Sparta Praha stadium.

3 Results of the experimental analyses

The first experiment was carried out at the home stadium of the AC Sparta Praha football club in Prague during the football match AC Sparta Praha vs. FK Teplice on March 21st 2015 that Sparta won 1:0. The only goal was scored in the 89th minute from the penalty.

The second experiment was realized at the home stadium of the SK Slavia Praha football club in Prague during the match SK Slavia Praha vs. FK Teplice on April 23rd 2016. The match ended in a draw 2:2.

Behaviour of the most active fans in the monitored grandstand area was recorded by camera and the time length of each cheering type was evaluated for both matches from these records. The results can be compared in the Table 1 and in the Table 2. The designations of the basic behaviour types were assumed from the reference [1]. However, other additional types were detected and distinguished in the course of the matches. The brief description is mentioned below. Eleven main types of fans behaviour were categorized during both matches that affected the grandstand vibrations:

• static state - all fans passively watch the match and sit or stand statically,

• walking - this type of fans behaviour occurs during breaks when spectators are coming to their seats or when they are leaving their places,

• jumping of the 1st type – independently - the fans are jumping en masse asynchronously and independently,

• jumping of the 2nd type – rows vertically - the fans are joined together in separated lines and jump synchronously on the spot,

• jumping of the 3rd type – rows horizontally - the fans are joined together in separated lines and move horizontally by synchronous jumping from right to left and back again usually in opposite directions in adjacent lines,

• jumping of the 4th type – sit and jump – group of spectators sit down at first after that they concurrently jump up on a signal of their leader and begin continuous jumping promptly,

• bouncing - the fans movement similar to jumping but the feet of persons remain in contact with the grandstand surface,
• swaying – the fans sway forward to support their shouted slogan and the entire motion is intensified moreover by arm movement,
• hand-clapping,
• goal scoring,
• Mexican wave - the sitting spectators stand up and raise their arms in turn creating the appearance of a wave passing through the stadium.

Table 1. Types of fans behaviour during the match AC Sparta Praha vs. FK Teplice.

	Types of fans behaviour	First half		Second half		Other time		All time	
		Time		Time		Time		**Time**	
		[min]	[%]	[min]	[%]	[min]	[%]	**[min]**	[%]
1.	Static	11:53	26.7	7:26	15.4	2:08	27.6	**21:27**	**21.3**
2.	Walking	-	-	-	-	0:59	12.7	**0:59**	**1.0**
3.	Jumping - independently	2:23	5.3	2:46	5.7	-	-	**5:09**	**5.1**
4.	Jumping – rows vertically	-	-	0:05	0.2	0:46	9.9	**0:51**	**0.9**
5.	Jumping – rows horizontally	-	-	0:59	2.0	-	-	**0:59**	**1.0**
6.	Jumping – Sit and Jump	1:17	2.9	-	-	-	-	**1:17**	**1.3**
7.	Bouncing	5:23	12.0	5:59	12.4	1:56	25.0	**13:18**	**13.2**
8.	Swaying	12:11	27.3	9:07	19.0	0:27	5.8	**21:45**	**21.6**
9.	Hand-clapping	11:14	25.0	19:28	40.5	1:28	19.0	**32:10**	**32.0**
10.	Scored goal	-	-	2:13	4.6	-	-	**2:13**	**2.2**
11.	Mexican wave	0:22	0.8	0:05	0.2	-	-	**0:27**	**0.4**

Table 2. Types of fans behaviour during the match SK Slavia Praha vs. FK Teplice.

	Types of fans behaviour	First half		Second half		Other time		All time	
		Time		Time		Time		**Time**	
		[min]	[%]	[min]	[%]	[min]	[%]	**[min]**	[%]
1.	Static	13:32	29.74	19:50	39.79	7:03	26.87	**40:25**	**33.24**
2.	Walking	-	-	0:370	1.24	16:00	60.99	**16:37**	**13.67**
3.	Jumping - independently	2:25	5.31	3:54	7.82	-	-	**6:19**	**5:20**
4.	Jumping – rows vertically	-	-	-	-	-	-	-	-
5.	Jumping – rows horizontally	-	-	0:29	0.97	-	-	**0:29**	**0.40**
6.	Jumping – Sit and Jump	0:27	0.99	-	-	0:31	1.97	**0:58**	**0.80**
7.	Bouncing	7:00	15.38	4:47	9.60	0:10	0.64	**11:57**	**9.83**
8.	Swaying	7:32	16.56	10:13	20.49	0:16	1.02	**18:01**	**14.82**
9.	Hand-clapping	13:28	29.60	6:50	13.71	2:14	8.51	**22:32**	**18.53**
10.	Scored goal	1:06	2.42	1:45	3.51	-	-	**2:51**	**2.34**
11.	Mexican wave	-	-	1:26	2.88	-	-	**1:26**	**1.18**

The interesting examples of the response of the grandstands caused by type of fans behaviour No. 5 and the relevant frequency spectra are shown in Fig. 3 (AC Sparta stadium) and Fig. 4 (SK Slavia stadium).

DOI: 10.1051/matecconf/201710700054

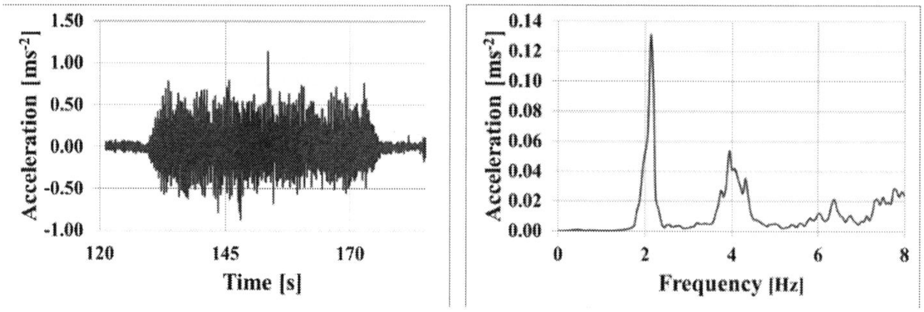

Fig. 3. The detail of the AC Sparta grandstand acceleration time record in vertical direction (left) and the relevant frequency spectrum that was measured on the middle girder during fans behaviour No. 5.

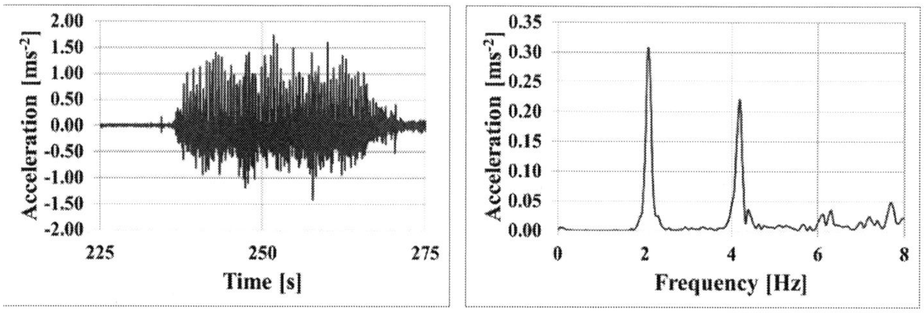

Fig. 4. The detail of the SK Slavia grandstand acceleration time record in vertical direction (left) and the relevant frequency spectrum that was measured in the middle of the horizontal beam during fans behaviour type No. 5.

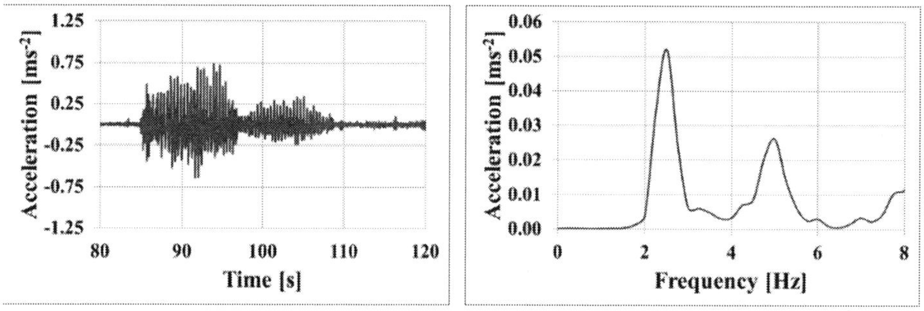

Fig. 5. The detail of the AC Sparta grandstand acceleration time record in vertical direction (left) and the relevant frequency spectrum that was measured on the middle girder during fans behaviour No. 10

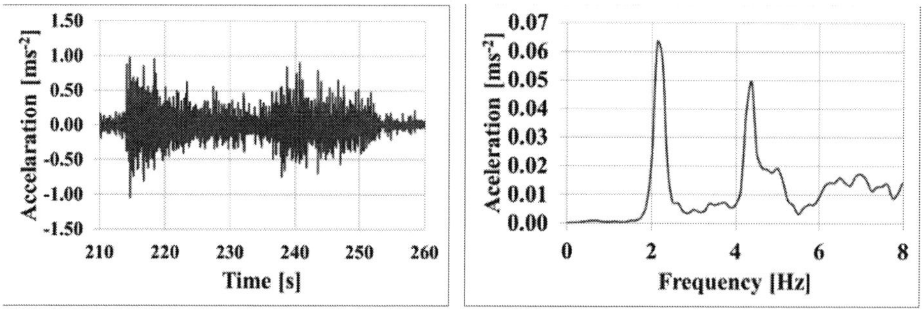

Fig. 6. The detail of the SK Slavia grandstand acceleration time record in vertical direction (left) and the relevant frequency spectrum that was measured in the middle of the horizontal beam during fans behaviour type No. 10.

4 Conclusions

The results of the fans behaviour monitoring are summarized in the Table 1 and in the Table 2. These tables show how the fans behaviour was changing during the observed matches and how their activity depends on quality of the match.

The most commonly used types of cheering in both matches were hand-clapping (type No. 9) or the static state (type No. 1) without cheering. The important parameter of the dynamic load caused by fans is not only time but also the percentage of active fans. This is why the largest vibration was caused during both matches when spectators support their team using Jumping of the 2nd, 3rd and 4th type because 100% of them were involved in these activities. The organization of cheering by the leader in the stadium sector where the most active fans are gathered caused almost always the involvement of all of the spectators present at the sector.

The highest levels of the grandstands vibrations were evaluated during the cheering type Jumping of the 3rd type (Table 1 – No. 5). The examples of evaluated accelerations for this type of cheering are shown in Fig. 3 and Fig. 4. The highest peak value of vertical acceleration 1.14 ms^{-2} was measured at the AC Sparta stadium and 1.54 ms^{-2} was measured at the SK Slavia stadium. The RMS values for these vibrations were 0.21 ms^{-2} at the AC Sparta stadium and 0.41 ms^{-2} at the SK Slavia stadium. For other types of cheering the vibration of these two stadia are lower, see Fig. 5 and Fig. 6. The frequency spectra show that the excitation frequencies caused by fans do not depend on the types of fans, on both stadia the spectra look similarly (Fig. 3 to Fig. 6 – right). Nor the panic criteria nor the comfort criteria for the passive part of spectators on stadiums according the ISO 10137 were achieved.

This work has been supported by the Czech Science Foundation project No. GA15-15728S.

References

1. C. A. Jones, P. Reynolds, A. Pavic, Vibration serviceability of stadia structures subjected to dynamic crowd loads: A literature review, J. Sound Vib. **330**, 1531 (2011)
2. M. Verner, M. Polak, T. Plachy, An Experimental Analysis of Grandstand Vibrations Caused by Crowd of Spectators during Two Football Matches, Appl. Mech. Mater **827**, 312 (2016)
3. M. Verner, T. Plachy, M. Polak, An Experiment Focused on Fans Behaviour and Induced Grandstand Vibrations during a Football Match, Appl. Mech. Mater **837**, 75 (2016)
4. M. Verner, M. Polak, T. Plachy, A Comparison of Spectators Induced Grandstand Vibrations and Fans Behavior during Two Football Matches between the Same Teams, Proceedings of 54th International Conference on EAN 2016, 6p. (2016)
5. P. Reynolds, A. Pavic, Vibration performance of a large cantilever grandstand during an international football match, J. Perform. Constr. Fac. **20**, 202 (2006)
6. ISO 10137:2007, Bases for design of structures – Serviceability of buildings and walkways against vibrations.

Dynamic analysis of an impact load applied to the composite wall structure

Maciej Major[1], *Krzysztof* Kuliński[2] and *Izabela* Major[3*]

[1] Technical Sciences, Department of Steel Constructions and Building Materials, Faculty of Civil Engineering, Częstochowa University of Technology, 42-200 Częstochowa, Poland
[2] Technical Sciences, Department of Concrete Constructions and Geotechnics, Faculty of Civil Engineering, Częstochowa University of Technology, 42-200 Częstochowa, Poland
[3] Technical Sciences, Department of Technical Mechanics and Engineering Graphics, Faculty of Civil Engineering, Częstochowa University of Technology, 42-200 Częstochowa, Poland

Abstract. In presented paper the composite wall structure made of concrete and Zahorski/Mooney Rivlin rubber cross pads subjected to impact loads was investigated. Mentioned rubber cross pads were assumed to be inserted or injected in specially formed holes during the production process. Moreover, two different impact loads were taken into further considerations - pressure and concentrated force. Such impact loads due to the small time step size allowed observation of mechanical wave propagation in analyzed structure in the form of stress plots. On the basis of composite wall comparison with the solid concrete wall it could be stated that the mechanical wave propagation was significantly damped via inserted/injected rubber cross pads. Discussed problem was solved with the use of finite element method using ADINA program.

1 Introduction

Composites due to their more preferable mechanical properties in comparison with individual materials are the subject of interest by many researchers. The most popular composite utilized in civil engineering are concrete with steel and steel with rubber. Utilization of composites allows mainly reduce the total mass of construction in comparison with the structure where individual material is used, reduce the influence of unfavourable effects, exploitation costs, improve forces redistribution etc.

Mechanical wave propagation is one of mentioned negative effects, which should be strongly avoided in newly designed structures. Such mechanical wave propagation in any considered material may be a result of earth movement vibrations, acoustic waves propagation, explosion blasts etc. The problem of wave propagation was discussed by Major and Major [1], where authors presented the possibility of numerical solutions with the use of finite element method in the ADINA program. However initial concrete or ceramic blocks acoustic/vibrations absorption property may be improved with the additional material utilization as a cover of host wall – such as polystyrene foam or mineral wool, the width dimension of such wall would also be increased. Such increase in structure

* Corresponding author: imajor@bud.pcz.czest.pl

DOI: 10.1051/matecconf/201710700055

dimensions is often undesirable from the designing point of view, especially in small structures with small interior area.

In this paper composite blocks made of concrete and rubber were proposed due to their significantly improved mechanical wave propagation damping properties in comparison with an individual solid concrete block. It is also worth to notice that discussed composite block allow transferring of significant compressive forces – load bearing capacity of such block in comparison with solid concrete block was only slightly reduced via specially prepared holes. It was assumed that the concrete block in the production process was embedded into special form and after its removal, rubber cross pads could be inserted or injected in mentioned holes. Moreover, three plane symmetry was assumed in order to eliminate possible assemblage mistakes. Two different material models representing rubber cross pads were adopted i.e. Mooney-Rivlin [2, 3] and Zahorski [4, 5], respectively. A similar analysis using the same materials was performed in the paper [6]. Additionally, two different impact loads were investigated – i.e. pressure and concentrated force. In order to determine the mechanical wave propagations resulting from applied loads, finite element method approach with the use of ADINA program was performed. Mechanical wave propagation phenomena were also the subject of interest of Coulson and Jeffrey [7], Kosiński [8] and Wesołowski [9]. Recently numerical approach is treated as an introduction to the expensive experimental tests or validation tool of already produced/constructed objects and structures (see [10]). As an example Cajka and Krejsa [11] validated the action of static and dynamic forces in repaired lightweight steel roof structure.

2 Methods

In order to perform numerical analysis of mechanical wave propagation, section of a wall made of composite blocks was adopted. Mentioned wall had following dimensions assumed – length equal 0.96 m, height 0.68 m and thickness of 0.25 m, respectively (see Fig. 1a). Each composite block had 0.47 m length, 0.25 m width and 0.22 m height. Inner mortars were assumed to be 1cm thick. In Fig. 1a whole wall was presented with the effective stress measurement points 1 – 4. It should be noted that points 1-4 on the rear wall side were introduced with the suffix letter "R" and were the reflection of points 1-4 from the presented frontal surface of the wall. Vertical and horizontal mortars with strength of 4 MPa were considered as 10 mm thick. Dimensions of all rubber cross pads were presented in Fig. 1b.

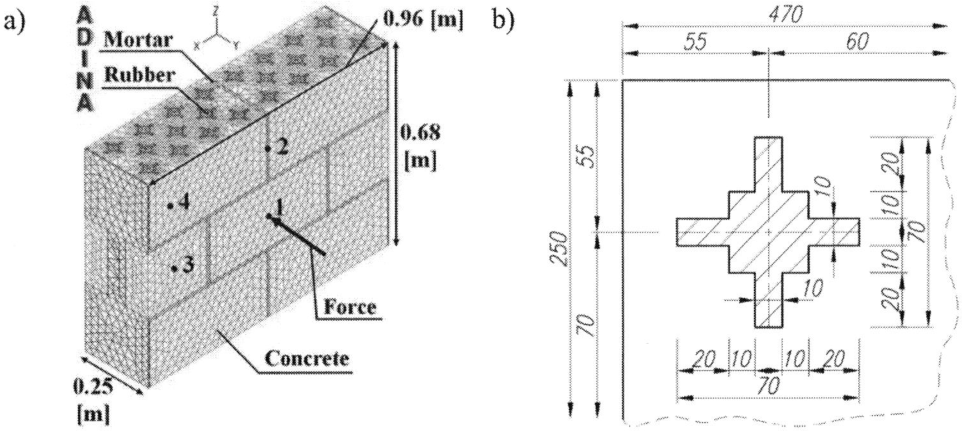

Fig. 1. (a) Wall numerical model made of composite blocks with concentrated force and measurement points 1-4, (b) rubber cross pads placement with respect to the left top corner of concrete block.

Dynamic pressure load with the value of 10000 Pa was applied on the whole frontal surface of the wall, whereas concentrated force with value of 5000 N was placed in the middle of front wall surface (see Fig. 1a). Both forces reached their maximum value at $t = 1\cdot10^{-5}$ s and after that time, force was removed – their value was equal 0. It should be noted that applied load values were treated only as a reference for comparative analysis between composite and solid concrete blocks. Three different time step values of obtained effective stress were taken into further considerations of obtained effective stress $t = 1\cdot10^{-5}$ s, $5\cdot10^{-5}$ s and $9\cdot10^{-5}$ s, respectively.

In presented model on exterior "YZ" and top "XY" plane surfaces of the wall X and Y - axis displacements were fixed, whereas on the bottom exterior "XY" plane surface of the wall X, Y and Z-axis displacements were fixed.

Discretization of considered wall model was performed with the use of 4-node 3D-solid finite elements (tetrahedrons), where size of each element in composite/concrete block was assumed to be ~0.02 m. Discretization of rubber and mortars was performed automatically with respect to the previously defined nodes in the composite/concrete blocks. Numerical model comprised of 26722 nodes, 89993 finite elements describing concrete block, 38578 elements describing rubber cross pads and 20163 elements describing mortars, respectively. In case of solid concrete blocks exactly the same mesh was used, whereas rubber elements were replaced with concrete finite elements. According to that 26722 nodes, 128571 finite elements describing concrete blocks and 20163 elements describing mortars were obtained.

Concrete blocks were made of C20/25 class concrete described via ADINA "Concrete" material module, mortars with strength of 4 MPa were described via ADINA "DF-Concrete", whereas rubber was represented via ADINA "Mooney-Rivlin" material module. In order to represent Zahorski material model description, the original ADINA "Mooney-Rivlin" material library was modified.

3 Numerical results

According to the fact that mechanical wave propagation transferred energy through analyzed model, effective stress redistribution was investigated. Obtained effective stress values in considered measurement points for both, wall made of composite blocks and wall made of solid concrete blocks were presented in Table 1. It should be noted that points with "R" suffix were the reflection of located on the front wall surface points 1-4 onto the rear side of the wall.

On the basis of Table 1 and performed numerical analysis one can state that for the Mooney-Rivlin and Zahorski material model obtained results were exactly the same. Differences between that two mentioned materials can be observed mainly at large deformations. According to the Fig. 2 it is clearly visible that the mechanical wave energy at $t = 5\cdot10^{-5}$ s was significantly absorbed via rubber pads and on the rear side of the composite blocks wall obtained effective stress was 37-65% lower for the pressure load and 55 to 97% lower for concentrated force compared to the wall made of solid concrete blocks. In case of concentrated force, it is worth to notice that the longer distance from the applied dynamic force to the measurement point and the more rubber cross pads were on the way of disturbance propagation the higher percentage damping was obtained (compare point 3R and 4R with 1R and 2R in Fig. 2). However, at the $t = 5\cdot10^{-5}$ s obtained effective stress values were lower in composites on the rear side of the wall compared to the solid concrete blocks, therefore at $t = 9\cdot10^{-5}$ s for example in point 2R or 4 for pressure load obtained stress was higher. Mentioned situation was connected with the mechanical wave refraction and reflection on the rubber material surfaces. According to that transferred energy overlapped providing greater values of discussed stress.

Table 1. Effective stress comparison in the given measurement points for composite wall and wall made of solid concrete blocks for two different loads at three different time steps.

Load	Point	Effective stress [Pa]					
		Composite wall (Mooney-Rivlin / Zahorski)			Solid concrete blocks		
		$t = 1\cdot10^{-5}$ s	$t = 5\cdot10^{-5}$ s	$t = 9\cdot10^{-5}$ s	$t = 1\cdot10^{-5}$ s	$t = 5\cdot10^{-5}$ s	$t = 9\cdot10^{-5}$ s
Pressure	1	5.34E+03	1.20E+03	1.30E+03	5.29E+03	8.93E+02	9.39E+02
	1R	4.41E-02	3.04E+02	1.70E+03	2.12E-01	4.81E+02	2.32E+03
	2	1.39E+03	3.08E+03	1.50E+03	1.36E+03	2.66E+03	1.78E+03
	2R	7.16E-02	4.11E+02	9.15E+02	9.69E-02	7.06E+02	3.97E+02
	3	5.72E+03	2.73E+03	1.52E+03	5.69E+03	2.60E+03	1.19E+03
	3R	2.35E-02	1.57E+02	1.88E+03	1.53E-01	4.50E+02	2.12E+03
	4	5.69E+03	4.85E+03	3.84E+03	5.75E+03	3.43E+03	1.72E+03
	4R	2.32E-02	1.72E+02	2.45E+03	1.30E-01	4.20E+02	2.63E+03
Concentrated force	1	2.12E+06	5.02E+05	3.25E+05	2.12E+06	4.97E+05	3.08E+05
	1R	7.66E-01	4.64E+03	3.61E+03	4.10E+00	1.38E+04	1.35E+04
	2	1.57E+00	4.80E+03	4.55E+04	1.40E+00	3.71E+03	3.25E+04
	2R	3.67E-03	1.40E+02	9.22E+03	1.21E-02	3.12E+02	1.27E+04
	3	1.15E-03	2.12E+01	1.57E+03	0.00E+00	3.08E+01	5.02E+02
	3R	0.00E+00	1.94E-01	3.24E+02	0.00E+00	7.14E+00	2.28E+03
	4	0.00E+00	4.01E-01	1.57E+02	0.00E+00	4.73E+00	6.11E+02
	4R	0.00E+00	3.94E-02	9.50E+01	0.00E+00	1.35E+00	7.33E+02

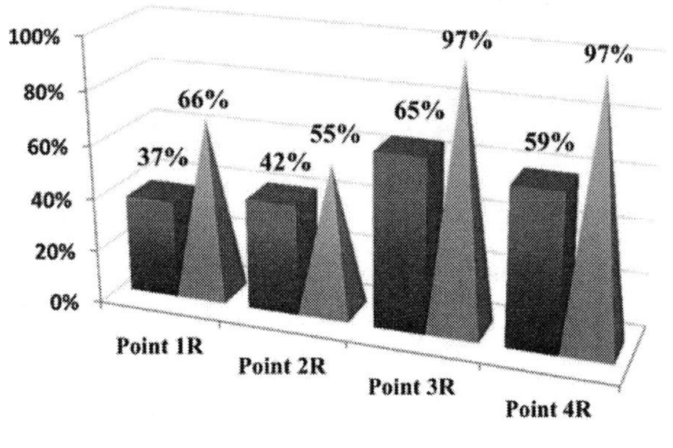

Fig. 2. Percentage damping of mechanical wave propagation for different measurement points and two types of load in composite wall related to the solid sandstone blocks wall at $t = 5\cdot10^{-5}$ s.

DOI: 10.1051/matecconf/201710700055

Moreover, value of effective stress in the composite block at the start of the analysis $t = 1 \cdot 10^{-5}$ s was slightly higher than in solid concrete blocks, which was connected with smaller overall volume of the concrete material – at first internal forces were distributed through material with higher rigidity, which in that case was concrete. Clearly visible mechanical wave propagation in the form of stress redistribution plots in the horizontal "XY" plane cross section passing through points 1 and 3 (see Fig. 1a) for the pressure impact load at $t = 5 \cdot 10^{-5}$ s was presented in Fig. 3 while for the concentrated force at $t = 7 \cdot 10^{-5}$ s was presented in Fig. 4.

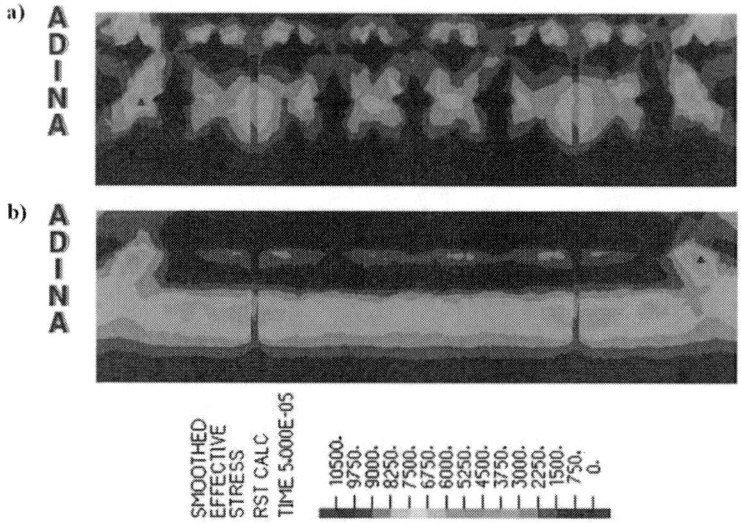

Fig. 3. Stress [Pa] redistribution plots from the pressure impact load at $t = 5 \cdot 10^{-5}$ s in the wall made of (a) concrete-rubber composite blocks, (b) solid concrete blocks.

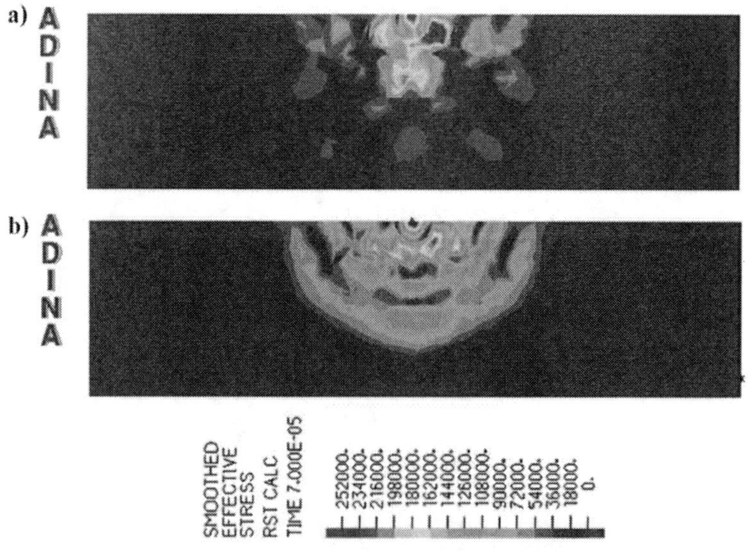

Fig. 4. Stress [Pa] redistribution plots from the concentrated force impact load at $t = 7 \cdot 10^{-5}$ s in the wall made of (a) concrete-rubber composite blocks, (b) solid concrete blocks.

4 Conclusions

In this paper mechanical wave propagation in two different types of wall was discussed. The reference numerical model was assumed as wall made of solid concrete blocks, whereas proposed new solution for the comprised of wall made of concrete blocks with specially formed holes in the production process, where rubber material could be inserted or injected. According to the performed numerical analysis it was shown that the mechanical wave propagation in wall made of composites could be significantly damped. Propagating wave on the rear side of the composite wall at $t = 5 \cdot 10^{-5}$ s, which was the time of wave passing through the block material from its front to rear surface was reduced in the range of 37 – 97% in comparison with the wall made of solid concrete blocks. Such high values of mechanical wave damping could be obtained due to the rubber mechanical energy absorption property i.e. rubber material elastic deformations. From the engineering and designing point of view presented composite could be easily utilized in civil engineering especially in areas exposed to small earth shakes, machines vibrations or to reduce acoustic waves. It should also be noted that due to the small area of the rubber cross pads the overall compressive forces bearing capacity was only slightly reduced in comparison with solid concrete block. Moreover, due to the three planes of symmetry the risk of incorrect block placement in the wall was utterly compromised. However, presented composite gave satisfactory results of mechanical wave damping via the numerical analysis, experimental tests should also be provided in order to estimate the actual load bearing capacity and other material properties unspecified in this paper.

References

1. I. Major, M. Major, Advanced Materials Research, **1020**, 171-176 (2014)
2. M. Mooney, Journal of Applied Physics, **11**, 582-592 (1940)
3. R.S. Rivlin, Philosophical Transactions of the Royal Society of London A 240, **822**, 459-490 (1948)
4. S. Zahorski, Archives of Mechanics, **5**, 613-617 (1959)
5. S. Zahorski, Rozprawy inżynierskie, **10** (1), 193-207, in Polish (1962)
6. M. Major, I. Major, Advanced Materials Research, **1020**, 165-170 (2014)
7. C.A. Coulson, A. Jeffrey, *Mathematics models,* in Polish (1982)
8. S. Kosiński, *Elastic waves in rubber-like layered composites,* in Polish (2007)
9. Z. Wesołowski, *Acoustics of elastic body,* in Polish (1989)
10. J. Melcer, Communication, **3**, 5-10 (2007)
11. R. Cajka, M. Krejsa, Applied Mechanics and Materials, **501**, 592-598 (2014)

Vibration tests of wall segments

Mariusz Żółtowski[1,*]

[1]UTP University of Science and Technology, Management Department, 85-790 Fordońska 430, Bydgoszcz, Poland

Abstract. In this study author described the degradation testing methodology with use of vibration on building structures to assess measures propagation of vibration energy in the masonry segments. Theoretical analysis and practical verification of susceptibility testing information of the proposed measures vibration processes indicate broad possibilities for their use. Recognizing the need to improve methods of testing machines and structures in order to assess their state of degradation - in this work author shows the important results in the field of effectiveness verification of proposed measures of masonry segments.

1 Introduction

Aging and consumption are inextricably linked with the existence of buildings, affecting destructive to their technical condition and leading inevitably to the possibility of observing their damage state. The combined impact of these degradation structure forms of the object leads to the development lesions, leading to the interruption of the structure correct functioning, and even the loss of physical cohesion of object.

The degradation of building materials and structures, can be track by finding changes in their degradation, and allows to predict damage, which determines the effectiveness of non-destructive testing and technical diagnostics.

Machines and structures requires to bring them (and discharge) respective types of energy, as well as a variety of material-technical products depending on the type of structure. Therefore it is needed to take into account the potential handler, which is used to rehabilitate the potential utility consumed during the existence of the building construction. Values parameters change characterize presented potentials, depends primarily on:

- the type of tasks performed by the structures,
- external conditions of existence (functioning) the construction,
- technical state of constructions (buildings)
- qualifications of the users and employees performing a repair works.[*]

The utility potential of the building structure can be defined as a material resource and energy to ensure its ability to use. This potential decreases during the existence of the structure while increasing its potential handler. ***Operating potential*** is interpreted as a material and energy resource needed to renew the potential utility of construction (buildings). These two types of potentials form an ***exploitation potential,*** which

[*] Corresponding author: mazolto@utp.edu.pl

characterizes the building structure as its ability to be used in accordance with the needs for which it has been enabled.

Against the background of the existing rules of previous studies in this paper author proposed research methodology, which represents a new approach in the study of building materials destruction. It uses a newer experimental research in the evaluation of changes in the degradation of the buildings structures, which is based on dispersed vibration energy and modal analysis methods which describes the degradation state of the tested object.

Dedicated algorithms methodology of degradation vibration testing in the construction industry is offend cooled the dynamic properties test procedure of the various components, machines and structures, including modeling, simulation studies and their verification in the research bench and supplies. It requires the association structure traits of evaluated object, with a set of measurement and evaluation processes of output. Dedicated algorithm for research of vibration energy analysis in the study of machines and structures construction products includes:
- formulate a research problem;
- acquisition and processing of vibration energy measurement;
- optimization the quality of vibration symptoms;
- models of degradation state binding measures;
- inference.

In general testing methodology of degradation contains detailed procedures of the study: the data acquisition system, processing and statistical inference.

The major issue of this work is to verify the effectiveness of the proposed above degradation testing methodology in the application of vibration energy propagation measures.

2 Masonry constructions research

The most reliable method for determining the wall resistance in the existing structure is to examine the strength of the samples taken from the wall out of its structure. However these wall elements can have relatively large dimensions and therefore is downloadable from the existing wall structure sample of this size is rarely possible. In the case of non-perforated wall masonry (full brick, cellular concrete blocks) can be cut into cylindrical samples, like in the case of concrete structures [1, 4]. Research masonry drilling can be conducted in different ways, which has an impact on the results. Correction factors samples (normalized strength of the wall) should be characterized by the trial elements.

The safety factors values of the wall are determined according to the category of production control and the category of the works on site. A distinction is made:
- category A of works - when work is performed by trained team under the supervision of master mason and apply grout produced in the factory. If the mortar are made on site, controlled dosing of ingredients and mortar strength and quality are controlled by a qualified person [2, 3].
- B categories of works - when the conditions of category A are not met, and the work quality monitor is performed by an authorized person, authorized by the contractor.

The decision to adopt the performance category is made by the designer according to the contractor confirmation by observations at the site. For exceptional design situations, regardless of the masonry production category and performance category, you can take: for wall - $\gamma_m = 1.3$; for anchoring of reinforcing steel - $\gamma_m = 1.15$; for reinforcing steel - $\gamma_m = 1.0$.

The basic information for the existing structure analysis is the knowledge of the wall elasticity modulus. The relationship between stress and strain is characterized by strain limits ε_L and ε_u, which is recommended to take according to the test results. Currently, there are technical possibilities of drilling in the walls of a diameter about 150 mm. Samples drilling can be carried out in a vertical or horizontal to the vertical plane of the wall. Based on the research it can be assumed that the relationship between the strength determined on samples cut from the wall, and his strength characteristic compressive strength for horizontal and vertical is about: $f_{k,\ mean} = (0.75\ /\ 0.83)\ f_{k,\ mean} \cdot m$, where $f_{k,\ mean}$ - standard compressive strength obtained from the wells, $m < 1$ - coefficient taken depending upon the masonry.

In many countries in order to evaluate the safety of existing masonry wall undertakes research on compression using pressure pads (*flat jacks*), placed in the horizontal joints of the wall. The different method over the classical evaluation of structural is that of interest is not the strength of the wall, and to securely transfer through the certain strength wall of the expected structure load.

Research of wall compression using pressure pads (*flat-jacks*), placed in the horizontal joints of the wall (Fig.1).

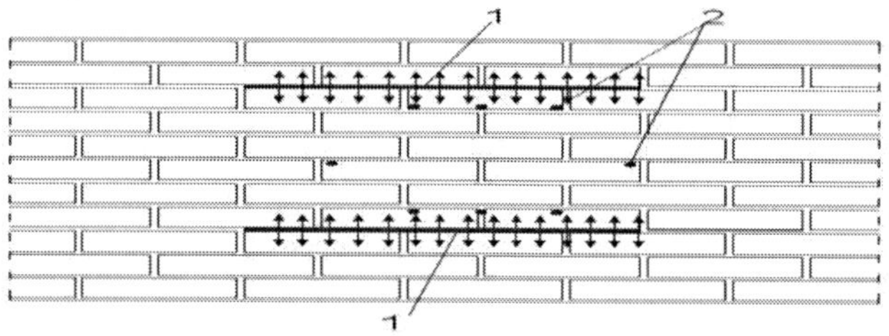

Fig. 1. The essence of the *"flat-jacks"* method.

Used in the pressure examination pad has a rectangular shape with aspect ratio of 1:2 or rounded to better fit the circular saw evacuated space in the joint. The pillows are made of stainless steel having a thickness of 0.8 to 1.0 mm and a thickness of about 10 mm. The surface area of cushion is about 80 000 mm^{2}. Horizontal welds, which have been introduced cushion pressure rupture of the mortar using a circular saw to a depth of not less than 2/3 of the wall thickness. The joints should be away from each other between 430 to 500 mm, which corresponds approximately to the length cushion. On the surface of the wall they are separated by a seam which is arranged by pressure cushions, glued suppositories for measuring (by means of a strain gauge socket) deformation of the wall under the effect of stress σ_p caused by the pressure pad. Both of the pressure pad are supplied by a single pump. Stress σ_p induced in the wall is determined by the formula: $\sigma_p = K_m K_P$, where K_m - constant value taking into account the characteristics of the pads and the rigidity of the welded joint metal pads on the perimeter (This value is determined by scaling the pad in press testing), K_a - ratio of the hollow joints (after mortar removal) into the bag, p - the pressure in the cushion caused by the supply pump.

Measurements are usually stress in the wall corresponding to the maximum load test piece construction in the foreseeable conditions of use of the object, or a little more. As a rule, it is a tension not exceeding 0.5 the estimated strength of the wall. Tests may also lead to stress short of strength, but it is not always necessary in view of the purpose of the investigation and with greater efforts of the - longer receive scratches on the wall.

The most reliable method, however, is not always possible to apply, to remove the item from the test masonry structures and to determine the value of f_B in accordance with the requirements of the standard. In the case of non-perforated masonry rather than remove the entire element can download into cylindrical sample of material from which the elements were walling and on that basis determine the strength f_B of masonry element. The diameter of the sample is typically 50 to 60 mm, and a correction factor of the sample converts the strength of the strength of the brick f_B 0.90 [1, 4]. Error in assessing the value of f_B - $\pm 10 \div 15$ %. The strength of the bricks you can explore by the Schmidt hammer [1, 3].

Compressive strength of mortar sets of rules "by feel", scraping it with a sharp instrument, or using research mortar adhesion to masonry - Using the *"pull-off"* (Fig. 2).

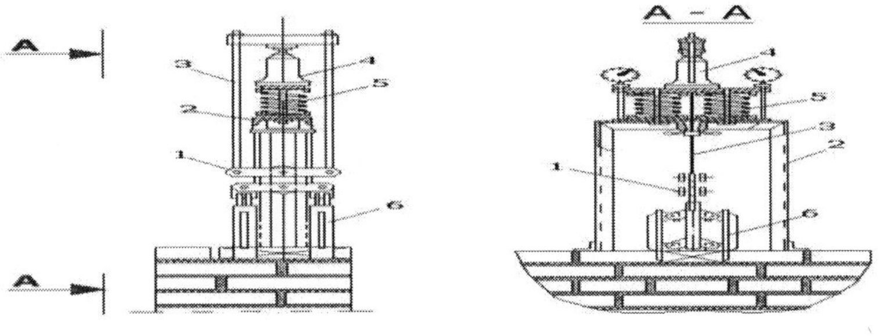

Fig. 2. Mortar adhesion research.

A different test method can safely transfers through the wall and a specific strength of the expected load of the structure is to subject the wall compression with flat pressure pads *(flat jacks),* placed in the horizontal joints of the wall, emptied the previously mortar – Fig. 3 [1, 4]. The presented methodology can be used also when the masonry compressive strength is known (e.g. the retrieved drilling of masonry) and the mortar strength (e.g. on the basis of a mason hammer sample), and - a necessary condition - the coefficient of variation v_x (e.g. ultrasonic method). The wall strength characteristic f_k can be determined using data from the PN-B-03002: 1999 constitute a generalization of the collected test results. Testing the strength of the wall shear along joints method - *"In situ"*.

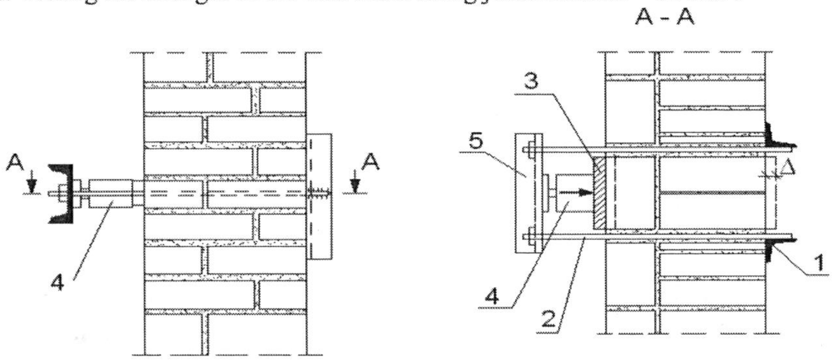

Fig. 3. The idea of the method of *"in-situ"*

When there is no data on homogeneity testing of the wall, it is usually identified with uniformity endurance masonry, as determined usually by the sclerometric testing [1, 3, 4].

3 Masonry segments research

Dynamic properties of building structures have a direct impact on the level of generated vibrations emitted noise, fatigue strength, controllability and stability of the structure. The analysis of the dynamic properties of the structure is identified based on structural models built on the basis of experiments or the real object.

Assessment of dynamic building structures using processes vibration requires the association structure traits evaluated object with a set of measurement and evaluation processes output. Introduced vibrations to the object are dynamic and keep the conditions of equilibrium between the state of inertia, elasticity, damping and extortion. Disturbances propagating from the source as a wave in a manner dependent on the physical properties

DOI: 10.1051/matecconf/201710700056

and the limits of the configuration, dimensions and shapes of buildings. This results, consequently, the power dissipation of the waves, the diffraction, reflection, and the overlap. As a result, the existence of the input and execution states representing the transition characteristics and properties of the structure is formed that can be measured by a number of symptoms characteristic of the processes included in the output [2, 5, 6, 12, 14].

For verification tests author selected wall segments made of ceramic elements, which for the purposes of this study were determined depending on the manner of their execution, as in Table 1. They were also given the destructive force the segment walled specified on the testing machine, available in a research laboratory.

Table 1. Characteristic of studied masonry segments.

Material	sample [cm]	reinforcement	sample sign	destructive force [kN]
ceramic brick	12×25	without	C1	290
		steel mesh	CD1	295
		Ruredil's mesh	CR1	185

Tested wall segment of ceramic bricks was made without reinforcement, with reinforcement steel mesh and reinforced Ruredil's mesh [9].

In the traditional research the strength subjected to a compression test to determine the limit of the destruction of the segment. In this study we determined waveform distortion depending on the load and the maximum value of the load, to destroy segment. Samples were subjected to an alternating load from 0 kN and the breaking force (breaking force values for individual segments of masonry are given in Table.1). Runs destruction from scratch and by buckling up to the destruction observed in the graphs σ- ε (Fig. 4). Courses of load changes, (kN) from 0 to the destruction force helped to predict the load points for each segment, for which further research has been carried out the destruction of the state of vibration.

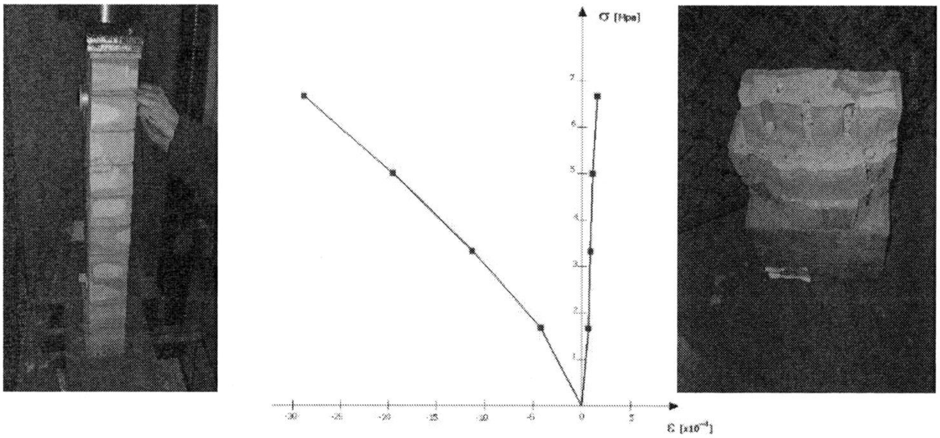

Fig. 4. Determination of breaking force and deformation σ- ε for sample C1.

Determined elasticity segment determines the ability of a material to return to its original form after removal of the external force that has caused deformation of the material. The elastic property of the material has a modulus of elasticity E is calculated by the following formula:

DOI: 10.1051/matecconf/201710700056

$$E = \frac{\sigma}{\varepsilon} \quad \text{[MPa]} \tag{1}$$

where: σ - the tension arising from (or tension) compression force F_n [kN] sample section A [cm^2]· appointed in [MPa], ε - elastic deformation caused by stress σ calculated from the relative change in length ΔL to its original length l (elongation caused by the stress) unscaled [-], however:

$$\sigma\,\varepsilon = E = F_n / A \quad \text{[MPa]; and: } \varepsilon = (\Delta l / l) \cdot 100\% \text{ [\%]} \tag{2}$$

including: F_n - squeezing force (tensile), [kN], Δl - Change of length [cm]
A - cross-sectional area of the sample [cm^2]· l - lengths original [cm].
The results of the study for all segments of brick are shown in Table 2 and Fig. 4.

Table 2. The results of sample C1 study.

C1 sample					
load [kN]	reading [mm]		tension [MPa]	s [×10^{-4}]	
	compression	extension		compression	extension
0	5.22	5.51	0.00	0.00	0.00
50	5.42	5.48	1.67	-4.28	0.64
100	5.75	5.47	3.33	-11.28	0.85
150	6.14	5.48	5.00	-19.57	1.08
200	6.58	5.44	6.67	-28.94	1.49
250	sample destruction		9.67		

In time of research also wall segments deformation was measured, which is shown in graph $\sigma(\varepsilon)$ to the appearance of cracks in the wall - a sign of the significant effort of the wall and usually causes interruption of research.

The traditional method of masonry structures testing is not always possible, and is often cumbersome to implement. The search for non-invasive way of testing the masonry leads directly to diagnostic testing vibration and modal methods proposed widely for many studies [4, 5, 6, 9]. The study segments of masonry was carried out in accordance with the proposed in many studies dedicated to the author's research methodology vibration in building structures, in accordance with the methodology shown in Fig. 5.

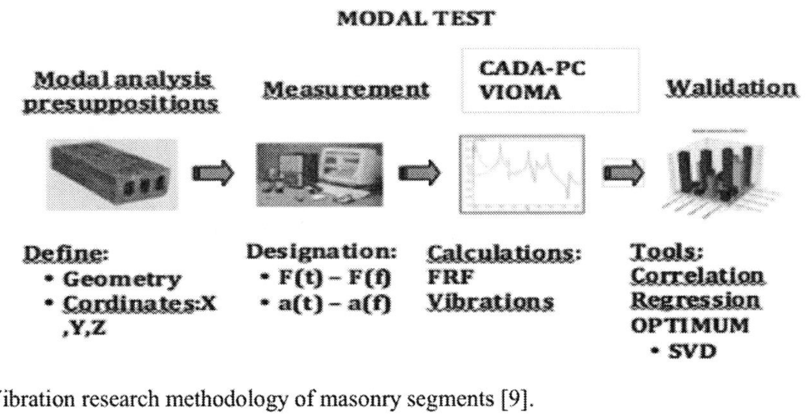

Fig. 5. Vibration research methodology of masonry segments [9].

The identification tests experiment was performed using a specialized signal analyzer, computer as control unit and a calculation interface equipped with special measuring software. Each segment tested was putted under the assumed load to extort vibration by modal hammer, and measure hardware recorded both signals the force putted from hammer and the answer at the extortion of the segment.

The basic method of signal processing used in the experiment identification is the ***frequency analysis***, which is carried out either by specialized analyzer, with use of specialized software implemented on the computer used to operate the experiment. Software that was used during the measurements is called LMS TestXpress [9].

The results of the research presents a stresses forces for distinguished states of segments load (Table 2), which provides the basis for further research, and the effectiveness of the developed vibration assessment procedure for masonry segments.

The following Fig. 6, 7, 8 shows the results of such treatment, presenting successively passes force, and the response, also the force amplitude of the vibration acceleration, vibration spectrum of the signal waveform FRF (real and imaginary parts), the transmittance waveforms (parts by real and imaginary), as well as the course of the coherence function.

The study was conducted for prepared 6 different masonry segments, with the extortion in two directions: along the x-axis and y. Presented results below are only some of these tests for a sample segment, along one axis.

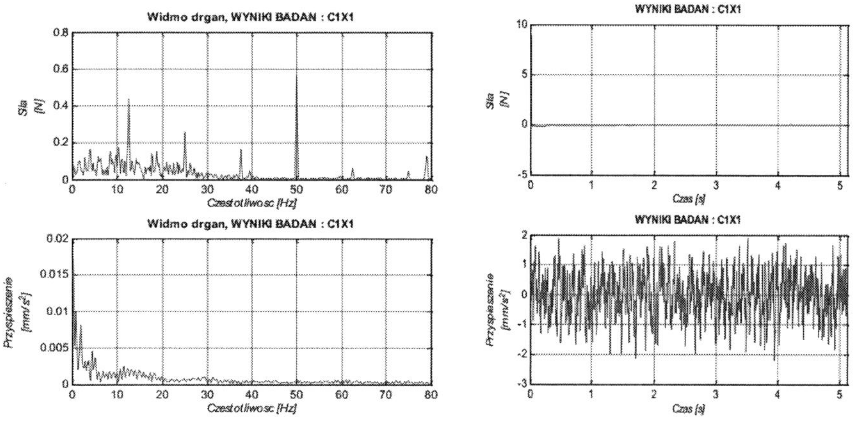

Fig. 6. Strength force and vibration acceleration as a response in the frequency and time domains.

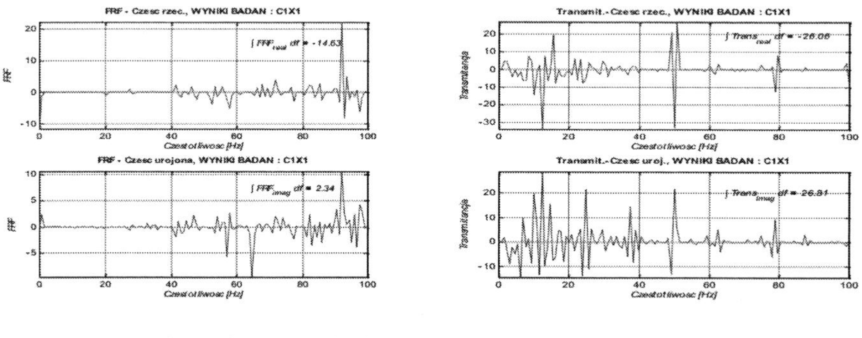

Fig. 7. Examples of waveforms transfer function and FRF *(Frequency Response Function).*

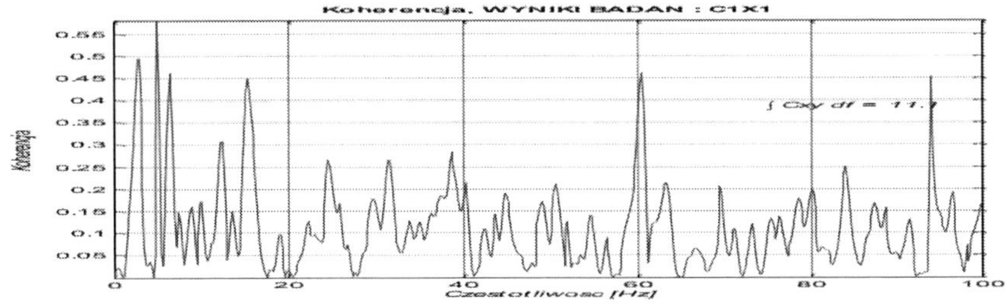

Fig. 8. The results of the coherence function for the segment C1.

Verification tests developed methods for the evaluation of destruction segments masonry was continued for three segments of masonry load taken from these studies: 0 kN, limit the force of destruction segment and the value of the average power load. Specific values of the force loads adopted for the study of individual segments of the masonry is included in Table 3.

Table 3. The figures measure the vibration signal for different load segments masonry.

	X data						*Y* data					
	kN	FRF_r	FRF_u	$H(f)_r$	$H(f)_u$	*coher*	kN	FRF_r	FRF_u	$H(f)_r$	$H(f)_u$	*coher*
C1	0	-14.63	2.34	-26.01	-26.80	11.11	0	15.47	5.27	-16.6	9.78	15.46
	150	6.67	12.06	-11.56	-58.72	14.97	150	-5.04	19.73	-11.08	50.94	10.59
	270	12.31	-6.18	39.83	13.16	12.67	270	-24.71	-9.81	-31.52	26.81	18.75
CD1	0	-21.61	2.01	43.87	12.72	15.83	0	28.64	4.17	-8.57	-24.88	12.37
	125	5.91	11.82	2.93	29.96	13.32	125	4.88	21.12	29.76	33.26	11.52
	295	40.93	-25.89	-14.28	77.32	14.76	295	4.87	6.52	-65.74	76.95	17.41
CR1	0	-1.23	7.45	13.02	32.04	14.83	0	-22.99	8.75	-16.62	-59.12	15.11
	150	-13.79	5.81	23.66	23.77	12.66	150	5.94	36.81	10.89	11.12	16.67
	305	-11.74	-3.72	-5.17	55.73	17.56	305	15.34	12.37	14.23	17.78	17.32

The test results of selected vibration estimators shown in Fig. 6, 7, 8 allows quality assessment, but do not allow convenient quantitative inferences about the change of destruction and impact load to a test segment destruction of masonry. All waveforms function has been further developed so (discretized) numerically and used during the inference quality using statistical methods proposed earlier [9]. Depending on measurement of vibration of the load conditions for the various segments of masonry are included in the regression models developed further.

4 Study of case-effect

Determining the relationship between measures of the state of destruction and load segments of masonry was carried out using polynomial regression. The table below presents regression models for segments made of ceramic bricks (with the beam cause - and effect) was determined for individual measures as well as for all measures simultaneously (Fig. 8, 9). Studies carried out indicated estimators segments brick for the construction of binding model are shown in Table 4.

Table 4. Results for the construction of polynomial regression model.

Parameters number	**FRFr**	**FRFu**	**Cohere**
1	-14.63	2.34	11.11
2	6.67	12.06	14.97
3	12.31	-6.18	12.67
4	-21.61	2.01	15.83
5	5.91	11.82	13.32
6	40.93	-25.89	14.76
7	-1.23	7.45	14.83
8	-13.79	5.81	12.66
9	-11.74	-3.72	17.56

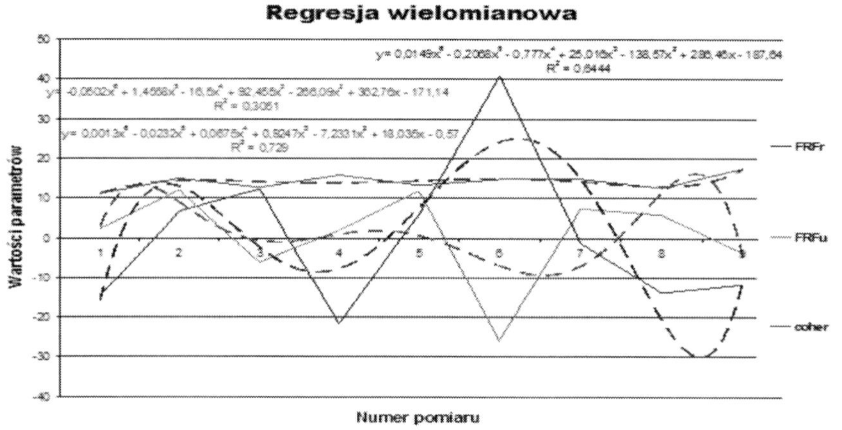

Fig. 8. Masonry segments regression models - axel X.

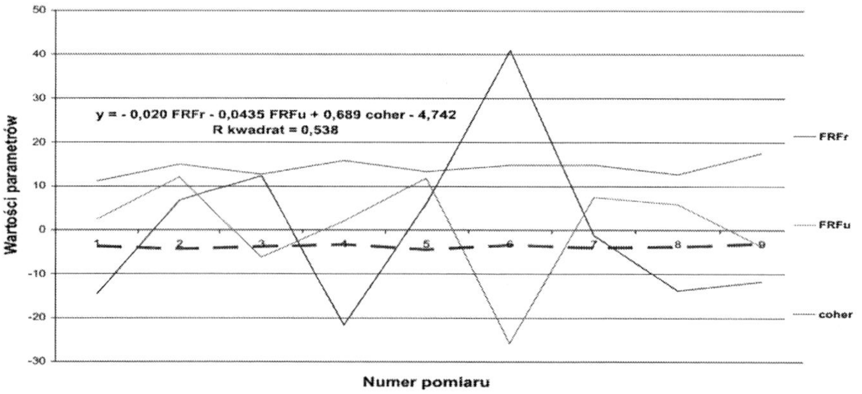

Fig. 9. Mathematic model of masonry segments destruction - axel X.

The drawings for the individual measurements and regression model show the relationships between the destruction of a segment in load values and measured vibration signal. The goodness of the models was assessed using the coefficient of determination R^2 assessor, which fitted the trend line to the measurement results.

Carried exemplification of methods on a group of masonry segments determines the model inference in terms of quality degradation of the studied masonry segments, and confirms the accuracy and efficiency of the vibration research methodology development for construction industry [9, 13, 14].

5 Summary

These considerations of this study confirm the usefulness of vibration measures and modal model parameters from experimental modal analysis to assess the destruction state of masonry structures. The following figures for individual measures and collective regression models shows the cause relationships - effect relationships between the values of segment destruction and the vibration signal. The goodness of the models was assessed using the coefficient of determination R^2, evaluating the fit of the trend line to the measurement results. Carried out in this work method exemplification confirmed the compliance and effectiveness of the proposed research methodology to the group of masonry segments as a method for the study of vibration in construction. Positively, therefore, we verified the assumptions developed methodology of research on the elements applied to the masonry segments, which will let author to study this methodology on bigger real objects in the future.

References

1. T. Chmielewski, Z. Zembaty, *Podstawy dynamiki budowli* (ARKADY Warszawa, 1998)
2. Z. Janowski, *Wymagania konstrukcyjne i wykonawcze murów*. (Konstrukcje murowe, 1998)
3. J. Kubica, *Własności wytrzymałościowe elementów murowych, zapraw i wykonanych z nich murów* (Wytrzymałość muru na ścinanie i rozciąganie,. Konstrukcje murowe, 1998).
4. W. Starosolski, R. Jasiński, A. Piekarczyk, *Impact-Echo - nowoczesna aparatura do nieniszczących badań konstrukcji betonowych i murowych*. (Konferencja Badań Nieniszczących, Zakopane 1999).
5. T. Uhl: *Computer-aided identification of mechanical structure models* (Scientific Technical Publishers, 1997)
6. B. Żółtowski, M. Łukasiewicz, *Diagnostyka drganiowa maszyn* (ITE-PIB Radom, 2012)
7. B. Żółtowski, M. Łukasiewicz, T. Kałaczyński, *Techniki informatyczne w badaniach stanu maszyn* (WU UTP Bydgoszcz, 2012)
8. B. Żółtowski, M. Żółtowski, *Vibration signals in mechanical engineering and construction* (ITE-PIB Radom, 2015)
9. M. Żółtowski, *Operacyjna analiza modalna w badaniach konstrukcji budowlanych* (WU UTP Bydgoszcz, 2012)
10. M. Żółtowski, *Informatyczne systemy zarządzania w inżynierii produkcji* (ITE-PIB Radom, 2011)
11. M. Żółtowski, *Opis drganiowy konstrukcji budowlanych*. Logistyka **6** (2014).
12. M. Żółtowski, *Investigations of harbour brick structures by using operational modal analysis*. Polish Maritime Research, **21,** 1 (2014)
13. M. Żółtowski, R.M. Martinod, *Technical Condition Assessment of Masonry Structural Components using Frequency Response Function (FRF)*. Masonry International Journal of the International Masonry Society **29**, 1 (2016)
14. M. Żółtowski, R.M. Martinod, *Quality identification methodology applied to wall-elements based on modal analysis* (Civil Engineering the Athens Institute for Education and Research, 2015)

Static and dynamic analysis of high-rise building with consideration of two different values of subsoil stiffness coefficients

Olga Ivankova[1,*], *Janka* Kovacikova[1] and *Lenka* Konecna[1]

[1]Slovak University of Technology in Bratislava, Faculty of Civil Engineering, Department of Structural Mechanics, Radlinskeho 11, SK-810 05 Bratislava, Slovakia

Abstract. This paper deals with the analysis of 21-storeyed cast in-situ reinforced concrete high-rise building. Two different 3D models were created, because of two considered values of subsoil stiffness coefficient - fixed structure (alt. 1) and the structure supported by elastic soil (alt. 2). For both alternatives of foundation of structure, required analyses (static and dynamic) were done and obtained results were compared in this paper. Short description of the structure, applied loads and other input parameters are also mentioned here. The main purpose of this analysis was to provide more information to planning engineers about the behaviour of structure exposed the wind load or seismic load when different soil conditions were considered.

1 Description of analysed structure

Investigated high-rise office building (Fig. 1) had 21 overground floors and 3 underground floors used as parking spots. The superstructure of the building was designed as a combination of cast in-situ reinforced concrete skeleton system with one stiffening core inside. Maximum dimensions of typical floor were (45.5×37.7) m. Total height of the building was 79.2 m over the ground. Under the structure, foundation slab with total dimensions (46.7×38.9) m and the thickness of 1.2 m was designed. In the place of stiffening core, the thickness of foundation slab was changed in the slope of 45°. Then, the final thickness was 1.7 m.

The thickness of reinforced concrete walls placed on the boundaries of underground floors was 300 mm. The walls of stiffening core had various thicknesses: 200, 250 and 300 mm. All columns had squared cross-sections with varied lengths of edge. From the 3rd underground floor up to the 6th overground floor, the length of edge was 900 mm. From the 7th to the 15th overground floor, it was 700 mm. From the 16th to last overground floor (21st), the length of edge was 550 mm. The slabs were designed as the combination of beamless slabs with the headers in the places of columns. The thickness was different in the case of underground floors (300 mm) and overground floors (270 mm). The thickness of headers was 100 mm. On the boundaries of slabs, the beams with the width 450 mm and the

* Corresponding author: olga.ivankova@stuba.sk

height of 650 mm were designed. They were used for the fixing of façade panels. Structural height was also different. The height of all underground floors was 3.3 m, but the height of 1st overground floor was 4.0 m. Other overground floors had the same structural height 3.6 m. The main considered material of the structure (reinforced concrete) was the combination of steel B500B with the classes of concrete C30/37, C35/45 and C45/55.

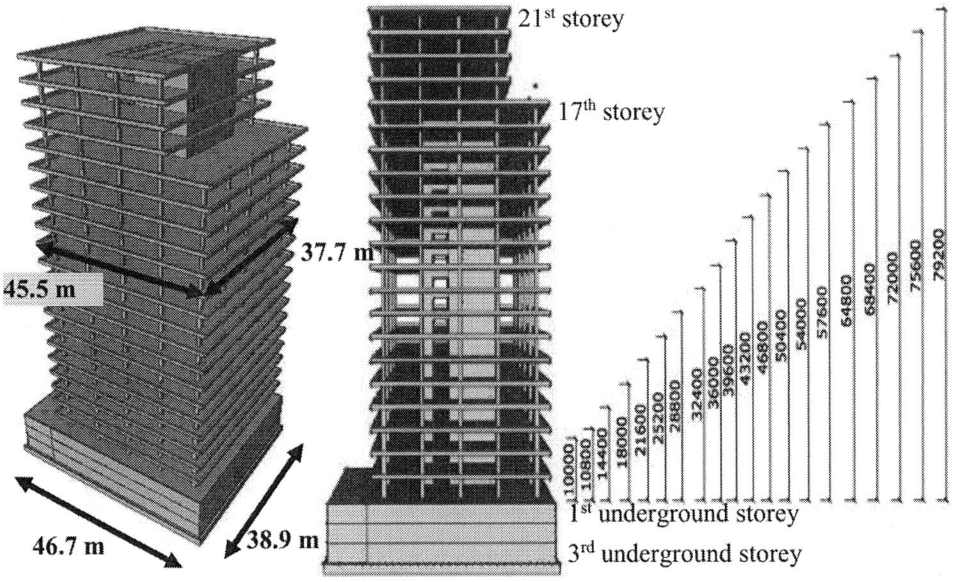

Fig. 1. The geometry. and total dimensions of analysed high-rise building.

2 Considered input parameters and the solution of the computing model

The following vertical and horizontal loads were applied on the structure. In the case of vertical loads, permanent loads (the self-weight of the structure, the weight of floor layers, the weight of roof layers) and variable loads (category B for office purpose and the category F for parting spots, partition walls) according to [1, 2] were taken into account. The snow load was calculated according to [3]. For the calculation of the wind load according to [4], the following input parameters were used: fundamental wind velocity $v_{b,0}$ = 26 m/s; the coefficient of season c_{season}= 1.0; the coefficient of prevailing wind c_{dir} = 1.0; the coefficient of orography c_o = 1.0 and the coefficient k_r = 0.234 for terrain category IV (built-up areas where the minimum 15 % of the area is covered by the high-rise buildings with total heights more than 15 m - centres of big cities). For the assessment of the structure, external and internal wind pressure coefficients were considered. Wind load was applied on the structure as horizontal load in x-direction and also in y-direction.

For the dynamic analysis, seismic load according to [5] was considered. Investigated structure was classified to the 2nd class of importance (the buildings with significant seismic response such as schools, halls, cultural institutions, etc.). Reference peak acceleration a_{gR} = 0.4 m/s^2 and the category of soil B were used in the calculation. It has to be noted, that the ground plan of the building was asymmetric. Also, the building was asymmetric in regard to its height (see Fig. 1).

Mentioned loads with adequate partial ratio were used in the combinations of the loads required for the assessment of the structure (Ultimate Limit State and Serviceability Limit State) [6].

As it was mentioned above, two models of the same structure were created because of considered different value of subsoil stiffness coefficient. In the 1st alternative, the structure was fixed and the value of subsoil stiffness coefficient k was ∞. In the 2nd alternative, the structure was supported by the elastic soil and the value of k_z [kN/m^3] was calculated using Eq. 1 (Winkler's one parametric hypothesis), where p is the contact stress in foundation gap [kPa] and s is the value of settlement of the structure [m].

$$k_z = p/s \tag{1}$$

Resultant value of subsoil stiffness coefficient for z-direction was k_z = 8 484 kN/m^3. According to recommendations given in [7], for other two directions (x and y), a half of this value can be considered. Hence, resultant value was $k_x = k_y$ = 4 242 kN/m^3. These values were used for static analysis.

However, for dynamic analysis, new value of subsoil stiffness coefficient had to be calculated using Eq. 2, where E_{dyn} is dynamic modulus of elasticity of the soil [kPa] (considered value was 150 000 kPa) and ν is Poisson's ratio of the soil [-] (considered value was 0.02).

$$k_z = E_{dyn}/(1-\nu^2) \tag{2}$$

Resultant values were: k_z = 21 347 kN/m^3 and $k_x = k_y$ = 10 178 kN/m^3. By the consideration of increased value of subsoil stiffness coefficients, the consolidation of the soil during the dynamic event can be expressed. In this case, the resultant value was 2.5 times larger than calculated value used for static analysis. The literature recommends the increasing of this value for dynamic analysis in range of 2.5-5.

For the solution of this structure, 3D model was created and solved by using the software Scia Engineer. The average size of the spatial/curved element was set as 1 000 mm in the case of static analysis and 2 000 mm in the case of dynamic analysis. Minimum length of beam element was set as 100 mm and maximum value was set as 100 000 mm.

3 Static analysis – wind effects on the high-rise building

In the case of static analysis, the influence of the wind on the stiffness of structure was analysed in this paper. From the most non-favorable combination of the permanent load, variable load and wind load were calculated maximum horizontal deflections. Wind load was applied on the structure in two directions (from the left side and also from the right side) in both axis x and y. Obtained maximum horizontal deflections are listed in Table 1.

Table 1. Maximum values of horizontal deflections due to wind load.

Direction of applied load	Subsoil stiffness coefficient [kN/m^3]		Maximum values of deflections [mm]		Increase of deflection [mm]
	Alt. 1	Alt. 2	Alt. 1	Alt. 2	
max. X	∞	k_z = 8 484 and $k_x = k_y$ = 4 242	15.2	35.1	19.9
min. X			14.9	28.5	13.6
max. Y			13.6	23.6	10.0
min. Y			3.9	5.5	1.6

 DOI: 10.1051/matecconf/201710700057

These values were compared with the limit value calculated using Eq. 3.

$$\Delta H \leq u_{lim} H \tag{3}$$

Where ΔH is calculated maximum horizontal deflection of the structure [m], H is total height of analysed structure measured from the foundation to the top [m] (in our case $H = 90.3$ m), and u_{lim} is the limit value defined in [8] (in our case this value was 1/2000). Resultant value $u_{lim} H$ was 45.15 mm.

All calculated values were less than the limit value. From this result is evident that if the stiffer subsoil (k equalled to ∞) is considered horizontal deflections are smaller. Hence, it is very important to consider right value of subsoil stiffness coefficient for the calculation because it affects the assessment of the structure.

3 Dynamic analysis – seismic effects

For this assessment, the seismic design combinations of the loads were used. It means, that the permanent loads (self-weight and other permanent loads) with the partial ratio equalled to 1.0 were considered. Variable loads (considered according to utilization of internal spaces – the category B for office and the category F for parking spots) with partial ratio equalled to 1.0 and the factor defining representative value of variable actions for combination value ψ_2 equalled to 0.3 (for B) and 0.6 (for F) were used. Snow load and wind load were not to be considered. Seismic load was considered in all three directions (x, y and z) with using the factor 1.0 or 0.3 according to requirements in [5].

Calculated maximum horizontal deflections from the most non-favorable combination of the loads are listed in Table 2. It is evident from the comparison of the values, that elastic soil caused the larger values of the horizontal deflections.

Table 2. Maximum values of horizontal deflections due to seismic load.

Direction of applied load	Subsoil stiffness coefficient [kN/m³]		Maximum values of deflections [mm]		Increase of deflection [mm]
	Alt. 1	Alt. 2	Alt. 1	Alt. 2	
max. X	∞	$k_z = 21\ 437$ and $k_x = k_y = 10\ 718$	18.40	22.5	15.9
min. X			35.00	44.1	11.1
max. Y			24.70	28.8	20.9
min. Y			29.80	38.8	25.8

Maximum values of horizontal deflections listed in Table 2, were compared with the limit value calculated using Eq. 3 and u_{lim} defined in [9] ($u_{lim} = (1/500)H$). Resultant value was 180.6 mm.

Also, maximum horizontal deflection between two slabs was compared with the limit value defined as 1% of structural height (4.0 for 1st overground floor or 3.6 m for other overground floors). In the case of alt. 1 (fixed structure), the maximum deflection between two slabs was 2.8 mm. In the case of alt. 2 (elastic soil), it was 3.9 mm. Limit value used for the assessment was 36 mm, because mentioned values were calculated between last two slabs.

4 Modal analysis – eigen frequencies

For the modal analysis, it was required to satisfy two conditions:

- the sum of effective mass considered eigen modes of vibration had to be equalled to 90 % of total mass of structure or more,
- all eigen modes with effective mass larger than 5 % had to be taken into account.

In this case, these conditions were not to be satisfied, therefore other two conditions (Eqs. 4 and 5) were considered.

$$k \geq 3\sqrt{n} \tag{4}$$

where k is considered number of eigen modes and n is the number of floors of the structure over the ground. Hence, $k \geq 3\sqrt{21} \Rightarrow k \geq 13.75$.

$$f_n \geq 5 \text{ Hz} \tag{5}$$

where f_n is eigen frequency of nth eigen modes.

Calculated eigen frequencies for both considered alternatives (fixed structure and the structure placed on elastic soil) are compared in Table 3. The values of effective mass for selected eigen modes are listed in Table 4. From the comparison of eigen frequencies is evident that the stiffness of soil had no significant influence on the change of the value of eigen frequencies.

Table 3. Calculated eigen frequencies – alt. 1 and alt. 2.

n	Alt. 1	Alt. 2	n	Alt. 1	Alt. 2	n	Alt. 1	Alt. 2	n	Alt. 1	Alt. 2
	f [Hz]	f [Hz]		f [Hz]	f [Hz]		f [Hz]	f [Hz]		f [Hz]	f [Hz]
1	0.69	0.539	**9**	3.896	3.656	**17**	5.726	5.66	**25**	6.779	6.768
2	0.7	0.544	**10**	3.977	3.784	**18**	5.905	5.77	**26**	7.016	6.839
3	1.253	1.234	**11**	4.038	3.9	**19**	6.021	5.816	**27**	7.041	6.942
4	2.47	2.389	**12**	4.172	4.168	**20**	6.318	5.981	**28**	7.123	7.034
5	2.924	2.71	**13**	4.379	4.348	**21**	6.427	6.325	**29**	7.137	7.067
6	3.065	2.821	**14**	4.51	4.375	**22**	6.479	6.34	**30**	7.186	7.12
7	3.629	3.16	**15**	4.942	4.966	**23**	6.632	6.557			
8	3.722	3.581	**16**	5.28	5.218	**24**	6.768	6.742			

Table 4. Effective mass – alt. 1 and alt. 2.

Alt. 1					Alt. 2				
n	f	W_{xi}/W_{xtot}	W_{yi}/W_{ytot}	W_{zi}/W_{ztot}	n	f	W_{xi}/W_{xtot}	W_{yi}/W_{ytot}	W_{zi}/W_{ztot}
	[Hz]	[%]	[%]	[%]		[Hz]	[%]	[%]	[%]
1	0.69	**50.5**	2.8	0.0	1	0.54	0.0	**54.8**	0.0
2	0.70	2.8	**50.0**	0.0	2	0.54	**54.7**	0.0	0.0
6	3.07	0.0	15.0	0.4	4	2.39	0.6	0.4	13.4
7	3.63	0.1	0.0	15.2	5	2.71	16.3	0.2	0.1
9	3.90	0.1	0.1	7.5	6	2.82	0.2	15.5	1.2
12	4.17	0.0	0.0	8.2	7	3.16	0.0	0.1	**63.2**
22	6.48	0.1	0.0	**20.5**	22	2.39	0.6	0.4	13.4

5 Conclusions

From the obtained results is evident, that considered value of subsoil stiffness coefficient have a significant influence on the behaviour of analysed structure. Hence, it is very important to take into account real soil parameters when the structure like this (high-rise building) is modelled and analysed. From these parameters, the subsoil stiffness coefficient is calculated. Its value affects horizontal deflections of structure caused by wind load or seismic load. These values have to be compared with the limit values defined by standards or recommendations in literature. If too large value of subsoil stiffness coefficient is used, horizontal deflections are smaller and the structure can satisfy the limit values, but it does not match with reality. On the other hand, if the value of subsoil stiffness coefficient is small and calculated horizontal deflections are too large, but we can suggest the possibilities how to make a structure stiffer. By this solution, we can prevent failures of the structure (and also damaging of non-bearing structures) in the future.

It has to be noted, that the value of subsoil stiffness coefficient is not the same during the whole life of structure. By the consolidation of the soil, its value is changed and the soil becomes stiffer. Winkler's one parametrical model of soil, used also in this paper, is only simplified calculation and it gives the value of subsoil stiffness coefficient in the beginning of building of structure. Therefore, for the detailed analysis of behaviour of structure, it should be useful to calculate changed value of subsoil stiffness coefficient with respect to the time (lifetime of structure).

The change of the value of subsoil stiffness coefficient did not have significant influence on the value of eigen frequencies. In both alternatives, calculated eigen frequencies had to be similar.

This paper was supported by the Grant Agency of Ministry of Education, Science, Research and Sports of the Slovak Republic - the grants No. 1/0544/15 and No. 1/0265/16. Authors thank to the Slovak University of Technology for the support - grant No. 1651.

References

1. STN EN 1990. Eurocode: Design of structures, SUTN, Bratislava, 2011.
2. STN EN 1991-1-1. Eurocode 1: Actions on structures. Part 1-1: General actions. Densities, self-weight, imposed loads for buildings, SUTN, Bratislava, 2007.
3. STN EN 1991-1-3. Eurocode 1. Actions on structures. Part 1-3: General actions. Snow loads, SUTN, Bratislava, 2012.
4. STN EN 1991-1-4. Eurocode 1: Actions on structures. Part 1-4: General actions. Wind actions, SUTN, Bratislava, 2010.
5. STN EN 1998-1. Eurocode 8: Design of structures for earthquake resistance - Part 1: General rules, seismic actions and rules for buildings, SUTN, Bratislava, 2005.
6. J. Bilcik, L. Fillo, V. Benko, J. Halvonik, Concrete structures: designed according to EN 1992-1-1, second ed., Publishing House of STU in Bratislava, Bratislava, 2008. ISBN 978-80-227-2940-6. (in Slovak)
7. N. Jendzelovsky, *Modelling of foundations by FEM* (in Slovak), (Publishing House of STU in Bratislava, Bratislava, 2009)

DOI: 10.1051/matecconf/201710700057

8. I. Harvan, *Concrete structures. High-rise buildings, design according to European standards* (in Slovak), (Publishing House of STU in Bratislava, Bratislava, 2015)
9. STN 73 1201. Design of concrete structures, SUTN, Bratislava, 1986

Foundation plate on the elastic half-space, deterministic and probabilistic approach

Katarína Tvrdá[1,*]

[1]Slovak University of Technology, Faculty of Civil Engineering, Radlinského 11, 81005 Bratislava, Slovakia

Abstract. Interaction between the foundation plate and subgrade can be described by different mathematical - physical model. Elastic foundation can be modelled by different types of models, e.g. one-parametric model, two-parametric model and a comprehensive model - Boussinesque (elastic half-space) had been used. The article deals with deterministic and probabilistic analysis of deflection of the foundation plate on the elastic half-space. Contact between the foundation plate and subsoil was modelled using contact elements node-node. At the end the obtained results are presented.

1 Introduction

Currently, the most widely used methods for the analysis of building structures are: finite element method (FEM), boundary element method and finite volumes method. Analytical solution of building structures gradually recedes. Most scientific as well as civil engineering software products are based on FEM deformation variant.

Each building structure is rested on the Earth´s surface, so it is necessary to analyse these structures in interaction with subsoil. The subsoil represents in fact an infinite area which, however, in the final model has its limits. From the historical viewpoint, the most used model is the Winkler model. Later two-parametric models subsoil had been introduced, implementing the effect of shear forces and impact of subgrade around the foundation into the calculation. Other models are based on the theory of elastic half-space where the different types of subsoil are described by physical properties (elastic modulus, Poisson's ratio, density). Using FEM introduced a lot of advantages to modelling structures in the interaction with the subsoil. It allows the inhomogeneous subsoil (layered), anisotropy to be implemented into the calculation.

Recently, numerous authors have been dealing with the interaction of the subsoil with structures, whether in statics, dynamics, or deterministic and probabilistic analyses [1-4].

2 The structure – subgrade interaction

Inseparable part of the building structure is groundmass in which the structure is built. This

* Corresponding author: katarina.tvrda@stuba.sk

massive area is infinite, but in modelling structures in the interaction with the subsoil it must have its limits. In modelling, it is important to choose the right mathematical - physical model to describe the subsoil.

Elastic foundation can be modelled with these types of models:
- one-parametric model (Winkler),
- two-parametric model (Pasternak),
- elastic half-space model (Boussinesque).

The theory of continuous elastic half-space implies that the subgrade forms a continuous, homogeneous and perfectly elastic body of infinitely large size, limited from above by the plane on which lies the foundation structure. Sometimes it is referred to as the Boussinesque model of soil. The model is characterized by two physical constants derived from experimental measurements (E - modulus of elasticity and ν - Poisson constant). To solve contact problems (foundation - elastic half-space) numerical methods have been primarily used to appropriately extend the closed form solutions while offering the possibility of a more realistical capture the interaction characteristics with the subsoil building structure for different boundary conditions.

Elastic foundation is modelled by spatial (3D) finite elements. In modelling of elastic half-space it is necessary to create a sufficiently large volume of groundmass, which to some extent depends on supporting and the boundary conditions. For the analysis of interaction of the foundation with the subsoil (static bond on the contact surface), two basic models can be defined: computational model with bilateral bond - continuous model, and calculation model with unilateral bond - discrete model.

3 Assessment of the reliability of structures

Reliability can be defined as the probabilistic measure of assurance of performance with respect to some prescribed conditions (ultimate limit state or serviceability limit state). Reliability of the structures can be determined by probabilistic methods (Monte Carlo, Latin Hypercube Sampling, Stratified Sampling, Adaptive Sampling, Directional Sampling, Responce Surface Method and others). Most problems concerning the reliability of structures are defined by comparison of two stochastic values, loading effects E and the resistance R, depending on the variable material and geometrics of the building structures. For such an assessment, performance criteria are first defined on the basis of functional relationship between the n-input variables, referred to as the basic random variables X_i.

Reliability function RF, security, usability, reliability function reserves (1) can be generally expressed as a function of stochastic parameters $X_1, X_2, \ldots, X_n$

$$RF = g(X_1, X_2, \ldots, X_n) \quad (1)$$

where the functional dependence $g(X_1, X_2, \ldots, X_n)$ is a computing model (i.e. idealization of reality). The failure functions $g(X_i)$ represent the condition of reliability (reserve) and can be expressed as a function of stochastic parameters. They may be defined as simple (e.g. for one section) or as complex structures, for more cross-sections (e.g. entire finite-element model). The probability of failure can be defined as:

$$P_f = P[R<E] = P[(R-E)<0] \quad (2)$$

where: P_f is given by the general expression

$$P_f = \iint \ldots \iint_{g(X)<0} f_X(X_1, X_2, \ldots, X_n)\, dX_1\, dX_2 \ldots dX_n. \quad (3)$$

4 Foundation plate in interaction with the subsoil

4.1 Deterministic approach

The analysed foundation plate is located under the residential building of the point type. It is made of concrete class C 25-30 with a thickness of 0.5 m. The material properties are - modulus of elasticity $E = 31$GPa, Poisson's ratio of 0.2. The subsoil consists of gravel G2 to a depth of 16.1 meters, then the rock continues. Standard gravel soil characteristics were taken, $E = 180$ MPa, $\upsilon = 0.2$.

Dimensions of the foundation plate are 12 × 12 m. Groundmass is modelled in sufficient width, length and depth of 36 × 36 × 16.1 meters. Loading of the foundation plate from the top of the building is in the point of columns, and it is calculated according to the relevant area and transfers application of load of eight floors (Fig. 1).

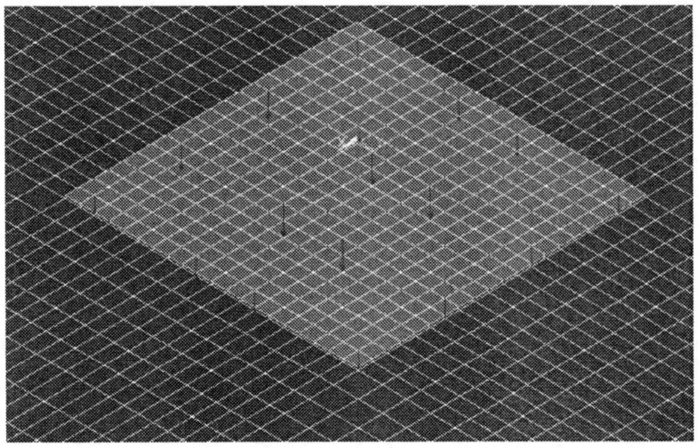

Fig. 1. Detail of loading of the foundation plate.

Structural model was created in the ANSYS software application. The plate was modelled from SHELL element 181, groundmass of elements Solid 185.

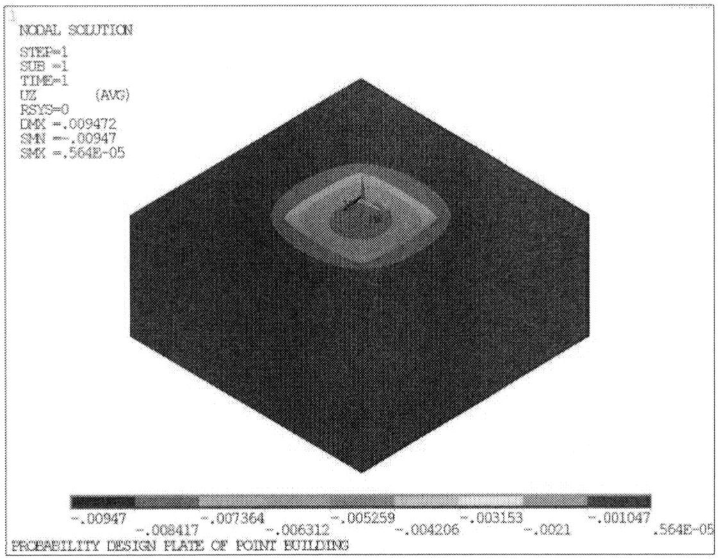

Fig. 2. Vertical deformation of groundmass and foundation plate.

Contact between the foundation plate and underlying groundmass in ANSYS can be defined in several ways: a) contact point - point b) point - area, c) surface-area. In this case, the contact surface was modelled through bilateral bond by element Contac52 with the stiffness of the surface in contact 6500 kN/m, coefficient of friction 0.5.

Normal stiffness is calculated from the stiffness of the blinding concrete, which depends on the modulus of elasticity, on the thickness and the area of the priming concrete. The foundation plate is assessed for the second limit state. Maximum deflection from determistic solution was 9.47 mm in the corner of the foundation plate, (Fig. 2).

4.2 Probabilistic approach

When applying probabilistic approach, the individual input values vary according to the tab. 1. Changing the geometrical characteristics of the plate is defined through $H = 0.5$ m, (plate thickness) and Hvar (change in thickness of the plate). The stiffness plate is determined through the modulus of elasticity $E_d = 31$ GPa, $\upsilon = 0.2$ and variable factor Evar_ and Mivar. The load is determined through uniformly distributed load q = 12 kN/m and forces F from 554 to 988 kN and qvar_ and Fvar_ variables.

Table 1. Probabilistic model of input variable.

Material	Variable parameter	Histogram	Mean value or Min. value	Stand. Dev. or Max.value
Concrete plate	EVAR_D	Gaus	1	0.05
	MIVAR_	Gaus	1	0.05
	HVAR_	Unif	0.99	1.01
Soil	EVAR_P	Gaus	1	0.05
	MIVAR	Gaus	1	0.05
Load	QVAR	Logl	1	0.1
	FVAR	Logl	1	0.1

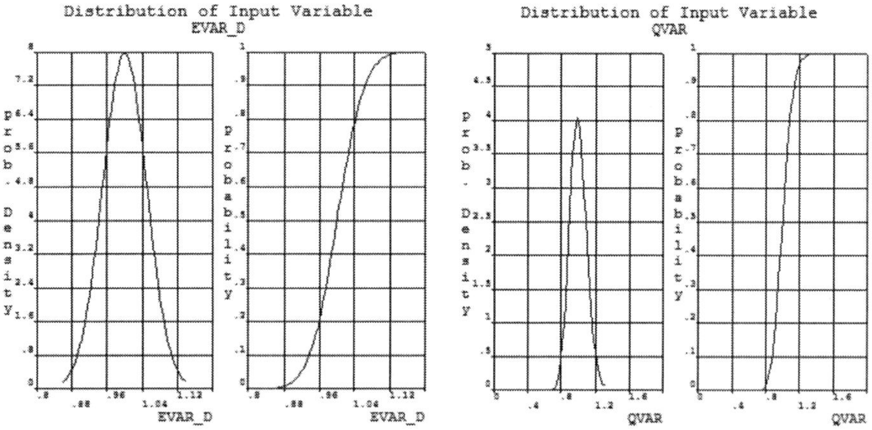

Fig. 3. PDF and CDF of Input Random Variable EVAR_D and QVAR.

The probability density functions - PDF and cumulative distribution functions- CDF of some input random variables are shown in Fig. 3 to Fig. 4.

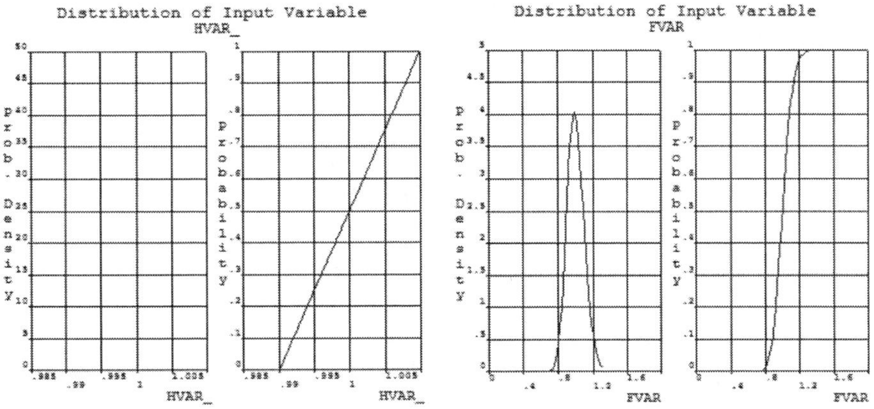

Fig. 4. PDF and CDF of Input Random Variable HVAR_ and FVAR.

During a probabilistic analysis, ANSYS executes multiple analysis loops to compute the random output parameters as a function of the set of random input variables. Histogram of ouput random variable PRIEH is shown in the Fig. 5.

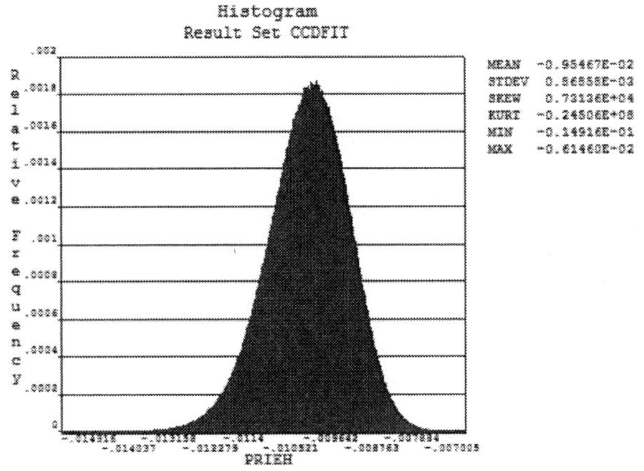

Fig. 5. Histogram of ouput random variable PRIEH.

Resulting from variability of input quantity, 45 simulations on the base of RSM method were realized. The probability of exceeding the limit deflection of plate structures was calculated using 5 million Monte Carlo simulations for 45 simulations of approximation method RSM on the structural FEM model. The probability that PRIEH is smaller than -1.1 mm is 5.456e-2. The probability is interpolated between:

PRIEH=-1.1000002 which has rank 272800 out of 5000000 samples,

PRIEH=-1.0999997 which has rank 272801 out of 5000000 samples.

The evaluation of the probabilistic sensitivities is based on the correlation coefficients between all random input variables and a particular random output parameter. Either Spearman rank order correlation coefficients or Pearson linear correlation coefficients may be used based on user's specifications. To plot the sensitivities of a certain random output parameter, the random input variables are separated into two groups: those that are significant (important) and those that are insignificant (not important) for the random output parameter. The sensitivity plots will only include the significant random input variables. The dominant impact to reliability of foundation displacement plate has the input parameter FVAR and QVAR (Fig.6).

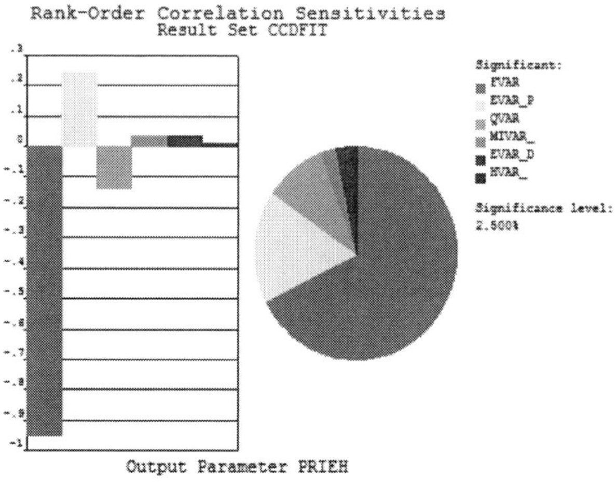

Fig. 6. Sensitivity result of ouput random variable PRIEH.

Conclusion

This article presented the static analysis of the residential building of point type rested on an elastic half-space, based on a deterministic and probabilistic analysis. Maximum deflection in the foundation gap obtained from deterministic analysis was 9.47 mm. The uncertainties at the level loads, geometry and material characteristics of the calculation model and numerical methods have been taken into account in the Monte Carlo simulation method based on the RSM. The probability of failure of the foundation plate in the ultimate deflection in the foundation gap is 5.456e-2 for maximum allowed deflection of 11 mm.

This contribution is the result of the research supported by the grant from VEGA Slovak Grant Agency Project No. 1/0544/15.

References

1. M. J. Tomlinson, *Foundation Design and Construction* (Pearson Education Ltd, England, 2001)
2. K. Kotrasova, E. Kormanikova, A case study on seismic behavior of rectangular tanks considering fluid - Structure interaction, International Journal of Mechanics **10**, (2016)
3. P. Marek, J. Brozzetti, M. Gustav, Probabilistic Assessment of Structures Using Monte Carlo Simulation Background, Exercises and Software, ITAM CAS, Prague, Czech Republic, (2003)
4. Martin Krejsa and Juraj Kralik, Probabilistic Computational Methods in Structural Failure Analysis Multiscale Modelling 6: 1550006 DOI:http://dx.doi.org/10.1142/ S17 56973715500067, (2015)

MATEC Web of Conferences **107**, 00059 (2017) DOI: 10.1051/matecconf/201710700059
DYN-WIND'2017

Design of the uniaxial shaker with variable stiffness

Ondrej Chlebo[1,*], *Martin* Sivý[1], *Miloš* Musil[1] and *Michal* Čekan[1]

[1] Slovak University of Technology in Bratislava, Nam. Slobody 17, 812 31 Bratislava, Slovakia

Abstract. During the vibration testing, various types of vibration drivers of different structures are used. Electrodynamic drivers are generally used for smaller excitation forces whereas for bigger excitation forces hydraulic drivers are used. The main disadvantage of drivers can be found in their size. One of the possibilities for size reduction may be a suitable engineering design. The paper deals with the design of a small uniaxial driver which uses an inertial driver as a source of the excitation force. The structure considers the requirements for stiffness changes of the whole system with the aim of available frequency range tuning.

1 Introduction

Every kinematic excitation system is composed of three basic elements. The first is the amplifier which is responsible for transferring sufficient energy into the working element, which is coincidentally the second part of the system. The third element is the fixing structure which is responsible for fixing the studied part and connecting it to the working element which must allow for motion in at least one dimension. The paper deals with a uniaxial kinematic driver. This driver is a system with one degree of freedom which is excited by the harmonic force $F_0 \sin\omega t$. This system is possible to represent by a simple motion equation in the form [1-4]:

$$m\ddot{x} + b\dot{x} + kx = F_0 \sin\omega t \quad (1)$$

where m is the mass of the system, b is the damping matrix, and k is the stiffness matrix. Equation (1) can be used to describe the natural frequency of a damped system, provided that the damping is relatively low, by:

$$\omega_d = \sqrt{\frac{k}{m} - \left(0.5\,\frac{b}{m}\right)^2} \quad (2)$$

From this equation (2) it can be seen that the natural frequency is dependent on three parameters. Other than the stiffness and damping properties of the system, mass has a significant effect on the frequency of the system. However unlike the other two parameters, the mass of the system is affected by the mass of the studied object as well. Therefore, it is desirable to account for system tuning within the structure of the driver.

* Corresponding author: ondrej.chlebo@stuba.sk

2 The structural design of the driver

The system TIRAvib TV 51165-IN was used as a source of kinematic excitation. This system is composed of a high-performance amplifier and inertial electrodynamic air cooled driver. This driver was fixed horizontally to the right side of the mounting table. Although the table must be fixed, it should allow for motion in a particular direction. One possibility is to use a pre-stressed linear bearing, this however does not allow for tuning of the system stiffness. Another possibility is to mount the vibrating table on springs such as leaf springs as seen in Fig. 1. Doing this allows for stiffness tuning by changing the length of the leaf spring such that the natural frequency is shifted away from the desired working frequencies of the system [5].

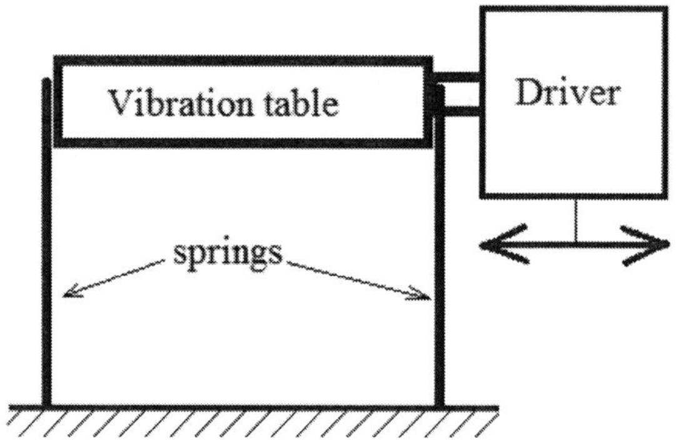

Fig. 1. Schematic model of the structural driver.

The driver is composed of mainly the vibrating table which is a welded steel ribbed plate with mounting grooves. Leaf springs are bolted to the table and fixed to a steel base. The leaf springs allow for significant movement in one direction which will correspond to the first resonant frequency of the vibrating system. Changing the length of the leaf springs changes the stiffness of the excited system.

3 FEA of the system

The first step was to design and analyse the vibrating plate of the driver. T80 profile was welded at its end to steel sheet. Through finite element analysis the structure was favourably reinforced to push the natural frequencies away from the measurement range. The final configuration of the table, as well as its corresponding natural frequencies, can be seen in Fig. 2. The first natural frequency of the vibrating table was 446 Hz. This frequency is varying depending on the number and placement of test piece mounting brackets as well as the character of the test piece itself.

 DOI: 10.1051/matecconf/201710700059

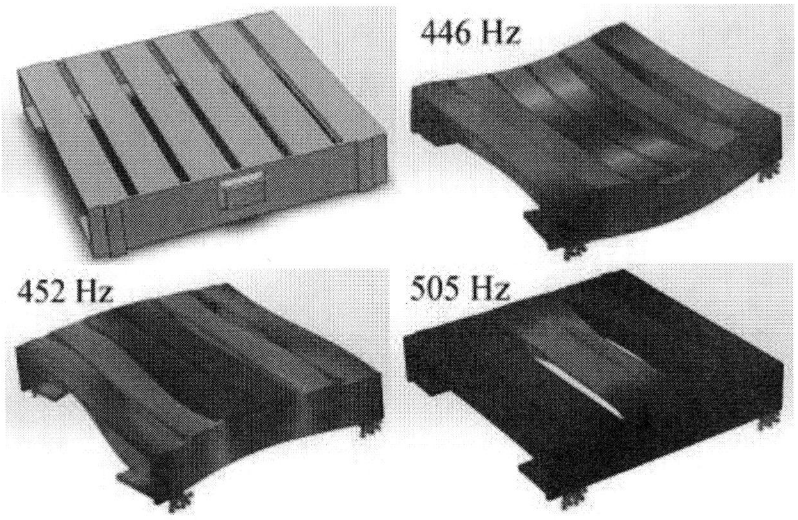

Fig. 2. Model of the vibration table and the first three modes.

The table was then mounted onto leaf springs. The natural frequency, and usable range, of the system is dependent on the length of these springs because they represent the least stiff member of the system. For this reason, analysis was performed with the table at the maximum and mid height of the springs. The results can be compared in Fig. 3. In this figure the effects of the leaf springs on the natural frequency of the system is clear, where the table acts as a rigid member at the first few natural frequencies of the system. This was done purposefully to satisfy the measurement range up to 100 Hz.

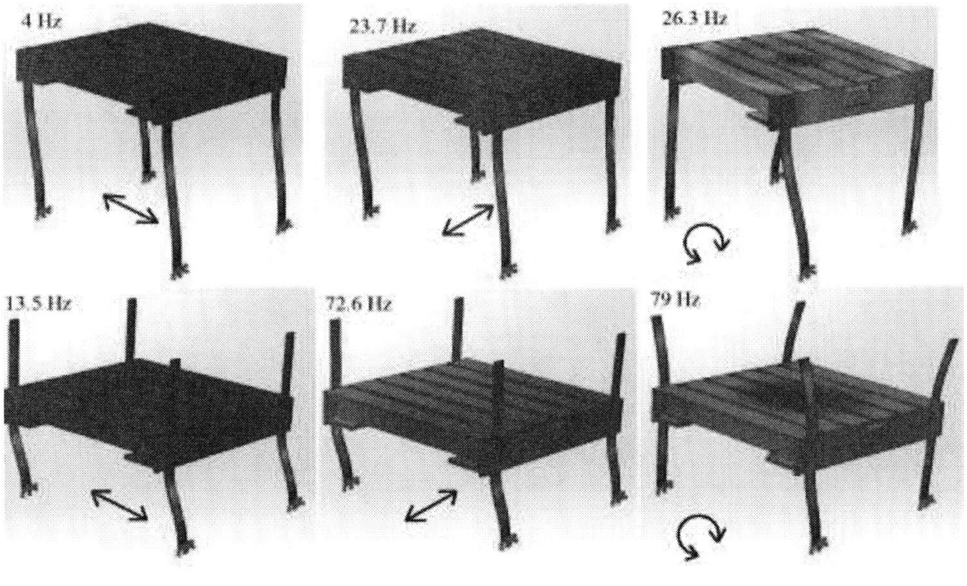

Fig. 3. Mode shapes of the vibrating system for the first three natural frequencies, top row of low stiffness and medium stiffness is the bottom row.

4 Experimental verification of the natural frequencies

After analysis, the vibrating table was built and tested. Modal analysis was performed to verify the numerical results (detail of the analysis is given in [6]). The modal analysis was performed using the measuring card B&K PULSE 3053-B-120 with accelerometer B&K 4534-B-001. The source was generated using an impact hammer PCB 086c03. Measured data was processed in the PULSE 19 software. Results can be seen in Fig. 4.

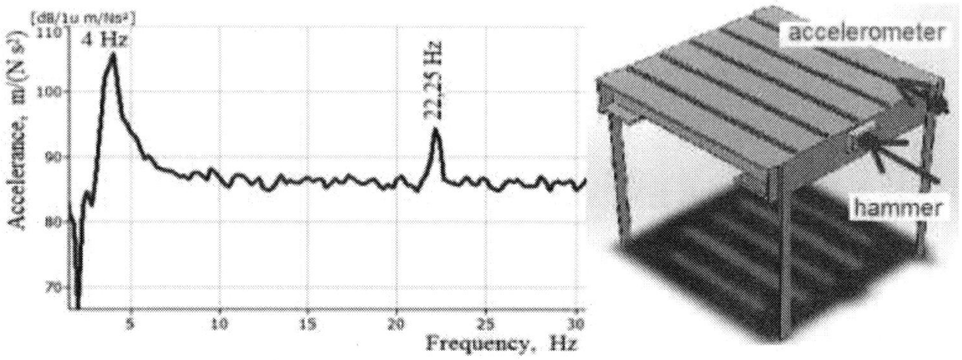

Fig. 4. Natural frequencies of the vibration system, first and third natural frequencies (the second natural frequency cannot be measured in this direction).

5 Conclusions

The purpose of this contribution was to design a vibrating system which could be utilized as a source of inertial excitation. During the design, the main requirement was to account for changing stiffness in order to accommodate the desired range of measurement. Based on numerical analysis, the table was optimized and built. Natural frequencies were confirmed using experimental modal analysis. In the low stiffness configuration, the first natural frequency was found at 4 Hz which corresponds to the numerical analysis. The third natural frequency was measured at 22.25 Hz whereas the numerical analysis computed the natural frequency at 23.7 Hz. This slight discrepancy, amongst other things, may be due to small variations in the mounting to the leaf springs.

This work was supported by the STU Grant scheme for Support of Young Researchers entitled *Design and Optimization of the Measuring Equipment for Vibration Tests*.

References

1. S. Braun, D. Ewins, S. S. Rao, *Encyclopedia of Vibration* (Academic Press, New York, 2002)
2. R. Brepta, L. Pust, F. Turek, *Mechanické kmitání* (Sobotáles, Praha, 1994)
3. M. Musil, *Základy dynamiky strojov s Matlabom* (STU, Bratislava, 2013)
4. M. Musil, M. Zuščík, *Aktívne a semiaktívne odpruženie vozidla* (STU, Bratislava,2012)
5. M. Naď, *Modification of Modal Characteristics of Vibrating Structural Elements*, (Scientific Monographs, Köthen,2010)
6. S. Žiaran, V. Kompiš, V. Mišun, J. Zajac, *Technická diagnostika* (STU, Bratislava, 2013)

The modification of a node-to-node algorithm for the modelling of beam connections in RFEM and SCIA using the explicit method

Hynek Štekbauer[1,*] and *Zbyněk* Vlk[1]

[1]Brno University of Technology, Faculty of Civil Engineering, Institute of Structural Mechanics, Veveří 331/95, 60200 Brno, Czech Republic

Abstract. The "node-to-node" (NTN) algorithm is one of the simplest and oldest contact discretization methods [1]. It is, therefore, used in cases where a more sophisticated method is not needed. Beam connections between 1D elements are a great example of an area of application of NTN discretization. Each connection is modelled by a hinge at the end of every connected 1D element. The explicit method is important with regard to many linear and non-linear dynamics problems. The central difference method is usually used in the Finite Element Method to solve explicit schemes. This and similar methods suffer significant errors if the time step exceeds the critical time step, which is directly derived from the element stiffness and tends towards zero as connected nodes get closer to each other. This is very inconvenient for NTN contacts, as with them there is usually only a small distance (or even zero distance) between the nodes and a rigid connection in a particular direction. This paper presents a modification that enables hinges that are rigid in a particular direction to be calculated using the explicit method.

1 Introduction

1.1 The central difference method

The basic equation of motion for dynamics in the central difference method at time t_n is

$$\mathbf{M}\ddot{\mathbf{u}}_n = \mathbf{P}_n - \mathbf{F}_n - \mathbf{C}\dot{\mathbf{u}}_n, \qquad (1)$$

where $\mathbf{M}$ is the diagonal mass matrix, $\mathbf{P}_n$ is the vector of external forces, $\mathbf{F}_n$ is the vector of internal forces, and $\mathbf{C}_n$ is the diagonal damping matrix [2]. Vectors $\mathbf{P}_n$ and $\mathbf{F}_n$ are usually represented by $\mathbf{R}_n$

$$\mathbf{R}_n = \mathbf{P}_n - \mathbf{F}_n. \qquad (2)$$

Acceleration $\ddot{\mathbf{u}}_n$ can be expressed as

$$\ddot{\mathbf{u}} = \mathbf{M}^{-1}(\mathbf{R}_n - \mathbf{C}\dot{\mathbf{u}}_{n-1}). \qquad (3)$$

* Corresponding author: stekbauer@seznam.cz

Due to the diagonal matrices $\mathbf{M}$ and $\mathbf{C}_n$, the system of equations can be decomposed into the solution of N independent equations for every degree of freedom in every node

$$\ddot{u}_n^j = (R_n^j - C_n^j \cdot \dot{u}_{n-1}) / m^j \tag{4}$$

$$\Delta\dot{u}^j = \ddot{u}_n^j \cdot \Delta t_n , \tag{5}$$

$$\dot{u}^j_{n+\frac{1}{2}} = \dot{u}^j_{n-\frac{1}{2}} + \Delta\dot{u}^j , \tag{6}$$

$$u^j_{n+1} = u^j_n + \dot{u}^j_{n+\frac{1}{2}} \cdot \Delta t_{n+\frac{1}{2}} , \tag{7}$$

where

$$\Delta t_{n+\frac{1}{2}} = \frac{\Delta t_n + \Delta t_{n+1}}{2} \tag{8}$$

and j takes values from 1 to N. This approach is fast and easy to parallelize. Rigidly supported nodes have prescribed deformations. The critical time step can be calculated from wave propagation velocity in beam c using the modulus of elasticity E and the density ρ

$$c = \sqrt{\frac{E}{\rho}} . \tag{9}$$

The Courant-Friedrichs-Levy stability criterion for the time needed for the wave to propagate through a beam of length l

$$\Delta \mathrm{t} \le \frac{l}{c} . \tag{10}$$

1.2 NTN model of a hinge

NTN discretization is a simple and stable discretization method for conforming meshes [3] and is used in RFEM and SCIA for hinge modelling. An NTN contact consists of two nodes (slave and master), and a transformation matrix that characterizes hinge direction (Fig. 1).

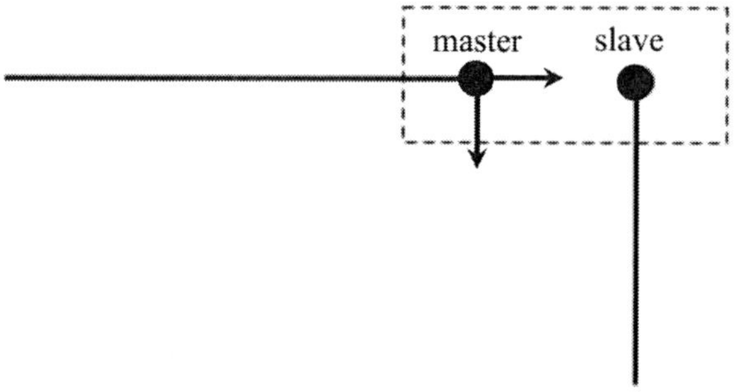

Fig. 1. The NTN contact (slave and master) and a transformation matrix that characterizes hinge direction.

The stiffness matrix of a hinge using the penalty stiffness method is given by

$$\mathbf{K} = \begin{bmatrix} +W & & & & & & -W & & & & & \\ & +W & & & & & & -W & & & & \\ & & +W & & & & & & -W & & & \\ & & & +W & & & & & & -W & & \\ & & & & +W & & & & & & -W & \\ & & & & & +W & & & & & & -W \\ & & & & & & +W & & & & & \\ & & & & & & & +W & & & & \\ & & & & & & & & +W & & & \\ & & sym. & & & & & & & +W & & \\ & & & & & & & & & & +W & \\ & & & & & & & & & & & +W \end{bmatrix} \tag{11}$$

where W is the penalty weight for a rigid directions of a hinge. The penalty weight W can be replaced in a required degree of freedom

$$\mathbf{K} = \begin{bmatrix} +A & & & & & & -A & & & & & \\ & +W & & & & & & -W & & & & \\ & & +W & & & & & & -W & & & \\ & & & +W & & & & & & -W & & \\ & & & & +W & & & & & & -W & \\ & & & & & +W & & & & & & -W \\ & & & & & & +A & & & & & \\ & & & & & & & +W & & & & \\ & & & & & & & & +W & & & \\ & & sym. & & & & & & & +W & & \\ & & & & & & & & & & +W & \\ & & & & & & & & & & & +W \end{bmatrix} \tag{12}$$

where A is a stiffness characteristic for direction u_x. The transformation matrix of a hinge

$$\mathbf{T}_R = \begin{bmatrix} \mathbf{T} & \mathbf{0} \\ \mathbf{0} & \mathbf{T} \end{bmatrix} \tag{13}$$

is composed of direction rotation matrices $\mathbf{T}$ between the global and local coordinate system

$$\mathbf{T} = \begin{bmatrix} C_{x,X} & C_{x,Y} & C_{x,Z} \\ C_{y,X} & C_{y,Y} & C_{y,Z} \\ C_{z,X} & C_{z,Y} & C_{z,Z} \end{bmatrix} \tag{14}$$

where $C_{i,J}$ means the cosine between axes i and J. The distance between nodes is taken into account in the transformation matrix extended by the eccentricity matrix $\mathbf{E}$

$$\mathbf{T}_{RE} = \begin{bmatrix} \mathbf{T} & \mathbf{TE} \\ \mathbf{0} & \mathbf{T} \end{bmatrix}. \tag{15}$$

Eccentricity matrix

$$\mathbf{E} = \begin{bmatrix} 0 & e_z & -e_y \\ -e_z & 0 & e_x \\ e_y & -e_x & 0 \end{bmatrix} \tag{16}$$

consists of eccentricities and is skew symmetric.

Generally, there are three types of hinges that are used in structures. First, there is a rigid hinge that simulates the infinitely stiff connection of two nodes. A rigid hinge alone has no meaning in the Finite Element Method because this type of connection can easily be modelled by elements with a common node, but it is needed if only some selected degrees of freedom are not rigid. The second type is a linear hinge with given stiffness in a particular direction. The third and most general type is the nonlinear hinge, whose behaviour can be described by a function or by the dependence of one direction on another direction (i.e. friction).

2 The modification of the NTN algorithm

In this section, the modification needed for the presented NTN hinge model to work properly in the explicit method is described. Upper script j is used to label the first and second node of the contact and takes the values 1 and 2.

$$\mathbf{R}_G^j = \begin{pmatrix} r_{ux} \\ r_{uy} \\ r_{uz} \\ r_{\varphi x} \\ r_{\varphi y} \\ r_{\varphi z} \end{pmatrix} \tag{17}$$

$$\mathbf{R}_L^j = \mathbf{T}_R \cdot \mathbf{R}_G^j \tag{18}$$

$$\mathbf{R}_{DUM}^j = \mathbf{R}_L^j \tag{19}$$

For all rigid directions rd of $\mathbf{R}_L^j$

$$\begin{aligned} \mathbf{R}_L^1(rd) &= \mathbf{R}_L^1(rd) + \mathbf{R}_{DUM}^2(rd) \\ \mathbf{R}_L^2(rd) &= \mathbf{R}_L^2(rd) + \mathbf{R}_{DUM}^1(rd) \end{aligned} \tag{20}$$

where rd can take values between 1 (for r_{ux}) and 6 (for $r_{\varphi z}$) inclusive.

$$\mathbf{R}_G^j = \mathbf{T}_R^T \cdot \mathbf{R}_L^j \tag{21}$$

Similar operations are performed for masses

$$\mathbf{M}_L^j = \begin{bmatrix} m_{Lx} & & & & & \\ & m_{Ly} & & & & \\ & & m_{Lz} & & & \\ & & & m_{L\varphi x} & & \\ & & & & m_{L\varphi y} & \\ & & & & & m_{L\varphi z} \end{bmatrix} \tag{22}$$

$$\mathbf{M}_L^j = \mathbf{T}_R \cdot \mathbf{M}_G^j \cdot \mathbf{T}_R^T . \tag{23}$$

Matrix $\mathbf{M}_L^j$ is not unconditionally diagonal, which is why it needs to be diagonalized if needed.

$$\mathbf{M}^{j}_{DUM} = \mathbf{M}^{j}_{L} \tag{24}$$

For all rigid directions rd of $\mathbf{M}^{j}_{L}$

$$\begin{aligned} \mathbf{M}^{1}_{L}[rd,rd] &= \mathbf{M}^{1}_{L}[rd,rd] + \mathbf{M}^{2}_{DUM}[rd,rd] \\ \mathbf{M}^{2}_{L}[rd,rd] &= \mathbf{M}^{2}_{L}[rd,rd] + \mathbf{M}^{1}_{DUM}[rd,rd] \end{aligned} \tag{25}$$

where rd can take values between 1 (for m_{ux}) and 6 (for $m_{\varphi z}$) inclusive. The masses are then transformed back to the global coordinate system and diagonalized if needed

$$\mathbf{M}^{j}_{G} = \mathbf{T}_{R}^{T} \cdot \mathbf{M}^{j}_{L} \cdot \mathbf{T}_{R} \,. \tag{26}$$

After such a modification, deformations are calculated using the central difference method (Eq. 4-8). Up until this point, only some of the necessary modifications have been presented. These are modifications which are good for hinges that are linear or rigid and without rigid support in any direction. A special treatment needs to be added for supported hinges and nonlinear hinges that should act in a rigid manner under certain circumstances. Deformations calculated by the central difference method are transformed to the local coordinate system of the hinge and examined for these two special cases.

$$\mathbf{U}^{j}_{G} = \begin{pmatrix} u_x^{\,j} \\ u_y^{\,j} \\ u_z^{\,j} \\ \varphi_x^{\,j} \\ \varphi_y^{\,j} \\ \varphi_z^{\,j} \end{pmatrix} \tag{27}$$

$$\mathbf{U}^{j}_{L} = \mathbf{T}_{R} \cdot \mathbf{U}^{j}_{G} \tag{28}$$

$$\mathbf{U}^{j}_{DUM} = \mathbf{U}^{j}_{L} \tag{29}$$

At first, the identification of supported hinge nodes in a particular direction is performed. Then, deformations from the rigidly supported node can be assigned to the other node in every rigid direction rd of a hinge. If the second node is rigidly supported

$$\mathbf{U}^{1}_{L}(rd) = \mathbf{U}^{2}_{DUM}(rd) \,, \tag{30}$$

and vice versa if the first node is rigidly supported

$$\mathbf{U}^{2}_{L}(rd) = \mathbf{U}^{1}_{DUM}(rd) \,. \tag{31}$$

After that, nonlinear hinges described by a function are evaluated. The user can define the limit stiffness which is taken as rigid.

A rigid branch of a function that is stiffer then the limit stiffness can be seen in Fig. 2. If any displacement d between contact nodes exceeds the limit value $d_{\lim}$ ($d_{\lim}$ = 1 mm for Fig. 2), where d is the x-th element of a vector $\mathbf{D}$

$$\mathbf{D} = \mathbf{U}^{2}_{L} - \mathbf{U}^{1}_{L} \,, \tag{32}$$

then local displacements are recalculated

$$\begin{aligned} \mathbf{U}^{1}_{L}(x) &= \mathbf{U}^{1}_{L}(x) \cdot \frac{d_{\lim}}{d} \\ \mathbf{U}^{2}_{L}(x) &= \mathbf{U}^{2}_{L}(x) \cdot \frac{d_{\lim}}{d} \end{aligned} \,, \tag{33}$$

and transformed back to the global coordinate system

$$\mathbf{U}^{j}_{G} = \mathbf{T}_{R}^{T} \cdot \mathbf{U}^{j}_{L} \tag{34}$$

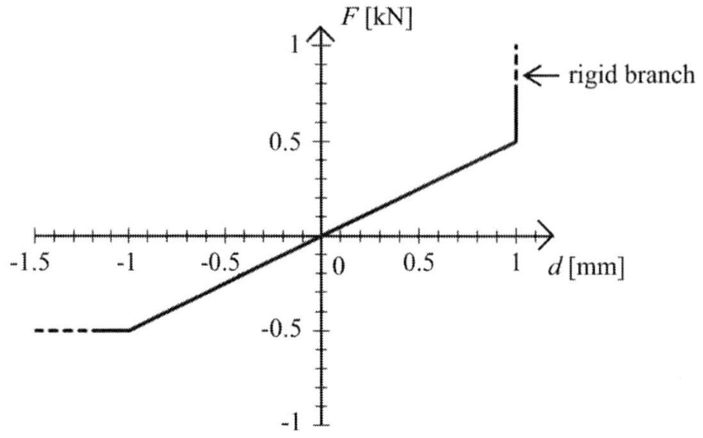

Fig. 2. Nonlinear hinge function with a rigid branch (a branch that is stiffer than the limit stiffness) showing the dependency of force F on a certain displacement u. The similar dependency of moment on rotation can be prescribed.

Accelerations and velocities need to be recalculated for modified displacements $\mathbf{U}_G^j$ using Eq. 5-7 backwards.

$$\dot{u}^j_{n+\frac{1}{2}} = \frac{u^j_{n+1} - u^j_n}{\Delta t_{n+\frac{1}{2}}} \tag{35}$$

$$\Delta\dot{u}^j = \dot{u}^j_{n+\frac{1}{2}} - \dot{u}^j_{n-\frac{1}{2}} \tag{36}$$

$$\ddot{u}_n^{\ j} = \frac{\Delta\dot{u}^j}{\Delta t_n} \tag{37}$$

3 Conclusions

The modification of an NTN algorithm for the modelling of beam connections in the explicit method is proposed. The modification is divided into two sections, the first of which involves the recalculation of nodal forces and masses based on the rigid local directions of a contact before the solution of equations. The second part is needed for the supported contacts and nonlinear contacts that act in a rigid manner under certain circumstances. Deformations calculated by the central difference method are transformed to the local coordinate system of the contact and modified to fulfil contact conditions and prevent solution instability at the same time.

This research was partially supported by the project FAST-J-17-4720.

References

1. G. Zavarise, L. De Lorenzis, Comput. Method Appl. M. **198**, 3428 (2009)
2. K. J. Bathe, *Finite Element Procedures* (Prentice Hall, New Jersey, 2009)
3. V. A. Yastrebov, *Numerical Methods in Contact Mechanics* (ISTE/Wiley, Oxford, 2013)

MATEC Web of Conferences **107**, 00061 (2017) DOI: 10.1051/matecconf/201710700061
DYN-WIND'2017

Influence of different types of trams on human perception of vibrations in buildings

Alicja Kowalska-Koczwara[1,*] and *Krzysztof* Stypuła[1]

[1] Cracow University of Technology, Institute of Structural Mechanics, 31-155 Kraków, Warszawska 24 St., Poland

Abstract. In an increasingly urbanized areas transport vibration source are closer and closer to the receiver which is a building. At the same time residents requirements for comfort in rooms are higher. In this paper influence of transport vibrations from various types of trams on human perception was tested and analysed. Measurements were made in Cracow on 11 buildings located near tramway. Human perception were investigated according to ISO and Polish standard. Basic RMS method and additional VDV method were applied because of high crest factor of recorded signal. Results of both methods were compared.

1 Introduction

Buildings, especially in city centers, are subjected to dynamic influences from many dynamic sources: transport, mining activity [1] or seismic [2]. Human perception of vibrations is defined as the effect of mechanical vibration on human body. During normal daily live human body is exposed to different types of vibrations: hand-arm vibrations, whole body vibrations and vibrations transmitted through the soil to the building structure. The first two types of vibration are mainly connected with working area for example with such occupations as bus drivers - effect of whole body vibration [3, 4] or builders, carpenters, joiners – effect of hand-arm vibration [5]. The third one is connected with transport vibrations especially in the city centers [6, 7]. In this article tramway vibration transmitted through the soil will be considered and influence of different type of trams on human perception of vibration will be investigated. There are two evaluation methods mainly used in the subject of human exposure to vibrations in buildings: root mean square method (RMS) and vibration dose value method (VDV). ISO standard [8] defines RMS method as a basic method of assessment but notes that sometimes additional methods are required. One of additional method which could be considered according ISO standard [8] is VDV method. VDV method is recommended in situations were high value of crest factor exist. Crest factor is defined as follows:

$$CF = \frac{|a_w|_{PEAK}}{a_{wRMS}} \tag{1}$$

* Corresponding author: akowalska@pk.edu.pl

where: a_{wPEAK} – peak value of weighted acceleration function,
a_{wRMS} – RMS value of weighted acceleration function.

According to ISO standard [8] when CF is equal 9 or higher VDV and RMS method should be both considered, because VDV method is more sensitive to peaks in recorded signal. Below this CF value RMS method is good enough. In Australian standard [9] regulations to the crest factor are more stricter. If CF is below 6 RMS method should be considered, if CF is in the interval of 6 and 9 both VDV and RMS method should be applied and when CF value is higher than 9 only VDV value should be calculated. British guidelines abandoned using RMS method at all and in Great Britain only VDV method is considered regardless to CF value [10].

VDV sensitivity for peaks in recorded signal comes from differences in both formulas for RMS and for VDV method. RMS method averages acceleration values in duration time and is defined as:

$$a_w = \left[\frac{1}{T} \int_0^T a_w^2(t)dt \right]^{\frac{1}{2}} \tag{2}$$

where: $a_w(t)$ – is weighted acceleration as a function of time [m/s^2];
T – is the duration time of measurement [s].

VDV method uses fourth power instead of the second power as is used in RMS and it is defined as follows:

$$VDV = \left[\int_0^T a_w^4(t)dt \right]^{\frac{1}{4}} \tag{3}$$

In this paper both methods are used for evaluation of human perception to vibrations and results of these methods are compared.

2 Measurement methodology

Vibration measurements were carried out before the modernization of the tramway in Krakow city centre. For measurements 11 representative buildings were selected in the so called zone of vibration influences. Ranges of zones of vibration influences are summarized in Table 1.

Table 1. Range of zones of dynamic influences.

Source of vibration	Range of zone [m]
Railway	25-50
Tramway, Road	15-25
Line of a shallow underground	40

Buildings were chosen according to following criteria: building structure, rooms destination, the distance of the building from the track and the nearest road, possibility of access to the

DOI: 10.1051/matecconf/201710700061

building and to the rooms localized as close as possible to the source of vibration. Measurement and analysis of the results was carried out using specialized sets of measurement equipment adapted to the measurement and analysis of low frequency vibrations. During measurements vibration acceleration was recorded in the range of 0.15 Hz to 100 Hz (range of vibrations influence on people in buildings).

Vibration measurement took place simultaneously at all measurement points located in a single building. Acceleration was recorded in three directions: two horizontal (x, y) and one vertical (z). Measurement points were located according to Polish standards [11, 12] as follows:

- one measurement point located on the foundation wall at the ground level,
- at least one measuring point located in the middle of the floor of room intended for human residence (office or residential)

Example of location of measurement point for evaluating the influence of vibration on humans (located in the middle of the floor) is shown in Fig. 1.

Fig. 1. Location of measurement point for evaluating the influence of vibration on humans.

3 Analysis of recorded signal

Analysis of influence of vibration on humans in buildings were made using two evaluation methods: RMS and VDV according to formulas (2) and (3). RMS analysis was made in one third octave bands and is mainly presented as bar graph as on example shown in Fig. 2.

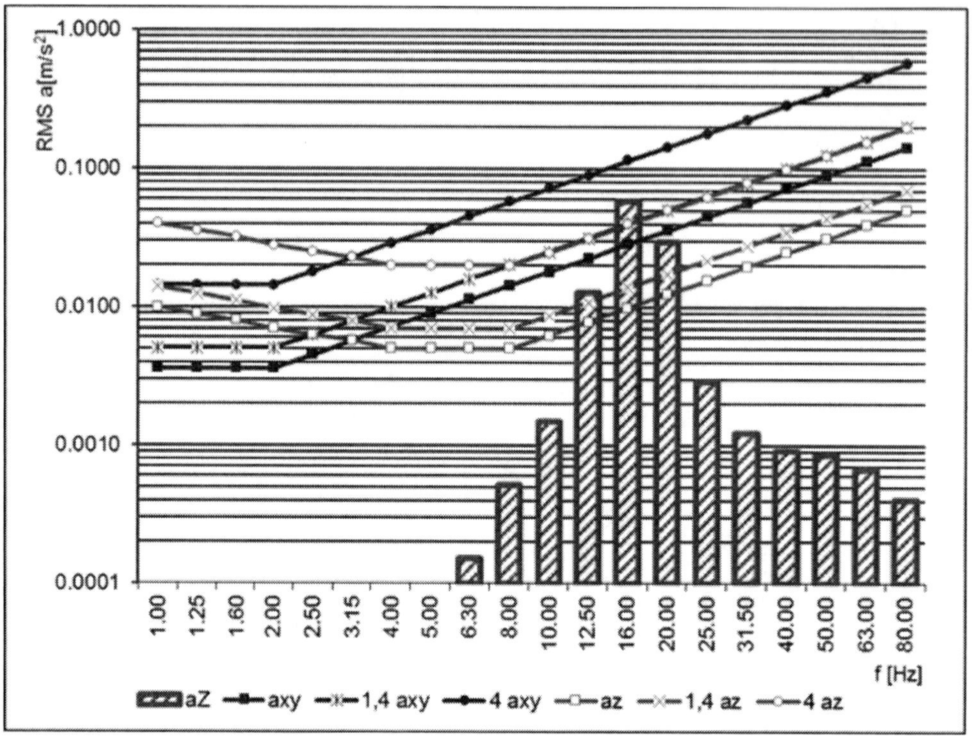

Fig. 2. The most unfavourable results of analyses of the influence of vibration on humans.

On the above figure there are some reference lines which according to ISO standard [8] and Polish standard [12] are interpreted as follows:

- axy – human perception threshold of vibration in x and y directions,
- 1.4 axy – comfort level during a night in residential buildings in x and y directions,
- 4 axy - comfort level during a day in residential and office buildings in x and y directions,
- az – human perception threshold of vibration in z direction,
- 1.4 az – comfort level during a night in residential buildings in z direction,
- 4 axy - comfort level during a day in residential and office buildings in z direction.

As a result of VDV procedure one value is obtained. This value should be than compare with table of thresholds (Fig. 3 acc. [10]).

Table 2. Ranges of VDV values.

Room destination	**Low probability of adverse comments**	**Adverse comments possible**	**Adverse comments probable**
Residential buildings – 16h day	0.2 – 0.4	0.4 – 0.8	0.8 -1.6
Residential buildings – 8h night	0.1 – 0.2	0.2 – 0.4	0.4 – 0.8
Office buildings – 16h day	0.4 - 0.8	0.8 – 1.6	1.6 – 3.2
Workshops – 16h day	0.8 – 1.6	1.6 – 3.2	3.2 – 6.4

Listed in Table 2 *VDV* values refer to whole day or whole night human exposure to vibrations, not to one dynamical episode. That is why having VDV_i values for one episode, whole day or night exposure for N episodes should be calculated according to formula (4):

$$VDV_{total} = \left(\sum_{i}^{N} VDV_i^4 \right)^{\frac{1}{4}} \tag{4}$$

To compare RMS and VDV analysis results, to RMS analysis so called *WODL* coefficient (in English, the human vibration perceptivity ratio – HVPR), proposed in [13] should be applied:

$$WODL = \max(\frac{aRMS}{az}) \tag{5}$$

where:

aRMS – acceleration RMS value obtained from analysis;

az – acceleration RMS value equivalent to the threshold for the perception of vibration in a z-direction in the same 1/3 octave band in which aRMS was obtained.

This coefficient is very useful because it describes if and how much human perception threshold is exceeded.

4 Results of analysis

Analysis was made using RMS and VDV method, for comparison VDV_{total} and *WODL* values were calculated. To show differences between different types of trams the results of analysis are divided not only to methods of evaluation but also to various trams. Episodes of maximum values were chosen for comparison. Because measurements were made in daytime only daily exposure to vibration was compared. In VDV analysis according to timetable of trams some assumption was made during daytime tram passes for every 5 minutes, which gives 192 tram passages. Summary of results of VDV analysis is listed in Table 3.

Table 3. Results of VDV analysis.

No.	Building type		Tram type				
			105 N	EU8N	GT8S	E1-C3	GT6
1	masonry four storey residential building	*WODL* max		5.87 (*f*=16Hz)			
		VDV	0.343	**0.793**	0.268	0.127	-
2	masonry two storey residential building	*WODL* max		10.97 (*f*=8 Hz)			
		VDV	**0.457**	0.167	0.340	0.216	0.219
3	Reinforced concrete 10-storey office building	*WODL* max	1.95 (*f*=10Hz)				
		VDV	**0.228**	0.070	0.072	0.211	0.058
4	masonry five storey residential building	*WODL* max					1.34 (*f*=50Hz)
		VDV	0.442	0.875	-	0.762	**1.159**

5	masonry two storey residential building	*WODL* max	3.22 (f=20Hz)				
		VDV	**0.750**	0.455	-	0.452	0.339
6	masonry three storey residential building	*WODL* max	7.61 (f=10Hz)				
		VDV	**0.776**	0.363	0.293	0.463	0.425
7	Reinforced concrete three storey office building	*WODL* max	1.40 (f=20Hz)				
		VDV	**1.098**	0.452	0.317	0.475	0.355
8	Reinforced concrete 8-storey residential building	*WODL* max	1.92 (f=12.5Hz)				
		VDV	0.199	**0.264**	-	0.117	0.068
9	masonry two storey residential building	*WODL* max	2.51 (f=8Hz)				
		VDV	**0.685**	0.102	0.223	0.366	-
10	masonry two storey residential building	*WODL* max				14.36 (f=10Hz)	
		VDV	0.907	1.188	0.287	**1.952**	0.359
11	masonry 6-storey residential building	*WODL* max	0.81 (f=20Hz)				
		VDV	**0.291**	-	-	0.209	0.134

As could be seen from above Table 3 the highest values of VDV_{total} are obtained for almost the same type of tram for each building excluding 8 storey reinforced concrete residential building (No. 8) in which differences in highest values of *VDV* are close to each other and excluding 2 storey masonry residential building (No. 2).

RMS analysis suggests that for almost every building (excluding the last one, No. 11) there is although one episode for which human perception threshold is exceeded. Meanwhile in VDV analysis similarly as in RMS analysis for only one building threshold is not exceeded, but this time it is building No. 3. There is simple explanation of this situation. In RMS analysis threshold is not connected with destination of the room, only comfort level is. In VDV analysis all ranges are connected to the destination of the room.

Analysing comfort level for both analysis there could be seen some similarities and differences. For building No. 1 and 10 results of both analysis are the same – comfort level is exceeded or is on the edge. But there are differences for buildings No. 2, 4 and 6. The highest difference is for 2 storey masonry building No. 2. *VDV* vale is in the middle range while *WODL* value is equal 10.97 which means that comfort level highly exceeded. The reverse situation occurred for building No. 4 in which *VDV* suggests that comfort is exceeded while *WODL* value is equal 1.34. For building No. 6 *VDV* is close to comfort level so both analysis can be interpreted as close to each other.

5 Conclusion

The article presents two methods of assessment and their application in analysis of influence of vibration on humans. It describes the procedures which should be used in such evaluations.

Differences and similarities between both methods of assessment are described and are shown on an exemplary analysis.

Analysis presented in this paper is based on measurements results that were carried out in 2013 in Krakow. Mostly measurements of human exposure to transport vibration concern on passanger or drivers comfort not on vibrational comfort for ground born vibration transmitted to buildings that can influence on humans in buildings. An interesting conclusion can be found in [14] in which RMS analysis and KB_F analysis (from German standard [15]) are compared, unfortunately conclusions are made based on numerical results.

Regarding results from analysis of influence of different types of trams which is based on measurements in this article some conclusion can be drawn:

- results from evaluation of human perception to vibration made using RMS and VDV analysis are mostly close to each other, but there are some vibrational episodes in which differences are significant,
- there are some situation especially of high value of crest factor in which is good to made evaluation of human exposure to vibration using basic RMS method and one of additional methods (for example VDV) given in ISO standard [8],
- in most analyzed dynamical episodes the highest RMS and WODL values occurred for tram type 105N, which should be expected because this is an old type of tram and very used,
- the highest value both RMS and WODL of all measured episodes occurred for tram type E1-C3 which is also an old tram called « nuremberg » but nowadays this type of tram is phased out from Krakow tram network,
- exceeding occurred also for tram type EU8N - there are upgraded wagons withdrawn from use on the light metro line U6 in Vienna.

References

1. T. Tatara, F. Pachla, P. Kuboń. Bull Earthquake Eng (2016)
2. L.A. Padrón, J.J. Aznárez, O. Maeso, Soil Dyn and Earthquake Eng, **29**, (2009)
3. R.P. Blood, J.D. Ploger, M.G. Yost, R.P. Ching, P.W. Johnson, J. Sound Vib. **329** (2010)
4. O. Thamsuwan, R. P. Blood, R. P. Ching, L. Boyle, P. W. Johnson, Int J Ind Ergonom **43** (2013)
5. K. T. Palmer, M. J. Grifin, H. Syddall, B. Pannett, C. Cooper, D. Coggon. Am J Ind Med **39** (2001)
6. D. C. Waddington, J. Woodcock, E. Peris, J. Condie, G. Sica, A. T. Moorhouse, A. Steele, The J. of the Acous. Soc. of Am. **135** (2014)
7. H. Hao, T.C. Anga , J. Shen. Build Environ **36** (2001)
8. ISO 2631-1, Guide to the evaluation of human exposure to whole body vibration. Part 1- General requirements, (2003)
9. AS 2670.2: Evaluation of human exposure to whole-body vibration - Continuous and shock-induced vibration in buildings (1 Hz to 80 Hz), (1990)
10. BS 6472-1, Guide to evaluation of human exposure to vibration in buildings, Part 1: Vibration sources other than blasting, (2008)
11. PN-85/B-02170: Ocena szkodliwości drgań przekazywanych przez podłoże na budynki / Evaluation of the harmfulness of building vibrations due to ground motion, (1985)
12. PN-88/B-02171: Ocena wpływu drgań na ludzi w budynkach / Evaluation of vibrations influence on people in buildings, (1988)

13. J. Kawecki, K. Stypuła *Ensuring vibrational comfort for people in buildings exposed for transport impact* (in Polish). (Publishing House of Cracow University of Technology, Cracow 2013)
14. G. Kouroussis, N. Pauwels, P. Brux, C. Conti, O. Verlinden, Sci Total Environ 482-483 (2014)
15. DIN 4150-2, Structural vibration, Part 2: Human exposure to vibration in buildings, (1999)

MATEC Web of Conferences **107**, 00062 (2017) DOI: 10.1051/matecconf/201710700062
DYN-WIND'2017

Identification of signal characteristics in the analysis of free-field vibration caused by trains

Barbara Kożuch[1,*] and *Tadeusz* Tatara[1]

[1]Institute of Structural Mechanics, Cracow University of Technology, Poland

Abstract. The paper deals with a selection of free-field vibration measurement results. The study is based on the vibration measurements. Vibration acceleration records and their dependence on the position of the accelerometer are shown. The study presents the results obtained in time domain and frequency domain (FFT). Nevertheless, the main aim of the paper is to compare the results of analysis which use the time – frequency domain. A Short-Term Fourier Transform (STFT) and Continuous Wavelet Transform (CWT) are used. The use of new methods to assess the dynamic characteristics is to validate the usefulness of proposed technique for non-stationary signals.

1 Introduction

Trains passages generate ground vibration, and the review of the worldwide literature (e.g. [1-5]) indicates that these vibrations are increasing engineering challenge and become a global concern. The problem is not unknown in the well-developed countries. Therefore, transport plays an important role in shaping the quality of the environment.

For this reason, large scale national testing of railway vibration took place in Poland at the railway line No. 4. For the use of the newly purchased Pendolino trains, homologation tests were conducted. Selected for testing train Pendolino was running at predetermined speeds. Pendolino and InterCity (the train with the locomotive ES64U4 Husarz) excited free-field vibrations. Detailed information about the study can be found at [6].

The paper presents a selection of free-field vibration measurement results in the selected point (in three directions, defined as P-04x for horizontal (perpendicular to the track), P-05y (parallel to the track) and P-06z vertical downward). A PCB Piezotronics piezoelectric transducers and a SCADAS Mobile LMS International analyser were used to perform in situ measurements. Accelerometers were fixed on the ground surface, 2.3 m below the railhead level, next to the railway embankment. The schematic arrangement of the sensors is shown in Fig. 1. Results of free-field vibrations caused by two types of trains running at the speed of 160 km/h were analysed to make comparison that is more tangible.

* Corresponding author: kozuchbm@gmail.com

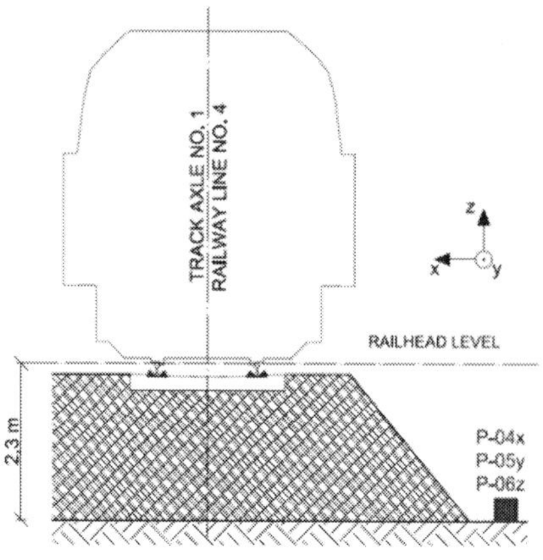

Fig. 1. Measuring polygon – schematic arrangement of sensors.

2 Experimental analyses

2.1 Fast Fourier Transform

The transition from the time domain (time histories ground surface acceleration) to the frequency domain was performed using Fast Fourier Transform (FFT). Transform was implemented for frequencies of up to 200 Hz. Fig. 2 presents each FFT of the acceleration record at ground surface (the directions x, y, z), located at 2.3m below railhead level for InterCity train. The data for Pendolino are shown in Fig. 3, in analogy. The largest amplitude of the vibrations for InterCity train occurs in the y direction (P-05y) and it is 9.5cm/s^2 at frequency 68Hz. The significant frequency band for the horizontal components (P-04x and P-05y) is frequency band of 50 – 115 Hz. For the vertical component the significant frequency band is narrower and it is in the range of 50 – 90 Hz. In the horizontal plane, the Pendolino train induces slightly lower parallel vibrations (P-05-y) but higher perpendicular ones (P-04x). The dominant frequency band is similar to the previously discussed one.

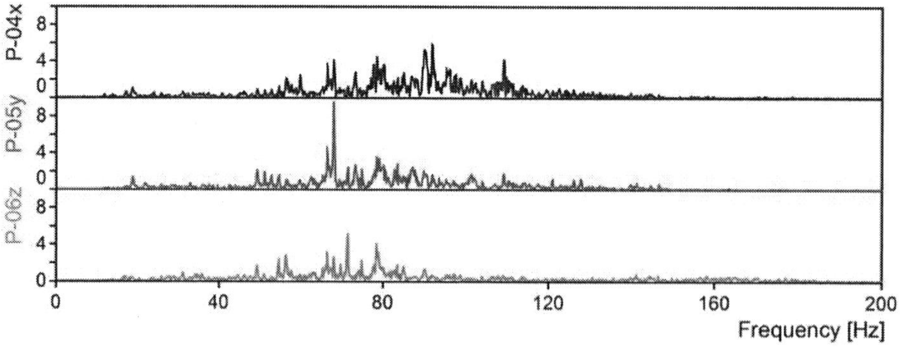

Fig. 2. The value of FFT for InterCity train – sensors P-04x, P-05y, P-06z.

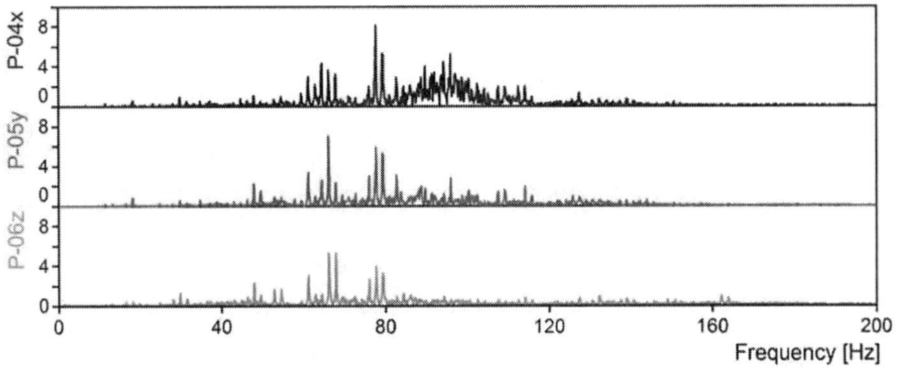

Fig. 3. The value of FFT for Pendolino train – sensors P-04x, P-05y, P-06z.

2.2 Analysis in time and frequency domain

FFT plays an important role in vibration analysis. It gives us the messages about the nature of the excited vibration, but the main drawback is that FFT does not include changing in time duration. Proceeding from the time domain to the frequency domain, we lose the information about the exact duration of the event and the beginning of individual excitations of vibrations during the course, which is particularly important in the decline of non-stationary vibration sources. Therefore, in further elaboration of the results it was decided to use the Short-Term Fourier Transform (STFT). The advantages of introduce analyzes which obtain data in the time - frequency domain are shown in reference [7].

The next step in the signal analysis is the Continuous Wavelet Transform (CWT). CWT is relatively new method and it was developed to improve the performance of STFT. The information and proposal to apply this analysis are available in references [8-10].

Signal analysis using STFT and CWT are presented for transducers P-04x, P-05y, P-06z in Figs. 4-6, respectively. It is used a scale - normalized dB to compare both analysis. The value of the amplitudes [cm/s^2] generated by the different runs was obtained from the FFT. In this case, the share of component in time - frequency domain was interesting, and therefore we decided on such a scale. Zero indicates maximum obtained amplitude. Other values are related to the reference value. In each figures the first columns are graphs obtained by applying the CWT analysis.

The first row in Fig. 4 shows the analysis for the Pendolino train. The eight points of excitation is clearly visible during the passage of a train. It has been associated with eight groups of wheels' sets. Schematic arrangement of bogies is shown in Fig. 7. The first and the last group contain half the numbers of wheels, therefore induced vibrations are reduced there. The discussed distribution in time domain is shown better in CWT graph. However, with such selected parameters, distribution in frequency domain is better illustrated in the STFT graph. The bottom row presents graphs for InterCity train. This is a typical arrangement for that train type - locomotive plus carriages. We can observe the largest amplitude for the first element (locomotive), and much less for the rest of the train (carriages). As above, in this case, CWT better represents characteristic of acceleration record in time domain but STFT better refers to frequency domain.

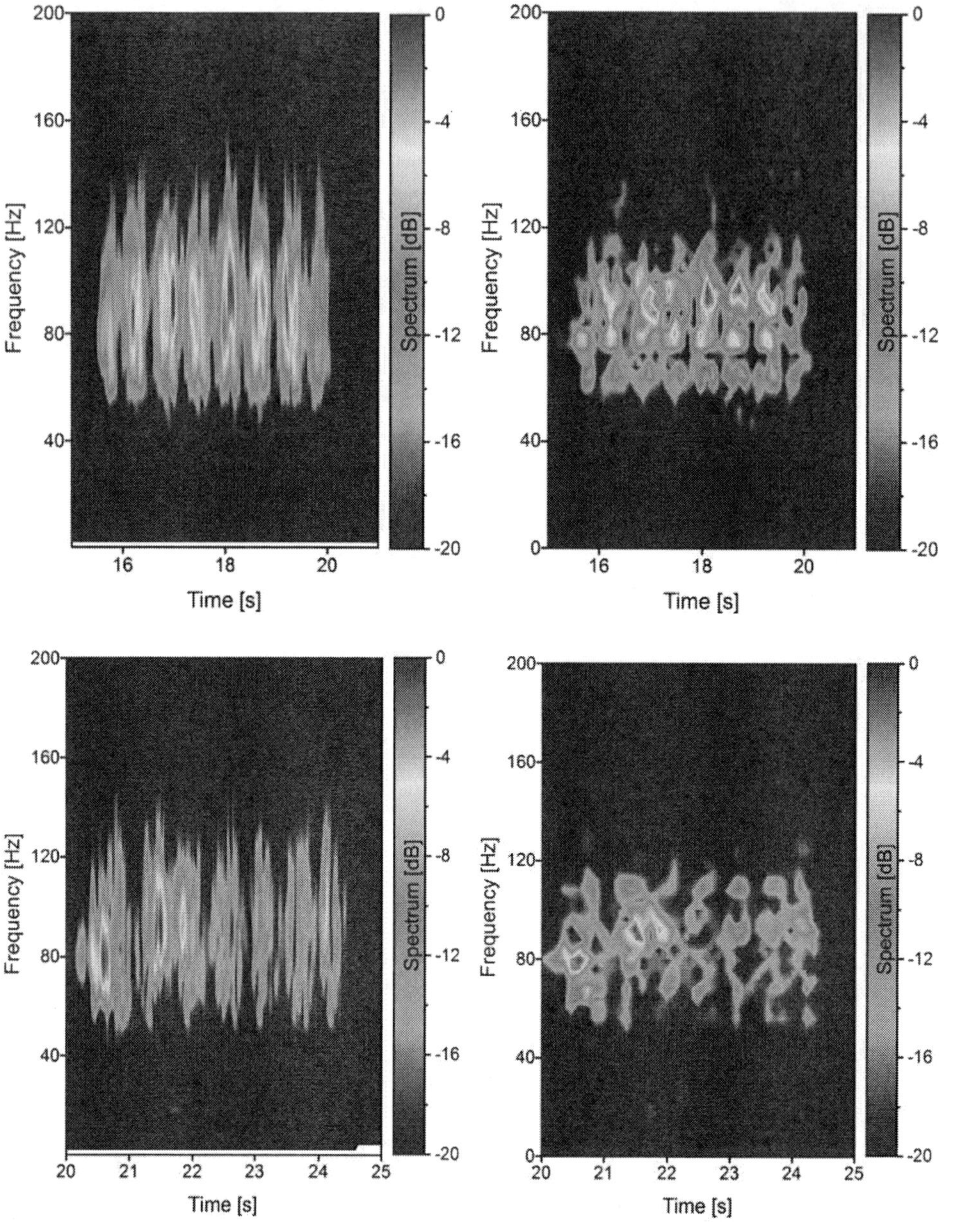

Fig. 4. Sensor P-04x - the value of CWT (the left column) and STFT (the right column) for Pendolino train (the upper row) and the value of CWT (the left column) and STFT (the right column) for InterCity train (the bottom row).

Graphs obtained for horizontal vibrations parallel to the axis of the railway track (P-05y) are shown in Fig. 5. The analyzes of the vibration records for this point (next to railway embankment) reveal a very similar nature of the two horizontal components of vibrations (x and y).

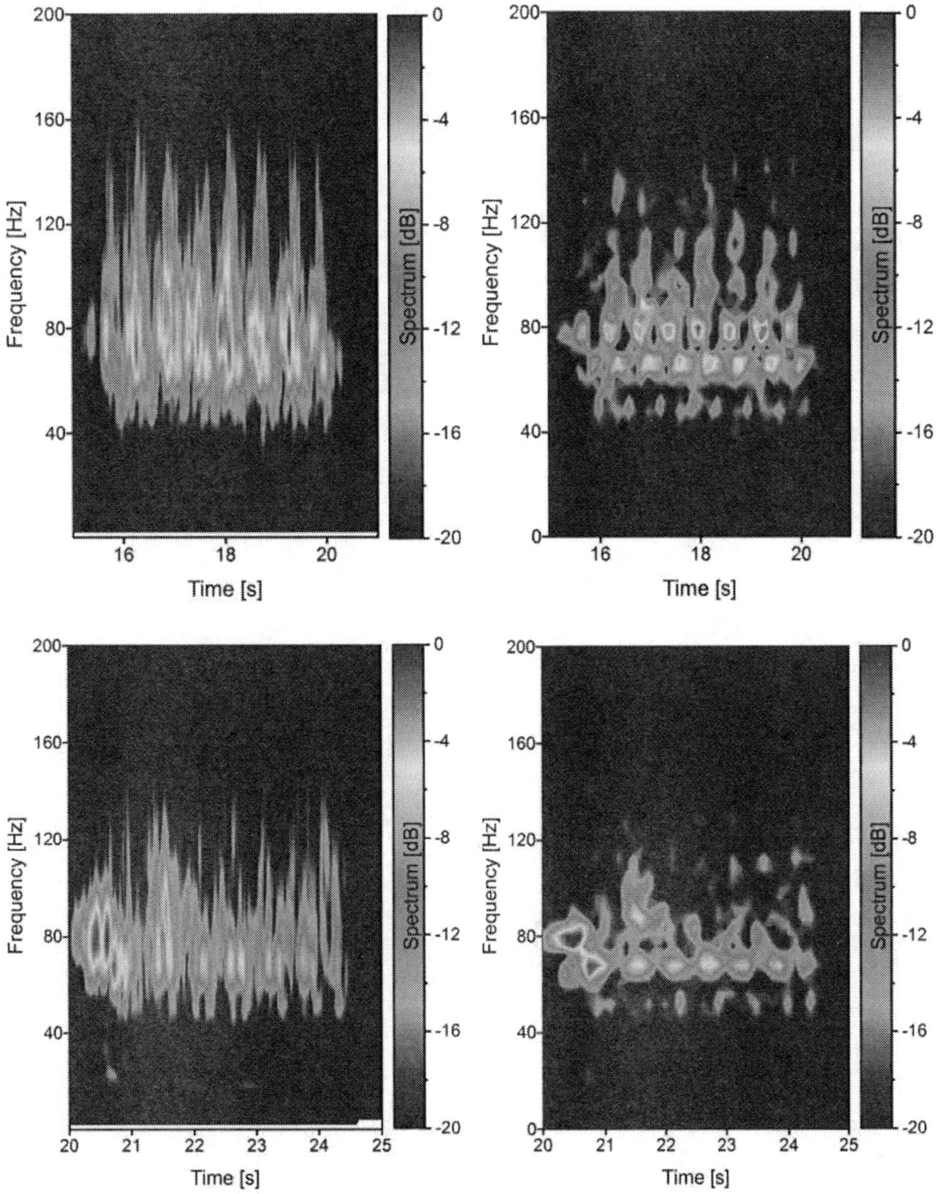

Fig. 5. Sensor P-05y - the value of CWT (the left column) and STFT (the right column) for Pendolino train (the upper row) and the value of CWT (the left column) and STFT (the right column) for InterCity train (the bottom row).

The values for the vertical component of vibration (P-06z) are a bit different. During the passages of Pendolino trains, vibration for the first and last group of wheels is higher in comparison to the other groups than in the previous cases. In the example of the InterCity train, the greatest strength of the signal occurs in the first second of movement. In both events, it can also be seen not negligible bandwidth vibration frequencies of about 160Hz, which does not exist for horizontal vibration. The same as for horizontal vibration, the representation of the time axis is better introduced by CWT but the frequency axis is more visible in STFT.

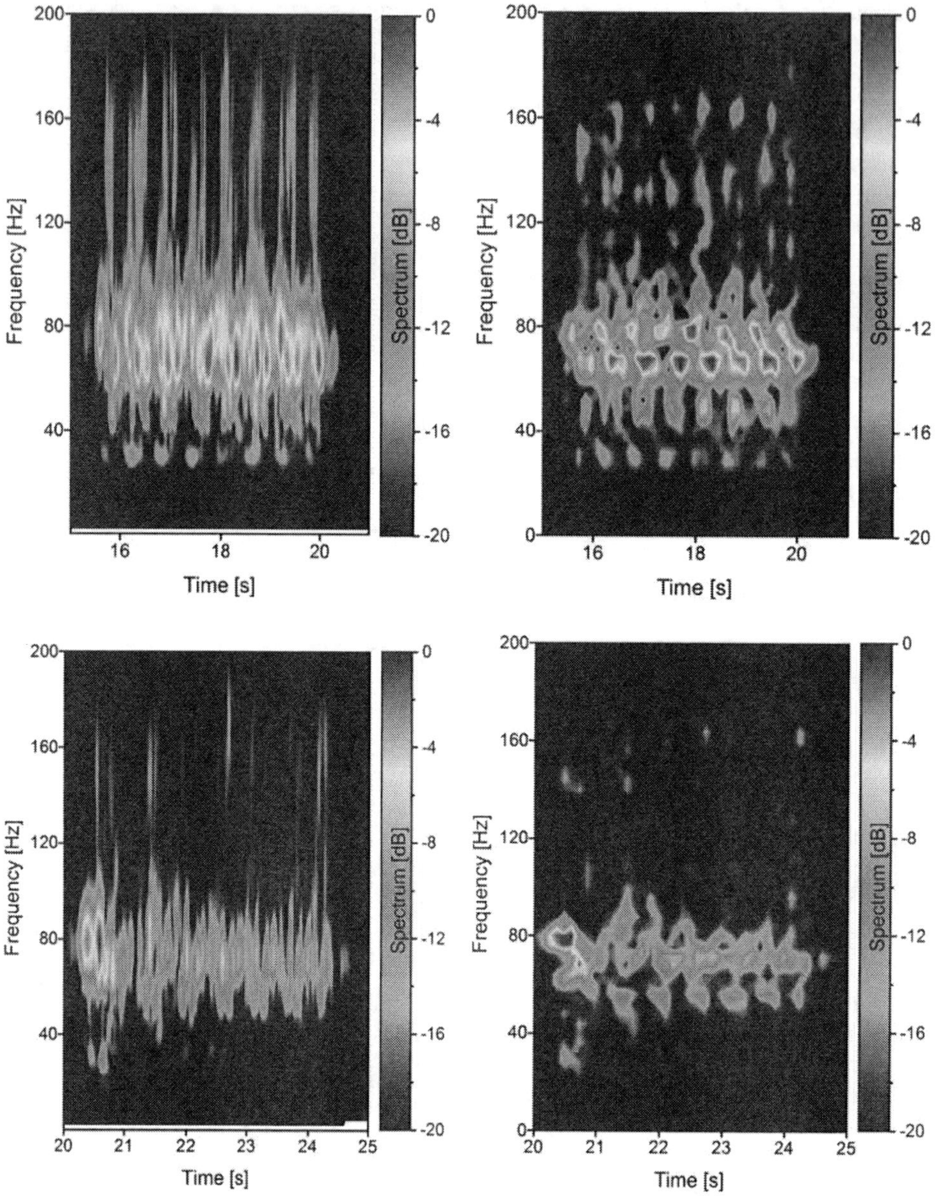

Fig. 6. Sensor P-06z - the value of CWT (the left column) and STFT (the right column) for Pendolino train (the upper row) and the value of CWT (the left column) and STFT (the right column) for InterCity train (the bottom row).

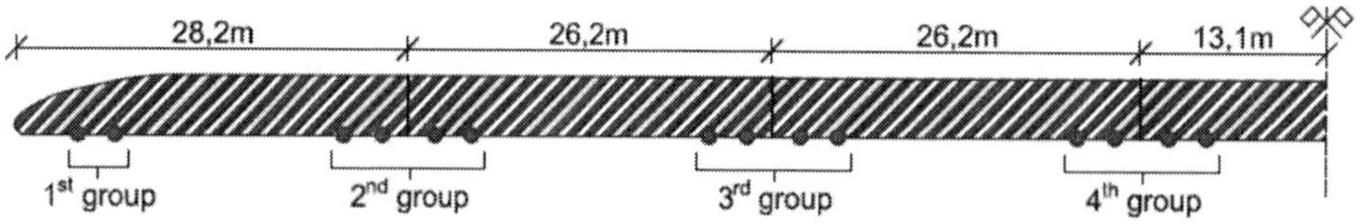

Fig. 7. Pendolino train – schematic arrangement of wheels.

3 Conclusions

Using world literature in the field of mechanical vibration it can be concluded that the vibration excitation of rolling stock is an important problem of the modern world. Therefore, the dynamic studies are necessary for each railway investments. This applies to both the design and construction of new lines and the purchase of new rolling stock.

The results of homologation tests, provides that there is no contraindication for release Pendolino train (Electric Multiple Unit – EMUs 250) for use by the Polish Railway Lines.

This study is a part of broader research aimed at the use of analysis giving an answer in frequency - time domain which should be a supplement for the analysis in the frequency or in the time domain.

As mentioned in [11-13] STFT brings significant limitations in the analysis of non-stationary signals. A large width of the analysis window provides high resolution in the frequency domain, while a low time resolution. A small width of the analysis window ensures high time resolution and a low frequency domain. It causes difficulty in the separation of certain spectral components in the analysis of broadband signals. That CWT's functions are characterized by specific properties - a good location in time (space) and the limited bandwidth in the frequency domain. Therefore, the wavelets are ideal for approximation signals varying in time. Important feature of the wavelet representation is that it includes the information about the signal at different levels of detail.

A time - frequency domain analysis provides more relevant information than classical FFT. Such verification data makes it possible to use vibration damping with special emphasis on the element which induces the largest vibrations. E. g. for vibroisolation design, which reduce vibration induced in the relevant frequency band or for changing the bogies that make the higher amplitudes.

All analysis and calculations were made using FlexPro 7.0 code.

The authors acknowledge the support of research funded by the Ministry of Science and Higher Education – project L-4/349/2016/DS-M.
In situ investigations of vibrations were performed by the Laboratory of Structural Mechanics at Cracow University of Technology. The authors would like to thank the team of the laboratory.

References

1. D.P.Connolly, G.P. Marecki, G. Kouroussis, I. Thalassinakis, P.K. Woodward, The growth of railway ground vibration problem – A review, Sci Total Environ **568**, 1-7, Elsevier (2015)
2. D. P. Connolly, G. Kouroussis, O. Laghrouche, C. Ho and M. Forde, Benchmarking railway vibrations – Track, vehicle, ground and building effects, Constr Build Mater **92**, 64-81, Elsevier (2015)
3. L. S. G. Degrande, Free field vibrations during the passage of a Thalys high-speed train at variable speed, J Sound Vib **247 (1)**, 131-144, Academic Press (2001)
4. X. Sheng, C. J. C. Jones, M. Petyt, Ground vibration generated by a load moving along a railway track, J Sound Vib **228 (1)**, 129-156 (1999)
5. A. Tesar, J. Melcer, D.Kucharova, Monitoring of small channel slabs in railway track science, Building Research Journal **61**, 15 – 24 (2014)
6. T. Tatara, K. Stypuła, Analysis of the results of measurements of ground vibration and some elements of the surface during the test of Pendolino train and other trains on the line CMK in agreed locations in the track No. 1, section Psary - Góra Włodowska, Cracow University of Technology, Cracow (in Polish) (2013)

7. S. Ervin, I. Djurowić, J.Jiang, Time – frequency feature representation using energy concetration: An overview of resent advances, Digital Signal Procesing **19**, 153-183 (2009)
8. T. Basiewicz, Wavelets and approximations, WNT, Warsaw (in Polish) (2004)
9. D. Cantero, M. Ülker-Kaustell, R. Karoumi, Time–frequency analysis of railway bridge response in forced vibration, Mech Syst Signal Pr **76-77**, 518-530 (2016)
10. J. Korzeb, Application of wavelet analysis in evaluation of vibrations propagation in transport infrastructure impact zones. Transport XXI century, Department of Transportation Warsaw University of Technology, Białowieża (in Polish) (2010)
11. A. Zacniewski, Comparative analysis of chosen transforms in the context of de-noising harmonic signals, Polish Naval Academy in Gdynia, Bulletin of WAT (in Polish) **LXIV**, 3 (2015)
12. M. Gonera, W. Kiciński, Time-frequency methods of signal analysis, The work of the Institute of Industrial Electronics (in Polish) **149**, Warsaw (2004)
13. S. Mallat, A Wavelet Tour of Signal Processing, Academic Press (2008)

MATEC Web of Conferences **107**, 00063 (2017) DOI: 10.1051/matecconf/201710700063
DYN-WIND'2017

Evaluation of the traffic impact on residential building

Damian Beben[1,*], *Wojciech* Anigacz[1] and *Piotr* Bobra[1]

[1]Faculty of Civil Engineering and Architecture, Opole University of Technology, 45-061 Opole, Poland

Abstract. The paper presents an example of evaluation of the traffic impacts on the residential building. Residents noticed numerous damage to the building, which, in their opinion, were caused by the movement of trucks from the nearby transport base of agricultural equipment's. The control seals have been glued on all cracks of the building and elements close to the road, in order to determine the damages propagation. The building and ground vibrations were established during the study. The vibration accelerations in the frequency function at the selected points of building were measured. Taking into account the dynamic impact scales (DIS), the vibration influences on the building were specified. Measurements of accelerations and vibration velocities inside a building were also conducted in order to determine the harmfulness of vibration in relation to the people. As a result of vibration accelerations test in different places of building, it was found that building is located in the first zone of the DIS, which vibrations are classified as imperceptible by the building. The harmfulness of occurring vibrations to the people staying in a residential building at the analysed road was also not found.

1 Introduction

Constant increase in traffic has negative effect on technical condition of residential buildings and engineering structures, as well as the comfort of life of the residents [1], [2]. It is a complex problem related to the dynamics of a structure, acoustics (influence of noise), wave propagation in ground, technical condition of a structure as well as a man's individual response to psychophysical stimuli [3]. The effect of traffic generated vibrations has already been discussed in literature on many occasions [4-7]. However, almost each individual case needs to be given individual attention because of slightly different ground and environmental conditions and, above all, of different structure and different vibrations source. Therefore the best way to evaluate the dynamic impact upon a given structure and/or people is direct measurement of acceleration (or velocity) and frequency of vibrations.

The aim of the research was: (i) to define the effect of vehicles driving along a county road on a residential building (identification of vibrations in the building and evaluation if their possible harmful effects), (ii) determination of possible propagation of damages

* Corresponding author: d.beben@po.opole.pl

 DOI: 10.1051/matecconf/201710700063

(cracks) by means of fixing control seals on cracks (iii) evaluation of technical condition of the structure of both the residential building and the road, (iv) evaluation of traffic intensity along the road near the building (measurement of traffic). Additionally effect of vibration on people staying in the considered building has been determined.

The discussed tests were undertaken in response to damage revealed in the structure of the building, caused, according to its residents, by heavy load traffic from the nearby base of vehicles and machinery used in agriculture.

2 Description the building structure and the damages

The discussed residential building was built before World War II and is located by a county road. Its 0.65 m thick walls are made of brick. The ceiling structure is made of wood. It is a storey building with no basement. It underwent reconstruction in the 90 s of the 20th century in order to improve its functioning – the floor was then lowered by 0.20 m at the ground level and the wooden ceiling structure over the storey was elevated to reach the minimal height for usable floor space, i.e. 2.20 m for the ground floor and 2.40 for the first floor. Extensions of the building (without basement) are a vestibule on the ground floor and an orangery on the first floor. The building is located at a distance of 12.0 m from the road axis. Fig. 1 shows and overview of the building at the time of a passage of a lorry.

Fig. 1. The residential building at the time of vibration measurements (passage of a HGV).

In order to determine the influence of traffic on the building, a detailed inventory of the damage identified in it was undertaken. Fig. 2 shows schematic distribution of control seals on identified cracks on the building. A total number of ten control seals were attached to the damages (cracks) of the building and on elements in the proximity of the road (fence foundation), the aim of which was to determine possible propagation of damages. The control stripes (seals) of plaster were 80×30 mm and 5 mm thick. The seals were secured with a transparent glass slab, at least as big as the seals. The cracks were generally classified as short and therefore the seals (1 or 2) were symmetrically fixed at the 1/3 and 2/3 of the crack height. Monitoring of the seals condition was held once a week over the period of two months. Fig. 3 shows example control seals on the building.

Within the period of monitoring of the seals condition, it was revealed, that the damages identified in the residential building were of non-structural nature, i.e. cracks and scratches of the coatings and plasters, relaxation of door and window mountings in the walls. These sorts of damages were observed on the first floor on the wall on the road side. Cracks on the fence foundation and the concrete floor around the building were observed too. Their dimension did not exceed 1.5 mm. Over the period of two months of monitoring of the seals no propagation of the cracks was observed on the floor of the building while the seals

 DOI: 10.1051/matecconf/201710700063

attached outside the building (the fence foundation, the concrete floor around the building) show a minor tendency of the cracks to increase (of no more than 2 mm).

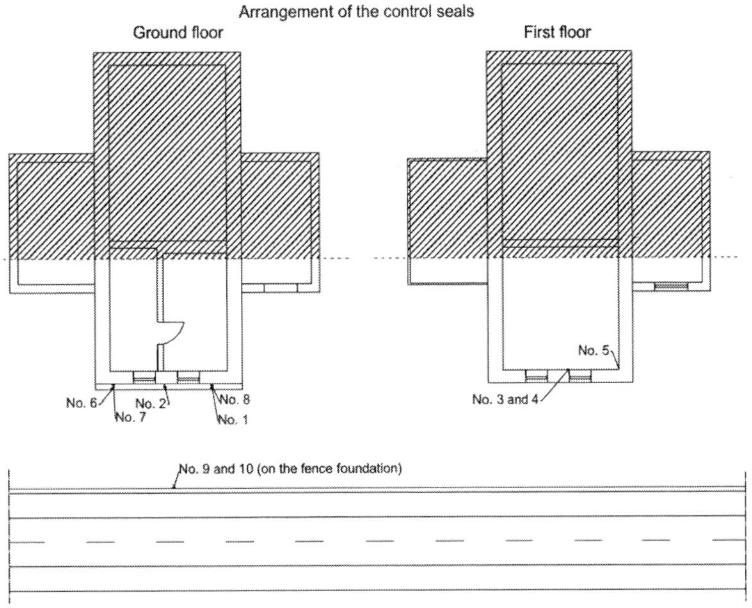

Fig. 2. Schematic distribution of control seals.

a) b)

Fig. 3. The sight of control seals a) nos. 6 and 7 on concrete floor around the building (at ground level) and b) nos. 3 and 4 (in a room on the first floor of the building). The sensor to measure vibrations, located on the wall, is also visible.

In order to determine the impact of traffic upon the residential building, an inventory of damages on the road in the vicinity of the building was held (ones with potential to increase dynamic impact on the building at the time of passage of vehicles). As a result numerous examples of damage and uneven asphalt layer (caused by previous minor maintenance repairs) were identified. Fig. 4 shows typical damage and uneven surface of the considered road.

At the time of inspection of the road the following were found: (i) a number of cases of uneven surface all along the road, (ii) cavities of the surface, (iii) collapse drainage and sewage wells, (iv) lots of pollutants (sand, gravel) and vegetation near pedestrian sidewalks that hindered drainage of the road, (v) lack of cross and longitudinal slopes of roadway (or being insufficient).

Fig. 4. Damages of the road pavement: a) cavities in asphalt filled with water (poor drainage), b) a collapsed drain.

The condition of the roadway was classified in accordance with GDPRaM (General Directorate of Public Roads and Motorways) directives [8] as at warning level - class C – unsatisfactory condition i.e. roadway with major damages requiring renovation. The warning level defines a condition requiring at least a detailed examination in order to improve the condition of the roadway.

As a result, it was found that the cavities and uneven surface of the considered road as well as damages to the wells could cause increased dynamic impact. Therefore additional passages of vehicles along the damaged areas were planned.

3 Characteristics of the traffic on the road

Within a couple of days a complete monitoring of traffic was held to find out about its structure and to estimate its impact on the building and its residents. The most intense traffic of heavy goods vehicles (HGV) was observed around 8.00 a.m. (vehicles leaving the transport base of a nearby company) and from 3 p.m. to 4 p.m. (vehicles coming back). It was observed that the total number of 21 HGVs drove along the road within the period of observation. Other agricultural equipment (tractors with attached trailers and farming machinery) were hardly ever seen (2-4 per day). In the majority of cases the road was used by personal cars of the inhabitants of the neighbouring villages.

It should be noticed that the road is of specific location i.e. it is relatively short (approx. 180 m) and it ends with double sharps which make drivers slow down to a considerable extent. This means that vehicles driving along this road do not exceed the speed of 30 km/h.

Complete traffic measurements were held twice. Categorisation of vehicles was done in accordance with GDPRaM directives [9]. The method of direct manual measurement was applied by the observers (the measurement point was located at the considered building). In accordance with GDPRaM directives for traffic measurement [9] the discussed road was classified as a type „W" section (i.e. a section that undergoes direct traffic measurement within a limited number of hours). Therefore in this case the time of measurement was limited to 8 hours (8:00 a.m. – 4:00 p.m.) on each of the days of measurement.

As a result of measurement of traffic structure on the discussed road it was found that:
- the traffic is light most of the time – nearly 97% of all traffic and it takes place mainly between 11.00 a.m. and 3.00 p.m.,
- heavy traffic can hardly ever be observed i.e. 8-10 passages, mainly from 7.00 a.m. to 8.00 a.m. and 3.00 p.m. – 4.00 p.m.,
- all-day measurement (8.00 a.m. – 4.00 p.m.) revealed 202 passages.

4 Measurements of the building and ground vibrations near the road

4.1 Impact of vibrations on the building structure

4.1.1 Measurement methodology

In accordance with the Polish Norm PN-85/B-02170 [10] vibrations of a building need to be characterised by displacements (or speed or acceleration) in characteristic points of the building and a relevant frequency of vibrations. Measurement of vibrations acceleration in the function of frequency was held in order to determine the impact of vehicles driving along the discussed road onto the residential building. The measurement considered the following three passages:
(i) a car of 2.0 t of weight driving at 20 and 40 km/h,
(ii) a combine harvester weighing 13 t and driving at z 15 km/h (maximum speed),
(iii) a heavy goods vehicle of 41.2 t driving at 35-40 km/h.

In fact only a couple of lorries, most of them unloaded (<20 t of weight), drive by during a day. A HGV loaded with corn (41.2 t of total weight) driving at 35-40 km/h (more than normally) was used for tests. At the time of tests research team members informed the driver regarding a possibility to drive safely with a given speed.

For short-time vibrations i.e. the total duration of which does not exceed 3 mins/24 hours (as in the discussed case) a record was held of a chosen parameter within at least three cycles of loading i.e. chosen vehicles drove by at least three times for each measurement point. The symbols *X* and *Y* for horizontal measurement directions, and *Z* for the vertical one were adopted in accordance with the Cartesian coordinate system. In the discussed case the measurement point (*X*, *Y*) was located on the wall from the side of vibration source, in surrounding ground level. The signals were analysed within the scope of vibrations from 0.5 Hz to 100 Hz.

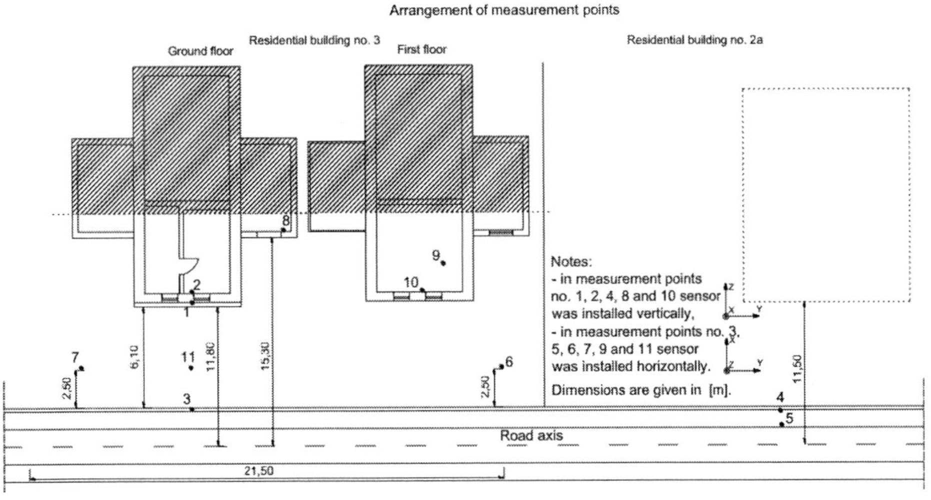

Fig. 5. Distribution of measurement points.

The measurements were held in the following test points (Fig. 5):
a) inside the building:

- on the bearing wall at ground level (on external facade of the building - Fig. 6a) measurement point No. 1,
- on the bearing wall at ground level (inside the building from the side of the source of vibrations – measurement points nos. 2 and 8,
- on the first floor on the bearing wall (measurement point No. 10) and on the floor (measurement point No. 9) – tests related to determining the impact of vibrations on the people staying in the building,

b) on elements outside the building on the way of the wave from the source of vibrations to the entry point (the building):

- on the fence foundation (measurement points 3 and 4),
- on the pavement (measurement point no. 5) at number 2a house,
- on a special tripod embedded in the ground (measurement points nos. 6, 7 and 11) – Fig 6b.

a) b)

Fig. 6. Measurement points: a) no. 1 located on the wall, b) no. 6 in ground.

A *FastTracer* vibration sensor by *Sequoia* was used. It allows to measure vibrations in three independent directions at the same time. Technical data of the sensor are as follows: maximum acceleration of +/- 5 g, frequency range of 0 – 2500 Hz, resolution of 0.0025 m/s^2, noise of 0.075 m/s^2, sampling of 8192 samples/s.

Each measurement point was cleaned prior to installation of the sensor. The sensor was fixedly attached by means of four magnets to a metal plate, which was consequently fixed to the measurement point with use of a double-sided sticky tape. A FFT (Fast Fourier Transform) analyser was used to analyse the signals from the sensor.

4.1.2 Approximate ways to check dynamic impact on the building

For approximate estimation of the impact of vibrations transmitted by the ground to certain types of buildings it is allowed to apply the dynamic impact scales (DIS). The DIS scales can be used in case of masonry elements (i.e. elements for manual construction of a wall). The DIS-I scale Fig. 7) refers to a building of compact shape and of small external dimensions of its plan (below 15 m), one- or two-storey one, and of height not exceeding any of the plan dimensions.

According with the PN-85 B-02170 [11], the DIS scales contain five zones limited with four limit lines (A, B, C and D). The lines have been shown in Fig. 7 within the coordinate system of vibration frequency (*f*) and acceleration (*a*). The coordinate values need to be determined from measurement in the points located on the structure at ground level. The following criteria were adopted for division into the zones of detrimental effect:

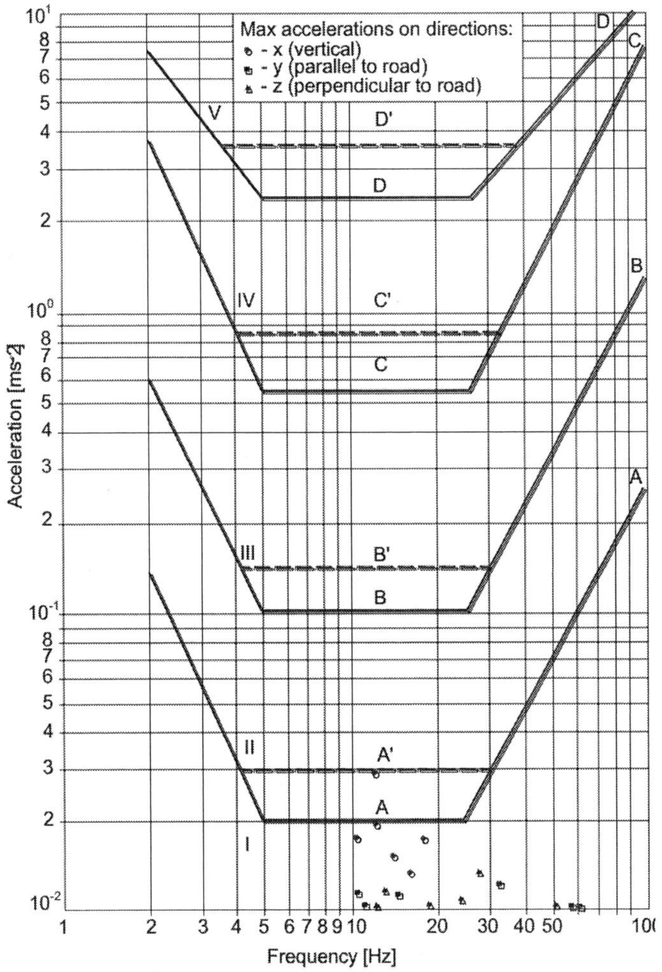

Fig. 7. Dynamic impact scales with measurement results (vibrations of the building).

a) zone I – vibrations not perceived by the structure (limit A – a lower limit for reception of vibrations by a building and a lower limit for considering dynamic impact. In case of vibrations below this limit, dynamic impact does not need to be considered).
b) zone II – vibrations perceptible for the building but of no harm to the structure; advanced wear of the building, first cracks in plaster and coatings etc. (limit B – the limit of the building stiffness, the lower limit for cracking and scratches in structural elements).
c) zone III – vibrations harmful to the building, they cause local scratches and cracks and therefore weaken the structure of the building and lower its load carry capacity as well as its resistance to further dynamic impact; coatings and plasters may fall off (limit C — the strength limit of individual elements of the building, the lower limit of severe construction damages).
d) zone IV – vibrations highly detrimental for the building, hazardous for people. Numerous cracks, local cases of destruction of walls and individual elements of the building are observed. Hanging objects may fall down, ceiling coating may peel off, ceiling beams may slide out from bearings etc. A possibly immediate removal of vibration source or reduction of their impact is required (limit D – the limit of a

structure stability, a lower limit of the whole structure collapse; vibrations exceeding this limit can cause construction disasters and hazards to human safety and life).

e) zone V – vibrations cause breakdown through collapse of walls, falling off ceilings etc.; total hazard to human life and safety; in case of possibility for such vibrations to appear, buildings of this type must not be used.

Zone limits have been presented in two variations, according to evaluation of the building condition, type of ground and type of vibrations. To classify a case within a specific category, a prevailing number of relevant characteristics specified in the Norm is taken into account [11].

4.1.3 Vibration test results and analysis

a) Vibrations of the residential building:

Due to very low values of vibration acceleration of the building structure measured in experiments, only the maximum courses of vibration acceleration were presented. In order to illustrate the level of vibrations in the building caused by passages of different vehicles, the obtained maximum acceleration results have been presented in Fig. 7 showing scales of dynamic impact (DIS). Whereas Fig. 8 shows a chosen course of vibration acceleration in the function of frequency and considers vibrations caused by a passage of a vehicle of 41.2 t of weight at the speed of 35-40 km/h (measurement point no. 1).

Table 1. Comparison of maximum results of vibration acceleration in the building.

Measurement point	Frequency range [Hz]	Amplitude of acceleration from directions *Y* and *Z* [m/s^2]	Amplitude of acceleration from direction *X* [m/s^2]
1 (see Fig. 9)	10 – 25	0.009	0.027
2 (see Fig. 10)	10 – 25	0.006	0.017
8 (see Fig. 11)	10 – 25	-	0.014
	50 – 60	0.004	-

Note: the following indications on the axis have been adopted: *Z* – perpendicular direction to the carriageway axis, *Y* – parallel direction to the carriageway axis, *X* – vertical direction.

Following analysis of test results it can be noticed that the signal is dominated by components of frequency ranging from 10–25 Hz (Table 1), and the highest amplitude of acceleration from horizontal directions (*Y, Z*) amounts at 0.009 m/s^2. It was registered at the time of passage of a HGV of 41.2 t of weight at the speed of 35-40 km/h. The impact of other vehicles was at a considerably lower level i.e. acceleration did not exceed 0.003 m/s^2. In line with the PN-85/B-02170 [11] and Fig. 7, the examined building is located within

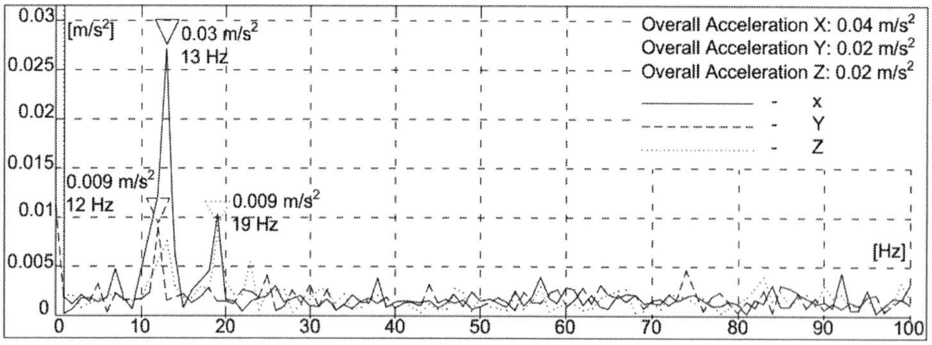

Fig. 8. Vibration accelerations course at measurement point no. 1 (foundation of the house).

zone I of the DIS-I scale, meaning the vibrations are not perceptible to the structure. Limit A means the lower limit of vibration perceptibility for the building. Dynamic impact can be neglected when vibrations fall below this limit. However it needs to be underlined that the analysis did not consider measurement from the vertical direction, because residential buildings are designed for vertical loads (dead load, snow load, service load).

b) Measurement points located outside the building:

Figs 9-10 show extreme courses of vibration acceleration at points located outside the residential building on the way of the wave from the vibration source to the receipt point of vibration (fence foundation and at ground – see Fig. 5) caused by passage of a HGV of 41.2 t of weight with the speed of 35-40 km/h. The maximum values of vibration acceleration of the measurement points located outside the building have been presented in Fig. 7. Table 2 shows extreme values of vibration acceleration for the three tested dimensions in relation to the vibration source.

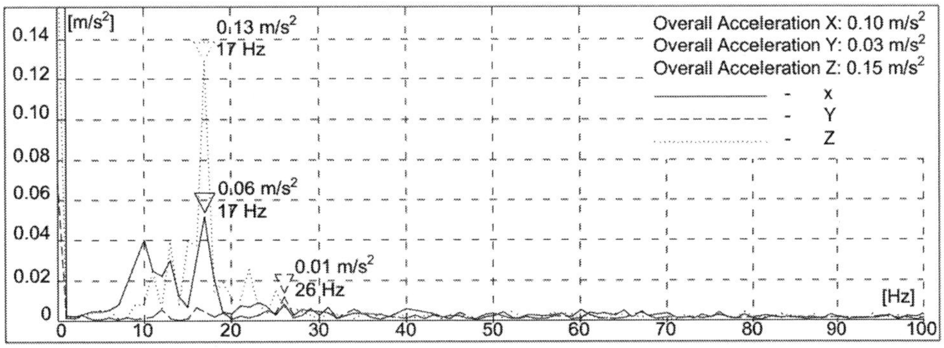

Fig. 9. Vibration accelerations course at measurement point no. 4 (foundation of the fence) (axes: *Z* – direction perpendicular to the carriageway; *Y* – direction parallel to the carriageway; *X* – vertical direction).

Analysis of graphs showing spectrums of vibration acceleration from points 6, 7 and 11 (located in ground) proves that they represent a very similar level, which means that vibrations propagate in soil in a similar way. This proves that soil in the area of tests is homogeneous. Moreover, in most cases, the obtained horizontal acceleration components of vibrations in ground are smaller than 0.005 g (g – gravitational acceleration). It should be noticed that it is not required to consider loads to a structure from vibrations transmitted through even in case of newly designed buildings.

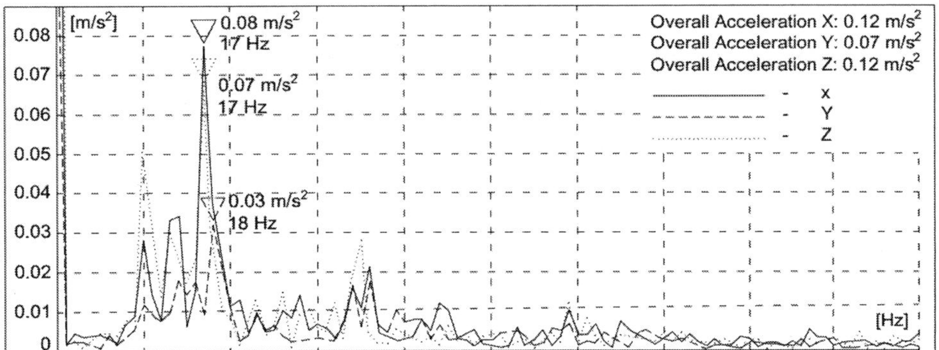

Fig. 10. Vibration accelerations course at measurement point no. 7 (in ground) (axes: *X* – direction perpendicular to the carriageway; *Y* – direction parallel to the carriageway; Z – vertical direction).

Acceleration of vibrations of measurement points located on the fence foundation (points 3 and 4) also do not show increased values. However, it needs to be underlined that the foundation structure is not well known and can contain the eigenvalues.

Table 2. Comparison of maximum values of vibration acceleration of elements located outside the building.

Measurement point number	Frequency range [Hz]	Amplitude of vibration acceleration from directions *X* (perpendicular) and *Y* (parallel) to the road [m/s^2]	Amplitude of acceleration from direction *Z* (vertical) [m/s^2]
3	10 – 15	0.025	0.075
4	10 – 25	0.128	0.051
4	10 – 25	0.070	0.085
5	10 – 20	0.045	0.111
5	10 – 30 50 – 65	0.030 0.030	0.070 0.021
6	10 – 25	0.044	0.021
6	10 – 25 80 – 85	0.022 0.036	0.020 0.015
7	10 – 20 30 – 40	0.077 0.021	0.068 0.027
11	10 – 25	0.021	0.022

4.2 Influence of vibrations on people

4.2.1 General remarks

Additional measurements of vibration acceleration inside the building were held in order to determine harmfulness of vibrations related to traffic along the nearby road to people, in accordance with the Polish Norm PN-88 B-02171 [10]. Vibration velocity values inside the building were determined on the basis of integration of the measured acceleration values.

The tests were held in places where people usually stayed (on the first floor) in points of vibration transmission from the structure to a man (on the floor – measurement point no. 9 and on the wall from the side of vibration source - measurement point no. 10 (see Fig. 5)). Due to the fact that a man can change position, measurements were held also in three directions: vertical Z and two horizontal (*X* and *Y*) marked by eutectic system related to geometry of the place of measurement. Vibrations were measured at the time of passage of a HGV of 41.2 t of weight at 35-40 km/h. Measurements were carried out in the least advantageous conditions in the whole evaluated frequency band for a man influenced by them. In accordance with the norm the observed vibrations were classified as intermittent i.e. of total duration time not exceeding 30 min/24 hours.

4.2.2 Allowable values in accordance with the norm

The allowable corrected value of vibration acceleration towards the vibration recipient needs to be determined on the basis of the following formula (1):

$$a_{\mathrm{k\,allo.}} = a_{\mathrm{k1}} \times n, \tag{1}$$

where:
a_{kl} – value of acceleration relevant for vibration noticeability threshold for a man (0.005 ms^{-2} for direction *Z*; 0.0036 ms^{-2} for directions *X* and *Y* of vibration reception [10]),
n – a factor dependent upon the type of vibrations and purpose of a place in a building (as it

has been assumed $n = 32$ – for flats and occasional vibrations of multiplicity not exceeding 10 per 24 hours [10]).

The allowable corrected acceleration value, for the Z direction of reception of vibrations by a man, amounts at (in the daytime): $a_{\text{k allo. (Z)}} = 0.005 \times 32 = 0.16$ m/s^2, and for directions X and Y of reception of vibrations by a man, the corrected allowable value of acceleration amounts at (in the daytime): $a_{\text{k allo. (X, Y)}} = 0.0036 \times 32 = 0.1152$ m/s^2. Whereas the corrected allowable values of vibration velocity towards the reception point of vibrations can be determined on the basis of the following formula (2):

$$V_{\text{k allo.}} = V_{\text{kl}} \times n, \tag{2}$$

where:
V_{kl} – the value of velocity corresponding with noticeability threshold in a man (0.0001 ms^{-1} for direction Z; 0.00029 ms^{-1} for directions X and Y of vibration reception [10]),
n – factor accepted as in formula (1).

The corrected allowable value of vibration velocity for direction Z of reception of vibration by a man amounts at (in the daytime): $V_{\text{k allo. (Z)}} = 0.0001 \times 32 = 0.0032$ ms^{-1} = 3.2 mms^{-1}, and for direction X and Y of reception of vibrations by a man, the corrected allowable value of vibration velocity (in the daytime), amounts at: $V_{\text{k allo. (X, Y)}} = 0.00029 \times 32 = 0.00928$ ms^{-1} = 9.28 mms^{-1}.

4.2.3 Measurements results and analysis

Fig. 11 and Table 3 show maximum vibration acceleration obtained for measurement point no. 9 located on the first floor of the building, on the floor from the vibration source side.

Table 3. Maximum vibration acceleration values at measurement point no. 9 (the floor).

Frequency [Hz]	The highest amplitude of acceleration from perpendicular direction X and the direction Y parallel to the carriageway [m/s^2]	The highest amplitude of acceleration from the vertical direction Z [m/s^2]
6		0.010
10	0.005	0.005
11	0.007	
13		0.008
18		0.007
20		0.007
82		0.009
84		0.024
15	0.006	0.015
17		0.009
67		0.008
88		0.009
93	0.005	0.020
95		0.010
79	0.008	0.054
9		0.008
67		0.028
69		0.022

Whereas the maximum values of vibration velocity have been presented in Table 4. They were observed at the measurement point no. 9 located on the first floor of the residential building (on the floor surface) by the discussed road.

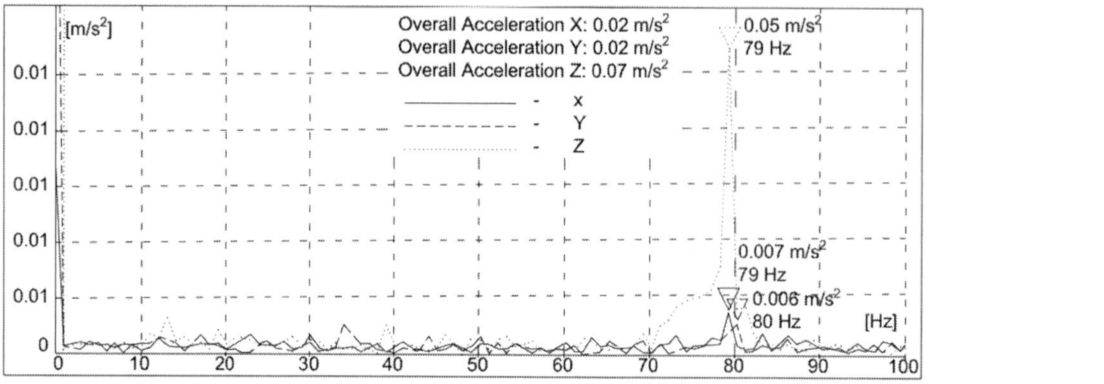

Fig. 11. Vibration accelerations course at measurement point no. 9 (the floor) – passage of a HGV over a collapsed drain.

Table 4. Maximum values of vibration velocity at the measurement point no. 9 (the floor).

Load description	Amplitude of vibration velocity from perpendicular direction *X* and direction *Y* parallel to the carriageway [mm/s]		Amplitude of vibration velocity from vertical direction *Z* [mm/s]	
	Extreme value	Effective value of the signal RMS	Extreme value	Effective value of the signal RMS
(normal passage of a vehicle of 41.2 t of weight	0.399 0.333	0.269 0.242	0.632	0.540
(passage of a vehicle of 14 t of weight over a drain)	0.354 0.294	0.207 0.178	0.461	0.288
(passage of a vehicle of 41.2 t of weight over a drain)	0.309 0.239	0.189 0.163	0.439	0.324
(passage of a vehicle of 41.2 t of weight over a drain)	0.239 0.292	0.142 0.179	0.441	0.287

In the process of analysis of the obtained results of vibration acceleration and velocity, the following has been found:

- for the vertical direction (*Z*) of vibration reception by a man, the corrected allowable vibration acceleration value amounts at 0,16 m/s^2, and is in all cases higher than one received in tests. Similarly the allowable value of vibration velocity being 3.2 mm/s was not exceeded in any of the case of passage of a loading vehicle. It needs to be underlined that a HGV of 41.2 t of weight was used in the tests, which, along with standard passages, did an additional passage over a collapsed drain (constituting a so called impact, i.e. increased dynamic influence).
- for perpendicular and parallel directions (*X* and *Y*) of vibration reception by a man , the corrected allowable value of vibration acceleration amounts at 0.1152 m/s^2, and is, in all discussed cases, higher than maximum results obtained from tests. The allowable value of vibration velocity of 9.28 mm/s also was not exceeded in any of the considered instances of vehicle passage.

5 Conclusions

Based on the conducted tests and observations, the following conclusions have been drawn:

1. As a result of examination of vibration acceleration in different parts of the residential building, it was found that the building falls within zone I of the DIS-I scale, which

means that vibrations are classified as unnoticeable for the structure (in line with Polish Norm PN-85/B-02170 [11]). It needs to be mentioned that the extreme values of vibrations were observed during passage of a loaded HGV of 41.2 t of total weight. Passages of other cars and heavy vehicles caused smaller values of vibrations.

2. As a result of additional measurements of vibration acceleration of soil, it was found that the values represent a similar level, which proves homogeneity of soil and therefore similar propagation of vibrations within. The values of vibration acceleration in majority of cases do not exceed 0.005 g, which is a condition for newly designed buildings.
3. No harmful effects of vibrations on people residing in the house by the discussed road have been proved. The obtained values of acceleration and velocity were smaller than permitted ones. It needs to be mentioned that a HGV of 41.2 t of weight was used during the tests. What is more, one of its passages was over a collapsed drain, causing an increased dynamic impact.
4. The damages found in the residential building were classified as non-structural, i.e. scratches and cracks in plasters and coatings. Monitoring of propagation of cracks held within the period of two months did not prove the damages would have a tendency to develop. The road pavement condition was classified at warning level in line with GDPRaM directives [8]. Numerous cases of surface roughness, cavities in pavement, collapsed drainage and sewage wells were observed.
5. It is recommended that the considered road is renovated, with special attention to drainage and sewage wells and cavities in the asphalt pavement, and with modelling of relevant cross and longitudinal downgrades. In the light of tests and analyses, it does not seem to be advisable to limit load of vehicles moving along the road. Moreover, it is suggested to undertake tests of noise level which probably causes poor psychophysical sensations in the residents.

References

1. B. Xia, P.B. Wei, Y.M. Cao, Traffic-induced vibrations of ground environments and buildings, *Environmental vibrations: Prediction, monitoring, mitigation and evaluation* (Taylor & Francis Group, London, 529–539, 2005)
2. O. Hunaidi, *Traffic vibrations in buildings* (National Research Council, 2000)
3. Z. Engel, *Environmental protection against vibrations and noise* (PWN, Warsaw 2001)
4. O. Hunaid, M. Tremblay, Traffic-induced building vibrations in Montreal, Can. J. Civ. Eng. **24**, 736–753 (1997)
5. R. Ciesielski, E. Maciąg, *Road vibrations and their effects on buildings* (WKŁ, Warsaw, 1990)
6. H. Hao, T.C. Ang, J. Shen, Building vibration to traffic-induced ground motion, Build. Environ. **36**, 321-336 (2001)
7. M. Petronijević, M. Nefovska-Danilović, M. Radišić, Analysis of frame structure vibrations induced by traffic, Gradevinar, **65**, 811-824 (2013)
8. M. Radzikowski, *The evaluation of the surface SOSN – application guidelines* (GDPRaM, Warsaw, 2010)
9. *Guidelines for measuring traffic*, Ministry of Infrastructure, Department of Roads and Highways, Warsaw (2009)
10. PN-88 B-02171, *Evaluation of vibration influence on people in buildings* (Polish Committee for Standardization of Measurement and Quality, Warsaw, 1988)
11. PN-85 B-02170, *Evaluation of the harmfulness of building vibrations due to ground motion* (Polish Committee for Standardization of Measurement and Quality, Warsaw, 1985)

Dynamic response analysis of multi-storey building to a non-uniform excitation

Pawel Boron[1,*] and *Joanna* Dulinska[1]

[1]Cracow University of Technology, Institute of Structural Mechanics, 31-155 Cracow, Poland

Abstract. In this paper the dynamic behaviour of multi-storey steel building is presented. The structure was subjected to a real, strong mining shock. During the analysis the time history analysis and the spectrum method were taken into account. The time history analysis concerns the uniform and non-uniform model of ground motion. A wave velocity of 300m/s was taken under consideration. The kinematic excitation was applied on the structure as support accelerations. The stress at selected points of structure were compared for all methods. The region nearby the connections was analysed particularly. The state of stress for connection zones was recognized.

1 Introduction

The mining shock was the typical dynamic load in the area of mining activity. The mining tremors may occurs very often and may be characterized by big energy. There are two main activity areas in Poland: Upper Silesian Coal Basin and Legnica-Głogow Copper Belt. In many cases the mining activity induce damage in buildings. The dynamic response analysis for buildings located in mining region are needed. The analysis allows to estimate the stress and displacements of structural elements. It is the method which helps to protect building against damage. Another issue is the wave passage effect. The non-uniform excitation indicated different object response than the uniform excitation. Some building shows vulnerability for wave propagation [1-4]. The time history analysis and response spectrum analysis are used to assessment the stress [5].

In this paper the multi-storey building was analysed. The building was subjected to uniform and non-uniform excitation. The dynamic response spectrum was calculated. The obtained results were compared with results from response spectrum method. The analysis allows to estimate the stress in elements during mining shock excitation. The analysis was conducted in Abaqus software [6].

2 Description of analysed construction

The five storey building was chosen to the dynamic analysis. The building was the steel structure on the plan of the rectangular. The building consist of five frames spaced in equal

* Corresponding author: pboron@pk.edu.pl

distance of 6 m. Total length of building was 24 m and the width 12 m. The total height of building was 17.8 m. The height of storey was not the same for each tier. The height of first storey was 4 m, the top storey 3.3 m and the rest of storeys 3.5 m. The main frames were created as a double span frames. The length of span equals 6 m. The elements of frame were made of typical section- double T sections. The external columns were made of HEB260; interior columns of HEB320 (it means the height of sections were 26 cm and 32 cm respectively). The lower beams (beams from 1st to 3rd floor) were made of IPE400 sections (40 cm height) and the rest of beams were made of IPE360 (height of 36 cm).

The beams and columns were joined together by rigid connections. That kind of connection allows to resist a bending moment.

The building was designed with a floor system consisting of a metal deck with a 15 cm lightweight concrete slab. It should be noted that the slabs were the same on each storey. Whole construction was supported by foundation slab.

The dimensions of analysed structure is presented in Fig. 1. The type of applied sections were also marked.

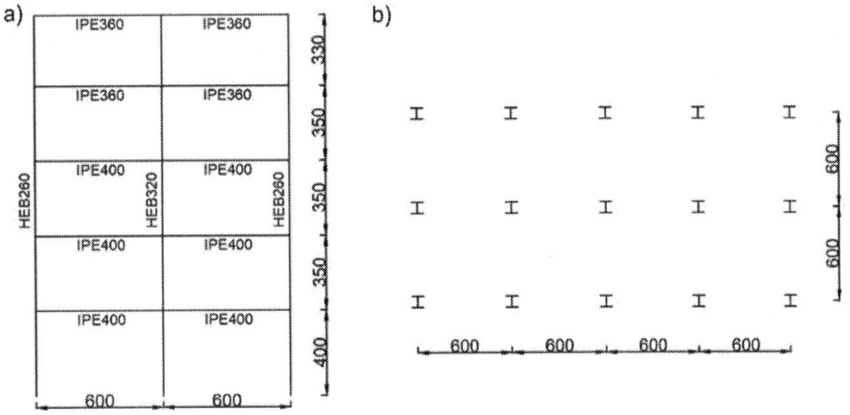

Fig. 1. Main dimension of analysed building: a) main frame, b) plan.

3 Numerical model

To carry out the dynamic analysis of structure the numerical model was created. The model represents whole structure. The building was presented as a multi storey frame. In the 3D model main dimensions like height, span length, spacing of frame and the real shape of cross sections were taken into account. In case of complex shape of cross section and numerous elements connections, in the numerical model some simplifications were applied. The model is shown in Fig. 2.

The main frames, except one of them, were modelled using beam elements (see Fig. 2a). Decreasing numbers of finite elements in whole model was possible by using beam elements. That allow accelerate the calculation without deterioration the quality of results. As mentioned, one of the frame was modelled by different than beams elements. The frame in the middle of building was represented by shell elements. The different kind of elements allow to calculate the stress distribution and deformation of middle frame. The cross section of girders and columns were made of plates which form a double T section (see Fig.3). The shell thickness- correspond to a real dimensions of cross section. The middle frame is presented in Fig. 2b.

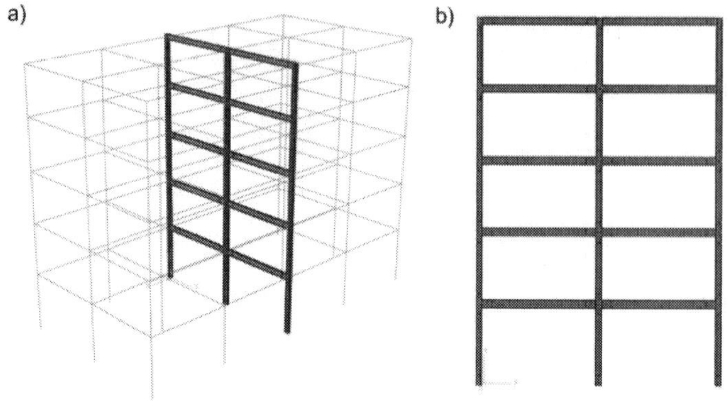

Fig. 2. Numerical model of building: a) isometric view b) main frame- front view.

In the numerical model the rigid, moment resistant joints were assumed. The connections were modelled as a welded connection. The exemplary connection between columns and beams is presented in Fig. 3. The elements were tied (there was compatibility of element displacements between column sides and beams ends). The rest of structural elements (modelled by beams) were coupled to the middle frame. The translations and also rotations were transmitted from beams to shells.

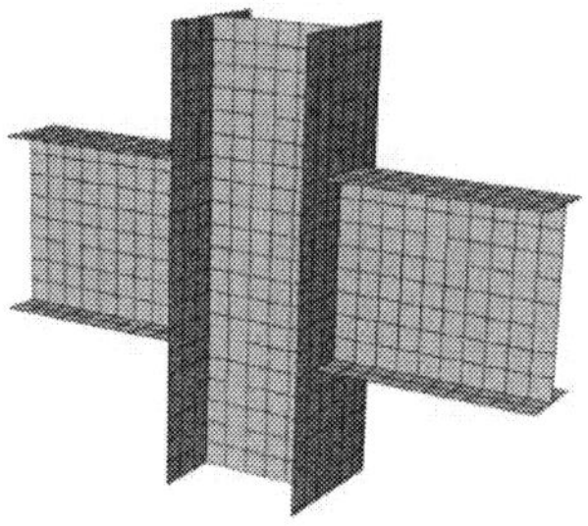

Fig. 3. Details of the frame structural node.

To represent the actual construction behaviour, the mass of non-structural elements were taken into account. The mass of the secondary elements were added to structure as inertiae masses. The masses were evenly distributed along each beams. The mass of concrete slab and 40% of live load were included into inertia.

The numerical model was supported at the end of columns. The fixed (each displacements equals 0) boundary conditions were used. The numerical model was created with Abaqus software [6]. The mesh consisted of 4092 beam elements (B31- 3D beam with 1st order interpolation) and 36504 shell elements (S4R - 4 nodes element with reduced integration). Used elements are available and describe in Abaqus element documentation. The size of elements were taken from convergence analysis.

4 Data of kinematic excitation

The dynamic response analysis of multi-storey building was carried out for a mining shock. The real, strong mining shock was taken into account. The shock was registered near Szombierki, in the Upper Silesian Coal Basin in Poland [7]. The region where the mining tremor was registered is the main region of mining activity in Poland. The shock was represented by the time history of ground accelerations in three directions: two horizontal-NS, WE and vertical. The accelerations record are presented in Fig. 4.

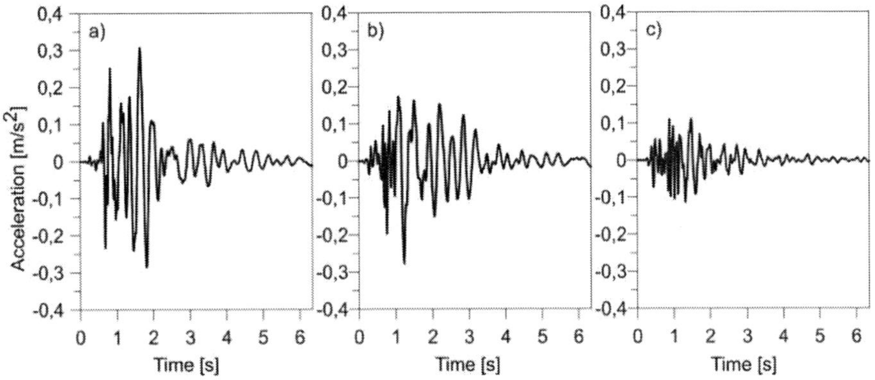

Fig. 4. Time histories of accelerations of the mining shock in: a) NS, b) WE, c) vertical direction.

Based on the figures 4a and 4b it can be seen that the maximum amplitudes in horizontal direction reach 0.3 m/s^2 , whereas the amplitude in vertical direction equal 0.12 m/s^2. In the analysis the mining shock was scaled up by 4. The maximum horizontal acceleration inputted into numerical model equals 1.2 m/s^2 and vertical 0.5 m/s^2. The Fourier analysis of the signal indicated that the dominant frequencies were located within the range of 2-5 Hz for all directions.

The time histories of acceleration were input to the structure as the kinematic excitation of supports. In the numerical model the varying ground motion model were taken into account [8, 9]. The uniform and non-uniform ground motion were applied. The non-uniform excitation model takes into consideration wave passage effect with defined wave speed- the building supports repeat the same motion with a delay depend on the wave velocity. In this study the wave speed of 300 m/s was assumed. It correspond to clayey sands ground. In the uniform excitation model, each supports move in the exactly the same way during analysis time.

5 Response Spectrum Analysis

The dynamic analysis presented in this paper include also the dynamic response spectrum method. This analysis based on spectral curve, specific for Upper Silesian Coal Basin region. The acceleration response spectrum is presented in Fig. 5. The spectrum curve was determined for damping ratio equals 5% and for ground type C (sand) [7]. The spectrum method allows to estimate the maximum stresses in structural elements. The complex simulation (like time history analysis) was not needed in response spectrum method.

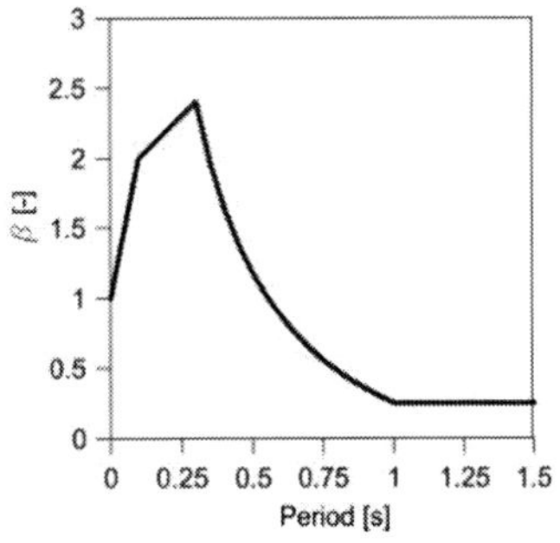

Fig. 5. The acceleration spectral curve.

6 Results

6.1 Uniform excitation

The dynamic response of analysed building was evaluated by the time history analysis. The Hilber-Hughes-Taylor direct integration method was used for the solution of equations of motion. The Rayleigh model of mass and stiffness proportional damping was applied. In the dynamic calculations the dead load and also a part of live load were taken into consideration. During analysis the stress in many points of structure were calculated. In this paper the results for representative points were presented only. The localization of chosen points is presented in Fig. 6. Point P1 and P2 were located in the middle of the span on 1st and 5th floor respectively. Point P3 was located in frame connection.

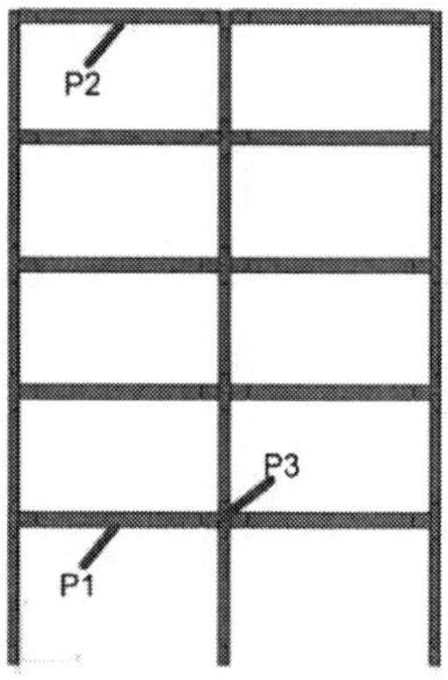

Fig. 6. Representative points chosen during analysis.

The first step of analysis was the calculation of stress for uniform excitation. The time history of stress obtained during calculation was presented in Fig. 7. Fig. 7a presents the stress distribution in point P1; Fig. 7b shows stress in point P2, and Fig. 7c represents results for point P3.

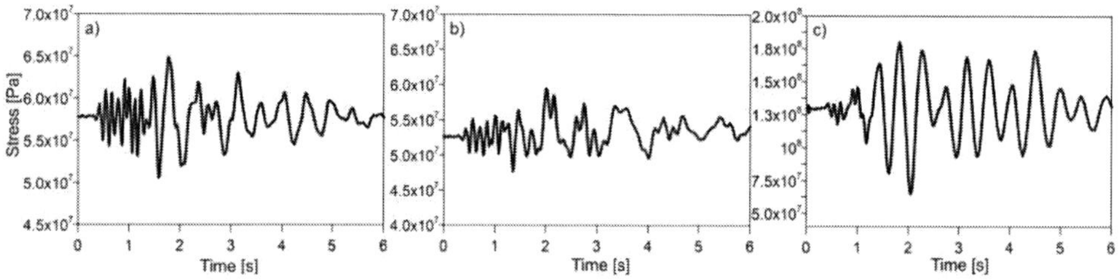

Fig. 7. Time history of stress in point: a) P1, b) P2, c) P3.

In Fig.7 the maximal principal stress is presented. The analysis of stress in point P1 shows that the values of stress oscillated around the constant stress. This stress comes from the dead load of structure. In case of point P1, the dead load stress equals 58 MPa. Based on the presented time history of stress, two regions of stress intensification can be identified. The first range occurred around 2nd second of excitation time and the another was set between 3-3.5 s of shock. Both intensification ranges occurred in the time of local increasing of shock acceleration.

The maximum value of stress in point P1 reach 65 MPa, and appears in 1.9 s of shock. It is the time of maximum peak ground acceleration. The peak stress obtained in point P1 during mine shock excitation is about 15% greater than stress achieved due to the dead load. The minimum value of stress equals 51 MPa.

The stress distribution presented in Fig. 7b (for point P2) shows similar variability in time as stress in point P1. The course of stress indicated two intensification regions too. It is worth to mentioned that this zones appeared at the same moment for points P1 and P2. The maximum stress in point P2 achieved almost 60 MPa, whereas minimum 47 MPa. It was clearly to see that the stress oscillated about value 52 MPa which comes from the self-weight of structure. The stress which occurred in the second intensification region, also reach the significant value. The peak stress (in 3.5 s time of shock) reach 57 MPa and it set 110% of initial stress (stress for dead load).

In addition the stress in frames connection was analysed. The results were presented in Fig.7c. The stress indicate the different variability than in other points. The stress oscillating with lower frequency and show steadily value of stress during whole excitation time. The stress in point P3 achieved maximum value in 2nd second of shock equals 180 MPa. The peak value of stress was almost 40% greater than stress for self-weight.
To complete the frame analysis, the stress distribution in connection was analysed. The map of stress in lower connection was shown in Fig. 8. The figure presents the Misses stress distribution in time of the maximal stress appearance.

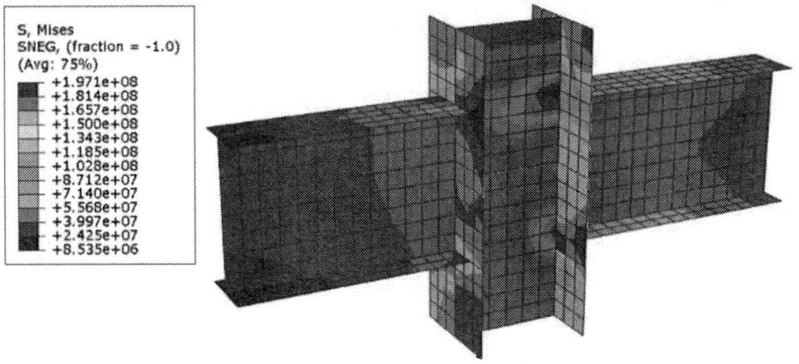

Fig. 8. Mises stress distribution in 2nd second time of shock.

Based on the Fig. 8 it can be noticed that the most charged elements occurred in column. Maximum stress appeared near the place where the beams joint to the column. It was caused by bending moment occurred in connection. The maximum value of stress equals about 200 MPa and was located nearby upper flange plate of girder. The stress in beams were significant lower than in column. The peak value of stress was 120 MPa and it was almost 1.7 times lower than maximum stress in connection.

6.2 Comparison of results obtained for uniform and non-uniform excitation

In the next step, the analysis for non-uniform excitation was conducted. The stress analysis was executed for the same points as previous calculation. The results obtained during both steps of analysis were compared. The comparison were presented in Fig. 9.

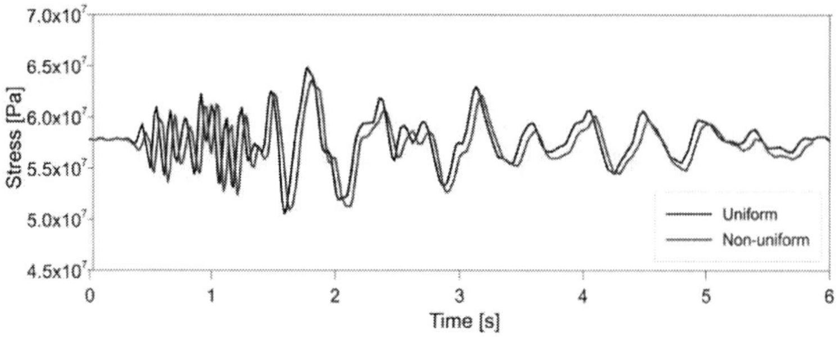

Fig. 9. Comparison the time history of stress in point P1.

Based on the results presented in Fig. 9 it can be seen that the stress obtained for uniform and non-uniform excitation were similar. The time history of stress indicated identical variability in time for both variants of excitation. The stress value for each points were also similar: the differences for stress in points P1 reach 1 MPa, so they did not exceed 10%. The maximum difference occurred form 1-2 s of excitation time. It is worth to mention that the distribution of stress were lagged in time about constant value during the whole shock. The translation was the effect of mining shock wave speed. Hence, the difference in stress values obtained in this point, for uniform and non-uniform excitation were unnoticeable. Only the time delay appeared. Similar situation occurred in points P2 and P3.

6.3 Estimation of results by response spectrum analysis

The last part of analysis was the response spectrum analysis. The calculation for chosen points of frame was conducted. The obtained results were compared to the value of stress from previous calculations. The stress comparison were presented in Table 1.

Table 1. Comparison of peak stress for THA and RSM.

Point	THA stress [MPa]	RSM stress [MPa]	Difference [%]
P1	65	72	10
P2	60	65	15
P3	180	209	17

The results summarized in Table 1 indicate that stress comes from response spectrum method (RSM) were greater than from time history analysis (THA). In each case the spectrum method brought the significant higher stress than the THA. The greatest difference equals 17% and it was reached for point P3 (the point in connection). For the other points the differences equals 10% for point P1 and 15% for point P2. The presented results indicated that the spectrum analysis provide to the secure stress level.

7 Conclusions

The dynamic response analysis of a multi-storey building to a mining shock allowed the formulation some conclusions.

The dynamic analysis for uniform excitation demonstrates, that the strong mining shock caused a significant increase of stresses in structural elements. The greatest increasing of stress was observed near connection of structure. The achieved stress were about 40%

greater than stress comes from the dead load. The maximum stress in connection reach 190 MPa. The stress which appeared in beams, indicated slight increase.

The observation of stress distribution achieved during analysis with uniform and non-uniform excitation, shows that the value of stress was similar for both cases. The analysis with non-uniform excitation indicate a unnoticeable difference in comparison with uniform excitation. The peak value of stress in structure was identical in each case of ground motion model. It shows that the analysed building- compact in plane with short sides- was invulnerable to wave passage effect.

The comparison of results from time history analysis and response spectrum method revealed that the spectrum method allow to estimate the solution. The maximum stress obtained during spectrum analysis were greater than stress from full time analysis. In that case the response spectrum method was the upper assessment of stress. The difference between results from THA and spectrum method were about 10-17%.

References

1. R. Jankowski, Eng Struct. **37**, 99-105 (2012)
2. T. Sugiyama, T. Ishii, M. Kaneko, Comput Geotech, **17**, 547–564 (1995)
3. A. Fathieh, O. Mercan, Eng Struct. **122**, 83-92 (2016)
4. L. Tian, Y. Wang, Z. Yi, H. Qian, Scientific World J. **2014**, 9 (2014)
5. J. Dulińska, AGH Journal of Mining and Geoengineering, **36/1**, 133-143 (2006)
6. ABAQUS 2012. User Manual V.6.12-2, Dassault Systems Simulia Corp., Providence
7. T. Tatara, *Odporność dynamiczna obiektów budowlanych w warunkach wstrząsów górniczych,* (2012)
8. J. Dulińska, D. Jasińska, AMM, **459**, 608-613 (2013)
9. A. Zerva, *Spatial variation of seismic ground motion*, (2009)

Systematization modern devices for the protection of buildings and structures due to seismic impacts

Natalia Kolosova[1], *Evgenii* Kolosov[1] and *Kamil* Agishev[2]

[1] Peter the Great Saint-Petersburg Polytechnic University, Institute of Civil Engineering, 195220 St.Petersburg, Russia
[2] TÜV International RUS, 190000 St.Petersburg, Russia

Abstract. The various types of structures and devices used for seismic isolation and protection of buildings and structures are considered in this work. In this paper the authors consider and analyze both traditional and modern design and protection means of buildings under seismic conditions with a view to their classification. As a result of this research the authors suggest modern classification of device for seismic protection and seismic isolation. According to the authors, this classification will allow to take the necessary constructive solution for the construction of facilities in the earthquake area.

1 The basic systematization of seismic protection methods

Every year, there are about a million earthquakes throughout the world, but most of them are so small and go unnoticed. Really strong earthquake capable of causing extensive damage occur on the planet about once every two weeks. [1].

Since seismic hazard is very important problem for the construction industry, many scientists have tried to systematize the methods and ways to protect structures from earthquakes. All used methods of protection buildings and structures are divided into traditional and special [2].

Traditional methods of protection are the methods based on increasing the carrying capacity and the sustainability of the building through the use of various design and architectural - planning decisions.

The first such systematization was presented in the form of a diagram, in which all methods of protection are divided into active (special) and passive (traditional) methods of protection [3-5].

In the future, to protect structures from earthquakes have applied combinations of active and passive methods [6-8] (Fig.1).

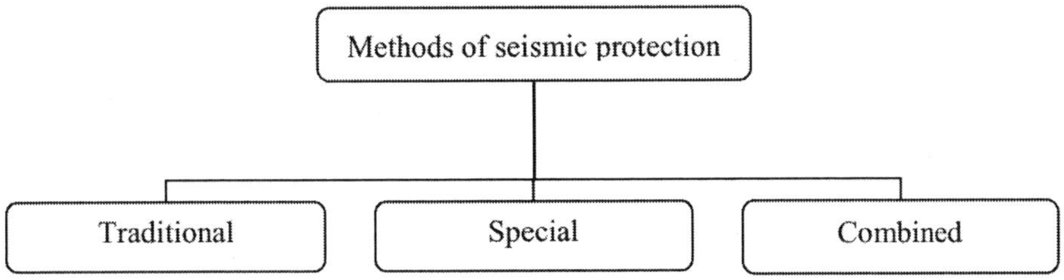

Fig. 1. Initial classification.

1.1 Traditional methods of seismic protection

As noted above, traditional passive methods of protection of buildings and structures are based on the principle of absence of special additional devices.

During design necessary to use a special constructive solutions:

- centrally symmetrical construction,
- elements of bearing structures should be equal in strength,
- to ensure the uniformity and monolithic character of construction,
- the architectural form of the building in the form of a circle, square, symmetric polygon,

Passive method of seismic protection generally leads to an increase in volume of the used materials, which in turn leads to an increase in weight and the inertial efforts.

With the development of the construction industry and the emergence of more complex building structures the above methods could not always be realize, required a new recommendation.

For example, for the construction of high multi-story facilities recommended to make a massive deep foundation; the long and not symmetrical buildings must be equipped with anti-seismic joints, providing the cutting of the building into separate pieces; in the presence of weak or fractured soils is recommended to carry out a partial or complete replacement [9].

Of course, these methods have led to positive results in some cases. However, as the experience of studying the consequences of strong earthquakes in Japan, Italy, China, Venezuela and other countries to ensure the full safety of buildings and structures using these recommendations is not possible [10].

In connection with this circumstance, the alternative (special) methods and means of increasing earthquake resistance intensively develop for the last 3-4 decades along with traditional methods of providing earthquake resistance.

1.2 Special methods of seismic protection

The ideas to create a special non-traditional method of providing seismic resistance based on well-known principles which include:

- natural frequencies of the structure compared to the predominant frequencies of seismic influence,
- foundation without a rigid connection with the construction,
- using dynamic vibration absorbers of different types, etc.

Special methods of seismic protection allow reducing the dynamic loads on the structure during the oscillation process by adjusting dynamic characteristics, thus reducing the resonance effect on the structure.

This effect can be achieved with the use of special seismic protection devices.

For the first time the seismic protection device was used in 1909 by J. A. Calanterients. His system played the role of seismic isolation based on the principle of material sliding along material. As seismic isolation material, the author recommended the use the mica, talc or sand [5].

Also, should be noted the use of sliding basements, moving rollers, balls of different diameters as early experiments of special non-traditional devices [3-7].

In the 70 years of the twentieth century with the development of science, there was a rubber used as seismic isolation material in Japan [10]. This material was used in the construction of basements.

Further, was developed a new seismic protection element by vulcanization of the steel plate with rubber material. This element increases the vertical stiffness of the shock absorber.

In the last 20-25 years the use of nontraditional methods of seismic isolation has been widely developed. Currently there are more than 100 patented designs of seismic isolation in the world. Consider some in detail.

1.2.1 Active seismic isolation systems

The principle of operation of such systems based on the use of automatic control of power elements, which creating the dynamic effect [11].

The effectiveness of such trade is very high, and the system can be versatile enough to allow rapid changeover the virtually any seismic effects.

A good example of active systems trade is the latest development of the Japanese researchers called Japanese Levitating House System design by the company "Air Danshin Systems Inc" [10-14].

The Japanese designers reasoned that if the earthquake is affecting you, only when you stand on the ground, therefore, in a couple of centimeters from the ground is already quite safe. This concept formed the basis of the new "flying houses". A normal-looking house set on concrete earthquake resistant foundations and structure is placed around the system mechanisms: a sensor that detects the tremors and sends a signal to the compressor, that it began to pump under the building a large enough volume of air. To keep the house maintained its stability during levitation, the amount of air is regulated by means of the special valve.

As a result, the building gradually rises above the base, three centimeters and waits in this state, the cataclysm. Then it's returned to safely to his seat. Such a futuristic house has already acquired 88 private homes throughout Japan, and in the meantime, a development company is hoping for wider adoption of its technology.

Also, the installation of frames earthquake resistant foundation designed by scientists from Stanford is one of the latest developments. Heavy multi-store structures difficult to get to fly, but they can move, while still standing on the quaking earth.

Greg Deierlein and his colleagues have developed a special frame of steel, which can be strengthened on the walls like old buildings and those that are just being built [15]. This durable frame connects all the floors with basement – but not very strongly, and so that during an earthquake the walls can deviate in different directions, up or down.

The steel frame is prudently covered with network wires that return the walls to original position. In fact, the house moves to the beat with underground vibrations and earthquake energy is extinguished. The protection of buildings and structures under seismic impacts is devoted to the many works of domestic and foreign experts.

Currently, the number of objects constructed using various means of seismic isolation and seismic protection in industrialized countries amounts to many hundreds, while these include structures with a high degree of responsibility such as the reactor compartment of nuclear power plants, large bridges, high-rise buildings and others [16-22].

Taking into account the diversity of seismic isolation devices we have necessity to systematize them. The authors analyzed the available literature about spatial seismic isolation devices and created the first attempt of their systematization, taking into account the latest developments (Fig. 2).

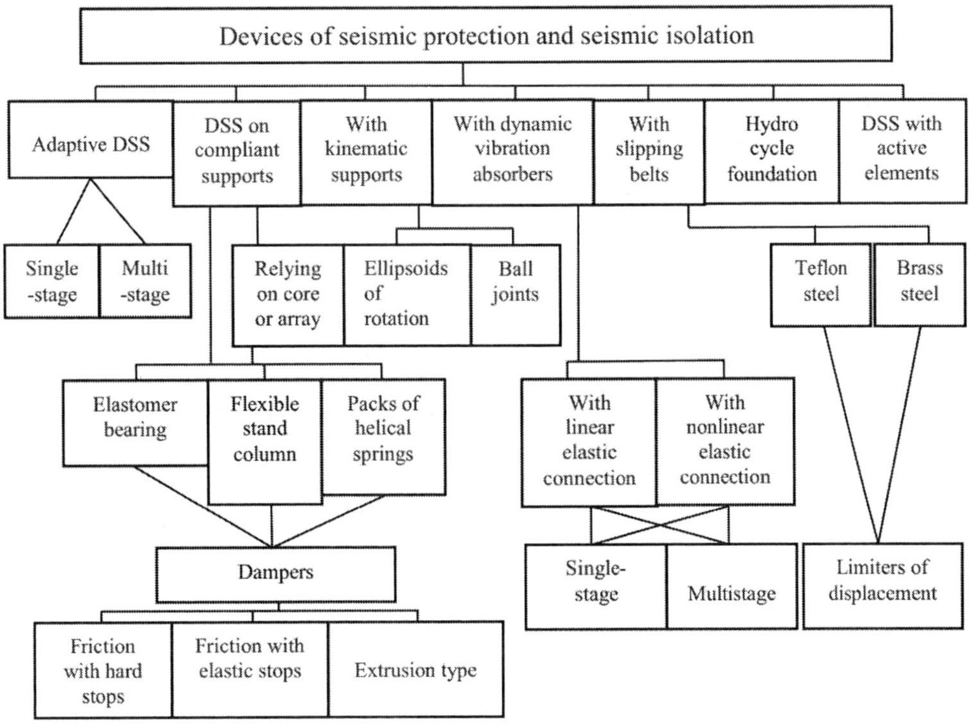

Fig. 2. Classification of devices of seismic protection and seismic isolation.

1.3 Combined methods of special seismic protection

Considering the different possibilities of traditional and special methods of protection of buildings and structures against seismic loads is also advisable to use a joint combination of traditional and special methods.

The Significant progress in the application of combined methods of protection was achieved by the American and Japanese scientists. They have achieved a reduction of destructive effects on building structures. [6, 10, 16]

Combining a set of different activities, traditionally recommended by the existing regulatory documents allows improve the bearing capacity of the building structure; significantly reduce the seismic loads on the structure, changing the dynamic stiffness or natural frequency of the building. Regulation of dynamic parameters allows to avoid resonance increasing of the vibration amplitudes of the structures and to reduce resonance effects.

As special devices you may use methods of moving belts, various dampers, kinematic foundations, braced systems with flexible diaphragms, elastomeric bearings, spatial-plastic dampers and others.

In other words, we have to use a combined method of protection. However, it should be noted that this combination does not mean using only one of these methods.

In the case of the combined approach and if necessary it is possible to use several methods, both traditional and special techniques at the same time.

2 Conclusion

The aim of this work was to study and analyze the existing methods and devices used in the protection of buildings and structures against seismic loads and their systematization. In our opinion, systematization is an area of research which has own scientific value.

In the literature the systematization of seismic protection methods is shown separately either in the form of short circuits, either in the form of small textual descriptions. After analyzing the available material, the authors of this article offer a more complete version of the systematization of modern methods and devices of seismic protection and seismic isolation of structures.

This work is aimed at practical application and will help participants of the construction process to orient correctly in the variety of available funds and choose the optimal method for protecting buildings from seismic effects in certain conditions. The authors intend to continue the study of methods and devices for seismic protection.

References

1. K.M. Shedlock, L.C. Pakiser, *Earthquakes. In science for changing world* (2007), available on internet: http:// Pubs.Usgs.gov/gip/earthquakes
2. A. R. Arutyunyan, *Modern methods of seismic isolation of buildings & structures*, MAG CIVIL ENG, **3 (13)**, (2010)
3. J. M. Eisenberg, A. I. Neumann, A. D. Abakarov, etc, *Adaptive system for seismic protection of structures* (Moscow, 1978)
4. V. S. Polyakov, *The question about the effectiveness of the dynamic damper of oscillations under seismic forces,* Construction mechanics and calculation of structures, **5**, 49-53 (1980)
5. A. M. Uzdin, *Basic theory of seismic & earthquake-proof construction of buildings & structures*, 176 (St. Petersburg, 1993)
6. R.I. Skiner, W.H. Robinon, G.H. McVerry, *Introduction to seismic isolation* (New Zealand, 1993)
7. J. M. Eisenberg, V. I .Smirnov, etc, *Effective system of seismic isolation. Research, design & construction,* SB. VNIINTPI, Earthquake-resistant construction. Safety of structures, **1**, 31-37 (2002).
8. V. A. Lyadsky, *Earthquake-resistant buildings,* Abstracts of scientific-research works of students and postgraduates of SPbSPU (2008)
9. J. M. Eisenberg, *Seismic isolation of tall buildings*, Earthquake-resistant construction. Safety of structures. **4**, 41-43 (2007)
10. A. Kato, S. Moro, M. Morishita, *Development Program of Three- Dimensional Seismic Isolation for Advanced Reactor System in Japan*. SMIRT-17, K 10-1, 1-8 (2003)

11. O. A. Savinov, *Dynamic problems of construction machinery,* St. Petersburg (1993)
12. J. D. Cherepansky, *Comparative analysis of the seismic isolation foundations reference type,* Earthquake-resistant construction. Safety of structures, **5**, 31-34, (2004)
13. G. E. Avidan, E. A. Carlina, *Vibration characteristics of buildings with seismic isolation foundations*, Earthquake resistant construction. Safety of structures, **1**, 42-44 (2008)
14. D. Rashid, E.S. Kolosov, N.B. Kolosova, T.N. Soldatenko, *Evaluation's methods of the seismic stability of buildings and structures in the Gaza Strip* (2016) http://www.crcnetbase.com/doi/book/10.1201/9781315393827
15. H. Burton, G. Deierlein, *Simulation of Seismic Collapse in Nonductile Reinforced Concrete Frame Buildings with Masonry Infill*, J STRUCT ENG, **8**, 140, (2014)
16. J. Barr, *The seismic safety of bridges: A view from the design office*, 12th European Conference on Earthquake Engineering (Oxford, UK, 2002)
17. A. D. Abakarov, *Study of the seismic response & evaluation of rational parameters of seismic isolation systems with sliding zone under seismic loading*, Earthquake-resistant construction. Safety of structures, **6**, 31 -34 (2003)
18. A.S. Scalat, *Evaluation of existing building in Israel for seismic hazard*, Wiley Inter Science. Earthquake Engineering Dynamics, *6*, Issue 3, 317- 325 (2007)
19. Wiggins, D.F. Noran, *Earthquake safety in the city of Long Beach* (California 2008), available on internet: www.amazon.com/Earthquake
20. A.S. Hokmabadi, B. Fatahi, B. Samali. *Assessment of soil-pile-structure interaction influencing seismic response of mid-rise buildings sitting on floating pile foundations*. Computers and Geotechnics, 172-186 (2014)
21. V.S. Vorob'yev, E.I. Valkner. *Features of designing zero-cycle constructions for multistorey block of flats in seismic regions*, All-Russian Scientific Conference of Students, PhD Students and Young Scientists, 204-206 (Novokuznetsk, SGIU, 2015)
22. H. Qu, R. Li, H. Hu, H. Jia, J. Zhang, *An approach of seismic design for sheet pile retaining wall based on capacity spectrum method.* Geomechanics and Engineering, 309-323 (2016)

Explicit and implicit method in nonlinear seismic analysis

Ivan Němec[1,*], *Hynek* Štekbauer[1], *Adéla* Vaněčková[1] and *Zbyněk* Vlk[1]

[1]Brno University of Technology, Faculty of Civil Engineering, Institute of Structural Mechanics, Veveří 331/95, 60200 Brno, Czech Republic

Abstract. The paper deals with suitability of use of the explicit and implicit method in heavily nonlinear seismic analysis. Considering typical time of duration of earthquakes from several seconds to approximately twenty seconds, it would seem that use of the implicit method is definitely more suitable. The shape of accelerograms, however, requires quite short time steps. For the explicit method, on the other hand, very short time steps are required in order to obtain a stable solution. Both methods must be compared with respect to a heavily nonlinear response, typical for seismicity. Conclusions of the paper are made on the basis of detailed numerical study and may be very useful for practice.

1 Numerical methods of direct integration of equation of motion

In the following text let us introduce the numerical methods used in the study. A system of motion equations of a discrete model of a structure subjected to a dynamic load can be written

$$\mathbf{Ma}(t) + \mathbf{Cv}(t) + \mathbf{Ku}(t) = \mathbf{F}(t) \tag{1}$$

Numerical methods of direct integration solve system (1) in a finite number of time instants $t_0, t_1, \ldots, t_m$. The distance between individual time instants $\Delta t_i = t_i - t_{i-1}$ is called the length of the integration step. The lengths of integration steps Δt_i influence the accuracy, stability and speed of the solution. Defined initial conditions are an integral part of system (1). The time $t = 0$ is considered to be the starting point at which $\mathbf{u}(t_0) = \mathbf{u}_0$, $\mathbf{v}(t_0) = \mathbf{v}_0$. System (1) can be thus written as

$$\mathbf{Ma}_i + \mathbf{Cv}_i + \mathbf{Ku}_i = \mathbf{F}_i \tag{2}$$

1.1 Explicit methods

In explicit methods we make use of the assumption about the distribution of motion characteristics $\mathbf{u}$, $\mathbf{v} = \dot{\mathbf{u}}$, $\mathbf{a} = \ddot{\mathbf{u}}$ in interval $\langle t_i, t_{i+1} \rangle$ and the knowledge of these characteristics

* Corresponding author: nemec@fem.cz

at time instant t_i, and we calculate vectors $\mathbf{u}_{i+1}$, $\mathbf{v}_{i+1}$, $\mathbf{a}_{i+1}$ from (2). Neither triangulation nor modification of the stiffness matrix is performed in explicit methods.

1.1.1 Method of central differences

The numerical integration of differential equations uses the substitution of the derivative of the independent variable with respect to time. If we replace the derivatives in (2) by

$$\mathbf{v}_i = \frac{1}{2\Delta t_i}\left(\mathbf{u}_{i+1} - \mathbf{u}_{i-1}\right) \tag{3}$$

$$\mathbf{a}_i = \frac{1}{\Delta t_i^2}\left(\mathbf{u}_{i+1} - 2\mathbf{u}_i + \mathbf{u}_{i-1}\right) \tag{4}$$

we get a recurrent formula for $\mathbf{u}_i$

$$\left(\frac{1}{\Delta t_i^2}\mathbf{M} + \frac{1}{2\Delta t_i}\mathbf{C}\right)\mathbf{u}_{i+1} = \mathbf{F}_i - \left(\mathbf{K} - \frac{2}{\Delta t_i^2}\mathbf{M}\right)\mathbf{u}_i - \left(\frac{1}{\Delta t_i^2}\mathbf{M} - \frac{1}{2\Delta t_i}\mathbf{C}\right)\mathbf{u}_{i-1} \tag{5}$$

The method has all the advantages of explicit methods as long as $[\mathbf{C}] = [\mathbf{0}]$ or $[\mathbf{C}] = \alpha[\mathbf{M}]$. Its application is most effective for diagonal mass matrix. However, the method is only conditionally stable. The length of the integration step must meet the condition

$$\Delta t_i \leq \frac{T_n}{\pi} \tag{6}$$

where T_n is the smallest vibration period.

1.2 Implicit methods

Implicit methods are based on system (2) at time instant t_i. The numerical integration of the system is carried out step by step using the following formula

$$\mathbf{a}_i = f\left(t_i, \mathbf{F}_i, \mathbf{u}_{i-1}, \mathbf{v}_{i-1}, \mathbf{a}_{i-1}\right) \tag{7}$$

with the necessity – in order to be able to start with the solution – to evaluate the acceleration at the beginning of the motion at time t_0 directly from system (2)

$$\mathbf{M}\mathbf{a}_0 + \mathbf{C}\mathbf{v}_0 + \mathbf{K}\mathbf{u}_0 = \mathbf{F}_0 \tag{8}$$

1.2.1 Newmark methods

The basic formula of the Newmark method that specifies the relations between displacement, velocity and acceleration vectors have the following form

$$\mathbf{u}_i = \mathbf{u}_{i-1} + \Delta t_i \mathbf{v}_{i-1} + \left(\frac{1}{2} - \beta\right)\Delta t_i^2 \mathbf{a}_{i-1} + \beta \Delta t_i^2 \mathbf{a}_i \tag{9}$$

$$\mathbf{v}_i = \mathbf{v}_{i-1} + (1-\gamma)\Delta t_i \mathbf{a}_{i-1} + \gamma \Delta t_i \mathbf{a}_i \tag{10}$$

where β and γ are what is termed *Newmark's parameters*. As formulas

$$\mathbf{u}_i = \mathbf{u}_{i-1} + \Delta\mathbf{u}_i,\ \mathbf{v}_i = \mathbf{v}_{i-1} + \Delta\mathbf{v}_i,\ \mathbf{a}_i = \mathbf{a}_{i-1} + \Delta\mathbf{a}_i \tag{11}$$

holds we can write the following formula for the vector of acceleration increments and velocity increments

$$\Delta\mathbf{a}_i = \Delta\bar{\mathbf{a}}_i + \frac{1}{\beta h^2}\Delta\mathbf{u}_i, \quad \Delta\mathbf{v}_i = \Delta\bar{\mathbf{v}}_i + \frac{\gamma}{\beta h}\Delta\mathbf{u}_i \tag{12}$$

where
$$\Delta\bar{\mathbf{a}}_i = \left(-\frac{1}{\beta\Delta t_i}\mathbf{v}_{i-1} - \frac{1}{2\beta}\mathbf{a}_{i-1}\right), \quad \Delta\bar{\mathbf{v}}_i = \left(1 - \frac{\gamma}{2\beta}\right)\Delta t_i \mathbf{a}_{i-1} - \frac{\gamma}{\beta}\mathbf{v}_{i-1} \tag{13}$$

The total increments of the displacement (and analogous increments of the velocity, acceleration and force vectors) can be written as

$$\Delta\mathbf{u}_i = \sum_{k=1}^{n_{ITER}} {}^k\Delta\Delta\mathbf{u}_i \tag{14}$$

Using the substitution according to formulas above into (2) and the obtained relation is modified we get

$$\left(\frac{1}{\beta\Delta t_i^{\,2}}\mathbf{M} + \frac{\gamma}{\beta\Delta t_i}\mathbf{C} + {}^k\mathbf{K}_{\mathrm{T},i}\right)\cdot {}^k\Delta\mathbf{u}_i = {}^k\Delta\mathbf{F}_i - \mathbf{M}\Delta\bar{\mathbf{a}}_i - \mathbf{C}\Delta\bar{\mathbf{v}}_i \tag{15}$$

The bracket on the left hand side of (15) represents what is termed modified stiffness matrix, which can be denoted $\hat{\mathbf{K}}_i$. The presented formula can be written in a similar form as

$${}^k\Delta\Delta\mathbf{u}_i : \begin{cases} \hat{\mathbf{K}}_i\,{}^k\Delta\Delta\mathbf{u}_i = {}^k\Delta\Delta\mathbf{F}_i - \mathbf{M}\Delta\bar{\mathbf{a}}_i - \mathbf{C}\Delta\bar{\mathbf{v}}_i\ ; & \text{for } k = 1 \\ \hat{\mathbf{K}}_i\,{}^k\Delta\Delta\mathbf{u}_i = {}^k\Delta\Delta\mathbf{F}_i\ ; & \text{for } k > 1 \end{cases} \tag{16}$$

Using this relation (16), it is possible to calculate partial increments of displacement.

2 Comparison of the used numerical methods

For this numerical study a wall of six floor building was used. The structure was subjected to the seismic load due to the accelerogram from Umbro-Marchigiana, Italy.

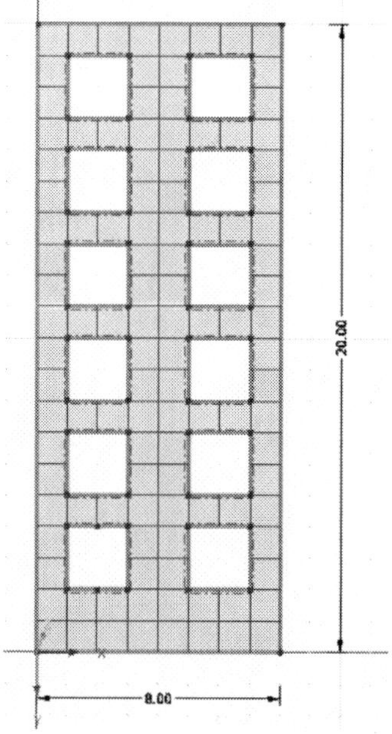

Fig. 1. The analysed structure.

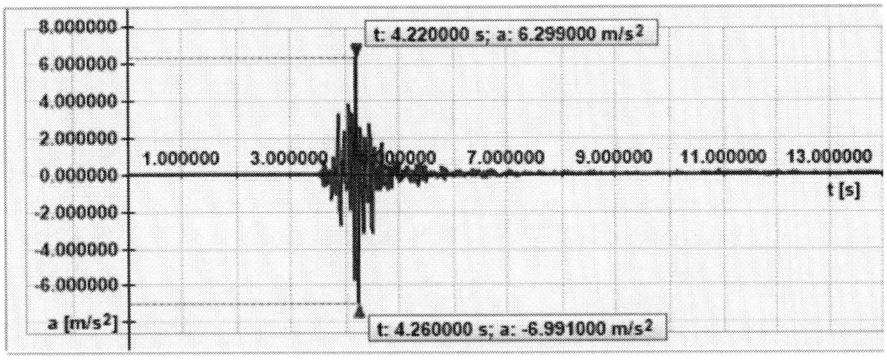

Fig. 2. Accelerograms Umbro-Marchigiana, station Colfiorita-Casermette.

The important aim of the paper is to compare the suitability of use of the explicit method and the Newmark's implicit method in the seismic analysis of structures.

It is commonly known that the explicit method (described e.g. in [1], [2] and [3]) is suitable for actions with very short duration, such as explosions or the impacts of vehicles against obstacles, i.e. actions generally studied via transient dynamic analysis. With the explicit method, the time step necessary from the point of view of computational stability is determined by the shortest time in which sound overcomes the distance between two arbitrary nodes of the structure of the given material. This time step must then be decreased to an even lower duration by a certain safety factor, let's say 0.7, to be sure that the calculations will be stable. The maximal possible time step then depends on the Young's modulus of the material, the density of the material (the square root of the ratio between the Young's modulus and density is the speed of sound in the material) and on the shortest distance between arbitrary nodes. It can thus be seen that refinement of the mesh will demand a pertinent decrease in the maximum allowed time step. Of course, this will mean an increase in the amount of computational time needed. It is clear from the above conditions that the explicit method requires an extremely short time step. For the analysis of building structures it is usually in the order of 10^{-5} s. Only in the case that the above-mentioned conditions are fulfilled is it possible to avoid the necessity of solving a system of linear equations at each time step, and instead solve only one equation of motion for each deformation parameter at each time step. In the case of transient dynamic analysis this is not a substantial disadvantage as, for actions like impacts or explosions, the speed at which changes to the external forces occur also requires very short time steps to be used, so the necessity for such time steps due to computational stability requirements is not an issue.

Implicit methods of solving a set of differential equations (described e.g. in [1], [4], [5] and [6]) are characterized by the fact that a system of linear equations must be solved at each time step. In the case of the dynamic solution of structures, dynamic equilibrium equations are employed. The Newmark implicit method thus ensures the equilibrium conditions of the structure are fulfilled at each time step. This method does not demand such a short time step as is needed in the explicit method, and therefore the implicit method is typically suitable for the solution of dynamical problems of a duration longer than several seconds.

In the case of dynamical tasks like the impact of a vehicle against an obstacle, or the effects of explosions, it is necessary to use a very short time step due to the high speed of load changes. Such a situation also occurs when investigating the response of structures to seismic load. Accelerograms commonly take the form of polygons with the time distance between extremes of acceleration being in the order of thousandths of seconds. The computation of such an accelerogram with sufficiently precise results requires the selection of a suitably small time step. This would be an argument for the use of the explicit method. On the other hand, the duration of seismic loading, which varies between seconds and tens of seconds, is a fact supporting the choice of an implicit method, which does not demand so short a time step. Of course, the computational demand for each time step is several times greater compared to that of the explicit method.

Recently, the development of a nonlinear dynamics program module for the RFEM program was completed and this module has now been introduced on the market. Regarding seismic analysis using the direct integration of equations of motion, a substantial improvement compared to competitors has been implemented in the module. Originally, at each time step the acceleration value was obtained by interpolation from the accelerogram and applied in the numerical solution. A study was performed which showed that this approach is not accurate enough because the speed and displacement obtained from numerical time integration from those discrete acceleration values are not accurate enough. A substantial improvement to the algorithm was suggested and implemented. Today, the RFEM program performs precise double time integration right at the beginning of computation in order to obtain the exact time course of displacement from the accelerogram. Additionally, during time step processing, the exact values of displacement in the nodal support are considered instead of the discrete values of acceleration which were used in the previous version of the program. These improvements to the program have been shown to provide a substantial improvement in the accuracy of seismic analysis. This is particularly noticeable in the case of the implementation of Newmark's implicit method in the program. As far as the explicit method is concerned, where the time steps are much shorter than in the implicit method for stability reasons, the improvement of accuracy is not as great as with the implicit method. Nevertheless, some improvement in accuracy has been achieved.

To compare the suitability of the used numerical method for the seismic analysis of a typical building wall, a real accelerogram from Italy was used along with the Drucker-Prager material model. The comparison of numerical methods was performed with regard to accuracy and computational performance. A time step of 0.0001s was applied in the case of the explicit method due to stability requirements. The same time step was also chosen for the implicit method in order to compare the time requirements of both methods for the same number of time steps, and also to obtain a very precise base solution for accuracy comparisons.

From the following graphs showing the course of the horizontal displacement of the upper right corner of the building it can be seen that there is very good concordance between the results of the explicit and implicit method. This can be regarded as proof that the results of the explicit method are good. The comparison between the implicit solutions from the RFEM and ANSYS programs shows that the results of both programs are practically identical.

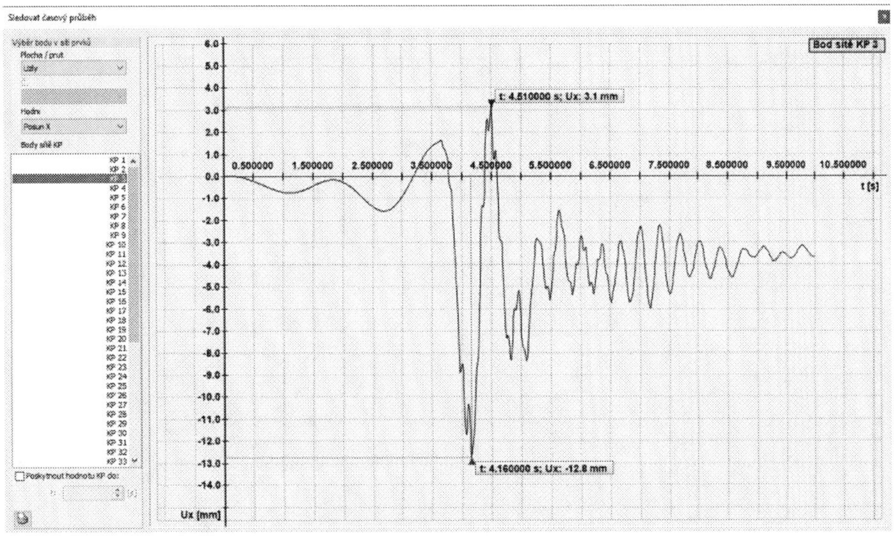

Fig. 3. Explicit method - Graph of the horizontal isplacement of the upper node for the time step 0.0001 s.

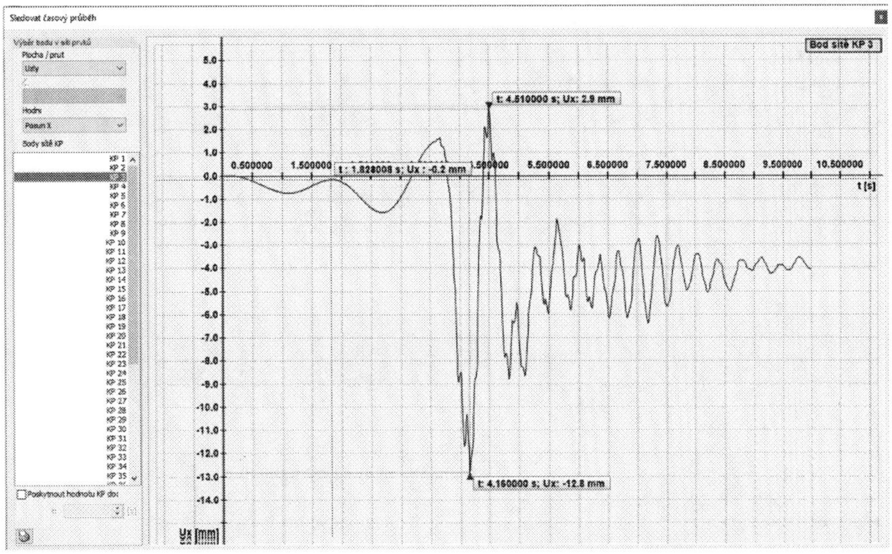

Fig. 4. Implicit method - Graph of the horizontal isplacement of the upper node for the time step 0.0001 s.

3 Conclusion

In accordance with expectations, the calculation performed by the implicit method for the same time step as with the explicit method was several times (about 5x) slower than when carried out by the latter method. The time step for the implicit method was then increased fivefold. In this case, the computational time for both methods was practically identical. The results for both solutions were still practically the same (only the magnitudes of the horizontal displacements of the upper right corner were compared). When the time step for the implicit method was increased further, the results provided by the method remained accurate enough until the time step was twenty times greater than the time step for the explicit method.

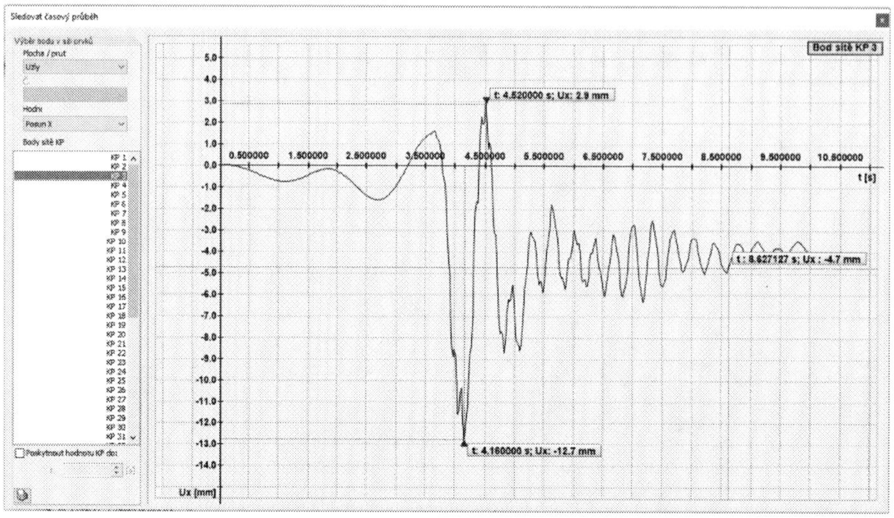

Fig. 5. Implicit method - Graph of the horizontal isplacement of the upper node for the time step 0.002 s.

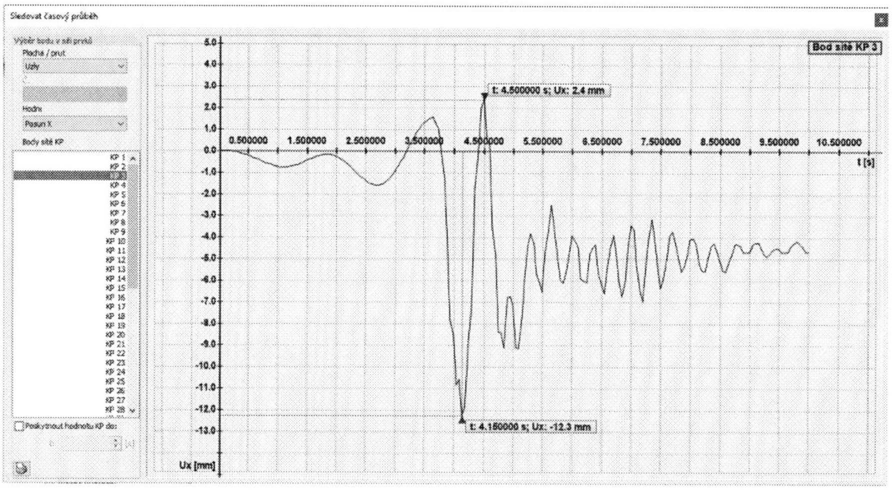

Fig. 6. Implicit method - Graph of the horizontal displacement of the upper node for the time step 0.005 s.

The extreme value of the displacement was only 1% smaller even in this case, which still can be regarded as acceptable accuracy. Good results were thus obtained from the implicit method for a computational time that was four times lower than that used with the explicit method. Taking into account the fact that it is not possible to estimate the highest acceptable time step for Newmark's implicit method accurately enough without additional effort, it can be accepted that both methods are comparable when used for seismic analysis, though the implicit Newmark method is perhaps the preferred option. The estimation of the highest time step value for the implicit Newmark method is possible with the help of Fourier transformation (in our case the same value was achieved by both of the above tests). However, the process is time-consuming and in practice it will be often easier to use a lower time step to be on the safe side, which partially decreases the advantage of the implicit method.

An interesting conclusion of this study is the discovery that the usage of both basic numerical methods for the direct integration of equations of motion, namely the explicit method and Newmark's implicit method, are both competitive in their seismic analysis capabilities. They are also both suitable and comparable as regards practical use.

From the results of numerical analysis it can be seen that there is excellent concordance between the results of the ANSYS and RFEM programs for the implicit method. The concordance demonstrated by the results for the two programs for the explicit method is partially lower, but still acceptable. ANSYS needed some mesh refinement to achieve comparable results.

This research was partially supported by the project FAST-J-17-4720.

References

1. I. Němec, V. Kolář, I. Ševčík, Z. Vlk, J. Blaauwendraat, J. Buček, B. Teplý, D. Novák, V. Štembera: *Finite Element Analysis of Structures, Principles and Praxis* (Shaker Verlag, 2010)
2. Shen R. Wu, Lei Gu, *Introduction to the Explicit Finite Element Method for Nonlinear Transient Dynamics* (John Wiley & Sons, 2012)
3. J. Rodriguez, G. Rio, J.M. Cadou, J. Troufflard, *Numerical study of dynamic relaxation with kinetic damping applied to inflatable fabric structures with extensions for 3D solid element and non-linear behavior*, Thin-Walled Structures **49** (11), 1468-1474, Elsevier, (2011)
4. M. Brio, *Numerical Time-Dependent Partial Differential Equations for Scientists and Engineers* (Academic Press, 2010)
5. K. J. Bathe, *Finite element procedures* (Prentice Hall, New Jersey, 1996)
6. T. Belytschko, T. J. R. Hughes, *Computational methods for transient analysis* (Elsevier science publishers B.V, 1983)
7. R. Courant, K. O. Fredrichs, H. Lewy, *On the partial difference equations of mathematical physics,* IBM Journal of Research and Development **11,** (2), 215–234 (1967)
8. J.Alamatian, *A new formulation for fictitious mass of the Dynamic Relaxation method with kinetic damping*, Computers & Structures **90-91**, 42–54, Elsevier (2012)
9. S. Oller, *Nonlinear Dynamics of Structures* (*Lecture Notes on Numerical Methods in Engineering and Sciences*), Springer (2014)

Safety assessment of the seismic resistance of nuclear power plant technology

Juraj Králik[1,*]

[1]Faculty of Civil Engineering, STU Bratislava, Radlinského 11, 813 68 Bratislava, Slovakia

Abstract. This paper gives the methodology to get the floor response spectra of the buildings on the base of the results of the probabilistic assessment. On the base of the geophysical and seismological monitoring of locality the peak ground acceleration and the uniform hazard spectrum of the acceleration was defined for the return period 10^4 years. The methodology of the seismic resistance of the cable support structures is described.

1 Introduction

One from the most importance technology segments in nuclear power plant (NPP) are the cable way structures. The effective supported structures were the prefabricated steel structures of OBO Bettermann Company. This paper presents the methodology and the results of the numerical and experimental verification of the seismic safety of these structures used in Mochovce NPP buildings.

The earthquake resistance analysis of NPP buildings in Mochovce were based on the recommends of International Agency of Energy Atomic (IAEA) in Vienna [1, 2] to get international safety level of nuclear power plants. Three logical possibilities of the source zones were defined for Mochovce site – contact of Eastern Alps and Western Carpathians, Dobra Voda and alternative fault. The seismic load was defined by PGA (Peak Ground Acceleration) and local seismic spectrum in dependence on magnitude and distance from source zone of earthquake [2]. Two principal methods are appropriate for assessing the seismic safety of facilities, the seismic margin assessment (SMA) method and the Seismic Probabilistic Safety Assessment (SPSA) method. The review level earthquake (RLE) is very effective method to reevaluation of NPP safety then the new seismic hazard is defined [2, 3, 4]. The SMA methodology is based on the RLE determination. The first seismic reevaluation of the seismic hazard of the Mochovce site was established on the deterministic method and seismic-tectonical consideration of the Mochovce site was analyzed on the base of probabilistic method. The design response spectra were prepared based on results of the probabilistic seismic hazard analysis (PSHA) study for the NPP Mochovce site developed by GFÚ SAV [2, 5].

Basic parameters of the design ground response spectrum (GRS) are as follows:

- Spectral shape MDE (SL-2) (Maximum Design Earthquake);
- It corresponds to 84% NEP (Non Exceedence Probability) (i.e. median + 1 sigma);

* Corresponding author: juraj.kralik@stuba.sk

- Seismic event with annual recurrent frequency 10^{-4};
- PGA (Peak Ground Acceleration) for horizontal direction $PGA_H = 0.150g$;
- PGA for vertical direction $PGA_V = 0.100g$;
- Vertical response spectra defined as 2/3 of horizontal response spectra.

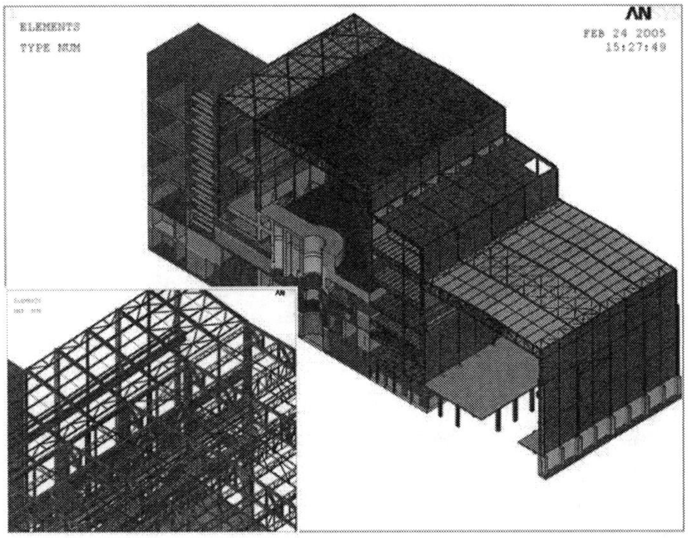

Fig. 1. Calculation model of NPP V2.

Firstly the value of PGA was defined at 1994 ($PGA_{RLE} = 0.1g$) follow in accordance of the results of seismological monitoring this locality at 2003 ($PGA_{UHS} = 0.142g$ and $PGA_{HS} = 0.143g$). The last PGA for horizontal direction was defined as $PGA_H = 0.150g$ on base of the seismic-tectonic investigation of Mochovce NPP site.

Mochovce NPP structures are typically the same as the NPP structures for reactor VVER 440. Around of the reinforced concrete structures of the box SG and bubbler tower are the steel structures of the reactor hall, longitude and transversal electrical buildings are built. The calculation model of NPP structures (Fig. 1) created by VUT Brno and modified by STU Bratislava has 161 856 elements and 135 629 nodes [2, 6]. The foundation of the reactor building is embedded into the rock subsoil

2 Response spectrum compatible accelerogram

To provide input excitations to structural models for sites with no strong ground motion data, it is necessary to generate artificial accelerogram. It has long been established that due to parameters such as geological conditions of the site, distance from the source, fault mechanism, etc. different earthquake records show different characteristics [2, 3, 5 - 13]. Based on Kanai's investigation regarding the frequency content of different earthquake records, Tajimi proposed the following relation for the spectral density function of the strong ground motion with a distinct dominant frequency [14]

$$S(\omega) = \frac{\left[1 + 4\xi_g^2 \left(\omega/\omega_g\right)^2\right]}{\left[1 - \left(\omega/\omega_g\right)^2\right] + 4\xi_g^2 \left(\omega/\omega_g\right)^2} S_0 \tag{1}$$

DOI: 10.1051/matecconf/201710700067

Here ξ_g and ω_g are the site dominant damping coefficient and frequency, and S_0 is the constant power spectral intensity of the bed rock excitation. The Kanai-Tajimi power spectral density function may be interpreted as corresponding to an "ideal white noise" excitation at the bedrock level filtered through the over-laying soil deposit at site. The generalized no stationary Kanai-Tajimi model is represented by the following equation:

$$\ddot{u}_f + 2\xi_g\omega_g\dot{u}_f + \omega_g^2 u_f = y(t), \qquad \ddot{u}_g = -(2\xi_g\omega_g\dot{u}_f + \omega_g^2 u_f).e(t) \qquad (2)$$

where u_f is the filtered response, ω_g is dominant ground frequency, ξ_g is the effective ground damping coefficient, $\ddot{u}_g$ is the output ground damping acceleration, and e(t) is the amplitude envelope function. After numerical integration of eq. (2) can be evaluate the ground damping acceleration $\ddot{u}_g$.

On the base of the methodology defined in [14] the software COMPACEL was created for the modified iteration processes and requirements of various standards [2].

In the case of the NPP structures the requirements of the standard ASCE 4/86 [3] are acceptable by Slovak nuclear authority. These requirements are following:

1. The mean of the zero-period acceleration (ZPA) values shell equal or exceed the design ground acceleration.
2. In the frequency range 0.5 to 33 Hz, the average of the ratios of the mean spectrum to the design spectrum, where the ratios are calculated frequency by frequency, shall be equal to or greater than 1.
3. No one point of the mean spectrum (from the time histories) shall be more than 10% below the design spectrum.
4. The three components of motion in the orthogonal directions shall be statistically independent (with mean correlation smaller than 0.3).

To provide input excitations to structural models for sites with no strong ground motion data it is necessary to generate more than one response spectrum compatible accelerogram. On the base of the US standards [3] requirements three synthetic design accelerograms compatible with the GRS response spectrum were generated by software COMPACEL [2].

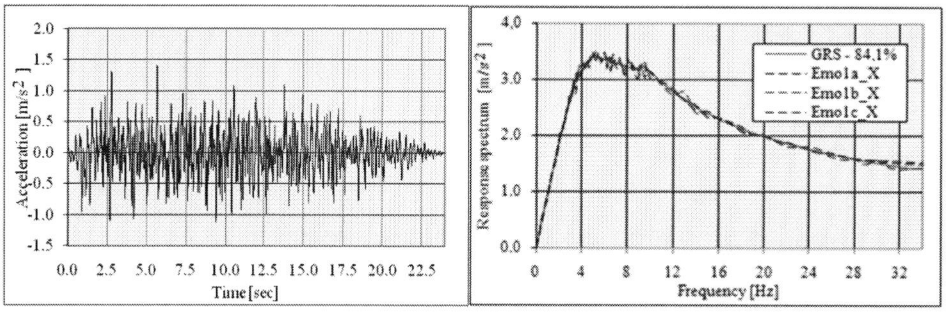

Fig. 2. The synthetic design accelerograms compatible with the response spectrum.

The comparison of the synthetic acceleration spectrum and GRS spectrum in the case of three synthetic accelerograms is showed in Fig. 2.

3 Seismic evaluation methodology

The methodology of structure resistance verification follows the requirements of IAEA [1] and acceptable US standards [3, 4]. This methodology is elaborately described in [2]. There are illustrated the procedures, requirements and criterion of calculation models and methods for design of structure reliability. There are two principal methodology available for seismic design of NPP structures – deterministic (SMA) and probabilistic (SPRA). The

CDFM method is very similar to the design procedure followed in the industry, except that the parameter values have been liberalized. The objective of seismic margin assessment (SMA) is to determine for a nuclear power plant the high-confidence-of-a-low-probability-of-failure (HCLPF) capacity for a pre-selected seismic margin earthquake (SME), which is always chosen higher than the design basis input. In probabilistic terms, the HCLPF is expressed as approximately a 95% confidence of about a 5% or less probability of failure.

4 Floor response spectrum

The equipment and interior structures of NPP are designed using floor response spectrum (FRS) as seismic loads. This spectrum is calculated from the time-history motions resulting from the dynamic analysis of the supporting structures in accordance of requirements of nuclear guide RG 1.122 [15]. The values of FRS are determined with the uncertainties of calculation methods and soil and structure properties. The FRS can be calculated from the in-structure spectra on the base of deterministic and semi-probabilistic methods [2]. The results from the calculation of FRS in all NPP buildings show, that in the case of the floor with the higher variability of distributed masses and slab stiffness, the FRS values are more conservative as the envelope of maximum spectrum values in various points of floor [2].

Deterministic method to generate FRS is defined in NUREG 0800 [16] and IAEA rep. No. 28 [1]. The response spectrum in the points of floor is calculated using the transient analysis of the structure from synthetic 3D accelerograms (median+sigma). One or three 3D accelerograms can be used as input loads. The material properties are calculated with median values (best estimation). The damping values are requirement in ASCE 4/98 [3] (max. 7% for steel and 10% for concrete structures and rock soil). The floor response spectrum may be calculated as envelope of maximum of mean spectrum values in each typical point (minimum 5 points are recommended).

5 Acceptable evaluation methods

For seismic analyses of structures and technological equipment following methods are acceptable [15]:

- Response Spectrum Modal Analysis Method – RSMAM,
- Methods of direct time integration of the motion equations system (Linear Time History Method – LTH and Non-linear Time History Method – NTH are discerned),
- Equivalent Static Method – ESM (or also "approximate equivalent static seismic loads method").

The equivalent static seismic loading method is used in cases when a structure or an equipment component is, in the viewpoint of the mechanics, a simple single-degree of freedom (SDOF) or 2-degree-of-freedom (2-DOF) system and when the intention of the analysis is to determine only the seismic forces and moments in the point of anchorage to the civil structure or to perform the stress analysis in some critical cross-section (when such a simple substitution is obviously sufficient).

The spectral analysis is based on linear behavior of the structures, while damping takes into account the plastic de-formations and the over thrust of friction-type connections. The response spectrum will be obtained by a combination of a response magnitudes for the individual shapes by the CQC method (considering the proximity) as follows [14]:

$$R_{CQC} = \left\{ \left| \sum_{i=1}^{N} \sum_{j=1}^{N} k.\varepsilon_{ij}.R_i.R_j \right| \right\}^2, \quad k = \begin{cases} 1\ ak\ i = j \\ 2\ ak\ i \neq j \end{cases}, \quad r = \omega_j/\omega i, \tag{3}$$

where $\varepsilon_{ij} = \dfrac{8\left(\xi_i \xi_j\right)^{1/2}\left(\xi_i + r\xi_j\right).r^{3/2}}{\left[1-r^2\right]^2 + 4\xi_i \xi_j r\left(1+r^2\right) + 4\left(\xi_i^2 + r\xi_j^2\right).r^2}$ and ξ_i is the modal damping.

Overall response R_k in the cross-section k has been obtained by the envelope of combinations of responses along-side the individual excitation directions in the direction of the axes X, Y and Z [3].

$$R_k = \max\{R_{xk} + 0.4R_{yk} + 0.4R_{zk}\ ;\ R_{xk} + 0.4R_{yk} + 0.4R_{zk}\ ;\ R_{xk} + 0.4R_{yk} + 0.4R_{zk}\} \qquad (4)$$

where R_{xk}, R_{yk} and R_{zk} are the responses in the cross-section k from excitation by the acceleration spectra in the direction X, Y and Z. The combination rule in (4) is more conservative than in Eurocode 1998.

6 Experimental testing

The correct numerical model of the technology structures depends on the real behavior of the structures verified by the experimental tests [2, 9, 11, 17 - 19]. The static and dynamic resistance of the typical connections of the cable trays structures were tested in the certified laboratory VOP-026 Šternberk,s.p., division VTÚPV Vyškov [18].

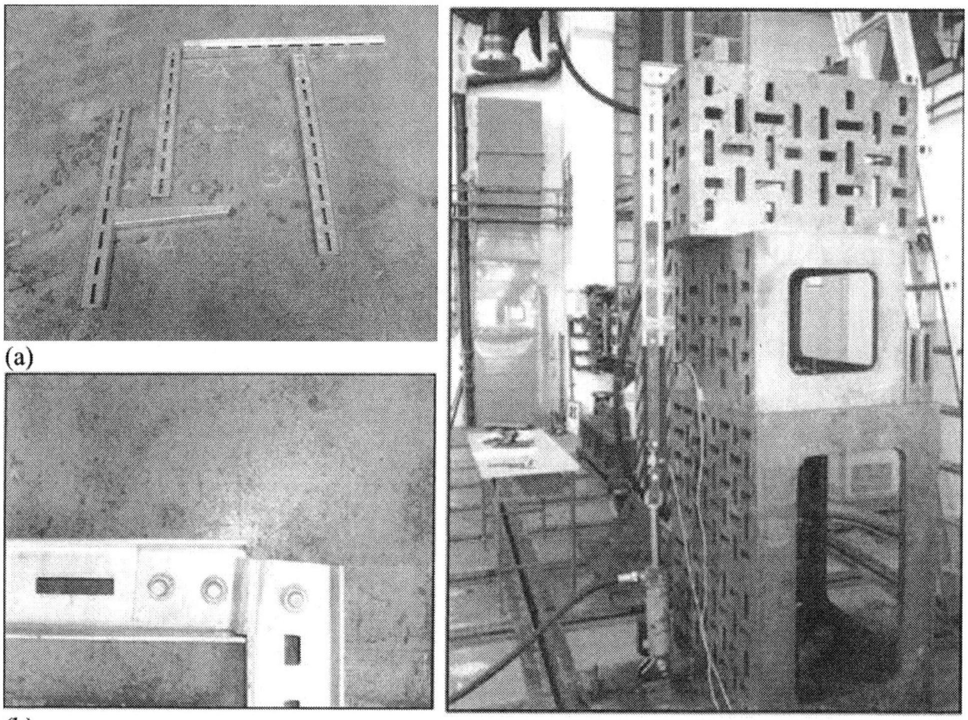

Fig. 3. The photo-documentation - a) the tested segments 1A, 2A and 3A, b) the connection after test, c) the view of the test segment˘.

Three segments were tested (Fig. 3):

1. Segment 1A - steel column IS8 (No.6337252) of 3680mm with console of AS 30/41 (No. 6418813) for cable masses of 90 kg.
2. Segment 2A - steel column and beam IS8 (No.6337252) connected with joint AHIS 8 (No.6019064) and bolts FRS10×25 (No.6407560) for cable masses 270 kg.

3. Segment 3A - steel column US7/160 (No.6337252) connected with joint KU7 NOX (No. 6349056) for cable masses 600 kg.

Dynamic resistance of the tested segments was determined for the cyclic loads in horizontal and vertical direction (Table 1). All tested segments were satisfied for the static and dynamic loads defined. In the case of the tested connection OBO2A the plastic deformation of 9 mm occurred, but this connection was not damaged. This connection was not used in the final cable trays structures in NPP finally.

Table 1. Recapitulation of the experimental tests of OBO Bettermann segments.

Test segment	Static resistance [N]	Dynamic resistance experimentaly determined				
		Horizontal force [N]		Vertical force [N]		Ratio [%]
		Min	Max	M in	Max	
1A	3 000	0	1040	450	810	35
2A	12 000	0	6370	2700	4864	53
3A	11 000	0	12530	5400	9730	88

7 Numerical analysis of the seismic resistance

The seismic resistance of the various typical cable support structures based on OBO Bettermann segments were numerically analyzed in software ANSYS (Fig. 4) [9]. The principal structures was made from the columns 5×I8/3680, beams 6×US7/1500, consoles 60×AS30/41/410 and the cable segments. These structures were used with the cable trays 48×WKSG140/1500 (OBO1Z) and the cable ladders 44×LG640N/1500 (OBO1R).

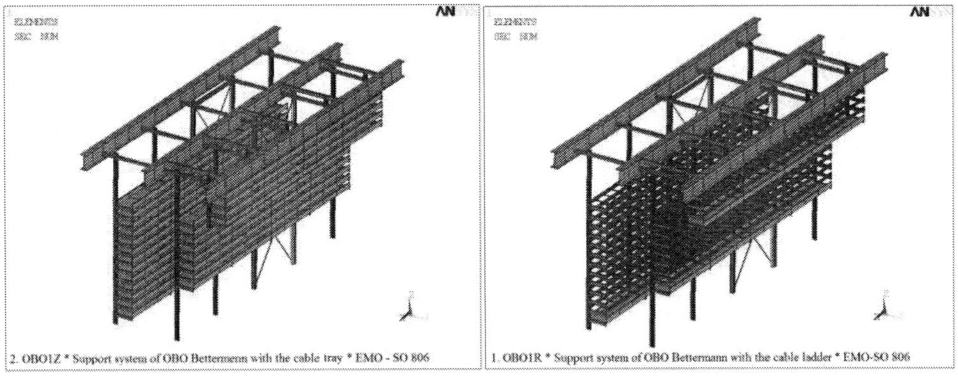

Fig. 4. Calculation models of the support structures with the cable tray EKS640 (OBO1Z) and the cable ladder LG640 (OBO1R).

The seismic response was solved by the method of the seismic response spectra from the floor spectra envelope in the horizontal direction as well as vertical direction at the level of the ceiling +5.4 m; upon considering 7% damping [2] (Fig. 5).

The value of the HCLPF parameter depends on the equipment structure or component resistance (R) and the corresponding effect of action (E) using elastic or inelastic behavior. Generally it follows

$$\text{HCLPF (CDFM)} = (FS)_{ep} \cdot PGA_{RLE} = \text{SL-2} \tag{5}$$

The optimal configuration of the support systems of the cable trays was determined by the acceleration peaks of FRS, which are in intervals 5-10Hz [9]. The seismic effective structures are with the principal modes out from this inter-val. There are presented two different conceptions of the structures. The principal frequencies of model OBO1Z and

OBO1R are out of this critical interval in direction X and Y (Table 2). The total masses of the structure OBO1Z is equal to 19 719 ton and OBO1R is equal to 22 176 ton.

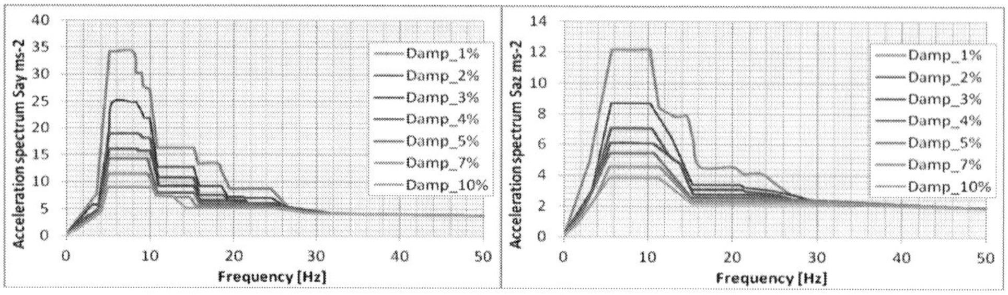

Fig. 5. FRS for horizontal and vertical direction at the level of +5.4m of SO805/806.

Table 2. Principal eigenvalues of the OBO Bettermann structures.

Model	f_0[Hz] in direct. X	Part. factor [%]	f_0[Hz] in direct. Y	Part. factor [%]	f_0[Hz] in direct. Z	Part. Factor [%]
OBO1Z	10.27	84.8	3.88	84.3	9.64	47.3
OBO1R	0.85	58.5	2.98	76.5	8.99	67.4

Table 3. Parameter HCLPF for the segments of OBO Bettermann structure.s

Model	Cable masses [kg/bm]	Column [g]	Beam [g]	Console [g]	Bracing [g]
OBO1Z	390	0.33	0.31	5.79	6.85
OBO1R	390	0.18	0.29	0.19	3.70

The seismic resistance of these support structures is determined by the lowest resistance of the structure segments (Table 3). The minimal HCLPF parameter of the structure OBO1Z is 0.31g and the structure OBO1R is 0.18g.

8 Conclusion

This paper presented the methodology of the seismic reevaluation of NPP in Mochovce due to new results from the geological and seismic-tectonic monitoring of this site. The deterministic methodology of seismic design based on CFDM methodology is presented. The synthetic spectrum compatible accelerograms generated in program COMPACEL (created by Králik) are presented in comparison with requirements IAEA and US standards [1, 3, 4, 15 and 16]. The methodology of the seismic resistance of the cable support structures is described. This methodology was used for the optimal design of the cable support systems from the structural elements of OBO Bettermann experimentally tested.

The project was performed with the financial support of the Grant Agency (VEGA 1/0265/16) and Ministry of Education, Science, Research and Sport of the Slovak Republic at the framework of the project "University Science Park of STU Bratislava", ITMS 26240220084.

References

1. IAEA No.28. *Seismic Evaluation of Existing Nuclear Power Plants* (IAEA Vienna, 2003)
2. J. Králik, *Safety and Reliability of Nuclear Power Buildings in Slovakia. Earthquake-Impact-Explosion*. (Ed. STU Bratislava, 307 pp., 2009)

3. ASCE 4/98. *Standard, Seismic Analysis of Safety - Related Nuclear Structures and Commentary on Standard for SASRNS - approved* 1998 (New York, ASCE 1998)
4. NUREG/CR-6926, *Evaluation of the Seismic Design Criteria in ASCE/SEI Standard 43-05 for Application to Nuclear Power Plants* (U.S. NRC Washington, DC 20555-0001, 2007)
5. P. Labák, P. Moczo, *Determination of the Review Level Earthquake (RLE) Characteristics for the Mochovce Nuclear Power Plant Site* (Contract No.370/96, Geophysical Institute SAV Bratislava, 1998)
6. V. Salajka, P. Hradil, J. Kala, Assess of the nuclear power plant structures residual life and earthquake resistance, In: Proc. *The Second International Conference on Engineering and Technology Innovation* (*ICETI* 2012), Kaohsiung, Taiwan, November 02-06, pp.4 (2012)
7. J. Králik, M. Šimonovič, Earthquake response analysis of nuclear power plant buildings with soil-structural interaction. *Mathematics and Computers in Simulation 50*. IMACS/Elsevier Science B.V. pp. 227-236 (1999)
8. J. Králik, Comparison of Probabilistic and Deterministic Assessments for Evaluation of Seismic Safety of Nuclear Power Plants in Slovakia. In: *First European Conference on Earthquake Engineering and Seismology*. 3-8 September, Geneve, p. 422 (2006).
9. J. Králik, *Design of the Seismic Resistance of the Connections. OBO Bettermann's Supporting Structures for the Electric Line in EMO34* (Rep. 04-130-088/OBO1a-2011, FCE STU Bratislava, 2011)
10. J. Králik, Risk-Based Safety Analysis of the Seismic Resistance of the NPP Structures, In: Proc. of the *8th International Conference on Structural Dynamics, EURODYN 2011*, Leuven, Belgium, 4-6 July, Vol.2, p. 292-299, ISBN 978-90-760-1931-4 (2011)
11. K. Kotrasová, E. Kormaníková, A case study on seismic behavior of rectangular tanks considering fluid - Structure interaction. In: International Journal of Mechanics, **10**, p. 242-252 (2016)
12. W. T. Lin, M. H. Hsieh, Y. C. Wu, C. C. Huang, Seismic Analysis of the Condensate Storage Tank in a Nuclear Power Plant. Journal of Vibroengineering, **14** (3), pp. 1021-1030 (2012)
13. A. Tesár, J. Melcer, Dynamic identification of fractal structures. In: Int. J. Numer. Meth. Eng. **71**, p. 1321-1341 (2007)
14. R. W. Clough, J. Penzien, *Dynamics of Structures* (Mc Graw-Hill, Inc. 1993)
15. RG 1.122. *Development of FRS for seismic design of floor-supported equipment or components* (U. S. NRC. Washington DC, Revision 1, February, 1978)
16. NUREG-0800. *Standard Review Plan for the Review of Safety Analysis Reports for Nuclear Power Plants, LWR Edition* (U.S. NRC Washington, DC 20555-0001, 2007)
17. J. Protivínsky M. Krejsa, Using scaled physical model for assessment of mechanical damping of power plant boiler structure, In: *Perspectives in Science* 7, p. 287-291, Ed. Elsevier, (2016)
18. J. Štuchal, et al. *Protocol of the experimental tests. Force measurements. Assembled joints segments of OBO Bettermann*. (No 6430-81/2008. VOP-026 Šternberk, s.p., divize VTÚPV Vyškov, June, 2008)

Response of endlessly long shipping channel due to earthquake

Kamila Kotrasová[1,*] and *Eva* Kormaníková[1]

[1]Department of Structural Mechanics, Institute of Structural Engineering, The Technical University of Košice, Faculty of Civil Engineering, Vysokoškolská 4, 042 00 Košice, Slovakia

Abstract. Ground-supported tanks are used to store a variety of liquid. The fluid develops hydrodynamic pressure on walls and bottom of tank during an earthquake. This paper provides theoretical background for specification of impulsive and convective actions of fluid in rectangular endlessly long shopping channel container by using analytical methods. This article contains description of methods for solution of response of vertical containment with fluid on the soil, which is subject to horizontal seismic loads.

1 Introduction

Ground-supported tanks are used to store a variety of liquids, e.g. water for drinking and firefighting, petroleum, chemicals, and liquefied natural gas. When subjected to external excitation like earthquake, liquid-containing structures are challenging to design due to sloshing effects. Indeed, fluid-structure interaction is the source of free surface fluctuation and hydrodynamic pressure loads that can cause unexpected instability or even failure of these structures, [2-6, 11, 21-25].

The seismic analysis and design of liquid storage tanks is, due to the high complexity of the problem, in fact, really complicated task. Number of particular problems should be taken into consideration, for example: dynamic interaction between contained fluid and tank, sloshing motion of the contained fluid; and dynamic interaction between tank and sub-soil. Those belong to wide range of so called fluid structure interactions (FSI). Tank-soil interaction could under specific conditions have a significant effect on seismic response of the tank [8-10, 12,].

The knowledge of forces, pressures acting onto walls and the bottom of containers and dynamic response of liquid storage tanks during an earthquake plays essential role in reliable and durable design of earthquake resistance structure/facility – tanks [13-20].

2 Hydrodynamic pressures on the wall and bottom of tank

The complete solution of the Laplace equation for the motion of fluid contained in a rigid containers can be expressed as the sum of two separate contributions, called "rigid impulsive", and "convective", respectively. The "rigid impulsive" component of the

* Corresponding author: kamila.kotrasova@tuke.sk

solution satisfies exactly the boundary conditions at the walls and the bottom of the tank (compatibility between the velocities of the fluid and the tank), but gives (incorrectly, due to the presence of the waves in the dynamic response) zero pressure at the original position of the free surface of the fluid in the static situation. The "convective" term does not alter those boundary conditions that are already satisfied, while fulfilling the correct equilibrium condition at the free surface.

Let us to consider a rectangular ground supported container, having the width $2L$ as shown in Fig. 1. Walls have uniform thickness. Rectangular container is filled with fluid to the height H and the container is excited by a horizontal acceleration $\ddot{u}_o$ in the X - direction.

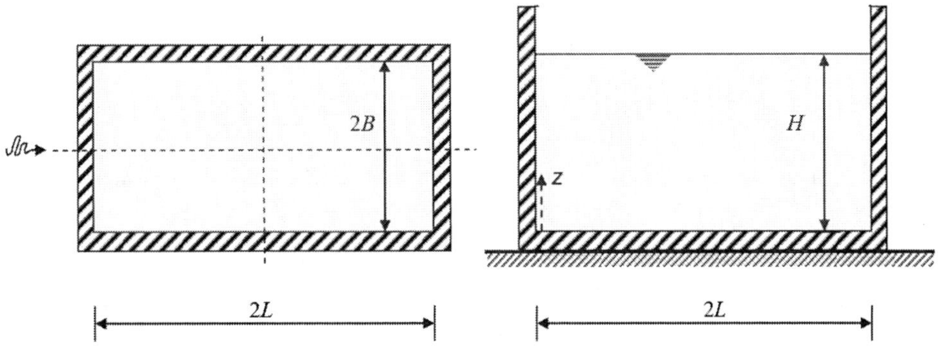

Fig. 1. Geometry of rectangular tanks a) plan, b) sectional elevation.

For rectangular container whose walls can be considered as rigid, a solution of the Laplace equation due to the horizontal excitation, can be obtained in the form (1). The total pressure is given by the sum of an impulsive and a convective contribution using the absolute summation rule [26]

$$p(z,t) = p_i(z,t) + p_c(z,t). \tag{1}$$

The impulsive component is expressed by

$$p_i(z,t) = q_0(z)\rho L A_g(t), \tag{2}$$

where L is the half-width of the tank in the direction of the seismic action, ρ is mass density of the liquid.. The function $q_0(z)$ [26] gives the variation of $p_i(\cdot)$ along the height. The time-dependence of the pressure p_i in eq. (2) is given by the function $A_g(t)$. $A_g(t)$ is the time-history dependent ground acceleration in free - field motion of the ground. $p_i(\cdot)$ is constant in the direction that is orthogonal to the seismic action.

The convective pressure component is given by the summation of modal terms (sloshing modes), each one having a different variation in time. The dominant contribution for rectangular tanks is the fundamental mode

$$p_{c1}(z,t) = q_{c1}(z)\rho L A_1(t), \tag{3}$$

where the function $p_{c1}(z)$ is given in [26], together with the contribution $p_{c2}(z)$ of 2nd mode. $A_1(t)$ is the acceleration response function of the simple oscillator having the frequency of the first mode, the appropriate value of the damping, and subjected to an input acceleration $A_g(t)$.

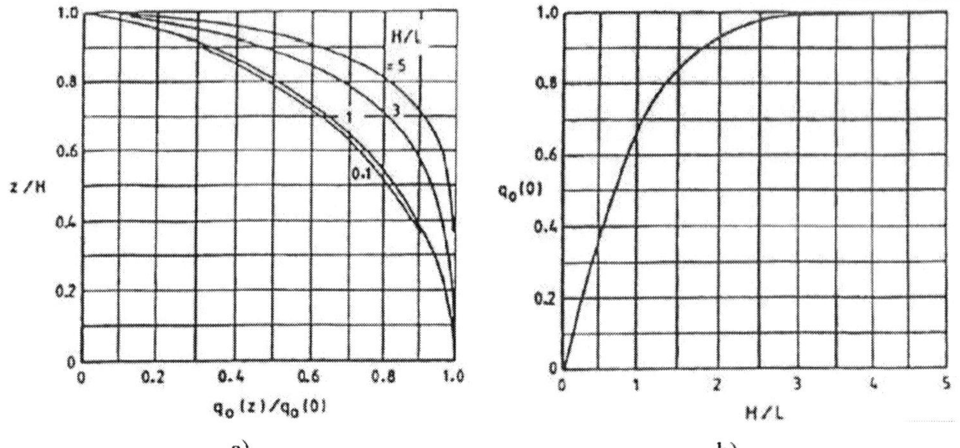

Fig. 2. (a) Dimensionalless impulse pressures at wall of rectangular reservoirs vertical at the direction of seismic excitation; (b) Maximum value of non - dimensional impulse pressures at the direction of seismic excitation [26].

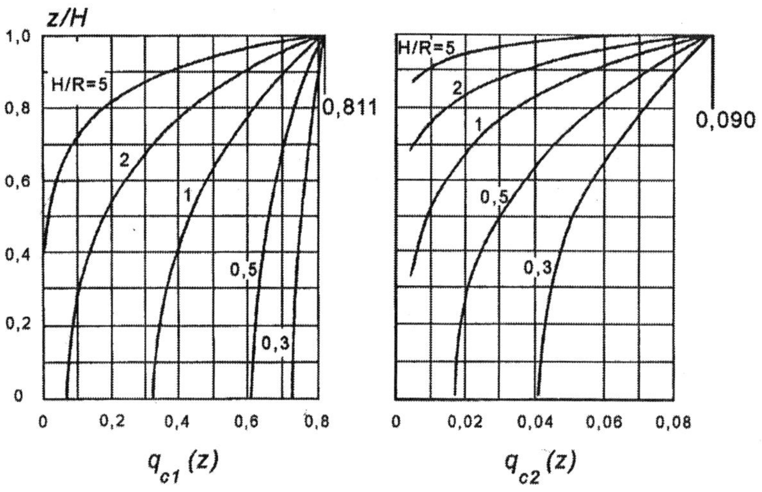

Fig. 3. Dimensionless convective pressures on rectangular tank wall which is perpendicular to the component of the seismic action [26].

The period of oscillation of the first sloshing mode is

$$T_{c1} = \sqrt{\frac{L/g}{\frac{\pi}{2}\tanh\left(\frac{\pi}{2}\frac{H}{L}\right)}}, \tag{4}$$

and the impulsive period is

$$T_i = 2\pi\sqrt{\frac{d_f}{g}}, \tag{5}$$

where g is gravity acceleration, d_f is the deflection of the flexible wall on the vertical centre-line and at the height of the impulsive mass, when the wall is loaded by a load uniform in the direction of the ground motion and of magnitude $m_i g/4BM$. The width of the tank $2B$ is perpendicular to the direction of the seismic loading.

A more rough procedure, summarized below, may be adopted. The procedure operates by changing separately the frequency and the damping of the impulsive rigid and the impulsive flexible pressure. In particular, for the rigid impulsive pressure components, whose modified natural period is

$$T_i^* = 2\pi\sqrt{\frac{m_i + m_0}{k_x \cdot \alpha_x} \cdot \frac{m_i \cdot h_i'}{k_\theta \cdot \alpha_\theta}}, \tag{5}$$

and modified damping value is

$$\xi = \xi_s + \frac{\xi_m}{\left(T^*/T\right)^3}. \tag{6}$$

3 Numerical experiment results and discussion

As an example case we will assume the ground supported rectangular endlessly long concrete shipping channel, with the length $2L = 5$ m and the height $H_w = 3$ m, where $E = 3.4 \cdot 10^{10}$ Pa, density $\rho = 2\,540$ kg/m^3. The endlessly long channel surrounding walls have the uniform thickness of 0.25 m. The channel base slab is $h = 0.4$ m thick. Shipping channel is filled with water to the height of 2.6 m, where water density is $\rho_w = 1\,000$ kg/m^3. There is no roof slab structure covering the shipping channel. This water filled endlessly long concrete shipping channel is grounded on sub-soil 30 MNm^{-3}. As the excitation input we consider horizontal earthquake load given by the accelerogram of the earthquake in Loma Prieta, California (18.10.1989). We use the horizontal excitation of this accelerogram in $2L$ - direction of rectangular channel for the seismic analysis.

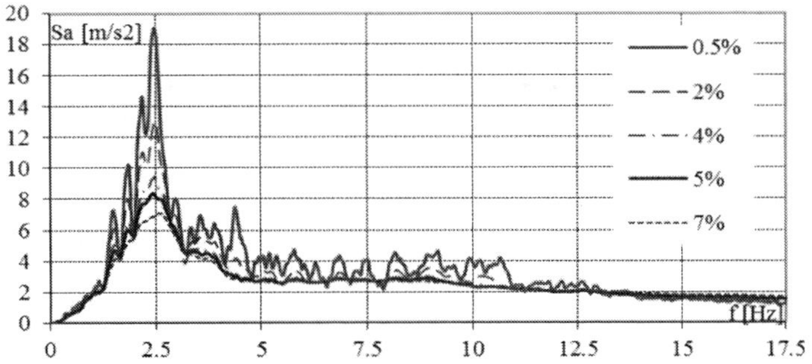

Fig. 4. Elastic response spectrums for accelerogram Loma Prieta

Dynamic response of concrete open top rectangular chipping channel was performed by using Finite Element Method (FEM). The solid walls and the base of the shipping channel was modelled by using 2D SOLID finite element under plain strain condition. The effect of fluid interaction was simulated under the quasi-static conditions. The hydrostatic and hydrodynamic components of the fluid pressures in addition with inertial effects of container were applied as the static load acting onto the walls and the bottom of the tank. The hydrodynamic pressure was given by Eq. (1) – (6). The elastic response spectrums for accelerogram Loma Prieta, Fig. 4, was used for the simulation of earthquake. Values of the elastic response spectrums were calculated with regards to the damping ratio 5% and 0.5%. Value of the elastic response spectrums with the damping ratio 5% was used for calculation of hydrodynamic impulsive pressure and inertial effect of tank solid. Values of the elastic response spectrums with damping ratio 0.5% was used for calculation of hydrodynamic convective pressures.

Fig. 5 show:

- the vertical displacement of ground supported rectangular endlessly long concrete full fluid filled shipping channel bottom in [m] - result of seismic analysis (red line),
- the vertical displacement of ground supported rectangular endlessly long concrete fluid empty shipping channel bottom in [m] - result of static analysis (blue dash line).

The bottom of tank is 5.5 m long. The vertical displacement from static analysis of empty shipping channel bottom is symmetric. The vertical displacement of full fluid filled shipping channel bottom as result of seismic analysis is asymmetric, it is due to considered horizontal earthquake load.

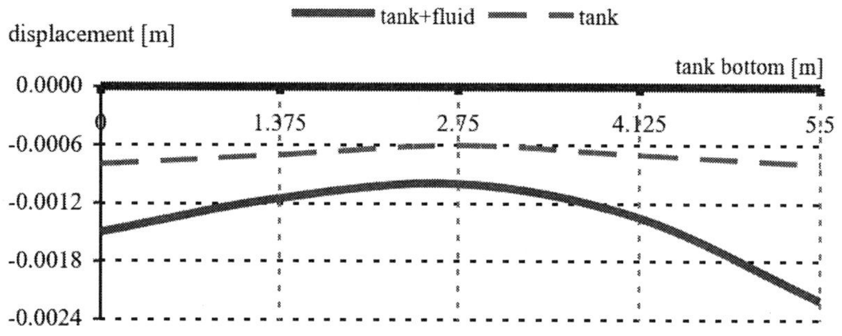

Fig. 5. Vertical displacement of ground supported rectangular endlessly long concrete shipping channel bottom in [m].

This work was supported by the Scientific Grant Agency of the Ministry of Education of Slovak Republic and the Slovak Academy of Sciences under Project VEGA 1/0477/15.

References

1. G.K. Batchelor, *An introduction to fluid dynamics* (Cambridge University Press, Cambridge, 1967)
2. Bathe, K., J., Zhang, H., jr, S., *Computer & Structures*, **72,** 2 (1999)
3. K.J. Bathe, H. Zhang, M.H. Wang, *Computer & Structures*, **56,** 2 (1995)
4. K.J. Bathe, H. Zhang, *International journal for numerical methods in engineering*, (2004)
5. A.Di Carluccio, G. Fabbrocino, E. Salzano, G. Manfredi, *ICSV18: 18th The World Conference on Earthquake Engineering,* October 12 – 17, Beijing, China (2008)
6. A. H. N. Chegini, G. Pender, *WSEAS Transactions on Environment and Development.* **8** (2012)
7. G.W. Housner, *Earthquake pressures on fluid containers* (California institute of technology, Pasadena California, 1955)

8. G.B. Jacobs, W.S. Don, T. Dittmann, *Theoretical and Computational Fluid Dynamics*, **26,** 1 (2012)
9. N. Jendzelovsky, L. Balaz, *Applied Mechanics and Materials*, **617,** 148-151 (2014)
10. E. Kock, L. Olson, *International Journal for Numerical Methods in Engineering*, **31,** 3 (1991)
11. K. Kralik, J. Kralik jr., *Advanced Materials Research*, **715,** 929-936 (2013).
12. M. Krejsa, P. Janas, V. Krejsa, *International Journal of Mathematics and Computers in Simulation,* **8,** 1 (2014)
13. G. Lajčáková, J. Melcer, *Transport and Telecommunication,* **16** (2015)
14. P.K. Malhotra, T. Wenk, M, Wieland, *Structural Engineering International*, **3** (2000)
15. J. Melcer, *Journal of Vibrational Engineering and Technologies, vibration engineering & technologies*, **3,** 2 (2015)
16. J. Michel, M. Mihalikova, *Acta Metallurgica Slovac*a, **2,** 108-115 (2000)
17. S. Rugonyi, K.J. Bathe, *CMES*, **2,** 2 (2001)
18. H. Sezen, R. Livaoğlu, A. Doğangün, *Engineering Structure,* **30,** 794-803 (2008)
19. M. Safari, *WSEAS Transactions on Mathematics*, **12** (2013).
20. E. Sávio De Góes Maciel, *WSEAS Transactions on Fluid Mechanics*, **8,** 1 (2013)
21. O. Sucharda, J. Brozovsky, *International Journal of Mechanics*, **7,** 3 (2013)
22. V. Michalcova, S. Kuznetsov, S. Pospisil, *International Journal of Mathematics and Computers in Simulation*, **8,** 1 (2014)
23. X. Wang, *Fundamentals of Fluid – Solid Interactions. Analytical and Computational Approaches* (Eselvier, Linacre House, Oxford OX2 8DO, 2008)
24. M. Zmindak, I. Grajciar, *Computers and Structures*, **64,** 5-6 (1997)
25. Manual ADINA (71 Elton Ave, Watertown, MA 02472, USA, ADINA R&D, Inc., wn, October, 2005)
26. STN EN 1998-4. Design of structure for earthquake resistance - Part. 4: Silos, tanks and pipelines (2006)

Sloshing effects in tanks containing liquid

Martin Sivý[1,*], *Miloš* Musil[1], *Ondrej* Chlebo[1] and *René* Havelka[1]

[1]Institute of Applied Mechanics and Mechatronics, Faculty of Mechanical Engineering, Slovak University of Technology, Námestie slobody 17, 812 31 Bratislava, Slovakia

Abstract. Dynamic analysis of a tank containing liquid is a complex problem involving fluid-structure interaction. The tank-liquid system is simplified by an equivalent model in which the total liquid mass is divided into two zones – impulsive and convective. The paper is primary focused on the behavior of the free liquid surface (the convective portion of liquid) subjected to the dynamic loading which may result in liquid spilling or tank wall damage. Therefore, the sufficient freeboard must be required to design. The paper deals with the seismic design of the open cylindrical liquid storage tank with the aim to determine convective dynamic properties (natural frequencies and modes of oscillation), maximum vertical displacements over tank radius and overall response of the liquid to an earthquake. The analysis is performed analytically by applying procedures for the determination of convective effects based on simplified equivalent spring-mass model, numerically response spectrum, and method of motion integrating equations utilizing ANSYS Multiphysics.

1 Introduction

Cylindrical tanks or pressure vessels are commonly used in the power and processing industries. They can be employed as storage vessels for various liquids (from non-flammable, nontoxic liquids to highly volatile, toxic chemicals with explosive nature). The ordinary operation of tank-liquid systems may be influenced by seismic loading which can threaten the safety and integrity of these structures. During a seismic event, time-dependent hydrodynamic forces, pressures and stresses are induced by the liquid on the tank walls and bottom. Knowledge of hydrodynamic effects is essential since they influence the response of the tank and must be considered in the design. The procedures for evaluation of these effects are covered in international, national standards and/or guidelines (e.g. API 650, Eurocode 8).

Earthquake damage to steel storage tanks can come in several forms. One of the phenomena observed at the free liquid surface is a sloshing effect of the upper (convective) portion of the liquid. Oscillation of the convective liquid in the storage tanks may result in negative effects such as deformations and ruptures of the tank walls due to impact [1]. In addition, when open tanks are subjected to ground motions, sloshing may lead to liquid spillage. Therefore, the sufficient freeboard (clearance) between the free surface and the top of the tank or the roof must be provided.

* Corresponding author: martin.sivy@stuba.sk

DOI: 10.1051/matecconf/201710700069

2 Basic concept

Past earthquakes and their destructive consequences led to a huge effort to describe the response of the tank-liquid systems when subject to dynamic excitation resulting to provisions and recommendations for earthquake resistance to prevent negative effects to people, environment, operation, etc. Based on analytical and experimental studies of G. W. Housner [2], A. S. Veletsos [3], P. K. Malhotra [4] and others, procedures for the evaluation of seismic characteristics for liquid storage tanks were formulated.

In calculations of the seismic effects, simple equivalent mechanical models are used to investigated tank-liquid systems. The most widely used model adopted in various international codes (e.g. in [5]) is the one based on spring-mass modeling proposed for rigid tanks by G. W. Housner [2] which was later modified and extended for flexible tanks.

In the simplified model (Fig. 1a) the total liquid is divided into two parts – the impulsive liquid, located near the tank base which moves with (and deforms) the tank walls, and the convective liquid which represents sloshing of the free surface and oscillates independently of the tank wall. The impulsive portion of the liquid is represented by a respective mass m_{i} rigidly attached to the tank wall, whereas the convective part is described by an infinite number of masses m_{cn} flexibly attached by a spring of appropriate stiffness k_{cn}. Each of the convective masses represents another slosh mode of the free liquid surface. In addition to spring-mass models, the convective portion of liquid can be replaced by another system, e.g. a system of simple pendulums (Fig. 1b), each of mass m_{cn} and length l_{cn}. All necessary parameters for the description of equivalent models can be found in [5, 6].

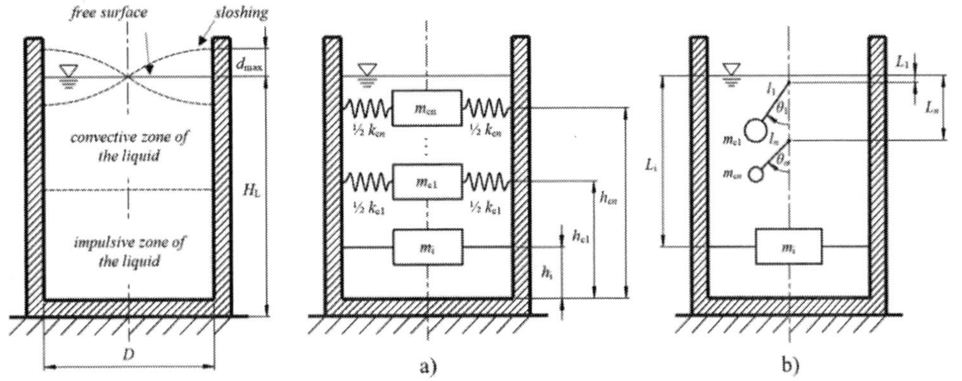

Fig. 1. Equivalent simplified spring-mass and pendulum mechanical models.

3 Liquid sloshing effects

In this chapter, we will focus on the determination of dynamic properties (natural frequencies, modes of oscillation and sloshing wave height) analytically using procedures introduced in [3-5] and by seismic response of the convective liquid portion to earthquake load utilizing numerical analysis based on finite element method in ANSYS Multiphysics.

The investigated model is a flexible (steel) ground-supported liquid storage tank fixed to the rigid foundation of a circular cross-section of radius R (5 m), height H (5 m) and uniform wall thickness t (0.005 m). The tank is filled with water to a height of H_L (4 m).

In the FEM formulation, a tank is modeled by structural shell elements (SHELL181); liquid is represented by fluid elements based on Lagrange's formulation (FLUID80). At the interface between structural and fluid elements it is necessary to define the coupling of tank and liquid displacements.

3.1 Convective natural frequencies and modes of oscillation

The motion of the convective portion of liquid can be expressed as a linear combination of the corresponding natural modes of oscillation. Each respective mode is associated with vertical displacements and is described analytically by the Helmholtz equation. The solution for each slosh mode of a circular liquid storage tank with radius R is proportional to

$$S_{n,m} = J_n\left(\lambda_{n,m}\frac{r}{R}\right)\cos n\theta \tag{1}$$

which varies towards the circumference by $\cos n\theta$. J_n represents a Bessel function of the first kind and n^{th} order and $\lambda_{n,m}$ are values for which the first derivative of J_n' are zero.

The natural frequency of the convective portion of liquid for the slosh mode of oscillation can be calculated as

$$f_{c(n,m)} = \frac{1}{2\pi}\sqrt{\lambda_{n,m}\frac{g}{R}\tanh\left(\lambda_{n,m}\frac{H_L}{R}\right)} \tag{2}$$

Using Eq. (1) and Eq. (2), it is possible to calculate any convective mode of oscillation at the free liquid surface and its respective natural frequency. In international standards, only the vertical displacements antisymmetric about the axis of rotation (one nodal diameter) can be calculated, i.e. the Bessel function of the first order is employed.

Fig. 2 presents the selected convective modes of oscillation for investigation of the tank model determined by Eq. (1) at frequencies calculated by Eq. (2). Based on results from seismic analyses of liquid storage tanks introduced in other publications of the authors, e.g. [7, 8], a good correlation of natural convective frequencies and modes of oscillation between analytical and numerical calculations (using FEM in ANSYS Multiphysics) is shown.

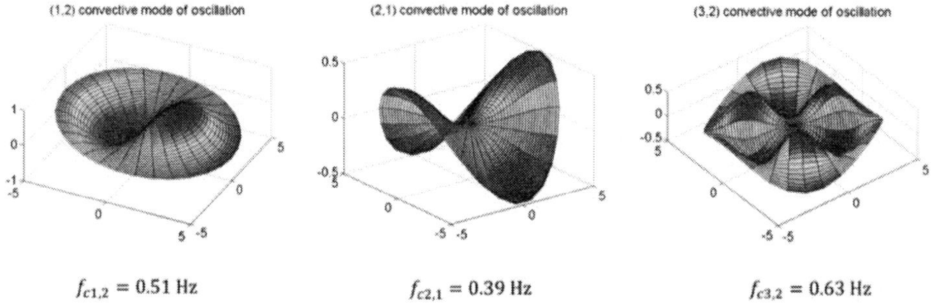

Fig. 2. Convective modes of oscillation and respective natural frequencies.

3.2 Sloshing response to seismic excitation

The natural frequencies of convective liquid are low in comparison with the impulsive portion of the liquid whose modes of oscillation are characterized by higher frequencies. Thus, the convective liquid is imparted to low acceleration but experiences high deformations, whereas impulsive liquid experiences high acceleration and low deformations during ground motion [9]. The convective liquid causes the sloshing motion and it does not contribute to a significant pressure component to the total hydrodynamic pressure.

For the seismic input, data recorded from the Parkfield California earthquake with a PGA of 4.25 m/s^2 at 7.52 s was used. Acceleration time-history of the earthquake and its response spectrum for proportional damping of 0.5% (liquid), 2% (steel or pre-stressed concrete structures) and 5% (concrete and masonry structures) are presented in Fig. 3.

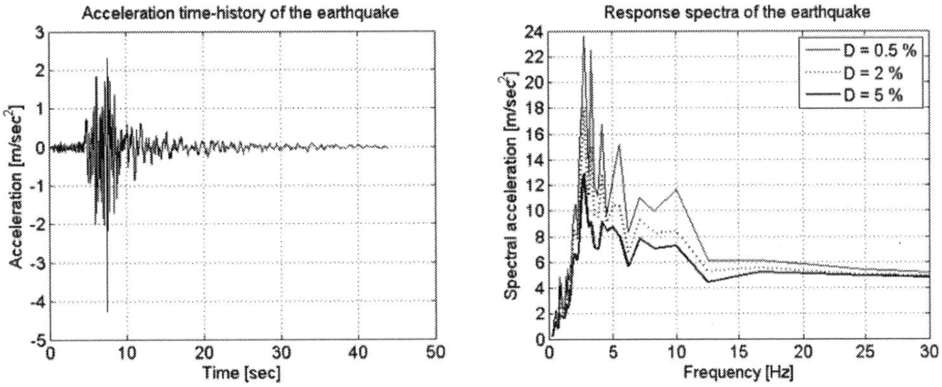

Fig. 3. Parkfield California earthquake.

3.2.1 Sloshing wave height

Introduced by A. S. Veletsos [3], the vertical displacement of free liquid surface due to sloshing may be analytically calculated from the following expression

$$d(r,\theta,t) = R\sum_{n,m=1}^{\infty}\frac{2}{\lambda_{n,m}^{2}-1}\frac{J_n\left(\lambda_{n,m}\frac{r}{R}\right)}{J_n(\lambda_{n,m})}\frac{S_e(f_{c(n,m)})}{g}\cos n\theta \tag{3}$$

where $S_e(f_c)$ is a convective spectral acceleration for a mode of oscillation corresponding to a natural frequency from a specific elastic response spectrum.

In codes aimed at tank-liquid systems design for earthquake resistance, the calculation of the sloshing wave height is reduced for specifying the maximum height at the edge by assuming only the first mode of oscillation. Fig. 4 presents the maximum vertical displacements of the tank model along the radius assuming only the (1,1) natural mode. Using Eq. (3), the peak value of the sloshing wave at the edge is 0.37 m.

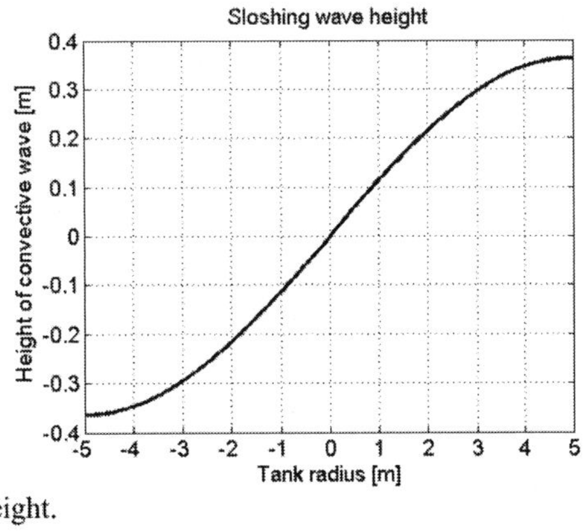

Fig. 4. Sloshing wave height.

3.2.2 Response spectrum method

One of the most common methodologies for describing the behavior of a linear structural system subject to seismic excitation is by performing a response spectrum modal analysis. It is often used instead of a time-history analysis to determine the response to time-dependent loading.

The equation of motion for the MDOF system subject to earthquake ground acceleration $\ddot{u}_g(t)$ is written as

$$\mathbf{M}\ddot{\mathbf{u}}(t) + \mathbf{C}\dot{\mathbf{u}}(t) + \mathbf{K}\mathbf{u}(t) = -\mathbf{M}\mathbf{d}\ddot{u}_g(t) \tag{4}$$

where $\mathbf{M}$, $\mathbf{C}$ and $\mathbf{K}$ are mass, damping and stiffness matrices; $\mathbf{u}$, $\dot{\mathbf{u}}$, $\ddot{\mathbf{u}}$ are the displacement, velocity and acceleration vectors and $\mathbf{d}$ is the excitation direction vector (direction cosines). Using modal transformation

$$\mathbf{u}(t) = \sum_{i=1}^{n} \boldsymbol{\phi}_i \mathrm{w}_i(t) \tag{5}$$

multiplying by $\boldsymbol{\phi}_j^{\mathrm{T}}$ and application of appropriate orthogonality conditions, Eq. (4) can be written as several uncoupled equations of motion representing a set of SDOF systems

$$\ddot{\mathrm{w}}_i(t) + 2\omega_i\xi_i\dot{\mathrm{w}}_i(t) + \omega_i^2\mathrm{w}_i(t) = -\gamma_i\ddot{u}_g(t) \tag{6}$$

in which γ_i represents a participation factor for a given excitation direction

$$\gamma_i = \boldsymbol{\phi}_i^{\mathrm{T}}\mathbf{M}\mathbf{d} \tag{7}$$

This procedure is applied only when the time histories of all the modal displacement vectors in all the modes are known.

The overall relative displacement for each mode of oscillation can be expressed by

$$\mathbf{u}_{i\,\max} = \frac{\gamma_i S_{\mathrm{A}i}}{\omega_i^2}\boldsymbol{\phi}_i \tag{8}$$

where $S_{\mathrm{A}i}$ represents the spectral acceleration for the ith mode obtained from the response spectrum of a recorded earthquake. For evaluation of the overall response, individual responses are combined by a proper combination method (e.g., SRSS, GRP, CQC, etc.).

For computation of response of the free liquid surface, single-point response spectrum analysis is performed. The tank-liquid system is excited with a corresponding response spectrum for 0.5% proportional damping of the Parkfield California earthquake record (Fig. 3) in the x-direction. Based on results from the modal analysis, SRSS method is used for the combination of maximum modal responses. This combination rule is used due to closely spaced modes of oscillation. Fig. 5 presents the overall response of the tank-liquid system to the response spectrum. The maximum response of the liquid-structure system is 0.349 m.

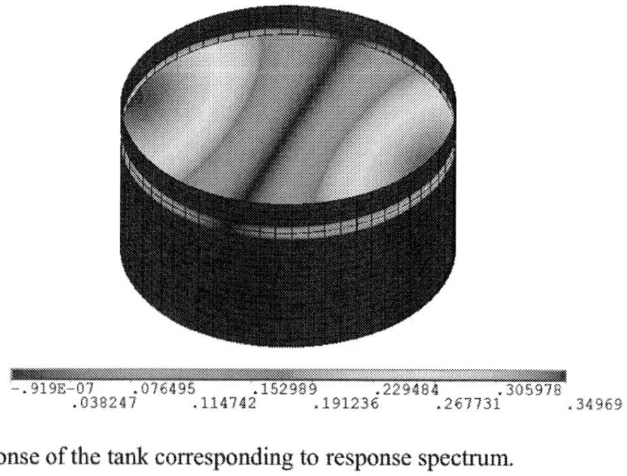

Fig. 5. Sloshing response of the tank corresponding to response spectrum.

3.2.3 Method of integrating of motion equations

Integration of motion equations is a method used in a time-history analysis. This analysis is a universal technique used to determine the response of the structures subjected to any time-dependent loading. The commonly used numerical integration method for evaluating the dynamic response of a structures is the Newmark's implicit method (more e.g. in [10, 11]).

The system of equations of motion of a MDOF system is

$$\mathbf{M}\ddot{\mathbf{u}}(t) + \mathbf{C}\dot{\mathbf{u}}(t) + \mathbf{K}\mathbf{u}(t) = \mathbf{F}(t) \tag{9}$$

where $\mathbf{F}(t)$ is the applied external load. In the case of the seismic excitation, the force $\mathbf{F}(t)$ is replaced in Eq. (9) by $-\mathbf{M}\ddot{u}_g(t)$.

Applying the Newmark's procedure, Eq. (9) is transformed into its static equivalent

$$\tilde{\mathbf{K}}\mathbf{u}_{t+\Delta t} = \tilde{\mathbf{F}}_{t+\Delta t} \tag{10}$$

where $\tilde{\mathbf{K}}$ represents the effective stiffness and $\tilde{\mathbf{F}}_{t+\Delta t}$ the effective load vector which are

$$\tilde{\mathbf{K}} = \frac{1}{\alpha\Delta t^2}\mathbf{M} + \frac{\delta}{\alpha\Delta t}\mathbf{C} + \mathbf{K} \tag{11}$$

$$\tilde{\mathbf{F}}_{t+\Delta t} = \mathbf{F}_{t+\Delta t} + \mathbf{M}\left(\frac{1}{\alpha\Delta t^2}\mathbf{u}_t + \frac{1}{\alpha\Delta t}\dot{\mathbf{u}}_t + \left(\frac{1}{2\alpha} - 1\right)\ddot{\mathbf{u}}_t\right) +$$

$$+ \mathbf{C}\left(\frac{\delta}{\alpha\Delta t}\mathbf{u}_t + \left(\frac{\delta}{\alpha} - 1\right)\dot{\mathbf{u}}_t + \frac{\Delta t}{2}\left(\frac{\delta}{\alpha} - 2\right)\ddot{\mathbf{u}}_t\right) \tag{12}$$

in which α, δ represent Newmark's integration parameters which influence accuracy and stability of the algorithm (average acceleration method, linear acceleration method).

Eq. (10) can be easily solved. To achieve correct results, the choice of the integration time step is important and depends on the integration method. If seismic excitation is applied, the integration step should not be shorter than the interval of the input acceleration data [11].

To evaluate the time-dependent sloshing response of the convective liquid, the acceleration time-history of the Parkfield California earthquake (Fig. 3) in the *x*-direction is applied. Fig. 6 presents the maximum sloshing response of the tank storing water. This maximum vertical displacement of 0.3397 m occurs at the edge of the tank at time 10.16 sec.

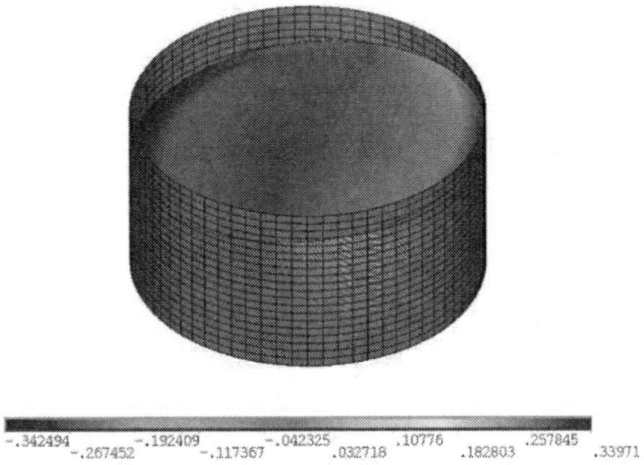

Fig. 6. Sloshing response of the tank to earthquake excitation.

4 Conclusions

The intention of this contribution was to perform a seismic analysis of the liquid storage tank focusing on the sloshing response to a given earthquake loading. Generating of sloshing waves during ground motions represents unfavourable conditions which impacts the top and the walls within closed tanks or can result in the liquid spilling when open tanks are considered. A minimum freeboard is desired to prevent these negative effects. Insufficient freeboard leads to increase of impulsive mass due to the constraining action of the roof and upward load due to impacts generating the forces, which can break the connections between the tank and head. It may lead to buckling of the tank at the base if the loading from the additional impulsive mass was not considered in the design.

The convective response was determined analytical with respect to international standards, subsequently the response was calculated by FE software in ANSYS Multiphysics in which response spectrum and time-history analysis were performed. The results obtained between each solution of the dynamic analyses represented good conformity.

This work was supported by the grant from the Grant Agency of VEGA no. 1/0742/15 entitled *Analysis for Seismic Resistance of Liquid Storage Tanks with Nonlinear and Time-Dependent Parameters*, by the Slovak Research and Development Agency under the contract no. APVV-15-0630 *Extension of the Validity of the Computation Standards for the Seismically Resistant Liquid Storage Tanks, in terms of Safety at NPPs and Other Industrial Areas* and by the STU Grant scheme for Support of Young Researchers entitled *Design and Optimization of the Measuring Equipment for Vibration Tests*.

References

1. K. Malhotra, M. Eeri, Earthquake Spectra **21**, 4 (2005)
2. G. W. Housner, *Bulletin of the Seismological Society of America* **47**, 1 (1987)
3. A. S. Veletsos, *Seismic Response and Design of Liquid Storage Tanks, Guidelines for the Seismic Design of Oil and Gas Pipeline Systems* (Technical Council on Lifeline Earthquake Engineering ASCE, 1984)
4. P. K. Malhotra, T. Wenk, M. Wieland, Struct. Eng. Int. **10**, 3 (2000)
5. EN 1998-4, *Eurocode 8 – Design of Structures for Earthquakes Resistance, Part 4 – Silos, Tanks and Pipelines* (2006)
6. R. A. Ibrahim, *Liquid Sloshing Dynamics Theory and Applications* (Cambridge University Press, Cambridge, 2005)
7. M. Sivy, M. Musil, Procedure for Seismic Analysis of Liquid Storage Tanks using FEM Approach and Analytical Models, *Advances in Mechanism Design II: Proceedings of the XII International Conference on the Theory of Machines and Mechanisms* (Springer, Liberec, 2017)
8. M. Sivy, M. Musil, Stroj. cas. – J. of Mech. Eng. **66**, 2 (2016)
9. P. K. Malhotra, P. Nimse, M. Meekins, Struct. Eng. Int. **24**, 4 (2014)
10. A. K. Gupta, *Response Spectrum Method in Seismic Analysis and Design of Structures* (Blackwell Scientific Publications, Boston, 1990)
11. M. N. Fardis, E. C. Carvalho, P. Fajfar, A. Pecker, *Seismic Design of Concrete Buildings to Eurocode 8* (CRC Press, Boca Raton, 2015)

Noise annoyance from different sources in young adults

Lubica Argalasova[1*], *Alexandra* Filova[1], *Katarina* Hirosova[1], *Diana* Vondrova[1], *Martin* Samohyl[1], *Ludmila* Sevcikova[1] and *Jana* Jurkovicova[1]

[1]Institute of Hygiene, Faculty of Medicine, Comenius University, Spitalska 24, 813 72 Bratislava, Slovakia

Abstract. The aim of the study is to evaluate the adverse effects of community noise on health and the response to noise exposure in residential areas of Bratislava. The study quantifies the exposure to different environmental noise sources as well as the social noise exposure in the sample of university students. The validated methodology according to ICBEN and the Ohrkan study was used. The measurement of ambient noise levels was done using hand-held sound level analyzer. There were 837 university students (256 males), average age 23.09 ± 2.23, enrolled into the study so far; 267 in the housing facility exposed to road traffic noise (L_{Aeq} = 67.6 dB) and 570 in the control one (L_{Aeq} = 53.4 dB). From the total sample 79.2% students reported the use of personal music players (PMP) in the last week in the average time of 293 minutes. There was a significant difference between the duration and frequency of listening to PMP between the exposed and the control group. There was not significant difference between the duration of time spent at activities with high noise exposure. The results of the study showed the importance of road traffic noise and the social noise for health as well and the need for a preventive action.

1 Introduction

Community noise constitutes an integral part of environmental pollution and has a great negative impact, especially in populations living in urban areas. Although the physiological and non-physiological adverse effects of noise are well documented nowadays, there is a need of more research and solutions, as noise pollution continues to rise [1].

Community noise is considered to be emitted from all the sources, apart from industrial facilities. The major sources of community noise affecting daily human activities and well-being are road traffic, railway and neighborhood noise. During the last decades, environmental studies and research being progressively carried out are undoubtedly proving the undesirable impact of noise in health and quality of life [1, 2, 3, 4].

* Corresponding author: lubica.argalasova@fmed.uniba.sk

In the European Union, about 56 million people (54%) living in areas with more than 250 000 inhabitants are exposed to road traffic noise of more than average L_{DEN} (day-evening-night) 55 dB per year, which is thought to be risky to health [5]. WHO estimated that in high-income western European countries (population about 340 million people), at least 1 million healthy life years are lost every year (DALY - disability-adjusted life-years) because of environmental noise [6].

Hearing impairment, annoyance, sleep disturbance and cardiovascular diseases are among the most common adverse health effects of community noise, posing a great growing concern. In spite of the numerous scientific studies and counter-measures to control and eliminate noise pollution, this public health problem continues to grow, presumably due to the advent of a more modernized way of life nowadays [4].

The aim of this study is to evaluate the adverse effects of community noise on health and the response to noise exposure in residential areas of Bratislava, Slovakian capital. The study quantifies the exposure to different environmental noise sources as well as the social noise exposure (personal music players, events with high noise exposure) in the sample of university students aged 19-25 years.

2 Material and methods

The research was conducted at the Institute of Hygiene Medical Faculty Comenius University in Bratislava. The validated methodology according to ICBEN and the Ohrkan study was used [7, 8]. The measurement of ambient noise levels was done using hand-held sound level analyzer.

2.1 Exposure assessment

Maximal, minimal and equivalent sound levels were assessed for both the control and exposed groups living in the Slovakian capital Bratislava by hand-held analyzer Brüel&Kjaer type 2250, with sound level meter software BZ-7222 and frequency analysis software from Brüel&Kjaer. Exposed housing facility is a student dormitory, situated near the major transportation route, the main thoroughfare with railway transport; control housing facility is a student dormitory in a quiet area with surrounding greenery. All measurements were recorded according to the valid legislation during the time intervals from 17.00-18.00 and from 20.00-21.00 in the exposed and at the same time in the control area. This time interval was chosen to record the afternoon traffic peak and to detect the time most annoying for students and for their activities (studying, watching TV, talking, and falling asleep). Measurements were recorded during spring period at working days (Tuesday) two times on each site. Road traffic flow composition was assessed as well.

2.2 Sample

There were 837 university students (256 males), average age 23.09 ± 2.23, enrolled into the study so far; 267 in the housing facility exposed to road traffic noise and 562 in the control one in Bratislava. Students significantly did not differ by gender, but they differed by age (older in the control area), flat location in relation to noise exposure, position of a flat in the floor height, length of stay in the given area, windows orientation, windows types and satisfaction with flat surrounding. The respondents have been living in the given area at least for four years.

2.3 Noise annoyance questionnaire

Subjective response was assessed by the authorized "Noise annoyance questionnaire", the different sources of environmental noise were quantified [9].The validated 5 grade noise annoyance verbal scale (Not at all; Slightly; Moderately; Very; Extremely), was developed and recommended by experts from the noise research ICBEN (The International Commission on the Biological Effects of Noise) team [7].

The questionnaire comprised personal (age, gender, education), behavioral (smoking, coffee and alcohol consumption), and questions focused on the characteristics of residential environment (localization, construction and surrounding of residential buildings, the location and amenities of the apartment, window orientation to quiet and noisy streets and the length of stay in the apartment). It also included questions on possible non-auditory health effects (noise annoyance from different sources, interference with various activities and sleep disturbance) and subjective assessment of health troubles (headache, nervousness and irritability, difficulties falling asleep, the use of different types of medications, the presence of cardiovascular diseases and overall assessment of the health status). A special section is dedicated to traffic noise annoyance and its psychological and physiological effects as well as the social noise exposure (personal music players (PMP), events with high noise exposure). The intensity and duration of exposure was subjectively assessed, as well as the type of headphones. At the end, we ask whether respondents have complained on noise annoyance and disturbance and about the corrective actions implemented in the house to reduce noise exposure. Information from respondents was obtained by personal interview. Response rate of the questionnaire was 90%.

Social noise exposure was quantified by subjective assessment of intensity and frequency of listening to personal music players and the duration of time spent at activities with high noise exposure - playing the music instrument, visiting cinema, classical, rock, pop, jazz concerts, discotheques, entertainment facilities, sport events, noisy household and garden work). The authorized methodology was based on the Ohrkan study [8]. The duration and frequency of PMP listening were assessed in the course of one week, as well as the loudness level and the type of headphones (headset, earphones, earbuds). The duration of time (hours and minutes per week or month) spent at activities with high noise exposure - entertainment facilities, discotheques, concerts of different music styles (classic, rock, pop, jazz), sport activities, playing the music instrument and noisy household and garden work was assessed subjectively as well.

2.4 Statistical analysis

Statistical evaluation comprises the methods of descriptive statistics, the relationship between continuous variables were examined by bivariate analysis, t-test, analysis of variance (ANOVA) and correlation coefficients. Relationships between categorical data were evaluated by contingency tables, chi-square test and stratified analysis. Statistical packages Epi Info™, Version 7.1.1.1, 2013 and SPSS, 2016 were implemented.

3 Results

The monitoring of sound levels in the exposed area showed the levels above the national and international limits in the afternoon and in the evening time interval (17.00-18.00 and 20.00-21.00) [3, 5, 10]. Sound levels in the control area were significantly lower

($p < 0.001$). The higher sound levels in the evening interval could be due to other noise sources (e.g. entertainment facilities) (Table 1, 2).

In the composition of the traffic flow, the number of passenger cars and trams, which are considered to be particularly annoying, was predominant. In the composition of the traffic flow in the control area the number of passenger cars was also predominant, however; not in such numbers as in the exposed area; there were buses and lorries, but no trams (Table 1, 2).

The indicators estimated from Bratislava strategic noise map were $L_{DEN} = 66 \pm 2$ dB vs $L_{DEN} = 56 \pm 4$ dB ($p < 0.05$) (http://www.laermkarten.de/bratislava/) [11].

Table 1. Sound levels in the exposed housing facility, April 2016.

Time intervals	Sound level $L_{A\min}$ (dB)	Sound level $L_{A\max}$ (dB)	Sound level L_{Aeq} (dB)	Road traffic flow composition
17.00-18.00	58.6	75.5	67.5	A 5460, B 36, L 60, T 72
20.00-21.00	52.1	81.3	66.9	A 4644, B 12, L 12, T 60

Legend: A - automobile, B - bus, L - lorry, T - tram

Table 2. Sound levels in the control housing facility, April 2016.

Time intervals	Noise level $L_{A\min}$ (dB)	Noise level $L_{A\max}$ (dB)	Noise level L_{Aeq} (dB)	Road traffic flow composition
17.00-18.00	41.5	68.3	50.4	A 108, B 12, L 0, T 0
20.00-21.00	41.7	69.9	53.6	A 60, B 12, L 0, T 0

Legend: A - automobile, B - bus, L - lorry, T - tram

Students in the exposed housing facility were significantly more annoyed by road traffic noise ($OR_{MH} = 3.14$, 95% CI = 2.66-3.71), by railways noise (trams) ($OR_{MH} = 1.43$, 95% CI = 1.19-1.72), noise from industry ($OR_{MH} = 1.72$, 95% CI = 1.45-2.03), noise from neighborhood ($OR_{MH} = 1.45$, 95% CI = 1.24-1.69). There was no significant difference concerning noise annoyance from entertainment facilities ($OR_{MH} = 0.90$, 95% CI = 0.73-1.12), house construction and aircraft noise (Table 3).

Table 3. Annoyance risks from different community noise sources (year 2016).

Noise annoyance (type of noise)	Risks in 2016 *OR* (95 % *CI*)
Road traffic	$^{+}3.14$ (2.66-3.71)***
Neighborhood	$^{+}1.45$ (1.24-1.69)***
Entertainment facilities	$^{+}0.90$ (0.73-1.12)
House construction	$^{+}0.99$ (0.80-1.22)
Railways	$^{+}1.43$ (1.19-1.72)***
Aircraft	0.91 (0.71-1.16)
Industry	$^{+}1.72$ (1.45-2.03)***

Legend: $^{***}p < 0.001$, + Mantel-Haenszel weighted odds ratio
CI = confidence interval; *OR* = odds ratio

From the total sample of respondents, 658 (79.2%) students reported listening to PMP in the last week for the average time of 293 minutes. There was a significant difference in PMP use between the exposed (85.3%) and the control group (76.5%) ($p = 0.003$) and also in the duration of listening to PMP in minutes (333 ± 390 vs 272 ± 362) ($p < 0.05$), but it was not significant between genders ($p = 0.21$).

More than 14% of students listen to the music on the loudness level 4 (they cannot hear the speech or even the traffic) and more than 80% (85.14) use earbuds. There was not significant difference between the loudness level of PMP or in the duration of time spent at most events with high noise exposure between the exposed and control group.

The significant difference was in the type of headphones; earbuds are more often used by students from the exposed area (more than 90% of students) ($p = 0.02$). Ear bud insert phone types are more harmful according to SCENIHR (2008) and increase the sound level by 7-9 dB [12].

The presence of subjective hearing impairment of PMP users was significantly higher (25%), than in the non-PMP users (18 %) ($p < 0.05$).

From activities with high intensity of noise exposure most students took part in household and garden work ($n = 438$), where they spent on average 447 ± 689 minutes. The second most frequent activity was visiting a cinema ($n = 374$), where students spent on average 194 ± 379 minutes per month. The third one was visiting discotheques ($n = 300$) where they spent on average 457 ± 661 minutes per month. The fourth most preferred activity was the visit of rock concerts ($n = 146$), where respondents spent on average 244 ± 368 minutes per month. Most of the time, subjects ($n = 143$) devoted to sport events, where they spent on average 479 ± 760 minutes per month. Students spent 459 ± 642 minutes playing musical instrument ($n = 125$), 129 ± 107 minutes visiting classical music concerts ($n = 59$) and 303 ± 350 minutes in sport shooting ($n = 11$) (Fig. 1).

Students from the exposed group to road traffic noise spent at the cinema about 244 ± 650 minutes per month, compared with subjects in the control group who spent in the cinema on average 170 ± 98 minutes per month ($p = 0.08$). There was no significant difference between the duration of time spent at the other events with high noise exposure between the exposed and the control group.

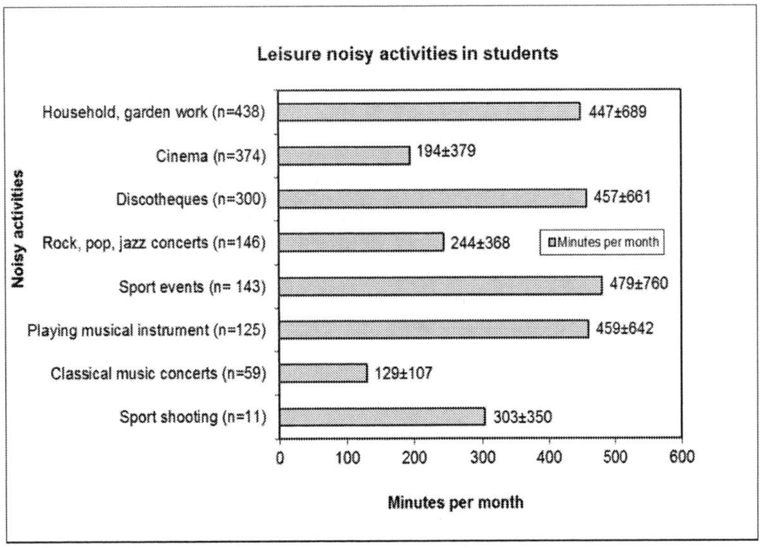

Fig. 1. Leisure noisy activities in the sample of university students.

4 Discussion

Our findings confirmed that students in the exposed housing facility were significantly more annoyed by road traffic noise, by railways noise (trams), but also noise from the other sources such as industry and neighbourhood. No significant difference concerning noise annoyance from entertainment facilities was probably due to similar exposure (and very high) in both student dormitories.

Results are consistent with the findings of other studies reporting 88-90% of teenagers and young adults listening to PMPs through earphones [13]. A number of papers published recently estimated the risk of hearing loss due to the use of PMPs, as well as the actual incidence of hearing loss and tinnitus in the exposed populations. The studies were carried out in German, American, Italian, Dutch, Slovenian, Brazilian, and Malaysian teenagers and young adults [14, 15, 16, 17, 18, 8, 19].

The percentage of teenagers and young adults at risk of developing noise induced hearing loss (NIHL) was estimated to be between 17% and almost 29% [13].

In our study, university students in the group exposed to road traffic noise were listening to PMP more often and with higher duration than students from the control group (possibly trying to mask the effect of this environmental noise source). However, the level of loudness in PMP listening was not significant between groups. The duration of time spent at the other events with high noise exposure was not significant between groups. These results are difficult to compare with the results of the other studies, because we are not aware of a study which has been studying such associations by now.

In the future, we would like to enlarge our study sample and to add a population group of adolescents in the age group of 15-19 years. In cooperation with the ENT experts we would like to perform audiometric, tympanometric and other objective examinations to determine hearing loss and to objectify the effects of social noise exposure on the hearing organ. Within the quantification of the environmental noise effects, we will explore the risk of chronic diseases in more detail in teenagers and in the young adults.

5 Conclusion

Results confirmed risk effect of road traffic noise on the young population. In our study 79.2% of respondents reported listening to personal music players (PMP) in the last week for the average time of 293 minutes. There was a significant difference between the frequency and duration of listening to PMP in the groups of respondents with a different road traffic noise exposure. There was no significant difference between the duration of time spent at events with high noise exposure in the exposed and in the control group.

The results of the study showed the importance of road traffic noise and the social noise as well for health and well-being. After the finalization of the study results, we would like to formulate the proposals and interventional procedures and effectively target the preventive measures (education, the use of noise-cancelling headphones for PMP users) in the vulnerable groups of teenagers and young adults and their parents and teachers as well.

The research was partially supported by a grant Y.A.B.S. (Youth and Parents Behavioral Survey in Slovakia) O-15-101-/0001-00

References

1. E. Murphy, E.A. King, *Environmental noise pollution. Noise mapping, Public health, and Policy* (Elsevier, 2014)
2. S. Žiaran, *Low-frequency noise and vibration* (Slovak Technical University, Bratislava, 2016)
3. B. Berglund, T. Lindvall, DH Schwela DH (eds), *Guidelines for Community Noise* (WHO, Geneva, 2000)
4. L. Fritschi, L.A. Brown, R. Kim, D. Schwela, S. Kephalopoulos (eds), *Burden of disease from environmental noise. Quantification of healthy life years lost in Europe* (WHO, Geneva, 2011)
5. Ch. Hurtley (ed), *Night noise guidelines for Europe* (WHO, Geneva, 2009)
6. M. Basner, W. Babisch, A. Davis, M. Brink, C. Clark, S. Janssen, S. Stansfeld, Lancet **383**, 9925 (2014)
7. J.M. Fields, R.G. de Jong, T. Gjestland, I.H. Flindell, R.F.S. Job, S. Kurra, P. Lercher, M. Vallet, T. Yano, R. Guski, U. Felscher-Suhr, R. Schumer, J. Sound Vib. **242**, 4 (2001)
8. D. Twardella, C. Perez-Alvarez, T. Steffens, G. Bolte, H. Fromme, U. Verdugo-Raab, Noise Health **15**, 67 (2013)
9. L. Sobotova, J. Jurkovicova, J. Volekova, L. Aghova L, IJOMEH **14**, 2 (2001)
10. Slovak Ministry of Health Decree No. 549/2007
11. EUROAKUSTIK: Web page Strategic noise maps, accessed on 6 March 2017: http://www.hlukovamapa.sk/strategicka-hlukova-mapa-bratislavskej-aglomeracie.html
12. SCENIHR, *Potential health risks of exposure to noise from personal music players and mobile phones including a music playing function* (Scientific Committee on Emerging and Newly Identified Health Risks, European Commission, Brussels, 2008)
13. M. Basner, M. Brink, W. Bristow, Y. de Kluizenaar, L. Finegold, J. Hong, S.A. Janssen, R. Klaeboe, T. Leroux, A. Liebl, T. Matsui, D. Schwela, M. Sliwinska-Kowalska, P. Sörqvist, Noise Health, **17**, 75 (2015)
14. I. Vogel, H. Verschuure, C.P.B. van der Ploeg, J. Brug, H., Am J. Public Health, **100**, 6 (2010)
15. R.R. Figueiredo, A.A. Azevedo, P.M. Oliveira, S.P. Amorim, A.G. Rios, V. Baptista, Braz J Otorhinolaryngol, 77, 3 (2011)
16. C.D.F. Portnuff, B.J. Fligor, K.H. Arehart, Int. J. Audiol. **52** (2013)
17. E. Pellegrino, C. Lorini, G. Allodi, C. Buonamici, G. Garofalo, G. Bonaccorsi, Ann. Ig. **25**, 5 (2013)
18. D.C. Colon, U.Verdugo-Raab, C.P.Alvarez, T. Steffens, S.C. Marcrum, S. Kolb, C. Herr, D. Twardella, Noise Health **18**, 85 (2016)
19. S. Jeram, N. Delfar, 11th International Congress on Noise as a Public Health Problem (ICBEN) (Nara, JAPAN, 2014)

Strategy for tram noise reduction

Eva Panulinová[1,*] and *Slávka* Harabinová[2]

[1,2]Technical University of Košice, Faculty of Civil Engineering, Institute of Structural Engineering, Slovakia

Abstract. When there are many sources it is important to determine the sound power and directivity of each to determine their relative contributions to the noise problem. Any noise problem may be described in terms of a source, a transmission path and a receiver and noise control may take the form of altering any one or all of these elements. The aim of this paper is to suggest ways of reducing tram noise by technical and technological means.

1 Introduction

Noise is sound which is unwelcome but the annoyingness depends on the individual. Noise is phenomenon, contaminating the environment. Noise from traffic specifically from tram traffic is unwanted sound and is among the most pervasive pollutants today. The problem with noise is not only that it is unwanted, but also that it negatively affects human health and well-being. For implementing the sustainability goals formulated in the EC 2011 Transport White Paper and the Greening of Transport package, the environmental impact of tramway operations needs to be minimised.

The aim of this paper is to suggest ways of reducing noise from trams by technical and technological means.

2 Sources of tramway noise

Rail noise is sound emissions arising from the operation of trains and trams. There are a wide variety of sources and causes of rail noise, such as wagons braking, squeal noise in curves, vibration from rail corrugation and out-of-round wheels, etc.

Many studies and publications exist concerning sources of rail noise. The Working Group Railway Noise of the European Commission published its Position Paper on the European strategies and priorities for railway noise abatement in 2003 [1]. The International Union of Railways (UIC) published its "Environmental Noise Directive Development of Action Plans for Railways" in April 2008 [2]. Both studies identify three dominant sources for railway noise:

- Rolling noise.
- Power equipment noise.
- Aerodynamic noise.

* Corresponding author: eva.panulinova@tuke.sk

Train speed is a major influence parameter for noise emission. Aerodynamic noise is important for high-speed trains. Tramways as a kind of urban railway transport, operating in densely populated areas. Trams run therefore speeds up to 50 km/h. At medium speed the dominant source is rolling noise [3]. The rolling noise depends strongly on the surface condition (roughness) of wheels and rails. According [4] the noise source analysis was indicated that most sources are located on the lower part of the vehicles. The area centre is located either at 0.5 - 1 meter height for ground.

The tram is often seen as environmentally friendly, namely non-polluting and silent. However complaints from residents living along the lines prove that it may sometimes lead to annoyance.

3 Effects of noise

WHO published in 2011 a study about the burdens of disease from environmental noise [4]. One result of the study is that, about 1 million years of healthy life are lost in the EU every year due to noise reasons.

While the human ear can perceive an increase in sound volume as sound energy increases, the relationship is logarithmic. If two identical 10 dB noise sources are placed together, the perceived increase is not a doubling of the volume but rather a 3 dB increase. As such, a sound level increase from 45 dB to 55 dB may not look like much on paper, but it represents a ten-fold increase in sound energy and its impact on human health. Humans are usually able to sense a change of 3 dB in sound level, which corresponds to a limit of sensitivity. Measures to reduce noise levels by less than 3 dB would, by themselves, be of no real value. Sound can also be transmitted as vibration through the ground and directly into the body, and this is also a form of noise pollution [5, 11, 12].

The faintest audible sound is at 0 dB; the pain threshold is about 120 dB. If it is louder than 120 dB, there is a risk of injury. At a detonated blast of 150 dB the eardrum can rupture. Noise exposure during sleep such as night flight noise is regarded as particularly critical. So night noise causes health hazards already at individual levels below 45 dB, if the difference between the individual level and the background noise is more than 3 dB. Noise above 55 dB is considered as noise pollution. If noise above this level lasts for an extended period of time, the efficiency and well-being of a person will be reduced. Noise in the range 65 to 75 dB causes stress to the body. This can lead to arterial hypertension (high blood pressure), cardiovascular disease and myocardial infarction (heart attack). Noise can also provide for a reduction of gastric secretion and be the cause of stomach ulcers [4-6, 9].

4 Analysis of noise regulation possibilities

One of the possibilities to place the tramway is its location in the common corridor with road traffic. This is so even in most of the localities in Kosice. The various types of permanent way in the city of Kosice are presented on photos in Fig. 1.

When we analyze the acoustic conditions it is important to identify the main source of the noise and to determine the dominant producer of noise. It is necessary, to be effective measures for noise reduction [13, 14].

Decibels are logarithmic "units", they may not be added linearly like other figures. In most cases we will add uncorrelated signals as noise. If a sound source creates a sound level of L_1 = 60 dB and another source with L_2 = 60 dB is added, then it is not the level of 120 dB, but gives an incoherent (noncoherent) summing of the signal level of 63 dB. If both values are equal, it is easy [4].

Adding of two incoherent (noncoherent) sound pressure levels we must use the relationship:

$$L = 10 \log (10^{L_1/10} + 10^{L_2/10}) \text{ dB} \tag{1}$$

Fig. 1. Photos of tram permanent way located in the common corridor with road traffic in the city of Kosice.

4.1 Example

The analysis of a noise event caused by a tram travelling along the road in Kosice, which is one of the busiest tram lines are operated by three regular and one irregular line. It is a densely built-up area where prevail residential houses. Tramway track is lined on both sides by four-lane road. The tram rails are located on concrete sleepers in the ballast (a classical open permanent way). The dominant sound source in a busy location, as published in the paper [7, 8] is a tram transport. Measurements were detected value of the equivalent noise level L_1 = 57 dB (at night without trams) in the monitored area. Especially noise from trams $L_2 = L_{Aeq,1h}$ (A-weighted equivalent sound level in dB measured over a period of 1 hour) was determined by theoretical calculations by Liberko [10].
As has been already shown in:

$$L_2 = 10 \log X + 40 = 10 \log 534 + 40 = 67.3 \text{ dB} \tag{2}$$

$$X = 44.5 \cdot F_5 \cdot m = 44.5 \cdot 0.50 \cdot 24 = 534 \tag{3}$$

where:

- F_5 factor depends on the average speed of the tram, for the classical open permanent way and speed of 30 km/h is valid F_5 = 0.50.
- m is the number of trams that pass the given section of the tramway per one hour. Value was found in the early night time m = 28 vehicles per hour.

The determination of the equivalent level of tramway noise is given by (2), $L_2 = 67.3$ dB. When we have the incoherent addition of two sounds, resulting by (1) is:

$$L = 10 \log (105.7 + 106.7) = 67.7 \text{ dB} \tag{4}$$

The result is almost equal to the noise level, which generates tram transport. So the noise pollution from road traffic will have no almost effect on the final value of the equivalent noise level. The phenomenon is called sound masking.

This will occur if the difference between the levels is large (almost more than 10 dB), then a higher level suppress entirely the effect of the source with lower levels. On the basis of this phenomenon can be confirmed that it is effective and efficient to focus the draft measures on noise pollution of the dominant source. Noise levels in decibels are not placed with greater accuracy than one decimal place because the resolution of the sound level meter is not greater than 0.1 dB. The resulting values are rounded to whole dB, because the change of less than 1 dB will be imperceptible to an average listener.

4.2 Recommendations

Noise mitigation is a set of strategies to reduce noise pollution or to reduce the impact of that noise. Therefore, developing a regulation scheme for a staged process towards low-noise tram cars must be the heart of a noise abatement strategy for tramways.

But let us return back to the technical anti-noise measures used to reduce tram noise, which is closely related to the strategy of reducing noise around of tram tracks.

Summary of recommendations:

- When designing the tram line in the town peripheries should be consider the design of the track to a separate body. Location on a separate track body has operational and maintenance advantages over the line in the pavement.
- Construction of tramway permanent way should be chosen such that the reduction of noise and vibration. It is therefore appropriate to choose the construction of contactless track using an elastic rail fastening systems and make other structural adjustments, which would lead to a reduction of negative impacts on the environment.
- To limit the emission of noise and vibrations from the surface of the track structure into free space. Most of these structural modifications is performed by inserting a flexible cushioning material, usually by rubber. One of the most commonly used adjustments is filler block profile. They are made in various lengths and shapes for all types of rails and track constructions. The use of rail dampers significantly reduces the noise emissions and vibrations caused by the operation of rail transport.
- To use the construction of the permanent way with grassing.
- To apply the layer rubber mat used to improve drainage and for sound and vibration damping.
- Among the most effective noise definitely passive measures include noise barriers. An excellent architectural and functional solution from an architectural and functional is a low concrete barrier.
- The crucial and costly difficult task played by restoring the tram car fleet of operator.
- Lastly, there are the acoustic treatment of windows and facades

5 Conclusion

In conclusion it can be stated that the equivalent level of noise arising from the operation of the tram line may be more than 10 dB higher than the equivalent noise level from the rest of noise load (cars, buses), which means that the resulting equivalent noise level (at action of the two sources at the same time) not higher become.

Recommendations for any further tramlines extensions is, to identified the dominant source of noise in the study area before designing the anti-noise measures and its follow-up noise emissions to minimized.

The preparation of the article was supported by the Scientific Grant Agency of the Ministry of Education of Slovak Republic and the Slovak Academy of Sciences under Project No.1/0477/15.

References

1. Working Group Railway Noise of the European Commission: Position Paper on the European strategies and priorities for railway noise abatement, Brussels (2003)
2. Directive 2002/49/EC relating to the assessment and management of environmental noise. Official Journal of the European Communities L 189/12 -25, 25 June (2002)
3. WHO JRC 2011, WHO European Centre for Environment and Health, Bonn Office, WHO Regional Office: Burden of disease from environmental noise Quantification of healthy life years lost in Europe; ISBN 978 92 890 0229 5 (2011) http://www.euro.who.int/__data/assets/pdf_file/0008/136466/e94888.pdf, last visit: January 3rd 2017.
4. Directorate general for internal policies policy department b: structural and cohesion policies transport and tourism reducing railway noise pollution study Brussels, © European Union (2012) http://www.europarl.europa.eu/RegData/etudes/etudes/join/2012/474533/IPOL-TRAN_ET(2012)474533_EN.pdf, last visit January, 30rd, (2017)
5. WHO 2009, World Health Organization: Night noise guidelines for Europe, WHO Regional Office for Europe Scherfigsvej 8, DK-2100 Copenhagen Ø, Denmark, (2009) http://www.euro.who.int/_data/assets/pdf_file/0017/43316/E92845.pdf, last visit February, 13rd (2017)
6. Ľ. Argalášová, J. Jurkovičová, Ľ. Ševčíková, Z. Štefániková, K. Hirošová, J. Babjaková, A. Kánovicsová, A. Filová, Applied Mechanics and Materials, **617** (2014)
7. E. Panulinová, S. Harabinová, Applied Mechanics and Materials, **617** (2014)
8. E. Panulinová, S. Harabinová, *SGEM 2013*, **1** (2013)
9. A. Fyhri, G.M. Aasvang, Science of the Total Environment, **408** (2010)
10. M. Liberko, Guidelines for calculation of traffic noise levels, **1** (1991)
11. J. Smutny, L. Pazdera, Insight, **46**, 10 (2004)
12. A. Pultznerová, L. Ižvolt, S. Hodas, The Transportation as the Factor of the Socio-economic Development of the Regions, **1** (2012)
13. J. Melcer, Transport and Telecommunication, **16**, (2015)
14. V. Valašková, J. Melcer, G. Lajčáková, Procedia Engineering, **111** (2015)

Measuring of noise emitted by moving vehicles

Tomáš Skrúcaný[1,*], *Branislav* Šarkan[1], *Tomasz* Figlus[2], *František* Synák[1] and *Ján* Vrábel[1]

[1]University of Žilina, Faculty of Operation and Economics of Transport and Communications, Univerzitná 8215/1, 010 26 Žilina, Slovakia
[2]Silesian University of Technology, Faculty of Transport, Krasinskiego St. 8, 40-019 Katowice, Poland

Abstract. The article aims to measure the intensity of the exterior noise of a vehicle in motion. It provides the results of the measurements of the external noise of selected vehicles in motion and the impact of selected factors on the sound level of driving. There are done two measurements in the paper. Results from the first one are comparing noise level of 9 passenger cars according to the Directive 71/157/EEC. The second one shows the road surface influence on the exterior noise of moving vehicle where the sound level was measured by a road whose surface was made of slightly degraded concrete, and at a different place of the same road, where the surface was renovated by applying asphalt mix onto it.

1 Introduction

The noise is defined as an unwanted sound which has an unpleasant and annoying effect. It is classified by the World Health Organisation as the second most serious environmental cause that threats the health of the population. The road transport is a major source of the environmental noise. The results of the European Environmental Agency's report show that one in four Europeans is exposed to the sound level above the upper limit laid down in European directives. The noise from the transport afflicts approximately 20 million of Europeans and approximately 8 million experience sleep disturbance. It is also responsible for nearly 43 000 hospitalisations, 900 000 cases of hypertension and 10 000 cases of premature deaths per year [1, 2].

Based on the reasons referred to above, the issues have been implemented into the legislation. On the European level these comprise the Council Directive 70/157/EEC, the Directive 2002/49/EC of 25.06.2002 relating to the Assessment and Management of environmental noise, and the Regulation 540/2014 of the European Parliament and of the Council relating to the Sound level of motor vehicles. On the national level these comprise the Act No. 355/ 2007 Coll. on the Protection, Support and Development of the Public Health, Decree No. 549/2007 Coll. establishing details for the noise limits and the others.

The factors affecting the noise which is caused by the driving of a vehicle can be divided into two groups [2, 3].

The first group involves the noise from the vehicle engine and the noise arising from the driving due to air flowing around the vehicle and through its cooling and ventilating

* Corresponding author: tomas.skrucany@fpedas.uniza.sk

system. The important factors affecting the noise caused by the engine are the engine construction, rotations, technical condition, and tightness of the exhaust manifold.

The second group involves the noise arising from the motion of tyres at the road surface. The noise intensity is mainly influenced by the tyre construction, its size, type and pattern condition, building materials and technical conditions of the road surface, and speed of a vehicle [4, 5].

The aim of the article is to measure the exterior noise arising from the driving of selected vehicles and to assess the effect of the vehicle speed and building and technical conditions of the road surface on such a noise.

2 Method of measurement

2.1 Exterior driving noise of chosen passenger cars

The first part of the measurement was focused on the measuring the exterior noise of selected vehicles in motion. The list of selected vehicles with their performances is given in the Table 1.

Table 1. Used vehicles during the practical measurement.

Vehicle brand	Trade name	Engine capacity [cm^3]	Production year	Maximum engine performance [kW/min]	Fuel type	Exterior noise of a vehicle in motion (methodology) [dB]
Skoda	Octavia	1896	2008	77/4000	Diesel	69
VW	Passat	1896	1996	81/4150	Diesel	72
Renault	Thalia	1149	2007	55/5500	Petrol	70
Opel	Zafira	1995	2004	74/4000	Diesel	72
Opel	Vectra	1598	1996	74/6200	Petrol	73
Mazda	323	1840	2000	84/6000	Petrol	73
Peugeut	206	1868	1998	51/4600	Diesel	72
Skoda	Felicia	1289	2001	50/5000	Petrol	73
Renault	Clio	1390	1997	55/5750	Petrol	75

The measurement was performed in accordance with Council Directive 71/157/EEC. According to this standard, A microphone must be placed at a distance of 7.5 m +/- 0.2 m from a reference line of the roadway and 1.2 m +/- 0.1 m above the ground. Its axis of maximum sensitivity must be horizontal and perpendicular to the vehicle roadway. When the vehicle front reaches the specific line, the accelerator fully opens as quickly as possible and is held in the fully opened position, until the rear vehicle intersects another specific line. Then, the accelerator closes as quickly as possible. The measurement conditions are depicted in the Fig. 1 [6].

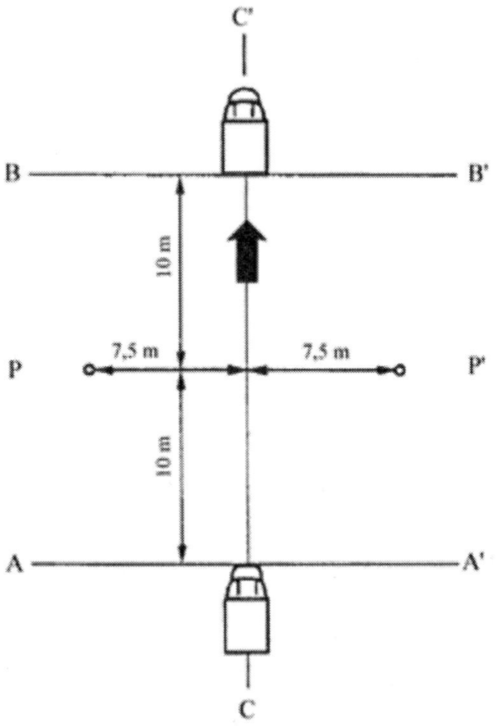

Fig. 1. The measurement of the noise of a vehicle in motion.

2.2 Exterior driving noise influenced by road surface type

The second part of the measurement was focused on the measuring of the impact of vehicle speed and road surface on the exterior vehicle noise. The measurements of traffic noise were performed using a Hyundai i40 Wagon car with a 1.7 litre engine. The measurement was performed at having constant engine speed of 2.200 rpm. The vehicle reaches the speed of 52 kmh^{-1} at 3rd gear, 75 kmh^{-1} at 4th gear and 98 kmh^{-1} at 5th gear.

The traffic noise tests were performed by road I/65 between the towns: Martin and Turčianske Teplice in Slovakia. The measurements were taken at two sections of the road. The surface of the first section was made of slightly degraded cement concrete, and the surface of the other section was previously renovated by applying a layer of asphalt mix onto the existing concrete layer [7 - 9].

The measurements were taken using the modified SPD method, in two directions of vehicle traffic, as shown in Fig. 2, at a controlled constant car speed. During the experiment, at least 3 runs of the vehicle, meaningful in terms of the measurement, were performed in each traffic direction.

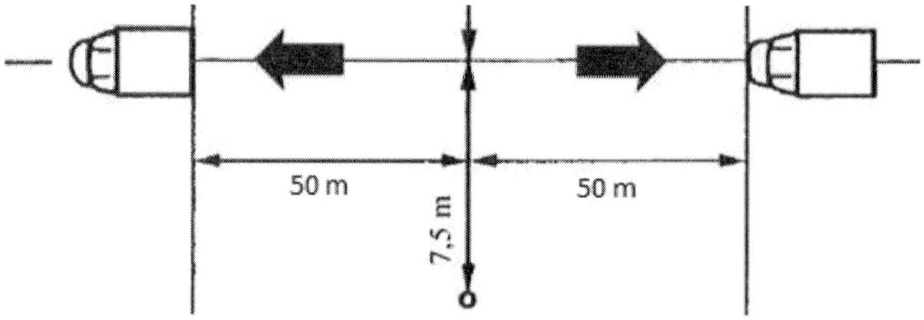

Fig. 2. Method of measurement.

3 Measuring devices

For both measurements it was used the Bruel&Kjaer 2250 digital sound analyzer with anti-wind attachments. Measurements were performed in the frequency range from 3 Hz to 20 kHz, and in 1/3 octave bands. The maximum sound level, L_{AFmax}, and the equivalent sound level, L_{Aeq}, were recorded, and the results were adjusted by means of correction curve A [10, 11].

Fig. 3. View of the Bruel&Kjaer 2250 sound analyzer during measurement.

3 Results and Discussion

The Table 2 shows the exterior noise measured value of the selected vehicles in motion and these are compared with the "Methodology."

Table 2. Noise emission limits during the test of the vehicle in motion.

Vehicle	Exterior noise from moving vehicle						Metho-dology dB	Highest measured value dB	Absolute difference dB	Relative difference [%]
	1st measurement		2nd measurement		3rd measurement					
	Left side dB	Right side dB	Left side dB	Right side dB	Left side dB	Right side dB				
Skoda Octavia	69	69.2	68	68.5	68.5	68.7	69	69.2	+0.2	0%
Mazda 323	70.1	69.9	70.7	70.5	71	69.9	73	71	-2	-3%
Renault Thalia	70	69.8	70.5	70.3	71	71.2	70	71.2	+1.2	2%
Skoda Felicia	71.9	71.9	71.8	71.7	72	71.9	73	72	-1	-1%
Opel Zafira	72.3	72.4	72	72.1	71.8	72	72	72.4	+0.4	1%
VW Passat	72.8	72.6	72.8	72.6	71.8	72	72	72.8	+0.8	1%
Peugeot 206	71.5	71.9	72.9	72.6	72.3	72.4	72	72.9	+0.9	1%
Renault Thalia (2)	70.1	69.9	72.9	72.6	72,3	72.2	70	72.9	+2.9	4%
Renault Clio	74.5	74.4	74.9	74.8	73.5	73.7	75	74.9	-0.1	0%
Opel Vectra	77	77.1	76.9	76.8	76.9	77	73	77.1	+4.1	6%

The lowest noise value was measured within the Škoda Octavia vehicle comprising 69.2 dB. The highest value was reached by Opel Vectra vehicle with 77.1 dB where also the highest positive difference between the permitted and measured value was. The seven out of nine tested vehicles reached higher noise values than those defined in their Registration certificate in accordance with 540/2014 [12, 13].

The results of the second part of the measurement are shown in the Fig. 4 and 5. The Figures present the results of measurements of the maximum sound level L_{AFmax}, and of the equivalent sound level L_{Aeq}, recorded in the entire frequency band at different car speeds [7].

DOI: 10.1051/matecconf/201710700072

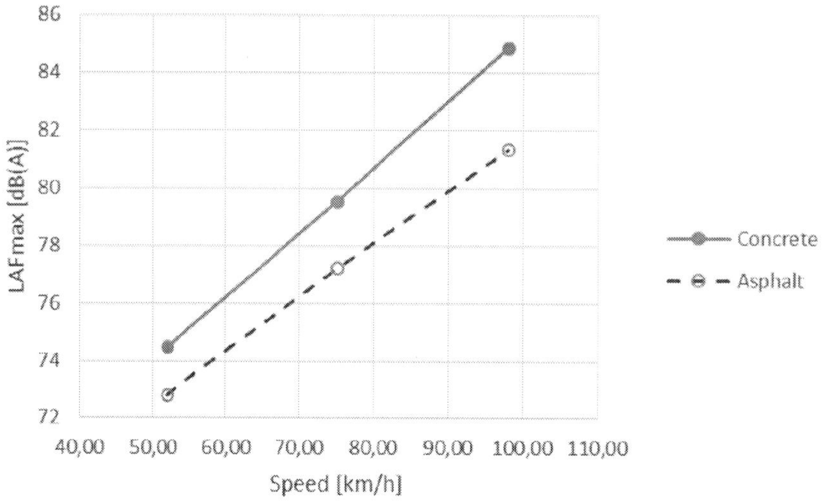

Fig. 4. The impact of driving speed and road surface on the L_{AFmax}.

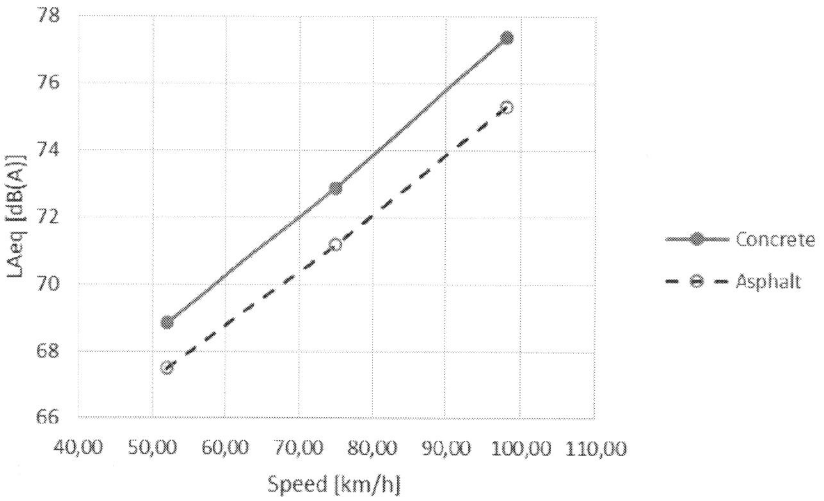

Fig. 5. The impact of driving speed and road surface on the L_{Aeq}.

The recorded values of the maximum sound level strongly depend on the car speed; on the cement concrete surface they reach values from 74.4 dB at a car speed of 52 kmh^{-1}, through 79.5 dB at 75 kmh^{-1}, up to 84.8 dB at 98 kmh^{-1}. It can be noticed that an increase in traffic speed by ca. 23 km/h induces an increase in the maximum traffic noise level near the road by ca. 5 dB. Comparing the recorded values of the maximum sound level for a car driving on a cement concrete surface with those recorded by the road with an asphalt mix surface, the differences amount to ca. 1.5-3 dB. Each time the sound levels recorded on the road with the asphalt mix surface are lower, and the differences in the values depend on the speed of the vehicle. Smaller differences are observed at lower traffic speeds [6].

While analyzing the existing traffic noise levels, it can be noticed that each time they exceeded the instantaneous values of 70 dB, and at the highest traffic speed, 80 dB. Such a sound level of sound is a nuisance to the environment. A change of the type of surface on the road under consideration and additionally, a reduction of the car speed may considerably contribute in this case to a reduction of the maximum sound level. For example, a change of the surface and a simultaneous car speed reduction from 98 to 75 kmh^{-1} may decrease the traffic noise by as much as 7.6 dB.

For comparison purposes, an analysis was made of the equivalent sound level recorded during the running of the tested vehicle at a distance of ca. 50 m in front of and behind the measurement point. The sound level values in this case are lower than 70 dB only at a traffic speed of 52 kmh^{-1}. At higher traffic speeds, they exceed 70 dB each time, the maximum recorded value of 77.4 dB for the road with a cement concrete surface [6].

4 Conclusion

Noise emitted by moving vehicles is a problematic part of traffic emissions. Also measuring and evaluating the results is a complicated thing. It is still hard to evaluate the real noise influence on the human hearing and health despite the fact that there are directives, standards and regulations to solve the problematic of noise. Reached and described results in this paper reflect the real state of the noise level in traffic [14, 15, 16]. The results from the first part of measurements, passenger cars comparison, show that real noise level is almost higher than the value declared in the vehicle technical specification. It can be caused by different tires (depreciation, changed type, mark), technical condition of the engine, vehicle drive train or other technical parts [17, 18]. The second result evaluation testifies the road surface influence on the noise level measured by one vehicle. The surface made from older and depreciated concrete panels is noisier than the renewed surface where the old surface was equipped by a new layer or asphalt. So not only the vehicle (type, technical condition, tires) but also the road surface is influencing the noise level of moving road vehicles.

References

1. J. Paslawski, Int. J. Environ. Pollut. **35**, 275 295 (2008)
2. T. Figlus, A. Wilk, Š. Liščák, Ł. Gozdek, Doprava a Spoje, 473-480, (2012)
3. W. Gardziejczyk, Transport Res. D-Tr E, **44**, 93 – 104, (2016)
4. R. Rajamani, *Vehicle dynamics and control* (New York: Springer, 2012)
5. J. Y. Wong, *Theory of ground vehicles* (Ottawa, 2001)
6. T. Figlus, J. Gnap, T. Skrúcaný, P. Szafraniec, K. Żuradzki, *CMDTUR* (CMDTUR, Žilina, Slovakia, 2016)
7. B. Sarkan, O. Stopka, L. Chenguang, (to be published)
8. Z. Matuszak, M. Jaśkiewicz, K. Ludwinek, Z. Gawęcki, *Selected Problems of Electrical Engineering and Electronics* (WZEE'2015, Kielce, Poland, IEEE, New York, USA, 2015)
9. T. Figlus, J. Gnap, T. Skrucany, B. Sarkan, J. Stoklosa, Entropy, **18**, 253 (2016)
10. P. Fiala, P. Drexler, D. Nespor, Z. SzabO, J. Mikulka, J. Polivka, Entropy, **18**, 443 (2016)
11. J. Caban, P. Drozdziel, D. Barta, S. Liscak, Diagnostyka - Diagnostics, **15**, 11 – 14 (2014)

12. Regulation (EU) No. 540/2014, 16 April (2014)
13. Ordinance of The Transport Minister (Poland) No. 951, 22 August (2013)
14. S. Kubikova, A. Kalasova, L. Cernicky, *14th Transport Systems Telematics Conference* (TST, POLAND, 2014)
15. J. Gnap, V. Konecny, *9th International Scientific Conference on Business Economics and Management* (BEM, Izmir, TURKEY, 2015)
16. J. Ondrus, L. Cernicky, Communications : scientific letters of the University of Zilina. **18**, 118 – 123 (2016)
17. V. Rievaj, L. Mokričková, Transport and communications, scientific journal, **1**, 22 -26, (2016)
18. O. Stopka, I. Simkova, V. Konecny, Nase More, **62**, 126 - 130 (2015)

Research of noise emission sources in railway transport and effective ways of their reduction

Peter Zvolenský[1*], *Juraj* Grenčík[1], *Alžbeta* Pultznerová[2] and *Ľubomír* Kašiar[1]

[1]University of Žilina, Faculty of Mechanical Engineering, Univerzitná 8215/1, 01026 Žilina, Slovakia
[2]University of Žilina, Faculty of Civil Engineering, Univerzitná 8215/1, 010 26 Žilina, Slovakia

Abstract. In the EU conditions attention is systematically paid to noise reduction on the railways. Because TSI rules systematically tighten limits for noise emissions from railway vehicles, noise research must be addressed by all Member States, as the main technical solutions for railway vehicles and construction technological aspects of railway operations can result in lower noise exposure of affected areas or objects. The paper focuses on theoretical investigation of sources and paths of sound propagation, possibilities of noise reduction both on vehicles and by infrastructure and experimental measurements of the situation in transport practice. Methodology for reducing railway noise around tracks has been presented, too.

1 Introduction

Environmental aspects of design, operation and maintenance of vehicles in recent years have gained particularly important significance. Under the conditions of the EU for long time systematically attention has been paying to rail noise reduction. Because TSI rules systematically have been tightening limits for noise emissions produced from railway vehicles, noise research must be addressed by all Member States. The main technical solutions for rolling stock and technological aspects of railway operations should result in lower noise exposure of affected areas or objects. Scientific approach in this area can be defined at several levels: a theoretical investigation of the source and paths of sound propagation, simulation and modelling of noise fields and their impact on the environment, verification of the theory and computer modelling by experimental research and verification of the situation in the transport practice, elaboration of procedures and methods for reduction of noise in the application for the design of specified railway vehicles as well as construction of railway tracks.

Noise reduction on the railway requires a systematic approach therefore continuous research of the aforementioned interconnected levels being the system components. In the transport practice, tasks associated with noise reduction usually appear in two equal time phases. Either as an urgent acute noise reduction, this usually arises from the measured situations - noise exposure, especially surveyed by the state regulatory authority, for example noise in railway hump yards affecting residential areas - or as a conceptual task

* Corresponding author: peter.zvolensky@fstroj.uniza.sk

resulting from the EU legislation and EU environmental policy of country concerned, for example concept of train formation and international transit so called silent trains [1].

Many designs of railway vehicles are on their design limits in terms of noise emissions and possibilities of their reduction. Equally solutions of the secondary noise reduction (acoustic barriers) are relatively well developed and parameters of sound reduction index and sound absorption are well-defined. So every additional noise reduction, even of only a few decibels is welcome, because for different acoustic conditions the maximum limits permitted by norms are often exceed just by few decibels. At the same time the use of materials and technologies that effectively encourage recycling has special significance Rapid transfer of scientific - research findings and results, particularly experimental research in the field of noise emissions from railway vehicles in transport practice, bring new tools and approaches to effective improvement of environmental quality

Development and modernization of railways in the Slovak Republic leads to targeted integration into European transport networks. This naturally means that high level of quality of transport, structures and vehicles is desired. When evaluating the rolling stock, among others, the emitted noise and vibration are assessed. The most dominant source of noise in the railway vehicle operation is a rolling of the wheel on the rail. We talk about so called combined noise resources, which are caused by roughness wheel and rail roughness and the produced by dynamic shocks of tracks and vehicles. Braking of the vehicles is another important noise source. When measuring and evaluating various parameters, it is important to comply with the provisions of the TSI. [2]

In the reduction of noise and vibration in transport it is most efficient to focus on their source and generally by active design modifications. Only when these primary possibilities cannot be use it is possible to use passive protection. If there is an urgent need to protect the environment, system of verification of compliance with the parameters of noise and vibration must be properly created. Their reduction represents a large and complex approach and proposals for solutions. Therefore the effects of noise and vibration are analysed separately, because they influence in different way the staff and a passengers on vehicles, and in different way people living in residential areas.

2 Noise emitted by railway vehicles

Sources of external noise from railway operations can be divided into several categories.
Primary noise – noise during movement of rolling stock

- basic – caused by rolling of wheel on rail
- other – noise from traction and auxiliary machines, air turbulence, noise produced by running gear, from pantographs and certainly vibrations of individual parts of vehicles.

Secondary – signals form locomotive, noise from shunting operations.

The basic source of noise is the wheel running on a rail - contact surface between wheel and rail. At this contact point, irregularities on the running surface of the wheel and so called transverse and longitudinal waves on the rail head produce oscillation of track and the vehicle and vibrations are generated, which are emitted to the air in the form of noise, or transmitted in the rail and wheel. From there the noise is propagated further in the construction or is radiated into the surrounding outer space. Sound waves are then propagated form a vehicle to surroundings, while the shape of the acoustic field is affected by:

- directional characteristics of noise source radiation,
- acoustic properties of the environment through which the emitted energy propagates,
- barriers in the path of noise propagation.

2.1 Noise reduction measures on railway vehicles

Active noise protection measures are such modifications of noise sources that will reduce the overall noise, such as structural modifications of railway vehicles, optimization of track alignment and structural arrangement of railway lines and the like.
For railway vehicles, whether traction or wagons, it is possible to reduce noise emissions by the following design measures [1]:

- Traction vehicle (locomotive) – where besides noise produced by rolling of wheel on rail, noise generated by traction and auxiliary aggregates is added. In general, electric traction is less noisy than diesel traction. Diesel engine noise is usually reduced using exhaust silencers of diesel locomotives.
- Railway wagon - differences between the running noise passenger and freight wagons are attributed to more efficient damping adjustments on coaches as flexible suspension of car body on bogies, introduction of new bearings with lower rolling resistance, and the like.
- Appropriate type of brakes - trends is replacement tread brake for disk brakes with an anti-skid device. In case of tread brakes used on freight wagons, new materials of brake shoes are being developed, such as composite that replaces classical cast iron material. Brake shoes made from composite substantially reduce squeal sound during braking.
- Optimisation of wheel shape – it is possible to achieve certain minimum noise and vibration emissions of a wheel. It is possible to achieve this also by wheel profile and axial symmetry of the wheel.
- Using wheel noise absorber – which reduces not only noise radiation yet it transforms it into higher frequency range and reduces noise level compared with conventional wheel by about 6 dB
- Shading cover – mounted on car body, it can be placed also o bogies, which contributes to reduction of emitted noise.
- Aerodynamic shape of car body - at higher velocities the aerodynamic shape of car body is important. It reduced remarkably irregularities on outer surface of car body and excess edges. By this uncontrollable turbulence on car body surface is reduced.
- Suitable pantograph type design – electric current pantograph is significant noise source at higher operational velocities.

3 Railway noise around railway tracks

Because most of the railway lines were built in the territory of the Slovak Republic in the 19th and the beginning of the 20th century, the question of the noise was not solved in their design and construction and therefore no arrangements for its reduction were implemented. Currently on the railway lines of the main corridor Košice - Žilina - Bratislava modernization of railway track, connected with the transition to a qualitatively new, higher standard is underway. This enables trains rung on upgraded lines at speed up to 160 km/h. These changes, however, will have impact on the environment. Among the biggest impacts of rail transport on the environment belong noise and vibrations and their impact on people living near the railway line.

3.1 Noise from the aspect of railway track alignment

Natural protection against traffic noise is longer distance of the noise source from residential buildings, which can be accomplished only in case of new lines construction.

For the design of new railway lines, especially high-speed, light rail and rapid lines, it is possible to reduce the noise load by using the following options [2, 3]:

- Tunnel line – most suitable for densely built-up area from the environmental protection point of view; near the tunnel line, however, buildings may be affected by vibrations (i.e. underground in large cities, eventual connecting lines of suburban and long-distance passenger traffic through the city centre).
- Deep cut – attenuates noise,
- Line above terrain level on embankment body – requires building-up an embankment with moderate slope gradient with noise protection berries and growing vegetation of bushes and trees which creates natural noise insulation and non-violent separation of traffic on the route from city life is achieved.
- Line construction on elevated lines – is noisier and more difficult solution for inclusion into countryside or residential area.

The most suitable technical solution of communication from the acoustic aspect is routing the line in a cut or tunnel. The least suitable is construction on the terrain level. It is very suitable in case of re-buildings, eventually new-built constructions, to locate road communication and railway line parallel in so called transport corridor and adjacent area protect by earth walls and higher greenery.

3.2 Noise from the track construction aspect

From the perspective of a railway line, it is possible to reduce noise levels by optimum directional and altitude alignment, sufficiently deformation resistant structure of railway substructure, by applying modern structures of the superstructure - such as continuous welded rail, resilient fixation of rails, anti-vibration mats and flexible pads, suitable design of the rails, applying modern types of switches, application rail absorbers, by suitable design of bridges and last but not least by regular and continuous maintenance of the superstructure [4].

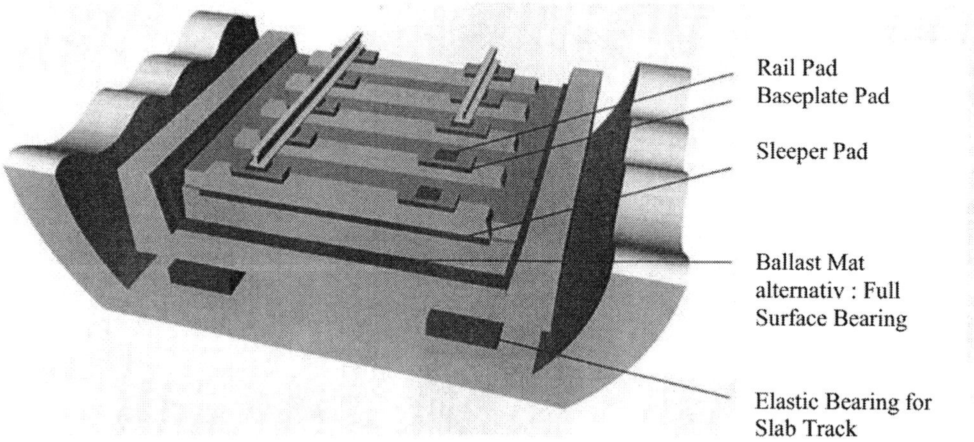

Fig. 1. Resilient railway fastenings [5].

4 Possibilities of noise reduction on railway tracks

Noise reduction generated from rail transport on already built-up track can be done by direct design solutions for the railway track – so called active noise abatement solutions or by constructing of sufficiently noise reducing barriers - so called passive noise solutions.

4.1 Active noise abatement solutions

Noise barriers in many cases help to significantly reduce the load on citizens by noise from traffic, but this solution is not always possible to be used to achieve maximum success. If necessary, preserving free sight, aesthetics and needs only for minor additional noise reduction, active noise solutions are preferred. Their advantage is that they prevent the spread of noise to the surrounding area. An option is using a rail absorber and so-called rail track absorber (rail mat).

Rail damper (absorber) is fixed to the rail web. Thanks to its design composition it has high absorption properties by which it absorbs vibrations arising at the wheel / rail interface and thereby prevents the generation of noise in the rail. In the case of a train run, in the contact between wheel and rail noise with variable spectrum is generated. Rail damper works on the system of vibrations absorption at frequencies with the greatest noise levels. For low and high frequencies, which correspond to lower noise levels, the structure of the rail damper is designed so that the damper had no effect on the noise generated. On the other hand, in the case of frequencies that are characteristic for most of the noise generated, the rail damper is the most efficient [6]. Rail damper that prevents the propagation of the vibration wave also facilitates its better attenuation.

There are several types of rail dampers that are different in shape and design structure. One of models manufactured by Vossloh company is shown in Fig. 2 and 3.

Fig. 2 Rail damper of Vossloh [6].

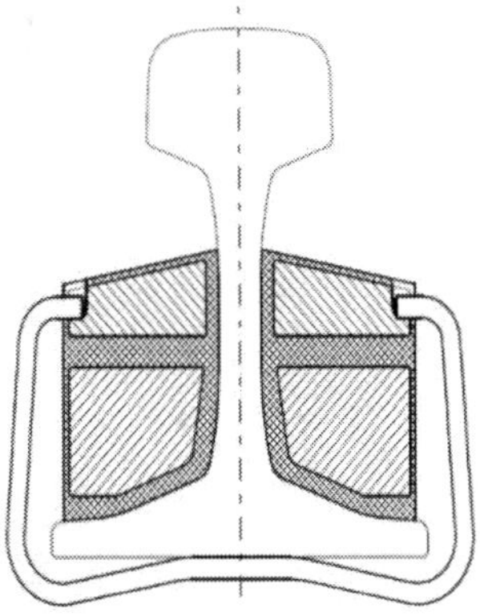

Fig. 3 Cross-section of rail damper [6].

Rail dampers reduce noise generally by about 1-3 dB, when well designed by 4-7 dB [6]. Their main advantage is that they can be mounted even after the construction of the section in critical locations for improving the noise parameters after putting construction into operation, there are no restrictions on track maintenance, they are maintenance-free and the rail damper does not interfere with the overall appearance of the track [4].

Track absorber is a system mats made from wooden-cemented parts Duris, or recycled synthetic material STERED® (Fig. 4)a thickness of 50 to 150 mm, and its design follows the corresponding shape of the tracks, whether it is for new construction or reconstruction of existing railway lines. [7] It consists of 5 types of absorbers, which are placed between the rails and next to them, while the combination significantly reduces noise from rail traffic (Fig. 5). Based on measurements by accredited laboratory, the compact system components in case of the standard audio spectrum in rail transport, achieves absorption up to 18 dB [8].

Fig. 4. Completed track absorber BRENS® with synthetic recyclate STERED® and artificial grass on the [7].

Fig. 5. Track noised absorber [8].

4.2 Passive noise reduction solutions

The purpose is to build an acoustically sufficient impermeable noise barrier built on the propagation path of sound waves, while a sound shadow of sound pressure levels is created behind the barrier. Such barrier may be a physical object, earth walls, vegetation and the most commonly used anti-noise walls.

Noise barriers nowadays are the most common structure utilized to reduce the noise already generated. They are constructed in parallel with modernization of railway corridors by comparing the permissible noise limit values according to [9] with the noise emission values determined by noise studies, detailed calculations of isophones (noise exposure contours) and verification measurements. Their elaboration is required by the proposed increase of line speed on upgraded sections of railway track to 160 km/h.

The efficiency of a noise barrier depends on structural elements, materials, shape (Fig. 6 and 7), height, length, location on the area and the quality of the construction corresponding to needs of the particular situation.

Fig. 6 Low noise barrier [10].

Fig. 7 track barrier [11].

5 Methodology for noise reduction on railways

For consistent design of noise prevention measures, especially location, dimensions and acoustic properties of noise barriers and measures on the buildings themselves, it is appropriate to make predictive mathematical modelling of sound propagation in the area of interest. This requires the methodology as follows: [12]

1. Specification of the area of interest - a user-defined polygon, which delimits the assessed area. Area of interest is defined by a strip (region) with a width of at least 100 m from the route of the railway line.
2. Creation of a digital terrain model - definition of terrain and objects (buildings) is formed on the basis of data from aerial photography.
3. Definition of railway communications - size, shape, character of trains' run, quality of the superstructure and the like.
4. Design of noise barriers - at this stage 3D dimensions of all objects, acoustic properties of the walls and the height of the walls are determined.
5. Modelling the proposed route of the railway line - using software for predicting noise immissions in the exterior, all elements of the proposed routes - bridges, overpasses, cuts, embankments, elevated highways, and so on are modelled.
6. Input of the expected number and types of trains, their speeds, lengths, run character and so on.
7. Calibration of the mathematical model - based on calibration measurements, the model is calibrated so that to ensure the necessary accuracy of modelling results.
8. Mathematical modelling - calculation of noise immissions in the area of interest. The objective is to develop a noise map of the area of the railway section using methods of mathematical modelling, using a methodology for traffic noise modelling in accordance with the valid Slovak legislation and the relevant norms. Example of noise map after application of the anti-noise wall is in Fig. 8.
9. Optimization of noise barriers - designed noise barriers will be optimized so that the permissible values determining the quantities of noise in accordance with applicable legislation have not been exceeded on the area of interest.
10. Proposal of other technical arrangements - arrangements for construction projects

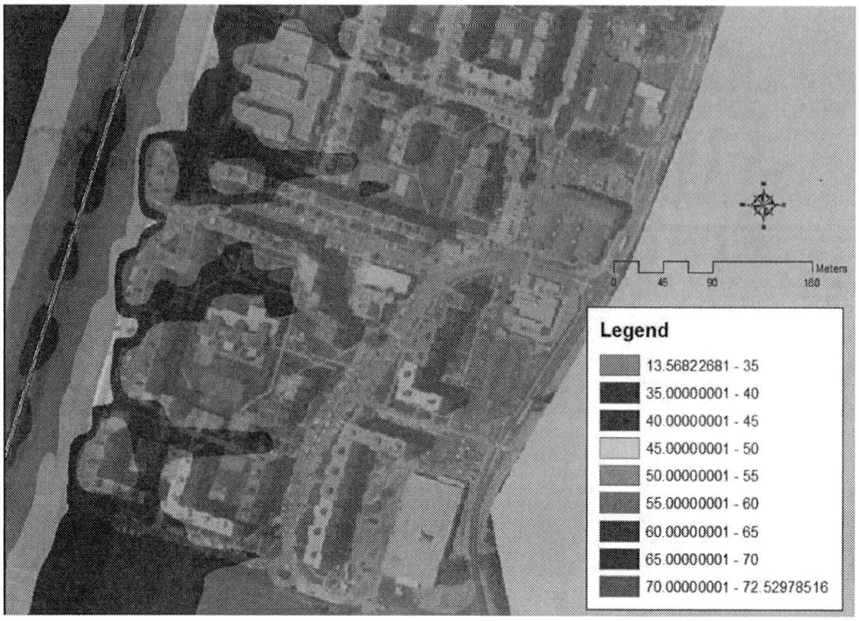

Fig. 8. Visualization of continuous sound propagation generated from a modernized railway line the specified area with highlighting of polygons in 3D (own research).

6 Practical example - noise analysis at hump marshalling yard

Particularly unfavourable acoustic situation arises in formation of freight train using technology of shunting at so-called hump marshalling yards using track retarders for braking. The braking performance of these retarders is generated automatically by clamping special jaws on the railway vehicle wheel according to the weight and the running speed of shunted freight wagons, while extremely high noise levels are generated, often exceeding 120 dB(A). Rather extensive research is focused on this problem area, both on the experimental analysis of noise conditions, computer simulations of noise propagation fields and propagation paths of noise, and also the proposals for noise control solutions. As an example acoustic situation during braking of shunting freight tank wagons series Zaes with use of track retarders (Fig. 9, 10, 11).

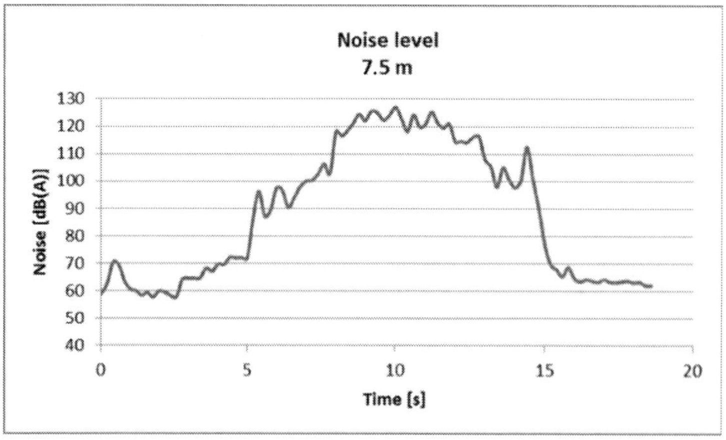

Fig. 9. Noise level at track retarder in the distance 7.5 m from track centre line (own research).

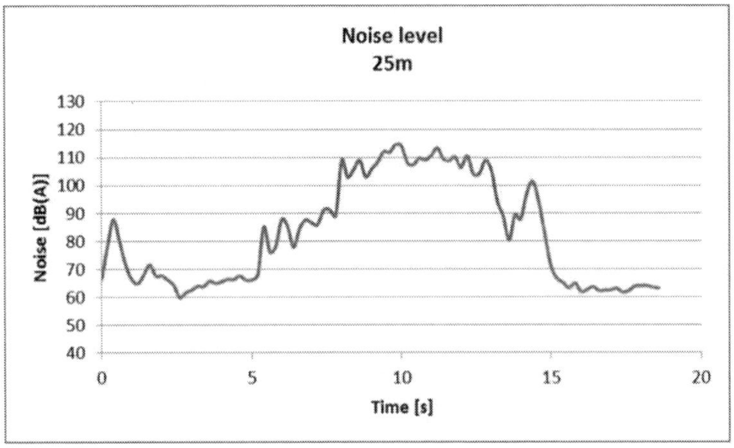

Fig. 10. Noise level at track retarder in the distance of 25 m from track centre line (own research).

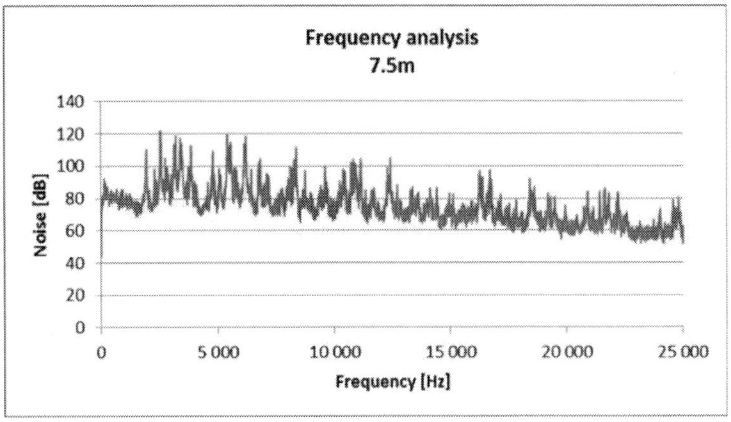

Fig. 11. Frequency analysis of noise at track retarder in the distance 7.5 m from track centre line (own research).

7 Conclusions

From a series of the measurements it implies that the different levels at the distance of 7.5 m from track centre line reached enormous values more than 127 dB(A). From the frequency analysis of noise measurements carried out it implies that up to 5000 Hz there are dominant tonal frequencies that exceed 120 dB. There were also dominant noise frequencies found, which are: 1548 Hz, in which the noise level was measured 124 dB, 1624 Hz with 123 dB, 3288 Hz with 122 dB. It is also interesting that with increased body weight of shunted wagons in the range of 66-78 tons, during the breaking by track retarder the dominant frequency range are moving from the higher to the lower frequencies.

Measurements have shown that the action of track retarders significantly creates noise burden over fairly large affected area. For this reason, in the program NoiseAtWork v3.32 we have simulated the propagation of the measured values of noise levels in the surrounding area. The simulation results show that the measured noise level with a value of 127 dB(A) reaches 122 dB(A) at the distance of 1000 m from the track retarder. This characterizes particularly unfavourable acoustic situation for wide surroundings of the hump marshalling yard technologies using track retarders.

There are partial results of the grant VEGA 1/0766/15" Research of noise emissions sources in the railway transport and methods of their effective reduction" in the paper.

References

1. U. Clausen, C. Doll, F. J. Franklin, G. Vasic Franklin, H. Heinrichmeyer, J. Kochsiek, W. Rothengatter, N. Sieber, *Reduction of noise pollution from rail transport*, study, European Parliament (2012)
2. U. Degn, U. Danneskiold-Samsøe, *A Study of European Priorities and Strategies for Railway Noise Abatement*, Report II, Retrieval of System Cases, Bruxelles, EU Commision, Directorate – General for Energy and Transport, Belgium, (2001)
3. C. Esveld, *Modern railway track* (Delft University of Technology, The Netherlands, 2001)
4. L. Bavlna, A. Pultznerová, P. Zvolenský, *Possibilities of railway traffic noise reduction depending on the railway structure and construction of the rail vehicle*, Logistika 4 (2015)
5. https://www.getzner.com/de/ueber-uns/wer-wir-sind (10.3.2017)
6. D. Benton, *Railway noise in urban areas, possible source noise reduction measures*
7. http://www.stered.sk/novinka/kolajove-absorbery-hluku-brens-dalsi-prirastok-do-rodiny-efektivnych-protihlukovych-rieseni-pre-zeleznice (10.3.2017)
8. http://www.leier.sk/hlukove-absorbery-a-nastupistne-hrany-durisol (10.3.2017)
9. Decree of MZ No. 549/2007 Z.z. stating the details on the permissible levels of noise, infrasound and vibration and the requirements on objectification of noise, infrasound and vibration in the environment of 16th August 2007
10. https://www.google.sk/search?q=n%C3%ADzke+protihlukov%C3%A9+steny&hl=sk&rlz=1T4MXGB_skSK563SK575&tbm=isch&tbo=u&source=univ&sa=X&ei=T4CZU8TqHYnY0QWC-YCYBA&ved=0CFsQsAQ&biw=1120&bih=824 (10.3.2017)
11. https://www.google.sk/search?q=protihlukov%C3%A9+steny&tbm=isch&tbo=u&source=univ&sa=X&ei=Qn2ZU76gLJOM4gTcsIDIAg&sqi=2&ved=0CCsQsAQ&biw=1119&bih=824 (10.3.2017)
12. E. Lumnitzer, M. Biľová, *Methodology for design of anti-noise measures in the vicinity of railway lines to ensure the acoustic quality of interiors,* Nehnuteľnosti a bývanie, **III**, 1 (2015)

MATEC Web of Conferences **107**, 00074 (2017) DOI: 10.1051/matecconf/201710700074
DYN-WIND'2017

Important moments in the history of structural aerodynamics

Jozef Melcer[1,*]

[1]University of Žilina, Faculty of Civil Engineering, Univerzitná 8215/1, 010 26 Žilina, Slovakia

Abstract. There were some important facts in the past that have influenced the development of structural aerodynamics very strongly. The first important event was the breakdown of the Tacoma Narrows Bridge (USA) in the year 1940. The second moment represents the disaster of the cooling towers at Ferrybridge (England) in 1965. In the year 2010 the first time in history the interesting aerodynamic phenomena was observed on the new-built bridge in Volgograd (Russia). It is likely that this event will also stimulate the additional development of the bridge aerodynamics.

1 Introduction

There were some important events in the past that have influenced the development of structural aerodynamics very strongly. They can be marked as faulty moments in the development of structural aerodynamics and in the approach of structure design on the wind load. The first such important event was the breakdown of Tacoma Narrows Bridge (USA) in the year 1940 [1]. The fatal event was the collapse of power plant cooling towers at Ferrybridge (England) in 1965 due to the wind load [1, 2]. In the year 2010 the demonstration of aerodynamic instability was observed on the new-built bridge crossing the river Volga in Volgograd (Russia) [3]. This phenomenon on such type of bridge structure was observed for the first time in the history of bridge engineering. It is likely that this event will also stimulate the additional development of the bridge aerodynamics and it will bring new hitherto unseen knowledge.

2 Collapse of Tacoma Narrows Bridge

In November 7, 1940 the collapse of Tacoma Narrows Bridge (USA) was observed. The native professor of architecture filmed the unique shots of this disaster. Leon Mojsejev was the bridge designer. The bridge was built within two years at the cost of 6. 4 millions US dollars in that time. Mojsejev died of the heart attack three years after the bridge disaster. The bridge was aerodynamically unstable. The loss of aerodynamic stability of the bridge and its collapse occurred under 68 km/h wind conditions, Fig. 1.

* Corresponding author: jozef.melcer@fstav.uniza.sk

 DOI: 10.1051/matecconf/201710700074

Fig. 1. Collapse of Tacoma Narrows Bridge, November 7, 1940.

What is the significance of this event for bridge engineering? The public blamed Mojsejev for the crash of Tacoma Narrows Bridge, but hurt him unjustly. Until then, no one ever thought that the problems of aerodynamic instability could occur also in relation to the bridge structure and no one would ever dream that the wind could cause the collapse of some bridge construction. Until then there were neither practical nor theoretical instructions how to avoid such situations.

The basic aerodynamic instability phenomena were already known at that time. Phenomena such as buffeting -caused by alternating impacts of wind on both sides of the structure due to turbulent air flow, galloping - movement generated flowing around curved and sharp-edged profiles resulting from the uneven tearing the air stream at different edges of diagonal wrap oscillating object, diffraction torsional flutter - resulting from the laminar flowing around the slim profile in the presence of inertial coupling, Strouhal effect – vibration of circular sections perpendicular to the wind direction caused due to alternate in the roll vortices on both sides of the structure, engineers already knew. But no one expected that these phenomena could be significantly applied to bridge constructions. Tacoma Narrows Bridge accident initiated intense theoretical and mainly experimental aerodynamic bridge construction research. Therefore this moment can be considered as a faulty moment.

3 Group effect

Another event that can be regarded as a faulty moment in the development of structure aerodynamics took place on November 1, 1965 in England at Ferrybridge, Fig. 2. The conclusions which were drawn from this event have direct impact on bridge construction as well. The cooling towers crash at Ferrybridge brought engineers to the conclusion that the isolated building structure behaves under wind loads diametrically opposed to that of the same object in a group of other objects (i.e., a group effect).

Fig. 2. Collapse of cooling towers in Ferrybridge, November 1, 1965.

 DOI: 10.1051/matecconf/201710700074

This fact significantly influenced experimental research in the wind tunnel. The engineers have realized that it is necessary to simulate the environment in the vicinity of the tested object.

4 New aerodynamic reality on the bridge in Volgograd

In October 2009 seven km long section of the bridge over the river Volga in Volgograd was put into operation. The bridge has a total length of 29.833 km. Seven kilometer stretch was built within 13 years. Circa 57 millions EUR were invested. This is a type of continuous beam composite steel-concrete construction. In May 2010 an unusual phenomenon was observed on the bridge. The section of the bridge around 1 km long was oscillating with amplitudes of 1 m, Fig. 3. These oscillations were caused by wind.

Fig. 3. Vibration of the Bridge in Volgograd, May 2010.

This event is interesting since it was the first ever observed case where the bridge with continuous beam on solid supports was aerodynamically unstable and vibrated due to the wind load in the same way as suspension bridges. This phenomenon is new to engineers, until now unprecedented aerodynamic reality. It is therefore very likely that this event also significantly affects the further development and exploration of bridge aerodynamics and it will bring us new previously unrecognized knowledge. Since then bridge designers have to consider the fact that even the bridge on solid support may under some circumstances be aerodynamically unstable and can oscillate due to the wind load as suspension bridges.

This work was supported by the Grant National Agency VEGA of the Slovak Republic, project number 1/0005/16.

References

1. O. Fischer, V. Koloušek, M. Pirner, *Aeroelasticity of Structures* (ACADEMIA, Prague, 1977, in Czech)
2. J. D. Holmes, *Wind loading of structures* (Spon Press, London, New York, 2001)
3. https://en.wikipedia.org/wiki/Volgograd_Bridge

Pressure field in measurement section of wind tunnel

Jakub Hnidka[1,*] and *Dalibor* Rozehnal[1]

[1] University of Defence, Department of Air Force and Aircraft Technology, Kounicova 65, 662 10 Brno, Czech Republic

Abstract. The University of Defence in Brno has a new low-speed wind tunnel. In order to confirm the quality of the wind inside of the measurement section, several measurements of the dynamic pressure have been performed with the Pitot-static tube. The pressure fields are then analysed and quality of the field is evaluated. Measurement of a pressure drop on the body of a standing helicopter was conducted.

1 Introduction

A new low-speed wind tunnel of return-flow type with an open measurement section was built at the University of Defence. The dimensions of the measurement section (MS) are 830×570×1370 mm (width × height × length). The maximum velocity is currently M = 0.09 (30 m/s); nonetheless, the fan currently operates in a restricted mode and the velocity will be increased in the future. Due to the relatively low values of Reynolds number achieved, the wind tunnel is ideal for e.g. testing the flight performance of small UAVs, which can be measured in scale of 1:1.

Investigation of the flow characteristics of the wind tunnel is a necessary step in the optimization of the wind tunnel [1]. E.g. in propeller testing, the aerodynamic performance of the propeller can be expressed as its efficiency. At a constant rotational speed of the propeller, a 1% change of the freestream velocity in the plane of rotation near the propeller take-off design point can alter the measured efficiency of the propeller by 0.5% and power absorbed by propeller by 1.3% [2]. If the wind tunnel is to be used to measure flight performance of the small UAVs, which almost exclusively relay on the propeller as a source of thrust, non-uniform velocity along the freestream could significantly influence the measured results.

The wind tunnel is equipped with vanes in all corner sections of the tunnel. Two corners upstream from the nozzle feature water-cooled vanes. Temperature stability of the freestream in the MS was investigated as well.

The flow fluctuation will be investigated in the close future.

* Corresponding author: jakub.hnidka@unob.cz

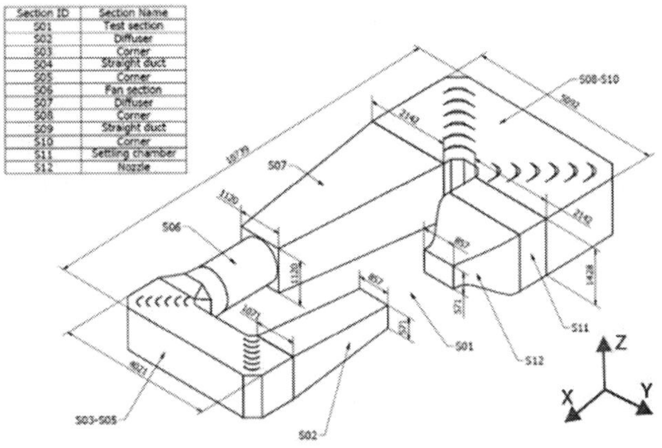

Fig. 1. Wind tunnel geometry with marked sections [3]. In the right bottom corner is shown the orientation of axes in the MS. *X*-axis is in the direction of wind flow.

2 Measurement instrumentation

The measurement of the velocity was performed with Pitot-static probe mounted on the traverse system. The system is capable of moving the probe in three axis to any destination in the MS. The system is controlled by three stepper motors, which are wired to the NI-7330 Motion Control PCI Module.

A differential pressure sensor with analogue output measured the dynamic pressure. The signal was sampled by NI 9215 module with 16-bit A/D converter, 100kS/s/ch maximum sampling speed and four BNC terminals [4]. This module was fitted to cDAQ-9184 chassis, which were connected to the PC via Ethernet bus. During the measurement, the application was testing the measurement capabilities of the system and it performed FFT analysis of the raw data. The purpose of this analysis was to search for any regular occurrences, which could indicate disruption of the freestream, sourced in e.g. screens, or blades of the fan. The differential pressure sensor was mounted outside of the traverse system in order to prevent any influence of the data by vibrations caused by the wind tunnel. The tubes connecting the sensor with the Pitot-static tube are therefore of a considerable length. Pressure measurement assumes the incompressibility of the air in the connecting tubes and this criterion is met for static events. Tested hypothesis was that if there is a significant disruption of the freestream, it could be measured by the pressure sensor despite the damping in the connecting tubes. Results of this analysis are not part of this paper – nonetheless, in order to observe higher frequencies, the sampling rate was set to 5 kHz. Occurrence of low frequencies was investigated separately, because the measurement of dynamic pressure distribution is time-consuming. The MS is divided into finite number of points in selected cross sections, where each node of the mesh is measured. The system implements a waiting procedure, where before initializing the measurement, it waits for a certain period of time (typically 1 second) and then starts the measurement. This procedure has a major impact on the overall duration of the measurement and because the wind tunnel is temperature sensitive device, the measurement time was reduced to acceptable minimum of 2 seconds. Stationarity of the freestream was evaluated separately, where the probe performed measurement in few selected nodes with measurement time up to 120 seconds. Average values of dynamic pressure measured in both cases are well comparable and measurement time of 2 seconds is long enough to correctly estimate the dynamic pressure in the measured point.

During the measurement, the ambient pressure and temperature of the freestream were measured separately. In total, six RTD sensors Pt1000 were used to measure temperature in five different sections of the wind tunnel and ambient temperature. Two four-channel NI 9219 modules with 24-bit A/D converter [5] were used to sample resistance measured at the sensors. Time-synchronization had to be solved, because NI 9219 modules used the same sampling clock as NI 9215 (pressure). Reference temperature used to calculate the velocity was measured at the nozzle output.

An application was written in LabVIEW, which both controls the pressure and temperature measurement and the motion of the system [6].

Data were evaluated in LabVIEW with implementation of Advanced Plotting Toolkit [7].

Fig. 2. Measurement of dynamic pressure distribution in X = 0.

3 Dynamic pressure distribution measurement

Dynamic pressure distribution along the MS is one of the key parameters of the wind tunnel. It is necessary to precisely map its changes with relation to the velocity achieved in the MS. Typically, the velocity in aerodynamics is expressed as a Mach number, or the ratio between the actual velocity v and speed of sound a.

$$M = \frac{v}{a} \tag{1}$$

The flowfield investigation was conducted at Mach numbers of 0.030, 0.045, 0.060, 0.075 and 0.090 (10, 15, 20, 25, 30 m/s) in four different cross sections of the MS, distanced from the nozzle output by 0, 200, 400 and 600 mm downstream. The length of the MS is 1 370 mm – the model is typically mounted 400 mm downstream from the nozzle output and the cross sections further downstream were not investigated yet. The results were also compared to a measurement conducted several years ago before the internal parts of the wind tunnel were tuned.

Measured data were evaluated in accordance with [2].

3.1 Procedure of data evaluation

Pitot-static probe measures total and static pressure – their difference is equal to dynamic pressure q_i. The velocity v_i can be expressed:

$$v_i = \sqrt{\frac{2q_i}{\rho_i}} \tag{2}$$

Due to the temperature instability, the air density ρ changes during the measurement. The density is not measured directly, but as a function of the ambient pressure $p_{amb.}$ and temperature of the air in the MS T_i:

$$\rho_i = \frac{p_{amb.}}{r \cdot T_i} \tag{3}$$

The specific gas constant of the air r is equal to 287.1 $Jkg^{-1}K^{-1}$. The ambient pressure is measured separately and it is considered to be constant during the measurement.

Distribution of both velocity and dynamic pressure is calculated. The ratio of dynamic pressures is more severe criterion than velocity – the velocity is proportional to the square root of the dynamic pressure (2). The dynamic pressure distribution ratio $\bar{q}$ is calculated according to formula:

$$\bar{q} = \frac{q_i}{q_{ref}} \tag{4}$$

The dynamic pressure q_i is related to the reference pressure q_{ref}, which is measured at the geometrical centre of the measured cross section area. Similarly, the velocity ratio $\bar{v}$:

$$\bar{v} = \frac{v_i}{v_{ref}} \tag{5}$$

The change of the dynamic pressure in the geometric centre of each cross section measured along the MS is calculated in accordance with (4).

3.2 Temperature instability during the measurement

The precision of the measurement is related to the temperature stability inside of the MS during the measurement. Based on the equations (2) and (5), 2% change of the temperature can cause almost 1% change of the velocity measured.

During the measurement, the air inside of the wind tunnel is heated by friction until a certain equilibrium is reached. Due to relatively small Mach numbers reached, any significant change of the temperature could be avoided by accelerating the wind tunnel to higher velocity to pre-heat the air inside and then decelerating to the desired value of velocity.

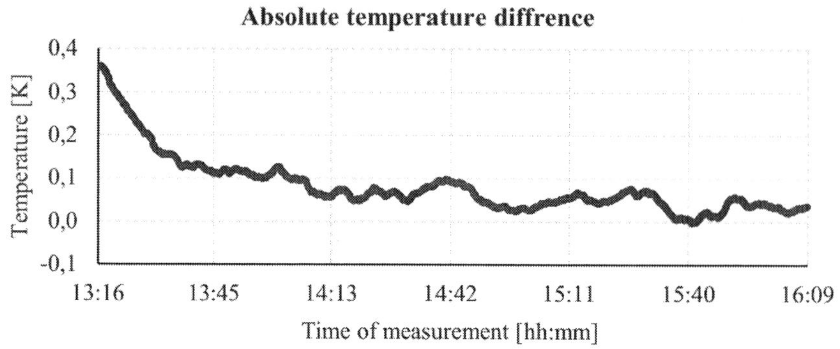

Fig. 3. Absolute temperature difference during the measurement at M = 0.03 (10 m/s).

The absolute temperature difference during the measurement at the M = 0.03 is shown in the Fig. 3. The temperature difference significantly drops in the first phase of the measurement. The pre-heating process causes this phenomenon. The temperature is stable during the measurement.

The ambient temperature does change during the measurement and it can affect the measurement. It is crucial to perform any measurement in the wind tunnel during as stable ambient conditions, as possible. Fig. 4. shows record of temperature during three-hour long measurement around noon – the temperature inside of the wind tunnel *Temp* changes only by 0.2 K. *Tamb.* is the ambient temperature.

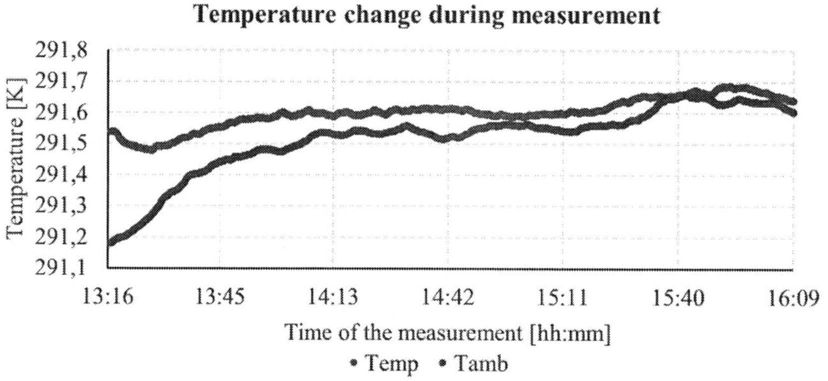

Fig. 4. Temperature change during a measurement at M = 0.03 (10 m/s).

Measured velocity change is proportional to the square root of temperature change:

$$\Delta v \sim \sqrt{\Delta Temp} \tag{6}$$

According to (6) in an example demonstrated in Fig. 4, temperature change from 291.5 K to 291.7 K requires 0.034 % correction of the measured velocity. Temperature screening during the measurement is used to confirm the temperature stability in the MS and temperature corrections are used only if the temperature change is significant.

The controller of the wind tunnel maintains a constant pre-set frequency of the fan. The temperature change is then inversely proportional to the pressure delivered by the fan. This phenomenon further increases the impact of the ambient temperature change to the precision of the measurement.

3.3 Flowfield measurement results in the cross section of MS

The investigation of the dynamic pressure and velocity distribution along the measurement section confirms uniform profile of the flowfield at all Mach numbers currently achievable in the wind tunnel. The Fig. 5-8 focus on the area, where the change of dynamic pressure is within ±1% or change of velocity within ±0.5%.

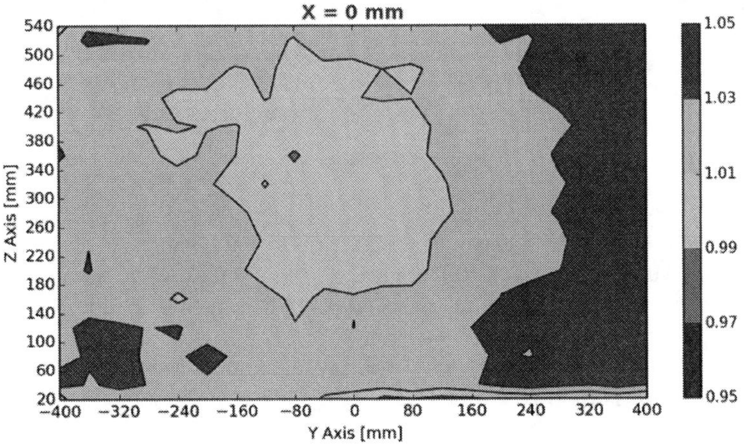

Fig. 5. Dynamic pressure at the output of the nozzle at M = 0.075 (25 m/s).

Worst parameters were observed at the output of the nozzle, due to its significant contraction (more than 6:1) as shown in Fig. 5. There is a noticeably asymmetrical profile caused by the setup of the vanes in the corner of the wind tunnel upstream of the nozzle.

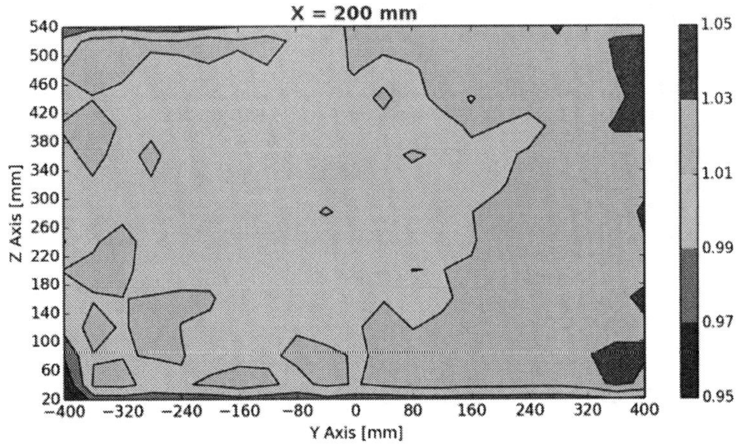

Fig. 6. The dynamic pressure distribution in the MS at M = 0.075 (25 m/s).

200 mm downstream the dynamic pressure distribution is almost entirely within ±1% change as shown in the Fig. 6.

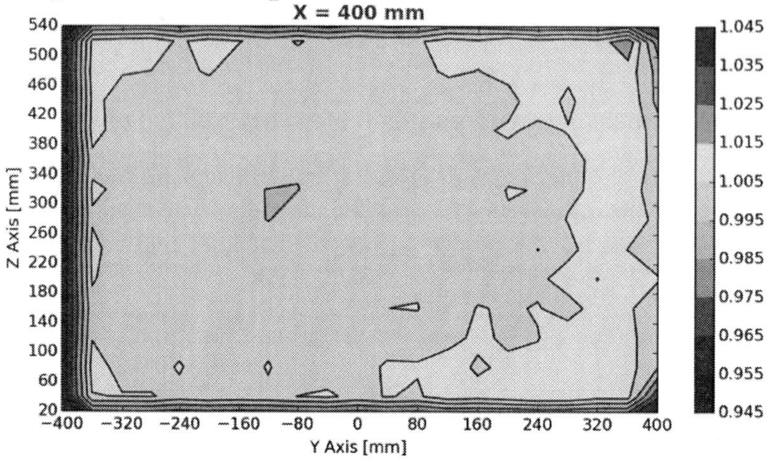

Fig. 7. Velocity distribution at M = 0.075 (25 m/s).

The flow in the open MS expands, which tends to smooth out any differences in the pressure and velocity. The X = 400 mm is the cross section, where model is typically mounted. The velocity change in the cross section is mostly within a ±0.5% of a change.

Fig. 8 confirms uniform velocity profile 600 mm downstream of the nozzle. The velocity distribution is within ±0.5% of a change in the large area of the cross section.

Similar parameters were observed for all Mach numbers measured.

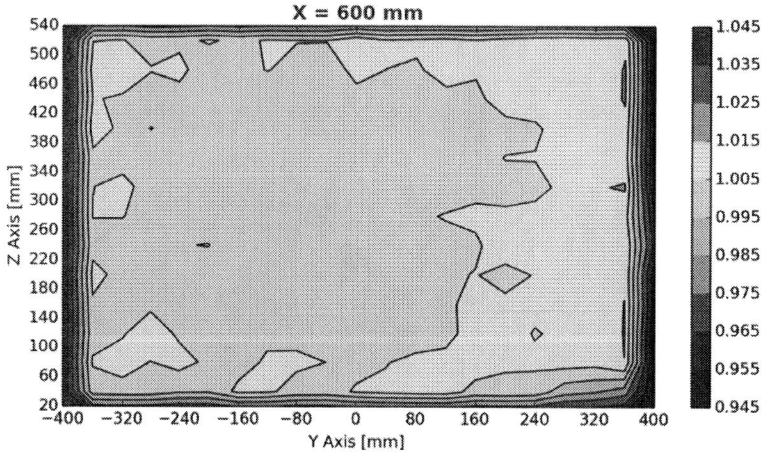

Fig. 8. Velocity distribution at M = 0.075 (25 m/s).

3.4 Reference pressure distribution along the MS

The reference pressure/velocity was obtained as an average measured at the geometric centre of the measured cross section. This pressure is not constant in the downstream direction.

The contraction of the nozzle is considerable, causing the stream to accelerate in the MS. However, at a certain distance, the freestream starts expanding, causing the pressure drop. This pressure drop is more significant at higher Mach numbers.

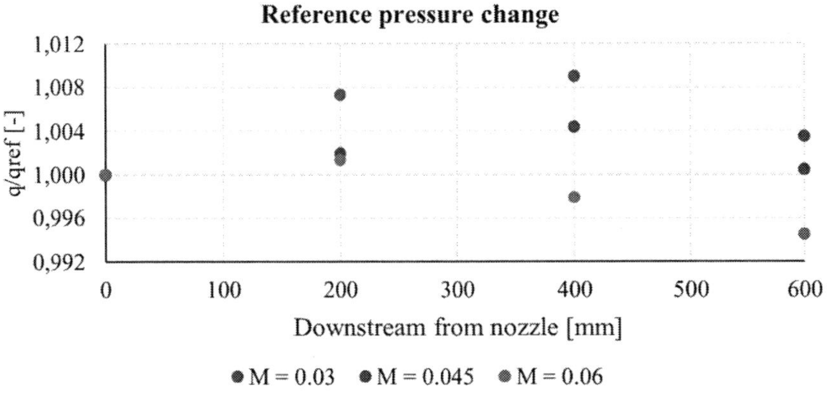

Fig. 9. Reference pressure change in the downstream direction in the MS.

3.5 Comparison of a flowfield parameters in the MS

During the development of the wind tunnel, numerous times were the flowfield parameters (i.e. dynamic pressure and corresponding freestream velocity) investigated – mostly to confirm the positive impact caused by any alteration caused to the wind tunnel configuration.

The very first measurement was performed manually – the traverse system was installed later. In addition, the wind tunnel was not properly configured. Yet, the measurement showed promising results, as shown at the left side of Fig. 10. The unnatural symmetry was achieved by mirroring the results, because the instrumentation at the time could not reach other side of the wind tunnel. In the right side of the Fig. 10 is the current state of the flowfield parameters.

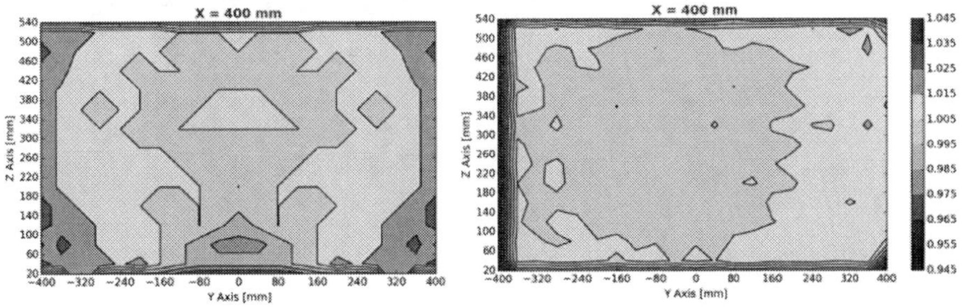

Fig. 10. Comparison of flowfield parameters in the MS along the time at M = 0.045.

The symmetry in the original data was only assumed – the measurement instrumentation measured only right half of the MS and the results were extrapolated to the left half of the MS. The alterations to the inside of the wind tunnel – among others described in [3] – considerably improved the flowfield parameters of the wind tunnel.

4 Measurement of the dynamic pressure loss on the model of a helicopter

The traverse system can be used not only to map the velocity profiles in the empty MS, but it can also be used to investigate the pressure drop on a model mounted in the MS.

If the flowfield parameters are constant, then the drop of dynamic pressure between cross section upstream and downstream from the model is proportional to its drag. Therefore, it is not necessary to implement a balance system to measure the drag directly. However, typically a balance system is more preferable, because as such, it is a much less time-consuming method.

Measurement of a dynamic pressure drop was performed at the model of the Mi-24 helicopter in order to investigate the dynamic pressure downstream from the model. Purpose of this measurement was to test the traverse system and its ability to map dynamic pressure behind standing aircraft, which is subjected to a significant cross- or headwind. The pressure loss was determined by measurement of the dynamic pressure in the cross section upstream and downstream from the model – the cross sections were as close to the model as possible. The pressure drop $q(\%)$ was then evaluated as difference of dynamic pressure upstream $q(ups, y, z)$ from the model and dynamic pressure downstream $q(dws, y, z)$ from the model – in each node with the same coordinates (y, z).

$$q(\%) = \frac{q(dws,y,z) - q(ups,y,z)}{q(ups,y,z)} 100\% \tag{6}$$

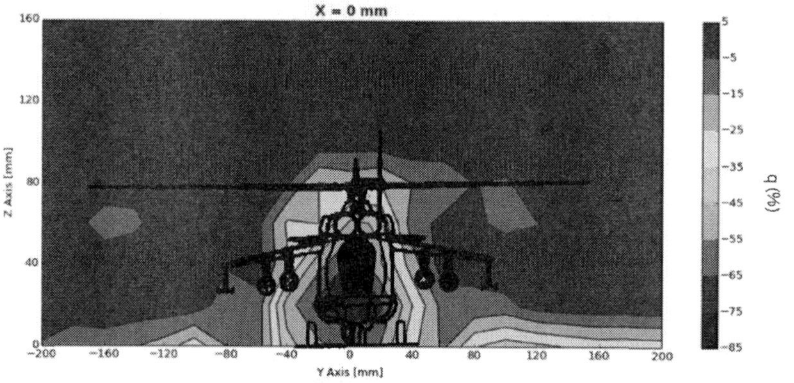

Fig. 11. Dynamic pressure change in the cross section downstream from a model at M = 0.06 (20 m/s).

The results of measurement of headwind are shown in the Fig. 11. The pressure drop was significant and the pressure field marks the silhouette of the model of the helicopter. The traverse system mapped precisely the dynamic pressure.

5 Conclusion

The investigation of flowfield parameters (i.e. dynamic pressure and velocity) in MS of the wind tunnel proved that the velocity distribution along the MS is uniform. With exception of the nozzle output, where there are still prevailing distortions caused by a considerable contraction of the nozzle, the velocity ratio is in the most parts of the MS within ±1% for all Mach numbers measured up to M = 0.09 (30 m/s).

The wind tunnel currently is not featuring a heat exchanger and the temperature inside of the MS cannot be controlled. Known issue with a return-flow type wind tunnel is the heat surplus caused by the friction. Temperature changes are monitored and temperature increase was investigated. Due to the relatively low Mach numbers achieved, the temperature change during the measurement can be minimized by pre-heating the wind tunnel by accelerating to higher velocity then desired and then decelerating. Because of the considerable duration of the measurement, a correlation between ambient temperature and temperature in the wind tunnel can be observed.

The wind tunnel delivers a uniform air flow in the MS and as such is ideal e.g. for measurements of a small UAV, where the velocity achieved is not large and as such typically operate at very low Reynolds numbers. The wind tunnel was used also to measure the impact of the crosswind and headwind to the standing aircraft. In this case, a model of a Mi-24 was used and a pressure drop was recorded.

This work was completed as a result of a cooperation with the company ENERGOKLASTR.

References

1. J. B.Barlow, W. H. Rae and A. Pope. *Low-speed wind tunnel testing* (3rd ed. New York: Wiley, c1999)
2. CH. Hughes, *Flowfield Measurements in the NASA Lewis Research Center 9- by 15-Foot Low-Speed Wind Tunnel: NASA Technical Memorandum 100883* [online]. (Lewis Research Center Cleveland, Ohio, 1989) [quot. 2017-03-05]. Available from: https://ntrs.nasa.gov/archive/nasa/casi.ntrs.nasa.gov/19890011631.pdf
3. V. Havránek, *CFD Analysis of Wind Tunnel Components* [quot. 2017-03-05]. (Diploma thesis. University of Defence. Supervisor doc. Ing. Dalibor Rozehnal, Ph.D. Brno, 2015)
4. *NI 9215 Datasheet* [online]. National Instruments, 2016 [quot. 2017-03-05]. Available from: http://www.ni.com/pdf/manuals/373779a_02.pdf
5. *NI 9219 Datasheet* [online]. National Instruments, 2014 [quot. 2017-03-05]. Available from: http://www.ni.com/datasheet/pdf/en/ds-198
6. J. Hnidka, *Design of traverse 3D system for low-speed wind tunnel* (Diploma thesis. University of Defence. Supervisor Doc. Ing. Dalibor Rozehnal, Ph.D. Brno, 2015)
7. *Advanced Plotting Toolkit* [online]. National Instruments [quot. 2017-03-05]. Available from: http://sine.ni.com/nips/cds/view/p/lang/cs/nid/213166

Determination of performance parameters of vertical axis wind turbines in wind tunnel

Van Bang Nguyen [1,*] and *Dalibor* Rozehnal [1]

[1]University of Defence, Department of Airforce and Aircraft Technologies, Brno, Czech Republic

Abstract. The paper deals with the determination of the performance parameters of a small vertical axis wind turbines (VAWT), which operate by the utilization of drag forces acting on the blades of the turbine. The performance was evaluated by investigating the electrical power output and torque moment of the wind machine. Measurements were performed on the full-scale model and the experimental data are assessed and compared to other types of wind turbines, with respect to its purpose.

1 Introduction

The energy industry today has increased interest in the use of renewable energy sources, which are more environmentally friendly. Among renewable ecological energy sources are undoubtedly wind turbines [1]. One goal of this article is to determine the power and torques characteristics of a very simple wind turbine with a vertical rotation axis VAWT by measurements in the wind tunnel. This model changes the kinetic energy of airflow into mechanical energy of the generator shaft and then it is converted to electricity. The expected output of this measurement is performance characteristics of the wind turbine depending on the velocity conditions of the wind tunnel.

On the basis of the measured results on the VAWT, characteristics were performed to its dimensionless state and a compared with the value obtain by theoretical assumptions. By applying dimensional analysis, it is possible to subsequently predict performance characteristics of a geometrically similar wind turbine, which will be produced in the desired size. From the results obtained by the measurement, and subsequent conversion to the parameters of the real model, it is possible to determine the power requirements for the connected electrical rotating machine - a generator which converts mechanical energy of VAWT to resulting electrical energy. The generator is an integral part of the entire set of wind turbines [Fig. 2b]. Its output parameters enable us to design and dimension the parameters of other devices connected to VAWT. The evaluation process also included the results of the performance measurement for an air flow velocity in the measurement section of the wind tunnel in a range from 8 m/s to 14 m/s.

* Corresponding author: vanbang.nguyen@unob.cz

DOI: 10.1051/matecconf/201710700076

2 The drag type of VAWT

2.2 The principle

VAWT can be classified from many perspectives, e.g. by the working principle of producing forces on the blade of the wind turbine. From this perspective, there are two types of wind engines: lift type (Darrieus) and drag type (Savonius) [2]. The drag type utilizes the differential drag force of a suitably shaped body during bypass of the wind flow around bodies from opposite directions. The typical example is a cup-type wind-speed sensor (anemometer), which is used as wind speed meter. The Fig. 1 illustrates this case - the cup on the right side of the picture produces much less of the drag force than a cup on the left side. The ratio of these drag force at the same wind speed is up to about 1:4 [3].

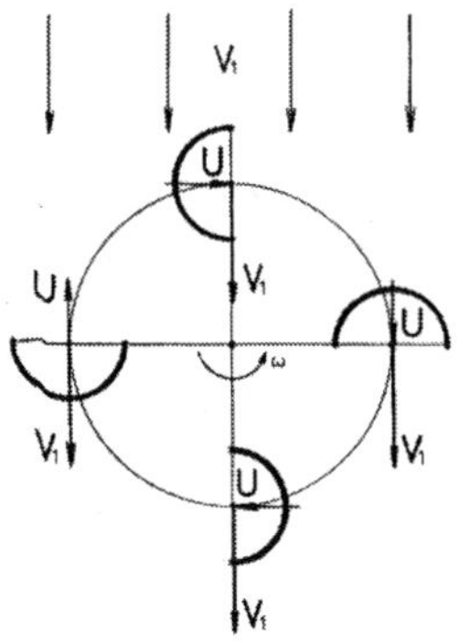

Fig. 1. Cup-type wind-speed sensor / Simple vertical drag type of wind turbine.

The previous kinematic scheme shows that the right cup has a low drag and this effect is further increased because the wind speed V_1 is added to the peripheral speed of rotation U. Also, the left cup generates more drag, because the wind speed is subtracted by the peripheral speed. Therefore, the resulting differential drag force decreases proportionally with the increasing value of angular velocity ω. The circumferential speed must be in all cases smaller than the wind speed in order to produce torque.

The drag of body depends on the square of the relative speed, the drag coefficient, the front surface area of the body and the specific weight (density) of the air. If the circumferential speed is limited, the absolute and differential value of drag force is limited too. Therefore, from the overall energy passing through the rotor, it can process only a few percent (about 5%) and the efficiency of these types of wind turbines is generally very low. The rest of the wind flow energy remains unused or it is transformed into a wake and heat.

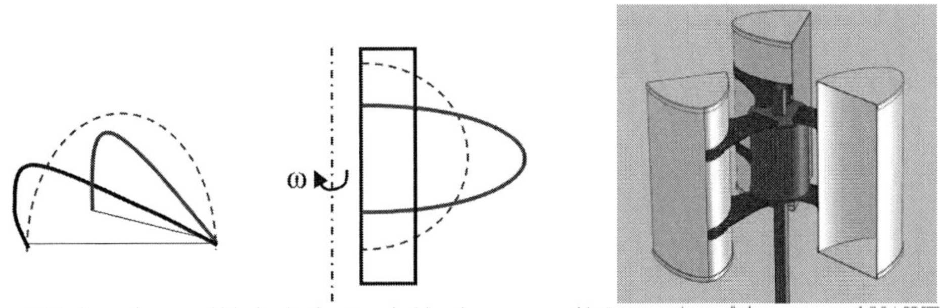

a) Various shapes of blades in front and side view b) Conception of the measured VAWT

Fig. 2. Construction modification of basic cup type and the measured model.

This disadvantage can be decreased by some special modifications of the construction. In this particular case, the rotor is designed with the various number and aerodynamic shapes of blades. The aim was to confirm and find the most efficient configuration – i.e. find configuration, where the differential value of drag coefficient and frontal area of the left and the right side is largest.

2.2 The mathematical description

Mathematical description of the wind turbine is based on the basic laws of mechanics, laws of conservation of mass, momentum, and energy [4, 5, 6]. The basic idea is that said wind turbine is put into the so-called flow tube, see Fig. 3. The mass air flow along the length of the tube is constant. Far ahead of the wind turbine, the speed of the flow in the tube is equal to the freestream velocity V. Around the wind turbine, it is diminishing and it is noted V_1. Far behind of the wind turbine the velocity in the flow tube is V_2.

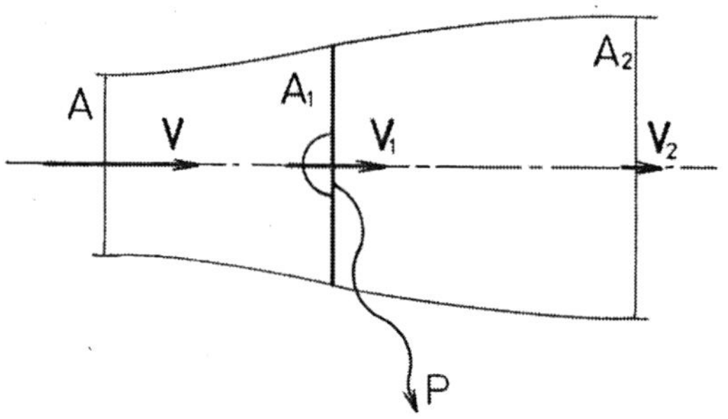

Fig. 3. Power take-off from the stream flow tube.

We selected three cross sections of the flow tube to write an equation describing the law of conservation of mass continuity

$$m_s = \rho \cdot V \cdot A = \rho \cdot V_1 \cdot A_1 = \rho \cdot V_2 \cdot A_2 \tag{1}$$

The density of the air ρ is a constant; A, A_1, A_2 are cross-sectional areas perpendicular to the axis of flow tube; V, V_1, V_2 are the air speed in the individual areas.
Ideal performance, which can be taken to the shaft of wind turbine, can be expressed as the difference between input and output power of the fluid flowing through the flow tube

$$P = m_s \cdot \frac{V^2 - V_2^2}{2} \tag{2}$$

The efficiency of the wind turbine is given by the ratio of the power P for the total performance of the fluid at the entrance of the flow tube. After some calculations, it can be shown that the theoretical efficiency η_t can be calculated only from speeds V_1 and V_2.

$$\eta_t = 1 - \frac{V_2^2}{V^2} = 4 \cdot \left(\frac{V_1}{V} - \frac{V_1^2}{V^2} \right) \tag{3}$$

Maximum efficiency can be achieved in the operating mode of the wind turbine, if V_2 is zero, and the ratio V_1/V is 0.5. In this case, the theoretical efficiency achieves value $\eta_t = 1$. Practically, this value is unachievable, because the air far behind the wind turbine must flow along the tube and speed V_2 cannot be zero. Because of the complex calculation of the actual mass flow, we estimate the maximum efficiency. The true efficiency η can be then calculated by the formula:

$$\eta = \frac{\rho \cdot A_1 \cdot V_1}{\rho \cdot A_1 \cdot V}\left(\frac{V^2 - V_2^2}{V^2}\right) = 4 \cdot \frac{V_1}{V}\left(\frac{V_1}{V} - \frac{V_1^2}{V^2}\right) \tag{4}$$

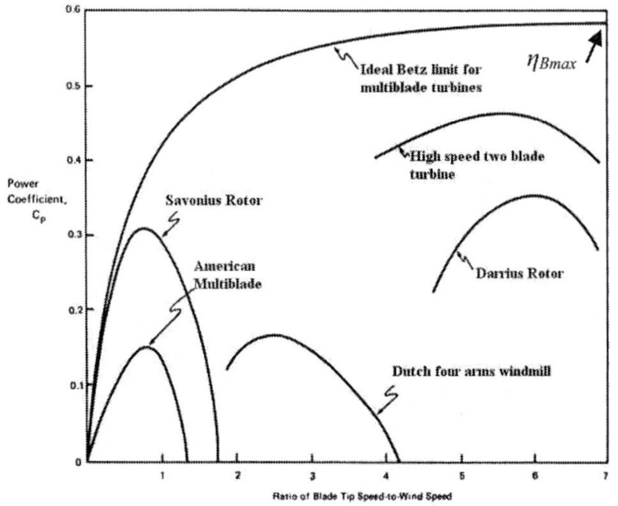

Fig. 4. Typical performances of wind machines. The power coefficient C_p is equal to the true efficiency η [6].

Maximum true efficiency η_{Bmax}, calculated as an ideal Betz limit for multiblade wind turbines, is achieved when the ratio $V_1/V = 2/3$. Its maximum value will be (after substituting 2/3 for V_1/V in equation (4)):

$$\eta_{B\max} = 16/27 \approx 0.593 \tag{5}$$

Maximum efficiency from the perspective of design parameters can be achieved with a ratio of tip speed and the wind speed V about 2.5 to 5 and more [6]. The tip speed and angular velocity depend on the product of the number of blades, width of the blades and lift force generated by the blades. The optimal angular velocity is inversely proportional to this product. This is a reason why multi-blade propellers and rotors optimized for lower angular velocity feature high torque, and they generally operate very well in the breeze. Multi-blade propellers are used mostly for water pumping [6].

The performance can be precisely estimated from known area A_1 of the turbine and the wind speed V:

$$P = \eta \cdot \rho \cdot A_1 \cdot V^3 / 2 \approx 0.3 A_1 \cdot V^3 \tag{7}$$

The constant 0.3 in the formula can be achieved if the efficiency η is equal to 0.5 and the air density is 1.2 kg·m^{-3}, which corresponds to the MSL ISA. This formula is valid for both the propeller and the rotor types of wind machines. The value of the constant 0.3 is the upper possible limit of the wind turbine of propeller type. For rotor type of wind turbine, the limit of efficiency ranges typically from about 0.10 to 0.12 and less [6]. This relation can be observed in the Fig. 4.

3 Test setup

Test equipment used to measure the unsteady aerodynamic characteristics of the wing section is shown in the Fig. 5. Model of VAWT was mounted in the test section of the wind tunnel, which was designed by company ENERGOKLASTR and is based at the University of Defence in Brno. It is a low-speed return-flow wind tunnel with an open test section. The wind flow's velocity is generated by a fan driven by a 260 kW electric motor. The dimension of the outlet nozzle is 2.4×2.4 m and the length of measurement section is 6 m.

a. The measured full-scale model b. Savonius VAWT full-scale model

Fig. 5. Placement of VAWT models in the wind tunnel.

In the test section of the wind tunnel was installed a system, onto which was mounted the measured model (Fig. 5 a.). The mechanical power of the VAWT was transformed into electricity produced by the generator, which was then consumed in a resistor. The investigated parameters were the angular velocity of VAWT, the force on the disc brake, and consumed the electric output of the generator. By changing the resistance of the resistor, the load on the generator changed as well and it was possible to continuously determine the torque of VAWT. In order to convert the measured values into their dimensionless form, it is necessary to know the ambient pressure, temperature, and velocity of the flow in the test section of the tunnel. All measured parameters are converted into electrical quantities, which are then processed and evaluated. The speed of wind flow in test section was gradually increased from 8 m/s to 14 m/s.

4 Results

4.1 The power characteristic

The aim of the test was to determine the performance of the wind turbine for 7 values of the speed of the wind flow. The Fig. 6 presents a measured electrical output power generated by turbines of VAWT versus RPM of its rotor shaft.

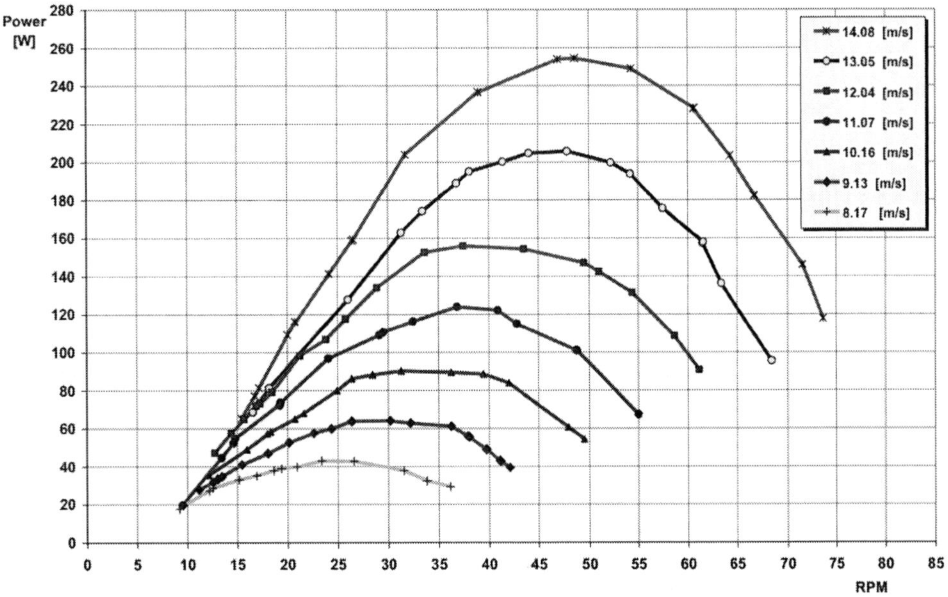

Fig. 6. Output power of the measured VAWT versus RPM.

The performance of the wind turbine is proportional to its size, effectiveness of the rotor and the parameters of the flowing fluid, which flows through the rotor. The measured performance parameters can be converted into dimensionless form accordingly:

$$C_p = \frac{2 \cdot P}{\rho \cdot A_1 \cdot V^3} \tag{8}$$

The aim of this transformation is to eliminate the influence of geometrical dimensions of the turbine, as well as the influence of ambient conditions under which tests were conducted.

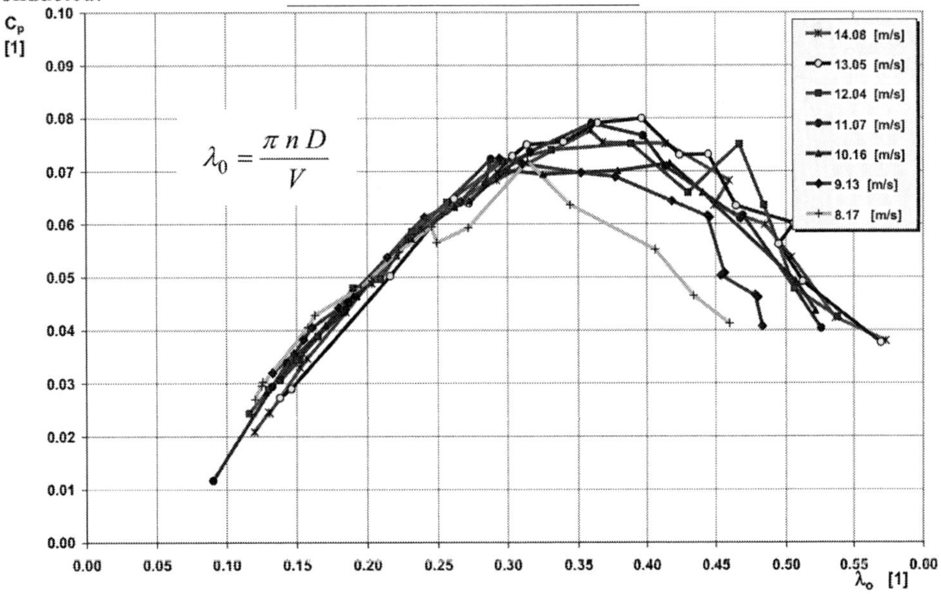

Fig. 7. Power coefficient of the measured VAWT versus tip speed ratio.

The value of dimensionless performance coefficient C_p versus dimensionless speed (tip speed ratio λ_o) allows us to perform a qualitative assessment of the VAWT. Graphical diagrams evaluating the dimensionless coefficients of VAWT are plotted in Fig. 7 and Fig. 9.

4.2 The power characteristic

The torque moment M_k of VAWT can be calculated from the measured output power and revolutions.

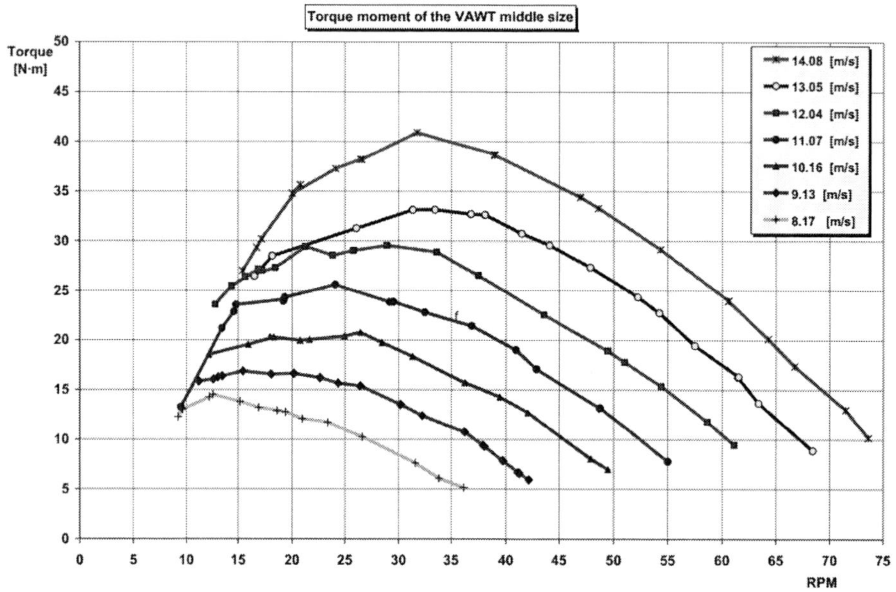

Fig. 8. Output torque moment of the VAWT versus RPM.

The values of dimensionless performance coefficients C_m can be converted into dimensionless form (9).

$$C_m = \frac{4 \cdot M_k}{\rho \cdot D \cdot A_1 \cdot V^2} \tag{9}$$

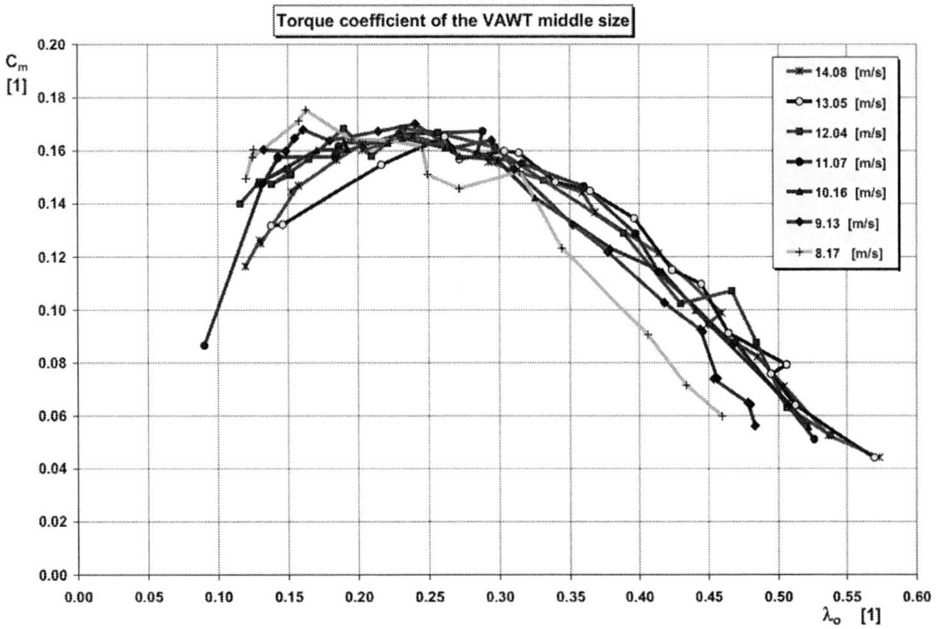

Fig. 9. Torque coefficient of the VAWT versus tip speed ratio.

5 Discussion

The performance parameters observed were the power and the torque moment generated by the VAWT and corresponding dimensionless coefficients, such as C_p and C_m. These coefficients can be used to precisely compare different types of wind turbines and assess their performance in different atmospheric conditions.

The obtained data show that the maximum coefficient of performance C_p is relatively low and corresponds to the theoretical predictions. If we only evaluated the mechanical performance of the VAWT without the generator, the investigated efficiency would be for about 20% higher. The efficiency of the generator, which is a part of the structure of VAWT, reached in the measured operating range of rpm and load about 80%. The measured characteristics differ with the velocity of the wind flow, as seen in Fig. 7 and Fig. 9. The efficiency of the generator is not constant. The revolutions of the generator are proportional to the wind velocity and with the increase of the velocity in the operating range, its efficiency grows to 80%. This effects the entire wind machine, which operates at a lower efficiency if its revolutions are lower. The measured results contain other factors such as mechanical losses in bearings of VAWT, thermal processes in the generator.

The advantage of this type of wind turbine is its high torque moment generated in relatively small revolutions – the maximum torque achieved is actually when the wind turbine is not rotating at all. However, the generator produces zero power if it is not rotating. The actual electrical power output generated by the VAWT, therefore, depends also on the type of the generator used. It is necessary to measure both the characteristics of the VAWT and the generator and they have to be tuned together. Coupling the generator with its maximum efficiency achieved in high revolutions with VAWT type of wind turbine can turn the wind machine extremely inefficient.

The paper has been prepared thanks to the support of a project SV16-FVT-K205-PET "Improvement of the experimental methods in aircraft technologies", the University of Defence in Brno, Czech Republic and the same thanks to supporting of ENERGOKLASTR. All tests were conducted in VAWT recently launched a return-flow type wind tunnel, which is owned by ENERGOKLASTR and located at the University of Defence in Brno. The operation of this facility is an example of successful cooperation between academic work and private association.

References

1. B. Koč, *Šance pro vítr* (EkoCentrum Brno, 1996)
2. A. Balmaceda, C. Maynez, *Wind turbines: types, design, and efficiency* (John Wiley & Sons, 2010)
3. F. W. RIEGELS, *Aerodynamische Profile: Windkanal-Meßergebnisse; theoretische Unterlagen* (Oldenbourg,1958)
4. M. O. L. Hansen, *Aerodynamics of wind turbines* (Routledge, 2015)
5. V. Rychetník, J. Janoušek, J. Pavelka, *Větrné motory a elektrárny* (ČVUT Praha, 1997)
6. G. L. Johnson, *Wind energy systems* (2006)

MATEC Web of Conferences **107**, 00077 (2017) DOI: 10.1051/matecconf/201710700077
DYN-WIND'2017

CFD simulation of dimpled sphere and its wind tunnel verification

Vojtěch Spálenský[1,*] and *Dalibor* Rozehnal[1]

[1]University of Defence, Department of Air Force and Aircraft Technology, Kounicova 65, 662 10 Brno, Czech Republic

Abstract. Paper deals with problems of CFD simulating airflow over a dimpled spherical surface and its verification by the wind tunnel measurement. The low-cost simulation approach was applied to be run on a common PC using the commercial software ANSYS CFX. The wind tunnel testing has been performed in the laboratory of aerodynamics at the Department of Air Force and Aircraft Technology of the University of Defence. Measured results of the drag coefficient versus the Reynolds number for smooth and dimpled spheres were compared and discussed. Presented simulation corresponds adequately to the experimental results. It can be stated that the CFD simulation is suitable for simulating the flow over the dimpled surfaces similar to sphere.

1 Introduction

The surface structure influences the air flow over objects of various shapes. Due to reducing the aerodynamic drag, the dimpled surface structure may be used. The typical example of using the dimpled surface is a golf ball. Dimpled golf balls have been used for almost a century. Empirical observations had shown that the dimpled balls travel farther due to reduction in pressure drag.

Dimpled surfaces reducing the aerodynamic drag are often applied in the design of sport equipment. There is a large potential in using of dimpled surfaces for the aerodynamic drag reduction of various objects, where the laminar-turbulent transition is inducted [1, 2]. Nowadays, the research of dimple's aerodynamics is also focused on the study of increasing the heat transfer rate [3, 4].

Most of the current CFD simulations of dimpled surfaces use the high demand turbulence models like DNS (direct numerical simulation), or LES (large eddy simulation) [3, 4]. These numerical models require a very fine mesh causing high computational demands of these applications.

The paper is focused on the study of the possibility of the low-cost, three-dimensional CFD simulation of flow around a dimpled surface. A common PC and commercial software ANSYS CFX were used. Simulation model was tested on dimpled sphere and was verified by a wind tunnel measurement.

* Corresponding author: vojtech.spalensky@unob.cz

2 Wind tunnel testing

For the validation of the simulation model, it is necessary to obtain experimental data for both smooth and dimpled spherical surfaces. The wind tunnel testing has been performed in the laboratory of aerodynamics at the Department of Air Force and Aircraft Technology of the University of Defence. The wind tunnel is the closed return one with a diameter of the testing section 600 mm and the range of free stream velocities from 2 to 45 m/s.

The wind tunnel model of the dimpled sphere has diameter $D = 216$ mm the surface of which is provided with 392 circular dimples of the diameter $d = 14.8$ mm and depth $h = 1.5$ mm, so the h/d ratio is about 0.1 (see Fig. 1).

Fig. 1. Model of the dimpled sphere in the wind tunnel section.

The aerodynamic force measurement system is based on two strain gauges. The analogue signals from them are processed by a computer. Measuring network is mostly build on National Instrument hardware and LAB-VIEW software.

The drag force was measured for various free stream velocity in the range of Reynolds numbers from 50 000 to 420 000. The lower limit of the measurement range is given by the capabilities of a force measuring instrument. The upper measurement range limit is given by the maximum velocity of the free stream in used wind tunnel.

3 Results of testing and analysis

Results of testing in the form of drag coefficient C_D versus the Reynolds number Re for dimpled and smooth spheres are given in Fig 2. Here, it is seen the substantial difference between drag curves for both observed objects.

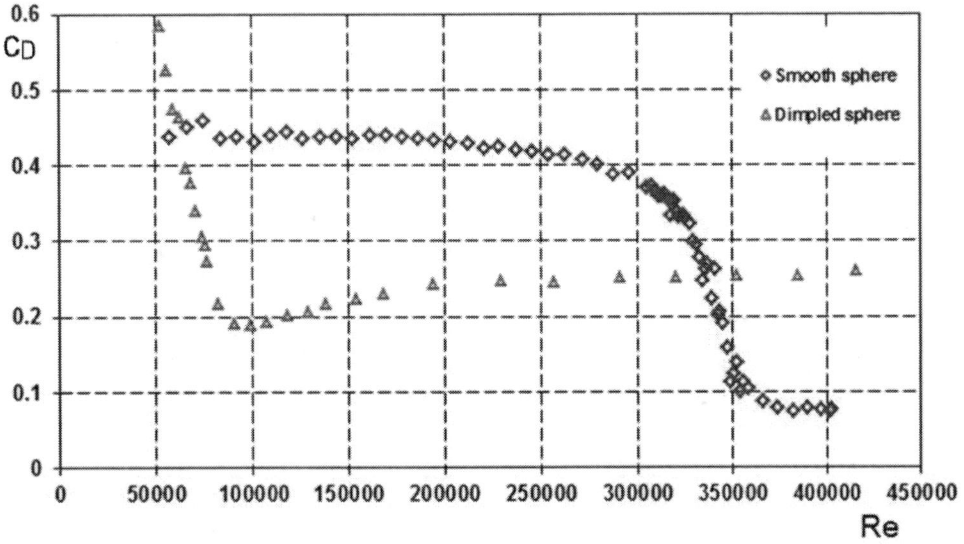

Fig. 2. Measured dependence of drag coefficient on Reynolds number for smooth and dimpled spheres.

In the case of smooth sphere the drag coefficient is nearly constant with the value about $C_D = 0.42$ in range of Re from 50 000 to 300 000, then there is a sudden decrease of C_D for Re between 300 000 and 350 000. The value of the drag coefficient drops to one quarter, i.e. under $C_D = 0.1$.

The same dependence of the drag coefficient on the Reynolds number for dimpled sphere is quite different. This decline in drag coefficient is between Re 50 000 and 100 000. The minimum value of the drag coefficient is $C_D = 0.195$. Then between Re 100 000 and 200 000 the drag coefficient slightly increases to $C_D = 0.27$. Almost constant value up to the maximum velocity corresponding to the Reynolds number 420 000 is observed next.

Finally, it can be stated that the drag coefficient for the dimpled sphere is lower than for the smooth one in the range of Reynolds numbers between 60 000 and 340 000. However, the smooth sphere has lower drag coefficient than the dimpled one for Reynolds numbers above 340 000.

Presented experimental results correspond very well with the results measured by [5] for smooth and dimpled golf balls.

4 Results of testing and analysis

The three-dimensional CFD simulation, based on Reynolds averaged Navier-Stokes equations by the CFD software ANSYS CFX 14.5 [6], was applied to simulate the flow over the dimpled and smooth spherical surfaces. The simulation was performed as a low-cost, on the standard notebook HP Pavilion g7 with the Intel core i7 processor and 8 GB operational memory. The computing time of the simulation for one flow of velocity took about twenty hours.

The tetra-prism mesh was used, because this one has proved to be most suitable for the simulation of flow over a single dimple [7] and also this kind of mesh is better for the simulation of a wake zone.

In the case of flow over the single dimple, with the increase in velocity of the free stream the flow character is changing from two symmetrical vortices to one asymmetric vortex. This change occurs between free stream velocities about 10 m/s and 20 m/s (see Fig. 3). That matched the observations [8] and also simulations [1, 2]. This change of the flow character is obvious, when streamlines in simulations that use tetra-prismatic mesh, as shown in Fig. 3, are compared. There are no further radical changes in a flow with the next velocity increase. An asymmetric vortex has two possible stabile orientations. Orientation of the vortex depends on the mesh parameters.

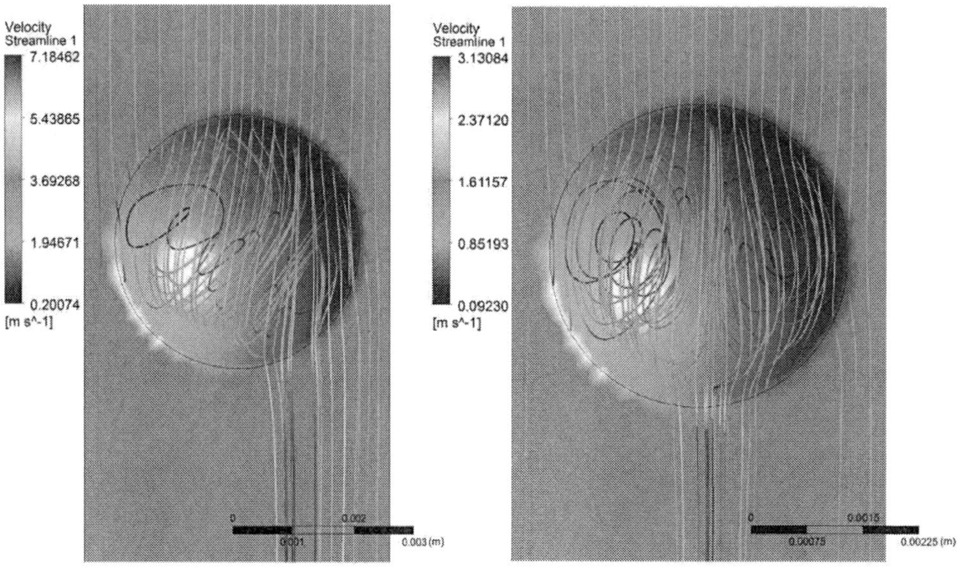

Fig. 3. Comparison of streamlines simulation of flow over single dimple for two free stream velocities: left 20 m/s and right 10 m/s. Tetra-prism mesh was used for simulation.

The geometry for the simulation of flow over the dimpled sphere is the same as the geometry of the wind tunnel model. Due to reducing the computational demands, the only a quarter of the dimpled sphere was simulated. The domain is chosen as the quarter of a cylinder of length 4 m and radius 1 m.

The position of the sphere center is in the distance of 1.5 m from the inlet section. The computational mesh was generated by using the mesh generator ICEM CFD [9]. The mesh consists of 8 000 000 cells and the overall view of the mesh is shown in Fig. 4.

A near surface region is composed by 32 prismatic layers. The surface mesh around the sphere is formed by 150 000 cells. The density of cells on the surface is constant. Detail of the computational mesh near the spherical surface is shown in Fig. 5.

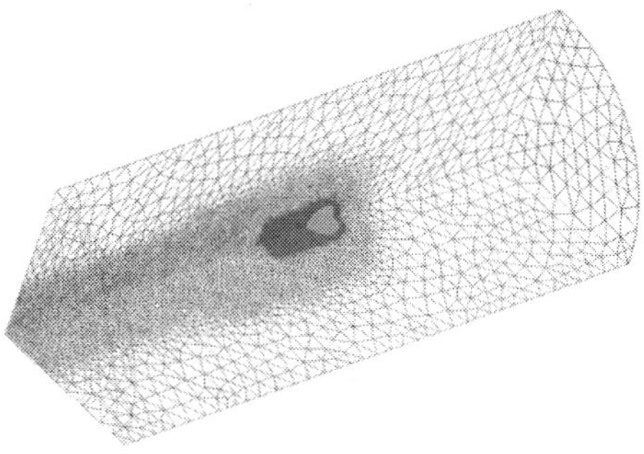

Fig. 4. Overall view of computational mesh.

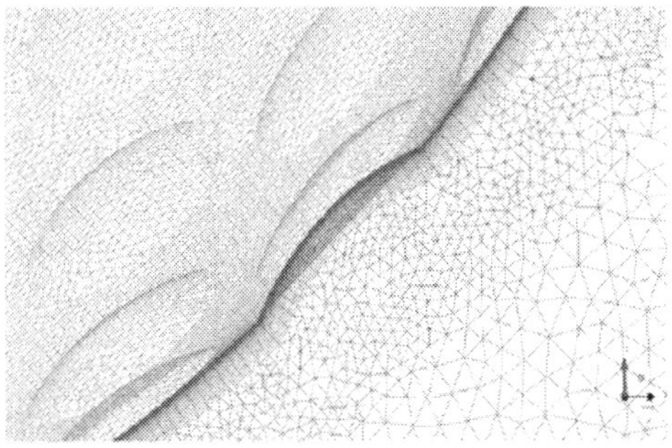

Fig. 5. Detail of computational mesh near surface.

The physical model and the boundary conditions of the CFD simulation were the following:

- •The steady state simulation, due to the lower computational demands. Although the flow over a single dimple is of unsteady state [7, 8], I try to examine whether such condition is acceptable.
- •The uncompressible fluid (air) of constant density, with no heat transfer.
- •The ambient air temperature is 25 °C.
- •The reference pressure is 101 325 Pa and the relative pressure on the outlet is 0 Pa.
- •Inlet free stream turbulence intensity is considered 5%
- •The used turbulence model is the shear stress transport model and the laminar-turbulent boundary layer transition onset follows the gamma-theta model [10].
- •The simulation was performed for seven free stream velocities: 5, 6.5, 7, 8, 9, 10, and 15 m/s.
- •The symmetry was set on sides of the domain, as I simulated only the quarter of observed object.
- •The surface of the sphere is considered as the smooth wall.
- •The domain outer cylindrical wall is set up as the free slip wall.
- •The convergence criterion was associated with a stabilization of the drag force.

5 Results of simulation and analysis

There are presented some examples of CFX POST outputs for smooth and dimpled spheres. The velocity distributions around the observed streamlined objects are shown in Fig. 7.

For the smooth sphere at high Reynolds number Re = 647 000, the turbulent separation takes place farther toward the rear of the sphere, at a position about 118.5° from the front stagnation point, as seen in Fig. 7a). This causes the sudden drop in the drag coefficient C_D, to a minimum of about 0.1 (see Fig. 2).

In the range of Reynolds numbers from 50 000 to 300 000, the drag coefficient of smooth sphere remains almost constant at about 0.4 (see Fig. 2). The flow separates at a position about 83° from the front stagnation point, and there is a fully developed turbulent wake, as seen in Fig. 7b). For the simulation of the flow around the smooth sphere, the same CFD computational model was applied.

In the case of the dimpled sphere, the flow pattern around this object is nearly the same (Fig. 7c) in the whole examined range of Reynolds numbers from 65 000 to 209 000 [7]. However, there is a significant difference between the flow patterns around the smooth and dimpled spheres. For the dimpled sphere at Re = 209 000, the flow separates at a position about 90° from the front stagnation point, as seen in Fig. 7c), and the drag coefficient is about 0.24.

Results of simulation for flow around the dimpled sphere in terms of integral characteristics as the dependence of drag coefficient versus Reynolds number are presented in Fig. 6. The comparison of calculated and measured values shows that the maximum difference is about 30%, but for Reynolds numbers over 130 000, the difference is lower than 10%.

These differences are probably caused by an inadequacy in the domain geometry when some dimples are divided. But the more significant is an inadequacy in the physical model, which does not allow the detailed simulation of transition processes in a zone of the sudden drop in the drag coefficient.

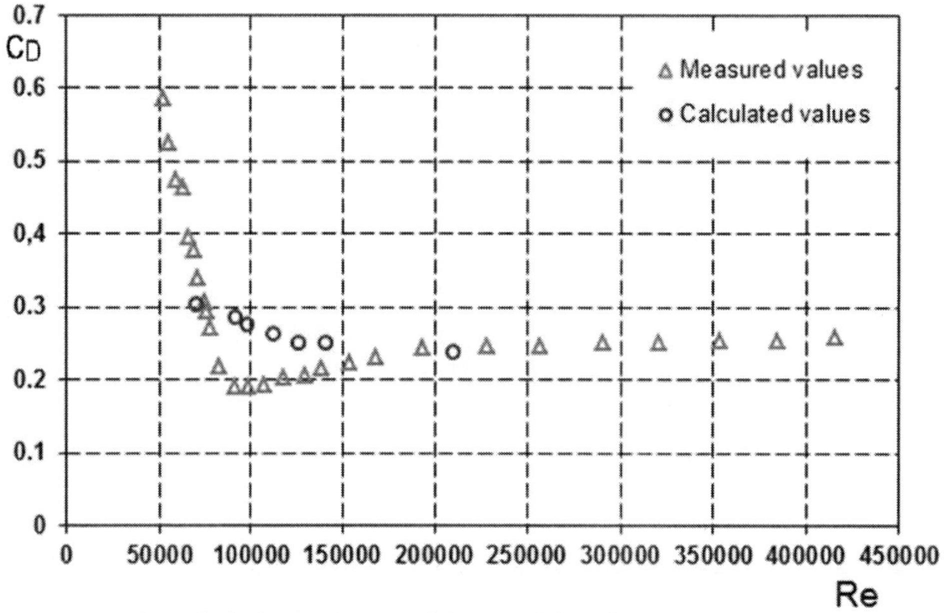

Fig. 6. Comparison of calculated and measured drag coefficients for dimpled sphere.

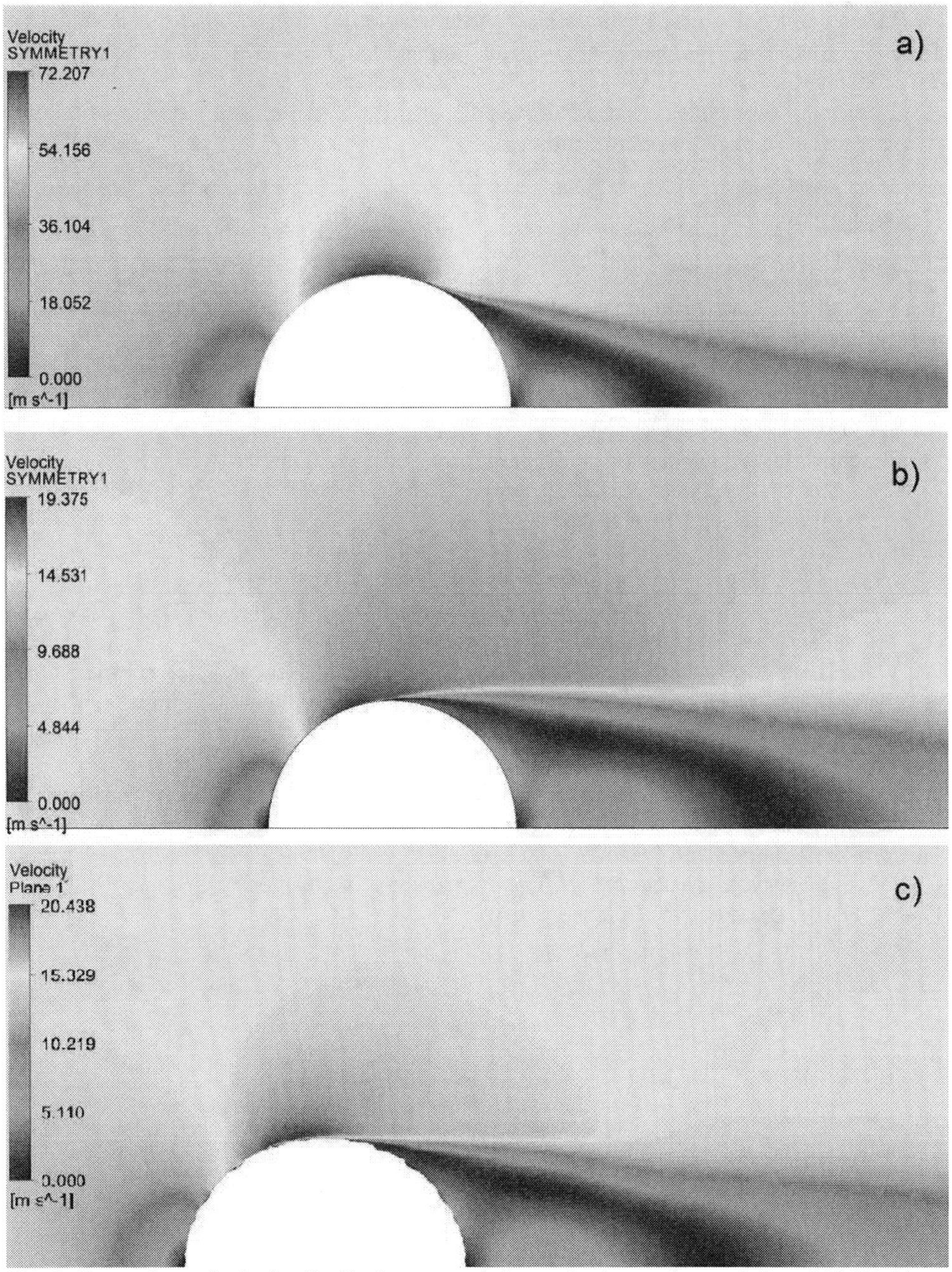

Fig. 7. Comparison of velocity distributions around observed spheres: a) smooth Re = 647 000, b) smooth Re = 194 000, c) dimpled Re = 209 000.

6 Conclusions

This easy-to-use CFD model has been developed for the simulation of a flow over the dimpled rounded surfaces. The simulation model is suitable for fast and low-cost calculations by using a standard PC.

Presented model is mainly suitable as a criterion for deciding whether to use smooth or dimpled surfaces and evaluate conditions and application areas of structured surfaces for a drag reduction.

Usability of this model has been already confirmed on the dimpled cylinder simulation, where it was used even for optimization [11].

References

[1] S. A. Isaev, N. V. Kornev, A. I. Leontiev, and E. Hassel, Influence of the Reynolds number and the spherical dimple depth on turbulent heat transfer and hydraulic loss in a narrow channel. int.j.heat mass tran **53,** 1–3, 178-197 (2010)
[2] J. Turnow, N. V. Kornev, S. A. Isaev, and E. Hassel, Vortex mechanism of heat transfer enhancement in a channel with spherical and oval dimples. int.j.heat mass tran **47,** 3, 301-313 (2011)
[3] H. Schlichting, *Boundary-Layer Theory* (New York: Mc Graw-Hill, 1979)
[4] C. E. Smith, *Direct Numerical Simulation of the Flow over a Golf Ball* [Dissertation], (Phoenix: Arizona State University, 2011)
[5] P. W. Bearman and J. K. Harvey, Golf Ball Aerodynamics, AeQ, 112-122 (1976)
[6] *ANSYS CFX, Reference Guide, Release 12.0*, ANSYS, Inc. (2009)
[7] V. Spálenský, *CFD Simulation of the flow over the dimpled surface* [Thesis, in Czech], (Brno, University of Defense, 2014)
[8] V. I. Terekhov, S. V. Kalinina, Y. M. Mshvidobadse, Heat transfer coefficient and aerodynamic resistance on a surface with a single dimple. j enhanc heat tranf. **4**, 131–145 (1997)
[9] *ANSYS ICEM CFD 11.0, Tutorial Manual*, ANSYS, Inc., January (2007)
[10] R. B. Langtry and F. R. Menter, Transition Modeling for General CFD
Applications in Aeronautics, AIAA Paper (2005-522), *43rd AIAA Aerospace Science Meeting and Exhibit, Reno, Nevada*: AIAA, (2005)
[11] V. Spálenský, *Optimization of dimples geometry on surface of cylinder* [Thesis, in Czech], (Brno, University of Defense, 2016)

MATEC Web of Conferences 107, 00078 (2017) DOI: 10.1051/matecconf/201710700078
DYN-WIND'2017

Wind tunnel tests of tent halls of different shape

Agnieszka Porowska[1,*], *Renata* Kłaput[1] and *Andrzej* Flaga[1]

[1]Wind Engineering Laboratory, Cracow University of Technology, Poland

Abstract. Aerodynamic investigations of wind pressure distribution on the surfaces of models of tent halls were carried out in the boundary layer wind tunnel at the Cracow University of Technology. Four types of objects of different shapes and construction were tested. Although tent halls are significantly vulnerable with respect to the wind action, there is no information about pressure distribution on objects of such type in standards, codes and normalization documents. Obtained results indicate that it is necessary to take into account different configurations of wind action while designing of the analysed structures.

1 Introduction

Four tent halls of different shape were tested in the Wind Engineering Laboratory at the Cracow University of Technology [1]. Tent halls are light buildings which are usually located in open space areas where wind velocities are high. For this reason they are vulnerable with respect to the wind action on them. Additionally, European standards do not provide enough information about this issue [2]. Construction of the roofs of tested halls does not fulfil a criterion of a gable or flat roof in accordance with [2]. Only one of them can be categorized as an arched roof, but [2] considers just the case of wind blowing perpendicular to the structure and the dependency on an angle of wind onflow is not taken into account. In this situation, wind tunnel investigations of uncommon structures are recommended.

Problem of wind pressure distribution on low-rise buildings was undertaken by many researchers, starting from pioneering works of Davenport [3, 4]. In these wind-tunnel tests models of different geometry and the influence of boundary layer and turbulence were investigated. Then the methods of codification of these results in the form of area-averaged pressure coefficient were elaborated [5]. However, there was still necessity to improve this area of knowledge, because of the development of the structures complexity and vulnerability. Therefore researchers studied influence of different parameters such as relative building height [6] or length [7] on wind pressure distribution on its surface. Also comparisons of full-scale and wind tunnel tests were already made [8]. Results of experiments carried out on models and calculations with accordance to the different standards were compared in [7, 9, 10]. Moreover, in many review papers [11, 12, 13] wind pressure distribution on buildings of varying shape was discussed. Up to now this issue is under consideration and there are still problems which need to be investigated thoroughly

* Corresponding author: agnieszkaporowska89@gmail.com

such as influence of roof shape or angle of wind onflow on wind pressure distribution on the external surfaces of low-rise buildings.

The aim of this work is to show that during the design process of atypical structures it is necessary to take into account different angles of wind attack and consider the most unfavourable ones. It is important, especially in the case of tent halls which are highly exposed to wind action.

2 Description of the research

2.1 Simulation of the boundary layer

The experiments on the models of tent-halls were carried out in the boundary layer wind tunnel. In the initial part of the investigations, a structure of wind flow was determined. Thermo-anemometers were used to measure mean as well as fluctuation of wind velocity in 6 points located in the working section of the wind tunnel on the height from 5 cm to 30 cm above the floor level. On the basis of the data received from measurements, mean wind velocity profile was obtained with use of the following power-low form:

$$V(z) = V_{ref}(z/z_{ref})^{a} \qquad (1)$$

where: $z_{ref} = 0.3$ m – reference height; $V_{ref} = 14.7$ m/s – reference velocity (mean wind velocity at the reference height); $a = 0.18$ – wind profile exponent dependent on the terrain roughness.

Obtained wind profile is shown in Fig.1a. Points mark values obtained in the wind tunnel tests and line marks function determined by least-square regression. This profile corresponds with 2nd terrain category with respect to [2] and complies with assumptions concerning wind conditions characteristic for destined localization of investigated tent halls. Turbulence intensity on the reference height was about 16%. Tests were carried out for 8 angles of wind attack with increment of 45°. Fig.1b shows orientation of the model with respect to the adopted wind directions.

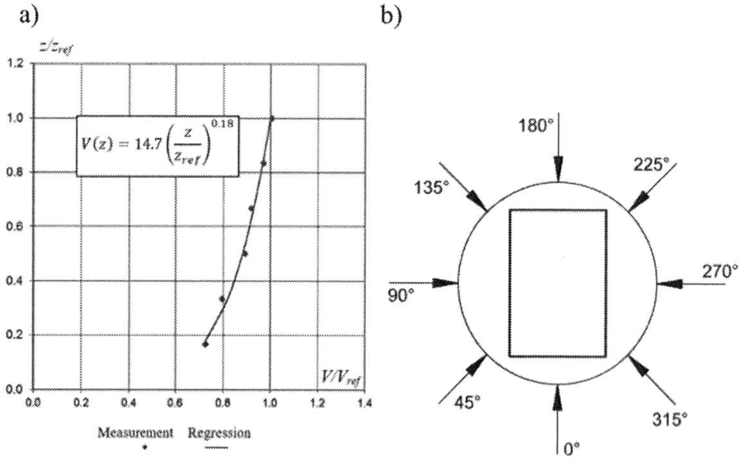

Fig. 1. Wind tunnel tests: mean wind velocity profile (a), angles of wind attack with respect to the tent hall orientation (b).

2.2 Description of the wind tunnel tests

Investigations were conducted on the models of four tent halls designated as H1, H2, H3 and H4 for the simplification. Fig. 2 shows schemes of all four models with the localization of the measurement points and cross-sections.

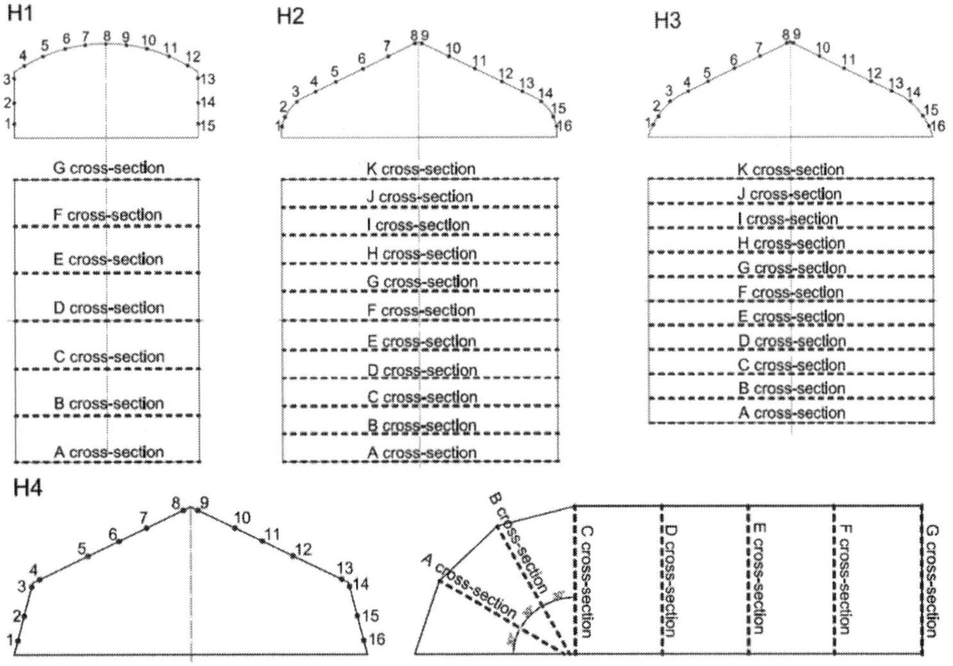

Fig. 2. Scheme of four tent halls H1-H4: distribution of the measurement points in the cross section and localization of the cross-sections on the scheme of the hall in the top view (H1-H3) or in the side view (H4).

Wind pressure was measured in points located in the cross-sections with use of the system of pressure electronic scanners of parallel type. Wind velocity reference pressure was determined in the reference point in the area of undisturbed wind flow in the front of the model. As a result external mean wind pressure coefficient was obtained, which was defined as:

$$C_{pe} = p_e / q_{ref} \quad (2)$$

where: p_e – mean wind pressure measured on the external surface of the object (positive values – positive pressure, negative values – negative pressure); q_{ref} – reference pressure.

3 Results of the wind tunnel tests

Following diagrams show results of the investigations as the mean external pressure coefficients determined in the measurement points on the cross-sections of four halls. For simplification of the data presentation, halls cross-sections were divided into three categories which are summarized in Table 1. Because of the model symmetry, it was decided that results will be presented only for angles of wind onflow in the following range: 0°-180°, but for the whole of the respective cross-sections.

Table 1. Cross-section categories adopted in the result presentation.

Name of hall	**H1**	**H2**	**H3**	**H4**
Cross-section category	**Name of a cross-section**			
Marginal	A, G	A, K	A, K	A, G
Intermediate	B, C, E, F	B-E, G-J	B-E, G-J	B
Middle	D	F	F	C-F

Results for one of marginal, intermediate and middle cross-sections are presented for each of the tent halls with exception of H4 –where two marginal cross-sections were taken into account because of specific geometry of the tent hall. In the analysis of the results, the angles of wind onflow for which the maximum values of mean external wind pressure coefficients were obtained are defined as unfavourable angles.

Results for H1 are shown in Fig. 3.

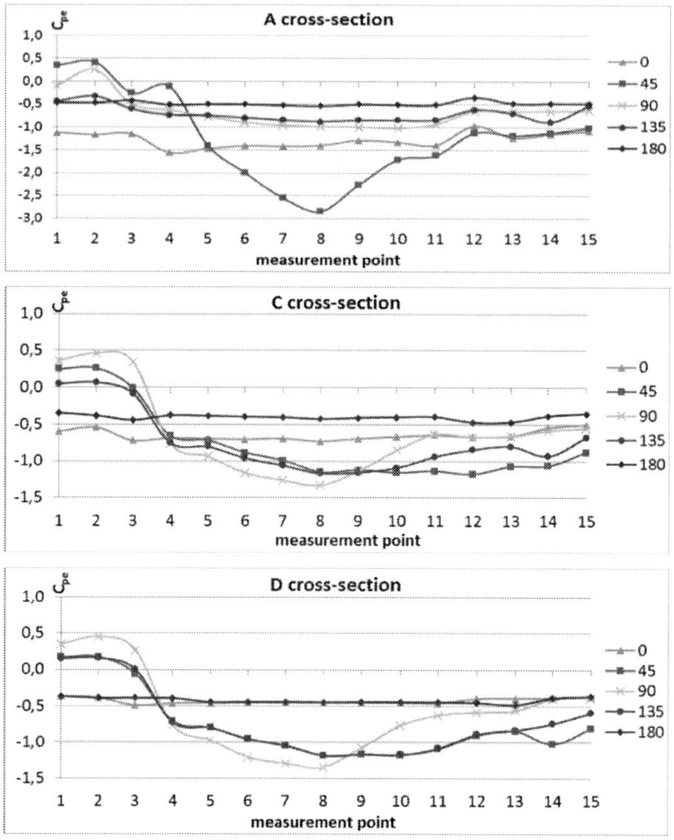

Fig. 3. Mean external wind pressure coefficients obtained for H1.

On the basis of the results for H1 one can draw the following conclusions:

1. The highest negative pressure appears in the middle part (points 7-8) of all of the cross-sections and maximum negative pressure occurs on the edge of the hall length (cross-sections: A, G);
2. The highest positive pressures appear on the side-walls of all of the cross sections (points: 1, 2) and regardless of the cross-section location on the length of hall maximum positive pressure is at the same level (C_{pe} about 0.5);
3. The most unfavourable angles of wind attack are as follows: marginal cross-sections – 45°; intermediate cross-sections – 90°, 45° and 135°; middle cross-section – 90° but for

the wall part of the cross-section on the leeward side (points 13-15) angles 45° and 135° are worse.

Diagrams for H2 and H3 are presented in fig.4 and Fig. 5, respectively.

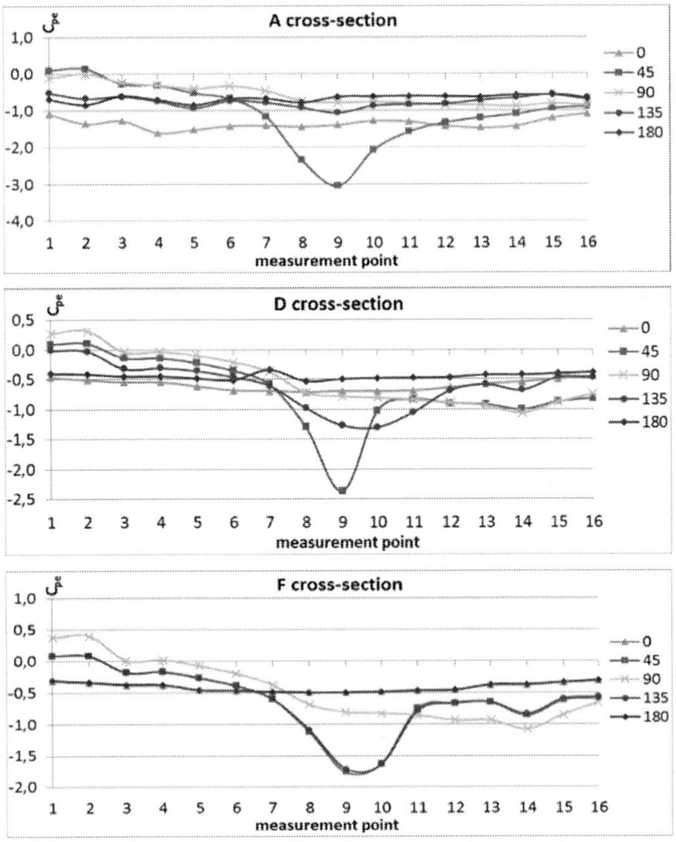

Fig. 4. Mean external wind pressure coefficients obtained for H2.

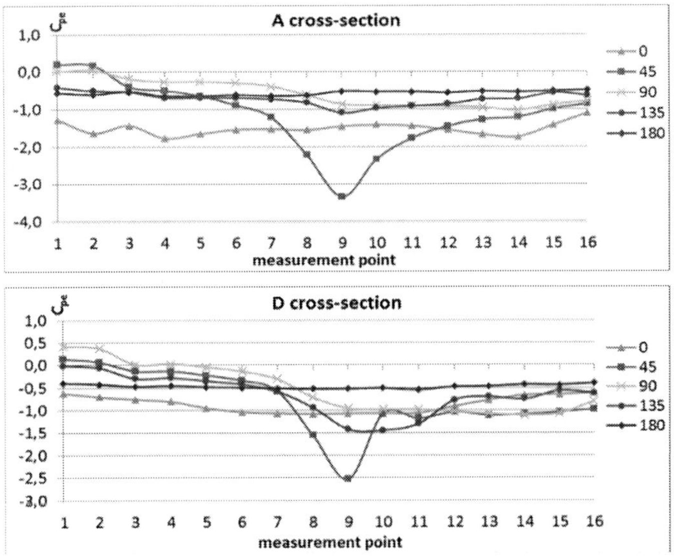

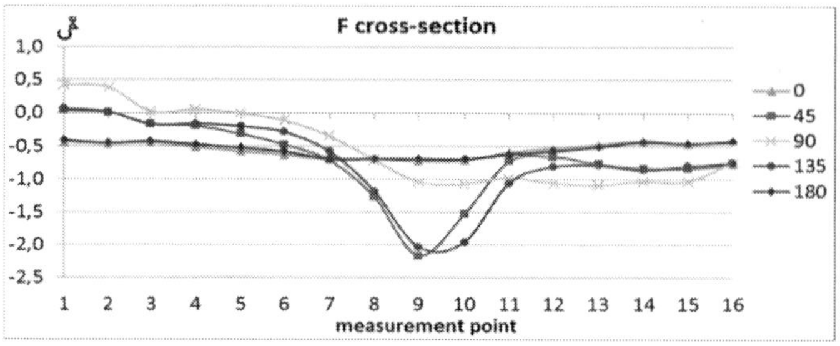

Fig. 5. Mean external wind pressure coefficients obtained for H3.

Results for H2 and H3 are similar, so conclusions for both of them are presented togheter:

1. The highest negative pressure appears in the middle (points 8-10) of all of the cross-sections and maximum negative pressure occurs on the edge of the hall length (cross-sections: A, K);
2. The highest positive pressures appear on the side-walls of all of the cross sections (points: 1, 2) and it is higher for cross-section in the middle of hall length (F cross-section);
3. The most unfavourable angles of wind attack are as follows: marginal and intermediate cross-sections – 45°, middle cross-sections – 45°, 135°.

Results for H4 are shown in Fig. 6.

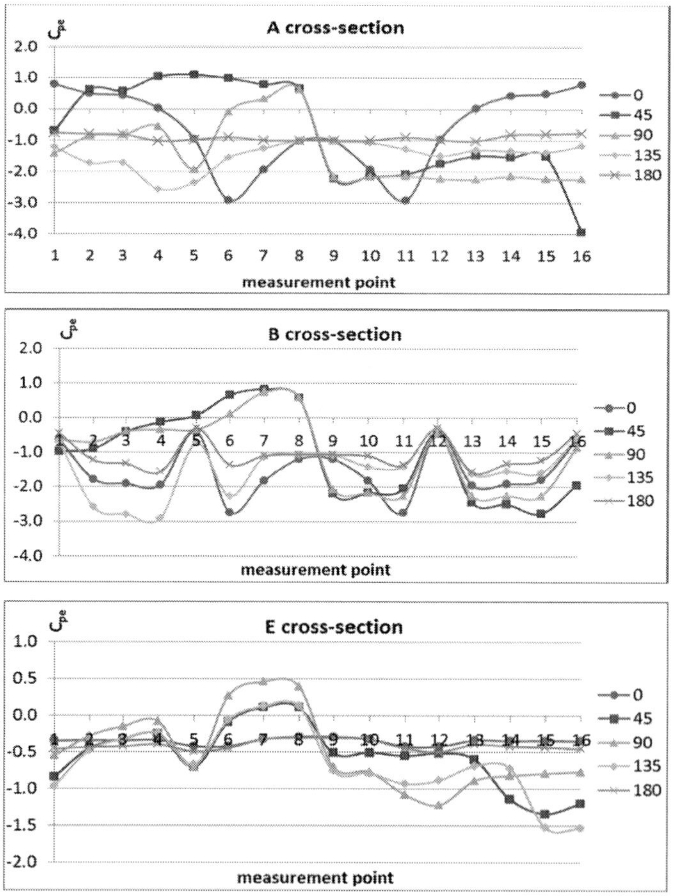

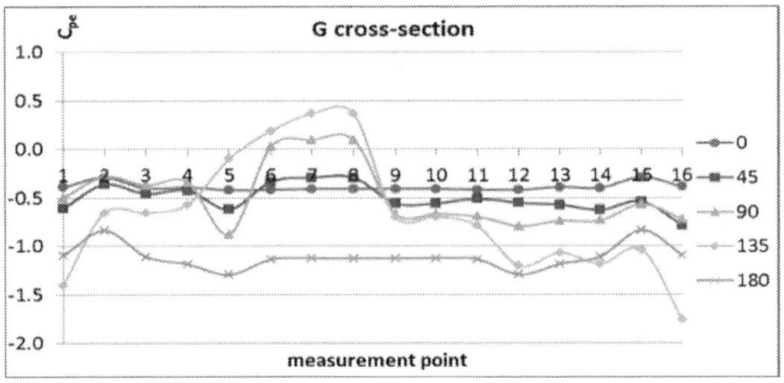

Fig. 6. Mean external wind pressure coefficients obtained for H4.

On the basis of the results obtained for H4, the following remarks can be presented:

1. Maximal negative pressures and positive pressures appear in the inclined cross-sections (A,B);
2. The highest negative pressures can be seen in the following parts of the cross-sections:

- A cross-section: high dependency on the angle of wind onflow,
- B cross-section: the middle and upper part of the wall (points 2-4 for the angle 135° and 13-16 for the angle 45°),
- C-G cross-sections: the lower part of the wall (points 1,2 and 15,16) for the angle 135°;

3. The highest positive pressures appear in the following parts of the cross-sections:

- A cross-section: high dependency on the angle of wind onflow,
- B-G cross-sections: in the middle and in the upper part of the roof (points 6-8),

4 General conclusions

The range of the mean external wind pressure coefficient for four halls is summarized in Table 2.

Table 2. The range of the mean external wind pressure coefficient C_{pe}.

Name of hall	**H1**		**H2**		**H3**		**H4**	
Cross-section category	$C_{pe,min}$	$C_{pe,max}$	$C_{pe,min}$	$C_{pe,max}$	$C_{pe,min}$	$C_{pe,max}$	$C_{pe,min}$	$C_{pe,max}$
Marginal cross-sections	-3.0	0.5	-4.0	0.3	-4.0	0.4	-4.0	1.0
Intermediate cross-sections	-1.5	0.5	-3.0	0.4	-3.3	0.45	-3.0	1.1
Middle cross-sections	-1.4	0.4	-2.0	0.45	-2.5	0.5	-2.0	0.5

It can be observed that maximum wind negative pressure appears in the marginal cross-section for all of the analysed halls. For three of them: H1-H3, maximum negative pressure occurs in the half of the cross-section (H1: points 7-9, H2, H3: points 8-10). H4 presents slightly different results because the highest negative pressures can be noticed in the lower part of the wall (points 1-4).

Mean external wind pressure coefficient has extreme values for the following angles of wind onflow for particular halls:

- H1: 90° for middle and intermediate cross-sections, but 45° for marginal cross-sections. It is worth to point out that values of C_{pe} in marginal cross-section are significantly higher than in the middle and itermediate cross-sections,
- H2, H3: 45° for all of the cross-sections,
- H4: 45° and 135° for inclined cross-sections (A, B), 90° and 135° for other cross-sections.

During the analysis of the results one can also notice differences appearing for C_{pe} for H2 and H3. These halls have similar geometry, but they differ only in the length, wherein H2 is longer than H3. Negative pressure for H3 is higher than for H2 and this trend is visible in results for all of the cross-sections. It can be concluded that length of hall has influence on mean external wind pressure coefficient: the smaller length, the higher value of C_{pe}. Especially, it is seen in marginal cross-section where the differnce is up to 0,5. It would be necessary to analyse this dependency thoroughly.

To summarise, in most of the analysed cases the extremal mean wind pressure coefficient appears at the heighest point of the roof of the marginal cross-section of tent halls. The shape of the hall cross-section has significant infuence on the value of this coefficient. The highest vallues of C_{pe} appears for 45° angle of wind onflow, so it is essential to take into consideration this information in the process of design of such structures.

This work was sponsored by Rekord Hale Namiotowe Sp. z o.o.

References

1. A. Flaga, R. Kłaput, A. Porowska, Ł. Flaga, M. Mańczyk. Aerodynamic investigations and design of the construction of four tent halls of the company REKORD HALE NAMIOTOWE Sp. z o.o., Cracow, (Technical Report, in polish, 2015)
2. PN-EN 1991-1-4, 2008. Eurocode 1 (2008)
3. A.G. Davenport, D. Surry, T. Stathopoulos. Wind Loads on Low-Rise Buildings. The University of Western Ontario, London, Ontario, Canada (Final Report on Phases I and II. BLWT-SS7, 1977)
4. A.G. Davenport, D. Surry, T. Stathopoulos. Wind Loads on Low-Rise Buildings. The University of Western Ontario, London, Ontario, Canada (Final Report on Phase III. BLWT-SS8, 1978)
5. D. Surry, T. Stathopoulos, J. Wind Eng. Ind. Aerodyn. **2,** 4, 385-397 (1978)
6. H.J. Gerhardt, C. Kramer, J. Wind Eng. Ind. Aerodyn. **43**, 1765-1773 (1992)
7. H. Alrawashdeh, T. Stathopoulos, J. Wind Eng. Ind. Aerodyn **147**, 212-225 (2015)
8. R.V. Milford, A.M. Goliger, J.L. Waldech, J. Wind Eng. Ind. Aerodyn. **43**, 1693-1704 (1992)
9. E. Gavanski, Y. Uematsu, J. Wind Eng. Ind. Aerodyn. **132**, 77-91 (2014)
10. L.M. St Pierre, G.A. Kopp, D. Surry, T.C.E. Ho, J. Wind Eng. Ind. Aerodyn. **93**, 31-59 (2005)
11. T. Stathopoulos, Eng. Struct. **6**, 119-135 (1984)
12. H.W. Tieleman, J. Wind Eng. Ind. Aerodyn. **91,** 1627-1649 (2003)
13. T.C.E. Ho, D. Surry, D. Morrish, G.A. Kopp, J. Wind Eng. Ind. Aerodyn. **93**, 1-30 (2005)

Measurements of dynamic wind pressures on gable roofs – comparison with ČSN EN 1991-1-4

Stanislav Pospíšil[1,*], *Michael* Macháček[1], *Stanislav* Hračov[1], *Sergey* Kuznetsov[1] and *Arsenii* Trush[1]

[1]ITAM AV ČR, v.v.i., Prosecká 76, Prague 9, 19000, Czech Republic

Abstract. The contribution is dealing with the wind tunnel measurements of dynamic pressures on the canopy roofs. It compares experimentally determined pressure coefficients C_p with the values in ČSN EN 1991-1-4 and former code ČSN 73 0035 as well. The pressures are expressed by the dynamic values. Thus, the averaged values are presented together with the fluctuating part, which significantly contributes to the overall loading. The measurement is carried out at two roof models, two basic directions, and two turbulence intensities.

1 Introduction

Knowledge of wind pressures and the pressure coefficients on the building roofs is basic to the accurate wind-resistant design of structures. There has been much investigation of the wind characteristics on the gable roofs in different regions, see, [1], [2]. The magnitude and distribution of the wind pressures are analyzed in [3]. The wind tunnel tests are often indispensable for the designers, because the current codes sometimes do not give adequate values with respect to some zones at wind exposure. The extremely high local suctions may occur in the leading edges of longitudinal wall and windward roof, roof corners, and roof ridge under strong wind. The aspect ratio of building has a certain effect on the mean wind pressure coefficients, and there exist the effect relates to wind attack angle as well. Also the type of the exposure (terrain) is important. The paper [4] discusses the phenomenon that the suburban exposure produces lower wind loads than those experienced in the open country and compares the local pressures on low buildings with gable roof. Another paper, [5] analyses the wind pressure on gable roof buildings, through comparisons with existing results and relative wind load codes. In [6], the pressure spectral values are compared for both wind tunnel and full-scale measurement. The coherence of pressures was identical in full scale and wind tunnel in many cases and is better than that predicted in quasi-steady theory. In the papers [7, 8], the background for the Eurocode EN 1991-1-4 of the pressure coefficients is discussed. The reliability of the codification for wind loading is treated in paper [9].

* Corresponding author: pospisil@itam.cas.cz

The present article is a preliminary study with emphasis on the understanding of wind pressure on gable roof buildings, through comparisons with existing related wind load codes. The wind tunnel measurements are used to compare pressure coefficients C_p with the values in ČSN EN 1991-1-4 [10] and in expired code ČSN 73 0035 [11] which in some zones give very different results. The pressures are expressed by the averaged values together with the fluctuating part, which significantly contributes to the overall loading. The measurement is carried out at two aerodynamic models with different roof inclination for two basic directions in both smooth and turbulent flow.

2 Measurement set-up

2.1 Model preparation

Two models made of plexiglass with plane dimensions $b \times d = 30 \times 30$ cm and height $h = 16$ cm ($\alpha = 5°$), respectively, $h = 30$ cm ($\alpha = 45°$) were produced. The legend for the gable roof and the exposed zones according to [10] is shown in Fig. 1. Scheme of the model and the photograph of the manufactured scaled exposed to the wind pressures are shown in Fig. 2. The size of the model was chosen so as to avoid the possibility of the flow blockage in the wind tunnel. The ratio between the model area A_m and the tunnel cross section A_t is $A_m/A_t = 2.5\%$ and 3.7%, respectively. Measurement of the pressure coefficient is thus not distorted; as the systematic error of 10% at the windward side of the model occurs when the blockage is higher than 5% while for the leeward side it is even less, see [3]. According to [4], the correction effect by blockage (calculated according for the model over the entire height of the tunnel) and used for the dimensions of the presented building model is only 6%. It can therefore be assumed that for the measured case in which pressure is determined on the roof it is not necessary to consider a correction.

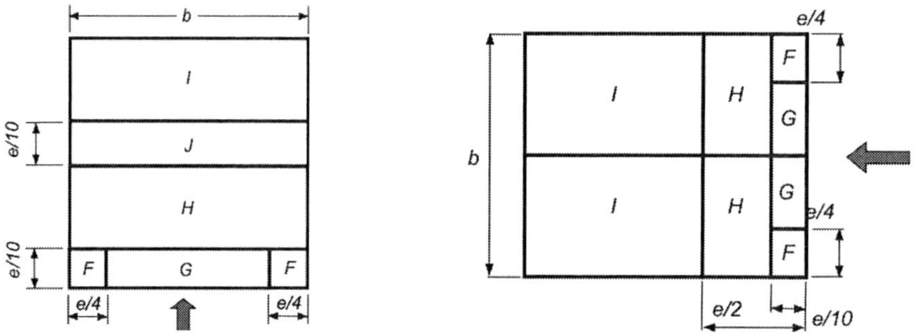

Fig. 1. Scheme of the pressure zones for the gable roof according to [10]. Left-wind direction 0°. Right-wind direction 90°. Dimension *e* is the lower one from the values *b* or *2h*.

The examined object models were equipped with the pressure probes; small holes on the roof. Silicone tubing connected these probes with the dynamic pressure sensor type Honeywell DC002NDR5. This is a sensitive differential pressure gauge with a range of about 500 Pa and the sensitivity 5 Pa in the full range. The objects were placed in a smooth flow with very low turbulence intensity $I_t = 1.4\%$ and then to a turbulent stream with modelled atmospheric boundary layer with turbulence intensity $I_t = 26\%$.

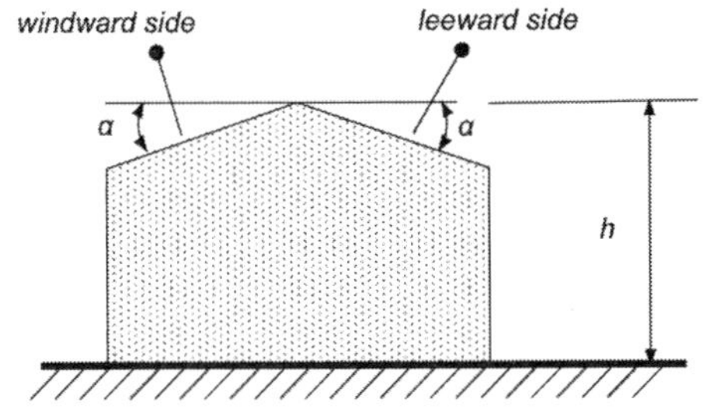

Fig. 2. Scheme of the gable roof and roof inclination. The photograph of the model with the pressure taps and sensors.

2.2 Model testing

The pressures at the smooth model gable roof were ascertained for five wind directions, i.e., the three main directions of $\theta = 0°$, $90°$, $180°$ and on the secondary incidence angles $\theta = 45°$ and $\theta = 135°$, which are not evaluated in this paper. The model was placed in the measuring part of the aerodynamic section of the tunnel, where the wind speed can be produced up to 50 m/s. Wind speed is measured by the Prandtl - Pitot tube in place of the undisturbed stream before the model, see photo in Fig. 3. The model was at first loaded with the speeds in the range of 10% - 70% of the fan power, which is 3.0 m/s to 20 m/s, in order to determine whether the pressure coefficient depends on the flow speed. Finally, the tests were carried out at a speed of about 16 m/s, i.e., in the part of the interval where the local pressures coefficients are independent of the wind speed. In other words, Reynolds number was in the range of $Re = 3.9 \times 10^5$ corresponding to the values in prototype scale. This value guarantees, see [3], the independence of the results at Re. Each partial measurement with 32 channels representing 32 spots on the roof surface took 30 seconds. This sampling time is sufficient with respect of the ergodicity and stationarity of the process.

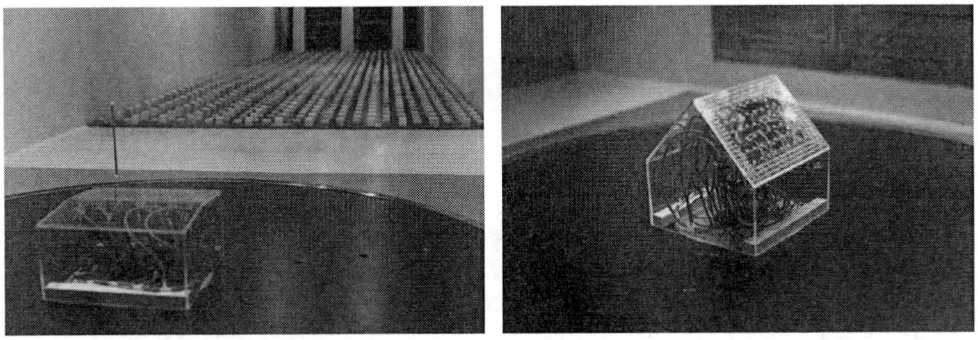

Fig. 3. Measured gable roof in the wind tunnel. The photograph is taken from the measurement in the turbulent atmospheric boundary layer.

Transfer of the results from the tunnel to the real measure is implemented through the pressure coefficient:

$$C_p = 2(p_x - p_r)/\rho V_r^2 \qquad (1).$$

Value p_x is pressure at the measuring points on the model; p_r is static (barometric) pressure at a reference point, i.e., at the site of the free stream; ρ is the density of air and V_r is the mean wind speed at the reference location. Differential pressure is measured directly by a pressure transducer and represents the wind pressure at the sampling site. Pressure on the real building is then obtained by multiplying the pressure coefficient by the value of dynamic pressure $p_d = 0.5\rho V_r^2$. In the case of the turbulent flow, the terrain category IV was selected. The wind speed profile fits very well the profile given in [10]. The turbulence intensity in the test section is $I_t = 26$ %. The normalized spectral density of the wind speed, showing the energy distribution of the fluctuation component in the frequency domain corresponds to the Von Karmán spectrum utilized to describe the dynamic effect of the wind. This is seen in Fig. 5 below.

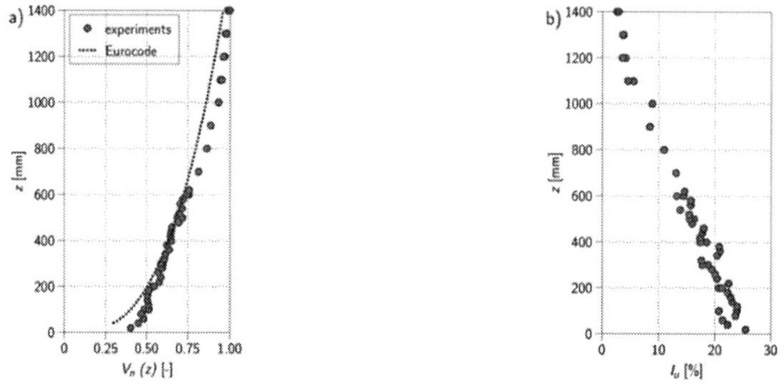

Fig. 4. a) Atmospheric boundary layer simulated in wind tunnel; b) turbulence intensity.

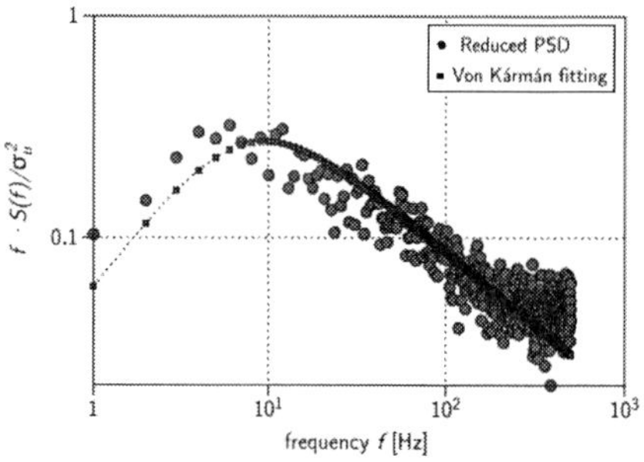

Fig. 5. Wind velocity power spectral density and the fitting to the theoretical values.

3 Results and discussion

The measurement results (non-dimensional pressure coefficients) are summarized in Fig. 6-13 and on Tables 1-2. The results are shown in the form of the maps of the local pressure coefficients over the roof plan. These values can be compared with the values in [10] as well as with the coefficients in the former Czechoslovak code [11]. Tables 1 and 2 give the average values of the coefficients and corresponding standard deviations (RMS) for each zone F, G, H, I and J as seen in Fig. 1. The methodology of division of the whole roof into the zones has been adopted for the [11] as well. For the evaluation of the data the authors selected relatively simple methodology expressed by the mean values pressure augmented by the standard deviation. The latter represents the fluctuating pressures which naturally create the additional loading at the roof. Thus, for the comparison with the codes, the effective maximum values of suction or pressure were selected. They are highlighted by red colour in Tables 1-2. Alike, the nearest values from the both codes are selected and accented.

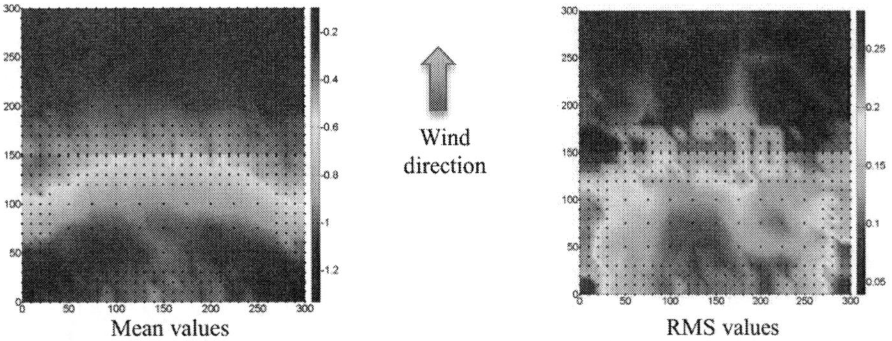

Fig. 6. Pressure coefficients for the roof with $\alpha = 5°$ in case of smooth flow and wind angle $\theta = 0°$.

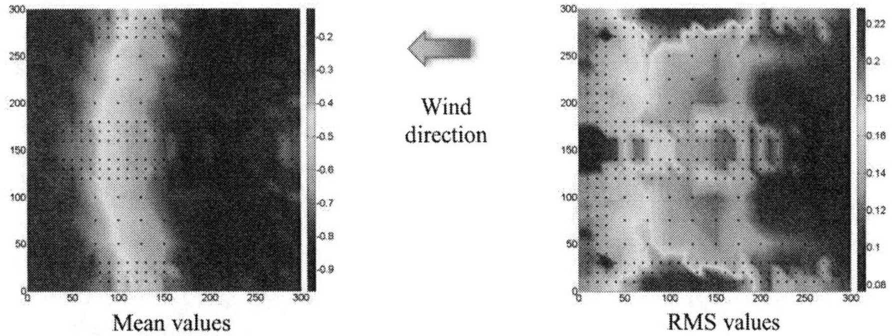

Fig. 7. Pressurs coefficients for the with $\alpha = 5°$ in case of smooth flow and wind angle $\theta = 90°$.

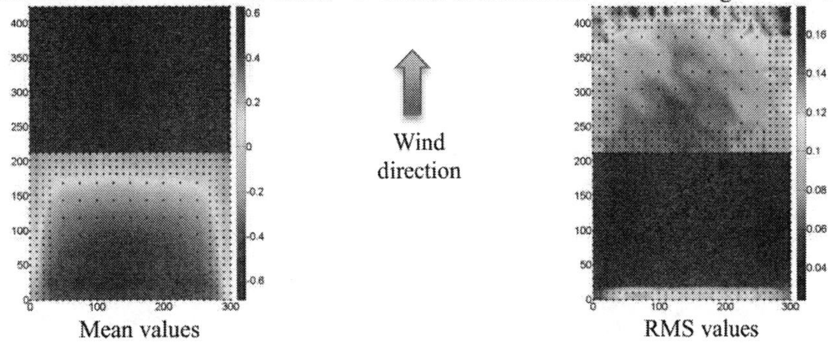

Fig. 8. Pressure coefficients for the with h $\alpha = 45°$ in case of smooth flow and wind angle $\theta = 0°$.

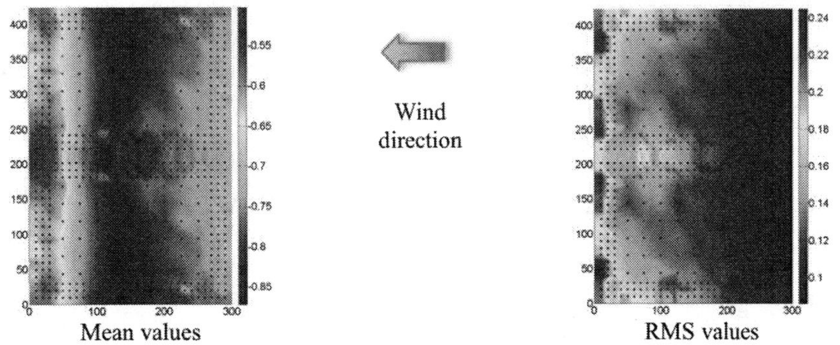

Fig. 9. Pressure coefficients for the with $\alpha = 45°$ in case of smooth flow and wind angle $\theta = 90°$.

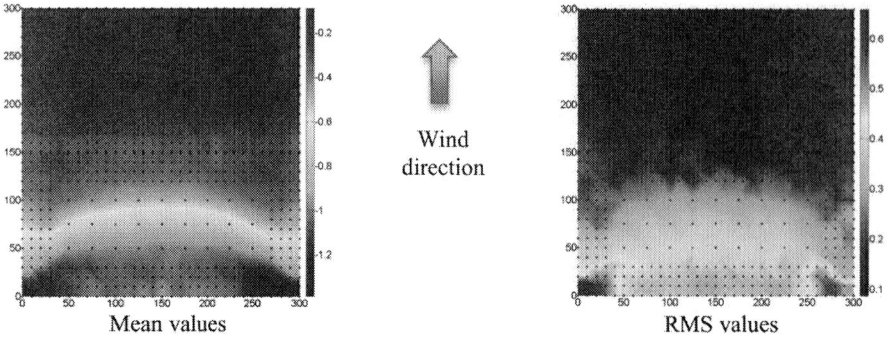

Fig. 10. Pressure coefficients for the roof with $\alpha = 5°$ in case of turbulent flow and wind angle $\theta = 0°$.

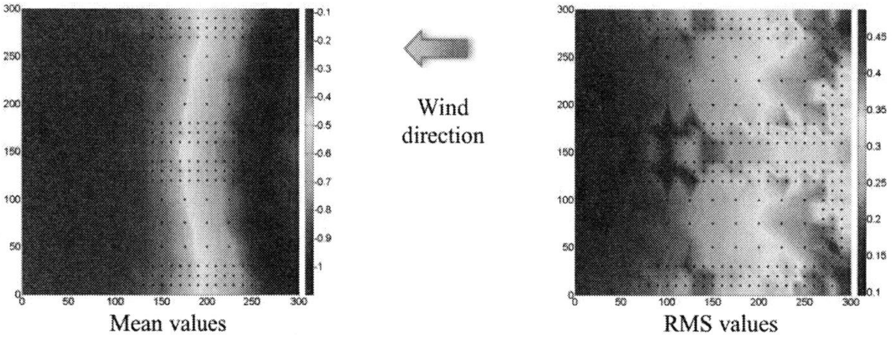

Fig. 11. Pressure coefficients for the roof with $\alpha = 5°$ in case of turbulent flow and wind angle $\theta = 90°$.

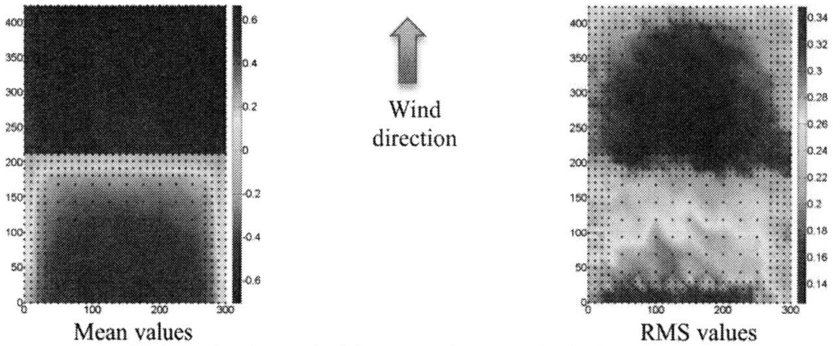

Fig. 12. Pressure coefficients for the roof with $\alpha = 45°$ in case of turbulent flow and wind angle $\theta = 0°$.

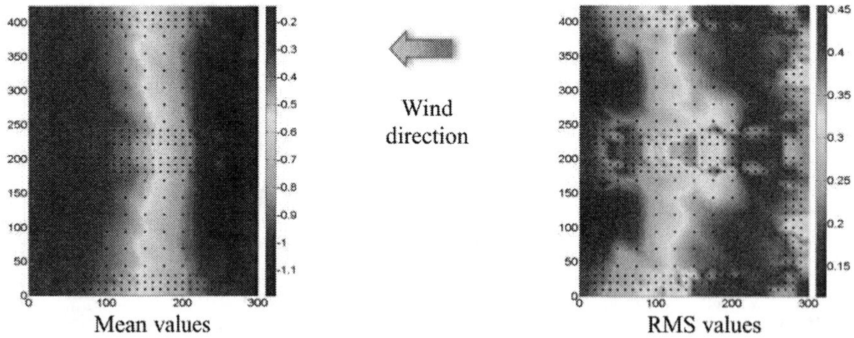

Fig. 13. Pressure coefficients for the roof with $\alpha = 45°$ in case of turbulent flow and incidence angle $\theta = 90°$.

Table 1. Measured local pressure coefficients for the gable with roof inclination α=5°.

θ	Zone	Code		Smooth flow				Turbulent flow			
		c_{pENV}	$c_{pČSN}$	c_p	σc_p	c_p-σc_p	c_p+σc_p	c_p	σc_p	c_p-σc_p	c_p+σc_p
0°	F	**-1.70**	-0.55	**-1.16**	0.16	-1.32	-1.00	**-1.12**	0.44	**-1.56**	-0.68
	G	**-1.20**	-0.55	**-1.02**	0.11	-1.13	-0.91	**-1.03**	0.36	**-1.39**	-0.67
	H	**-0.60**	-0.55	**-0.86**	0.15	**-1.01**	-0.71	**-0.60**	0.31	-0.91	-0.29
	J	**-0.60**	-0.40	**-0.41**	0.11	**-0.52**	-0.30	**-0.30**	0.16	-0.46	-0.14
	I	-0.60	**-0.40**	**-0.25**	0.08	**-0.33**	-0.17	**-0.18**	0.14	-0.32	-0.04
90°	F	**-1.60**	-0.70	**-0.87**	0.11	-0.98	-0.76	**-1.01**	0.40	**-1.41**	-0.61
	G	**-1.30**	-0.70	**-0.80**	0.09	-0.89	-0.71	**-0.92**	0.33	**-1.25**	-0.59
	H	**-0.70**	**-0.70**	**-0.85**	0.12	**-0.97**	-0.73	**-0.65**	0.32	**-0.97**	-0.33
	I	**-0.60**	**-0.50**	**-0.40**	0.15	**-0.55**	-0.25	**-0.17**	0.18	-0.35	0.01

Table 2. Measured local pressure coefficients for the gable with roof inclination α=45°.

θ	Zone	Code		Smooth flow				Turbulent flow			
		c_{pENV}	$c_{pČSN}$	c_p	σc_p	c_p-σc_p	c_p+σc_p	c_p	σc_p	c_p-σc_p	c_p+σc_p
0°	F	**+0.70**	0.30	**0.43**	0.06	0.37	0.49	**0.48**	0.29	0.19	**0.77**
	G	**+0.70**	0.30	**0.57**	0.06	0.51	0.63	**0.62**	0.30	0.32	**0.92**
	H	**+0.60**	0.30	**0.26**	0.03	0.23	0.29	**0.37**	0.25	0.12	**0.62**
	J	-0.30	**-0.40**	**-0.54**	0.07	-0.61	-0.47	**-0.50**	0.15	**-0.65**	-0.35
	I	-0.20	**-0.40**	**-0.59**	0.09	-0.68	-0.50	**-0.58**	0.18	**-0.76**	-0.40
90°	F	**-1.10**	-0.70	**-0.73**	0.09	-0.82	-0.64	**-1.04**	0.38	**-1.42**	-0.66
	G	**-1.40**	-0.70	**-0.71**	0.09	-0.80	-0.62	**-1.09**	0.36	**-1.45**	-0.73
	H	**-0.90**	-0.70	**-0.78**	0.11	-0.89	-0.67	**-0.87**	0.39	**-1.26**	-0.48
	I	**-0.50**	**-0.50**	**-0.73**	0.15	**-0.88**	-0.58	**-0.29**	0.25	-0.54	-0.04

The values of coefficients from Eurocode reflect the measurement with higher accuracy in most of the zones compared to former Czechoslovak code. This applies particularly for the windward side of the roof and mainly the zones F, G. The mean values for both types of flow do not differ much in most cases. Turbulence affects the RMS values of the pressures in both cases of the roof inclination. However, it is more significant for roof inclination $\alpha = 45°$.

4 Conclusions

The measurements of wind pressure coefficients at two gable roofs with different inclination were carried out in a wind tunnel. The results are presented in terms of the pressure maps in order to compare the results with the design codes where the roof is divided into several areas of exposition. The model has been placed in a smooth airflow, as well as into the turbulent one, which was used to model the atmospheric boundary layer above the urban terrain (category IV). Measurements were performed at Reynolds numbers that are already typical of the actual flow, so that the results are not affected by the nature of the flow.

The data of this study has been developed with sufficient versatility to allow comparison of wind load provisions of codes related to gable roofs with two different slopes. It provides instantaneous detailed pressure patterns over entire roof area. By comparison with the codes it can be concluded at first that the high tap areal density in the aerodynamically important zones may improve the capture of large suction loads or other important phenomena. Also, it turns out that there is still relatively high difference between the respective codes [10] and [11] in some zones. This discrepancy naturally reflects the historical contents and development of both codes. In the most cases Eurocode [10] gives more accurate results than [11], especially for the chosen methodology. This goes in hand with the probabilistic approach applied in recent wind engineering science.

It is our perspective to analyse the data with respect of its statistical parameters. The influence of other direction on the coefficients and deviation of the pressure coefficients from their coded values is envisaged too.

This research was financially supported by the project GAČR No. 17-26353J of the Czech Science Foundation and CET sustainability project LO12 (SaDeCET) of the Ministry of Education Youth and Sport of the Czech Republic.

References

1. J. D. Holmes, Journal of Wind Engineering and Industrial Aerodynamics **53**, 105-123, (1994)
2. M. Kanda, E. Maruta, Journal of Wind Engineering and Industrial Aerodynamics **50**, 173–182, (1993)
3. D. Meecham, D. Surry, A. G. Davenport, Journal of Wind Engineering and Industrial Aerodynamics **38**, 257–272, (1991)
4. P. C. Case, N. Isyumov, Journal of Wind Engineering and Industrial Aerodynamics **77-78**,107–118, (1998)
5. X. Jing, Y. Li, The Scientific World Journal **2013**, 11 pages, (2013).
6. C.P.W. Geurts, Journal of Wind Engineering and Industrial Aerodynamics **64**, 89-100, (1996)
7. C. Guerts, P. Blackmore, S.O Hansen, M. Hortmanns, G.,Sedlacek, P. Spehl, P. Staalduinen, B. Zimmerli. *Proceedings 3rd European & AfricanConference on Wind Engineering*, 165-172, (2001)
8. C. Geurts, C. van Bentum, *Wind Effects on Buildings and Design of Wind-Sensitive Structures*, 31-65, (2007)
9. M. Kasperski, C. Geurts, A. Goliger, Wind and Struct., An Int. J. 8 (4), 295-307, (2005)
10. ČSN EN 1991-1-4, *Zatížení konstrukcí – Část 1-4: Obecná zatížení-Zatížení větrem*
11. ČSN 73 0035, *Zatížení stavebních konstrukcí*
12. R. Meroney, Report University of Karlsruhe (1986)
13. ASCE: Wind Tunnel Studies of Buildings and Structures, Manual of practice no. **67**, (1999)

Investigation of Strouhal number of ice-accreted bridge cables at moderate flow turbulence

Piotr Górski[1,*], *Marcin* Tatara[1] , *Stanislav* Pospíšil[2] and *Sergej* Kuznetsov[2]

[1]Opole University of Technology, Faculty of Civil Engineering and Architecture, Department of Roads and Bridges, Katowicka 48, Opole 45-061, Poland
[2] Academy of Sciences of the Czech Republic, Institute of Theoretical and Applied Mechanics, Czech Republic

Abstract. This paper is concerned with the investigation of the Strouhal number (*St*) for a stationary iced cable model of cable-supported bridges using wind tunnel tests with respect to three principal angles of wind attack. The investigations were carried out in the Climatic Wind Tunnel Laboratory of the Czech Academy of Sciences in Telč. The *St* values were determined at the flow turbulence intensity of the order of 12%, within the range of the Reynolds number from $2.4 \cdot 10^4$ to $12.9 \cdot 10^4$, based on the dominant vortex shedding frequencies measured in the wake of the model. The obtained experimental *St* values of the iced cable of a cable-supported bridge will allow to determine the critical wind velocity as well as will be helpful for prediction of the cable response due to the vortex excitation phenomenon at moderate flow turbulence condition.

1 Introduction

It is well known that the vortex street is formed in the wake behind a body and the vortex-shedding frequency is characterized by the Strouhal number (*St*). The knowledge of the *St* value is necessary for determination of the critical wind velocity at which the largest amplitude of body vibration due to vortex excitation are observed. It is also known that *St* strongly depends on the body shape, its motion, flow turbulence intensity and may in some cases depends on the *Reynolds* number (*Re*). In the case of a body with the circular cross-section, e.g. a dry bridge cables, such dependences are well recognized [1]. It is noteworthy that previous wind tunnel studies, e.g. [2], show that the increase of the surface roughness of the cylinder as well as the increase of the turbulence intensity of free stream induce transition to the critical and transcritical regimes at a lower *Re*. Moreover, the greater flow turbulence may also produces the greater *St* values [3]. In the case of highly roughened cylinder the critical regime can be strongly reduced or may almost disappear.

In some weather conditions, the circular cross-section of a bridge cables may change due to ice accretion. In this case, the full-scale monitoring of slender bridge elements proved that ice accretion on the bridge cables has a significant influence on the flow field around the cables and their aerodynamics and can leads to a much larger amplitude of cable

* Corresponding author: p.gorski@po.opole.pl

vibrations under wind action than without the ice [4]. It should be noticed that the literature concerning the icing influence on aerodynamics of cables of cable supported bridges is still limited, e.g. [5-8]. Therefore it is very advisable to conduct further studies to quantify the cable aerodynamic variation due to ice accretion. In the existing literature, the investigation results of *St* of iced bridge cables are presented in the paper [9] at low flow turbulence intensity with the value of 3%. However, in most practical applications the flow field is usually turbulent with the turbulence intensity in the range between 10-20%. It means that the mentioned experimental results of *St*, received at low flow turbulence, cannot be applied in practice without any corrections.

This paper is concerned with the investigation of *St* for a stationary iced cable model of cable-supported bridges using wind tunnel tests with respect to three principal angles of wind attack at moderate flow turbulence. The investigations were carried out in the Climatic Wind Tunnel Laboratory of the Czech Academy of Sciences in Telč. One of the more frequent weather conditions in order to obtain the experimental icing of the inclined cable model was simulated. The ice shape of the cable was registered by a photogrammetry method. For the investigations of *St*, a new iced cable model was made at a scale of 1:1.6 using a 3D printing procedure. The St values were determined at the flow turbulence intensity of the order of 12%, within the range of the Reynolds number from $2.4 \cdot 10^4$ to $12.9 \cdot 10^4$, based on the dominant vortex-shedding frequencies measured in the wake of the model. The correctness of the *St* investigations was proved based on the *St* test of a circular smooth cylinder with the same experimental arrangement. Good agreement with the results reported in the existing literature was obtained. The obtained experimental *St* values of the iced cable of a cable-supported bridge will allow to determine the critical wind velocity as well as will be helpful for prediction of the cable response due to the vortex excitation phenomenon at moderate flow turbulence condition.

2 The ice accretion process on the cable model

The experimental icing process of a cable section model was conducted in a climatic section of the closed-return Climatic Wind Tunnel Laboratory of the Czech Academy of Sciences in Telč [10]. The 2.5 m long cable model as a pipe with a circular cross-section and 0.160 m diameter was inclined at an angle of 30° in the vertical plane, at an angle of 60° in the horizontal plane with respect to the wind velocity direction, and in this position was fixed to a special frame (see Fig. 1a). The model was made of polyvinylchloride (PVC), which surface is similar to the surface of a cable cover made of high-density polyethylene (HDPE). The icing process was performed at the average temperature about - 2 °C, the mean free stream velocity 2.8 m/s, the rainy conditions using rain sprinklers of the diameter heads 2.8 mm, and during a 40 min time period.

The final ice shape on the bottom side of the cable was obtained as the characteristic irregularly ice ribs with rounded edges and a relative surface roughness of 18% (Fig. 1b). On the upper part of the model the ice shape was similar to the circular shape with a surface roughness of 0.73%. The cross-section of the cable with ice became strongly nonsymmetrical with the dimensions of 0.192 m height and 0.181 m width.

Immediately after the icing process the shape of the iced cable model was registered by a photogrammetry method. After a numerical image analysis a three-dimensional (3D) numerical model of the iced cable was received (Fig. 1c). For the *St* investigations the new iced cable model, shown in Fig. 1d, was made of polylactide plastic at a scale of 1:1.6 using a 3D printing procedure. The outer dimensions of the model cross-section were 0.120 m in height and 0.113 m in width while the model length was 0.43 m. The descriptions of the wind tunnel, the icing process of the cable model in the climatic section, the final icing effect and preparation of the new iced cable model for the wind tunnel investigations are presented in the paper [9] in detail.

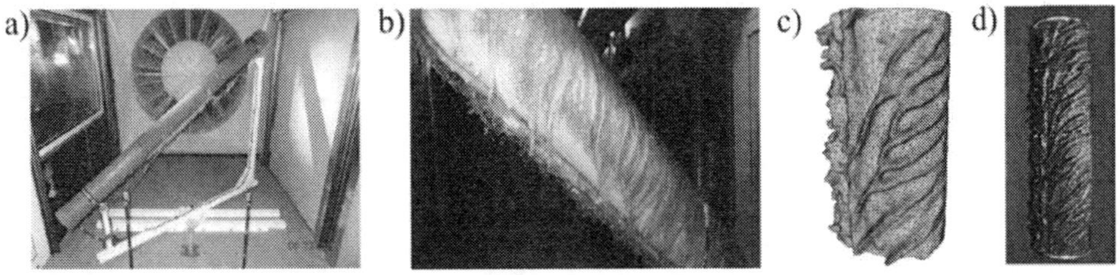

Fig. 1. (a) View of the cable model fixed to the special frame in the climatic section, (b) final icing effect of the cable model, (c) 3D numerical model of the ice shape of the cable model, and (d) new iced cable model for wind tunnel investigations made using a 3D printing method.

3 Experimental investigations of Strouhal number

The *St* investigations were carried out in the aerodynamic section of the wind tunnel. During the tests, the iced cable model was fixed horizontally in a special frame at a level of 69 cm above the floor of the aerodynamic section, crosswise to the airflow. Two sides of the frame were equipped with the plexi-glass end-plates to ensure two-dimensional flow around the model [11]. The height and the length of the plates were both 140.5 cm, the width was 17 cm and their spacing was 47 cm. The turbulent flow condition was modelled using a wooden grid with square bars placed 2.07 m upstream the model [12]. The width and the thickness of the grid bars was 5.5 cm and 4 cm, respectively, and their spacing was 18 cm. The experimental arrangement in the aerodynamic chamber is shown in Fig. 2.

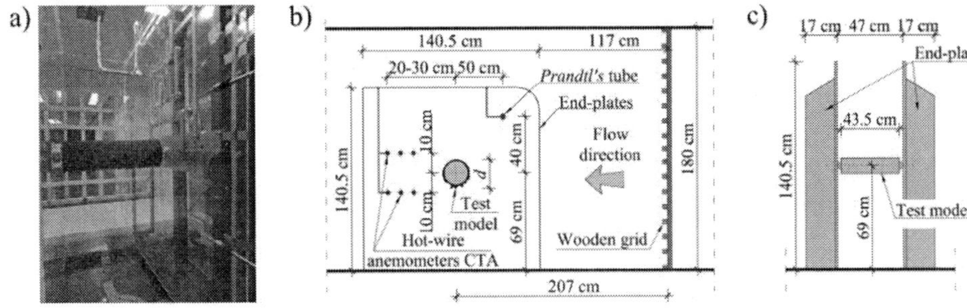

Fig. 2. (a) View of the model fixed to the special frame and the turbulence wooden grid, and sketch of the experimental arrangement (b) along and (c) across the aerodynamic section of the wind tunnel.

The *St* investigations were conducted under the assumption that the mean wind direction during the icing process is independent of the wind direction during the aerodynamic investigations. Thus, the *St* values were examined for three principal positions (configurations) of the ice cable model in relation to the flow direction which are presented in Fig. 3. In this figure the reference dimension *d* perpendicular to the airflow direction of each model configuration are also shown.

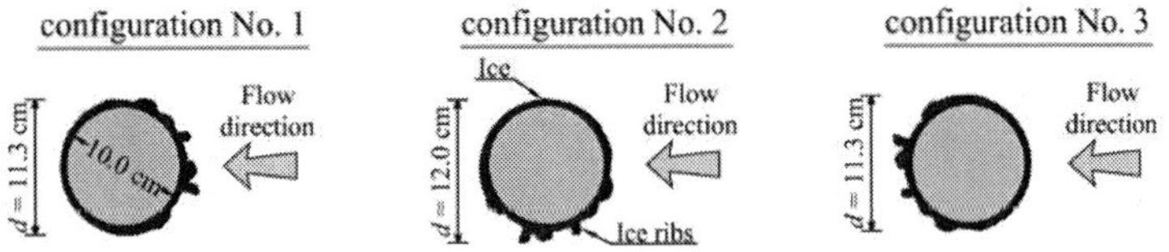

Fig. 3. Model configurations and their reference dimension *d* perpendicular to the airflow direction considered in the *St* investigations.

During the investigations the reference free stream velocity in front of the model u was measured by the *Prandtl's* tube while the horizontal along-wind velocity component, at a distance of $1.7 \cdot d$ to $2.6 \cdot d$ (20 to 30 cm) behind the model, was measured by two hot-wire anemometers CTA (Constant Temperature Anemometry) over a 30 s interval. The sampling rate of data acquisition was set as 1000 Hz. Two CTA sensors were fixed 10 cm above and below the longitudinal axis of the model (see Fig. 2b). The St values were determined based on the dominant vortex-shedding frequency f_s extracted from the Power Spectral Density (PSD) of the airflow velocity measured in the wake of the model according to the formula

$$St = \frac{f_s \cdot d}{\overline{u}} \tag{1}$$

where f_s is the vortex-shedding frequency, d is the reference dimension of the iced model perpendicular to the wind direction given in Fig. 3, and $\overline{u}$ is the mean free stream velocity in front of the model.

The Re values were evaluated according to the formula

$$Re = \frac{\rho \cdot d \cdot \overline{u}}{\mu}, \tag{2}$$

where ρ is the air density depends on the atmospheric pressure of the air and the air temperature, μ is the dynamic viscosity of the air depends on the air temperature.

The ρ and μ values were calculated based on the actual air temperature and the atmospheric pressure of the air inside the aerodynamic section measured continuously during the tests. It was found that the air temperature in the aerodynamic chamber was changeable within the range from 25.1 to 31.7 °C whereas the atmospheric pressure was 958.2 hPa. The tests were carried out at ten mean free stream velocities $\overline{u}$ in the range from 3.5 up to 18.0 m/s. Consequently, the corresponding Re number regime in the interval $Re = 2.4 \cdot 10^4$ to $12.9 \cdot 10^4$ was obtained.

The turbulence flow characterization in the test chamber was carried out using one CTA sensor. In this case, the measurements were performed in front of the model between the end-plates of the special frame revealing a longitudinal wind turbulence intensity upstream the center of the model. The turbulence intensity T_i calculated as a ratio between the standard deviation of free stream velocity to mean value of the velocity was found about 12% and was classified as a moderately turbulent wind. The comparison of the turbulence spectra measured in front of the model between the end-plates of the special frame in a smooth flow, i.e. without the turbulence grid ($T_i = 3\%$), and a moderate turbulent flow, i.e. with the turbulence grid, is depicted in Fig. 4. Due to the presence of the turbulence grid in the test chamber the free stream velocity decreased about 25% in relation to the same power of a ventilator which propels the airflow in the wind tunnel.

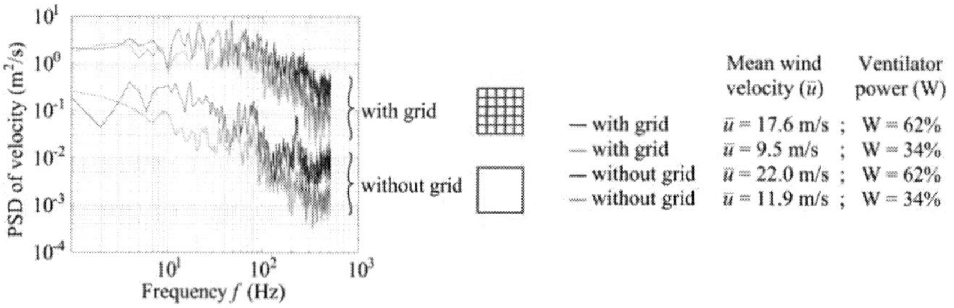

Fig. 4. PSDs of the airflow velocity in front of the model between the end-plates of the special frame measured in a smooth and moderate turbulent flow conditions.

4 Experimental results and discussion

In order to verify of the adopted methodology, described in Section 3, as well as to examine the possibility of a tunnel blockage effect on the test results, the *St* value of a circular smooth cylinder was investigated in a calibration test with the same experimental arrangement that was used for the iced cable model. The diameter of the cylinder was 0.10 m. The mean free stream velocity was in the range between 3.6 and 17.0 m/s. The corresponding *Re* range was from $2.1 \cdot 10^4$ to $10.2 \cdot 10^4$.

According to the statement given in Section 1, one should be aware that the main effect of free stream turbulence for the smooth cylinder is to develop transition to so-called the critical and transcritical regimes at a lower *Re* values as well as to produce the greater *St* values than in the case of a smooth free stream. Fig. 5c shows the variation of *St* of the circular smooth cylinder with *Re* for various turbulence intensity T_i noted by Cheung and Melbourne [3] based on an experimental data.

Fig. 5a shows an example of the PSD of the airflow velocity in the wake behind the circular smooth cylinder corresponding to the mean free stream velocity of $\bar{u} = 4.2$ m/s. Fig. 5b depicts the variation of *St* of the smooth cylinder with *Re* for various T_i based on the experimental results obtained from the calibration test at the turbulence intensity of $T_i =$ 12% and provided from the paper [3], i.e. from Fig. 5c, within the entire *Re* range that was studied in this paper (the *Re* range was indicated in Fig. 5c by the gray field). All *St* values obtained in the calibration test at $T_i =$ 12% were changed from 0.184 to 0.265 and were matched well with those reported in the paper [3]. Therefore, it was proved that the adopted methodology of the investigations was suitable and any corrections for a blockage effect were not needed.

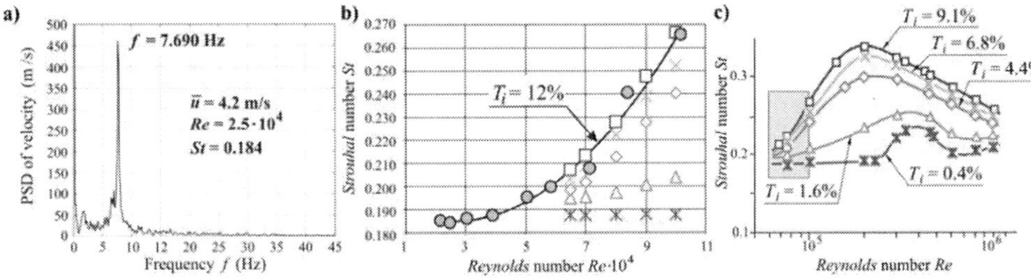

Fig. 5. (a) PSD of the airflow velocity behind the cylinder and dominant f_s corresponding to $\bar{u} = 4.2$ m/s, (b) variation of *St* with *Re* in the studied range for various T_i based on the experimental results from the calibration test at $T_i =$ 12% and the paper [3], and (c) variation of *St* with *Re* for various T_i reported in the paper [3].

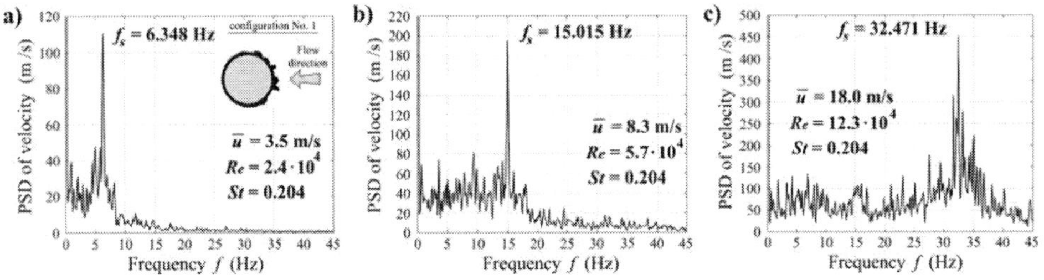

Fig. 6. Selected PSDs of the airflow velocity behind the iced cable model and dominant f_s for configuration No. 1 corresponding to: (a) $\bar{u}$ =3.5 m/s, (b) $\bar{u}$ =8.3 m/s, and (c) $\bar{u}$ =18.0 m/s.

Figs 6, 8 and 10 depict selected PSDs of the airflow velocity behind the iced cable model for configurations No. 1, 2 and 3, respectively, measured by CTA below the model at various $\bar{u}$ in the range of 3.5 m/s to 18.0 m/s. Figs 7a, 9a and 11a show the variation of

 DOI: 10.1051/matecconf/201710700080

f_s with $\bar{u}$, and Figs 7b, 9b and 11b depict the variation of St with Re for configurations No. 1, 2 and 3, respectively. All St values obtained for each configuration were approximated by a continuous line. The obtained results were related to the results for the smooth cylinder at T_i = 12% which were indicated as a dotted line in Figs 7, 9 and 11.

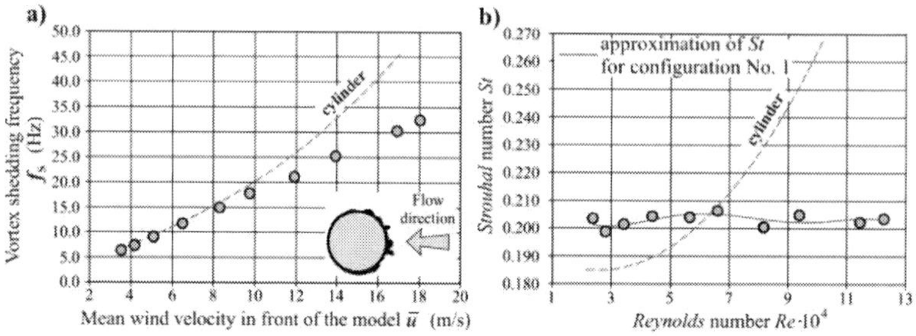

Fig. 7. (a) Variation of f_s with $\bar{u}$, and (b) variation of St with Re for configuration No. 1 with reference to the results for cylinder (dotted line).

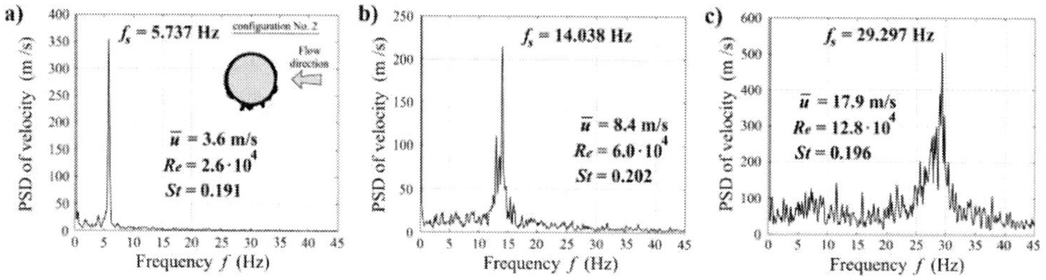

Fig. 8. Selected PSDs of the airflow velocity behind the iced cable model and dominant f_s for configuration No. 2 corresponding to: (a) $\bar{u}$ =3.6 m/s, (b) $\bar{u}$ =8.4 m/s, and (c) $\bar{u}$ =17.9 m/s.

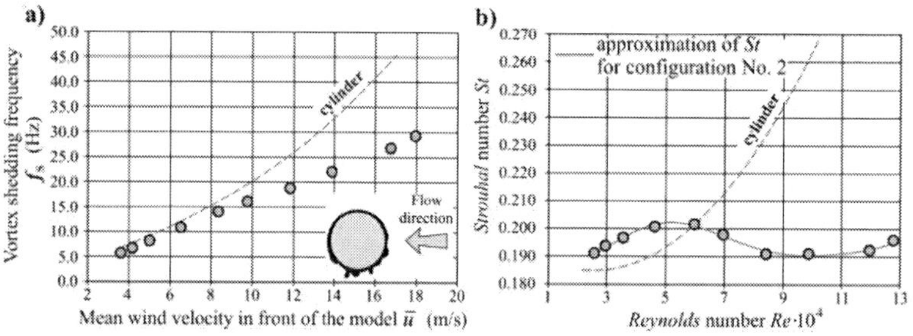

Fig. 9. (a) Variation of f_s with $\bar{u}$, and (b) variation of St with Re for configuration No. 2 with reference to the results for cylinder (dotted line).

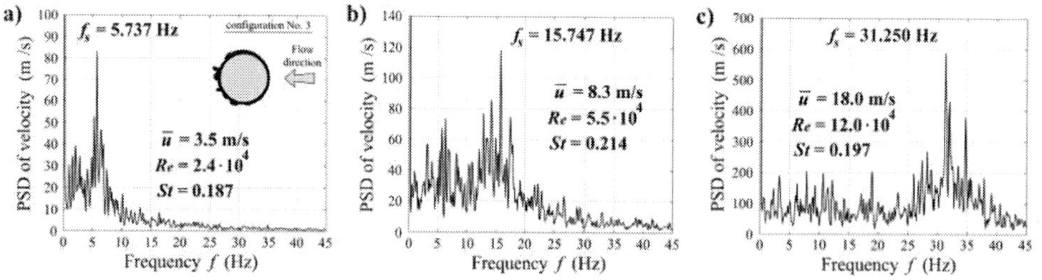

Fig. 10. Selected PSDs of the airflow velocity behind the iced cable model and dominant f_s for configuration No. 3 corresponding to: (a) $\bar{u}$ =3.5 m/s, (b) $\bar{u}$ =8.3 m/s, and (c) $\bar{u}$ =18.0 m/s.

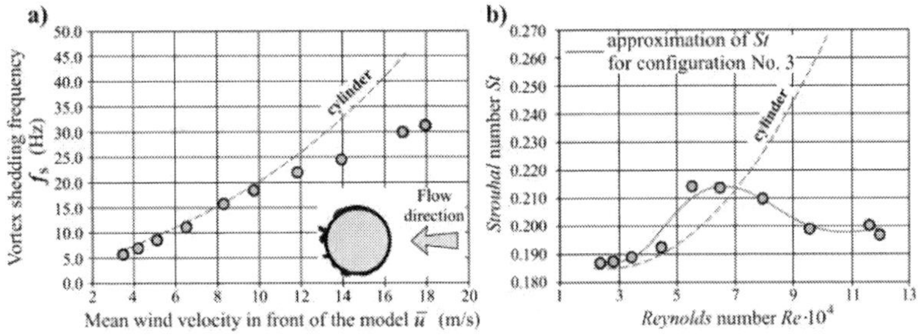

Fig. 11. (a) Variation of f_s with $\bar{u}$, and (b) variation of *St* with *Re* for configuration No. 3 with reference to the results for cylinder (dotted line).

As it was stated in the paper [9], the effect of cable inclination in the vertical plane on *St* in two principal practical cases should be considered in the following manner: (i) for the wind direction across the longitudinal axis of a bridge the cable inclination is unimportant and the *St* results obtained for configuration No. 2 are suitable, and (ii) for the wind direction along the longitudinal axis of a bridge the *St* results obtained for configuration No. 1 or 3 should be taken into account with the reference *Reynolds* number Re_{ref} related to the mean wind velocity in the horizontal plane, calculated according to the formula

$$Re_{ref} = \frac{Re}{sin(\alpha)} \tag{3}$$

where α is the angle of a cable inclination in the vertical plane.

5 Conclusion

The experimental icing of an inclined cable model of a cable-supported bridge was carried out in the Climatic Wind Tunnel Laboratory of the Czech Academy of Sciences in Telč. The *St* values were examined for the stationary iced cable model with respect to three principal angles of wind attack within the range of *Re* between $2.4\cdot10^4$ and $12.9\cdot10^4$ at the flow turbulence intensity of the order of 12%.

In order to ensure the correct interpretation of the test results for the iced cable model, in the first stage, the *St* values was investigated for the circular smooth cylinder with the same experimental arrangement. The *St* results obtained at the mean turbulence intensity about 12% matched well with those reported in the literature within the *Re* range that was studied in this paper, i.e. from $Re = 2.1\cdot10^4$ to $10.2\cdot10^4$. Therefore, the adopted methodology of the *St* investigations was confirmed as suitable.

St determined for configuration No. 1 has approximately a constant value, slightly varies in the range from 0.199 to 0.207 and seems to be independent of *Re* in the range that was studied (see Fig. 7b). For $Re < 6.6\cdot10^4$ the *St* values obtained for configuration No. 1 are greater than for the smooth cylinder. The maximum difference in this *Re* range was about 8%. For $Re > 6.6\cdot10^4$ the *St* values for the smooth cylinder are significantly greater than for configuration No. 1.

St determined for configuration No. 2 depends slightly on *Re* (see Fig. 9b). Within the range of $Re = 2.6\cdot10^4$ - $5.2\cdot10^4$ the *St* values increase from $St = 0.191$ to 0.202. Within the range of $Re = 5.2\cdot10^4$ - $9.9\cdot10^4$ *St* decreases to minimum value of $St = 0.191$, and for $Re > 9.9\cdot10^4$ *St* again increases to $St = 0.196$. Similarly to configuration No. 1, also in this case for $Re > 6.0\cdot10^4$ the *St* values obtained for configuration No. 2 are greater than for the smooth cylinder with the maximum difference of about 6%. For $Re > 6.0\cdot10^4$ the *St* values for the smooth cylinder are significantly greater than for configuration No. 2.

St determined for configuration No. 3 strictly depends on *Re* (see Fig. 11b). Within the range of $Re = 2.4\cdot10^4$ - $6.5\cdot10^4$ *St* increases nonlinearly from 0.187 to 0.214. Within the range of $Re = 6.5\cdot10^4$ - $10.1\cdot10^4$ *St* decreases from $St = 0.214$ to 0.198 and for $Re > 10.1\cdot10^4$ *St* has a nearly constant value $St = 0.199$. The *St* values obtained for configuration No. 3 are greater than for the smooth cylinder for $Re < 7.0\cdot10^4$ with the maximum difference of about 8%. For $Re > 7.0\cdot10^4$ the *St* values for the smooth cylinder are significantly greater than for configuration No. 3.

The differences between f_s obtained from the measurements using CTA located above and below the model did not exceed 6% for three considered configurations of the ice cable model.

Based on the experimantal data it was shown that the ice accreated on the bridge circular cables has a significant impact on the dependence of *St* with *Re* as well as on the *St* values within the entire *Re* range that was studied. Therefore, it is very advisable to conduct further such studies for various ice shapes accreted on a bridge cables in another weather conditions.

This work was supported from the funds for the research and development works and connected tasks serving the development of young researchers and PhD students financed within the internal competition under The Minister of Science and Higher Education No. DS-MN/40/WBiA/16 and Centre of Excellence Telč sustainability project LO1219 (SaDeCET) of the Ministry of Education Youth and Sport of the Czech Republic.

References

1. M.M. Zdravkovich, *Flow around circular cylinders* (Oxford Sciences Publications, 2007)
2. G. Buresti, J. Wind Eng. Ind. Aerod. **8**, 105-114 (1981)
3. J.C.K. Cheung, W.H. Melbourne, J. Wind Eng. Ind. Aerod. **14**, 399-410 (1983)
4. Z. Savor, J. Radic, G. Hrelja, Bridge Struct. **2**, 97-106 (2006)
5. H. Gjelstrup, C.T. Georgakis, A. Larsen, Wind Struct. **15,** 385-407 (2012)
6. C. Demartino, H.H. Koss, C.T. Georgakis, F. Ricciardelli, J. Wind Eng. Ind. Aerod. **138**, 98-119 (2015)
7. C. Demartino, F. Ricciardelli, J. Fluid Struct. **52**, 81–100 (2015)
8. H. Gjelstrup, C.T. Georgakis, A. Larsen, A., Wind Struct. **15**, 385-407 (2012)
9. P. Górski, S. Pospišil, S. Kuznetsov, M. Tatara, A. Marušić, Wind Struct. **22**, 253-272 (2016)
10. S. Kuznetsov, S. Pospíšil, R. Král, Technical Transactions **12(2-B)**, 303-316 (2015)
11. S. Pospíšil, A. Buljac, H. Kozmar, S. Kuznetsov, M. Macháček, R. Král, J. Bridge Eng. (to be published)
12. A. Trush, S. Pospíšil, S. Kuznetsov, H. Kozmar, J. Bridge Eng. (to be published)

DOI: 10.1051/matecconf/201710700081

Determination of the wind pressure distribution on the facade of the triangularly shaped high-rise building structure

Norbert Jendzelovsky[1,*], *Roland* Antal[1] and *Lenka* Konecna[1]

[1]STU in Bratislava, Faculty of Civil Engineering, Radlinskeho 11, Bratislava 810 05, Slovakia

Abstract. Methods for determination of the wind pressure distribution on the facade of the atypically shaped high-rise building structure will be shown. The standard Eurocode 1 [1] does not mention recommendations for calculation of wind pressure distribution on the facade for the most cases of buildings having an irregular cross-section. Analysed object has a shape of the equilateral triangle. For the determination of the pressure distribution two methods were used. Experimental analysis was carried out in the boundary layer wind tunnel and second approach was performed by numerical simulation of wind flow using computational fluid dynamics (CFD) software based on the finite volume method. Both methods used scaled model of the structure. For the wind tunnel experiment, 16 points of interests around an object were monitored in steady wind flow and from those points, results of external pressure coefficients from 2 directions of wind flow were obtained. In conclusion comparison of results of both methods will be shown.

1 Analysis of the wind pressure distribution on the equilateral triangle

High-rise structure was considered having a shape of an equilateral triangle (Fig. 1).

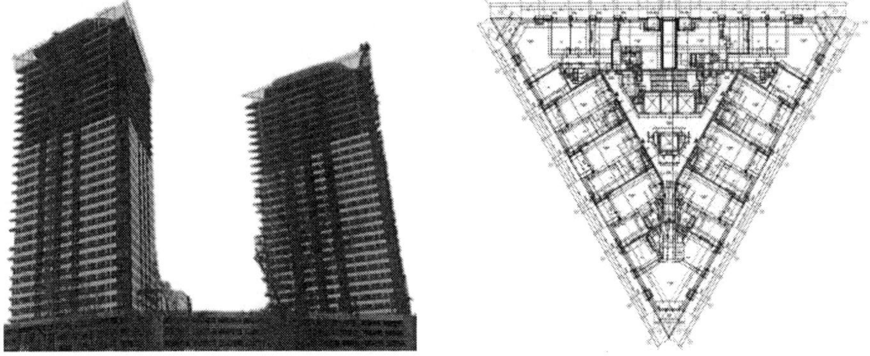

Fig. 1. Object of Panorama city buildings (Bratislava, Slovakia) and floor plan.

* Corresponding author: norbert.jendzelovsky@stuba.sk

DOI: 10.1051/matecconf/201710700081

Analysis of the wind pressure distribution was performed on the triangular cross-section of the scaled model. Length of the edge of the real size building is 51.34 m. Experimental and numerical model was created in scale 1:171.

1.1 Experimental study of pressure distribution in the boundary layer wind tunnel

Model for wind tunnel tests was created as a scaled simplification of mentioned high–rise building (Fig. 2a). Scale of this experimental model compared to the size of the real building is 1:172. Scale of the analysed object was set in regard to size of wind tunnel and previous measurements related to experimental investigations of an air flow and influences of walls and ceilings in the wind tunnel [2, 3].

Before positioning of sampling points, where wind pressure was measured, preliminary CFD simulations were done. Results of these numerical simulations gave us insight of how the pressure is distributed around an object, and where sampling points for experimental measurements should be positioned. 16 sampling points were created irregularly around the model (Fig. 2b). At these particular points pressure was measured by digital pressure scanner DSA 3217 Scanivalve from two directions of wind flow.

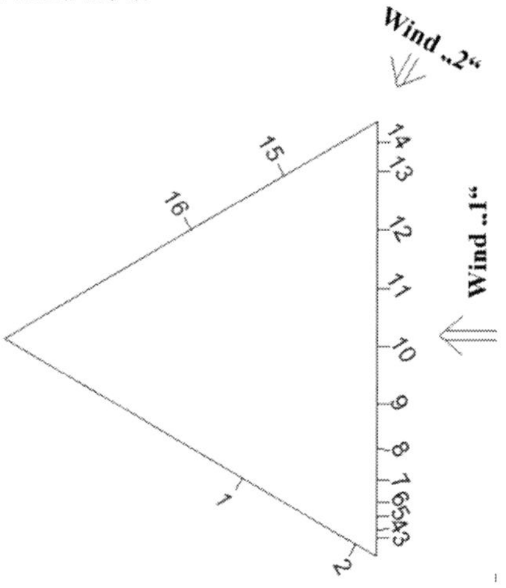

Fig. 2. a) Scaled model in wind tunnel and b) positioning of 16 sampling points.

For experimental measurements, opened, under-pressure, boundary layer wind tunnel was used. Tunnel has length of 26.3 m and its cross section has shape of rectangle with dimensions 2.6 m × 1.6 m. Tunnel consists of two main sections – front and rear space. Wind speed is adjustable in a range of 0.2 – 32 m/s. For turbulent flows, where turbulence of wind flow is required, rear space is used. For our purposes, front space of wind tunnel was used. Frontal area is used, when steady wind flow is required. Therefore, analysed object was tested in constant wind speed, without any significant turbulences. Properties of air measured on the day of tests are shown in Table 1.

Table 1. Boundary conditions and properties of air in the wind tunnel at the day of measurements.

Wind speed [m/s]	Temperature of air [°C]	Atmospheric pressure [Pa]	Density of air [kg/m^3]
12.08	22.8	99870	1.170502

Pressure was measured in 16 sampling point from both directions of wind flow twice, therefore averaged values of measured pressure were calculated. For better illustration, measured pressures were converted to external pressure coefficient, which is a dimensionless variable calculated by formula (1). Pressure coefficient demonstrates, how the pressure is distributed around an object depending on the direction of wind flow.

$$C_p = \frac{p - p_\infty}{\frac{1}{2} \cdot \rho_\infty \cdot v_\infty^2} \tag{1}$$

Where p is the pressure at the point at which pressure coefficient is being evaluated in [Pa],
p_∞ is the pressure in the free stream in [Pa],
ρ_∞ is the free stream fluid density in [kg/m3],
v_∞ is the free stream velocity of the fluid in [m/s].

In the following figure (Fig. 3) distribution of wind pressure around an object in the dimensionless external pressure coefficient for both directions of wind flow is shown.

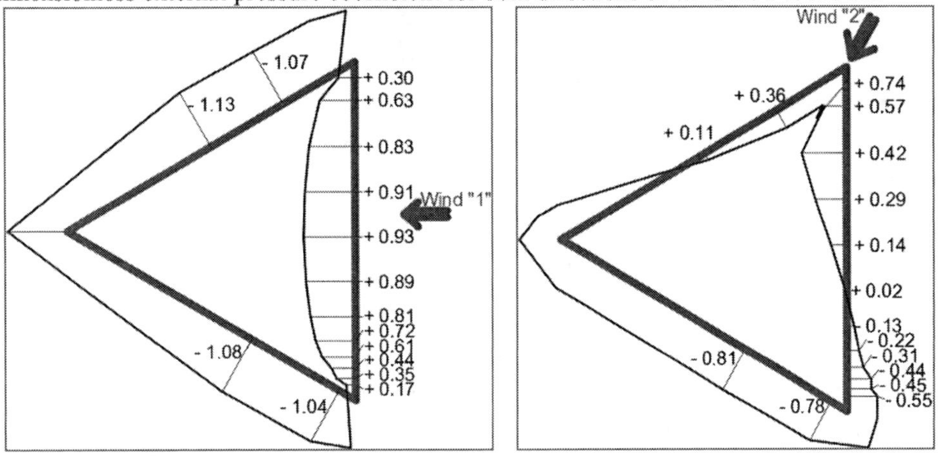

Fig. 3. External pressure coefficient *Cp* [-] for both analysed wind flow directions.

1.2 Computational fluid dynamic simulation of wind flow

Computational model was analysed by software based on finite volume method. Additional software systems for geometry, mesh and results were used. Size of analysed boundaries for an air flow simulation were created as follows (Fig. 4): Length of boundary is 9.4 m, width of boundary is 6.6 m and height of boundary is 5.6 m.

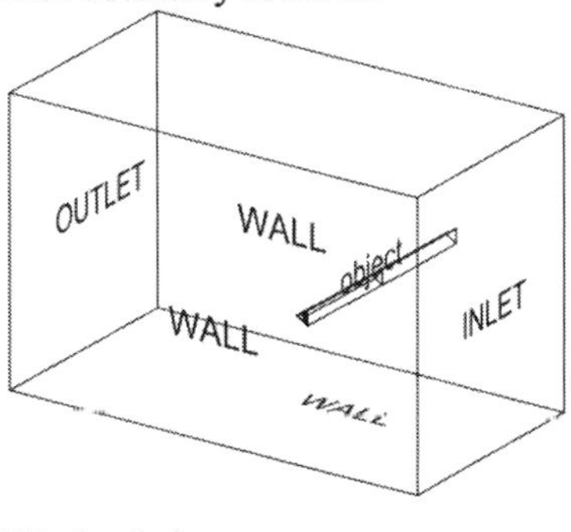

Fig. 4. Analysed boundaries for CFD simulation.

Mesh generator was used to create following mesh: Element type was set as Hexahedra, which were refined closer to the object of interest. Near analysed model, first boundary layer was created 0.01 mm from object edge. 40 other fine mesh layers were created gradually around an object with growth rate 1.2 (Fig. 5). Largest elements with size of 400 mm were created in the remote surroundings where turbulences of air does not significantly affect flow around an object.

For computational simulations, parameters were set according to data measured during wind tunnel tests. Input parameters used for CFD simulation are listed in (Table 1) Parameters and properties, which were not measured (roughness of object, small turbulence of wind flow etc.) were set simplistically by estimation.

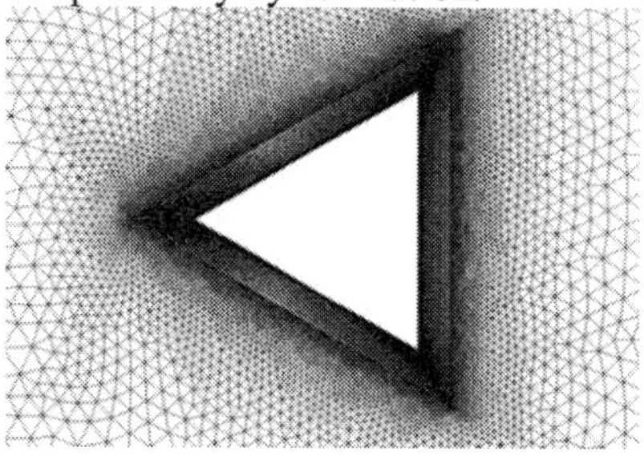

Fig. 5. Hexahedral mesh near object.

Flow simulation was carried out by CFD software module. Selected module calculates variables by finite volume method based on elements – EbFVM, with a cell vertex formulation.

For the given flow simulation, shear stress transport (SST) mathematical model was selected. It is the two-equation eddy-viscosity model, which combines two mostly used k-ω and *k-ε* models. The formulation of the SST model is based on wind tunnel experiments and it attempts to predict solutions to typical engineering problems. Equations used in *k-ω* model work best for near wall treatment through the viscous sub-layers. In the free stream flow, where near wall treatment is not essential, SST model automatically switches to k-ε model and hence, it avoids common *k-ω* problem, which is too sensitive to the inlet free-stream turbulence properties. Other information about mentioned mathematical models can be found in [4, 5, 6].

Process of calculation was iterative, therefore residuals, imbalances and pressures were monitored until satisfying results were reached. After 90 iterations, results of pressure didn't change significantly and iteration process was complete and solution was found.

Results of pressure coefficients for both wind flows are shown in (Fig. 6). For graphical representation wind flows from left to right in both cases.

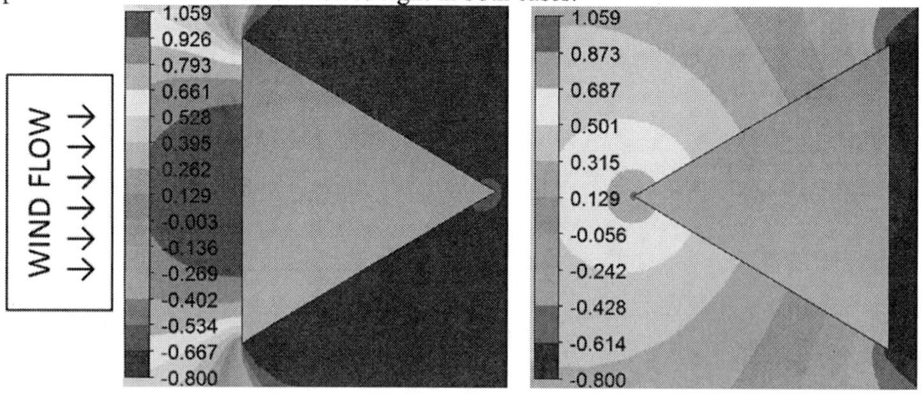

Fig. 6. Results of calculated pressure coefficient *Cp* [-] for both wind flows.

1.3 Comparison of wind tunnel measurements with CFD simulation

Demonstration of 2 methods used for calculations of wind pressure distribution was a matter of interest here. Results of external pressure coefficient measured in wind tunnel are compared with CFD simulation.

Comparison of results are shown in figures (Fig. 7) and (Fig. 8), where values of calculated pressure coefficient in sampling points are compared for both methods and both wind flow directions. On the horizontal axis, numbers of sampling points are shown. On the vertical axis, values of calculated external pressure coefficient are shown.

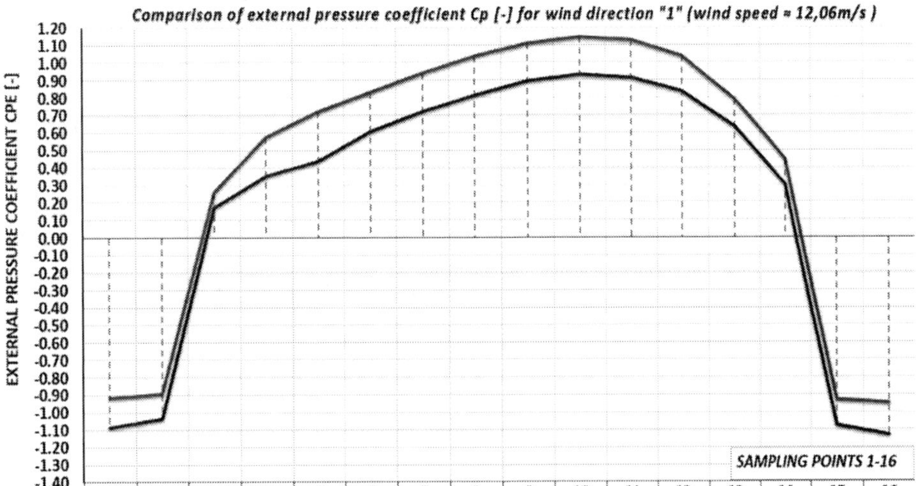

	1	2	3	4	5	6	7	8	9	10	11	12	13	14	15	16
—Wind tunnel	-1.08	-1.04	0.17	0.35	0.44	0.61	0.72	0.81	0.89	0.93	0.91	0.83	0.63	0.30	-1.07	-1.13
—CFD Simulation	-0.91	-0.90	0.26	0.58	0.72	0.83	0.94	1.03	1.11	1.15	1.13	1.04	0.79	0.45	-0.93	-0.95

Fig. 7. Comparison of calculated pressure coefficient *Cp* [-] for wind flow "1".

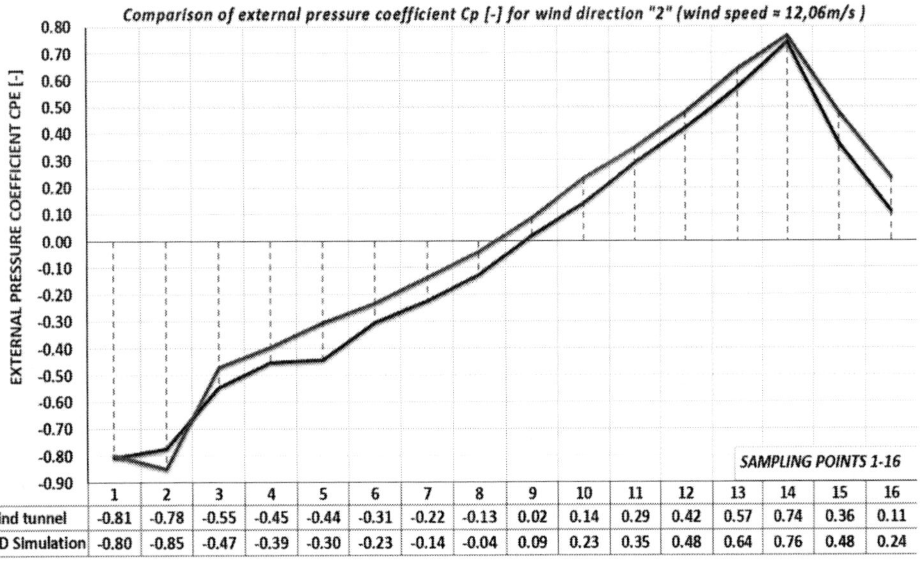

	1	2	3	4	5	6	7	8	9	10	11	12	13	14	15	16
—Wind tunnel	-0.81	-0.78	-0.55	-0.45	-0.44	-0.31	-0.22	-0.13	0.02	0.14	0.29	0.42	0.57	0.74	0.36	0.11
—CFD Simulation	-0.80	-0.85	-0.47	-0.39	-0.30	-0.23	-0.14	-0.04	0.09	0.23	0.35	0.48	0.64	0.76	0.48	0.24

Fig. 8. Comparison of calculated pressure coefficient *Cp* [-] for wind flow "2".

As we can see in both figures (Fig. 7) and (Fig. 8), results measured in the wind tunnel are slightly different to those, which were calculated by CFD simulation.

As a conclusion, it can be said, that both methods are capable to calculate variables needed to design a high–rise building structure. CFD simulation where SST mathematical model was used, proved, that it is very good and relatively accurate tool for calculation of many variables, even for those, which cannot be measured directly in the wind tunnel.

Differences between numerical simulation and experiment could be caused by many factors. Main factors responsible for differences in similar cases are: a) small differences between geometry of wind tunnel model and the model used in computational flow simulation – just a small difference in geometry in that scale could cause abnormality in results; b) position of model in wind tunnel during air flow – even very small rotation or movement of model can cause different results compared to "ideal" conditions simulated by CFD software; c) Boundary size, mesh size and boundary conditions set in CFD software – changes in boundary size, mesh size, number and size of near wall layers and of course changes of properties of air, wind speed, turbulence intensity, will change results significantly.

These days, cooperation between wind tunnel tests with addition of CFD simulation is the best way, how to analyse wind effects on irregular shaped buildings and determine wind load for unusual structures.

This paper was supported by Grant Agency VEGA of Slovak Republic, project No. 1/0544/15.

References

1. EN 1991-1-4, Eurocode 1: Actions on structures-Part 1–4: General actions-Wind actions, (2007)
2. O. Hubová, P. Lobotka, Scientific-Technical Journal **1**, 2, (2014)
3. M. Jirsák, *Studie budov a konstrukcí ve vetrných tunelech*, (ČKAIT, Praha, 2009)
4. F. R. Menter, AIAA Journal, **32**, 1598, (1994)
5. J. D. Anderson, *Computational Fluid Dynamics: The Basics with Applications.* (McGraw-Hill Inc., 1995)
6. T. J. Chung, *Computational Fluid Dynamics - Second Edition* (Cambridge University Press, New York, 2010)

Numerical modelling of flow around thermally loaded object

Vladimira Michalcova[1,*], *Lenka* Lausova[1] and *Ivan* Kolos[1]

[1]VŠB - Technical University of Ostrava, Faculty of Civil Engineering, 708 33 Ostrava-Poruba, Czech Republic

Abstract. The paper focuses on numerical analysis of selected turbulent characteristics near flow around thermally loaded object. Changes in the flow and thermal fields are examined in response to the change of the reference values of velocity and temperature of the flow around body. The results are evaluated at different heights on the windward side, above the object and on the leeward side. The tasks are solved using computational fluid dynamics (CFD) software based on the finite volume method.

1 Introduction

The task of the flow around heated model of a hill is numerically modelled in cooperation with the department of experimental research CWT CET in Telč [1, 2]. The reason is the "mapping" of the vertical heat flux which affects the velocity profiles of the flow field of wind [3, 4]. The tasks are solved using CFD codes in software Ansys Fluent. From the viewpoint of numerical modelling this is an interesting issue in terms of flow characteristics [5]. The model of the solved hill has a height h = 200 mm (Fig. 1 on the left). There are considered two cases of the reference air velocity: v_0 = 0.5 m/s a v_0 = 1.5 m/s.

The described process corresponds to a flow of Reynolds number $Re = 6\cdot10^3 - 2\cdot10^4$. It is a flowing with the transition from the low turbulence at the beginning to the fully developed turbulence behind the obstacle [6-8]. A flow around and temperature loaded object contributes to a change of momentum and turbulent flow characteristics [9] that leads to a significant change in the momentum in the transition area of turbulent flow at low *Re* numbers.

The experimental measurements for both air velocities v_0 [10] were carried out for the temperature of the hill T = 150 °C. The results were evaluated from the wind tunnel and numerical simulations for above mentioned cases in [10], there have been achieved satisfactory conclusions. In the submitted paper the task is expanded by a numerical calculation for cases where the model temperature of flow around hill is reaching values T = 100 °C a T = 200 °C. The velocity profiles are monitored at four height levels (Fig.1 right), with the lowest of these (h = 0.15 m) intersects the model of the hill in the upper quarter of the height.

The analysis of the correctness for solution of numerical models suitable for described task was carried out in [11]. Based on the comparison of outputs of the numerical

* Corresponding author: vladimira.michalcova@vsb.cz

calculations and experimental measurements the best results were achieved using the Transition SST (Shear Stress Transport) model. This model was able to take into account the difference between high and low turbulence at the interface between the wake behind an obstacle and the free stream.

2 Numerical model and computational area

The standard k-ω model is an empiric model based on the solution of two transport equations for kinetic energy k (1) and dissipation of this energy ω (2):

$$\frac{\partial(\rho k)}{\partial t}+\frac{\partial(\rho k u_i)}{\partial x_i}=\frac{\partial}{\partial x_j}\left(\Gamma_k\frac{\partial k}{\partial x_j}\right)+G_k-Y_k+S_k \tag{1}$$

$$\frac{\partial(\rho\omega)}{\partial t}+\frac{\partial(\rho\omega u_i)}{\partial x_i}=\frac{\partial}{\partial x_j}\left(\Gamma_\omega\frac{\partial \omega}{\partial x_j}\right)+G_\omega-Y_\omega+D_\omega+S_\omega \tag{2}$$

where k is kinetic energy [m^2/s^2], ω is specific dissipation rate [1/s], G_k is generation of turbulence kinetic energy due to mean velocity gradients [m^4], G_ω is generation of ω [$kg/(m^3\cdot s^2)$], Γ_k, Γ_ω are effective diffusivity of k and Y_k, Y_ω are dissipation of k and ω due to turbulence [$kg/(m\cdot s^3)$], [$kg/(m^3\cdot s^2)$], S_k, S_ω are user-defined source terms and D_ω is cross-diffusion term [$kg/(m^3\cdot s^2)$].

Both equations describe the anticipated sheer flows well. The Transition SST model, in addition to the original k and ω equations, uses two transport equations: the transport equation for the intermittency γ (3) and the transport equation for the transition momentum thickness Reynolds number $\overline{Re}_{\Theta t}$ (4). By this, the model is able to take into account the difference between high and low turbulence at the interface between the wake behind the barrier and free flow:

$$\frac{\partial(\rho\gamma)}{\partial t}+\frac{\partial(\rho u_j\gamma)}{\partial x_j}=P_{\gamma 1}-E_{\gamma 1}+P_{\gamma 2}-E_{\gamma 2}+\frac{\partial}{\partial x_j}\left[\left(\mu+\frac{\mu_t}{\sigma_\gamma}\right)\frac{\partial \gamma}{\partial x_j}\right] \tag{3}$$

$$\frac{\partial(\rho\overline{Re}_{\Theta t})}{\partial t}+\frac{\partial(\rho u_j\overline{Re}_{\Theta t})}{\partial x_j}=P_{\Theta t}+\frac{\partial}{\partial x_j}\left[\sigma_{\Theta t}(\mu+\mu_t)\frac{\partial \overline{Re}_{\Theta t}}{\partial x_j}\right] \tag{4}$$

where μ is dynamic viscosity [$kg/(m\cdot s)$], μ_t is turbulent dynamic viscosity [$kg/(m\cdot s)$], $P_{\gamma 1}$, $E_{\gamma 1}$ are transition sources, which include: strain rate magnitude and empirical correlation that controls the length of the transition region, $P_{\gamma 2}$, $E_{\gamma 2}$ are sources for the destruction/relaminarization, which include: vorticity magnitude, a critical Re number and also necessary empirical correlation and $P_{\Theta t}$ is cross-diffusion term [$kg/(m^3\cdot s^2)$].

Near the surface of the thermally burdened and flow around object the buoyancy forces play an important role. The Boussinesq model, which describes the transfer of heat by natural convection and buoyancy, is included into the heat transfer equation. This model adds a source term describing the change in current density into the momentum equation. In such a case, it can be written the following formula for the real density of media:

$$(\rho-\rho_0)\cdot g\approx\rho_0\cdot\beta\cdot(T-T_0)\cdot g \tag{5}$$

where ρ_0 is the constant flow density [kg/m^3], g is gravitational acceleration [m/s^2], β is the thermal expansion coefficient [1/K], T_0 is the operating temperature [K] and T is current temperature [K].

Equation (5) is solved by Boussinesq approximation:

$$\rho = \rho_0 \cdot (1 - \beta \cdot \Delta T) \tag{6}$$

If real density changes are small, as it is in the described task, the approximate solution can be considered accurate.

The task is solved in a 3D computational area with the following dimensions b × h × l = 0.1 × 1.8 × 8 m. The axis of the hill is located at a distance of 2 meters from the input to the area. The remaining 6 meters behind the axis is sufficiently long for the possibility of generating eddies in the wake. Boundary condition (BC) on the input is the velocity inlet, turbulence intensity is 1%. BC on the output is the pressure outlet and on the side walls it is the symmetry. The height of the computational area corresponds to the wind tunnel space, so the top and bottom surfaces provides BC wall describing the limited space of the wind tunnel.

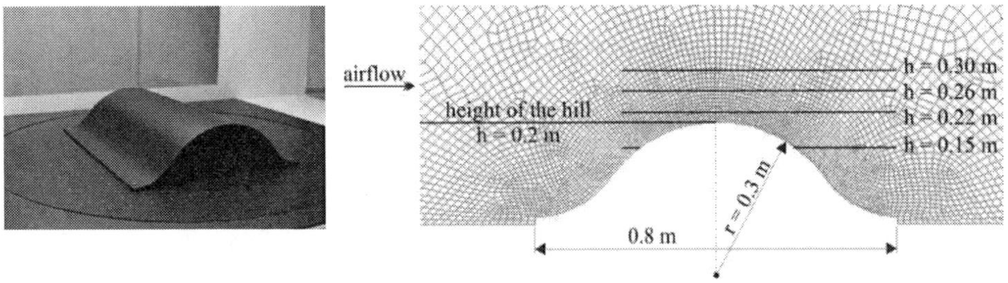

Fig. 1. Model of the hill in the wind tunnel (left) and scheme of the hill in the computational area marked with monitored height levels (right).

3 Results

Changes of velocity profiles for both reference velocities at given height levels are evident from Fig. 2. They are expressed as dimensionless ratios of the mean value of velocity in the direction of flow v_x to the velocity inlet (reference) v_0. The temperature change ± 50 °C does not significantly affect the velocity field, which can be seen in Figure 2. There is a distinct swirl on the leeward side (negative v_x) in the level of $h = 0.15$ m. Gradient decreases with increasing height velocity. A slight swirl can be seen at a height of 0.2 meters above the hill ($h = 0.22$ m).

Figure 3 describes the change of the temperature field in the surroundings expressed by dimensionless ratio of T/T_0, where T is the mean value of temperature at a given point and T_0 is the inlet temperature of air flow $T_0 = 20$ °C. The fact that the temperature gradient decreases with increasing height and simultaneously moves in the direction of air flow is confirmed in all three cases of temperature load of the model of the hill (100 °C, 150 °C, 200 °C). The gradient is lower at a higher velocity, and its shift slightly increases. This is also illustrated by isolines of temperature field for the two inlet velocities of the flow around model of the hill at temperature $T = 150$ °C (Fig. 4).

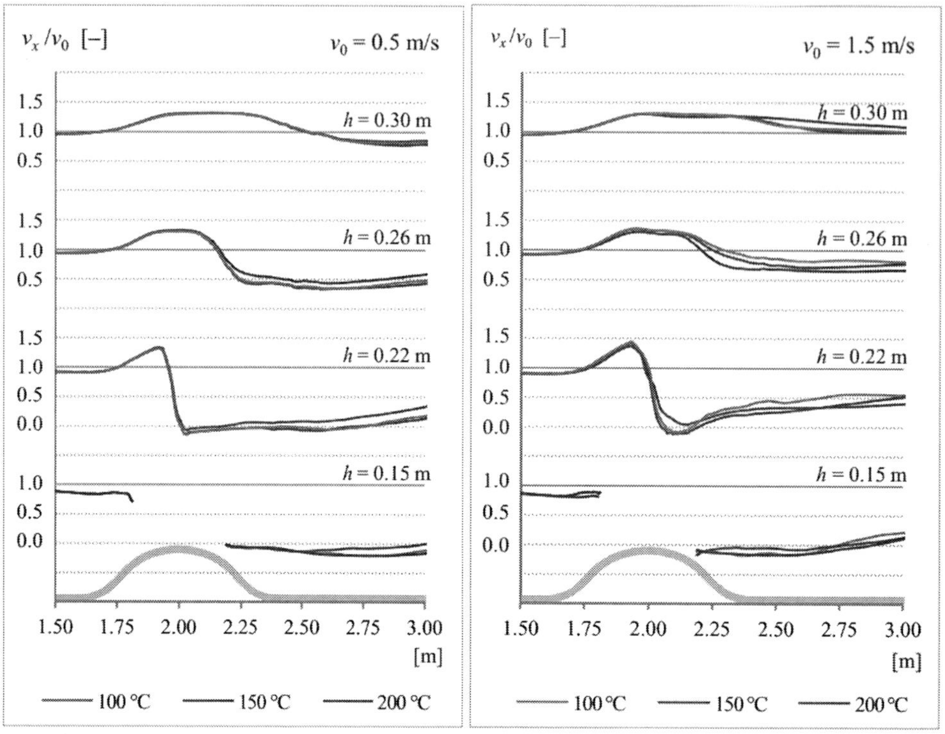

Fig. 2. Velocity field near the hill.

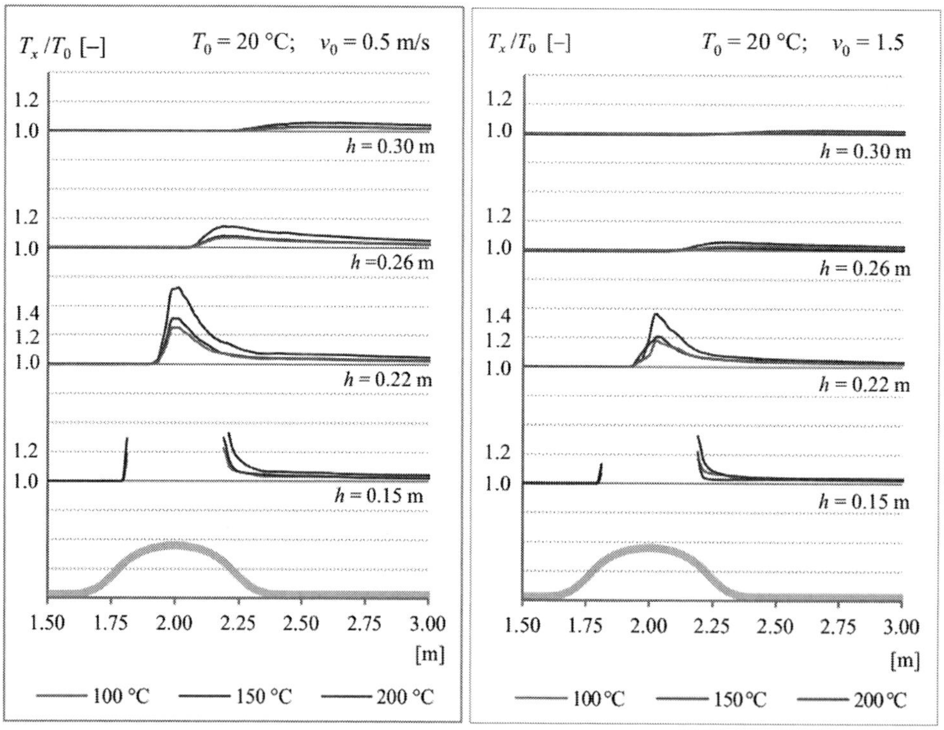

Fig. 3. Temperature field near the hill.

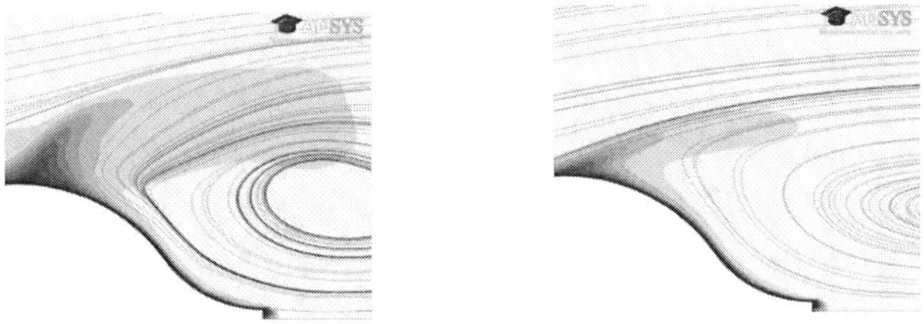

Fig. 4. Velocity isolines with thermal field for v_0 = 0.5 m/s (left) and v_0 = 1.5 m/s (right).

4 Conclusions

The paper presents the changes of the flow field around a thermally loaded object. The velocity and temperature fields are evaluated for the cases of temperature load changes of that object. The problem is solved numerically and builds on a task validated with experimental measurements. Those findings are presented in this work to verify correctness of that solution. It is possible to say that the results of the flow field in the case of temperature load change are satisfying. The following works of the authors will focus on a verification of new experimental measurements for detailed confirmation.

The paper has been supported by the project of "Conceptual development of science and research activities 2017" on the Faculty of Civil Engineering, VŠB-TU Ostrava.

References

1. V. Michalcova, S. Kuznetsov, J. Brozovsky, S. Pospisil, Appl. Mech. Mater. **617**, 275 (2014)
2. V. Michalcova, S. Kuznetsov, S. Pospisil, Math. Comput. Simulat. **8,** 135 (2014)
3. Y. Qu, M. Milliez, L. Musson-Genon, B. Carissimo, J. Wind. Eng. Ind. Aerod. **104-106,** 474 (2012)
4. S.S. Pillai, R. Yoshie, J. Wind. Eng. Ind. Aerod. **104-106,** 447 (2012)
5. J. Kralik, O. Hubova, L. Konecna, *Transactions of the VŠB – Technical University of Ostrava, Civil Engineering Series*, **15**, (2016)
6. J.E. Pieterse, T.M. Harms, J. Wind. Eng. Ind. Aerod. **121**, 82 (2013)
7. H. Nakayama, T. Takemi, H. Nagai, Adv. Sci. Res. **11,** 75 (2014)
8. K. Kotrasova, E. Kormanikova, Key Eng. Mat. **635,** 147 (2015)
9. Y. Ohya, H. Hashimoto, S. Ozono, J. Wind. Eng. Ind. Aerod. **67–68**, 793 (1997)
10. V. Michalcova, L. Lausova, I. Skotnicova, S. Pospisil, *Transactions of the VŠB – Technical University of Ostrava, Civil Engineering Series*, **16,** (2016)
11. V. Michalcova, L. Lausova, I. Skotnicova, S. Kuznetsov, Key. Eng. Mat. (to be published)

Full scale measurements and simulations of the wind speed in the close proximity of the building skin

Radoslav Ponechal[1,*] and *Peter* Juras[1]

[1] University of Zilina, Faculty of Civil Engineering, Department of Building Engineering and Urban Planning, Univerzitna 8215/1, 010 26 Zilina, Slovak Republic

Abstract. 36 meteorological stations are located on the facade of the Research Centre building´s from 2016. Weather stations measure basic climate parameters and wind velocity and direction using a powerful ultra-sonic anemometers. These are located on the walls of the building oriented to four cardinal points at different heights and positions on the façade. The location selections were made with use of CFD simulation which analysed flow around the building. Thus they give faithful image of wind flow near the façade of five-storey office building. Such detailed measurements make it possible to achieve high accurate calibration of CFD models and measurements in the wind tunnel. This paper describes these meteorological stations in detail. First outputs from measurement in autumn are published and analysed.

1 Introduction

Wind engineering deals with interaction between wind in the atmospheric boundary layer and human works on the earth surface. Good knowledge of fluid mechanics is the fundamental background necessary for the detailed understanding of interaction among wind flow and civil engineering structures of buildings.

1.1 Experimental methods in wind engineering

Wind caused loads on the buildings can be studied by different methods, such as wind tunnel measurements, experiments on the real buildings, which can be full or part scale and CFD (computational fluid dynamics) simulations. Aerodynamic boundary layer wind tunnels are simulating atmospheric flows in the lowest part of the atmosphere and are often used for the evaluation of wind loads on buildings.

Computational methods made big progress through the last decade and are often better, cheaper and more used compared to the ABL wind tunnel measurements [1]. Before the computer era, full-scale measurements were widely used. Nowadays, these measurements are expensive, time consuming and less used. But several distinctive features of the wind

* Corresponding author: radoslav.ponechal@fstav.uniza.sk

have been found from the full-scale measurements in the past [2]. Sometimes wind tunnel experiments have some limitations, which may be solved by full scale experiments. Such cases include the pressure difference over permeable facade or roof or the local conditions in small areas. Another reason, when the full scale experiments are still useful is validation and verification of wind tunnel experiment or other simulations.

For wind tunnel measurement and simulation especially, experienced supervisor for representation of the results is needed. On the other hand, full scale experiments are limited with the availability of sufficient wind conditions, large amount of time, effects of surrounding terrain, measuring devices and involved costs. Full scale measurements on city buildings, where the fluctuating characteristics of wind were subject of investigation, have been performed in many countries [3, 4, 5, 6].

1.2 Wind velocity and buildings

The unsteady character of wind regime, combined with the additional unsteadiness of the flow after the wind impacts on a building generates highly fluctuating velocities depending on the flow characteristics and the building configuration. The unsteadiness of the wind on the windward and leeward side of the façade is clear from the graph in Fig. 1.

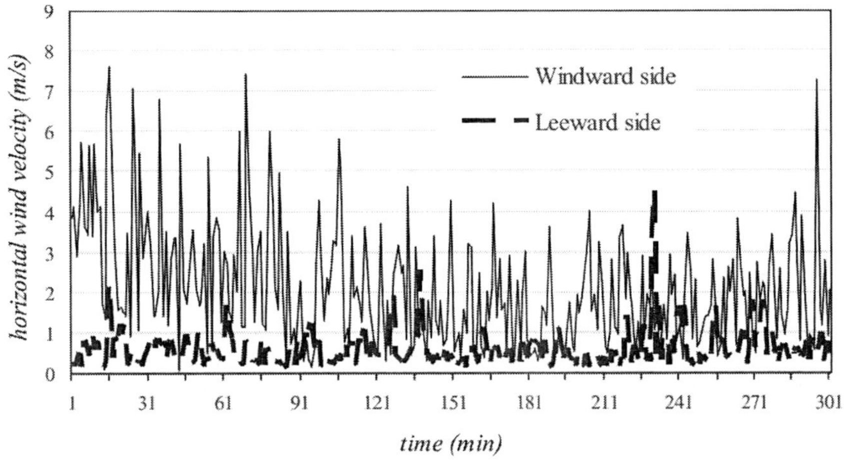

Fig. 1. Measured wind velocity on the façade of Research Centre building in Zilina.

The wind velocity can be considered as it consisted of mean wind speed and fluctuating component (turbulence). It may be caused by its own movement or ground roughness. The turbulence caused by ground roughness is higher in rougher terrain then in smoother terrain or flat open terrain and it decreases with height above the terrain. The turbulence is characterized by turbulence intensity $I\,(h)$, defined as:

$$I(h) = \sigma(h) \,/\, u_m(h) \tag{1}$$

where $u_m(h)$ is mean wind speed at elevation h and $\sigma(h)$ is standard deviation value of fluctuating velocity.

1.3 Wind structure

The mean wind speed changes in time, large fluctuations occur in a period of several hours or days. The variation of wind height is almost nil. The stationarity of the mean wind speed over periods of 10 minutes to many hours is the significant characteristic of the wind structure, since it forms the basic idea of wind tunnel testing.

Fig. 2 shows a record of wind velocity obtained from measurements at three different heights in the close proximity of the building skin.

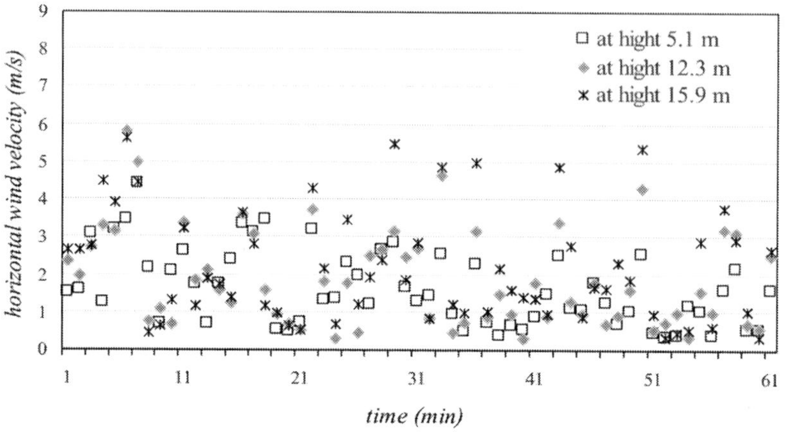

Fig. 2. Wind velocity from full scale measurements in three different heights from ground level on the north side of Research Centre building in Žilina.

2 Full scale measurement

2.1 Test side and the building

The building is situated north of the city center. The main part of building has dimensions: height 19.2 m, width 22.9 m and depth 16.4 m. The building has almost an exact east-west orientation so that the short facades are the north and south facade. Building consist of 5 storey administrative part and low rise laboratory part as can be seen in Fig. 3 from north-west side. Some surroundings in the prevailing wind directions consist of low-rise buildings (maximum height 20 m) and trees up to 25 m height. A view towards west side is in Fig. 4. The building has a partially aerated façade with tiles and partially ETICS with coating. Porous surface is produced by windows with external blinds.

Fig. 3. Picture of the Research Center building in Zilina, view from the west side.

Fig. 4. Picture of the surrounding building to the north-west direction- prevailing wind direction.

2.2 Meteorological parameters

The measured values can be compared with data sets from the meteorological station located at Zilina airport (is located approximately 16 km to the west, belongs to the Slovak Hydrometeorological Institute network) or the experimental detached weather station located on the roof of a nearby building [7]. Obviously, the weather station on the roof cannot be considered as perfect free standing wind-stream because of flow deformation.

2.3 Instrumentation

First station works since august 2016, full scale measurements with all 36 stations were conducted since October. Used measurement is from November 2016. 36 meteorological stations are fixed on each side of the building: 3 on the south side at three heights, 14 on the east side at four heights, 15 on the west side at four heights and 4 on the north side at four heights. The marked positions of the meteorological stations are shown in Fig. 5. The exact locations of the stations are summarized in the Table 1. The distances are from the building´s edges. Horizontal distance between the middle of the anemometer and facade is 270 mm. View on the façade meteorological station is in Fig. 6.

Table 1. Positions of meteorological stations in full scale, x in metres from the building edge, z in metres from the ground.

distance from the NE building edge						distance from the SW building edge					
station no.	x	z	station no.	x	z	station no.	x	z	station no.	x	z
1	14.3	8.5	10	1.1	11.3	16	18.3	4.1	25	6.3	4.1
2	10.4	7.7	11	6.4	14.9	17	14.6	4.1	26	1.3	7.7
3	14.3	12.1	12	1.1	14.9	18	10.0	4.1	27	6.3	11.3
4	10.4	11.3	13	2.6	5.1	19	18.3	7.7	28	1.3	11.3
5	14.3	15.7	14	2.6	12.3	20	14.6	7.7	29	6.3	14.9
6	10.4	14.9	15	2.6	15.9	21	10.0	7.7	30	1.3	14.9
7	6.4	4.1	34	14.6	1.9	22	18.3	11.3	31	2.8	8.7
8	1.1	4.1	35	10.8	1.9	23	14.6	11.3	32	2.8	12.3
9	6.4	11.3	36	2.7	1.9	24	10.0	11.3	33	2.8	15.9

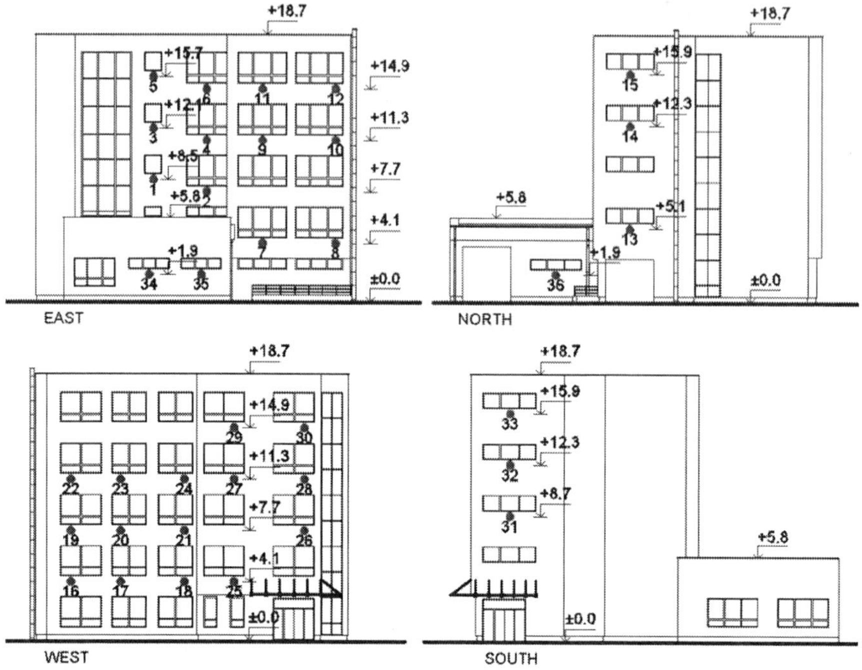

Fig. 5. Schematic view at the Research Centre building with marked and numbered positions of meteorological stations. Rows are marked from the top.

Measured values are collected wireless and transferred to the data storage server located in the building. The measured data are recorded in time steps of 60 seconds. The ultrasonic anemometer has a measuring range up to 40 m/s, accuracy +-0.01 m/s or 2% of reading, resolution 0.01 m/s, readout up-date interval 4 Hz and response time 250 ms. Measured values of wind velocity and other values are averaged in time, which is 60 seconds.

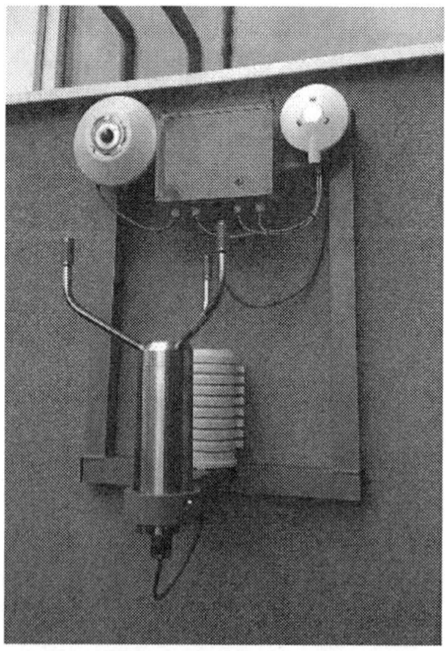

Fig. 6. Meteorological station on the façade consists of anemometer, pyranometer and pyrheliometer and sensor for temperature, relative humidity and air pressure.

3 Measurement results

Wind velocity and direction in the horizontal direction near the building are measured only a few months. However, from these results it is possible to select several days with higher wind speeds (up to 10 m/s) and the steady flow from one direction. Under such conditions we can observe the logical image of wind flow near the façade of five storey office building.

3.1 Wind speed

The measured wind speed in the horizontal line at approximately half height of the building (station No. 3, 4, 9 and 10) is documented towards to the distance from the north-east building edge (Fig. 7). Horizontal wind velocity at approximately same height is on Fig. 8.

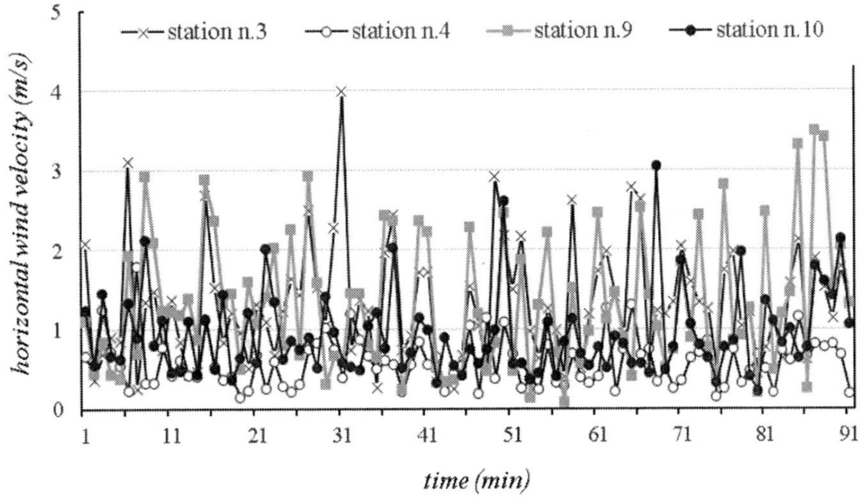

Fig. 7. Measured wind velocity on the east façade of Research Centre building.

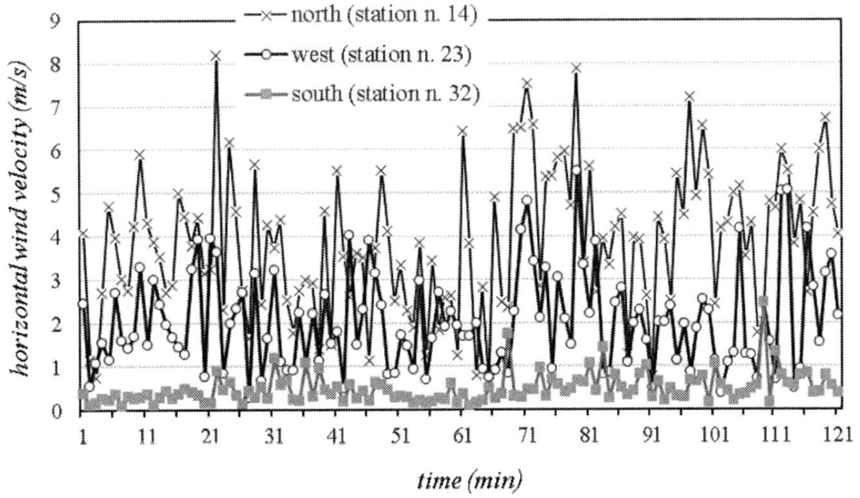

Fig. 8. Measured wind velocity on the façade of Research Center building at height 11.3 - 12.3 m.

3.2 Wind direction

The measured wind direction at approximately half height of the building is for illustration documented towards to façade orientation (Fig. 9). On the windward side (north) is the wind in almost the same direction. On the leeward sides (west and south) is the wind direction very changeable due to the high turbulence.

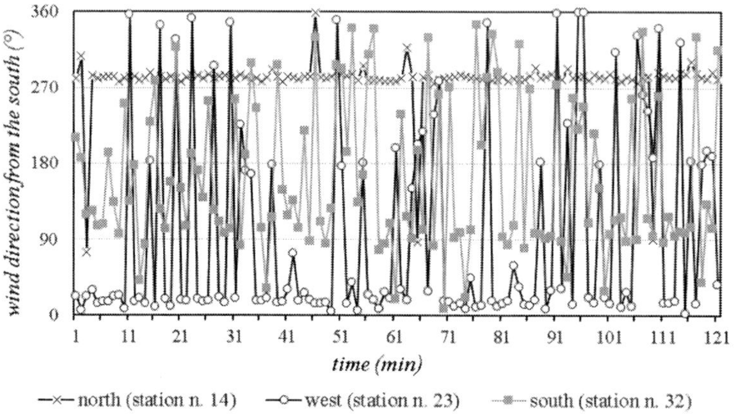

Fig. 9. Measured wind direction on the façade of Research Centre building at height 11.3 - 12.3 m.

3.3 Turbulence intensity

The graph in Fig. 10 describes measured values of wind speed and the turbulence intensity of the facade at one point on the windward side. While the air velocity is higher there is less intensity of turbulence and on the other hand, at the time when the speed is lower, the intensity of turbulence is greater.

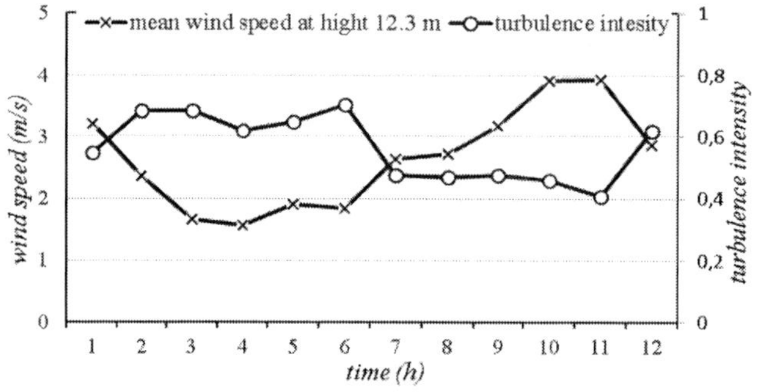

Fig. 10. Intensity of turbulence on the north façade (weather station No.14).

4 Simulation

Computational fluid dynamics (CFD) simulations were made before making the final test setup (decision where to fix the stations). Purpose of this simulation was not to compare exact values and comparison between simulation and measurement, because this simulation is far too simplified (surroundings, dimensions of the domain, etc.). But it can provide useful information of flow around the building of this geometry. This simulation was used as a draft for selection of good position on the façade for the meteorological stations.

For the simulation was used computer software OpenFOAM in version 2.2.0. It is an open-source, implicit, segregated and double precision solver. In the paper, an aerodynamic roughness length of 0.5 is chosen which represents a landscape totally and is quite regularly covered with similar-size large objects according to the Davenport roughness classification [8]. The inlet profile of the mean wind velocity (Fig. 11) is defined by the typical log-law expression [8]:

$$U(h) = \frac{u^*_{ABL}}{k} \ln\left(\frac{y + y_0}{y_0}\right) \tag{2}$$

The Aerodynamic boundary layer (ABL) friction velocity, u^*_{ABL}, is chosen to obtain the desired reference wind speed, U_{10}, at a height of 10 m (in this study 5 m/s). Height of the domain is 100 m with wind velocity U_{100} = 8.7 m/s. The inlet wind profile is shown in Fig. 11.

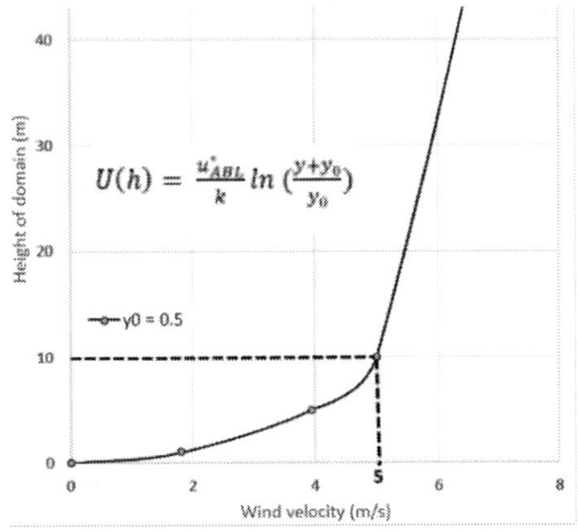

Fig. 11. Wind profile in the used inlet plane of the computation domain, wind velocity U_{10} = 5 m/s

The computational domain with dimensions 300×450×100 m consists of 3 727 272 cells. The cells are mostly hexahedra's with a few polyhedras. The mesh around the building is refined in 4 steps with use of snappyHexMesh tool. Computational domain and refined mesh around the building are in Fig. 12. The blockage ratio of the domain is 0.1% and the distances of the modelled building among the domain boundaries are according to the guidelines of Franke [9]. Simulated were more variants of wind direction, main flow direction is from south-west, which is represented by the geometry of domain (enough space behind the building for restoration the flow). However in this article is presented case with wind direction which represents west flow.

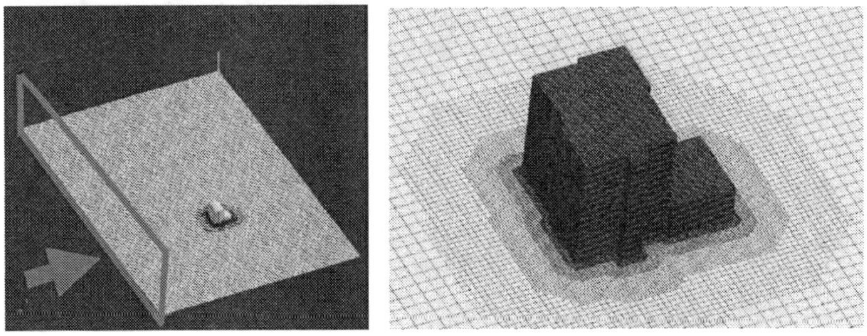

Fig. 12. Computational domain with dimension 300×450×100 m and marked inlet plane and wind direction, wind velocity U_{10} = 5 m/s. Refined mesh around the building on the right.

In Fig. 13, we can see wind flow field and trajectories through the vertical plane in the center of the high part of building. Fig. 14 shows wind flow field at mean wind velocity on the horizontal plane in high 11.3 m (4th story of the building, 2nd row of the meteorological stations from top).

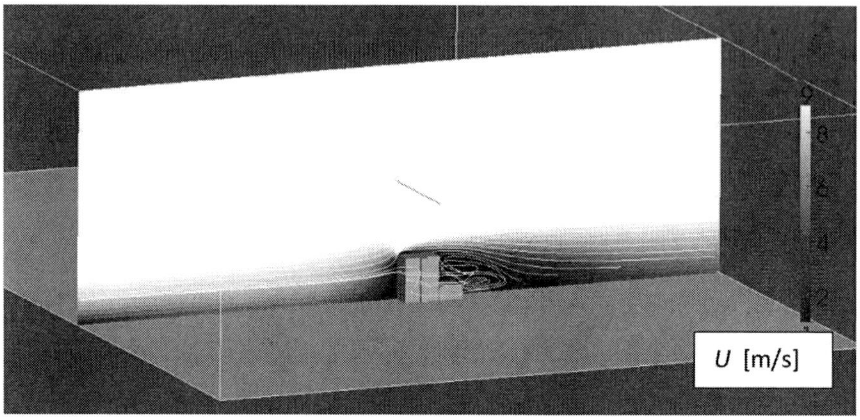

Fig. 13. Wind flow field through vertical plane. Reference wind velocity U_{10} = 5 m/s.

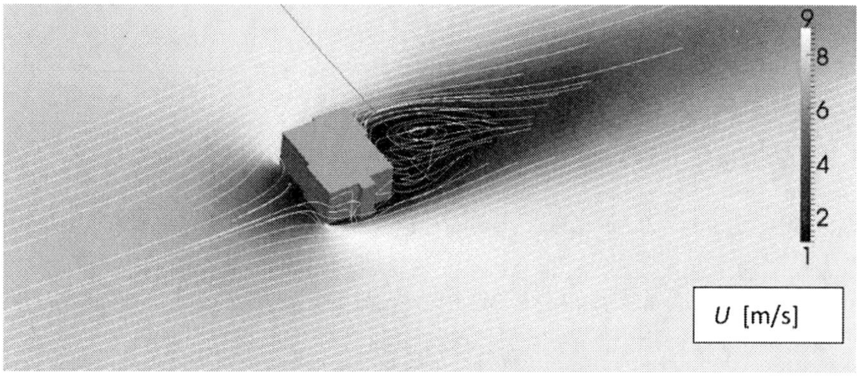

Fig. 14. Wind flow field through horizontal plane at height 11.3 m. Wind velocity U_{10} = 5 m/s.

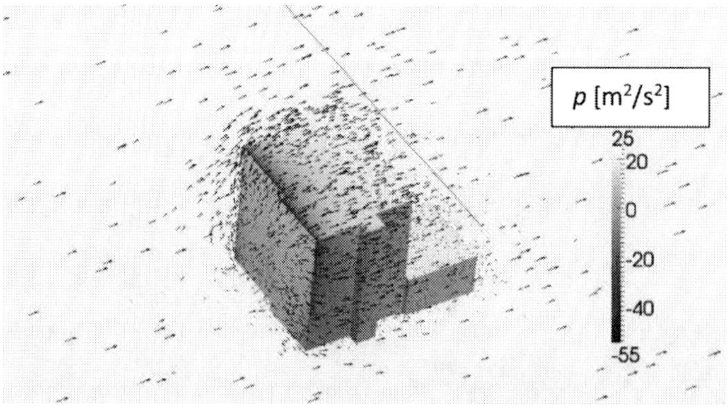

Fig. 15. Pressure on the building inducted by the wind flow and arrows representing wind flow around the building with wind flow from west. Induced pressures are in [m^2/s^2].

Influence of the wind flow on the building divide it to several sessions. According to the basic aerodynamics knowledge, approaching flow inducted positive pressure on the windward façade and negative on the other sides (Fig. 15). Typical flow features as the standing vortex, the down-flow from the stagnation point, large vortices behind the building, and flow restoration behind the building are visible (Fig 13 and 14). 2nd row of the meteorological stations on the façade is located in height 11.3 m above the terrain, which enable the possibility to analyze more precisely the change in the wind velocity from the unobstructed flow (measured generally at height 10 m above the terrain).

5 Results and Discussion

For the design of building structures, the wind loading in the range up to 15 and more years are relevant. This full scale experiment has much shorter period, however, it is important to perform these measurements under atmospheric circumstances with velocities, which are typically in 5 - 10 m/s range, for the wind tunnel experiment or other simulative validation. These tests have a limited applicability, because the reference conditions are often badly measured. Experience shows the relatively high differences between real measurements and wind tunnels are not so uncommon, therefore using the same full scale measurements may help to meet the more relevant values.

In this paper are summarized basic results of the conducted measurement on the façade of Research Centre and showed preliminary CFD simulation which helped with selection of good spots on the façade to set up the measurement. The measurement itself is planned as long-term one.

The research is supported by VEGA No. 1/0945/16.

References

1. B. Blocken, 50 years of Computational Wind Engineering: Past, present and future, J. Wind Eng. Ind. Aerod. **129**, p. 69-102 (2014)
2. V. D. Hoven, Power spectrum of horizontal wind speed in the frequency range from 0.0007 to 900 cycles per hour, J. Meteorology, **14**, (1957)
3. Ch. P. W. Geurts, *Wind-induced pressure fluctuations on the building Facades*, Technische Universiteit Eindhoven, (1997)
4. W. A. Dalgliesh, Comparison of model/full-scale wind pressures on a high-rise building, J. Ind. Aerodynam. **1**, (1975)
5. J. D. Holmes, Pressure fluctuation on a large building and along wind structural loadings, J. Indust. Aerodynam. **1**, (1976)
6. J. Kanda, Recent developments in full-scale wind pressure measurements in Japan, J. Wind E. Ind. Aerod. **33**, (1990)
7. D. Staffenova et al., Importance of using the right climate data sets for needs of energy analysis and choosing the accurate building materials, *Modern materials and technologies in sustainable building*, (2014)
8. B. Blocken et al., CFD Simulation of the Atmospheric Boundary Layer: Wall Function Problem, Atmos. Environ. **41**, (2007)
9. J. Franke et al., The Cost 732 Best Practice Guideline for CFD Simulation of Flows in the Urban Environment: A Summary. Int. J. Environ. Pollut. **44**, (2011)

DOI: 10.1051/matecconf/201710700084

Response of television tower to wind load

Pavol Marton[1]
[1] STU Bratislava, Faculty of Civil Engineering, Department of Structural Mechanics, Radlinského 11, SK-813 68 Bratislava, Slovakia

Abstract. This paper presents an analysis of the response of television tower to wind load. Computational model of tower is a model with the mass concentrated in the center of gravity of each part of the tower. The values of wind pressure were obtained using a record of wind speed.

1 Introduction

Reinforced concrete core of the tower consists of a rod with a square base 7×7 meters. External structure consists of a steel box columns that are above the platform and through steel beams connected with the core. In this amount of reinforced concrete construction continues and ends steel pipe, culminating in a laminate tube transmitters, which will consist of the tower. Individual platforms are woven into the reinforced concrete cores and corners hanging on a steel construction, Fig. 1.

Fig. 1. The television tower.

2 Computational model of supporting tower

The dynamic model of the tower is a system of concentrated masses in selected centers of mass of the tower [1]. Static computational model contains a reinforced concrete rod in the core of tower, steel platformy nad rods (box columns), steel and laminate tube.

To the level of 130 m are located platforms and reinforcing rod, Fig. 2a. Static computational model is shown in Fig. 2b, the dynamic Fig. 2c.

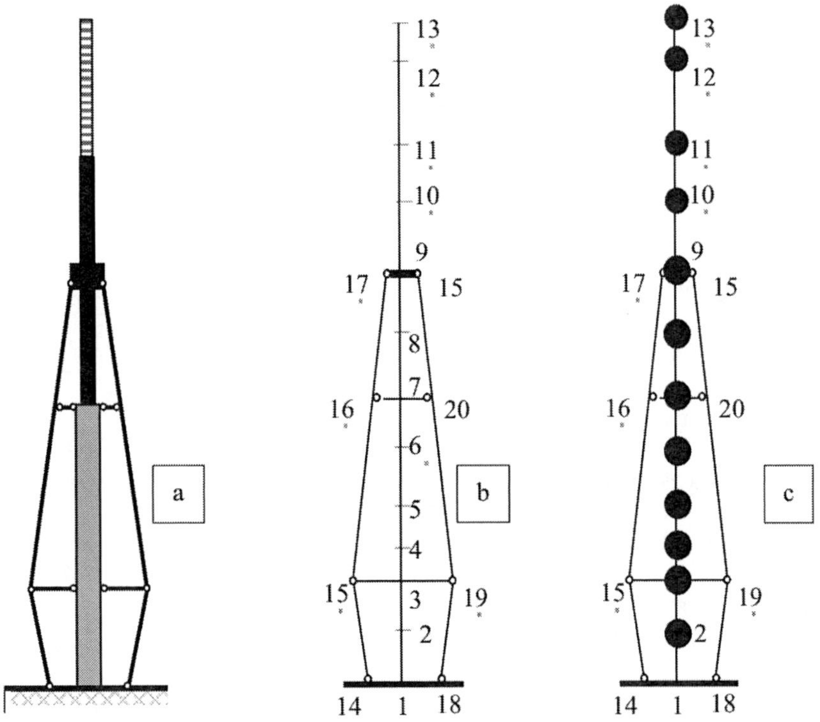

Fig. 2. Computational model of the tower.

3 Quasi-static calculation of the response of the tower

Modeling the wind speed is simulated by course with discrete time interval Δt. Wind pressure $w(t)$ is depending on the wind speed stated by the generally valid relation for laminar flow

$$w(t) = c\ A\ q(t) = c\ A\ \frac{v^2(t)}{16} \tag{1}$$

Displacement due to the load impulse is given by

$$u(t) = w(\tau)\,W(t-\tau)\,\mathrm{d}\tau \tag{2}$$

where $W(t)$ is an influence function that is the response of systems with one degree of freedom to the unit impulse

$$W(t) = \frac{e^{-\delta\,\omega_0\,t}}{m\sqrt{1-\xi^2}\ \omega_0}\sin\sqrt{1-\xi^2}\ \omega_0\,t \tag{3}$$

Displacement in time $\tau = k\ \Delta t$ and $t = j\ \Delta t$ is

 DOI: 10.1051/matecconf/201710700084

$$u_j = \sum_{k=1}^{k=j-1} w_k \, W[(j-k)\Delta t] \tag{4}$$

where $W(t-\tau) = W[(j-k)\Delta t]$

After the substitution, the elastic force in system with one degree of freedom after a single transformation is defined by the formula

$$F_j = 2\pi\kappa \sum_{k=1}^{k=j-1} w_k \, e^{-\delta\kappa(j-k)} \sin 2\pi\kappa(j-k) \tag{5}$$

where δ is the logarithmic decrement.

4 Methodology of processing record wind speeds

In methodology following assumptions are valid:

a) continuous chart recording wind speed is substituted by a discrete step system [2]

b) discrete values of wind speed shall be determined by discrete pressure values based on the unit area of wind $c{\cdot}A = 1$

c) the pressure values were averaged and represent the static value of wind pressure q_s

d) the relation $w_d = w - w_s$ determines the maximum value of the dynamic component

e) dynamic wind pressure will be standardized $q_d / q_{d,\max} \leq 1$

f) the relative value of dynamic load is defined by the relation

$$\frac{F_j}{F_{jk,\max}} = 2\pi\kappa \sum_{k=1}^{k=j-1} \frac{q_{dk}}{q_{d,\max k}} e^{\delta\kappa(j-k)} \sin 2\pi(j-k) \tag{6}$$

g) the dynamic coefficient of increase in wind pressure (dynamic factor) is defined by the relation

$$\beta = \max \frac{F_j}{F_{j,\max}} - 1 \tag{7}$$

h) impact factor is related to the maximum wind pressure resulting from the relation

$$\beta = 1 + \beta \frac{q_{d,\max}}{q_{\max}} \tag{8}$$

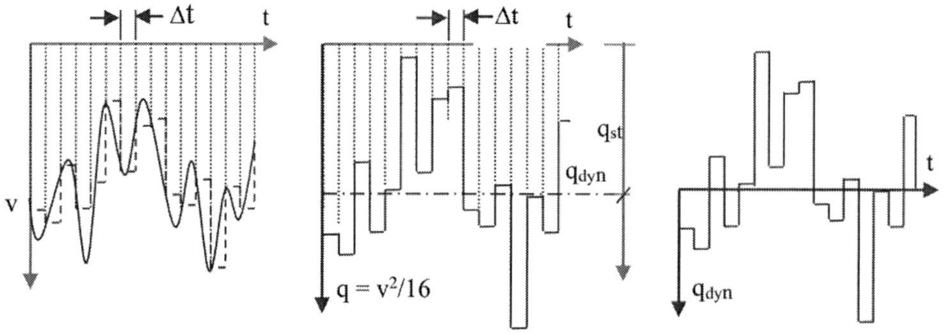

Fig. 3. Substitution of continuous recording of speed and wind pressure by the discrete one.

i) for producing the sum of equations (6) it is appropriate to limit the correlation length of t_w

$$t_w = t - \tau = (j - k)\Delta t = m\,\Delta t \tag{9}$$

4.1 Analysis of wind speed record

Wind speed record [3] during a severe storm recorded maximum wind speed v = 45 m/s. Wind speed record was taken in a period of about 9 minutes. The speed record of 30 values per minute corresponding to the time interval is in Fig. 4.

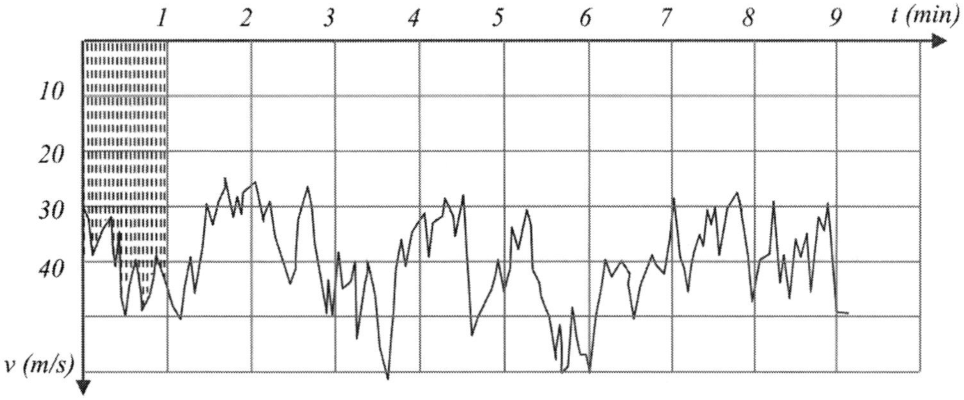

Fig. 4. Wind speed record.

a) Determination of mean and maximum dynamic components of the wind pressure Fig. 4 is in Table 1.

Table 1. Static and dynamic share stress.

q_s [kNm^{-2}]	$q_{d,\max}$ [kNm^{-2}]	$q_{\max}$ [kNm^{-2}]	q_s / $q_{\max}$	$q_{d,\max}$/ $q_{\max}$
0.588	0.458	1.046	0.56	0.44

b) Determination of mean and maximum dynamic components of the wind pressure is in Table 2. It indicates maximum load ratio for different values of logarithmic decrement and vibration period of the non-damped natural vibration.

Table 2. Determination of mean and maximum dynamic components.

δ	T_0 [s]				
	0.5	1.0	2.0	3.3	5.0
0.150	0.413	0.829	0.852	1.430	1.717
0.075	0.413	0.867	0.971	1.593	1.957
0.025	0.406	0.883	1.661	2.028	2.496

DOI: 10.1051/matecconf/201710700084

c) Determination of the gust dynamic coefficient
Gust factor φ is determined for $T_0 = 3.5$ s and the logarithmic decrement $\delta = 0.15$.

Table 3. Calculation of the gust factor φ.

φ	T_0 [s]				
δ	0.5	1.0	2.0	3.3	5.0
0.150	0.732	0.925	0.935	1.189	1.315
0.125	0.735	0.926	0.943	1.218	1.329
0.100	0.738	0.927	0.952	1.242	1.353
0.075	0.742	0.941	0.987	1.261	1.421
0.050	0.743	0.938	1.010	1.328	1.524
0.025	0.739	0.959	1.291	1.452	1.658

4.2 Numerical calculation

The load was calculated from the blast effects of wind calculated. Mechanical parameters and geometrical characteristics of the tower are shown in Table 4.

Table 4. Mechanical and geometrical characteristics of the tower.

No.	m^*	l	EI	No.	m	No.	l	EI
-	$kg \cdot 10^3$	m	$N \cdot m^2$	-	$kg \cdot 10^3$	-	m	$N \cdot m^2 \cdot 10^{10}$
1	496	32.8	$5.104 \cdot 10^9$	11	48	11	20.5	$0.025 \cdot 10^9$
2	1236	15.6	$6.268 \cdot 10^9$	12	27	12	10	$0.024 \cdot 10^9$
3	1100	5.4	$6.737 \cdot 10^9$	13				
4	838	9.4	$6.260 \cdot 10^9$	14				
5	1606	16.2	$5.555 \cdot 10^9$	15, 19				
6	1157	13.5	$4.853 \cdot 10^9$	16, 20				
7	326	21.6	$0.555 \cdot 10^9$	17, 21				
8	166	19.4	$0.158 \cdot 10^9$					
9	75	19.3	$0.036 \cdot 10^9$					
10	60	19.2	$0.021 \cdot 10^9$					

The calculation of load forces from the blast effects of wind is shown in Table 5.

Table 5. Load characteristics.

Sec.	l	b	$c_{f,0}$	c_f	$A_{ref}(z)$	$w_p(z)$	F_i	φ	F_i
1	16.2	18.0 4.0	15.6	1.19	73	161	644	1.2	773
2	15.6	25.7 4.0	15.6	1.19	408	161	644	1.2	773
3	5.4	32,7 4.0	5.4	1.24	343	161	564	1.2	677
4	9.4	30.5 4.0	9.4	1.28	226	151	459	1.2	551
5	16.2	27.6 4.0	16.2	1.33	353	156	490	1.2	588
6	13.5	22.6 4.0	13.5	1.42	335	167	602	1.2	722
7	21.6	17.6 3.0	21.6	1.52	309	179	269	1.2	323
8	19.4	11.5 3.0	19.4	1.0	236	94	100	1.2	120
9	19.3	5.4 3.0	19.3	1.2	58	14	336	1.2	403
10	19.2	3.0	19.2	1.2	58	94	100	1.2	120
11	20.5	2.6	20.5	1.2	59	14	168	1.2	202
12	10	2.6	10	1.2	46	14	84	1.2	101
13	5	2.6	10	1.2	13	14	41	1.2	50

5 Conclusion

The present method of determining the effects of blast wind upon structures is applicable to support systems, such as antenna masts, towers, bridges and civil buildings.

References

1. P. Marton, *Dynamic investigation of poles in the system of power lines* (ES SVŠT, 1979)
2. Ch. Petersen, *Anchored masts and chimneys* (W.E&S. 1970)
3. F. Steiger, Measurement of wind speed on the antenna mast, Stahlbau **34** (2008)

Random variation and correlation of the weather data series – evaluation and simulation using bounded histograms

Petr Konecny[1,*] and *David* Pustka[2]

[1]VSB – Technical University of Ostrava, Faculty of Civil Engineering, Department of Structural Mechanics, Ludvika Podeste 1875/17, 708 33 Ostrava-Poruba, Czech Republic
[2]VSB – Technical University of Ostrava, Faculty of Civil Engineering, Department of Structures, Ludvika Podeste 1875/17, 708 33 Ostrava-Poruba, Czech Republic

Abstract. The contribution is focused on the random variation and correlation of input parameters for the climate data description. The climate data such as ambient temperature, solar intensity, wind speed and direction are of the random nature. The description of ambient temperature can be based on the climate data time series in the form of climatic "load duration curve". Particular input parameters such as ambient temperature and solar radiation have a significant correlation. The attention is paid especially to the evaluation and simulation application of the histograms using Monte Carlo type process considering correlation of particular parameters.

1 Introduction

Numerical modelling of the wind load, thermal expansion related problems or building thermal balance is affected by the variability of input parameters such as ambient temperature, wind speed, direction and solar radiation.

The parameters for the analysis of the effect of thermal changes-induced deformation on structures such as bridges may benefit from data about the distribution of temperature since the design to most unlikely conditions is frequently too conservative. There are also applications when the wind speed and wind direction plays their role in the quasi static analysis. Then description of the variation of these parameters as well as correlation of wind speed and direction may be of interest in such cases, especially in cases of unsymmetrical objects.

Also the field of the energy performance of the building may benefit from description of variation of input parameters and their correlations. Thermal balance is generally based on the comparison of thermal losses with the capability of heating/cooling system. The system is frequently designed to meet the most demanding conditions. Those most demanding conditions can be rare, thus and design of a single system such as heat pump without supplementary heat source can be expensive. Typical analysis is based either on the deterministic evaluation of the performance in case of extreme summer and winter

* Corresponding author: petr.konecny@vsb.cz

temperatures or on the time-dependent analysis based on the typical year data in case of non-stationary processes (e. g. ground source heat pumps application [1]).

Since the input parameters are of random nature application of probabilistic reliability analysis can give an idea what is the probability that designed structural or thermal system does not meet the design condition with acceptable likelihood. The probabilistic assessment may be conducted using the probabilistic assessment [2-4]. The correlation between random inputs characteristics can be statistically described as well [5-8]. The suitable probabilistic approach in case of time-dependent parameters such as temperature and snowfall can be well described by time pulses (see e. g. [9]) or load-duration curves [2]. It may be worth explanation that load duration curve is a sorted history of loading formed as a frequency histogram.

Lot of such analyses often use the direct Monte Carlo method ([11], [2-4], [12-14]) which is robust, although it may be time consuming. Even though, enhanced probabilistic procedures that use computer power more effectively such as Latin Hypercube Sampling [15], Response Surface Method [16] or Direct Optimized Probabilistic Calculation [17], are not discussed herein. The paper is focused on the analysis of climate random input parameters variability and their mutual correlation necessary for the probabilistic building thermal losses analysis or thermal expansion related problems. This pilot study is based on the weather data time-series available from [7] that cover "so called" typical year.

2 Analysis of data

Time series of weather data hourly records that represents typical year in Ostrava are statistically analysed for mean value, standard deviation, minimum and maximum value. The frequency histograms are created as well. The relationship between the parameters is studied using the correlation index. Strong linear correlation is expressed if the correlation index approaches to value one either positive or negative. Variables are uncorrelated if the correlation index approaches to zero value. Since the weather data are seasonal with variation throughout a day the two time flow-related parameters are added. Parameter *time* represents hours from the beginning of the year while parameter *hours* represents hours from the beginning of the day.

3 Variation of climate data

The resulting statistical parameters are given in Table 1 and frequency histograms are shown in the Fig. 1-3. Statistics for the time parameters are not included in the Table 1.

Relationship between studied parameters expressed through correlation coefficient is given in Table 2. Graphical description of selected correlated parameters is shown in 2D plots Fig. 4 and 5.

Histogram of time shows some irregularities due to the number of days within even and odd months throughout the year (Fig. 1 on the left). This can be further refined. Histogram of the temperature has two peaks caused by the winter and summer (Fig. 1 on the right). Average value of temperature is 8.5°C. Histograms of extraterrestrial and global radiation (Fig. 2) show high occurrence of zero value that is caused by the night measurements with average value of 287 W/m^2 for extraterrestrial and 116 W/m^2 for global radiation. Maximum values are 1181 W/m^2 and 890 W/m^2 respectively. Wind direction is dominating from south west and from north (see Fig. 3 on the left). Wind speed average is 3.8 m/s with extreme value 19 m/s.

The correlation index between the time and temperature (*temp*) is almost zero even though there is a strong seasonal dependence as depicted at Fig. 4. The reason for not capturing the relationship between time and temperature is its periodic behaviour. Correlation index between radiation and temperature (*temp* is about 0.5 which is significant - see Fig. 4). Correlation between radiation extraterrestrial horizontal (*radia_eh*) and radiation global horizontal (*radia_gh*) around 0.9 illustrates strong influence of extraterrestrial radiation on the global radiation (see Fig. 5 on the left). Small correlation was found between wind speed and direction and between wind speed and extraterrestrial radiation (*radia_eh*).

Table 1. Statistical parameters of typical year weather data from Ostrava [7].

Description	Notation	Units	Mean	Std	Variation	Min	Max
Dry-bulb temperature	*temp*	[°C]	8.5	8.5	1.0	-14.1	32.9
Radiation extraterrestrial horizontal	*radia_eh*	[W/m²]	286.8	369.0	1.3	0.0	1181.0
Radiation global horizontal	*radia_gh*	[W/m²]	116.3	190.7	1.6	0.0	890.0
Wind direction	*wind_d*	[°]	165.8	104.9	0.6	0.0	360.0
Wind speed	*wind_s*	[m/s]	3.8	2.5	0.7	0.0	19.0

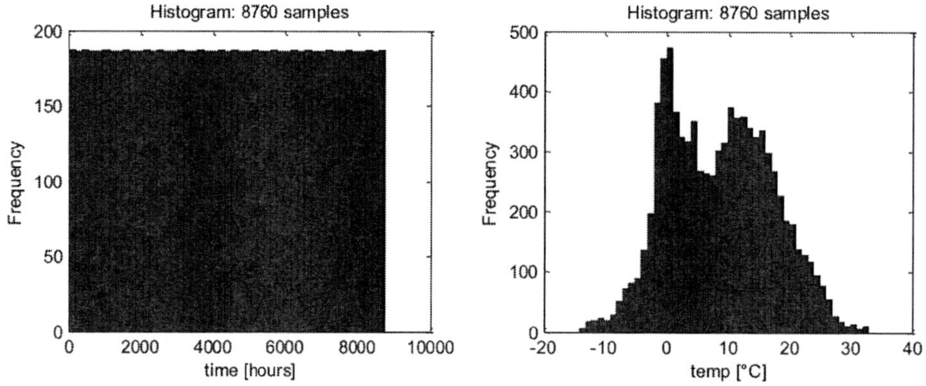

Fig. 1. Histograms of time /on the left/ and dry-bulb temperature [°C] /on the right/.

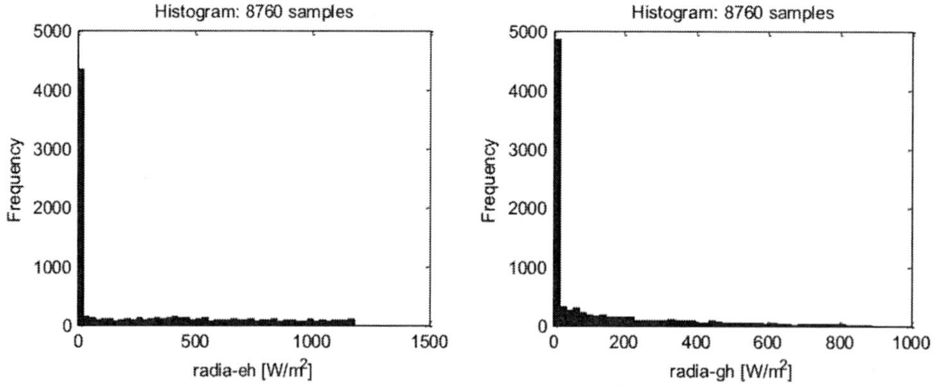

Fig. 2. Histograms of radiation extraterrestrial horizontal [W/m²] /on the left/ and radiation global horizontal [W/m²] /on the right/.

DOI: 10.1051/matecconf/201710700085

Table 2. Correlation indexes for parameters of typical year weather data from Ostrava [7].

	time	*hours*	*temp*	*radia_eh*	*radia_gh*	*wind_d*	*wind_s*
time	1.00	0.00	0.17	-0.05	-0.05	-0.07	-0.14
hours	0.00	1.00	0.10	-0.02	-0.02	0.03	0.03
temp	0.17	0.10	1.00	0.51	0.53	0.11	0.03
radia_eh	-0.05	-0.02	0.51	1.00	0.88	0.08	**0.24**
radia_gh	-0.05	-0.02	0.53	0.88	1.00	0.06	0.18
wind_d	-0.07	0.03	0.11	0.08	0.06	1.00	**0.26**
wind_s	-0.14	0.03	0.03	**0.24**	0.18	**0.26**	1.00

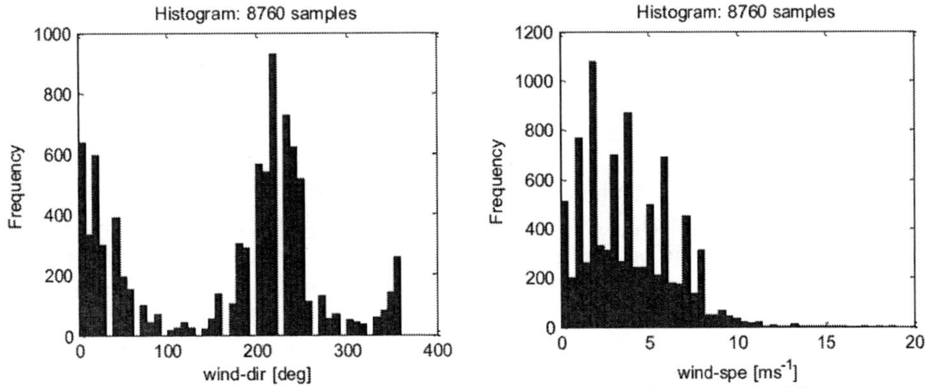

Fig. 3. Histograms of wind direction [°] /on the left/ and wind speed [m/s] /on the right/.

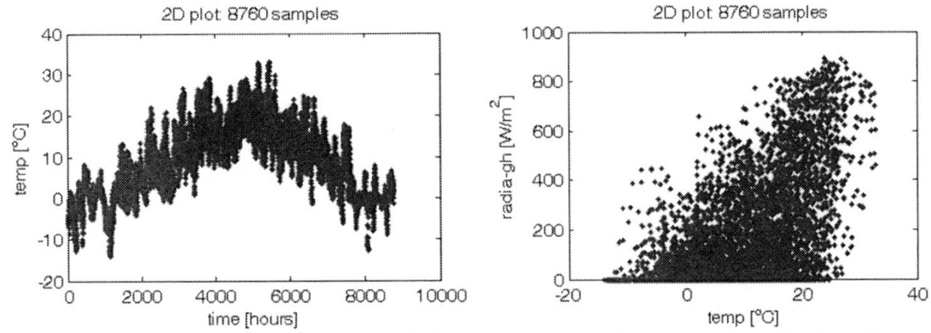

Fig. 4. 2D plot of time [hours] and dry-bulb temperature [°C] /on the left/; 2D plot of dry-bulb temperature [°C] and radiation global horizontal [W/m^2] /on the right/.

4 Monte Carlo simulation

Since the probabilistic simulation of observed climate phenomena is of interest considering their mutual correlation, the model for generation of correlated distributions is applied with 50 thousands simulation steps as described by Phoon et al. [6, 10] and applied also in [8]. The advantage of this procedure is direct generation of random occurrences without dependence on previous simulations. This approach is suitable for the direct Monte Carlo method. The disadvantage of this approach is the ambiguous description of the resulting joint distribution despite the quality of the marginal distributions being maintained. The correlation coefficient of the generated distributions is close to the specified value.

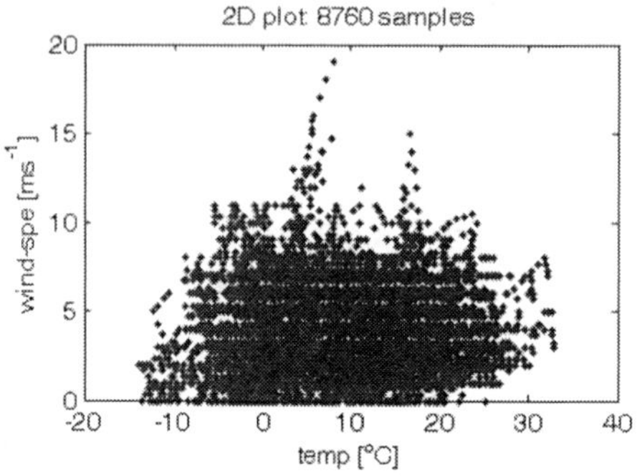

Fig. 5. 2D plot of temperature [°C] (horizontal axis) and wind speed [m/s] (vertical axis).

The simulation process using the Monte Carlo method can be enhanced using knowledge of the marginal distributions and the correlation between the parameters. The correlation matrix is modified to better preserve fractile correlation.

The first step in the simulation is to generate independent variables described by the normalized Gaussian distribution, which is adjusted by the Choleski decomposition into the correlated unit Gaussian distribution. The normal (Gaussian) distribution can be converted to a uniform distribution through its inverse transformation of the distribution function. The correlated uniform distribution is equivalent to the primary generator of random variables described by a frequency histogram.

Simulation steps are following:

- **Transformation of correlation matrix.** Due to differences in the normal distribution and the general distribution characterized, for example, by a histogram, it is appropriate to modify the obtained correlation index. The resulting correlation matrix R must satisfy the condition of positive definiteness to generate vectors of the correlated normal distribution.
- **Generation of correlated normal distributions.** The correlated Gaussian distributions may be, according to [5], [10] generated using either of two methods – Eigen decomposition, or Choleski factorization.
- **Transformation of a normal distribution into a uniform distribution.** Two generated occurrences of a normal distribution with a suitable correlation need to be transformed into a uniform distribution Y_F, using the inverse distribution function for the normal distribution.
- The probability density obtained by orthogonal transformation of a Gaussian random occurrence is a uniform distribution. It is suitable for generation of samples described by the frequency histogram.

Transformation of a uniform distribution into a general distribution (non-Gaussian process). A relevant general distribution is consequently obtained using the inverse transformation of the distribution function of the sought distribution.

- The probability density obtained by orthogonal transformation of a Gaussian random occurrence is a uniform distribution. It is suitable for generation of samples described by the frequency histogram.
- Transformation of a uniform distribution into a general distribution (non-Gaussian process). A relevant general distribution is consequently obtained using the inverse transformation of the distribution function of the sought distribution.

Since the resulting distribution follows the same marginal function described by bounded histogram (cumulative density function, CDF) as shown in [8]. It is not necessary to show the plots. The fractile correlation of respective CDFs that is more informative is shown in Table 3.

All parameters except of time show very good agreement with fractile correlation almost 1 (complete match). In case of the time, the difference is quite high here even though resulting histogram is approaching the original one on Fig 1 (on the left).

Table 3. Fractile correlation between original and simulated histograms.

time	-0.0890
hours	0.9994
temp	0.9991
radia-eh	1.0000
radia-gh	1.0000
wind-dir	0.9999
wind-spe	0.9995

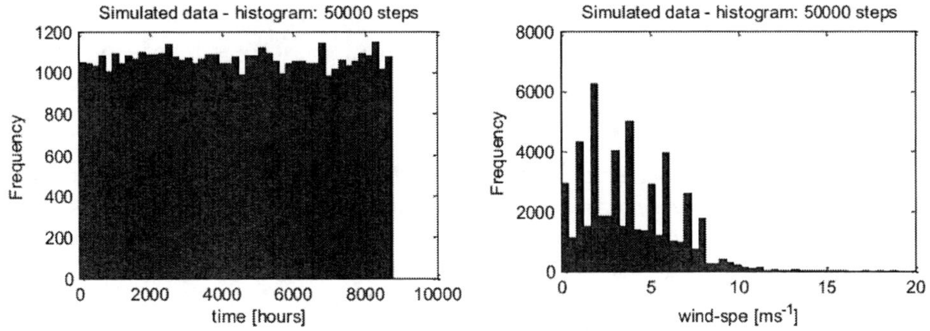

Fig. 6. Histograms of time [hours] /on the left/ and wind speed [m/s] /on the right/; both 50,000 simulations.

To verify the simulation process, the correlation matrix of randomly generated correlated occurrences is subsequently verified and compared with the original matrix. The Pearson's correlation coefficient of simulated correlation matrix is shown in Table 4.

Table 4. Correlation matrix of parameters simulated by Monte Carlo simulation.

	time	*hours*	*temp*	*radia_eh*	*radia_gh*	*wind_d*	*wind_s*
time	1.00	0.00	0.17	-0.04	-0.04	-0.07	-0.14
hours	0.00	1.00	0.10	-0.02	-0.02	0.03	0.03
temp	0.17	0.10	1.00	0.47	0.46	0.11	0.03
radia_eh	-0.04	-0.02	0.47	1.00	0.82	0.07	0.22
radia_gh	-0.04	-0.02	0.46	0.82	1.00	0.04	0.15
wind_d	-0.07	0.03	0.11	0.07	0.04	1.00	0.25
wind_s	-0.14	0.03	0.03	0.22	0.15	0.25	1.00

The resulting correlation indexes after simulation are similar to original ones as shows Table 4. The highest difference is in case of solar radiations and temperature as shows Table 5.

Table 5. Difference between original and simulated correlation matrices.

	time	*hours*	*temp*	*radia_eh*	*radia_gh*	*wind_d*	*wind_s*
time	0.00	0.00	0.00	-0.01	-0.01	0.00	0.00
hours	0.00	0.00	0.00	0.00	0.00	0.00	0.00
temp	0.00	0.00	0.00	0.04	0.08	0.00	0.00
radia_eh	-0.01	0.00	0.04	0.00	0.07	0.01	0.02
radia_gh	-0.01	0.00	0.08	0.07	0.00	0.01	0.02
wind_d	0.00	0.00	0.00	0.01	0.01	0.00	0.01
wind_s	0.00	0.00	0.00	0.02	0.02	0.01	0.00

5 Conclusion

The paper was aimed at the initial analysis of the typical year weather data from Ostrava [18] with respect to application of probabilistic structural and thermal balance assessment. It shows statistical parameters and frequency histograms of temperature, solar radiation, wind direction and speed. Correlation index between studied parameters and time is examined as well.

Significant correlation between solar radiation and temperature that would have to be considered in the probabilistic assessment is observed. Global radiation and extraterrestrial radiation are correlated very well. Correlation between time and temperature was not found due to the applied linear correlation coefficient. Suitable transformation shall be applied to adjust for the seasonal changes in temperature in order to simulate the relationship between season and the temperature in case of non-stationary processes such as ground source heat pumps application.

The analysis of typical year can illustrate the behaviour of selected parameters. There is a difference between typical year data used for the presented analysis and actual data sets with a decades of data. Storing a data in the form of histograms and correlation indexes can help to cover longer period of time with lower amount of storage space required. The suitable correlation field process shall be used in order to reproduce also the time history data variation related to seasonal changes relationship with sufficient quality and feasibility.

Moreover testing Monte Carlo simulation process is demonstrated. The process of simulation consists of preparation of a set of correlated normally distributed vectors. These vectors are transformed into two correlated uniform distributions that are used to produce a general distribution (a histogram). The histograms are created by a standard procedure, i.e., by the inverse transformation of a uniform distribution.

Good agreement between the input data and the results generated using the Monte Carlo method was observed. The correlation coefficient of the simulated values approached the required degree of correlation, and the prescribed marginal distributions were almost identical to the simulated distributions.

Financial support from VŠB-Technical University of Ostrava by means of the Czech Ministry of Education, Youth and Sports through the Institutional support for conceptual development of science, research and innovations for the year 2017 is gratefully acknowledged.

References

1. J.D. Spitler. Ground-source heat pump system research - Past, present, and future, HVAC&R Research, Vol: **11,** 2, 165-167 (2005)
2. P. Marek, M. Guštar, T. Anagnos, *Simulation-Based Reliability Assessment for Structural Engineers* (CRC Press, Inc., Boca Raton, Florida, 1995)
3. R.E. Melchers, *Structural reliability analysis and prediction* (2nd ed., John Wiley, Chichester, New York, 1999)
4. P. Marek, J. Brozzetti, M. Guštar, P. Tikalsky (ed.), *Probabilistic Assessment of Structures using Monte Carlo simulation. Basics, Exercises, Software,* (2nd edition, CD ROM attached, ITAM Czech Academy of Sciences, Prague, 2003)
5. M. Grigoriu, Simulation of non-Gaussian translation processes. J. Engng. Mech, ASCE, **124,** 2, 121-6 (1998).
6. K.K. Phoon, S.T. Quek, H. Huang, Simulation of non-Gaussian processes using fractile correlation, Probabilist. Eng. Mech., **19**, 287-292 (2004)
7. M. Vořechovský, D. Novák, Correlation control in small-sample Monte Carlo type simulations in a simulated annealing approach, Probabilist., Eng., Mech. **24,** 3, 452-462 (2009)
8. P. Konečný, Correlation of Steel Yield Stress and Ultimate Strength – Monte Carlo Simulation via Non-Gaussian Random Variables, Mater. Sci. Forum. (to be published in 2018)
9. M. Sýkora, *Probabilistic Analysis of Time-Variant Structural Reliability* (Doctoral thesis, CTU Prague, 2005)
10. K.K. Phoon, S.P. Huang, S.T. Quek, Simulation of second-order processes using Karhunen - Loeve expansion, Comput. Struct., **80**, 12, 1049-60 (2002)
11. N. Metropolis, S. Ulam, S. The Monte Carlo Method. Journal of American Statistical Association, **44**, 247, 335-341 (1945)
12. H.J. Pradlwarter, G.I. Schuëller, Assessment of low probability events of dynamical systems by controlled Monte Carlo simulation, Probabilist. Eng. Mech. **14**, 3, 213-227 (1999)
13. P. Konečný, J. Brožovský, P. Ghosh, Evaluation of Chloride Influence on the Cracking in Reinforced Concrete Using Korozeeneck Software, Transactions of the VŠB – Technical University of Ostrava, Civil Engineering Series, **11**, 1, 1-7 (2011)
14. P.S. Marsh, D.M. Frangopol, Reinforced concrete bridge deck reliability model incorporating temporal and spatial variations of probabilistic corrosion rate sensor data, Reliab. Eng. Syst. Saf. **93,** 3, 394-409 (2008)
15. R.L. Iman, W.J. Conover, A Distribution-Free Approach to Inducing Rank Correlation among Input Variables. Communications in Statistics Part B-Simulation and Computation, **11**, 3, 311-334 (1982)
16. Kralik, A RSM Method for Probabilistic Nonlinear Analysis of Reinforced Concrete Bubbler Tower Structure Integrity. In: *Proceedings of European Safety and Reliability Conference (ESREL)*, Reliability, Risk and Safety: Theory and Applications, 2010, 1-3, 1369-1372. Taylor & Francis (2009)
17. M. Krejsa, P. Janas, R. Čajka, Using DOProC Method in Structural Reliability Assessment, Appl. Mech. Mater. **300-301**, 860-869 (2013)
18. EnergyPlus Energy Simulation Software available on-line at: <http://apps1.eere.energy.gov/buildings/energyplus/> (2013)

AUTHOR INDEX

AUTHOR INDEX